T0318369

ANIMALS AND HUMAN SOCIETY

ANIMALS AND HUMAN SOCIETY

COLIN G. SCANES
University of Wisconsin–Milwaukee,
Milwaukee, WI, United States

SAMIA R. TOUKHSATI
Honorary Fellow, The University of Melbourne,
Parkville, VIC, Australia

ELSEVIER

ACADEMIC PRESS

An imprint of Elsevier

Academic Press is an imprint of Elsevier
125 London Wall, London EC2Y 5AS, United Kingdom
525 B Street, Suite 1800, San Diego, CA 92101-4495, United States
50 Hampshire Street, 5th Floor, Cambridge, MA 02139, United States
The Boulevard, Langford Lane, Kidlington, Oxford OX5 1GB, United Kingdom

Notices
Knowledge and best practice in this field are constantly changing. As new research and experience broaden
our understanding, changes in research methods, professional practices, or medical treatment may become
necessary.

Practitioners and researchers must always rely on their own experience and knowledge in evaluating and
using any information, methods, compounds, or experiments described herein. In using such information or
methods they should be mindful of their own safety and the safety of others, including parties for whom they
have a professional responsibility.

To the fullest extent of the law, neither the Publisher nor the authors, contributors, or editors, assume any
liability for any injury and/or damage to persons or property as a matter of products liability, negligence
or otherwise, or from any use or operation of any methods, products, instructions, or ideas contained in the
material herein.

Library of Congress Cataloging-in-Publication Data
A catalog record for this book is available from the Library of Congress

British Library Cataloguing-in-Publication Data
A catalogue record for this book is available from the British Library

ISBN: 978-0-12-805247-1

For information on all Academic Press publications visit our website at
https://www.elsevier.com/books-and-journals

Working together
to grow libraries in
developing countries

www.elsevier.com • www.bookaid.org

Publisher: Sara Tenney
Acquisition Editor: Kristi Gomez
Editorial Project Manager: Pat Gonzalez
Production Project Manager: Karen East and Kirsty Halterman
Designer: Miles Hitchen

Typeset by Thomson Digital

Contents

10. Animals in Entertainment
COLIN G. SCANES

11. Animals and Religion, Belief Systems, Symbolism and Myth
COLIN G. SCANES, PU CHENGZHONG

12. Animals as Companions
KATE MORNEMENT

13. Animals in Medicine and Research
TIFFANI J. HOWELL

14. Animals and Human Disease: Zoonosis, Vectors, Food-Borne Diseases, and Allergies
COLIN G. SCANES

15. Pest Animals
SAMIA R. TOUKHSATI, COLIN G. SCANES

List of Contributors

Pu Chengzhong Centre for the Study of Humanistic Buddhism, Chinese University of Hong Kong, Sha Tin, Hong Kong

Tiffani J. Howell La Trobe University, Bendigo, VIC, Australia

Katrina Kluss TC Beirne School of Law, University of Queensland, Brisbane, QLD, Australia

Kate Mornement Pets Behaving Badly, Melbourne, VIC, Australia

Clive J.C. Phillips Centre for Animal Welfare and Ethics, School of Veterinary Science, University of Queensland, Gatton, QLD, Australia

Colin Salter University of Wollongong, Wollongong, NSW, Australia

Colin G. Scanes University of Wisconsin–Milwaukee, Milwaukee, WI, United States

Samia R. Toukhsati Honorary Fellow, The University of Melbourne, Parkville, VIC, Australia

Preface

In this volume, we consider the complex interplay between animals and human society to explore the multiple ways that one has impacted the other in the past and our outlook for the future. We begin by exploring perceptual differences between humans and animals, such as four color vision in birds (including UV), and profoundly, the more sensitive and discriminating olfactory ability of dogs, and the ways in which these differences have been harnessed to benefit humans.

Animals have played a pivotal and benign role in human evolution, for instance, as sources of meat and leather. Human hunting of animals not only provided the high-energy foods necessary for the development of a large human brain, but also stimulated the development of both communication and tool making, paving the way for the development of complex human societies. With the domestication of livestock animals and poultry, humans were more able to control and predict their access to resources. In contemporary times, domesticated animals continue to be used by humans in myriad ways, such as for food (e.g., muscle/meat, milk, and later eggs), clothing (e.g., leather, wool, hair, and silk), transportation (e.g., horses, oxen, and camels), entertainment (e.g., horse racing), and to augment human health (e.g., service dogs, passive immunization with horse serum, and animal experimentation).

Human reliance on animals and the vast benefits we derive from them are not without costs and trade offs. The most injurious costs to humans are those caused by zoonotic diseases, such as AIDS and influenza. Moreover, multiple human diseases are spread by animals, for instance, Zika virus being transmitted by its vector, mosquitoes. Parasitic animals have deleterious effects on the health and well-being of people, particularly in developing countries. Moreover, sometimes our interests are in competition with animals, such as when insects and rodents consume or contaminate our food.

Human society has had injurious effects on many wildlife animals, pushing many species to extinction. We have enabled the spread of invasive species that disrupt ecosystems, for instance, rats have decimated sea bird populations on many islands and pythons are adversely affecting the Everglades in Florida. Human activities are causing habitat destruction, fragmentation, and degradation. For example, livestock, poultry, and aquaculture excreta, also known as animal waste, can have adverse effects on the environment, such as waterways. Anthropomorphic greenhouse gas is causing climate change with such diverse effects, including the degradation of corals and decreasing polar bear population. Ultimately, human activities are causing extinction of animal species; possibly leading to an uncontrolled extinction event.

We apply ethical theories to guide our exploration of the human–animal relationship, including animal welfare and animal rights perspectives. This volume is designed to promote discussion, debate, and research to optimize the quality of human–animal relationships and to encourage ethical, sustainable, and mindful interactions.

Animal Perception Including Differences With Humans

Colin G. Scanes

University of Wisconsin–Milwaukee, Milwaukee, WI, United States

1.1 INTRODUCTION

Humans and animals use their senses to perceive the environment to locate food, water, mates, shelter, dangers, such as predators, or to communicate between other members of their species. Animals have marked differences in their sensory systems depending on their evolution and environment. This chapter will consider the senses under the following sections.

- auditory perception in Section 1.2.
- light/visual perception (sight) in Section 1.3.
- chemoreception in Section 1.4.
- tactile perception or touch in Section 1.5.
- other senses in Section 1.6.

These are discussed in detail in succeeding sections. It should be noted that wild mammals do not appear to be able to detect many environmental stimuli that humans in the developed societies are able to, by using various artificial devices. For example, we can detect nonvisual spectra, such as infrared, radio waves, and gamma radiation using instruments. The relative paucity of human senses may be viewed as a detriment or as a means of freeing up neurons and allowing the brain to be used for other tasks, such as higher-level thinking. For instance, although Neanderthals and anatomically modern humans had similarly sized brains, Neanderthals had larger visual systems and presumably associated brain areas (Pearce et al., 2013).

1.2 HEARING OR SOUND DETECTION

Ears detect the auditory frequency of sound or acoustic waves in the air (or for aquatic mammals in water). Sound varies with wave frequency (pitch) and amplitude (loudness). Low notes have a relatively low frequency of sound waves per second (hertz or Hz) while high notes have a high frequency (high kilohertz or kHz or $Hz \times 1000$). The ear is made up of the following:

- The external ear consists of the pinnae or auricles and the ear canals (also named as the external auditory canals or external auditory meatus). The ear canals end at the tympanic membrane or eardrum.

- The middle ear with three bones (the auditory ossicles: malleus, incus, and stapes) in the air-filled tympanic cavity.
- The inner ear including the semicircular canals (responsible for balance) and cochlea (responsible for hearing).

The pinnae have important roles in transmitting sound (Ballachanda, 1997) acting to do the following:

- filter sound attenuating low-frequency sound
- amplifying middle frequencies of sound

People and animals can locate the direction of sounds based on various factors including the time differences in detection by the two ears, and amplification of different frequencies together with movement of the head and body. In some mammals, the pinnae rotate partially to aid in location (discussed in Section 1.2.1).

Sound is transmitted to the middle ear; first inducing vibrations of the tympanic membrane and then the auditory ossicles in the air-filled tympanic cavity. The vibrations are detected by sensory hair cells in the organ of Corti within the inner ear. This leads to depolarization of the auditory neurons (Uemura, 2015).

1.2.1 Auditory Differences Between Human and Animals

There are marked differences between the frequencies of sound detected by the ear in humans and animals (Table 1.1). Ultrasound is above the highest range of sound/vibrations of air normally detected by the human ear (above 20 kHz). Animals including cats, rats, and mice can detect ultrasonic sound with mice and rats communicating via ultrasonic vocalization (Carruthers et al., 2013).

The pinnae can be moved (rotate) in some animals including cats, mice, rats, some breeds of dogs, and bats. This facilities detection of the spatial location of the origin of the sound (cats, Musicant et al., 1990; Phillips et al., 1982; reindeer, Flydal et al., 2001). Moreover, this is used for echolocation in bats with the pinnae focusing certain frequencies of ultrasonic sound (Kuc, 2009). The evolution of echolocation is interesting. It is thought that echolocation based on laryngeal ultrasonic vocalization was developed in the ancestors of horseshoe bats and Old World fruit bats. It was subsequently lost in the ancestors of Old World fruit bats and then regained in some with tongue clicking ultrasonic vocalization (Jones and Teeling, 2006; Razak and Fuzessery, 2015; reviewed by Jones and Holderied, 2007).

TABLE 1.1   Range of Sound Frequencies Detected Between Humans and Different Animal Species

Species	Lower limit (Hz)	Upper limit (kHz)
Human	20	20
Dog	67	45
Cat	45	78
Mouse	1000	91
Bat	2000	110
Whale	1000	123

Different units in lower and upper limits.

Based on Elert, G., 2003. Frequency range of human hearing. Available from: http://hypertextbook.com/facts/2003/ChrisDAmbrose.shtml; Fay, R.R., 1988. Hearing in Vertebrates: A Psychophysics Databook. Hill-Fay Associates, Winnetka, IL; Heffner, R.S., Heffner, H.E., 1985. Hearing range of the domestic cat. Hear. Res. 19, 85–88; Heffner, H.E., Heffner, R.S., 2007. Hearing ranges of laboratory animals. J. Am. Assoc. Lab. Anim. Sci. 46, 11–13; Warfield, D., 1973. The study of hearing in animals. In: Gay, W. (Ed.), Methods of Animal Experimentation, IV. Academic Press, London, pp. 43–143.

There are some other differences in the external ear between humans and animals. For instance, there are no pinnae in birds. Similarly, there are either no or very small pinnae in marine mammals of the order *Cetacea* (dolphins, porpoises, and whales) and *Pinnipedia* (a clade within the order *Carnivora*). In Cetaceans, the auditory canal is not open to the tympanic membrane. Instead it may be narrow and plugged with wax. In Pinnipeds, a valve closes the auditory canal.

It is interesting to also consider hearing in human ancestors with other great apes. Based on the morphology of the auditory canal, there are differences in sensitivity to frequencies 1.5–3.5 kHz between humans and chimpanzees and also with the early hominin species, *Australopithecus africanus* and *Paranthropus robustus* (Quam et al., 2015).

> Dog whistles produce a sound in the ultrasonic range above the audible sound frequency range for humans (>23,000 Hz).
>
> Dogs are used as hearing dogs for the deaf.

1.3 VISION

The protein, opsin, is involved in photoreception in all animals. Opsin is an ancient protein found in insects and vertebrates (Yokoyama and Yokoyama, 1989) whose common ancestor lived about 950 million years ago (see Chapter 10, Section 10.1). In vertebrates at least, the opsin is a G protein-linked transmembrane protein. The collision of photon(s) with light is detected by the interaction of a photon with the 11-*cis* isoform of retinal and its transformation to an all-*trans* confirmation and then induction of conformation of the opsin molecule.

There are two photoreceptor cell types in the eye: (1) rods and (2) cones. There are over 20-fold more rods than cones. The rods detect

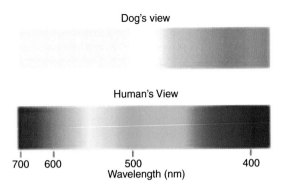

FIGURE 1.1 **Comparison of what dogs and people see.** *Source: Based on Coren, S., 2008. Can Dogs See Colors? Psychology Today. Available from: http://www.psychologytoday.com/blog/canine-corner/200810/can-dogs-see-colors.*

monochromatic light, even dim light, down to a few photons (Solovei et al., 2009). The cones are responsible for color vision and require bright light. The photopigment in rods is called rhodopsin. In the cones, the photoreceptor molecules contain forms of opsin linked to 11-*cis* retinal. In most mammals but not humans, there are two distinct photoreceptors in the cones detecting blue and red light (Okano et al., 1992) with the following absorption maxima in dogs (Neitz et al., 1989):

- blue pigment (absorption maxima 429 nm)
- red/yellow pigment (absorption maxima 555 nm)

Fig. 1.1 summarizes the approximation of a comparison of colors that a dog and a person would see.

1.3.1 Differences in Vision Between Humans and Animals

The number of rods and cones are similar at least between humans and other primates:

- There are 3.1 million cones in rhesus monkey (Wikler et al., 1990) and pigtail macaque (*Macaca nemestrina*) (Packer et al., 1989) compared 3.2 million in human (Jonas et al., 1992).

- There are 61.0 million rods in rhesus monkey and 60.5 million pigtail macaque (*M. nemestrina*) as compared to 60.1 million in human (Jonas et al., 1992).

However, there are differences between nocturnal versus diurnal species with larger eyes and greater density of rods in the retina of nocturnal species, such as cats, mice, rats, and deer, compared to diurnal species, such as cattle and humans (Solovei et al., 2009). Moreover, there are differences in the structure of the rods (Solovei et al., 2009). Thus, nocturnal species have much better night vision than diurnal species.

There are also differences in cones between humans and most mammals. In humans, there are three distinct photoreceptors detecting blue, green, and red light (Fig. 1.2) (reviewed by Yokoyama and Yokoyama, 1989).

- blue pigment (absorption maxima 420 nm)
- green pigment (absorption maxima 530 nm)
- red pigment (absorption maxima 560 nm)

The red and green pigments are found only within primates and specifically in Old World monkeys and great apes (clade *Catarrhini*). These pigments resulted from gene duplication about 35–40 million years ago in the common ancestor of Old World monkeys and great apes (Ibbotson et al., 1992; Yokoyama and Yokoyama, 1989). There appears to have been a parallel development of trichromatic vision

and independent gene duplication in the ancestors of the New World monkey—the howler monkey (Dulai et al., 1999). Moreover, the blue and red/green pigments resulted from a much earlier gene duplication about 250 million years ago (Yokoyama and Yokoyama, 1989). In contrast to the three photopigments in higher primates including humans, there are four in chickens and other birds with absorption maxima of the following: violet/ultraviolet, 415 nm; blue, 455 nm; green, 508 nm; and red, 570 nm (Okano et al., 1989, 1992) (Fig. 1.2).

1.3.2 Extraretinal Detection of Light (Pineal and Hypothalamus)

Unlike in humans and other mammals, light can be detected within the brain, and specifically the hypothalamus, of birds (reviewed by García-Fernández et al., 2015). It has been known since the 1930s that reproductive functioning is influenced by light detected in the hypothalamus (reviewed by Kuenzel et al., 2015). The opsin, vertebrate ancient opsin, present in the hypothalamus neuroendocrine cells (e.g., in the gonadotropin-releasing hormone expressing neuroendocrine cells) is different from that in the rods and cones (reviewed by Davies et al., 2012). In addition, there is light detection together with photosensitive proteins expressed in the avian pineal (Kubo et al., 2006; Okano et al., 1994).

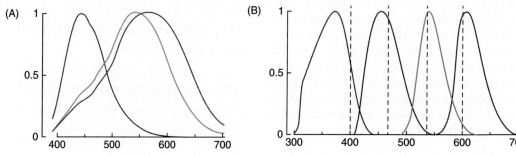

FIGURE 1.2   Comparison of spectra absorption by cone pigments between (A) human and (B) starling. *Source: Based on Osorio, D., Vorobyev, M., 2008. A review of the evolution of animal colour vision and visual communication signals. Vision Res. 48, 2042–2051. Elsevier and open access.*

1.4 CHEMICAL SENSES

Mammals have a least four distinct systems of chemoreception:

- olfactory chemoreception (sense of smell)
- vomeronasal chemoreception (detecting pheromones)
- taste or gustatory chemoreception (on tongue)
- trigeminal chemoreception (for noxious compounds, e.g., in cheek)

These will be considered next.

1.4.1 Olfactory Chemoreception

Humans and other mammals detect a vast number of volatile chemicals; these being perceived as odors or odorants. The olfactory system consists of the olfactory epithelium lining the posterior nasal cavity with neural processes to the olfactory bulb and other brain regions. The terminals of the olfactory sensory neurons are present in the olfactory epithelium. These contain specialized/modified cilia, which contain the olfactory receptors (OR) in the cell membrane. Binding of odorants to specific OR then leads to the generation of action potentials.

These contain OR. In most mammals, there are over 1000 different such receptor proteins encoded by the OR gene family with over 1000 different genes (reviewed by Buck, 2004).

1.4.1.1 Comparison of Olfaction Between Humans and Animals

Humans have about 350 functioning OR genes and over 600 OR pseudogenes (Gilad et al., 2003; reviewed by Buck, 2004). This is in contrast to the situation in most mammals where there are over 1000 OR (e.g., mice and pigs) (Gilad et al., 2003; Nguyen et al., 2012; reviewed by Buck, 2004) and over 1300 genes in dogs (Olender et al., 2004). Catarrhinine primates (Old World monkeys and the great apes) have increased numbers of OR pseudogenes (Gilad et al., 2003; Rouquier et al., 2000)

coincident with the acquisition of trichromatic vision (Gilad et al., 2004). There is evidence that the human versus the chimpanzee lineage had a higher rate of loss of OR genes (becoming pseudogenes) (Gilad et al., 2003, 2005). There are also major losses of the number of functional OR and the sense of smell in marine mammals (Cetaceans, such as whales and porpoises, and Pinniped carnivores, such as sea lions) and reptiles including sea turtles and sea snakes (Kishida and Hikida, 2010; Kishida et al., 2007). These are independent and parallel following the transition to a marine environment.

Dogs have an exquisite sense of smell due to the sensitivity of their olfactory system with over 1300 OR genes. There is evidence for polymorphisms in the OR genes with heterozygous dogs having greater detection abilities (Lesniak et al., 2008). This olfactory ability is used for sniffer dogs (see Chapter 2).

1.4.1.2 Odor Detection in Pigeons

It is perhaps surprising that the speed of return of homing pigeons was reported to be improved under smoggy polluted conditions in the Peoples Republic of China (Li et al., 2016). There is evidence that environmental odor plays a role in homing pigeons (Gagliardo et al., 2011) and that severing the olfactory nerve impairs homing (Papi et al., 1971).

1.4.2 Taste or Gustatory Chemoreception

The ability to taste ingested food and if necessary spit it out is an obvious advantage. If food or water contains a toxicant, it may be bitter and not swallowed. A sour taste may reflect spoiled food. If a food contains sugar and is sweet, it will be swallowed.

What can we or animals taste? The well-established tastes in humans and throughout the vertebrates are the following:

- sweet
- bitter

- sour (acidity or low pH)
- salty
- umami: savoriness/glutamate, such as in monosodium glutamate

Additional candidate or putative tastes include detectors for fatty acids/lipids and water (Bachmanov and Beauchamp, 2007).

Taste is mediated by taste buds or gustatory papillae on the surface of the tongue. Specific molecular receptors are responsible for taste. When a specific stimulus binds to the receptor, there is depolarization, which is transmitted to the taste receptor cell, and activation of a neuron [in branches of three cranial nerves: VII (facial), IX (glossopharyngeal), and X (vagus)] via the intervening synapses (reviewed by Bachmanov and Beauchamp, 2007; Purves, 2001). It was assumed that all molecular taste receptors span the cell membrane. This is true for the taste 1 and 2 families of receptors (T1R and T2R) are members of the superfamily of G protein-coupled receptors (Zhang et al., 2003; reviewed by Bachmanov and Beauchamp, 2007). These consist of extracellular, intracellular, and transmembrane domains, the latter across the plasma membrane. These form dimers and activate the G protein, gustducin. These receptors are in the apical microvilli of the taste cells (reviewed by Bachmanov and Beauchamp, 2007). However, some other taste stimuli including sodium, protons (pH), some bitter, and some sweet compounds enter the cell.

Sweet taste is primarily mediated by a heterodimer of two members of the type 1 taste receptor family, Tas1r2/Tas1r3; the heterodimer being two different molecules linked together as a functional unit (reviewed by Bachmanov and Beauchamp, 2007). The molecular taste receptors are the following:

- The principal umami taste receptor is a heterodimer of two members of the type 1 taste receptor family, Tas1r1/Tas1r3 (reviewed by Bachmanov and Beauchamp, 2007).

- The bitter taste receptor is mediated by the 25 members of the type 2 taste receptor families (Meyerhof et al., 2010).
- Salt is detected by epithelial-type sodium channels leading to depolarization of the receptor cell.
- Sourness is detected by proton (hydrogen ion) gated cation and chloride channels.

1.4.2.1 Is Taste Reception the Same Across Mammals and Birds?

In a number of species of the order *Carnivora* including cats, there is no taste reception to sugar and other sweeteners. This is due to the gene *Tas1r2* becoming inactive as a pseudogene (Jiang et al., 2012; Li et al., 2005, 2009). In an analogous manner, there is also loss of functional umami taste reception in bats (Zhao et al., 2012). Interestingly, based on genomic data, penguins appear to have lost the ability to taste bitter, sweet, and umami (Zhao et al., 2015). This may be explicable in terms of their diets of cephalopods and fish or their swallowing food immediately with no time in the mouth.

1.4.2.2 Presence of Molecular Taste Receptors in Other Organs

Molecular taste receptors are also present in the gastrointestinal tract. For instance, activation of umami receptors in the colon leads to peristalsis (Kendig et al., 2014) and activation of the sweet taste receptors in the small intestine leads to the release of glucagon-like peptide 1 (Jang et al., 2007) and the expression of sodium glucose cotransporters (Margolskee et al., 2007). The impact of gastrointestinal molecular taste receptors in livestock or poultry has received little attention.

1.4.3 Vomeronasal Chemoreception

The vomeronasal or Jacobson's organs (VNO) detect pheromones and other chemical signals by their binding to VNO chemoreceptor cells. [It should be noted that pheromones influence the VNO and olfactory epithelium with the response in pigs to the pheromone androstenone not mediated via the VNO (Dorries et al., 1997).] Activation of the VNO by specific pheromones leads to physiological or behavioral changes (reviewed by Dulac and Axel, 1995). There are two gene families of VNO chemoreceptors, V1R and V2R, with over 100 genes (Dulac and Axel, 1995). The V1R and V2R proteins have seven transmembrane domains and are linked to G proteins. The activation of the V1R or V2R proteins increases intracellular inositol 1,4,5-trisphosphate and hence with an ion channel of the transient receptor potential family (TRP) playing a critical role in signal transduction (reviewed by Keverne, 1999; Zhang and Webb, 2003).

The tubular VNO are found at the base of the nasal septum or in the roof of the mouth. The pheromones are detected by chemosensory neurons with axonal projections to the accessory olfactory bulb and other centers, such as in the amygdala and hypothalamus (reviewed by Dulac and Axel, 1995).

1.4.3.1 Differences Between Humans and Animals: Vomeronasal Chemoreception

There is still controversy as to whether or not adult humans even have vomeronasal organs and whether humans have functional vomeronasal chemoreceptors (reviewed by Meredith, 2001). Perhaps this is not surprising given that the existence of human pheromones is still not fully established (reviewed by Meredith, 2001). Similarly, there appears to be loss of vomeronasal signaling in bats (Zhao et al., 2011). Components of the VNO chemoreception system are not functional throughout the catarrhine primates (Old World monkeys and great apes) with V1R and TRP genes becoming nonfunctional or pseudogenes (Zhang and Webb, 2003).

It has been suggested that the reddening and swelling of the vulva at the time of estrus in Old World monkeys is a superior signal than pheromones and made the VNO dispensable (Zhang and Webb, 2003). Alternatively, the ability to discern differences between reds, oranges, and greens would have distinct advantages for fruit-eating animals during foraging (Surridge et al., 2003).

1.4.4 Trigeminal Chemoreception

Trigeminal nerve terminals can be activated by capsaicin, a chemical responsible for the sensation of spicy hot foods containing chili peppers or plants of the genus *Capsicum* (Liu and Simon, 2000). In addition, the trigeminal nerve terminals detect acid (e.g., in vinegar), acting via H^+-gated ion channels, and other noxious stimuli. These include air pollutants (e.g., sulfur dioxide and ammonia) and carbon dioxide (e.g., in cold sodas).

1.4.4.1 Animal–Human Differences in Trigeminal Responses

The trigeminal system of birds is not sensitive to capsaicin, the usual model for trigeminal chemoreception in mammals (Mason et al., 1991).

1.5 TOUCH

1.5.1 Animal–Human Differences: People Do Not Have Whiskers or Vibrissae

Vibrissae are stiff specialized hairs. They function as tactile or touch sensory organs. They can be either macrovibrissae (long whiskers) or microvibrissae (Prescott et al., 2014). Macrovibrissae are found on the face of most mammals including rodents, carnivores, and many primates (Fig. 1.3). Humans are one of the few mammals that are devoid of vibrissae. It has been suggested that vibrissae are particularly useful to provide an awareness of the environment in nocturnal species.

The movement of or shear forces in vibrissae are transduced to Merkel cells in the epidermis and then via a synapse-like process to adrenergic afferent neurons (Cha et al., 2011; Ikeda

FIGURE 1.3 **Vibrissae are sense organs in various animals, but have been lost in humans through evolution.** (*Top row*) Vibrissae by the nose of a dog. (*Middle row*) Vibrissae by the nose and mouth of a cat and rats. (*Bottom row*) Vibrissae on sea lion and walrus. *Source: From Wikimedia Commons.*

et al., 2014; Ma, 2014; Maksimovic et al., 2014). In rodents, vibration of the vibrissae plays a role in intraspecies communication (Mitchinson et al., 2011).

Based on a comparison of the human genome to that of the chimpanzee and other species, it was possible to identify deletions in the human genome responsible for the loss of the vibrissae enhancer from the human androgen receptor (McLean et al., 2011). There was also a loss of penile spines as a result of this deletion (McLean et al., 2011) leading to desensitization of the penis, longer duration of coitus, and presumably strengthening of bonding (van

Driel, 2011). While people do not have vibrissae, cancers of the Merkel cells is an uncommon but highly aggressive skin carcinoma in people (Sibley et al., 1980; Tilling and Moll, 2012) and dogs (Joiner et al., 2010).

1.6 OTHER SENSES

Birds have two magnetoreceptor systems: one based on magnetite in or near the beak, and the other in the retina based on putative magnetosensory molecules, the cryptochromes. These allow detection of the direction of the Poles and hence facilitate migration (reviewed by Mouritsen and Ritz, 2005).

References

Bachmanov, A.A., Beauchamp, G.K., 2007. Taste receptor genes. Annu. Rev. Nutr. 27, 389–414.

Ballachanda, B.B., 1997. Theoretical and applied external ear acoustics. J. Am. Acad. Audiol. l8, 411–420.

Buck, L.B., 2004. Olfactory receptors and odor coding in mammals. Nutr. Rev. 62, S184–S188.

Carruthers, I.M., Natan, R.G., Geffen, M.N., 2013. Encoding of ultrasonic vocalizations in the auditory cortex. J. Neurophysiol. 109, 1912–1927.

Cha, M., Ling, J., Xu, G.Y., Gu, J.G., 2011. Shear mechanical force induces an increase of intracellular Ca^{2+} in cultured Merkel cells prepared from rat vibrissal hair follicles. J. Neurophysiol. 106, 460–469.

Davies, W.I., Turton, M., Peirson, S.N., Follett, B.K., Halford, S., Garcia-Fernandez, J.M., Sharp, P.J., Hankins, M.W., Foster, R.G., 2012. Vertebrate ancient opsin photopigment spectra and the avian photoperiodic response. Biol. Lett. 8, 291–294.

Dorries, K.M., Adkins-Regan, E., Halpern, B.P., 1997. Sensitivity and behavioral responses to the pheromone androstenone are not mediated by the vomeronasal organ in domestic pigs. Brain Behav. Evol. 49, 53–62.

Dulac, C., Axel, R., 1995. A novel family of genes encoding putative pheromone receptors in mammals. Cell 83, 195–206.

Dulai, K.S., von Dornum, M., Mollon, J.D., Hunt, D.M., 1999. The evolution of trichromatic color vision by opsin gene duplication in New World and Old World primates. Genome Res. 9, 629–638.

Flydal, K., Hermansen, A., Enger, P.S., Reimers, E., 2001. Hearing in reindeer (*Rangifer tarandus*). J. Comp. Physiol. A 187, 265–269.

Gagliardo, A., Ioalè, P., Filannino, C., Wikelski, M., 2011. Homing pigeons only navigate in air with intact environmental odours: a test of the olfactory activation hypothesis with GPS data loggers. PLoS One 6 (8), e22385.

García-Fernández, J.M., Cernuda-Cernuda, R., Davies, W.I., Rodgers, J., Turton, M., Peirson, S.N., Follett, B.K., Halford, S., Hughes, S., Hankins, M.W., Foster, R.G., 2015. The hypothalamic photoreceptors regulating seasonal reproduction in birds: a prime role for VA opsin. Front. Neuroendocrinol. 37, 13–28.

Gilad, Y., Man, O., Glusman, G., 2005. A comparison of the human and chimpanzee olfactory receptor gene repertoires. Genome Res. 15, 224–230.

Gilad, Y., Man, O., Pääbo, S., Lancet, D., 2003. Human specific loss of olfactory receptor genes. Proc. Natl. Acad. Sci. USA 100, 3324–3327.

Gilad, Y., Przeworski, M., Lancet, D., 2004. Loss of olfactory receptor genes coincides with the acquisition of full trichromatic vision in primates. PLoS Biol. 2, E5.

Ibbotson, R.E., Hunt, D.M., Bowmaker, J.K., Mollon, J.D., 1992. Sequence divergence and copy number of the middle- and long-wave photopigment genes in Old World monkeys. Proc. R. Soc. Lond. B247, 145–154.

Ikeda, R., Cha, M., Ling, J., Jia, Z., Coyle, D., Gu, J.G., 2014. Merkel cells transduce and encode tactile stimuli to drive Aβ-afferent impulses. Cell 157, 664–675.

Jang, H.-J., Kokrashvili, Z., Theodorakis, M., Carlson, O., Kim, B.-J., Zhou, J., Kim, H., Xu, X., Chan, S., Juhaszova, M., Bernier, M., Mosinger, B., Margolskee, R., Egan, J., 2007. Gut-expressed gustducin and taste receptors regulate secretion of glucagon-like peptide-1. Proc. Natl. Acad. Sci. USA 104, 15069–15074.

Jiang, P., Josue, J., Li, X., Glaser, D., Li, W., Brand, J.G., Margolskee, R.F., Reed, D.R., Beauchamp, G.K., 2012. Major taste loss in carnivorous mammals. Proc. Natl. Acad. Sci. USA 109, 4956–4961.

Joiner, K.S., Smith, A.N., Henderson, R.A., Brawner, W.R., Spangler, E.A., Sartin, E.A., 2010. Multicentric cutaneous neuroendocrine (Merkel cell) carcinoma in a dog. Vet. Pathol. 47, 1090–1094.

Jonas, J.B., Schneider, U., Naumann, G.O., 1992. Count and density of human retinal photoreceptors. Graefes Arch. Clin. Exp. Ophthalmol. 230, 505–510.

Jones, G., Holderied, M.W., 2007. Bat echolocation calls: adaptation and convergent evolution. Proc.Biol. Sci. 274, 905–912.

Jones, G., Teeling, E.C., 2006. The evolution of echolocation in bats. Trends Ecol. Evol. 21, 149–156.

Kendig, D.M., Hurst, N., Bradley, Z.L., Mahavadi, S., Kuemmerle, J.F., Lyall, V., DeSimone, J.A., Murthy, K.S., Grider, J.R., 2014. Activation of the umami taste receptor (T1R1/T1R3) initiates the peristaltic reflex and pellet propulsion in the distal colon. Am. J. Physiol. 307, G1100–G1107.

Keverne, E.B., 1999. The vomeronasal organ. Science 286, 716–720.

Kishida, T., Hikida, T., 2010. Degeneration patterns of the olfactory receptor genes in sea snakes. J. Evol. Biol. 23, 302–310.

Kishida, T., Kubota, S., Shirayama, Y., Fukami, H., 2007. The olfactory receptor gene repertoires in secondary-adapted marine vertebrates: evidence for reduction of the functional proportions in cetaceans. Biol. Lett. 3, 428–430.

Kubo, Y., Akiyama, M., Fukada, Y., Okano, T., 2006. Molecular cloning, mRNA expression, and immunocytochemical localization of a putative blue-light photoreceptor CRY4 in the chicken pineal gland. J. Neurochem. 97, 1155–1165.

Kuc, R., 2009. Model predicts bat pinna ridges focus high frequencies to form narrow sensitivity beams. J. Acoust. Soc. Am. 125, 3454–3459.

Kuenzel, W.J., Kang, S.W., Zhou, Z.J., 2015. Exploring avian deep-brain photoreceptors and their role in activating the neuroendocrine regulation of gonadal development. Poult. Sci. 94, 786–798.

Lesniak, A., Walczak, M., Jezierski, T., Sacharczuk, M., Gawkowski, M., Jaszczak, K., 2008. Canine olfactory receptor gene polymorphism and its relation to odor detection performance by sniffer dogs. J. Hered. 99, 518–527.

Li, Z., Courchamp, F., Blumstein, D.T., 2016. Pigeons home faster through polluted air. Sci. Rep. 6, 18989.

Li, X., Glaser, D., Li, W., Johnson, W.E., O'Brien, S.J., Beauchamp, G.K., Brand, J.G., 2009. Analyses of sweet receptor gene (*Tas1r2*) and preference for sweet stimuli in species of Carnivora. J. Hered. 100 (Suppl. 1), S90–S100.

Li, X., Li, W., Wang, H., Cao, J., Maehashi, K., Huang, L., Bachmanov, A.A., Reed, D.R., Legrand-Defretin, V., Beauchamp, G.K., Brand, J.G., 2005. Pseudogenization of a sweet-receptor gene accounts for cats' indifference toward sugar. PLoS Genet. 1, 27–35.

Liu, L., Simon, S.A., 2000. Capsaicin, acid and heat-evoked currents in rat trigeminal ganglion neurons: relationship to functional VR1 receptors. Physiol. Behav. 69, 363–378.

Ma, Q., 2014. Merkel cells are a touchy subject. Cell 157, 531–533.

Maksimovic, S., Nakatani, M., Baba, Y., Nelson, A.M., Marshall, K.L., Wellnitz, S.A., Firozi, P., Woo, S.H., Ranade, S., Patapoutian, A., Lumpkin, E.A., 2014. Epidermal Merkel cells are mechanosensory cells that tune mammalian touch receptors. Nature 509, 617–621.

Margolskee, R.F., Dyer, J., Kokrashvili, Z., Salmon, K.S., Ilegems, E., Daly, K., Maillet, E.L., Ninomiya, Y., Mosinger, B., Shirazi-Beechey, S.P., 2007. T1R3 and gustducin in gut sense sugars to regulate expression of Na$^+$-glucose cotransporter 1. Proc. Natl. Acad. Sci. USA 104, 15075–15080.

Mason, J.R., Bean, N.J., Shah, P.S., Clark, L., 1991. Taxon-specific differences in responsiveness to capsaicin and several analogues: correlates between chemical structure and behavioral aversiveness. J. Chem. Ecol. 17, 2539–2551.

McLean, C.Y., Reno, P.L., Pollen, A.A., Bassan, A.I., Capellini, T.D., Guenther, C., Indjeian, V.B., Lim, X., Menke, D.B., School, B.T., Wenger, A.M., Bejerano, G., Kingsley, D.M., 2011. Human-specific loss of regulatory DNA and the evolution of human-specific traits. Nature 471, 216–219.

Meredith, M., 2001. Human vomeronasal organ function: a critical review of best and worst cases. Chem. Senses 26, 433–445.

Meyerhof, W., Batram, C., Kuhn, C., Brockhoff, A., Chudoba, E., Bufe, B., Appendino, G., Behrens, M., 2010. The molecular receptive ranges of human TAS2R bitter taste receptors. Chem. Senses 35, 157–170.

Mitchinson, B., Grant, R.A., Arkley, K., Rankov, V., Perkon, I., Prescott, T.J., 2011. Active vibrissal sensing in rodents and marsupials. Proc. Biol. Sci. 366, 3037–3048.

Mouritsen, H., Ritz, T., 2005. Magnetoreception and its use in bird navigation. Curr. Opin. Neurobiol. 5, 406–414.

Musicant, A.D., Chan, J.C., Hind, J.E., 1990. Direction-dependent spectral properties of cat external ear: new data and cross-species comparisons. J. Acoust. Soc. Am. 87, 757–781.

Neitz, J., Geist, T., Jacobs, G.H., 1989. Color vision in the dog. Vis. Neurosci. 3, 119–125.

Nguyen, D.T., Lee, K., Choi, H., Choi, M.K., Le, M.T., Song, N., Kim, J.H., Seo, H.G., Oh, J.W., Lee, K., Kim, T.H., Park, C., 2012. The complete swine olfactory subgenome: expansion of the olfactory gene repertoire in the pig genome. BMC Genomics 13, 584.

Okano, T., Fukada, Y., Artamonov, I.D., Yoshizawa, T., 1989. Purification of cone visual pigments from chicken retina. Biochemistry 28, 8848–8856.

Okano, T., Kojima, D., Fukada, Y., Shichida, Y., Yoshizawa, T., 1992. Primary structures of chicken cone visual pigments: vertebrate rhodopsins have evolved out of cone visual pigments. Proc. Natl. Acad. Sci. USA 89, 5932–5936.

Okano, T., Yoshizawa, T., Fukada, Y., 1994. Pinopsin is a chicken pineal photoreceptive molecule. Nature 372, 94–97.

Olender, T., Fuchs, T., Linhart, C., Shamir, R., Adams, M., Kalush, F., Keh, M., Lancet, D., 2004. The canine olfactory subgenome. Genomics 83, 361–372.

Packer, O., Hendrickson, A.E., Curcio, C.A., 1989. Photoreceptor topography of the retina in the adult pigtail macaque (*Macaca nemestrina*). J. Comp. Neurol. 288, 165–183.

Papi, F., Fiore, L., Fiaschi, V., Benvenuti, S., 1971. The influence of olfactory nerve section on the homing capacity of carrier pigeons. Monit. Zool. Ital. 5, 265–267.

Pearce, E., Stringer, C., Dunbar, R.I., 2013. New insights into differences in brain organization between Neanderthals and anatomically modern humans. Proc. Biol. Sci. 280, 20130168.

Phillips, D.P., Calford, M.B., Pettigrew, J.D., Aitkin, L.M., Semple, M.N., 1982. Directionality of sound pressure transformation at the cat's pinna. Hear. Res. 8, 13–28.

Prescott, T.J., Mitchinson, B., Grant, R.A., 2014. Vibrissal behavior and function. Scholarpedia 6, 6642.

Purves, D., 2001. Taste receptors and the transduction of taste signals. In: Purves, D., Augustine, G.J., Fitzpatrick, D., Katz, L.C., LaMantia, A.-S., McNamara, J.O., Williams, S.M. (Eds.), Neuroscience. second ed. Sinauer Associates, Sunderland, MA.

Quam, R., Martínez, I., Rosa, M., Bonmatí, A., Lorenzo, C., de Ruiter, D.J., Moggi-Cecchi, J., Conde Valverde, M.C., Jarabo, P., Menter, C.G., Thackeray, J.F., Arsuaga, J.L., 2015. Early hominin auditory capacities. Sci. Adv. 1, e150035.

Razak, K.A., Fuzessery, Z.M., 2015. Development of echolocation calls and neural selectivity for echolocation calls in the pallid bat. Dev. Neurobiol. 75, 1125–1139.

Rouquier, S., Blancher, A., Giorgi, D., 2000. The olfactory receptor gene repertoire in primates and mouse: evidence for reduction of the functional fraction in primates. Proc. Natl. Acad. Sci. USA 97, 2870–2874.

Sibley, R.K., Rosai, J., Foucar, E., Dehner, L.P., Bosl, G., 1980. Neuroendocrine (Merkel cell) carcinoma of the skin. A histologic and ultrastructural study of two cases. Am. J. Surg. Pathol. 4, 211–221.

Solovei, I., Kreysing, M., Lanctôt, C., Kösem, S., Peichl, L., Cremer, T., Guck, J., Joffe, B., 2009. Nuclear architecture of rod photoreceptor cells adapts to vision in mammalian evolution. Cell 137 (2), 356–368.

Surridge, A.K., Osorio, D., Mundy, N.I., 2003. Evolution and selection of trichromatic vision in primates. Trends Ecol. Evol. 51, 198–205.

Tilling, T., Moll, I., 2012. Which are the cells of origin in Merkel cell carcinoma? J. Skin Cancer 2012, 680410.

Uemura, E.E., 2015. The auditory system. In: Reece, W.O. (Ed.), Dukes' Physiology of Domestic Animals. thirteenth ed. Wiley-Blackwell, Hoboken, NJ, pp. 49–56.

van Driel, M.F., 2011. Words of wisdom. Re: human-specific loss of regulatory DNA and the evolution of human-specific traits. Eur. Urol. 60, 1123–1124.

Wikler, K.C., Williams, R.W., Rakic, P., 1990. Photoreceptor mosaic: number and distribution of rods and cones in the rhesus monkey retina. J. Comp. Neurol. 297, 499–508.

Yokoyama, S., Yokoyama, R., 1989. Molecular evolution of human visual pigment genes. Mol. Biol. Evol. 6, 186–197.

Zhang, Y., Hoon, M.A., Chandrashekar, J., Mueller, K.L., Cook, B., Wu, D., Zuker, C.S., Ryba, N.J., 2003. Coding of sweet, bitter, and umami tastes: different receptor cells sharing similar signaling pathways. Cell 112, 293–301.

Zhang, J., Webb, D.M., 2003. Evolutionary deterioration of the vomeronasal pheromone transduction pathway in catarrhine primates. Proc. Natl. Acad. Sci. USA 100, 8337–8341.

Zhao, H., Li, J., Zhang, J., 2015. Molecular evidence for the loss of three basic tastes in penguins. Curr. Biol. 25, R141–R142.

Zhao, H., Xu, D., Zhang, S., Zhang, J., 2011. Widespread losses of vomeronasal signal transduction in bats. Mol. Biol. Evol. 28, 7–12.

Zhao, H., Xu, D., Zhang, S., Zhang, J., 2012. Genomic and genetic evidence for the loss of umami taste in bats. Genome Biol. Evol. 4, 73–79.

Further Reading

Douglas, R.H., Jeffery, G., 2014. The spectral transmission of ocular media suggests ultraviolet sensitivity is widespread among mammals. Proc. Biol. Sci. 281, 20132995.

Pinc, L., Bartoš, L., Reslová, A., Kotrba, R., 2011. Dogs discriminate identical twins. PLoS One 6, e20704.

2

Animal Attributes Exploited by Humans (Nonfood Uses of Animals)

Colin G. Scanes

University of Wisconsin–Milwaukee, Milwaukee, WI, United States

2.1 INTRODUCTION

The nonfood uses of animals began with the use of antlers, bones, and so on as tools as early as before the emergence of the genus *Homo* in the development of hominids (see Chapter 5). With growing ingenuity, these tools became more sophisticated during the evolution of the various species in the genus *Homo*. The development of tools was followed by other uses of animals or animal products, such as leather and furs, for clothing almost a million years ago and the domestication of dogs aiding with hunting over 15,000 years ago (see Chapter 5). Particularly since the Neolithic, the uses of animals or their products greatly expanded and is still continuing.

This chapter will discuss the nonfood uses of vertebrate animals by people. Table 2.1 summarizes the extensive nonfood uses of animal tissues (ranging from clothing to biodiesel to biomedical uses) together with the animals themselves for transportation (horses, camels, donkeys, mules, and dogs), plowing, and other agricultural uses (oxen, donkeys, mules, and horses together with dogs for herding and guarding), and for hunting (see Section 2.6.1). In addition, animal excreta were extensively used (see Section 2.12). Uses for invertebrate animals are considered in Chapter 8.

2.2 USE OF ANIMAL SKIN: ANIMAL FIBERS (LEATHER, WOOL, FUR)

2.2.1 Introduction

Animal skins have long been valued particularly as raw materials for clothing (leather and wool), gloves (leather), shoes (leather), furniture (leather), blankets (wool), and other uses.

TABLE 2.1   Examples of Animal Tissues and Attributes Used by People Excluding for Food

Organ	Protein (other constituent)	Uses
Skin (epidermis)	Keratin	Shampoos and hair conditioners
Wool and hair	Keratin	Clothing
Hooves	Keratin	Fire extinguisher foam for planes
Skin (dermis)	Collagen	Leather
Skeletal muscle	Myosin and actin	Locomotion (e.g., horses, donkeys, oxen, and camels)
Short tendons	Collagen	Biomedical and cosmetic collagen
Adipose tissue		Triglyceride (livestock feed) Fatty acids (biodiesel, soap, candles, etc.) Glycerol (explosives)
Small intestine	Collagen in serosa	Violin strings (sheep gut but not cat gut); tennis racquet strings (cattle)
Blood	Hemoglobin	Blood meat as a slow-release fertilizer providing phosphate and nitrogen or animal feed
	Multiple	Binders
	Immunoglobulins or antibodies	Antitoxins
Bone*	Collagen	Hydrolyzed to gelatin
	Collagen	Ground to bone meal as a slow-release fertilizer providing phosphate and nitrogen or animal feed
Following ashing to bone ash (calcium phosphate)	Not applicable	Organic phosphate fertilizer and ingredient in bone china [e.g., bone ash (6 parts) used with feldspar (4 parts) and kaolin (3.5 parts)]
Sense organs	Multiple	Odor detection in dogs
Brain	Multiple	Behaviors useful to humans, such as for service, therapy, and working dogs
Endocrine organs	Multiple	Sources for hormones including insulin and thyroxine
Sex organs (primary, secondary, and characteristics)	Multiple	Hyaluronic acid for biomedical uses from male chicken combs
Heart valves	Collagen	Used for replacements in people

Leather in Protective Clothing

Leather is used as protective clothing for motorcyclists (bikers), such as protective jackets, gloves, pants, and boots. Leather was integral to armor for soldiers in the medieval period.

TABLE 2.2   Global Production of Animal Fiber-Based Products in 2013

Production	Production (million metric tons)
Sheep skins	9.19
Cattle hides	8.12
Wool, greasy	2.13
Goat skins	1.26
Buffalo hides	0.96
Horse hair	0.14

Data from FAOStats (United Nations Food and Agriculture Organization Statistics), 2015. Production/Livestock Primary. Available from: http://faostat3.fao.org/browse/Q//E.*

Table 2.2 summarizes the global production of hides and skin. Hides represent 6% of body weight in cattle and pigs but 9% for sheep (Jayathilakan et al., 2012). Hides or skins are made up of three layers:

- Epidermis: 1% of the skin with hair formed predominantly of the protein, keratin. The major product is wool or hair or feathers and related products.
- Dermis: 85% of the skin composed largely of the protein, collagen. The major product is leather after tanning.
- Subcutaneous layer (or flesh): 14% of the skin, which is removed during leather- or fur-making.

According the Economic Research Service of the US Department of Agriculture (USDA), "The drop value reflects the wholesale price that packers receive from the animal's byproducts that 'drop' off an animal's carcass when it is dressed, on a dollar per hundredweight basis" (ARS, 2017) and "Hides account for 30–75% of the byproduct drop value for cattle but very little of the drop value for hogs" (Marti et al., 2011).

2.2.2 Leather

Leather is produced from the hides or skins of animals. The sources of leather are the following:

- hides: cattle and water buffalo hides
- skins: sheep, goat, pig, and deer
- exotics: snake, alligator, crocodile, and ostrich

Leather is a flexible, strong (high tensile strength), moldable, supple, and durable product. It is also a good thermal insulator. Leather is resistant to tearing, piercing, abrasion, water, fire, and mold; it is permeable to water vapor. Leather is valued by humans for clothing and shelter and can be dyed and buffed to a shiny or velvety (in suede) appearance.

2.2.2.1 Hides and Skins

Hides and skins are considered either as a valuable by-product or coproduct in animal production (Table 2.2.). The major hide producing countries are the following:

1. China with 1.63 million metric tons in 2013 increasing from 1.33 million metric tons in 2003.
2. USA with 1.12 million metric tons of hides (predominantly from cattle) in 2013 compared to 1.09 million metric tons in 2003.
3. India with 1.04 million metric tons of hides (predominantly from water buffalo) in 2013 increasing from 0.94 million metric tons in 2003 (FAOStats, 2015).

The major goat- and sheepskin producing countries in 2013 were the following:

1. New Zealand with 7.33 million metric tons (sheepskin).
2. China with 755,000 metric tons (goat, 392,000; sheep, 364,000) (FAOStats, 2015).

2.2.2.2 *Tanning*

Definitions

1. Tanning is a "process of converting raw hides or skins into leather" (US EPA, 2015).
2. The grain side of the hide contains the hair and sebaceous glands.
3. The flesh side of the hide is thicker and softer.

Interesting Factoid: Dog Feces and Leather

Dog feces was valued and collected from the streets, for instance, in London. Dog feces was used, along with pigeon feces, in the process of tanning in the Middle Ages up to the 19th century.

Tanning is the process that chemically modifies the hides and skins (and specifically the collagen fibers) to preserve them and, in particular, to prevent the breakdown or decay of the proteins by the process of putrefaction (US EPA, 2015). After tanning, leather is durable, resistant to water, and supple. Tanning can be achieved by the following methods:

- Chromium tanning using salts of trivalent chromium (this is completely different from hexavalent chromium or chromium 6, which is highly toxic). This results in leather that is light blue until dying. Globally, 90% of leather is chromium tanned.
- Vegetable tanning using plant extracts (e.g., the bark of oak trees) containing tannin.
- Aldehyde tanning using formaldehyde or glutaraldehyde.
- Smoke tanning.
- Alum tanning.
- Synthetic chemical or syntans.
- Oil tanning with buckskin tanned using brain tissue.

The leather production process can be simplified as follows: Trimming, soaking, and washing → Fleshing → Unhairing → Baiting → Pickling → Tanning → Siding → Splitting (→ may require retanning) → Shaving → Coloring → Drying, conditioning, and milling → buffing and finishing (based on US EPA, 2015). Fig. 2.1 illustrates the traditional tanning in Fez, Morocco, upgraded leather production in Pakistan, and modern leather processing.

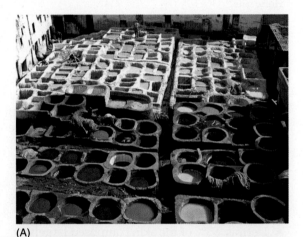

(A)

(B)

FIGURE 2.1 **Leather production.** (A) Traditional tanning in Fez, Morocco. (B) Leather production at the Leather Finishing Centre–Kasur tannery cluster in Pakistan developed in cooperation with the United Nations Industrial Development Organization. *Source: Part A: From Wikimedia Commons; Part B: Courtesy United Nations Industrial Development Organization.*

2.2.2.3 Leather Production and the Environment

According to the US Environmental Protection Agency (EPA), leather processing requires large amounts of water. Discharges from tanneries can be major sources for chemical pollution (including acids, alkalis, and tanning compounds) of waterways and hence effluent is regulated. In addition, there are gases released during processing including hydrogen sulfide and ammonia during "unhairing" and volatile organic compounds in several processes. In some developing countries, the environment regulations can be less rigorous or they are not well enforced. Moreover, hair may be disposed of by burning resulting in air pollution.

Unusual Uses of Animal Skins

1. Parchment: Parchment is made from the skin of a calf, sheep, or goat that has been limed but not tanned. Vellum is fine parchment made from the skin of a calf (Fig. 2.2). Prior to the use of paper in Europe, books and manuscripts were written on parchment because of the long life of the material (hundreds of years). The Gutenberg Bible was the first book printed with movable type in the 1450s. It was printed either on parchment or paper.

2. Gelatin: Gelatin is produced from the partial hydrolysis of collagen extracted from animal skins, ligaments, and bones and is readily digestible. Its gelling properties account for it being the key component in Jell-O, jellied meats, and canned hams. It is an excellent foam stabilizer and is used in bakery items, such as marshmallows, mousses, and cream fillings. Nonfood uses include adhesives and glues; ammunition testing (using gelatin blocks known as ballistic gelatin for examination of the performance of bullets) (Fig. 2.3); concrete foaming agent; cosmetics (e.g., lipstick); shampoos, conditioners, and other hair products; printing; rubber tires; and biomedical uses (such as gel pharmaceutical capsules and sterile sponge with these accounting for 6.5% of the total production).

3. Musical instruments: In Scotland, bagpipes are made with sheepskin with five pipes (blow pipe, chanter, bass drone, and two tenor drones); these connected to the holes remaining after the removal of the legs and head. In other parts of Europe, bagpipes are made of either sheep or goat skin. Leather or rawhide is used for the percussion surface of some drums, such as Native American and African drums.

4. Sports: Despite being referred to as pigskin, American footballs are made from leather from cattle.

2.2.2.4 Leather in Human Evolution

It is thought that human ancestors started using leather or at least animal skins about 780,000 years ago (Carbonell et al., 1999). This was based on the development of scrapers for hides (Carbonell et al., 1999). The resultant product was presumably used for shelters and/or clothes (Carbonell et al., 1999). It is argued that protection from the elements (e.g., rain) was important as by this time humans did not have body hair. Based on shifts to a gene encoding a receptor in the skin, it has been estimated that the loss of body hair in human ancestors occurred about 1.2 million years ago (Rogers et al., 2004). Parenthetically, it is noted that there is also a potential link between the

FIGURE 2.2   **Vellum deed from 1638.** *Source: From Wikimedia Commons.*

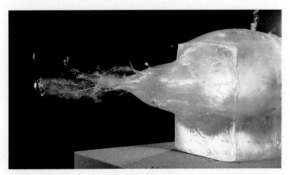

FIGURE 2.3   **Terminal fragmentation of a 0.243 projectile in ballistic gelatin.** *Source: From Wikipedia.*

loss of body hair and human success as hunters as the loss of hair facilitated dissipation of the large amounts of heat produced during hunting on the Savannah (Carrier, 1984). The loss of body hair was accompanied by the loss of the ectoparasites—lice on the bodies. It is estimated that human ancestors developed clothes over 70,000 years ago.

Body lice provide evidence for the origins of clothing. Humans are unique with the following three distinct populations of ectoparasites (most animals have one species of lice):

- Head louse *(Pediculus humanus capitis)*
- Body louse *(Pediculus humanus corporis)*
- Pubic louse (crabs) *(Phthirus pubis)*

Head and body louse are the same species but are genetically distinct with different ecologies and behaviors (Leo et al., 2005). It is thought that body lice developed from head lice once there was clothing as a refuge of the lice. Based on comparisons between the genomes of the head and body louse, the divergence time for head and body lice has been estimated. One study employed molecular clock analysis indicating that body lice originated about 72,000 ± 42,000 years ago (Kittler et al., 2003). In another study, it is estimated that the divergence of head and body louse occurred at least 83,000 years ago or as early as 170,000 years ago (Toups et al., 2011). Clothing provides advantages including insulation from high or low temperatures and from injuries (Gilligan, 2010). The development of needles about 40,000 years ago allowed clothing to be improved and much better fitting with improved insulation (Delson et al., 2000; Hoffecker, 2005).

The development of shoes was an important step in human development as they provide protection from injury, from hookworms, from extreme temperatures (e.g., frostbite), and from the adverse effects of forces of walking or running. Shoes are made from hides or other animal skins. Analysis of foot bones has provided evidence that people were wearing shoes as early as 32,000–40,000 years ago (Trinkaus, 2005; Trinkaus and Shang, 2008).

2.2.3 Wool

2.2.3.1 Overview

Different types of wool and hair used by human society are summarized in Table 2.3. Global production of wool is summarized in Table 2.2.

TABLE 2.3   Different Wools and Hairs (Natural or Bio-Based Fibers of Animal Origin Composed of Keratin) Used in Textiles

Name used in the USA	Species	Fiber diameter (micron or μm)	Major producer(s)
Alpaca	Alpaca	26	Peru, Bolivia, Chile
Angora (or mohair)	Angora goat	23–38	South Africa
Angora rabbit wool (or Angora wool or Angora fiber)	Angora rabbit	13	China
Camel hair	Bactrian camel	16–24	China, Mongolia
Cashmere wool	Cashmere goats	15	China, Mongolia
Horse hair	Horses	50–100	Argentina, Canada
Llama	Llama	20–40	South America
Merino	Merino sheep	<22	Australia
Mohair	Angora goat	23–38	South Africa
Vicuña	Vicuña	10–15	South America
Wool	Sheep	11–>40	China, Australia, New Zealand

Data from FAOStats (United Nations Food and Agriculture Organization Statistics), 2015. Production/Livestock Primary. Available from: http://faostat3.fao.org/browse/Q//E.*

Keratin

- Keratin is the major protein in tissues, such as the epidermis of the skin, hair, wool, feathers, horn, and nails. There are two principal forms: α keratin and β keratin. It is not readily digestible. Keratin from cattle hooves are used in fire extinguishing foams for planes. Feathers, a by-product of chicken production, are hydrolyzed chemically or by heat or by a bacterial enzyme, keratinase.

Terms Used for Wool

- Fleece is the wool sheared from a sheep.
- Greasy wool is shorn wool or wool following shearing.
- Clean wool is wool that has had the grease (lanolin) removed.
- Fiber diameter (wool or hair) is measured in microns or micrometers (μm, previously μ) (1 millionth of a meter).

- Coefficient of variation is the variation in wool fiber diameter (standard deviation × 100/mean). Wools and other animal fibers with high uniformity of fiber diameter or low coefficient of variation are desirable.

Crimp is the natural curl or bend or wave in an individual wool fiber. The frequency of crimps is expressed as number of crimps per inch. There can also be variation in the amplitude or depth of the crimp. In some breeds of sheep, such as merino, there can be as much as 16 crimps per inch. Wool with high regular crimp is spun into superior lighter yarn.

Flannel is wool or a wool–cotton blend or a wool–synthetic fiber blend used for clothing and bed sheets because of its ability to provide insulation.

Hank is the unit for the wool yarn and is equal to 560 yards of yarn.

- Shoddy is the technical term for recycled or remanufactured wool.
- Staple of wool is a cluster or lock of wool fibers.

Terms Used for Other Hair-Based Animal Fibers

- Bristles are hair from large pigs (boars).
- Cattle hair is used in camel-hair brushes.

- Horse hair uses the mane and tail hair of horses (usually used when combined with other fibers). It is used in fabrics, paint brushes, stuffing for upholstered furniture and mattresses, and in the strings of violin bows.
- Vicuña is the very fine hair from the Vicuña, a South American nondomesticated animal and the animal from which alpacas were domesticated. Hair is collected by a capture–shear–release system.

2.2.3.2 Production of Wool

Wool is derived from animals and is highly valued by humans. There are over 2 million metric tons of wool produced each year globally with 2.13 million metric tons of greasy wool produced in 2003 and 2013 (FAOStats, 2015). This represents only 3% of fiber production globally. The major wool-producing countries are the following:

1. China, with 471,000 metric tons produced in 2013 increasing from 338,000 metric tons in 2003.
2. Australia, with 361,000 metric tons produced in 2013 decreasing from 503,000 metric tons in 2003.
3. New Zealand, with 165,000 metric tons produced in 2013 decreasing from 230,000 metric tons in 2003.
4. Russia, with 54,600 metric tons produced in 2013 increasing from 44,600 metric tons in 2003.

In comparison, US production of wool is 20,000 metric tons.

2.2.3.3 Lanolin as a By-Product of Wool Production

Lanolin is a wax ester—a lanolin acid (mixture of 170 fatty acids) covalently linked to lanolin alcohol (e.g., sterol). Lanolin is similar in function to sebum that comes from sebaceous glands in the human skin. Sebum, like lanolin, acts to waterproof hair and skin. Because of its moisturizer or emollient (reducing water loss and itching) and antimicrobial actions, lanolin is used in personal care products, such as baby oil, diaper rash products, hemorrhoid medications, lip balm or salve for chapped lips, lotions and skin creams, medicated shampoos, makeup (lipstick, powder, foundation), nipple cream for lactating mothers, and shaving creams. Moreover, there are multiple industrial uses for lanolin including as a lubricant, leather production, textile additive as an emollient making textiles softer, in paints, varnishes, polishes, inks, and as waterproofing for concrete.

Natural is not necessarily safe. The ingestion of lanolin leads to lanolin poisoning (National Institutes of Health, 2015).

2.2.4 Fur

Definitions

- Fur is the soft hair (short and fine) together with skin stabilized by tanning and made into clothing.
- Pelage is all the hair of an animal.
- Pelt is the skin and hair.

Globally, 85% of fur traded is from fur farming with the rest from animals that are trapped. The major animals reared for their fur are the following:

- mink: America mink (*Neovison vison*) primarily but also European or Russian mink (*Mustela literal*) (in the family *Mustelidae* as are ferrets and weasels)
- fox
- chinchilla (a rodent)
- rabbit

Welfare and Fur

There are concerns about the welfare of farmed and wild animals used for fur. In the case of the former, the conditions for raising the animals may not be humane. Moreover, wild animals used for fur are not necessarily killed quickly.

Based on information for 2011, there are 52 million mink pelts sold each year (Kopenhagen Fur, 2015). In 2011, the major producers globally were the following countries and regions (Kopenhagen Fur, 2015):

1. Denmark 28%
2. China 25%
3. Netherlands 9%
4. Poland 9%
5. Commonwealth of Independent States and Baltic Countries 9%

2.2.5 Other Epidermal Products

2.2.5.1 Down

Down is the under-feather of ducks, geese, and other water fowl. Like hair, it is composed of the protein, keratin. It is used in clothing and bedding to provide insulation.

Horns, Antlers, and Tusks

Deer antlers are not true horns because when fully developed they are dead bone. Antlers are used for decoration and historically have been used as a substitute for ivory. In human prehistory, antlers were used for rooting in the soil.

Horns are boney projections of the skull surrounded by a keratin horn (similar to hooves, finger nails). Examples of the uses of cattle horns historically are as drinking horns or to carry gunpowder (powder horns).

Ivory is from tusks; these being elongated teeth (elephant and mammoth: incisors; walruses, narwhal (narwhale), and hippopotamus: canine teeth). Because of the killing and maiming of endangered animals and the illegal trade in tusks, the use of ivory has declined greatly but ivory is still illegally sought and traded. The use is banned under the Convention on International Trade in Endangered Species (CITES, 2017) (see Section 2.13).

Tortoise shells were used for ornaments and functional aspects of the material. The use is banned under CITES.

2.3 USE OF ANIMAL SKELETAL MUSCLE: ANIMALS FOR LOCOMOTION (HAULING, PLOWING, ETC.)

2.3.1 Overview

Horses have and still provide locomotive power for humans. Examples include pulling sleighs, carriages, and plows. Based on their speed and mass (inertia), horses were used by heavy cavalry in charges.

The powerful skeleton and muscles of animals are used for the following:

- Transportation: ridden or as pack animals carrying goods on their backs or pulling carts, carriages, chariots, wagons, and sleds (Figs. 2.4–2.7).
- Moving animals to markets, for instance, with cattle drives to railheads.
- Draught animals in agriculture, such as plowing, sewing, and harvesting crops (Figs. 2.4–2.7).
- Logging and moving wood (Fig. 2.7B).
- Moving water with animal traction powering pumps.
- Warfare, for example, cavalry (Figs. 2.4C and 2.6C).
- Pleasure (e.g., racing and pleasure riding).
- Historical shift from oxen to horses or donkeys (Figs. 2.4–2.6)

Horsepower

Horsepower is a unit of power, or rate of work, devised by James Watt in about 1805. It was based on the observation that a horse could lift 150 lb for 220 ft in 1 min. This leads to the horsepower being 33,000 (150 × 220) ft lb min^{-1}. The metric horsepower equals 4500 kg m min^{-1} (or 32,549 ft lb^{-1}). In contrast, cars have 150–300 horsepower and semitrucks have 500 horsepower.

The number of working or draught animals is estimated as the following (Guyo et al., 2015; Ramnaswamy, 1994):

- cattle (oxen), 246 million
- water buffalo, 60 million
- donkeys and mules, 50 million (90% in developing countries)
- horses, 27 million (60% in developing countries)
- camels, 10 million

In addition, dogs are used for pulling sleds, such as in the Iditarod Great Sled Race (Fig. 2.7C), while bantengs and yaks are used for plowing and other agricultural purposes, and llamas as pack animals transporting goods. Animals increase the ability to move goods and people in a faster manner.

Several historical examples of the use of animals for transportation are briefly covered:

1. The speed of horses is illustrated by the pony rider combinations of the Pony Express travelled 2000 miles (3200 km) in about 10 days carrying mail (Pony Express National Museum, 2015).
2. It has been estimated that 5.7 million cattle were moved in cattle drives from Texas to railheads in Kansas between 1866 and 1885 (Bronson, 1910). One trail was the Chisholm Trail running from San Antonio (Texas) to Abilene (Kansas). This was 600 miles (960 km) long with Fort Worth being an intermediate stop for provisioning. In each cattle drive, about 10 cowboys (each usually with 3 ponies) moved about 3000 cattle. Cattle drives ended when farmers had established barbed wire fences prevent free movement of cattle.
3. Horses made up to 80% of its transportation by the German army at the beginning of World War II with 3/4 million horses used following the attack of the then Soviet Union (Johnson, 2006). The use of horses predominantly slowed the advances and required 16,350 tons of feed per day (Dykman, 2015). The number of horses killed with the German army in World War II is estimated at 6.7 million (Dykman, 2015).

2.3.2 Uses of Animal Locomotion in Agriculture

The use of animal locomotion by people represented a significant advance over human

FIGURE 2.4 **Animals whose locomotive power has been employed by humans.** (A) Horse pulling a sleigh. (B) Plowing with horses. (C) Horses still being used by the US military and brewery. (D) Brewery horses pull barges on canals. Prior to the replacement of horses with engines, horses were required by breweries to walk in circles all day grinding the malt for brewing beer. (E) Pit ponies were used in coal mines in the United Kingdom with a peak number of 70,000 in 1913 together with Australia and the USA (Sunday, 2011). Ponies were replaced by mechanization. Prior to the use of ponies to haul coal, children were used! *Source: Part A–C: From Wikimedia Commons.*

(A)

(B)

FIGURE 2.5 **Donkeys have long been used to transport people and goods.** (A) Man riding a donkey. (B) Farmers with donkeys pulling two-wheeled carts still in use in the Middle East. *Source: Part A–B: From Wikimedia Commons.*

(A)

(B)

(C)

FIGURE 2.6 **Cattle locomotion has been used by humans.** (A) Farmer plowing with a donkey and cow in Near East in 1922. (B) Oxen pulling loaded cart in India. (C) Field gun being drawn by oxen in Serbia during World War I. *Source: Part A: G. Eric and Edith Matson Photograph Collection, Library of Congress Prints and Photographs Division; Part B: From Wikimedia Commons; Part C: Library of Congress Prints and Photographs Division.*

labor (person power) reducing drudgery, increasing production, and allowing increasing intensity of agriculture (Starkey, 1989). As late as the 1980s, increasing the use of draught animals was advanced as a means for economic development in sub-Saharan Africa (Onyango, 1987; Starkey, 1989).

"Quantities of draught power output...... are rarely quantified. Furthermore valuation is difficult ......" (FAO, 2011). However, it has been estimated that as late as the 1990s, draught animals were used on 52% of agricultural land in

(A)

(B)

(C)

FIGURE 2.7 **The locomotive power generated by the musculoskeletal system of animals is used by humans, for instance, camels, elephants, and dogs.** (A) Camel carrying loads. (B) Elephants working in Sri Lanka in 1907. (C) Sled dog race on snow in winter. *Source: Part A–C: From Wikimedia Commons.*

the world (Ramnaswamy, 1994) (Fig. 2.3B–C). Moreover, 82% of mechanical power in Africa is from draft animals (Ramnaswamy, 1994).

There have been marked changes with, for example, Europe, China, and North America essentially fully mechanized, but areas in Asia and Africa still employ draught animals substantially (Böttinger et al., 2014). Along with mechanization, there has been a dramatic decline in the number of draught animals with, for instance in the USA, the number of horses and mules decreasing from 26.5 million in the early 20th century to 9.2 million in 2003 (reviewed by Kilby, 2007). Moreover, along with the shift from animal power, there have been great reductions in the need for feed for the animals but increased use of petroleum products.

2.3.3 Welfare Issues and Draught Animals

There are multiple welfare issues on the use of draft animals. These include animal neglect, abuse such as beating, pain from ill-fitted harnesses, inadequate or poor nutrition, disease, parasites, and injured or diseased animals forced to work (Guyo et al., 2015; Ramnaswamy, 1994). Before mechanization, large numbers of ponies were used in coal mines (Fig. 2.1E). The 1911 Coal Mines Act (UK) or "Pit Pony's Charter" regulated the welfare of pit ponies. The conditions were undeniably harsh. Arguably, the ponies were treated much better than the miners.

2.4 USE OF ANIMAL SENSES: SNIFFER ANIMALS

2.4.1 Sniffer Dogs

Dogs have a much superior sense of smell than humans (see Chapter 1). The sensitivity of olfaction in dogs is illustrated by their ability to distinguish even between identical twins by odor (Pinc et al., 2011). In addition to their extensive history in hunting and trailing, sniffer dogs (Fig. 2.8) are used to detect the following:

- bed bugs (Pfiester et al., 2008)
- cadavers (Lasseter et al., 2003)
- diseases and/or their indicators
  - various cancers (Balseiro and Correia, 2006; Taverna et al., 2015)
  - hypoglycemia—a side effect of insulin treatment of people with type 1 diabetes (Hardin et al., 2005)
- drugs, with marijuana the easiest to detect, followed by amphetamine, cocaine, and then heroin (Jezierski et al., 2014)
- estrus in cattle based on the urine (Fischer-Tenhagen et al., 2015; Heuwieser, 2015)
- explosives (Mullis et al., 2015)

2.4.2 Sniffer Pigs

Sniffer female pigs have been traditionally used to detect truffles based on the presence of 5α-androst-16-en-3-one or androstenone. This compound is a pheromone in pigs, produced in the testes and affecting the behavior and physiology of the female. Female pigs are much more sensitive to androstenone than males (Dorries et al., 1995). However, it appears to be detected by the olfactory system rather than the vomeronasal organ (Dorries et al., 1997).

2.5 USE OF ANIMAL IMMUNE SYSTEM: PRODUCTION FOR IMMUNOGLOBULINS (SPECIFIC ANTIBODIES) FOR THERAPY

Passive immunity or the administration of animal-specific immunoglobulins is used to ameliorate acute problems of snake and other venoms, toxins from bacteria, and to immediately address possible infection after exposure, through either bites or saliva with breaks in the skin, to

rabid animals (Table 2.4) (also see Chapter 15). Animal antivenoms and antitoxins were first developed in the 1890s with the development of therapy with horse antidiphtheria toxin and antisera to snake venom (reviewed WHO, 2015c). The animal immunoglobulin employed can be in the form of antisera or as purified antibodies or as fragments of the antibodies.

2.5.1 What Are Animal Immunoglobulins Used to Treat?

2.5.1.1 Botulism

Botulism is caused by a neurotoxic protein produced largely by *Clostridium botulinum* but also by *Clostridium butyricum* and *Clostridium baratii*. It is a serious disease with the toxin being present in food or produced by bacteria present in wounds or the intestines in children (CDC, 2015).

2.5.1.2 Diphtheria

Diphtheria is a disease caused due to infection by the bacterium *Corynebacterium diphtheriae* with much of the problems due to the bacterial toxin. Until the widespread application of vaccination against the bacterial toxin, diphtheria was a serious disease and widespread with high mortalities. If someone has diphtheria, they can be treated with antisera against the bacterial toxin.

2.5.1.3 Gas Gangrene

Gas gangrene is treated by the administration of hyperimmune horse antiserum to *Clostridium septicum* toxin (NIBSC, 2015).

2.5.1.4 Rabies

Immunoglobulin against the rabies virus is administered postexposure when a possibly rabid animal bites a person or licks them when there is broken skin or by exposure to bats (WHO, 2015d). In this case, human immunoglobulin can also be used.

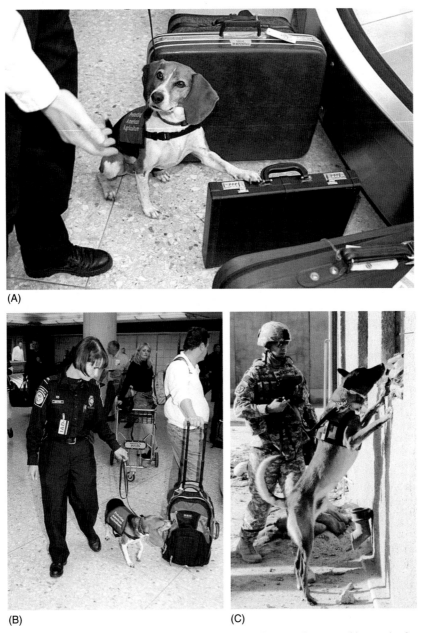

FIGURE 2.8 **The superior olfactory system of dogs is used to detect drugs, cadavers, and improvised explosive devices.**
(A) USDA's beagle brigade detects illegal importation of meats and vegetables. (B) Sniffer dog at airport. (C) Bomb sniffer dog and dog handler. *Source: Part A, C: From Wikimedia Commons; Part B: Courtesy USDA.*

TABLE 2.4   Examples of the Use of Animal Serum (Immunoglobulin) as Therapy for Toxins and Venom

Disease	Antisera raised against listed agent and organism
Toxin	
Botulism	Botulinum toxin produced by *Clostridium botulinum*
Diphtheria	Toxin from bacterium *Corynebacterium diphtheriae*
Gas gangrene	Toxin from bacterium *Clostridium perfringens*
Rabies	Virus (rabies virus)
Venom and other alien agents	
Venom inducing multiple symptoms	Venomous snakes
Venom inducing multiple symptoms	Venomous spiders
Venom inducing multiple symptoms	Scorpion stings
Anaphylaxis	Multiple organisms including bees

2.5.1.5 *Venomous Snakes*

According to the WHO (2015a), 2.5 million people are bitten by snakes and receive venom with over 100,000 people dying and over 300,000 people having limb amputations or other permanent disabilities as a result (see Chapter 16). To ameliorate the adverse effects of snake venom, antivenoms are administered. To produce antivenoms, venom is collected from poisonous snakes. Such snakes are maintained in captivity and "milked" on a regular basis to provide the source of immunogen to raise antisera. This is critically important and antivenom is included in the WHO List of Essential Medicines in places where there are venomous snakes (WHO, 2015a).

2.5.1.6 *Bee Stings and Other Allergens*

Animal antisera can be used to prevent anaphylaxic reactions against bee stings and other allergens.

2.5.2 What Animals Are Used as the Source of Antivenom, Antitoxin, or Antiorganism?

Horses, donkeys, sheep, and goats are predominantly used as a source of antisera for snake venom or antivenoms with the horse as the preferred species. This is because of their ability to produce a strong antibody response (hyperimmune), their size and the consequent ability to obtain large quantities of serum, the relative ease to maintain them, and their relative docility (WHO, 2015c). The use of sheep and goat has the disadvantages of the presence of prions if the animals are infected with prion diseases (for discussion of prion disease, see Chapter 15) and hence is discouraged (WHO, 2015c).

2.6 USE OF ANIMAL BEHAVIORS, INCLUDING SERVICE AND THERAPY ANIMALS

2.6.1 Inherited Behaviors Facilitating Utility of Working Dogs

Humans use working dogs for multiple purposes including herding sheep and cattle, hunting, retrieving, pointing, detection of odors (sniffer or tracker dogs), and as guard dogs, service and therapy dogs (discussed in Sections 2.6.2 and 2.6.4), police dogs, and sled dogs.

Working Dogs

1. Working with livestock
   a. Sheep dogs and herding dogs, such as the Border Collie, Australian Shepherd, and Shetland Sheepdogs.
   b. Cattle herding dogs, such as the Australian Cattle Dog.
   c. Livestock guardian dogs, such as the Great Pyrenees or Pyrenean Mountain Dog.
2. Hunting
   a. Pointing (e.g., German short-haired pointer).
   b. Flushing game birds using setters and spaniels, or flushing animals, such as foxes and rabbits, from burrows using terriers and Dachshunds.
   c. Retrievers including water dogs.
3. Others
   a. Search and rescue, police, and military dogs, such as the German shepherd dog.
   b. Racing dogs, such as Greyhounds and whippets.
   c. Guard dogs.
   d. Sled dogs, such as the Alaska or Siberian husky, Samoyed, Chinook, and Alaskan Malamute.

Some, possibly all, of the roles of working dogs exploit specific inherited canine behaviors or series of behaviors. Breeds of dogs were developed by breeders and fanciers in different locales predominantly between 1750 and 1900 (reviewed by Vaysse et al., 2011). Breeds exhibit marked differences in size, height, coat color, texture, and other characteristics, such as shape of ears (Akey et al., 2010; Vaysse et al., 2011). There are also tremendous breed differences in specific behaviors that are useful to humans, such as herding, pointing, tracking, and guarding (reviewed by Spady and Ostrander, 2008). As might be expected based on the development of herding breeds albeit by nonscientific breeding, behaviors related to this characteristics exhibit low to moderate heritability (Arvelius et al., 2013; Hoffmann et al., 2003). Another heritable trait is pointing (Akkad et al., 2015). Moreover, there are major differences in behaviors on the aggression–fearfulness scale with aggression either being increased in guard dogs or decreased in lap dogs (reviewed by van Rooy et al., 2014). Boldness is on the shy–bold spectrum (Starling et al., 2013).

2.6.2 Service Dogs

According to the Civil Rights Division of the US Department of Justice (US Department of Justice, 2015), service animals are defined as "dogs that are individually trained to do work or perform tasks for people with disabilities." Trained service animals must be allowed into areas where members of the public are allowed to go. They also state that "allergies and fear of dogs are not valid reasons for denying access or refusing service to people using service animals."

Service dogs are trained by accredited organizations and, at a minimum, must be "house broken" (i.e., trained to urinate and defecate outside) and respond to their owners' commands. Service animals assist people with disabilities as follows (Fig. 2.8A–B):

- Sight-impaired or blind people being guided by guide dogs or Seeing Eye dogs (Fig. 2.9).
- Hearing-impaired or deaf people being alerted by hearing or signal dogs.
- Movement/ambulation-restricted people requiring wheelchairs being pulled by service dogs.

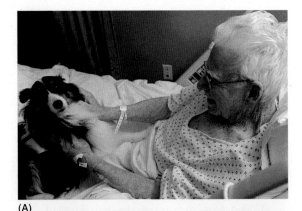

(A)

(B)

(C)

FIGURE 2.9 **Dogs can be trained as service dogs (e.g., Seeing Eye) and therapy dogs.** (A) Therapy dog. (B) Guide or Seeing Eye dog allowing a vision-impaired person to lead a fuller life. (C) Psychiatric service dog. *Source: Part A, C: From Wikimedia Commons; Part B: From Wikimedia Commons/ Shutterstock.*

- People prone to seizures being alerted and/ or protected by Seizure Response dogs.
- People with autism being interrupted from repetitive behaviors by sensory signal dogs or social signal dogs.
- People with posttraumatic stress syndrome being calmed during an anxiety attack.
- People with mental illness being, for instance, reminded to take medication or to stop self-utilization by psychiatric service dogs (Brennan and Nguyen, 2016).

2.6.3 Miniature Horses as Service Animals

Miniature horses have a shoulder height from 24 to 34 in. (61–86 cm) and weight between 70 and 100 lb (32–45 kg). Miniature horses can be considered as service animals working with people with disabilities, for instance, visual impairment. According to the Civil Rights Division of the US Department of Justice, "entities covered by the ADA must modify their policies to permit miniature horses where reasonable." Miniature horses have the advantage, or at least potential advantage, over dogs in their much longer lifespan and usefulness.

2.6.4 Therapy Animals

According to the American Veterinary Medicine Association, animals, predominantly dogs, are used in three ways to provide emotional support or comfort to people suffering from mental and physical diseases. These three approaches are the following (AVMA, 1982) (Fig. 2.8A):

- animal-facilitated therapy or animal-assisted therapy or "pet therapy"
- animal-assisted activities (AAA)
- animals resident in nursing homes

The human–animal bond is synergistic and aids the quality of life for both (reviewed by Beetz et al., 2012) irrespective of age including children (reviewed by Goddard and Gilmer, 2015) and older people with dementia (reviewed by Cherniack and Cherniack, 2014). These approaches have been reported to be effective in a number of situations but there is an absence of clinical trials. Animal therapy is proven useful in reducing anxiety in people hospitalized with psychotic disorders and mood disorders (Barker and Dawson, 1998). In children after surgery, animal-assisted therapy lowers the perception of pain (Calcaterra et al., 2015). Similarly, AAA is followed by decreases in depression and anxiety in patients receiving chemotherapy (Orlandi et al., 2007). Moreover, there is an increased oxygen saturation of the blood in patients receiving chemotherapy receiving AAA compared to those who do not (Orlandi et al., 2007).

The presence of animals in hospitals or nursing homes requires attention to mitigating the risk of increased infection and to assure welfare of the animals. The possibility of an animal-to-human transfer of pathogens can be greatly reduced by following guidelines (Lefebvre et al., 2008).

2.7 USE OF ANIMAL HORMONES

2.7.1 Insulin

Insulin from cattle and pigs has been used to ameliorate the symptoms of diabetes since the 1920s. Since the development and approval of biosynthetic human insulin for the treatment of diabetes (FDA, 1982), the use of pork/beef insulin has declined precipitously and is no longer available.

2.7.2 Premarin

Premarin is a highly prescribed orally administered estrogen. It is also available as a topical cream. This estrogenic product is isolated from the urine of pregnant horses (PREgnant MARes' urINe). It has been used for hormone replacement therapy for postmenopausal women in the USA since 1942. Premarin is manufactured by Wyeth–Ayerst. Over 9 million American postmenopausal women take Premarin to relieve the symptoms of menopause including "hot flashes," night sweats, and vaginal dryness together with preventing osteoporosis (FDA, 2009).

Premarin is a complex of sodium salts of predominantly three conjugated estrogens: estrone sulfate (about 50%) and two unique horse estrogens [equilin sulfate (3-hydroxyestra-1,3,5,7-tetraen-17-one) (25%) and eqilenin sulfate (15%)], together with the sodium salts of 17β-estradiol sulfate, 17α-estradiol sulfate, 17 β-dihydroequilin sulfate, 17α-dihydroequilin sulfate, equilenin sulfate, 17β-dihydroequilenin sulfate, 17α-dihydroequilenin sulfate, and δ-8-estrone sulfate. All are biologically active estrogens (reviewed by Bhavnani, 1998). In addition, they influence neurons, which may have implications to Alzheimer's disease (Zhao and Brinton, 2006). Generic versions of Premarin have not been approved in the USA as "precisely how these various estrogens (in Premarin) contribute to the drug's effectiveness has not been definitively determined" and thus demonstration of bioequivalence is not possible (FDA, 2009). There are thoughts of increased risks from therapy including coronary heart disease and breast cancer when Premarin is combined with progestin treatment (Rossouw et al., 2002).

Essential to the production of Premarin is a reliable source of pregnant mare urine. There are about 700 farms with 80,000 horses producing the urine. The mares have to be maintained such that the urine can be readily collected. There are concerns about the welfare of the horses (HSUS, 2015).

2.8 USE OF ANIMAL TISSUES AS XENOGRAFTS

2.8.1 Heart Valves

Valves from pigs and cattle continue to be used in surgery to replace problematic aortic and mitral valves (American Heart Association, 2015). This is because of their flexibility and strength. These xenografts normally function for 10–20 years before being replaced but problems can occur earlier (Gordon et al., 1980; Lawton et al., 2009; Lehmann et al., 2015).

2.9 USE OF ANIMAL MACROMOLECULES WITH SPECIAL CHARACTERISTICS

2.9.1 Collagen

There are multiple biomedical uses of collagen, such as drug delivery, wound dressings, and scaffolds. Biomedical-grade collagen can be purified from short tendons from cattle (Mokrejs et al., 2009; Ran and Wang, 2014). Collagens are also used in cosmetics due to their characteristic as a moisturizer (humectants) and their ability to adhere to the skin (reviewed by Peng et al., 2004). Partially hydrolyzed collagen or gelatin is also used in cosmetics.

2.9.2 Heparin

Among the animal macromolecules used in human medicine is heparin. Heparin is a sulfated glycosaminoglycan that is very useful as an anticoagulant. The raw materials for the production of heparin are usually either the intestinal mucosa of pigs or the lungs of cattle (reviewed

by Warda et al., 2003). There are concerns due to the religions of patients and risks from pathogens (reviewed by Warda et al., 2003). Alternative sources are being investigated including camel (Warda et al., 2003), turkeys (Warda et al., 2003), and mollusks (Saravanan and Shanmugam, 2010).

2.9.3 Hyaluronic Acid

Hyaluronic acid can be derived from animal tissue or bacterial fermentation. The comb of male chickens is a good source of hyaluronic acid. Intraarticular administration of hyaluronic acid has been demonstrated to have some positive effects on knee functioning in a metaanalysis of osteoarthritis (Lo et al., 2003). Hyaluronic acid is also used as a lip filler in cosmetic surgery and in eye surgery. Similarly, there were positive effects of postsurgery administration of hyaluronic acid into the joints of dogs with patellar luxation (Nganvongpanit et al., 2013).

2.9.4 Animal Products as Dietary Supplements

2.9.4.1 Overview

In the USA, the FDA regulates dietary supplements under the Dietary Supplement Health and Education Act of 1994. The FDA does not allow claims, such as reducing pain, or treats a disease for these food ingredients (FDA, 2017).

2.9.4.2 Chondroitin Sulfate

Chondroitin sulfate is a carbohydrate that is obtained from cattle cartilage. It is considered as a dietary supplement in the USA. The evidence that chondroitin sulfate is useful for people with osteoarthritis is mixed. In a multicenter, double-blind study, there were positive effects of glucosamine and chondroitin sulfate in people with osteoarthritis and moderate-to-severe pain but not lower levels of pain (Clegg et al., 2006). However, a metaanalysis of several studies in this area concluded that chondroitin sulfate was not effective (Reichenbach et al., 2007; Wandel et al., 2010). On the other hand, results from the treatment with glucosamine and chondroitin together with exercise were encouraging in addressing the pain and progression of osteoarthritis (Sterzi et al., 2016). In contrast, the available evidence, albeit limited, supports positive effects of two carbohydrates, glucosamine hydrochloride and chondroitin sulfate, in dogs. Combined treatment of osteoarthritic dogs with glucosamine hydrochloride and chondroitin sulfate was followed by reductions in pain and the severity of osteoarthritis together with improvements in weight-bearing ability (Gupta et al., 2012; McCarthy et al., 2007).

2.10 USE OF ANIMAL TOXINS AS DRUGS: EMPLOYMENT OF UNIQUE ANIMAL PROTEINS

There are animal toxins that are the basis of human drugs (reviewed by Utkin, 2015). Examples include the following:

- Prialt (Ziconotide), a neurotoxic peptide from cone snail venom used as an analgesic anticoagulants.
- Integrilin (Eptifibatide) from the venom of the southeastern pygmy rattlesnake *Sistrurus miliarius barbouri* is used as an anticoagulant.
- Aggrastat (Triofiban) from the saw-scaled viper *Echis carinatus*.

The latter two are used as an anticoagulant for thromboses (blood clots formation).

2.11 USES OF ANIMAL ADIPOSE TISSUE (FAT): NONFOOD USES

2.11.1 Overview

Animal fats are considered as by-products of animal production with fat representing 3%–4%

of body weight (Jayathilakan et al., 2012). There are food and nonfood uses of animal fat.

2.11.2 Rendering

Rendering converts "waste" animal tissue into useable products based either on protein or fat with 26.8 million metric tons of tissue rendered per year in the USA (National Renders Association, 2015). Rendering minimizes potential biohazards (Aberle et al., 2001) from animal waste from packing plants including offal and carcasses deemed unwholesome, fat trimmings from butcher shops, expired meat and meat products from grocery stores, restaurant grease, condemned carcasses, and dead animals. Protein products of rendering include meat and bone meals and these are used in livestock and poultry feeds and in pet food. Oil (fat, greases, and tallow) products include the following (Swisher, 2015):

- Products edible to humans and that have undergone considerable processing and purification:
  - Predominantly edible tallow (fat derived from cattle or sheep), 0.74 million metric tons produced in the USA in 2014.
  - Edible lard (pig derived), 62,300 metric tons produced in the USA in 2014.
- Products inedible to humans
  - Inedible tallow (cattle or sheep derived), 1.36 million metric tons produced in the USA in 2014.
  - Choice white grease (pig), 0.58 million metric tons produced in the USA in 2014.
  - Poultry fat, 0.50 million metric tons per year.
  - Greases from animals and vegetable oils (yellow grease), 0.93 million metric tons produced in the USA in 2014.
  - White grease, 0.60 million metric tons per year.
  - Tallow, 2.20 million metric tons per year.

Inedible oils are used to produce lubricants, soaps and detergents, candles, inks, personal care items; as a raw material for the production of biodiesel; or as a substitute No. 2 or No. 6 fuel oil or natural gas for the production of steam. About 35% of lard is employed for nonfood uses, such as production of paints, varnishes, resins, plastics, lubricants, and biodiesel (see Section 2.11.3) (USDA ERS, 2015). In the European Union, production of animal fat was 2.7 million metric tons in 2011 (Alm, 2015). Of this, 33% goes to biodiesel, 26% to feed, 22% to soaps and oleo, 11% to pet food, 4% to food, and 2% to other uses (Alm, 2015). One of the advantages of adding fats to animal feed is reducing dust.

Other Uses of Animal Fat

1. Fatty acids: Industrial uses for fats/fatty acids include soap, cosmetics, paints, industrial oil and lubricants, crayons, insecticides floor wax, creams and lotions, shaving cream, biodegradable detergents and hair conditioner, and other oleochemical products (McGlashan, 2006). Soaps are produced by the saponification of animal or vegetable triglycerides (from the Latin *Sapo* or soap):
   Esterified 3 fatty acids ~ glyceride + NaOH (or KOH) → fatty acid Na(K) salt + glycerol fat (triglyceride)
2. Glycerol: Glycerol (also known as glycerin or glycerine) can be used in livestock and poultry feed. For example, glycerol is used for the production of the explosive, trinitroglycerine or TNT. Possible future uses of glycerol include the production of hydrogen (McGlashan, 2006).

2.11.3 Biodiesel

In the USA, 1.27 billion gallons of biodiesel (B100) (4.8 billion liters) were produced in 2014 (US Energy Information Administration, 2015). Biodiesel can also be blended with conventional diesel from petroleum into B20 (with 20% biodiesel and 80% diesel from petroleum) or B5 (5% conventional diesel) or B2 (2% biodiesel). Biodiesel provides about 8% less per unit volume of fuel (Alternative Fuels Data Center, 2015).

Fatty acids + methanol → methylated fatty acid

Biodiesel is produced by the methylation of fatty acids in triglyceride of fats and oils. The source (i.e., feedstock) is predominantly from soybeans (USA) and rapeseed or sunflower seed (European Union). An additional and cost-effective feedstock is inedible animal fats (Nelson and Schrock, 1994) or surplus edible fats with about 13% of biodiesel produced in the USA being derived from animal fat (Centric Consulting Group, 2014). About a quarter of rendered fat is used for biodiesel (Brazil, 25%; European Union, 28%; USA, 25%) (Swisher, 2015). The use of inedible animal fat is supported based on energetics, economics depending on the price of oil (Nelson and Schrock, 1994), and sustainability (Mata et al., 2009). There are over 80% greenhouse gas savings from the use of waste animal fat for biodiesel and no land use changes (USDA FAS, 2015). Moreover, in the European Union, they do not count in the 7% cap on biofuels from crops (USDA FAS, 2015).

2.12 ANIMAL EXCRETA

2.12.1 Animal Excreta as Fertilizer

It has been long known that the application of animal excreta to land increases the yield of crops.

Indeed, Knowlton and Cobb (2006) clearly articulated that "animal manure can be a valuable resource for farmers, providing nutrients, improving soil structure, and increasing vegetative cover to reduce erosion potential." While undoubtedly important, globally, organic manure contributes less than a quarter of the nitrogen added to the soil for crop production (Smil, 1999).

In an analysis of 14 field trials, it was concluded that chemical fertilizers or manure were equally effective in increasing crop yield but manure was more effective in increasing soil organic matter together with potassium, phosphate, calcium, and magnesium (Edmeades, 2003). In addition, manure application was associated with higher porosity, hydraulic conductivity, and aggregate stability of the soil (Edmeades, 2003). Similarly, the application of manure in relatively arid areas increased crop yields but also improved efficient use of rainfall (Hati et al., 2006; Wang et al., 2016). The application of manure increases the carbon storage in the soil (Ren et al., 2014). With the intensification of livestock and poultry production, the amounts of excreta and distance from the field for application increasingly become a problem (Naylor et al., 2005).

2.12.2 Cattle Waste for Fuel

In the USA in the 19th century, dry buffalo and later cattle manure was used extensively for cooking and heat by the Plains Native Americans, people on trails to the West, trappers, and settlers (Welsch Dannebrog, 2007). Even today, dried cow feces is still used for cooking and heat by millions of farmers particularly in Africa and Asia with it being estimated that 300 million metric tons of dung is being used as fuel in India alone (Sansoucy, 2007). Cow feces are also used as plaster for buildings (Sansoucy, 2007).

2.13 ILLEGAL USES OF ANIMAL PRODUCTS

Despite the stringent restrictions under the CITES developed to protect endangered species, there is still an illegal trade in many animal products, such as tusks and horns. Bone, gall bladder, and other tissues from tigers together with rhinoceros horn and organs from multiple wild species are prized by some Asian folk medicine. The widespread poaching is threatening endangered species (Corwin, 2012). Illegal trading in animals and plants (predominantly trees) is estimated by the World Wildlife Federation to be a US$19 billion "industry" (WWF, 2012). This is the fourth largest criminal operation globally only exceeded by human trafficking, illegal drugs, and counterfeiting (WWF, 2012). Alternative sources for some of the animal tissues or products may be farming. However, while bears are farmed and fistulated to supply bear bile for Chinese traditional medicine, the conditions have been described as inhumane (Feng et al., 2009).

2.14 CONCLUSIONS

Humans have employed animal products and attributes for nonfood uses throughout human evolution. It might be argued that human uses of animals is at a peak or is declining due to mechanization (e.g., for agriculture and transportation) or technology. The latter is supported by the shift from leather and wool with the development of synthetic fibers. However, production of wool and leather continues to increase.

References

Aberle, E.D., Forrest, J.C., Gerrard, D.E., Mills, E.W., 2001. Principles of Meat Science, fourth ed. Kendall/Hunt Publishing, Dubuque, IA.

Akey, J.M., Ruhe, A.L., Akey, D.T., Wong, A.K., Connelly, C.F., Madeoy, J., Nicholas, T.J., Neff, M.W., 2010. Tracking footprints of artificial selection in the dog genome. Proc. Natl. Acad. Sci. USA 107, 1160–1165.

Akkad, D.A., Gerding, W.M., Gasser, R.B., Epplen, J.T., 2015. Homozygosity mapping and sequencing identify two genes that might contribute to pointing behavior in hunting dogs. Canine Genet. Epidemiol. 2, 5.

Alm, M., 2015. Animal fats. The American Oil Chemist's Society. Available from: http://lipidlibrary.aocs.org/Oils-Fats/content.cfm?ItemNumber=40320.

Alternative Fuels Data Center, 2015. Available from: http://www.afdc.energy.gov/fuels/biodiesel_blends.html.

American Heart Association, 2015. Available from: http://www.heart.org/HEARTORG/Conditions/More/Heart-ValveProblemsandDisease/Types-of-Replacement-Heart-Valves_UCM_451175_Article.jsp#.

ARS, 2017. USDA. Available from: https://www.ars.usda.gov/is/np/alwayssomethingnew/Animal32.pdf.

Arvelius, P., Malm, S., Svartberg, K., Strandberg, E., 2013. Measuring herding behavior in Border Collie-effect of protocol structure on usefulness for selection. J. Vet. Behav. 8, 9–18.

AVMA, 1982. Guidelines for animal-assisted activity, animal-assisted therapy and resident animal programs. Available from: https://ebusiness.avma.org/files/productdownloads/guidelines_AAA.pdf.

Balseiro, S.C., Correia, H.R., 2006. Is olfactory detection of human cancer by dogs based on major histocompatibility complex-dependent odour components?—A possible cure and a precocious diagnosis of cancer. Med. Hypotheses 66, 270–272.

Barker, S.B., Dawson, K.S., 1998. The effects of animal-assisted therapy on anxiety ratings of hospitalized psychiatric patients. Psychiatr. Serv. 49, 797–801.

Beetz, A., Uvnäs-Moberg, K., Julius, H., Kotrschal, K., 2012. Psychosocial and psychophysiological effects of human-animal interactions: the possible role of oxytocin. Front. Psychol. 3, 234.

Bhavnani, B.R., 1998. Pharmacokinetics and pharmacodynamics of conjugated equine estrogens: chemistry and metabolism. Proc. Soc. Exp. Biol. Med. 217, 6–16.

Böttinger, S., Doluschitz, R., Klaus, J. Jenane, C., Samarakoon, N., 2014. Agricultural development and mechanization in 2013. A comparative survey at a global level. In: Fourth World Summit on Agriculture Machinery. B%C3%B6ttinger-et-al_2013_Agricultural-Development-and-Mechanization-in-2013.pdf.

Brennan, J., Nguyen, V., 2016. Service Animals and emotional support animals: where are they allowed and under what conditions? The ADA National Network. Available from: https://adata.org/publication/service-animals-booklet.

Bronson, E.B., 1910. The Red-blooded Heroes of the Frontier. A.C. McClurg & Company. Available from: Googlebooks.

Calcaterra, V., Veggiotti, P., Palestrini, C., De Giorgis, V., Raschetti, R., Tumminelli, M., Mencherini, S., Papotti, F., Klersy, C., Albertini, R., Ostuni, S., Pelizzo, G., 2015. Post-operative benefits of animal-assisted therapy in pediatric surgery: a randomised study. PLoS One 10, e0125813.

Carbonell, E., Garcia-Anton, M.D., Mallol, C., Mosquera, M., Olle, A., Rodriguez, X.P., Sahnouni, M., Sala, R., Verges, J.M., 1999. The TD6 level lithic industry from Gran Dolina, Atapuerca (Burgos, Spain): production and use. J. Hum. Evol. 37, 653–693.

Carrier, D.R., 1984. The energetic paradox of human running and hominid evolution. Curr. Anthropol. 25, 483–495.

CDC (Centers of Disease Control), 2015. National Center for Emerging and Zoonotic Infectious Diseases. Available from: http://www.cdc.gov/nczved/divisions/dfbmd/diseases/botulism/.

Centric Consulting Group for the National Biodiesel Board, 2014. Biodiesel demand for animal fats and tallow generates an additional revenue stream for the livestock industry. Available from: http://biodiesel.org/docs/default-source/news---supporting-files/animal-fats-and-tallow-bd-demand-impact-report.pdf?sfvrsn=2.

Cherniack, E.P., Cherniack, A.R., 2014. The benefit of pets and animal-assisted therapy to the health of older individuals. Curr. Gerontol. Geriatr. Res., 623203.

CITES (Convention on International Trade in Endangered Species of Wild Fauna and Flora), 2017. Available from: https://cites.org/eng/disc/what.php.

Clegg, D.O., Reda, D.J., Harris, C.L., Klein, M.A., O'Dell, J.R., Hooper, M.M., Bradley, J.D., Bingham, 3rd, C.O., Weisman, M.H., Jackson, C.G., Lane, N.E., Cush, J.J., Moreland, L.W., Schumacher, Jr., H.R., Oddis, C.V., Wolfe, F., Molitor, J.A., Yocum, D.E., Schnitzer, T.J., Furst, D.E., Sawitzke, A.D., Shi, H., Brandt, K.D., Moskowitz, R.W., Williams, H.J., 2006. Glucosamine, chondroitin sulfate, and the two in combination for painful knee osteoarthritis. N. Engl. J. Med. 354, 795–808.

Corwin, J., 2012. Trafficking in tragedy: the toll of illegal wildlife trade. Available from: http://iipdigital.usembassy.gov/st/english/publication/2012/11/20121116138811.html#ixzz46HwN8kvc.

Delson, E., Tattersall, I., Van Couvering, J., Brooks, A., 2000. Encyclopedia of Human Evolution and Prehistory. Garland, New York, NY.

Dorries, K.M., Adkins-Regan, E., Halpern, B.P., 1995. Olfactory sensitivity to the pheromone, androstenone, is sexually dimorphic in the pig. Physiol. Behav. 57, 255–259.

Dorries, K.M., Adkins-Regan, E., Halpern, B.P., 1997. Sensitivity and behavioral responses to the pheromone androstenone are not mediated by the vomeronasal organ in domestic pigs. Brain Behav. Evol. 49, 53–62.

Dykman, J.T., 2015. The Soviet Experience in World War Two. The Eisenhower Institute, Gettysburg College.

Available from: http://www.eisenhowerinstitute.org/about/living_history/wwii_soviet_experience.dot.

Edmeades, D.C., 2003. The long-term effects of manures and fertilisers on soil productivity and quality: a review. Nutr. Cycl. Agroecosyst. 66, 165–180.

FAO, 2011. Animal Production and Health. Available from: http://www.fao.org/docrep/014/i2294e/i2294e00.pdf.

FAOStats (United Nations Food and Agriculture Organization Statistics), 2015. Production/Livestock Primary. Available from: http://faostat3.fao.org/browse/Q/*/E.

FDA (Food and Drug Administration), 1982. Human insulin receives FDA approval. FDA Drug Bull. 12, 18–19.

FDA, 2009. FDA Backgrounder on Conjugated Estrogens. Available from: http://www.fda.gov/Drugs/DrugSafety/InformationbyDrugClass/ucm168838.htm.

FDA, 2017. Available from: http://www.fda.gov/Food/DietarySupplements/.

Feng, Y., Siu, K., Wang, N., Ng, K.-M., Tsao, S.-W., Nagamatsu, T., Tong, Y., 2009. Bear bile: dilemma of traditional medicinal use and animal protection. J. Ethnobiol. Ethnomed. 5, 2.

Fischer-Tenhagen, C., Johnen, D., Le Danvic, C., Gatien, J., Salvetti, P., Tenhagen, B.A., Hardin, D.S., Anderson, W., Cattet, J., 2015. Dogs can be successfully trained to alert to hypoglycemia samples from patients with type 1 diabetes. Diabetes Ther. 6 (4), 509–517.

Gilligan, I., 2010. The prehistoric development of clothing: archeological implications of a thermal model. J. Archaeol. Method Theory 17, 15–80.

Goddard, A.T., Gilmer, M.J., 2015. The role and impact of animals with pediatric patients. Pediatr. Nurs. 41, 65–71.

Gordon, M.H., Walters, M.B., Allen, P., Burton, J.D., 1980. Calcific stenosis of a glutaraldehyde-treated porcine bioprosthesis in the aortic position. J. Thorac. Cardiovasc. Surg. 80, 788–791.

Gupta, R.C., Canerdy, T.D., Lindley, J., Konemann, M., Minniear, J., Carroll, B.A., Hendrick, C., Goad, J.T., Rohde, K., Doss, R., Bagchi, M., Bagchi, D., 2012. Comparative therapeutic efficacy and safety of type-II collagen (UC-II), glucosamine and chondroitin in arthritic dogs: pain evaluation by ground force plate. J. Anim. Physiol. Anim. Nutr. 96, 770–777.

Guyo, S., Legesse, S., Tonamo, A., 2015. A review on welfare and management practices of working equines. Global J. Anim. Sci. Livest. Prod. Anim. Breed. 3, 203–209.

Hardin, D.S., Anderson, W., Cattet, J., 2015. Dogs can be successfully trained to alert to hypoglycemia samples from patients with type 1 diabetes. Diabetes Ther. 6, 509–517.

Hati, K., Mandal, K., Misra, A., Ghosh, P., Bandyopadhyay, K., 2006. Effect of inorganic fertilizer and farmyard manure on soil physical properties, root distribution, and water-use efficiency of soybean in Vertisols of central India. Bioresour. Technol. 97, 2182–2188.

Heuwieser, W., 2015. Validation of bovine oestrous-specific synthetic molecules with trained scent dogs; similarities between natural and synthetic oestrous smell. Reprod. Domest. Anim. 50, 7–12.

Hoffecker, J.F., 2005. Innovation and technological knowledge in the Upper Paleolithic of northern Eurasia. Evol. Anthropol. 14, 186–198.

Hoffmann, U., Hamann, H., Distl, O., 2003. Genetic analysis of markers of the performance test for herding dogs. 1. Performance traits. Berl. Münch. Tierärztl. Wochenschr. 116, 81–89.

HSUS (Humane Society of the United States), 2015. Prescription for cruelty by refusing drugs like Premarin, women can end an industry that still holds thousands of mares captive. Available from: http://www.humanesociety.org/news/magazines/2015/03-04/premarin.html.

Jayathilakan, K., Sultana, K., Radhakrishna, K., Bawa, A.S., 2012. Utilization of byproducts and waste materials from meat, poultry and fish processing industries: a review. J. Food Sci. Technol. 49, 278–293.

Jezierski, T., Adamkiewicz, E., Walczak, M., Sobczyńska, M., Górecka-Bruzda, A., Ensminger, J., Papet, E., 2014. Efficacy of drug detection by fully-trained police dogs varies by breed, training level, type of drug and search environment. Forensic Sci. Int. 237, 112–118.

Johnson, P.L., 2006. Horses of the German Army in World War II. Schiffer Military History Book, Schiffer Publishing, Atglen, Pennsylvania, USA.

Kilby, E.R., 2007. The demographics of the U.S. equine population. In: Salem, D.J., Salem, D.J., Rowan, A.N. (Eds.), The State of the Animals IV: 2007. Humane Society Press, Washington, DC, pp. 175–205.

Kittler, R., Kayser, M., Stoneking, M., 2003. Molecular evolution of *Pediculus humanus* and the origin of clothing. Curr. Biol. 13, 1414–1417.

Knowlton, K.F., Cobb, T.D., 2006. ADSA Foundation Scholar Award: implementing waste solutions for dairy and livestock farms. J. Dairy Sci. 89, 1372–1383.

Kopenhagen Fur, 2015. Available from: http://furcommission.com/world-mink-output-climbs-for-second-straight-year/.

Lasseter, A.E., Jacobi, K.P., Farley, R., Hensel, L., 2003. Cadaver dog and handler team capabilities in the recovery of buried human remains in the southeastern United States. J. Forensic Sci. 48, 617–621.

Lawton, J.S., Moazami, N., Pasque, M.K., Moon, M.R., Damiano, Jr., R.J., 2009. Early stenosis of Medtronic Mosaic porcine valves in the aortic position. J. Thorac. Cardiovasc. Surg. 137, 1556–1557.

Lehmann, S., Merk, D.R., Etz, C.D., Oberbach, A., Uhlemann, M., Emrich, F., Funkat, A.K., Meyer, A., Garbade, J., Bakhtiary, F., Misfeld, M., Mohr, F.W., 2015. Porcine xenograft for aortic, mitral and double valve replacement:

long-term results of 2544 consecutive patients. Eur. J. Cardiothorac. Surg. 49 (4), 1150–1156.

Lefebvre, S.L., Golab, G.C., Christensen, E., Castrodale, L., Aureden, K., Bialachowski, A., Gumley, N., Robinson, J., Peregrine, A., Benoit, M., Card, M.L., Van Horne, L., Weese, J.S., Writing Panel of Working Group, 2008. Guidelines for animal-assisted interventions in health care facilities. Am. J. Infect. Control 36, 78–85.

Leo, N.P., Hughes, J.M., Yang, X., Poudel, S.K., Brogdon, W.G., Barker, S.C., 2005. The head and body lice of humans are genetically distinct (*Insecta: Phthiraptera, Pediculidae*): evidence from double infestations. Heredity (Edinb.) 95, 34–40.

Lo, G.H., LaValley, M., McAlindon, T., Felson, D.T., 2003. Intra-articular hyaluronic acid in treatment of knee osteoarthritis: a meta-analysis. J. Am. Med. Assoc. 290, 3115–3121.

Marti, D.L., Johnson, R.J., Mathews, K.H., 2011. Where's the not meat? By-products from beef and pork production. Economic Research Service. USDA. Available from: http://www.ers.usda.gov/media/147867/ldpm20901.pdf.

Mata, T.M., Cardoso, N., Ornelas, M., Neves, S., Caetano, N.S., 2009. Sustainable production of biodiesel from tallow, lard and poultry fat and its quality evaluation. Available from: http://www.aidic.it/CISAP4/webpapers/46Mata.pdf.

McCarthy, G., O'Donovan, J., Jones, B., McAllister, H., Seed, M., Mooney, C., 2007. Randomised double-blind, positive-controlled trial to assess the efficacy of glucosamine/chondroitin sulfate for the treatment of dogs with osteoarthritis. Vet. J. 174, 54–61.

McGlashan, S.A., 2006. Industrial and energy uses of animal by-products. In: Meeker, D.L. (Ed.), Essential Rendering: All About the Animal By-products Industry. National Renderers Association, Alexandria, VA, pp. 229–243.

Mokrejs, P., Langmaier, F., Mladek, M., Janacova, D., Kolomaznik, K., Vasek, V., 2009. Extraction of collagen and gelatine from meat industry by-products for food and non food uses. Waste Manage. Res. 27, 31–37.

Mullis, R.A., Witzel, A.L., Price, J., 2015. Maintenance energy requirements of odor detection, explosive detection and human detection working dogs. PeerJ. 3, e767.

National Institutes of Health, 2015. Lanolin poisoning. Available from: http://www.nlm.nih.gov/medlineplus/ency/article/002663.htm.

National Renders Association, 2015. Available from: http://www.nationalrenderers.org/environmental/.

Naylor, R., Steinfeld, H., Falcon, W., Galloway, J., Smil, V., Bradford, E., Alder, J., Mooney, H., 2005. Losing the links between livestock and land. Science 310, 1621–1622.

Nelson, R.G., Schrock, M.D., 1994. Energetics and economics of producing biodiesel from beef tallow look positive. Economic Research Service (USDA). Available from: http://www.ers.usda.gov/media/935863/ius3g_002.pdf.

Nganvongpanit, K., Boonsri, B., Sripratak, T., Markmee, P., 2013. Effects of one-time and two-time intra-articular injection of hyaluronic acid sodium salt after joint surgery in dogs. J. Vet. Sci. 14, 215–222.

NIBSC (National Institute for Biological Standards and Control) (UK), 2015. Available from: http://www.nibsc.org/documents/ifu/64-014.pdf.

Onyango, S.O., 1987. Reducing present constraints to the use of animal power in Kenya. FAO Corporate Document Repository. Available from: http://www.fao.org/wairdocs/ilri/x5455b/x5455b2b.htm.

Orlandi, M., Trangeled, K., Mambrini, A., Tagliani, M., Ferrarini, A., Zanetti, L., Tartarini, R., Pacetti, P., Cantore, M., 2007. Pet therapy effects on oncological day hospital patients undergoing chemotherapy treatment. Anticancer Res. 27, 4301–4303.

Pinc, L., Bartoš, L., Reslová, A., Kotrba, R., 2011. Dogs discriminate identical twins. PLoS One 6, e20704.

Peng, Y., Glattauer, V., Werkmeister, J.A., Ramshaw, J.A., 2004. Evaluation for collagen products for cosmetic application. J. Cosmet. Sci. 55, 327–341.

Pfiester, M., Koehler, P.G., Pereira, R.M., 2008. Ability of bed bug-detecting canines to locate live bed bugs and viable bed bug eggs. J. Econ. Entomol. 101, 1389–1396.

Pony Express National Museum, 2015. Available from: http://ponyexpress.org/history/.

Ramnaswamy, N.S., 1994. Draught animals and welfare. Rev. Sci. Tech. 13, 195–216.

Ran, X.G., Wang, L.Y., 2014. Use of ultrasonic and pepsin treatment in tandem for collagen extraction from meat industry by-products. J. Sci. Food Agric. 94, 585–590.

Reichenbach, S., Sterchi, R., Scherer, M., Trelle, S., Bürgi, E., Bürgi, U., Dieppe, P.A., Jüni, P., 2007. Meta-analysis: chondroitin for osteoarthritis of the knee or hip. Ann. Intern. Med. 146, 580–590.

Ren, T., Wang, J., Chen, Q., Zhang, F., Lu, S., 2014. The effects of manure and nitrogen fertilizer applications on soil organic carbon and nitrogen in a high-input cropping system. PLoS One 9, e97732.

Rogers, A.R., Iltis, D., Wooding, S., 2004. Genetic variation at the MC1R locus and the time since loss of human body hair. Curr. Anthropol. 45, 105–108.

Rossouw, J.E., Anderson, G.L., Prentice, R.L., LaCroix, A.Z., Kooperberg, C., Stefanick, M.L., Jackson, R.D., Beresford, S.A., Howard, B.V., Johnson, K.C., Kotchen, J.M., Ockene, J., 2002. Risks and benefits of estrogen plus progestin in healthy postmenopausal women: principal results from the Women's Health Initiative randomized controlled trial. J. Am. Med. Assoc. 288, 321–333.

Sansoucy, R., 2007. Livestock—a driving force for food security and sustainable development. Available from: http://www.fao.org/docrep/v8180t/v8180t07.htm.

Saravanan, R., Shanmugam, A., 2010. Isolation and characterization of low molecular weight glycosaminoglycans from marine mollusc *Amussium pleuronectus* (linne) using chromatography. Appl. Biochem. Biotechnol. 160, 791–799.

Smil, V., 1999. Nitrogen in crop production: an account of global flows. Global Biogeochem. Cycles 13, 647–662.

Spady, T.C., Ostrander, E.A., 2008. Canine behavioral genetics: pointing out the phenotypes and herding up the genes. Am. J. Hum. Genet. 82, 10–18.

Starkey, P., 1989. Animal traction for agricultural development in West Africa: production, impact, profitability and constraints. FAO Corporate Document Repository. Available from: http://www.fao.org/wairdocs/ilri/x5455b/x5455b0g.htm.

Starling, M.J., Branson, N., Thomson, P.C., McGreevy, P.D., 2013. "Boldness" in the domestic dog differs among breeds and breed groups. Behav. Processes 97, 53–62.

Sterzi, S., Giordani, L., Morrone, M., Lena, E., Magrone, G., Scarpini, C., Milighetti, S., Pellicciari, L., Bravi, M., Panni, I., Ljoka, C., Bressi, F., Foti, C., 2016. The efficacy and safety of a combination of glucosamine hydrochloride, chondroitin sulfate and bio-curcumin with exercise in the treatment of knee osteoarthritis: a randomized, double blind, placebo-controlled study. Eur. J. Phys. Rehabil. Med. 52 (3), 321–330.

Sunday, D.M., 2011. More about pit ponies. Available from: http://www.equiculture.org/1more-about-pit-ponies.aspx.

Swisher, K., 2015. Market report going down, down, down but upcoming uptrend is coming. Render Magazine, 10–16. Available from: https://d10k7k7mywg42z.cloudfront.net/assets/55281d9ec0d6715235004d2e/MarketReport2014.pdf.

Taverna, G., Tidu, L., Grizzi, F., Torri, V., Mandressi, A., Sardella, P., La Torre, G., Cocciolone, G., Seveso, M., Giusti, G., Hurle, R., Santoro, A., Graziotti, P., 2015. Olfactory system of highly trained dogs detects prostate cancer in urine samples. J. Urol. 193, 1382–1387.

Toups, M.A., Kitchen, A., Light, J.E., Reed, D.L., 2011. Origin of clothing lice indicates early clothing use by anatomically modern humans in Africa. Mol. Biol. Evol. 28, 29–32.

Trinkaus, E., 2005. Anatomical evidence for the antiquity of human footwear use. J. Archaeol. Sci. 32, 1515–1526.

Trinkaus, E., Shang, H., 2008. Anatomical evidence for the antiquity of human footwear: Tianyuan and Sunghir. J. Archaeol. Sci. 35, 1928–1933.

US Department of Justice, 2015. Available from: http://www.ada.gov/service_animals_2010.htm.

US Energy Information Administration, 2015. 1.27 billion gallons of B100 produced in 2014 in USA. Available from: http://www.eia.gov/biofuels/biodiesel/production/table1.pdf.

US EPA (Environmental Protection Agency), 2015. Leather tanning. Available from: http://www3.epa.gov/ttnchie1/ap42/ch09/final/c9s15.pdf.

USDA ERS (Economic Research Service), 2015. Oilseed, oilmeal, and fats and oils supply and use statistics. Available from: http://www.ers.usda.gov/data-products/oil-crops-yearbook.aspx.

USDA FAS (United States Department of Agriculture Foreign Agriculture Service), 2015. EU Biofuels Annual 2015. Available from: http://gain.fas.usda.gov/Recent%20GAIN%20Publications/Biofuels%20Annual_The%20Hague_EU-28_7-15-2015.pdf.

Utkin, Y.N., 2015. Animal venom studies: current benefits and future developments. World J. Biol. Chem. 6, 28–33.

van Rooy, D., Arnott, E.R., Early, J.B., McGreevy, P., Wade, C.M., 2014. Holding back the genes: limitations of research into canine behavioural genetics. Canine Genet. Epidemiol. 1, 7.

Vaysse, A., Ratnakumar, A., Derrien, T., Axelsson, E., Rosengren Pielberg, G., Sigurdsson, S., Fall, T., Seppälä, E.H., Hansen, M.S., Lawley, C.T., Karlsson, E.K., Consortium, L.U.P.A., Bannasch, D., Vilà, C., Lohi, H., Galibert, F., Fredholm, M., Häggström, J., Hedhammar, A., André, C., Lindblad-Toh, K., Hitte, C., Webster, M.T., 2011. Identification of genomic regions associated with phenotypic variation between dog breeds using selection mapping. PLoS Genet. 7, e1002316.

Wandel, S., Jüni, P., Tendal, B., Nüesch, E., Villiger, P.M., Welton, N.J., Reichenbach, S., Trelle, S., 2010. Effects of glucosamine, chondroitin, or placebo in patients with osteoarthritis of hip or knee: network meta-analysis. Br. Med. J. 341, c4675.

Wang, X., Jia, Z., Liang, L., Yang, B., Ding, R., Nie, J., Wang, J., 2016. Impacts of manure application on soil environment, rainfall use efficiency and crop biomass under dryland farming. Nat. Sci. Rep. 6, 20994.

Warda, M., Gouda, E.M., Toida, T., Chi, L., Linhardt, R.J., 2003. Isolation and characterization of raw heparin from dromedary intestine: evaluation of a new source of pharmaceutical heparin. Comp. Biochem. Physiol. C 136, 357–365.

Welsch Dannebrog, R.L., 2007. Encyclopedia of the Great Plains. In: Wishart, D.J. (Ed.). Available from: http://plainshumanities.unl.edu/encyclopedia/doc/egp.ii.007.

WHO (World Health Organization), 2015a. Available from: http://www.who.int/mediacentre/factsheets/fs337/en/.

WHO, 2015b. Available from: http://www.who.int/blood-products/snake_antivenoms/snakeantivenomguideline.pdf.

WHO, 2015c. International travel and health: rabies. Available from: http://www.who.int/ith/vaccines/rabies/en/.

WWF, 2012. Wildlife crime a threat to regional security: a new report on the crisis of illegal wildlife trafficking details its unprecedented scale and global impact. Available from: http://www.worldwildlife.org/stories/wildlife-crime-a-threat-to-regional-security.

Zhao, L., Brinton, R.D., 2006. Select estrogens within the complex formulation of conjugated equine estrogens (Premarin) are protective against neurodegenerative insults: implications for a composition of estrogen therapy to promote neuronal function and prevent Alzheimer's disease. BMC Neurosci. 7, 24.

Further Reading

Hidaka, S., Liu, S.Y., 2003. Effects of gelatins on calcium phosphate precipitation: a possible application for distinguishing bovine bone gelatin from porcine skin gelatin. J. Food Compos. Anal. 16, 477–483.

WHO, 2015. Available from: http://apps.who.int/iris/bitstream/10665/43858/1/9789241563482_eng.pdf.

3

Animal Products and Human Nutrition

Colin G. Scanes

University of Wisconsin–Milwaukee, Milwaukee, WI, United States

3.1 INTRODUCTION

Consumption of animal products in foods is increasing globally irrespective of whether it is meat, milk, eggs, or fish. This is reflected by large increases in global production of animal products for human consumption (Table 3.1) with, for instance, production of meat increasing 62.5% between 1993 and 2013. Animal production is growing at a faster rate than the human population growth (Table 3.1). The corollary to the increase in production is increased consumption per capita.

The increase in the per capita consumption of animal products reflects the following:

1. Improved living standards with economic development.
2. Reductions in poverty globally with the number of undernourished people declining (see Section 3.8).
3. Consumer demand or people's "desire" to eat animal products. According to The Economist, as people have more resources they eat more meat and shrimp not as a cheap source of calories but because they are tastier (Jensen and Miller, 2011).
4. Improved efficiency of agricultural production (discussed in Chapter 9) with concomitant price stability and/or decreases in inflation-adjusted moneys.

In the USA, people with an omnivorous diet consume 65% of their protein intake and 26% of their energy from animal products (Pimentel and Pimentel, 2003). In contrast, lacto-ovo-vegetarians consume 20% less protein of which 40% of their protein intake and 15% of their energy are from animal products (Pimentel and Pimentel, 2003). The omnivorous diet in the USA is similar to the diets of other industrialized (Western) countries with an average of about 25% of calories from animal products (FAO, 2003, 2017). In contrast in developing countries, per capita consumption of animal products is about a third of that in industrialized countries, but with considerable variation (FAO, 2003). There is a high consumption of animal products in the People's Republic of China and in Brazil, approaching

41

TABLE 3.1   Global Productions of Major Animal Food Products and its Growth Between 1993 and 2013

Animal commodity	Production (in million metric tons) (% increase in decade)		
	1993	2003	2013
Meat	191.0	244.9 (28.2)	310.4 (26.7)
Milk	520.0	618.0 (18.8)	768.6 (24.4)
Eggs	41.1	58.7 (42.8)	73.9 (25.9)
Human population	5.5	6.3 (14.5)	7.1 (10.9)

Population data from US Census Bureau, In Press. Available from: https://www.census.gov/population/international/data/idb/worldpoptotal.php; Production data from FAO Stats (United Nations Food and Agriculture Organization Statistics), 2015. Available from: http://faostat3.fao.org/browse/Q//E*

that of Western industrialized nations (FAO, 2003). In contrast, in Sub-Saharan Africa, consumption of animal product is low (~5% of energy) and is only beginning to increase (FAO, 2015a).

"The increase in animal-source food products has both positive and adverse health effects" (Popkin et al., 2012). The positive aspects of consumption of animal products are discussed in Sections 3.2 (overview), 3.3 (amino-acids and proteins), 3.4 (minerals), 3.5 (vitamins), 3.6 (lipids), 3.7 (carbohydrate), and 3.8 (nutrient deficiencies), while health considerations are considered in Sections 3.9 (human health aspects of red meat, poultry, fish, milk, and egg consumption), 3.10 (food safety with viral, bacterial, and protozoan considered in Chapter 19), and 3.11 (sustainability of production of meat and other animal products).

3.2 ANIMAL PRODUCTS AS A SOURCE OF NUTRIENTS

Table 3.2 summarizes the nutrient composition of a series of animal products. With the exception of dairy products, these have very low or no carbohydrate. Animal products have high contents of protein and fat, particularly saturated fat. Meat, fish, eggs, and dairy products are a valuable source of minerals including calcium, zinc, and iron together with vitamins, such as vitamin B_{12}.

3.3 ANIMAL PRODUCTS AS A SOURCE OF AMINO-ACIDS/ PROTEINS

Proteins are required in the diet with animal products being very good sources of protein to meet this requirement. The dietary needs are defined as the Recommended Dietary Allowance (RDA) for protein. The RDA is defined as covering 97.5% of the age" group (Institute of Medicine Panel for Macronutrients, 2002/2005). The RDA for protein is established as the following:

- 56 g day^{-1} men
- 46 g day^{-1} women

The RDA increases by 21 g day^{-1} during pregnancy and lactation (Institute of Medicine Panel for Macronutrients, 2002/2005). This is not surprising as approximately 0.9 kg protein is deposited during pregnancy into the fetus, uterus, placenta, breasts, and maternal blood (Joint FAO/WHO/UNU Expert Consultation, 2007b). Injuries and disease accentuate protein loss (Joint FAO/WHO/UNU Expert Consultation, 2007c) and therefore increase protein requirements. The protein/amino-acid/nitrogen balance in adult humans is summarized in Figs. 3.1–3.3.

Proteins differ in their digestibility (Table 3.3). Predominantly, proteins from animal products are more digestible than those from plants. There is an exception with collagen, with its glycine-X-proline or glycine-X-hydroxyproline

TABLE 3.2   Composition of Selected Animal Products

Species	Protein (g 100 g^{-1})	Fat (g 100 g^{-1})	Carbohydrate (g 100 g^{-1})	Cholesterol (mg 100 g^{-1})
Beef (grass-fed cattle meat) ground	19.4	12.7	0	62
Pork/pig meat (trimmed retail cuts)	20.2	4.9	0	64
Chicken (raw including skin)	18.6	15.1	0	75
Chicken (white meat without skin)	23.2	1.6	0	58
Chicken (dark meat without skin)	20.1	4.3	0	80
Eggs	12.6	9.5	0.7	372
Whole milk	3.3	3.7	4.6	10
Skim milk	3.4	0.1	5.0	2
Cheese (cheddar)	22.9	33.3	0.5	99
Cottage cheese	11.1	4.3	3.4	17
Yogurt (plain)	3.5	3.2	4.7	13
Salmon	19.8	6.3	0	55 (85*)
Shrimp	13.6	1.0	0.9	126 (123*)
Blue crab	18.1	1.1	0	78 (98*)

Species	Saturated fatty acids	Monounsaturated fatty acids	Polysaturated fatty acids	Trans-fatty acids
Beef (grass-fed cattle meat) ground	5.3	4.8	0.5	0.8
Pork/pig meat (trimmed retail cuts)	1.6	2.1	0.5	0.02
Chicken (raw including skin)	4.3	6.2	3.2	0.1
Chicken (white meat without skin)	0.4	0.4	0.4	0
Chicken (dark meat without skin)	1.1	1.3	1.1	0
Eggs	3.2	3.7	1.9	0.02
Whole milk	1.8	0.8	0.2	0
Skim milk	0.1	0.05	0.01	0
Cheese (cheddar)	18.9	9.2	1.4	0.9
Cottage cheese	1.7	0.8	0.1	0
Yogurt (plain)	2.1	0.9	0.1	0
Salmon	1.0	2.1	2.5	0
Shrimp	0.3	0.2	0.3	0.02
Blue crab	0.2	0.2	0.4	0

(Continued)

TABLE 3.2   Composition of Selected Animal Products (*cont.*)

Species	Calcium (mg 100 g^{-1})	Iron (mg 100 g^{-1})	Vitamin B$_{12}$ (μg 100 g^{-1})
Beef (grass-fed cattle meat) ground	12	2	2
Pork/pig meat (trimmed retail cuts)	11	0.9	0.67
Chicken (raw including skin)	12	1.0	0.36
Chicken (white meat without skin)	12	0.7	0.38
Chicken (dark meat without skin)	12	1.0	0.36
Eggs	56	1.75	0.89
Whole milk	113	0.03	0.45
Skim milk	122	0.03	0.5
Cheese (cheddar)	710	0.14	1.1
Cottage cheese	83	0.07	0.43
Yogurt (plain)	121	0.05	0.37
Salmon	12	0.8	3.2
Shrimp	54	0.2	1.1
Blue crab	89	0.7	9

Data from USDA (United States Department of Agriculture), 2017. National Nutrient Database for Standard Reference. Available from: http://ndb.nal. usda.gov/ndb/foods

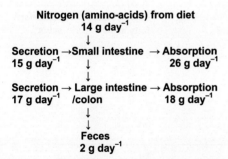

FIGURE 3.1 **Protein/amino-acid/nitrogen balance in adult humans.** *Source: Based on Jackson, A.A., 2000. Nitrogen trafficking and recycling through the human bowel. In: Furst, P., Young. V.R. (Eds.). Proteins, Peptides and Amino Acids in Enteral Nutrition (Nestlé Nutrition Workshop Series, Clinical & Performance Program, Volume 3). Vevey, Nestec Ltd./Basel, S. Karger: pp. 89–108.*

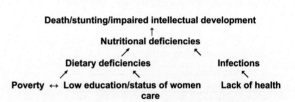

FIGURE 3.2 **Causes and impacts of nutritional deficiencies.** *Source: Based on Müller, O., Krawinkel, M. 2005. Malnutrition and health in developing countries. Can. Med. Assoc. J. 173, 279–286; Bailey, R.L., West, K.P. Jr., Black, R.E. 2015. The epidemiology of global micronutrient deficiencies. Annals Nutr Metab 66 (Suppl. 2), 22–33.*

repeated motifs. Collagen fibers are found in meats and are poorly digested (Harkness et al., 1978).

There are specific requirements for amino acids (Joint FAO/WHO/UNU Expert Consultation, 2007c). Table 3.4 shows the amino-acid composition of animal- and plant-based products. There is a substantial variation in amino acid content with methionine levels in legumes considered marginal (Joint FAO/WHO/UNU Expert Consultation, 2007c). The significance of this is considered in Box 3.1.

Amino-Acid Requirements in Cats

Just because we need or do not need a nutrient does not mean we can extrapolate this to other animals. Taurine and cats is a good example. Taurine is required in the diet of cats with it estimated that cats need a minimum of 35–56 mg day^{-1} of taurine (Burger and Barnett, 1982). The National Research Council (1986) recommended that feeds for cats should contain 400 mg kg^{-1} feed and 500 mg kg^{-1} during growth. It was later concluded that the requirement was between 1000 and 2000 mg kg^{-1} feed (Earle and Smith, 1994).

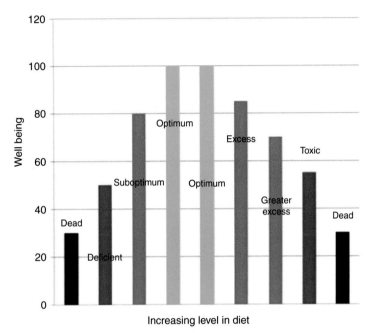

FIGURE 3.3 **Relationship between dietary level of nutrient and its effect.** Color code from left to right: black, death; red, deficiency; orange, suboptimal; green, optimal; orange, excessive; red, toxic; black, death. *Source: Based on Andersson, M., Karumbunathan, V., Zimmermann, M.B. 2012. Global iodine status in 2011 and trends over the past decade. J. Nutr. 142, 744–750; Bailey, R.L., West, K.P. Jr., Black, R.E., 2015. The epidemiology of global micronutrient deficiencies. Annals Nutr. Metab. 66 (Suppl. 2), 22–33.*

TABLE 3.3   Digestibility of Protein in Humans

Food	Digestibility (%)
Eggs	97
Milk/cheese	95
Meat/fish	94
Soy flour	86
Corn/maize flour	85
Beans	78

Based on Joint FAO/WHO/UNU Expert Consultation, 2007a. Protein quality evaluation. In: Protein and Amino-Acid Requirements in Human Nutrition. pp. 93–102, World Health Organization. Available from: http://whqlibdoc.who.int/trs/who_trs_935_eng.pdf

TABLE 3.4   Amino-Acid Composition of Major Animal Products Used as Human Food (Shown as % of Protein) (FAO. Amino-Acid Content of Foods)

	Bean (*Phaseolus vulgaris*)	Wheat (*Triticum* spp.) whole grain	Maize (*Zea mays*) whole grain	Chicken egg	Cattle meat or beef	Chicken meat	Cow's milk
Isoleucine	2.62	2.04	2.30	3.93	3.01	3.34	2.95
Leucine	4.76	4.17	7.83	5.51	5.07	4.60	5.96
Lysine	4.50	1.79	1.67	4.36	5.56	4.97	4.87
Methionine	0.66	0.94	1.20	2.10	1.69	1.57	1.57
Cysteine	0.53	1.59	0.97	1.52	0.80	0.82	0.51
Phenylalanine	3.26	2.82	3.05	3.58	2.75	2.50	3.36
Tyrosine	1.58	1.87	2.39	2.60	2.25	2.09	2.97
Threonine	2.48	1.83	2.25	3.20	2.87	2.48	2.78
Tryptophan	0.63	0.68	0.44	0.93	0.70	0.64	0.88
Valine	2.87	2.76	3.03	4.28	3.13	3.18	3.62
Arginine	3.55	2.88	2.62	3.81	3.95	3.48	2.05
Histidine	1.77	1.43	1.70	1.52	2.13	1.64	1.67
Alanine	2.62	2.26	4.71	3.70	3.65	2.13	2.17
Aspartate	7.48	3.08	3.92	6.01	5.62	5.73	4.81
Glutamate	9.24	18.66	11.84	7.96	9.55	9.38	13.90
Glycine	2.37	2.45	2.31	2.07	3.04	3.31	1.23
Proline	2.23	6.21	5.59	2.60	2.36	2.59	5.71
Serine	3.47	2.87	3.11	4.78	2.52	2.44	3.62
A/E[a]	47	68	55	100	80	79	66

[a] *Average divided by that in eggs (an ideal).*

3.4 ANIMAL PRODUCTS AS A SOURCE OF CARBOHYDRATES

Dairy products are the only animal product with significant carbohydrate content. The carbohydrate is the disaccharide, lactose. This is readily digested in infants and many adults. There is a high expression of lactase in the epithelial cells of the small intestine villi (enterocytes) in babies and other nursing mammals (Freund et al., 1990; Tivey et al., 1991). The lactase is released into the small intestine allowing the digestion of milk sugar, lactose.

$$\text{Lactose} \xrightarrow{\text{Lactase}} \text{Glucose} + \text{Galactose}$$

However, after weaning, expression and levels of lactase decline leading to lactose intolerance (see Box 3.2).

3.5 ANIMAL PRODUCTS AS A SOURCE OF LIPIDS

3.5.1 Introduction

Meat, eggs, and dairy products are relatively high in fat, particularly saturated fat. The overconsumption of saturated fats, particularly in Western countries, is a public health issue (Dietary Guidelines Advisory Committee, 2015). Fats in the diet can be used, for instance, for the

BOX 3.1 AMINO-ACIDS

Amino-acids are considered in the following categories (Institute of Medicine Panel for Macronutrients, 2002/2005; Joint FAO/WHO/UNU Expert Consultation, 2007a):

- indispensable (formerly essential)
- dispensable (formerly nonessential)
- conditional indispensable amino-acids (formerly categorized as nonessential)

Indispensable or Essential Amino-Acids

The indispensable amino-acids (formerly known as essential amino-acids) must be present in the diet at sufficient quantities, as the body cannot synthesize them at all or not in sufficient quantities. The nine indispensable or essential amino-acids are the following: (1) histidine, (2) isoleucine, (3) leucine, (4) lysine, (5) methionine, (6) phenylalanine, (7) threonine, (8) tryptophan, and (9) valine.

Dispensable or Nonessential Amino Acids

The dispensable or nonessential amino acids can be synthesized in sufficient quantities by the body. The dispensable amino acids are the following: alanine, asparagine, aspartic acid, glutamic acid and serine.

Conditional Indispensable Amino Acids (e.g., At Times of Stress and Disease)

The conditional indispensable amino acids are required at times, such as stress and disease.

They are the following: arginine, cysteine (but can be synthesized from methionine), glutamine, glycine, proline and tyrosine.

Amino-acids are not only essential precursors of proteins but also precursors for other critically important biological molecules. Examples of amino-acids as nonprotein precursors include the following (based on Institute of Medicine Panel for Macronutrients, 2002/2005):

- arginine to nitric oxide and ornithine (in the urea cycle),
- cysteine to glutathione and taurine,
- glycine to heme, bile acids, and nucleic acid bases,
- glutamine to GABA (gamma-aminobutyric acid),
- histidine as a methyl group donor,
- lysine to carnation,
- phenylalanine/tyrosine to dopamine, norepinephrine, epinephrine, thyroid hormones, and melanin, and
- tryptophan to serotonin, melatonin, and nicotinic acid.

These conversions are predominantly irreversible with consequent losses of the amino acids. Moreover, glucogenic amino acids (alanine, arginine, aspartate, asparagine, cysteine, glutamate, glutamine, methionine, proline, serine, and valine) can be either oxidized or converted to glucose as well as used for protein synthesis. Obviously the body needs supplies of amino-acids to replenish losses in amino-acids and those used in protein synthesis.

BOX 3.2 LACTOSE INTOLERANCE

The decline in lactase is seen in about 65% of humans with the decline starting at the age of about 2–3 years old (reviewed by Grand et al., 2003; Ingram et al., 2009; Swallow, 2003). In these people, there is lactose intolerance; digestion of milk sugars being greatly impaired. If someone with lactose intolerance or maldigestion consumes milk, the lactose will not be digested. This then passes to the colon where it is fermented by intestinal bacteria leading to gas/flatulence (hydrogen together with methane and carbon dioxide), discomfort, bloating, stomach cramps, and diarrhea. It can be diagnosed by the level of hydrogen on a person's breath. It should be emphasized that lactose intolerance is not a food allergy rather it is an intestinal disorder.

There are marked differences in the incidence of lactose intolerance in different human populations with very high incidences in Asian populations and very low incidences in populations in Northern Europe (de Vrese et al., 2001) (Table 3.5). The US population reflects a series of immigrations and forced migrations. It is not surprising, therefore that there are marked differences in the incidence of lactose intolerance in different ethnic groups (de Vrese et al., 2001) (Table 3.6). The incidence of lactase persistence is shown in Table 3.7.

The concentration of lactose is much reduced in processed dairy products, such as cheese and yogurt with added bacteria (*Lactobacilli*, such as *Lactobacillus acidophilus*) metabolizing the lactose to lactate.

Myth

If you are lactose intolerant, you can drink sheep or goats milk.

This is not true! Lactose is found in milk irrespective of whether it came from cows, sheep, goats or, in fact, human. You can drink sheep or goats milk if you have allergies to proteins in cow's milk.

Lactose Intolerance and Human Evolution

About 35% of people in the world have lactase persistence where there is much less or no decline in the level of lactase. This is an advantage as these people can digest milk sugar and hence consume milk without ill effects. Lactose persistence is essentially the reciprocal of lactose intolerance.

Lactose persistence varies considerably between different human populations/ancestries being very high in people of North European descent, decreasing in people whose ancestry is from around the Mediterranean Sea and being low in people whose ancestry was from East Asia, Africa, South India, and Latin America (Table 3.7) (Gerbault et al., 2011). It appears that the lactose persistence in North European populations is due to a single mutation that occurred within the last 10,000 years (Gerbault et al., 2011).

There are different mutations responsible for the lactase persistence in, for instance, the Western Sahara, Sudan, the Arabian Peninsula, Iran, and North India (Gerbault et al., 2011). Even within the same country, there can be marked differences by population. There is very high lactase persistence in pastoralists in Sudan compared to nonpastoralist neighboring populations.

Lactase persistence provides a strong selective advantage allowing the consumption of milk and milk products without discomfort and the concomitant nutritional advantages of high-quality protein together with calcium. Hand-in-hand, this would facilitate the development of dairying. There have been independent origins of lactase persistence in different regions of the world. It has been suggested that the single-point mutation seen in North European populations today was helpful in the development of agriculture in Northern Europe in the Neolithic era.

BOX 3.2 (cont'd)

TABLE 3.5   Differences in the Incidence of Lactose Intolerance in Different Human Populations

Human population	Lactose intolerance (%)
East Asian (Chinese, Japanese, Korean, Vietnamese)	90
Central Asia, Africa	80
South India/Latin America	70
Italy/Balkans/Southern France	55
North India	30
Germany/Austria/Finland/Northern France	15
British Isles	10
Scandinavia	5

Based on de Vrese, M., Stegelmann, A., Richter, B., Fenselau, S., Laue, C., Schrezenmeir, J., 2001. Probiotics—compensation for lactase insufficiency. Am. J. Clin. Nutr. 73 (Suppl. 2), 421S–429S.

TABLE 3.6   Differences in the Incidence of Lactose Intolerance in Adults of Different Ethnic Groups in the United States

Human population	Lactose intolerance (%)
Chinese-American	95
Native American	80
Ashkenazi (of European ancestry) Jews	70
African-American	80
Hispanic	53
Caucasian	20

Based on National Institutes of Health. Lactose intolerance Information for Health Care Providers. Available from: https://www.nichd.nih. gov/publications/pubs/documents/NICHD_MM_Lactose_FS_rev.pdf; de Vrese, M., Stegelmann, A., Richter, B., Fenselau, S., Laue, C., Schrezenmeir, J., 2001. Probiotics—compensation for lactase insufficiency. Am. J. Clin. Nutr. 73 (Suppl. 2), 421S–429S.

TABLE 3.7   Lactase Persistence in Different Human Populations in Asia, Africa, and Europe

Human population	Lactase persistence (%)
British Isles/Scandinavia	90–100
Germany/Northern France/BeNeLux countries/Finland	80–90
Western Sahara/Sudan/Arabian peninsula/Iran/North India	60–80
Iberian Peninsula/Southern France/Eastern Europe	50–70
Middle East excluding Arabian peninsula	30–50
South India	20–40
Southern Africa/East Asia	<20

Based on Gerbault, P., Liebert, A., Itan, Y., Powell, A., Currat, M., Burger, J., Swallow, D.M., Thomas, M.G., 2011. Evolution of lactase persistence: an example of human niche construction. Philos. Trans. R. Soc. Lond. B Biol. Sci. 366, 863–877.

synthesis of cell membranes or as a source of energy. The energy can be used immediately with fatty acid oxidation . Alternatively, energy is stored efficiently as triglyceride in adipose tissue. Fats are also important for facilitating the absorption of fat-soluble vitamins (A, D, and E).

3.5.2 Essential Fatty Acids

Essential fatty acids are polyunsaturated fatty acids and there are two of these:

- alpha-linolenic acid (ALA), an omega-3 $(\omega - 3)$ $[18:3n - 3]$ fatty acid
- linoleic acid (LA), an omega-6 $(\omega - 6)$ $[18:2n - 6]$ fatty acid

These are elongated and desaturated in the body to other fatty acids including:

- eicosapentaenoic acid (EPA) $(\omega - 3)$ $[20:5n - 3]$ fatty acid
- docosahexaenoic (DHA) acids $(\omega - 3)$ $[22:6 - 3]$ fatty acid
- arachidonic acid $(\omega - 6)$ $[22:6n - 3]$ fatty acid

Essential fatty acids and their derivatives are critical to the brain and nervous system with these representing 15%–30% of the dry weight of the brain (reviewed by Hallahan and Garland, 2005). Animal products, such as liver and eggs contain essential fatty acids. The fat from cattle fed grass-based diets has more omega-3 $(n - 3)$ fatty acids (Daley et al., 2010). Fish are a particularly rich source of essential fatty acids (Table 3.8) (also see Section 3.9.4). Vegetarian

diets are low in ALA, EPA, and DHA (Davis and Kris-Etherton, 2003) and it is not surprising that circulating concentrations of EPA and DHA are low in vegans and to a less extent vegetarians (Rosell et al., 2005). In the USA, intake of omega-3 fatty acids (EPA and DHA) does fully meet recommended levels (Papanikolaou et al., 2014).

3.5.3 Conjugated Linoleic Acid

Conjugated linoleic acids (CLA) are minor components in dairy products and meats from ruminants, such as cattle and sheep (reviewed by Whigham et al., 2007). The fat from cattle fed grass-based diets has more CLA (C18:2) isomers (Daley et al., 2010). There is evidence that CLA is effective in reducing body weight in humans (reviewed by Whigham et al., 2007).

3.5.4 Cholesterol

Cholesterol is a nutrient present in animal products and used in the cell membrane and the synthesis of steroid hormones. However, sufficient cholesterol is synthesized in the human liver and there is no evidence that cholesterol is required in the diet. Moreover, the Dietary Guidelines Advisory Committee (2015) in the USA considered that "cholesterol is not a nutrient of concern for overconsumption." Shellfish are high in protein and cholesterol but low in fat (Table 3.2). Issues of cholesterol are addressed later in Section 3.9.1.

TABLE 3.8 Omega-3 Fatty Acids Content in Various Seafood Products (Mozaffarian and Rimm, 2006)

Fish	Omega-3 fatty acids (mg 100 g^{-1})
Salmon (wild)	1043
Salmon (farmed)	2649
Swordfish	827
Trout	932
Shrimp	314

3.6 ANIMAL PRODUCTS AS A SOURCE OF MINERALS

Among the important minerals found in animal products at relevant levels are the following:

- calcium (Ca^{++}), essential for bone formation and muscular functioning,
- cobalt (Co^{++}), essential for vitamin B$_{12}$,
- iron (Fe^{++}), essential for the synthesis of erythrocytes,

- iodine (I^-), essential for thyroid functioning and synthesis of thyroxine,
- phosphate (PO^{4-}), essential for bone formation and metabolism, and
- zinc (Zn^{++}), essential for various facets of metabolism.

Deficiencies in calcium, iron, zinc, and iodine are discussed, respectively, in Sections 3.8.4.1–3.8.4.4.

3.7 ANIMAL PRODUCTS AS A SOURCE OF VITAMINS

3.7.1 Overview

Particular emphasis will be placed on vitamins B_{12} and D (see Sections 3.7.2 and 3.7.3 further). Among the important vitamins found in animal products at relevant levels are the following:

- vitamin A (found in liver and fish together with fortified milk), essential for eye functioning and as a precursor for retinoid acid and other retinols
- vitamin B_{12} or cobalamin (Section 3.7.2)
- vitamin D (found in liver, fish, and eggs together with fortified milk), precursor for hormone; 1,25-dihydroxy vitamin D (calcitriol) is essential for calcium metabolism and bone development
- vitamin K (found in eggs), essential for blood clotting
- biotin or vitamin B_7 (found in liver and eggs), essential cofactor for metabolism
- folic acid (found in liver), essential cofactor for metabolism
- niacin or vitamin B_3 (found in meat, fish, and eggs), essential cofactor for metabolism
- pantothenic acid or vitamin B_5 (found in meat), essential cofactor for metabolism
- pyridoxine or vitamin B_6 (meat), essential cofactor for metabolism
- riboflavin or vitamin B_2 (found in dairy products), essential cofactor for metabolism
- thiamine or vitamin B_3 (found in liver and eggs), essential cofactor for metabolism

3.7.2 Vitamin B_{12}

Vitamin B_{12} (cobalamin) is a water-soluble vitamin that is only found in animal-derived foods. Vitamin B_{12} has critically important roles in the biosynthesis of erythrocytes. Indeed, without it there is pernicious anemia. Moreover, vitamin B_{12} has important neurological roles. The sources of vitamin B_{12} include the following: shellfish (e.g., clams 84 µg per serving), liver (71 µg per serving), fish (~5 µg per serving), red meat (1.4 µg per serving), milk (1.2 µg per serving), cheese (0.9 µg per serving), eggs (0.6 µg per serving), and poultry (0.3 µg per serving) compared to the daily value of 6 µg (USDA, 2017). Deficiencies in vitamin B_{12} are addressed in Section 3.8.4.5.

3.7.3 Vitamin D

Vitamin D is a fat-soluble vitamin, which is hydroxylated to form the hormone, 1,25-dihydroxy vitamin D or 1,25-dihydroxy calciferol or calcitriol. This metabolite of vitamin D plays an important role in the control of bone development and calcium absorption in the small intestine. Animal products are superior sources of vitamin D. Excluding food items that are fortified with vitamin D, the top animal foods as sources of vitamin D are the following: sword fish (666 IU in 100 g), salmon (526 IU in 100 g), tuna (canned) (196 IU in 100 g), and egg (82 IU in 100 g) compared to the daily value of 400 IU (National Institute of Health, 2016; USDA, 2017).

3.8 MEAT AND NUTRIENT DEFICIENCIES

3.8.1 Prevalence of Hunger or Chronic Undernutrition

In 2000, the United Nations (UN) established the Millennial Goals. These included Goal 1, "to eradicate extreme poverty and hunger," with target 1.C, to "halve, between 1990 and 2015, the proportion of people who suffer

TABLE 3.9   Changes in the Prevalence of Chronic Undernourishment (Hunger or Food Deprivation) Between 1990–92 and 2014–16

	1990–1992	2014–2016
Africa	182	224
North Africa	6	4
Sub-Saharan Africa	176	220
Asia	742	512
Central Asia	10	6
East Asia	295	145
South Asia	291	281
Southeast Asia	138	61
Western Asia	8	19
Latin America	66	34
Developed countries	20	15
World	1010	765

Data from FAO, 2015b. The state of food insecurity in the world. Available from: http://www.fao.org/3/a-i4646e.pdf

from hunger." What is meant by suffer from hunger is addressed in the sidebar. Substantial progress has been made in reaching the target (Table 3.9). In the new "17 goals for sustainable development," the UN now seeks to "end hunger, achieve food security, and improved nutrition and promote sustainable agriculture" (Goal 2) (United Nations, 2015). The proportion of underweight children in developing countries has declined from 28% to 17% between 1990 and 2013 (FAO, 2015b; WHO, 2015a). Table 3.9

According to the United Nations Food and Agriculture Organization, hunger is equated to chronic undernourishment. In turn, undernourishment is "the population whose dietary energy consumption is less than a predetermined threshold… kilocalories required to conduct sedentary or light activities" (FAO, 2013).

summarizes the changes in hunger in the World and major regions between 1990–1992 and 2014–2016. Large decreases in chronic food deprivation occurred with large reductions in Southeast Asia (55.8%), East Asia (50.8%), and Latin America (48.5%). Unfortunately, a marked increasing number of people are still suffering from hunger particularly in Sub-Saharan Africa (up 25%).

3.8.2 Protein–Energy Malnutrition

There are two forms of severe protein–energy malnutrition: (1) marasmus, with severe wasting, and (2) kwashiorkor, with severe wasting and edema (Müller and Krawinkel, 2005). The major cause is deficiency of energy (Müller and Krawinkel, 2005). Over 19 million children suffer from severe malnutrition (Table 3.10). Protein–energy deficiency results in stunting and underweight children with over 175 million impacted (Black et al., 2008) (Table 3.10). Chronic protein energy malnutrition impairs cognitive development and impairs learning in children (Kar et al., 2008). Low birthweight impacts the effectiveness of education with low-birthweight-associated risks of failure at school (Martorell et al., 2010). Meat consumption markedly decreases stunting in children (Krebs et al., 2011).

3.8.3 Deficiencies in Protein and Specific Amino-Acids

Low levels of animal products in the diet can be associated with deficiencies of specific amino-acids. According to the Institute of Medicine Panel for Macronutrients (2002/2005) in the USA, "individuals who restrict their diet to plant foods may be at risk of not getting adequate amounts of certain indispensable amino-acids because the concentration of lysine, sulfur amino-acids, and threonine are sometimes lower in plant food proteins than in animal food proteins" (Table 3.4).

TABLE 3.10   Prevalence of Nutritional Deficiencies in Children and the Newborn

	Severe acute malnutrition/severely wasted (in millions)[a]	Stunted (in millions)[b]	Underweight (in millions)[c]	Low birth weight/intrauterine growth restriction (in millions)[d]
Africa	5.6	56.9	31.1	2.7
Asia	13.3	111.6	78.6	9.5
Latin America − South and Central America + Caribbean	0.3	9.2	2.7	0.6

[a] Weight-for-length or weight-for-height less than mean of healthy population minus 3 standard deviations (SD).
[b] Height-for-age less than mean of healthy population minus 2 SD.
[c] Weight-for-age less than mean of healthy population minus 2 SD.
[d] Less than 2.5 kg.
Based on Black, M.M. 2003. Micronutrient deficiencies and cognitive functioning. J. Nutr. 133 (Suppl. 2), 3927S–3931S; WHO, 2009. Child growth standards and the identification of severe acute malnutrition in infants and children. A Joint Statement by the World Health Organization and the United Nations for terminology and definitions.

3.8.4 Deficiencies in Micronutrients

Micronutrient deficiencies are widespread globally with the most widespread being the following (Bailey et al., 2015):

- iron
- iodine
- folate
- vitamin A
- zinc

Deficiencies in micronutrients result in 50% of deaths, directly and indirectly, in children and infants less than 5 years old globally. The most vulnerable populations are pregnant women and children under 5 years old (Bailey et al., 2015). The concentrations of critical micronutrients, such as vitamins A, vitamin B_{12}, and iodine are reduced depending on the nutrition of the mother (reviewed by Black et al., 2008).

3.8.4.1 Calcium

The Dietary Guidelines Advisory Committee (2015) in the USA considered that there is underconsumption of calcium in the country. Similarly, over 10% of omnivorous women and over 5% of men consume less than the recommended amount of calcium (Craig, 2010) (Table 3.11). The proportion of ovo-lacto-vegetarians who consume less than the recommended amount of calcium is greater than that in omnivores.

TABLE 3.11   Comparison of Intake of Various Nutrients (Daily Intake as a Percentage of Daily Recommended Intake) Between the Average US Diet (Omnivore) With Ovo-Lacto-Vegetarian and Vegans

Nutrient	Women			Men		
	Omnivorous	Ovo-lacto-vegetarian	Vegan	Omnivorous	Ovo-lacto-vegetarian	Vegan
Calcium	89.8	87.5	62.2	94.6	90.6	75.5
Zinc	>100	>100	>100	>100	93.4	>100
Iron	62.8	81.7	98.9	>100	>100	>100
Vitamin D	72	42	16	68	40	20
Vitamin B_{12}	100	87.5	41.7	>100	>100	41.7

Calculated From Craig, W.J. 2010. Nutritional concerns and vegetarian diets. Nutr. Clin. Practice 25, 613–620.

Consumption of sufficient calcium is markedly lower in vegans (Craig, 2010) (Table 3.11). In a study of children in New Zealand, children with very low milk intake were shorter and had a lower bone mineral content (Black et al., 2002).

3.8.4.2 Iron

Iron deficiency impacts 30% of the world's population (Bailey et al., 2015; de Benoist et al., 2008). Estimates of the prevalence of iron deficiency in the world vary from 500 million (Cook et al., 1994) to 2 billion people (Bailey et al., 2015; de Benoist et al., 2008). Iron deficiency is obviously associated with anemia but also pregnancy problems with low birth weight and premature births together with "impairment in psychomotor development and cognitive function in infants and preschoolers" (Cook et al., 1994).

Iron deficiency causes at least half of anemia even in areas where malaria is widespread (Black et al., 2008). The prevalence of anemia globally and in WHO regions is summarized in Table 3.12. Globally, 1.62 billion people are affected with 47.4% of preschool children, 41.8%

of pregnant women, 30.2% of nonpregnant women of child-bearing age, school-age children 25.4%, the elderly 23.9%, and men 12.7% (WHO, 2015a).

In Western countries, iron deficiency is still widespread (Table 3.11). There is insufficient iron consumption in the USA, and presumably other Western countries, particularly in adolescents, premenopausal women, and pregnant women (Dietary Guidelines Advisory Committee, 2015).

3.8.4.3 Zinc

Zinc is primarily found in animal products and seafood (Bailey et al., 2015). Low consumption of animal products is associated with zinc deficiency (Wessells and Brown, 2012). An estimate of zinc insufficiency is 17.3% globally (Wessells and Brown, 2012). Generally, there is sufficient consumption of zinc in Western countries, for instance, the USA (Table 3.11). One consequence of zinc deficiency is reduced growth with lack of zinc correlating with stunting in young children (5 years old) (Wessells and Brown, 2012).

TABLE 3.12 Prevalence of Anemia Globally and in World Health Organization (UN WHO) Regions (Joint FAO/WHO/UNU Expert Consultation, 2007c)

	Preschool children[a] number (prevalence %)	Nonpregnant women[b] number (prevalence %)	Pregnant women number (prevalence %)
African region[c]	83.5 million (67.6)	69.9 million (47.5)	17.2 million (57.1)
Americas region[d]	23.1 million (29.3)	39.0 million (17.8)	3.9 million (24.1)
East Mediterranean region[e]	46.7 million (46.7)	39.8 million (32.4)	44.2 million (46.7)
European region[f]	11.1 million (21.7)	40.8 million (19.0)	2.6 million (25.1)
Southeast Asia region[g]	115.3 million (65.5)	182.0 million (45.7)	18.1 million (48.2)
Western Pacific region[h]	27.4 million (23.1)	97.0 million (21.5)	7.6 million (30.7)
World	293.1 million (47.4)	468.4 million (30.2)	56.4 million (41.8)

[a] Less than 5 years old.
[b] From 15 to <50 years old.
[c] Africa except Egypt, Sudan, South Sudan, Tunisia, Morocco, and Somalia.
[d] North and South America, Egypt, Sudan, South Sudan, Tunisia, Morocco, and Somalia.
[e] North African countries (Egypt, Morocco, Sudan, Somalia, South Sudan, and Tunisia) plus Arab countries, Iran, Pakistan, and Afghanistan.
[f] Europe plus Asian Russia, and countries of the Commonwealth of Independent States.
[g] Bangladesh, Burma, India, Indonesia, Thailand, and Turkey.
[h] Australia, Cambodia, China, Japan, Laos, Malaysia, Oceana, and Vietnam.

3.8.4.4 Iodine

Globally, about 2 billion people are deficient in iodine (Bailey et al., 2015). The prevalence of iodide deficiency in school-age children is estimated at 30% or 241 million globally (Andersson et al., 2012). Iodine deficiency during prenatal development impairs cognitive development (reviewed by Black, 2003) with a 13.5 point reduction in IQ (Black et al., 2008). Vegans can have iodine deficiency as indicated by elevated circulating concentrations of thyroid-stimulating hormone, presumably due to reduced feedback from thyroid hormones (Appleby et al., 1999).

Nutrient Supplementation or Adequate Nutrition

It might be theoretically optimal for people to receive all nutrients in an adequate diet. However, food fortification has been implemented as a public health measure even in high-income countries with programs for public welfare. For instance, these public health policies in the USA have led to the fortification of foods and water including the following:

- folic acid in flour,
- fluoride in drinking water,
- iodide in salt, and
- vitamins A and D in milk.

In addition, soy-milk and orange juice are available that are fortified with calcium. Given the widespread deficiencies globally, there is a strong case for micronutrient supplementation. Iodide fortification has reduced iodine deficiencies in some countries.

3.8.4.5 Vitamin B_{12}

Pernicious anemia and severe vitamin B_{12} deficiency is defined as a serum concentration of vitamin B_{12} of less than 148 pmol L^{-1} or

(200 pg mL^{-1}) (Allen, 2009). Vitamin B_{12} deficiency is caused by the following:

- diets with insufficient vitamin B_{12} and
- malabsorption of vitamin B_{12}.

Vitamin B_{12} deficiency is a major problem across the world particularly in developing countries. Vitamin B_{12} deficiency may impair neural development (reviewed by Black, 2003). For instance, vitamin B_{12} deficiency and marginal depletion is observed at the following levels (reviewed by Allen, 2009):

- 40% children and adults in Latin America
- 70% children in Kenya
- 70% adults in India
- 80% children in India

There is little vitamin B_{12} deficiency in children in the USA (<1%) but it is observed in infants (3%) (reviewed by Allen, 2009). Vitamin B_{12} deficiency is a problem in developed countries particularly in the elderly. In the USA, 6% of people over 70 years old have a vitamin B_{12} deficiency and an additional 16% have marginal vitamin B_{12} depletion (reviewed by Allen, 2009). Similar levels of vitamin B_{12} deficiency are reported in the United Kingdom (reviewed by Allen, 2009). Moreover, vegans have low vitamin B_{12} intake (Table 3.11).

3.9 HUMAN HEALTH AND THE INTAKE OF ANIMAL PRODUCTS

3.9.1 Meat Consumption and Human Health

3.9.1.1 Overview

Consumption of meat meets the requirements for many nutrients including protein, zinc, vitamin B_{12}, and iron. Black et al. (2008) considered that globally "the major cause of iron deficiency anemia is low consumption of meat, fish, or poultry." Ostensibly, we would think that the diets of ovo-vegetarian and/or lacto-vegetarian and vegans might be deficient in specific nutrients. There is good information about the intake of these nutrients (Table 3.11). Perhaps surprisingly, vegetarians (particularly vegans) take in the required amounts of zinc and iron. This is presumably due to the intake of supplements and fortified foods. However, ovo-lacto-vegetarians and particularly vegans have much lower than recommended intakes of calcium and vitamin D. There is a lower bone mineral density in vegans but "the magnitude of the association is clinically insignificant" (Ho-Pham et al., 2009).

There are marked differences in the intake of major nutrients and serum concentrations of cholesterol between vegans, vegetarians, omnivores consuming fish, and meat-eating omnivores (Table 3.13). While there are no differences in energy intake, vegans consume higher levels of carbohydrates and lower levels of protein, fat, saturated fat, and cholesterol than meat-eaters with vegetarians intermediate. Moreover, vegans have lower serum concentrations of cholesterol and low-density lipoprotein (LDL) cholesterol.

TABLE 3.13   Difference in the Intake of Major Nutrients and Serum Concentrations of Cholesterol Between Vegans, Ovo-vegetarians and/or Lacto-vegetarians, Omnivores Consuming Fish, and Meat-Eating Omnivores

Parameter	Vegan (n = 114)	Vegetarian (no meat or fish but includes dairy and eggs) (n = 1550)	Omnivorous with fish but no meat consumed (n = 415)	Omnivorous with meat consumed (n = 1198)	
Intake					
Intake of energy, kcal (MJ)	2245 ± 24 (9.4 ± 0.1)	2292 ± 24 (9.4 ± 0.1)	2317 ± 24 (9.7 ± 0.1)	2257 ± 24 (9.45 ± 0.1)	N.S.
Protein as % of energy intake	11.7 ± 0.1%	12.3 ± 0.1%	13.1 ± 0.1%	15.0 ± 0.1%	$P < 0.001$
Carbohydrate as % of energy intake	51.9 ± 0.3%	47.0 ± 0.3%	43.3 ± 0.3%	43.1 ± 0.3%	$P < 0.001$
Fat as % of energy intake	34.8 ± 0.3%	36.3 ± 0.3%	39.3 ± 0.3%	38.4 ± 0.3%	$P < 0.05$
Saturated fat as % of energy intake	6.8 ± 0.4%	13.2 ± 0.6%	12.9 ± 0.7%	13.7 ± 0.6%	$P < 0.001$
Cholesterol (mg)	5 ± 0.05	234 ± 4	255 ± 4	286 ± 3	$P < 0.001$
Serum cholesterol					
Cholesterol, mg dL^{-1} (mMol L^{-1})	166 ± 5.4 (4.29 ± 0.14)	189 ± 3.9 (5.31 ± 0.10)	194 ± 4.2 (5.01 ± 0.11)	205 ± 3.9 (5.31 ± 0.10)	$P < 0.001$
HDL cholesterol, mg dL^{-1} (mMol L^{-1})	58 ± 1.9 (1.49 ± 0.05)	58 ± 1.4 (1.50 ± 0.03)	58 ± 1.4 (1.50 ± 0.03)	58 ± 1.9 (1.49 ± 0.05)	N.S.
LDL cholesterol, mg dL^{-1} (mMol L^{-1})	88 ± 4.9 (2.28 ± 0.13)	106 ± 3.5 (2.74 ± 0.09)	111 ± 3.8 (2.88 ± 0.10)	123 ± 3.5 (3.17 ± 0.09)	$P < 0.001$

N.S., Not significant.

Based on Appleby, P.N., Thorogood, M., Mann, J.I., Key, T.J., 1999. The Oxford vegetarian study: an overview. Am. J. Clin. Nutr. 70 (Suppl. 3), 525S–531S.

TABLE 3.14 Comparison of Adults Globally and in the USA who are Either Overweight or Obese

Year	Overweight, with BMI 25–29.9 (%)	Obese, with BMI >30 (%)
World (over 18 years old)[a]	Total, 1.3 billion	Total, 0.6 billion
	26	13
USA (over 20 years old)[b]		
1978	32.1	15.0
1991	32.7	23.2
2000	33.6	30.9
2008	34.2	33.8
2011	33.9	35.1

[a] Data from WHO (2015b).
[b] Data from the CDC (2015) and Ogden et al. (2014).

Meat intake was associated with increased mortality in men but not women in China (Takata et al., 2013). Moreover, there was an increase in deaths from cardiovascular disease with high intakes of meat in a study in Japan (Nagao et al., 2012). It is questioned whether red meat, poultry, or processed red meat have the same effects. This is discussed in Section 3.9.1.2.

The number of people who are overweight or obese is increasing in Western countries, for example, the USA and globally (Table 3.14). This is referred to as a global epidemic of obesity (WHO, 2015b). Among the underlying reasons for obesity appear to be the consumption of meat. For instance, the consumption of meat is associated with greater daily intake of energy and obesity using body mass index (BMI) or waist circumference (Wang and Beydoun, 2009) or weight gain (Vergnaud et al., 2010) as the

---

Body mass index (BMI)
BMI = body weight (in kilograms)/ (height in meters square)
Overweight and obesity
People with a BMI 25–29.9 are considered overweight and those with a BMI >30 are considered clinically obese.

---

determinants. Recently, it has been reported that Chinese people consuming a Western diet with high meat intake have increased obesity and BMI (Shu et al., 2015).

3.9.1.2 Red Meat and Processed Meat Consumption and Human Health

The intake of processed red meat and probably nonprocessed red meat have negative effects on health and mortality. In a study of about 500,000 people in the USA, the consumption of red and processed meat is associated with some increase in mortality and deaths from cancer and cardiovascular disease (Sinha et al., 2009). The consumption of red meat was associated with increased mortality in a European study but when the analysis was corrected for measurement error, the association was no longer present (Rohrmann et al., 2013). No relationship was found between nonprocessed red meat consumption and mortality in a metastudy of Europeans and Asians, but there was a relationship in US populations (Wang et al., 2015). In contrast, there was a strong association between consumption of processed red meat and mortality with it estimated that 3.3% of deaths could be eliminated if processed red meat intake was reduced to less than 20 g day^{-1} (Rohrmann et al., 2013). Furthermore, there was a relationship between processed red meat and mortality, together with mortalities from cancer and cardiovascular disease in a metaanalysis (Wang et al., 2015).

The intake of red meat and processed red meat is associated with increased obesity and BMI (metaanalysis by Rouhani et al., 2014). A metaanalysis showed a link between the intake of processed meat but not red meat and coronary heart disease (Micha et al., 2010). The consumption of red meat or processed red meat is associated with some increase in the risk of death in the Health Professionals Follow-up Study and Nurses Health Study in the USA (Pan et al., 2012). High consumption of red meat increases the risk of metabolic syndrome; encompassing risk factors for CVD (waistline), triglyceride, LDL, fasting glucose, and high blood pressure (Azadbakht and Esmaillzadeh, 2009).

There appears to be a link between the intake of processed meat but not red meat and incidence of type 2 diabetes mellitus (van Dam et al., 2002; metaanalysis by, Micha et al., 2010). Although the intake of fat and saturated fat is associated with an increased risk of type 2 diabetes, this is not independent of BMI (van Dam et al., 2002).

There may be a link between cancer and consuming red and processed red meat. A positive association between the consumption of red and processed red meat and colorectal cancer risk was reported possibly due to carcinogenic compounds produced during cooking (Norat et al., 2005). The high intake of well-done red meat has been reported to increase the risk of cancer (Zheng and Lee, 2009). However, a series of metaanalyses do not support a link between either red meat or processed red meat consumption and either prostate (Alexander et al., 2010a; Bylsma and Alexander, 2015) or colorectal (Alexander et al., 2011) or breast cancer (Alexander et al., 2010b).

The American Cancer Society Guidelines on nutrition and physical activity for cancer prevention include recommendations to consume a healthy diet, with an emphasis on plant foods and to limit the consumption of processed meat and red meat (Kushi et al., 2012).

3.9.1.3 Poultry Consumption and Human Health

Poultry meat meets multiple nutritional requirements and appears to have little or no detrimental effects of health. The consumption of poultry was not associated with either increased or decreased mortality in a European study (Rohrmann et al., 2013), but a decreased mortality in a study in the USA (Sinha et al., 2009). Risks of mortality are reduced by substituting poultry for red meat or processed red meat (Pan et al., 2012). No association between the consumption of poultry and colorectal cancer risk was reported (Norat et al., 2005).

3.9.1.4 Consumption of Fish, Shellfish, and Human Health

Seafood are important to the diet supplying high-quality protein, long-chain omega-3 fats, together with vitamin D and selenium (Harvard T.H. Chan School of Public Health, In press) with the American Heart Association recommending two servings per week (American Heart Association, 2017). The level of some of macronutrients in various seafood is summarized in Table 3.2 with levels of omega-3 polyunsaturated fatty acids shown in Table 3.8. A negative association between consumption of fish and colorectal cancer risk was reported (Norat et al., 2005).

Despite the high levels of cholesterol in shellfish (Table 3.2), their consumption is associated with either decreased or a tendency for decreased circulating concentrations of LDL (Childs et al., 1990). Moreover, cholesterol absorption was depressed (Childs et al., 1990). Moreover, mollusks are rich in noncholesterol sterols that may interfere with the absorption of cholesterol (Childs et al., 1987, 1990).

In the USA, government agencies have concluded that pregnant and breastfeeding women

TABLE 3.15 Mercury Contamination of Fish

Species	Mercury (ppb)
Scallop (3 ppb), shrimp (9 ppb)	<10
Oyster (12 ppb), tilapia (13 ppb), salmon (22 ppb), catfish (25 ppb), crawfish (33 ppb), crab (65 ppb), trout (71 ppb), and herring (84 ppb)	10–100
Cod (111 ppb), perch (150 ppb), lobster (93–166 ppb), snapper (166 ppb), halibut (241 ppb), and Chilean seabass (354 ppb)	100–400
Tuna (144–689 ppb), grouper (448 ppb), and orange roughly (571 ppb)	400–700
King mackerel (730 ppb), shark (979 ppb), swordfish (995 ppb), and tilefish (1450 ppb)	>700

Based on data from the FDA (Food and Drug Administration), In Press. Mercury levels in fish. Available from: http://www.fda.gov/food/foodborneillnesscontaminants/metals/ucm115644.htm

and young children were consuming very little seafood and recommended increased seafood intake (EPA and FDA, 2014). In addition, the recommendation was to avoid fish with high levels of mercury (tilefish, shark, swordfish, and king mackerel) (Table 3.15).

3.9.2 Milk Consumption and Human Health

Milk has high biological value protein, vitamin B_{12}, niacin, vitamin B_6, iron, zinc, phosphorus (phosphate), long-chain omega-3 polyunsaturated fats, riboflavin, pantothenic acid, selenium, together with vitamins A and D in fortified milk. Whole milk and cheese consumption is inversely related to weight gain (Rosell et al., 2006).

Gao et al. (2005) conducted a metaanalysis of studies on the effect of consumption of milk on the rates of prostate cancer. They concluded that "High intake of dairy products and calcium may be associated with an increased risk of prostate cancer, although the increase appears to be small." In a more recent study, there was a two-fold increase in prostate cancer with dairy products, but the effect was only observed with milk and not with other dairy products (Raimondi et al., 2010). The intake of dairy products was associated with a small increase in the risk of nonaggressive prostate cancer but not the aggressive form of the disease with the effects particularly observed with low-fat dairy products (Ahn et al., 2007).

3.9.3 Impact of Egg Consumption and Human Health

Eggs are an excellent source of protein, minerals, and vitamins. There are little or no negative effects of low to moderate intake of eggs. For instance, no linkage was reported between the intake of eggs, up to one per day, and serum concentrations of cholesterol or incidence of myocardial infarction, coronary heart disease, or stroke (Djousse and Gaziano, 2008; Hu et al., 1999; Nakamura et al., 2006; metaanalysis; Rong et al., 2013). In contrast, a higher intake of eggs increases total and LDL cholesterol. Serum cholesterol was high in people consuming two eggs per day compared to egg substitutes (Chenoweth et al., 1981). Moreover, high egg intakes (two per day) increased LDL and high-density lipoprotein (HDL) cholesterol in people with elevated serum cholesterol compared to egg substitutes (Knopp et al., 1997). Similarly, there were markedly higher concentrations of LDL and some increase in HDL between moderate egg intake (1 per day) and high (two or greater) (Weggemans et al., 2001).

3.9.4 Low-Carbohydrate Diets (Such as Atkins Diet)

Low-carbohydrate diets, such as the Atkins diet are effective in achieving weight loss (Gardner et al., 2007). These often contain high amounts of animal products. When a high protein/high fat low-carbohydrate diet was compared with a conventional low-fat diet, both were effective in weight loss and there were no differences in LDL (Foster et al., 2003). Moreover, low-carbohydrate and isocaloric-balanced diets achieve similar weight losses and decreases in circulating concentrations of cholesterol and LDL (Naude et al., 2014). In contrast, when low-carbohydrate and low-fat diets were compared as weight loss programs in a metaanalysis, the low-carbohydrate diet was more effective but was associated with an increase in LDL (Mansoor et al., 2016).

3.10 FOOD SAFETY AND ANIMAL PRODUCTS

3.10.1 Pathogen Contamination

Pathogens are found in many animal products. This is discussed in detail in Chapter 19 (specifically Section 19.8). To combat food-borne diseases, there are a number of approaches:

- Preharvest and processing to reduce the pathogen load.
- Hazard Analysis Critical Control Point analysis with the critical control being cooking for meat and pasteurization for milk and shelled eggs (FDA, 2015).

3.10.2 Toxicant Contamination

There is little or no contamination of animal products with toxicants. There is one major exception, mercury. Methyl mercury bioaccumulates in aquatic animals and when consumed has adverse effects including impaired neurological development (Bose-O'Reilly et al., 2010). Table 3.15 gives examples of the levels of mercury contamination in fish. With inspection, there is little or no contamination of animal products with antibiotics.

3.11 SUSTAINABILITY AND CONSUMPTION OF ANIMAL PRODUCTS

In the USA, food production uses 50% of the land area of the country together with 80% of its freshwater and 17% of its energy utilization (Pimentel and Pimentel, 2003). Except for range animals (predominantly cattle), livestock and poultry production takes a disproportionately high amount of land (to produce animal feed, such as corn, soybeans, and hay), water for irrigation and production and processing of livestock and poultry, and energy (e.g., for fertilizers, planting, and harvesting of plants for animal feed). Moreover, it has been calculated that producing a kilogram of beef protein requires 100-fold more water compared to a kilogram of wheat protein (Pimentel and Pimentel, 2003).

It is argued that agriculture in the USA and other Western countries is not sustainable with its high-energy inputs and water usage irrespective of whether an omnivorous (including meat) or a lacto-ovo-vegetarian diet is assumed (Pimentel and Pimentel, 2003). This will be exacerbated with the needs of a growing population, losses of agricultural land, soil loss, and restrictions on fossil energy (either by availability or price) with measures to mitigate climate change (Godfray et al., 2010). Parenthetically, in the USA, omnivorous people consume 111 g protein day^{-1}, which is about twice the RDA (Pimentel and Pimentel, 2003).

3.12 CONCLUSIONS

Rather than drawing conclusions, a series of questions are posed:

- How can we meet the nutrient requirements of the world's growing population with or without animal products?

- Does eating meat, milk, eggs, and fish and hence having livestock, poultry, and aquaculture species compete with people for food resources? If you consider the answer to be yes, what are the consequences?
  - Should we transition to a vegan diet?
  - What are the consequences for nutrition and health?
  - Should this be voluntary or mandatory? If mandatory, is this consistent with freedom and human rights?
  - Should ownership of companion animals be outlawed as they consume high protein or animal product diets?
  - To what extent do livestock and poultry provide invaluable fertilizer particularly in developing countries?

Are the questions setting up a series of false dichotomies? Do national and state parks, wildlife reserve forests, and large suburban yards use land that otherwise could (or would) be used for raising crops?

References

Ahn, J., Albanes, D., Peters, U., Schatzkin, A., Lim, U., Freedman, M., Chatterjee, N., Andriole, G.L., Leitzmann, M.F., Hayes, R.B., Prostate, Lung, Colorectal, and Ovarian Trial Project Team, 2007. Dairy products, calcium intake, and risk of prostate cancer in the prostate, lung, colorectal, and ovarian cancer screening trial. Cancer Epidemiol. Biomarkers Prev. 16, 2623–2630.

Alexander, D.D., Mink, P.J., Cushing, C.A., Sceurman, B., 2010a. A review and meta-analysis of prospective studies of red and processed meat intake and prostate cancer. Nutr. J. 9, 50.

Alexander, D.D., Morimoto, L.M., Mink, P.J., Cushing, C.A., 2010b. A review and meta-analysis of red and processed meat consumption and breast cancer. Nutr. Res. Rev. 23, 349–365.

Alexander, D.D., Weed, D.L., Cushing, C.A., Lowe, K.A., 2011. Meta-analysis of prospective studies of red meat consumption and colorectal cancer. Eur. J. Cancer Prev. 20, 293–307.

Allen, L.H., 2009. How common is vitamin B-12 deficiency? Am. J. Clin. Nutr. 89, 693S–696S.

American Heart Association, 2017. Eating Fish for Heart Health. Available from: http://www.heart.org/HEARTORG/GettingHealthy/NutritionCenter/HealthyEating/Eating-Fish-for-Heart-Health_UCM_440433_Article.jsp

Andersson, M., Karumbunathan, V., Zimmermann, M.B., 2012. Global iodine status in 2011 and trends over the past decade. J. Nutr. 142, 744–750.

Appleby, P.N., Thorogood, M., Mann, J.I., Key, T.J., 1999. The Oxford Vegetarian Study: an overview. Am. J. Clin. Nutr. 70 (Suppl. 3), 525S–531S.

Azadbakht, L., Esmaillzadeh, A., 2009. Red meat intake is associated with metabolic syndrome and the plasma C-reactive protein concentration in women. J. Nutr. 139, 335–339.

Bailey, R.L., West, Jr., K.P., Black, R.E., 2015. The epidemiology of global micronutrient deficiencies. Ann. Nutr. Metab. 66 (Suppl. 2), 22–33.

Black, M.M., 2003. Micronutrient deficiencies and cognitive functioning. J. Nutr. 133 (Suppl. 2), 3927S–3931S.

Black, R.E., Allen, L.H., Bhutta, Z.A., Caulfield, L.E., de Onis, M., Ezzati, M., Mathers, C., Rivera, J., Maternal, Child Undernutrition Study Group, 2008. Maternal and child undernutrition: global and regional exposures and health consequences. Lancet 371, 243–260.

Black, R.E., Williams, S.M., Jones, I.E., Goulding, A., 2002. Children who avoid drinking cow milk have low dietary calcium intakes and poor bone health. Am. J. Clin. Nutr. 76, 675–680.

Bose-O'Reilly, S., McCarty, K.M., Steckling, N., Lettmeier, B., 2010. Mercury exposure and children's health. Curr. Probl. Pediatr. Adolesc. Health Care 40, 186–215.

Burger, I.H., Barnett, K.C., 1982. The taurine requirement of the adult cat. J. Small Anim. Pract. 23, 533–537.

Bylsma, L.C., Alexander, D.D., 2015. A review and meta-analysis of prospective studies of red and processed meat, meat cooking methods, heme iron, heterocyclic amines and prostate cancer. Nutr. J. 14, 125.

CDC (Centers for Disease Control), 2015. Adult obesity facts. Available from: http://www.cdc.gov/obesity/data/adult.html

Chenoweth, W., Ullmann, M., Simpson, R., Leveille, G., 1981. Influence of dietary cholesterol and fat on serum lipids in men. J. Nutr. 111, 2069–2080.

Childs, M.T., Dorsett, C.S., Failor, A., Roidt, L., Omenn, G.S., 1987. Effect of shellfish consumption on cholesterol absorption in normolipidemic men. Metabolism 36, 31–35.

Childs, M.T., Dorsett, C.S., King, I.B., Ostrander, J.G., Yamanaka, W.K., 1990. Effects of shellfish consumption on lipoproteins in normolipidemic men. Am. J. Clin. Nutr. 51, 1020–1027.

Cook, J.D., Skikne, B.S., Baynes, R.D., 1994. Iron deficiency: the global perspective. Adv. Exp. Med. Biol. 356, 219–228.

Craig, W.J., 2010. Nutritional concerns and vegetarian diets. Nutr. Clin. Pract. 25, 613–620.

Daley, C.A., Abbott, A., Doyle, P.S., Nader, G.A., Larson, S., 2010. A review of fatty acid profiles and antioxidant content in grass-fed and grain-fed beef. Nutr. J. 9, 10.

Davis, B.C., Kris-Etherton, P.M., 2003. Achieving optimal essential fatty acid status in vegetarians: current knowledge and practical implications. Am. J. Clin. Nutr. 78 (Suppl. 3), 640S–646S.

de Benoist, B., McLean, E., Egli, I., Cogswell, M., 2008. Worldwide Prevalence of Anaemia 1993–2005: WHO Global Database on Anaemia. World Health Organization, Geneva.

de Vrese, M., Stegelmann, A., Richter, B., Fenselau, S., Laue, C., Schrezenmeir, J., 2001. Probiotics—compensation for lactase insufficiency. Am. J. Clin. Nutr. 73 (Suppl. 2), 421S–429S.

Dietary Guidelines Advisory Committee. 2015. Scientific report of the 2015 dietary guidelines advisory committee. Available from: http://health.gov/dietaryguidelines/2015-scientific-report/pdfs/scientific-report-of-the-2015- dietary-guidelines-advisory-committee.pdf

Djousse, L., Gaziano, J.M., 2008. Egg consumption and risk of heart failure in the Physicians' Health Study. Circulation 117, 512–516.

Earle, K.E., Smith, P.M., 1994. The taurine requirement of the kitten fed canned foods. J. Nutr. 124 (Suppl. 12), 2552S–2554S.

EPA and FDA (US Environmental Protection Agency and Food and Drug Administration), 2014 http://www2.epa.gov/choose-fish-and-shellfish-wisely

FAO, 2003. Global and regional food consumption patterns and trends (Chapter 3). Available from: http://www.fao.org/docrep/005/ac911e/ac911e05.htm.

FAO, 2013. An introduction to the basic concepts of food security. Available from: http://www.fao.org/docrep/013/al936e/al936e00.pdf

FAO, 2015a. World Agriculture: Towards 2015/2030. An FAO perspective. Available from: http://www.fao.org/docrep/005/y4252e/y4252e07.htm

FAO, 2015b. The state of food insecurity in the world. Available from: http://www.fao.org/3/a-i4646e.pdf

FAO, 2017. Fisheries and Aquaculture Department Statistics. Available from: http://www.fao.org/fishery/statistics/en

FDA, 2015. Hazard Analysis Critical Control Point (HACCP). Available from: http://www.fda.gov/Food/GuidanceRegulation/HACCP/

Foster, G.D., Wyatt, H.R., Hill, J.O., McGuckin, B.G., Brill, C., Mohammed, B.S., Szapary, P.O., Rader, D.J., Edman, J.S., Klein, S., 2003. A randomized trial of a low-carbohydrate diet for obesity. New Engl. J. Med. 348, 2082–2090.

Freund, J.N., Duluc, I., Foltzer-Jourdainne, C., Gosse, F., Raul, F., 1990. Specific expression of lactase in the jejunum and colon during postnatal development and hormone treatments in the rat. Biochem. J. 268, 99–103.

Gardner, C.D., Kiazand, A., Alhassan, S., Kim, S., Stafford, R.S., Balise, R.R., Kraemer, H.C., King, A.C., 2007. Comparison of the Atkins, Zone, Ornish, and LEARN diets for change in weight and related risk factors among overweight premenopausal women: the A TO Z Weight Loss Study: a randomized trial. J. Am. Med. Assoc. 297, 969–977.

Godfray, H.C., Beddington, J.R., Crute, I.R., Haddad, L., Lawrence, D., Muir, J.F., Pretty, J., Robinson, S., Thomas, S.M., Toulmin, C., 2010. Food security: the challenge of feeding 9 billion people. Science 327, 812–818.

Gao, X., LaValley, M.P., Tucker, K.L., 2005. Prospective studies of dairy product and calcium intakes and prostate cancer risk: a meta-analysis. J. Natl. Cancer Inst. 97, 1768–1777.

Gerbault, P., Liebert, A., Itan, Y., Powell, A., Currat, M., Burger, J., Swallow, D.M., Thomas, M.G., 2011. Evolution of lactase persistence: an example of human niche construction. Philos. Trans. R. Soc. Lond. B 366, 863–877.

Grand, R.J., Montgomery, R.K., Chitkara, D.K., Hirschhorn, J.N., 2003. Changing genes; losing lactase. Gut 52, 617–619.

Hallahan, B., Garland, M.R., 2005. Essential fatty acids and mental health. Br. J. Psychiatry 186, 275–277.

Harkness, M.L., Harkness, R.D., Venn, M.F., 1978. Digestion of native collagen in the gut. Gut 19, 240–243.

Harvard T.H. Chan School of Public Health, In Press. Available from: http://www.hsph.harvard.edu/nutrition-source/fish/

Ho-Pham, L.T., Nguyen, N.D., Nguyen, T.V., 2009. Effect of vegetarian diets on bone mineral density: a Bayesian meta-analysis. Am. J. Clin. Nutr. 90, 943–950.

Hu, F.B., Stampfer, M.J., Rimm, E.B., Manson, J.E., Ascherio, A., Colditz, G.A., Rosner, B.A., Spiegelman, D., Speizer, F.E., Sacks, F.M., Hennekens, C.H., Willett, W.C., 1999. A prospective study of egg consumption and risk of cardiovascular disease in men and women. J. Am. Med. Assoc. 281, 1387–1394.

Ingram, C.J., Mulcare, C.A., Itan, Y., Thomas, M.G., Swallow, D.M., 2009. Lactose digestion and the evolutionary genetics of lactase persistence. Hum. Genet. 124, 579–591.

Institute of Medicine Panel for Macronutrients, 2002/2005. Dietary reference intakes for energy, carbohydrate, fiber, fat, fatty acid, cholesterol, protein, and amino-acids. National Academies Press, Washington, DC. Available from: https://www.nal.usda.gov/fnic/DRI/DRI_Energy/energy_full_report.pdf

Jensen, R., Miller, N., 2011. A revealed preference approach to measuring hunger and undernutrition. Available from: publicaffairs.ucla.edu/robert-jensen summarized in the Economist, March 26th–April 1st, p. 88.

Joint FAO/WHO/UNU Expert Consultation, 2007a. Amino acid requirements of adults. In: Protein and Amino-Acid Requirements in Human Nutrition. pp. 135–159, World Health Organization. Available from: http://whqlibdoc.who.int/trs/who_trs_935_eng.pdf

Joint FAO/WHO/UNU Expert Consultation, 2007b. Energy requirements for pregnancy. Available from: http://www.fao.org/docrep/007/y5686e/y5686e0a.htm#TopOfPage.

Joint FAO/WHO/UNU Expert Consultation, 2007c. Protein and Amino Acid Requirements in Human Nutrition.

Available from: http://www.who.int/nutrition/publications/nutrientrequirements/WHO_TRS_935/en/.

Kar, B.R., Rao, S.L., Chandramouli, B.A., 2008. Cognitive development in children with chronic protein energy malnutrition. Behav. Brain Funct. 4, 31.

Knopp, R.H., Retzlaff, B.M., Walden, C.E., Dowdy, A.A., Tsunehara, C.H., Austin, M.A., Nguyen, T., 1997. A double-blind, randomized, controlled trial of the effects of two eggs per day in moderately hypercholesterolemic and combined hyperlipidemic subjects taught the NCEP step I diet. J. Am. Coll. Nutr. 16, 551–561.

Krebs, N.F., Mazariegos, M., Tshefu, A., Bose, C., Sami, N., Chomba, E., Carlo, W., Goco, N., Kindem, M., Wright, L.L., Hambidge, K.M., Complementary Feeding Study Group, 2011. Meat consumption is associated with less stunting among toddlers in four diverse low-income settings. Food Nutr. Bull. 32, 185–191.

Kushi, L.H., Doyle, C., McCullough, M., Rock, C.L., Demark-Wahnefried, W., Bandera, E.V., Gapstur, S., Patel, A.V., Andrews, K., Gansler, T., American Cancer Society 2010 Nutrition and Physical Activity Guidelines Advisory Committee, 2012. American Cancer Society Guidelines on nutrition and physical activity for cancer prevention: reducing the risk of cancer with healthy food choices and physical activity. Ca-A Cancer J. Clin. 62, 30–67.

Mansoor, N., Vinknes, K.J., Veierød, M.B., Retterstøl, K., 2016. Effects of low-carbohydrate diets v. low-fat diets on body weight and cardiovascular risk factors: a meta-analysis of randomised controlled trials. Br. J. Nutr. 115, 466–479.

Martorell, R., Horta, B.L., Adair, L.S., Stein, A.D., Richter, L., Fall, C.H., Bhargava, S.K., Biswas, S.K., Perez, L., Barros, F.C., Victora, C.G., Consortium on Health Orientated Research in Transitional Societies Group, 2010. Weight gain in the first two years of life is an important predictor of schooling outcomes. J. Nutr. 140, 348–354.

Micha, R., Wallace, S.K., Mozaffarian, D., 2010. Red and processed meat consumption and risk of incident coronary heart disease, stroke, and diabetes mellitus: a systematic review and meta-analysis. Circulation 121, 2271–2283.

Mozaffarian, D., Rimm, E.B., 2006. Fish intake, contaminants, and human health: evaluating the risks and the benefits. J. Am. Med. Assoc. 296, 1885–1899.

Müller, O., Krawinkel, M., 2005. Malnutrition and health in developing countries. Can. Med. Assoc. J. 173, 279–286.

Nagao, M., Iso, H., Yamagishi, K., Date, C., Tamakoshi, A., 2012. Meat consumption in relation to mortality from cardiovascular disease among Japanese men and women. Eur. J. Clin. Nutr. 66, 687–693.

Nakamura, Y., Iso, H., Kita, Y., Ueshima, H., Okada, K., Konishi, M., Inoue, M., Shoichiro Tsugane, S., 2006. Egg consumption, serum total cholesterol concentrations and coronary heart disease incidence: Japan Public Health Center-based prospective study. Br. J. Nutr. 96, 921–928.

National Institute of Health, 2016. Office of Dietary Supplements. Available from: https://ods.od.nih.gov/factsheets/VitaminD-HealthProfessional/

National Research Council, 1986. Nutrient Requirements of Cats. National Academy Press, Washington, DC.

Naude, C.E., Schoonees, A., Senekal, M., Young, T., Garner, P., Volmink, J., 2014. Low carbohydrate versus isoenergetic balanced diets for reducing weight and cardiovascular risk: a systematic review and meta-analysis. PLoS One 9, e100652.

Norat, T., Bingham, S., Ferrari, P., Slimani, N., Jenab, M., Mazuir, M., Overvad, K., Olsen, A., Tjønneland, A., Clavel, F., Boutron-Ruault, M.C., Kesse, E., Boeing, H., Bergmann, M.M., Nieters, A., Linseisen, J., Trichopoulou, A., Trichopoulos, D., Tountas, Y., Berrino, F., Palli, D., Panico, S., Tumino, R., Vineis, P., Bueno-de-Mesquita, H.B., Peeters, P.H., Engeset, D., Lund, E., Skeie, G., Ardanaz, E., González, C., Navarro, C., Quirós, J.R., Sanchez, M.J., Berglund, G., Mattisson, I., Hallmans, G., Palmqvist, R., Day, N.E., Khaw, K.T., Key, T.J., San Joaquin, M., Hémon, B., Saracci, R., Kaaks, R., Riboli, E., 2005. Meat, fish, and colorectal cancer risk: the European Prospective Investigation into cancer and nutrition. J. Natl. Cancer Inst. 97, 906–916.

Ogden, C.L., Carroll, M.D., Kit, B.K., Flegal, K.M., 2014. Prevalence of childhood and adult obesity in the United States, 2011–2012. J. Am. Med. Assoc. 311, 806–814.

Pan, A., Sun, Q., Bernstein, A.M., Schulze, M.B., Manson, J.E., Stampfer, M.J., Willett, W.C., Hu, F.B., 2012. Red meat consumption and mortality. Arch. Intern. Med. 172, 555–563.

Papanikolaou, Y., Brooks, J., Reider, C., Fulgoni, 3rd, V.L., 2014. US adults are not meeting recommended levels for fish and omega-3 fatty acid intake: results of an analysis using observational data from NHANES 2003–2008. Nutr. J. 13, 31.

Pimentel, D., Pimentel, M., 2003. Sustainability of meat-based and plant-based diets and the environment. Am. J. Clin. Nutr. 78 (Suppl. 3), 660S–663S.

Popkin, B.M., Adair, L.S., Ng, S.W., 2012. Now and then: the global nutrition transition: the pandemic of obesity in developing countries. Nutr. Rev. 70, 3–21.

Raimondi, S., Mabrouk, J.B., Shatenstein, B., Maisonneuve, P., Ghadirian, P., 2010. Diet and prostate cancer risk with specific focus on dairy products and dietary calcium: a case-control study. Prostate 70, 1054–1065.

Rohrmann, S., Overvad, K., Bueno-de-Mesquita, H.B., Jakobsen, M.U., Egeberg, R., Tjønneland, A., Nailler, L., Boutron-Ruault, M.C., Clavel-Chapelon, F., Krogh, V., Palli, D., Panico, S., Tumino, R., Ricceri, F., Bergmann, M.M., Boeing, H., Li, K., Kaaks, R., Khaw, K.T., Wareham, N.J., Crowe, F.L., Key, T.J., Naska, A., Trichopoulou, A., Trichopoulos, D., Leenders, M., Peeters, P.H., Engeset, D., Parr, C.L., Skeie, G., Jakszyn, P., Sánchez, M.J., Huerta, J.M., Redondo, M.L., Barricarte, A., Amiano, P., Drake,

I., Sonestedt, E., Hallmans, G., Johansson, I., Fedirko, V., Romieux, I., Ferrari, P., Norat, T., Vergnaud, A.C., Riboli, E., Linseisen, J., 2013. Meat consumption and mortality—results from the European Prospective Investigation into Cancer and Nutrition. BMC Med. 11, 63.

Rong, Y., Chen, L., Zhu, T., Song, Y., Yu, M., Shan, Z., Sands, A., Hu, F.B., Liu, L., 2013. Egg consumption and risk of coronary heart disease and stroke: dose-response meta-analysis of prospective cohort studies. Br. Med. J. 2013, 346.

Rosell, M., Håkansson, N.N., Wolk, A., 2006. Association between dairy food consumption and weight change over 9 y in 19,352 perimenopausal women. Am. J. Clin. Nutr. 84, 1481–1488.

Rosell, M.S., Lloyd-Wright, Z., Appleby, P.N., Sanders, T.A., Allen, N.E., Key, T.J., 2005. Long-chain $n-3$ polyunsaturated fatty acids in plasma in British meat-eating, vegetarian, and vegan men. Am. J. Clin. Nutr. 82, 327–334.

Rouhani, M.H., Salehi-Abargouei, A., Surkan, P.J., Azadbakht, L., 2014. Is there a relationship between red or processed meat intake and obesity? A systematic review and meta-analysis of observational studies. Obes. Rev. 15, 740–748.

Shu, L., Zheng, P.F., Zhang, X.Y., Si, C.J., Yu, X.L., Gao, W., Zhang, L., Liao, D., 2015. Association between dietary patterns and the indicators of obesity among Chinese: a cross-sectional study. Nutrients 7, 7995–8009.

Sinha, R., Cross, A.J., Graubard, B.I., Leitzmann, M.F., Schatzkin, A., 2009. Meat intake and mortality: a prospective study of over half a million people. Arch. Intern. Med. 169, 562–571.

Swallow, D.M., 2003. Genetics of lactase persistence and lactose intolerance. Annu. Rev. Genet. 37, 197–219.

Takata, Y., Shu, X.O., Gao, Y.T., Li, H., Zhang, X., Gao, J., Cai, H., Yang, G., Xiang, Y.B., Zheng, W., 2013. Red meat and poultry intakes and risk of total and cause-specific mortality: results from cohort studies of Chinese adults in Shanghai. PLoS One 8, e56963.

Tivey, D.R., Hilton, K.J., Dauncey, M.J., 1991. Compensatory increase in lactase expression by enterocytes of neonatal pigs on a low energy intake. Exp. Physiol. 76, 285–288.

United Nations (UN), 2015. 17 Goals for sustainable development. Available from: http://www.un.org/millenniumgoals/pdf/MDG_Gap_2015_E_web.pdf

USDA (United States Department of Agriculture), 2017. National Nutrient Database for Standard Reference. Available from: http://ndb.nal.usda.gov/ndb/foods

van Dam, R.M., Willett, W.C., Rimm, E.B., Stampfer, M.J., Hu, F.B., 2002. Dietary fat and meat intake in relation to risk of type 2 diabetes in men. Diab. Care 25, 417–424.

Vergnaud, A.C., Norat, T., Romaguera, D., Mouw, T., May, A.M., Travier, N., Luan, J., Wareham, N., Slimani, N., Rinaldi, S., Couto, E., Clavel-Chapelon, F., Boutron-Ruault, M.C., Cottet, V., Palli, D., Agnoli, C., Panico, S., Tumino, R., Vineis, P., Agudo, A., Rodriguez, L., Sanchez, M.J., Amiano, P., Barricarte, A., Huerta, J.M., Key, T.J., Spencer, E.A., Bueno-de-Mesquita, B., Büchner, F.L., Orfanos, P.,

Naska, A., Trichopoulou, A., Rohrmann, S., Hermann, S., Boeing, H., Buijsse, B., Johansson, I., Hellstrom, V., Manjer, J., Wirfält, E., Jakobsen, M.U., Overvad, K., Tjonneland, A., Halkjaer, J., Lund, E., Braaten, T., Engeset, D., Odysseos, A., Riboli, E., Peeters, P.H., 2010. Meat consumption and prospective weight change in participants of the EPIC-PANACEA study. Am. J. Clin. Nutr. 92, 398–407.

Wang, Y., Beydoun, M.A., 2009. Meat consumption is associated with obesity and central obesity among US adults. Int. J. Obes. 33, 621–628.

Wang, X., Lin, X., Ouyang, Y.Y., Liu, J., Zhao, G., Pan, A., Hu, F.B., 2015. Red and processed meat consumption and mortality: dose-response meta-analysis of prospective cohort studies. Public Health Nutr. 6, 1–13.

Weggemans, R.M., Zock, P.L., Katan, M.B., 2001. Dietary cholesterol from eggs increases the ratio of total cholesterol to high-density lipoprotein cholesterol in humans: a meta-analysis. Am. J. Clin. Nutr. 73, 885–891.

Wessells, K.R., Brown, K.H., 2012. Estimating the global prevalence of zinc deficiency: results based on zinc availability in national food supplies and the prevalence of stunting. PLoS One 7, e50568.

Whigham, L.D., Watras, A.C., Schoeller, D.A., 2007. Efficacy of conjugated linoleic acid for reducing fat mass: a meta-analysis in humans. Am. J. Clin. Nutr. 85, 1203–1211.

WHO, 2015a. Millennium Development Goals (MDGs). Available from: http://www.who.int/mediacentre/factsheets/fs290/en/

WHO, 2015b. Obesity and overweight. Available from: http://www.who.int/mediacentre/factsheets/fs311/en/

Zheng, W., Lee, S.A., 2009. Well-done meat intake, heterocyclic amine exposure, and cancer risk. Nutr. Cancer 61, 437–446.

Further Reading

Evershed, R.P., Payne, S., Sherratt, A.G., Copley, M.S., Coolidge, J., Urem-Kotsu, D., Kotsakis, K., Ozdoğan, M., Ozdoğan, A.E., Nieuwenhuyse, O., Akkermans, P.M., Bailey, D., Andeescu, R.R., Campbell, S., Farid, S., Hodder, I., Yalman, N., Ozbasaran, M., Biçakci, E., Garfinkel, Y., Levy, T., Burton, M.M., 2008. Earliest date for milk use in the Near East and southeastern Europe linked to cattle herding. Nature 455, 528–531.

Marangoni, F., Corsello, G., Cricelli, C., Ferrara, N., Ghiselli, A., Lucchin, L., Poli, A., 2015. Role of poultry meat in a balanced diet aimed at maintaining health and wellbeing: an Italian consensus document. Food Nutr. Res. 59, 27606.

Phillips, K.M., Ruggio, D.M., Exler, J., Patterson, K.Y., 2012. Sterol composition of shellfish species commonly consumed in the United States. Food Nutr. Res. 56, 18931.

Salque, M., Bogucki, P.I., Pyzel, J., Sobkowiak-Tabaka, I., Grygiel, R., Szmyt, M., Evershed, R.P., 2013. Earliest evidence for cheese making in the sixth millennium BC in Northern Europe. Nature 493, 522–525.

Hunter–Gatherers

Colin G. Scanes

University of Wisconsin–Milwaukee, Milwaukee, WI, United States

4.1 INTRODUCTION

Humans and their ancestors were hunter–gatherers for over 99% of the existence of the genus *Homo* (Cordain et al., 2000). However, "the history of the past 13,000 years consists of tales of hunter–gatherer societies becoming driven out, infected, conquered or exterminated by farming societies in every area of the world suitable for farming" (Diamond, 2002).

This chapter will discuss present-day and Paleolithic hunter–gatherer societies together with the Paleolithic diet and fishing. As opposed to following a chronological approach, present-day hunter–gatherers are considered first. This is due to our ability to study and quantify metrics in present-day hunter–gatherers. The caveat is that there are marked differences between individual societies of present-day hunter–gatherers. Second, present-day hunter–gatherers trade with agricultural communities. For instance, the Agta people of the Philippines trade meat for plant foods and consumer goods with adjacent agricultural communities (Konner, 2005).

4.2 PRESENT HUNTER–GATHERER SOCIETIES

4.2.1 Animal Products in the Diets of Existing Hunter–Gatherer Societies

Presently, hunter–gatherer societies rely on over 60% animal products for energy across regions geographically (Table 4.1) (Cordain et al., 2000). There is considerable variation with some such societies using plant products for the majority of their energy requirements, but most using animal products for the majority of their energy needs (Table 4.2) (Cordain et al., 2000). A different analysis indicated that Paleolithic hunter–gatherers and presumably present hunter–gatherers consumed 65% of their energy requirements from plants and 35% from animals (Eaton et al., 1997).

The diet of present hunter–gatherers has been compared to Western diets (Table 4.3). Even with the high range between different societies, there is consistently greater consumption of protein in hunter–gatherers (Table 4.3).

TABLE 4.1 Subsistence in Present-Day Hunter–Gatherer Societies

Environment	Dependency on plant foods (%)	Dependency on hunted animal foods (%)	Dependency on fished animal foods (%)
Tundra	10.5	40.5	50.5
Temperate	27.3	31.9	22.7
Subtropical	40.5	43.8	17.2
Tropical	41.6	30.5	29.5

Calculated from Cordain, L., Miller, J.B., Eaton, S.B., Mann, N., Holt, S.H., Speth, J.D., 2000. Plant-animal subsistence ratios and macronutrient energy estimations in worldwide hunter-gatherer diets. Am. J. Clin. Nutr. 71, 682–692.

TABLE 4.2 Present Hunter–Gatherer Societies Consuming More Than 50% of Energy From Animals or Plants

Source	Percentage of hunter–gatherer societies (%)
>50% Animal foods	73
>50% Plant foods	13.5

Calculated from Cordain, L., Miller, J.B., Eaton, S.B., Mann, N., Holt, S.H., Speth, J.D., 2000. Plant-animal subsistence ratios and macronutrient energy estimations in worldwide hunter-gatherer diets. Am. J. Clin. Nutr. 71, 682–692.

TABLE 4.3 Comparison of Dietary Constituents of Present Hunter–Gatherer Societies With Western Diets

Dietary constituents	Present hunter–gatherers (%)	Western diet (%)
Carbohydrate	22–40	49
Protein	19–35	15.5
Fat	28–58	34
Alcohol	—	3.1

Data from Cordain, L., Miller, J.B., Eaton, S.B., Mann, N., Holt, S.H., Speth, J.D., 2000. Plant-animal subsistence ratios and macronutrient energy estimations in worldwide hunter-gatherer diets. Am. J. Clin. Nutr. 71, 682–692.

4.2.2 Energetics and Hunter–Gatherers

There have been studies of the time spent working by hunter–gatherers. In a "classical" study, the mean time spent working was reported as 202 min day^{-1} or 23.6 h week^{-1} across hunter–gatherer societies in Australia, Africa, the Philippines, South India, Alaska, and South America (Belovsky, 1987). The time expenditure to gather meat and plant foods have been analyzed in the !Kung San people of the Kalahari in Southern Africa:

- Meat requires 0.34 min of work per gram cooked meat (including time for cooking and repairing tools).

- Mongongo nuts, the main gathered plant food, requires 0.12 min of work to gather and transport g^{-1} nuts (Belovsky, 1987).

With the differences in gender roles, women provide 69% of the food by foraging, but work about 85% more time than men in the !Kung San (see Fig. 4.1) (Bentley, 1985).

There have been intensive studies of the Hadza people in Tanzania (East Africa). They have markedly lower body fat than people on Western diets and subsistence farmers. It was reported that their total daily energy expenditure (TEE) (kcal day^{-1}) was no different (tended to be lower) than in hunter–gatherers but with more

FIGURE 4.1 **Hunter–gatherer Khoisan group in Namibia, Southern Africa.** *Source: Reproduced from https://pixabay.com/en/bushman-indigenous-people-509238/. Public domain.*

physical activity than either Western peoples or subsistence farmers (Pontzer et al., 2012), or TEE corrected for fat-free body weight (Pontzer et al., 2015). The amount of walking daily has been quantified:

- Men: 11.4 km day^{-1} (11% of TEE)
- Women: 5.8 km day^{-1} (7% of TEE) (Pontzer et al., 2012)

Meat represents about 37% of the food brought back to camps by the Hadza (Pontzer et al., 2012).

It is assumed that in hunter–gatherer societies, children contribute little to foraging. This is the case with the !Kung in Southern Africa (Blurton-Jones et al., 1994). In contrast, in the Hadza in north-central Tanzania, young children spend about 26 min day^{-1} foraging while older children spend no time (<1 min day^{-1}) foraging per

day (Hawkes et al., 1995). In hunter–gatherer societies, children are taught by adults and older children (Konner, 2016).

Many hunter–gatherer societies are nomadic moving their camps to locales at regular intervals to have access to new foods. Movement to a new camp requires time and energy for surveying for optimal locations, planning, and movement (Hamilton et al., 2014). Consequences include the energy costs and the lack of the ability to store food (Hamilton et al., 2014). Some hunter–gatherer societies are sedentary with long established camps in locales that are highly productive (Diamond, 2002).

4.2.3 Gender Roles and Hunter–Gatherers

Hunter–gatherer societies are assumed to have strong gender roles with women predominantly the gatherers of plant foods (fruits, seeds, nuts, tubers, etc.) and men, the hunters. For instance, Hadza men in Tanzania hunt singly and bring meat back to the camps. In contrast, women, in groups, gather tubers and other plant foods (Pontzer et al., 2012). Similarly, in the !Kung San of the Kalahari (Fig. 4.1), there is a division of labor with men hunting and women gathering (Bentley, 1985). The assumption of defined gender roles is challenged by the Agta, a hunter–gatherer people of the Philippines where women are responsible for about half the game killed (Konner, 2005).

4.2.4 Population Stability and Reproduction in Hunter–Gatherer Societies

Hunter–gatherer populations exhibit little growth. As noted by Malthus (1798), "population must always be kept down to the level of the means of subsistence." For hunter–gatherers, there is less food gathered by foraging expressed per person as the population foraging increases. This is illustrated in Fig. 4.2 (adapted from Weisdorf, 2005). This reflects the carrying

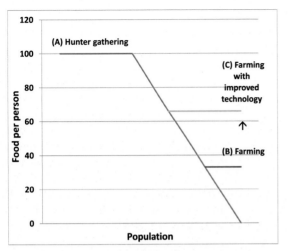

FIGURE 4.2 **Conceptual view of changes in food gathered per person in hunter–gatherer societies as human population increases.** (A) There is initially a linear relationship *(blue line)* with the inflection point being at the carrying capacity. Then, as populations rise, there is less food gathered per capita. (B) In contrast, there is a linear relationship between food per person raised in subsistence agriculture as populations rise *(red line)*, and then (C) with improved technology, such as improved genetics as quantal "jump" *(green line)*. Agriculture has the advantage that many of such quantal jumps increase food production per capita.

capacity of the locale. There is a considerable survival pressure for hunter–gatherer societies to restrict the number of births and hence the population. Infanticide is thought to be found in hunter–gatherers to increase the probability of survival of the other children. The rate of infanticide is reported to be 1% of births of the !Kung people (Howell, 1979). In addition, infants need to be carried during migrations.

Hunter–gatherer societies have low fertility with, for instance, 4.7 births to !Kung San women (Bentley, 1985). The low birth rate is due to the following:

- Delayed menarche compared to farming communities, for example, 16.6 years old for !Kung (reviewed by Konner, 2005). Menarche is also thought to have been earlier in Paleolithic hunter–gatherers than after the development of agriculture in the Neolithic (Gluckman and Hanson, 2006; Patton and

Viner, 2007). The time from weaning to sexual maturity is thought to be much longer in humans than in great apes (Konner, 2005).
- Prolonged nursing/delayed weaning at 3–4 years old in the !Kung and 2.5 years old in the Hadza. Birth intervals are the following: 4 years for !Kung, 3.6 years for the Aka (Pygmy) hunter–gatherers of Central Africa, and 3 years for the Agta people of the Philippines (reviewed by Konner, 2005).
- Low fertility for nonlactating women. For example, Hadza women have body fat percentages of 21% (Pontzer et al., 2012). This compares to the requirement of 22% for the maintenance of reproductive activities in women (Frisch, 1987). Thus, it would appear that a significant number of Hadza women have reduced fertility.

4.2.5 Nutrition and Health in Present Hunter–Gatherer Societies

It is generally thought that hunter–gatherers had periods of nutritional deprivation intermittently. However, the incidence and severity of these periods appears to be less adverse for hunter–gatherers than for subsistence farmers (Berbesque et al., 2014; Cordain et al., 1999).

While 65% of energy in hunter–gatherers is from animal sources, there is a lack of cardiovascular and other diseases associated with a Western lifestyle (Cordain et al., 2002; Eaton and Konner, 1985). This is despite the similar level of fat consumption between Western diets and hunter–gatherer diets. This apparent paradox may be explained by one or more of the following: differences in the types of fat consumed, the amount of exercise (Cordain et al., 2002), consumption of sugar, and other reasons.

4.2.6 Nutrient Deficiencies in Present Hunter–Gatherer Societies

Deficiency diseases exist in present-day hunter–gatherers. For example, there are vitamin C deficiencies, subacute scurvy, and low plasma concentrations of ascorbic acid in Inuit and Eskimos (Geraci and Smith, 1979; Heller, 1949; Nutrition Canada, 1975). This is not surprising given that the diets traditionally have been predominantly marine (e.g., seals) and terrestrial mammals (e.g., caribou, Arctic hare, and muskox), fish, and some plants in the summer. It has been suggested that the low level of degenerative eye disease in hunter–gatherers is related to the consumption of phytonutrients (London and Beezhold, 2015).

HUNTER–GATHERERS AS PROFESSIONAL TRACKERS

Native American Scouts

Native American scouts have been used by the military including the United States Army Indian Scouts since the Colonial era in what is now the United States of America (prior to 1776). One group was the Apache Scouts who helped pursue those Apaches, including those led by Geronimo, who did not accept various treaties in the USA and Mexico. Between 1916 and 1917, multiple units of the US Army under Brigadier General John J. Pershing moved into the sovereign nation of Mexico pursuing Pancho Villa, who is described either as a revolutionary or a bandit. This incursion was authorized by President Wilson and categorized initially as Punitive Expedition, US Army but is now known as the Mexican Expedition or the Pancho Villa Expedition (US Army Campaigns, 2015; Yockelson, 1997). Accompanying the US troops were 39 Apache Scouts (Huachuca Illustrated, 2017). The unit was dissolved in 1920. In a similar manner, the US Army employed Montagnards, subsistence farmers, in the Vietnam War, and the British Special Forces acquired tracking knowledge from the Iban peoples, also subsistence farmers, in Borneo in World War II (Moreira, 2010). The Montagnards and Iban peoples were extensive hunters.

The US Immigration and Customs Enforcement (ICE) agency employ Native Americans from the Tohono O'odham Nation in a tactical patrol known as the Shadow Wolves. This group uses their tremendous tracking skills to find people smuggling drugs and those crossing the border without documentation (US Immigration and Customs Enforcement, 2007).

U.S. Immigration and Customs Enforcement

Ju/'hoansi-San as Professional Trackers

Professional Ju/'hoansi-San trackers from Namibia (Southern Africa) have been used by teams of archaeologists to examine human footprints from 15,000 to 18,000 years ago in caves in present-day France. The trackers provided new insights including identifying footprints previously overlooked and identifying different individuals who made the tracks with sexes and ages (Pastoors et al., 2015, 2016). The critical contribution of the three Ju/'hoansi-San trackers was recognized by their inclusion as authors in the research papers (Pastoors et al., 2015, 2016).

4.3 PALEOLITHIC HUNTER–GATHERER SOCIETY

4.3.1 Animal Products in the Diets of Paleolithic Hunter–Gatherer Societies

Hunter–gatherers have high-energy expenditures and hence require high-energy diets (Leonard and Robertson, 1997). Based on a series of assumptions, it was estimated that the late Paleolithic diet was 35% derived from animals and 65% derived from plants but with animals accounting for 76% of protein intake (Eaton and Konner, 1985). An alternate view, based on the diet of present-day hunter–gatherers, is that the diet of Paleolithic hunter–gatherers was 65% derived from animals and 35% derived from plants (Cordain et al., 2000). Moreover, analysis of Paleolithic skeletal remains supports a strongly carnivorous diet (see Chapter 5 for details). It should be noted that the mineral intake of Paleolithic hunter–gatherers was estimated to be greater for calcium, iron, and potassium but not sodium (Table 4.4).

4.3.2 Nonmeat Animal-Derived Items in the Diet of Paleolithic Hunter–Gatherer Societies

Insects are likely to have made a significant contribution to the overall nutrition of Paleolithic hunter–gatherers, contributing to the needs for micronutrients (vitamins and trace minerals), together with protein and energy (Raubenheimer et al., 2014).

Fish, including shellfish, contributed to the diet of Paleolithic hunter–gatherers particularly in coastal areas, around lakes, and along rivers. For example, there is evidence that Paleolithic hunter–gatherers had a ready source of protein in molluscan shellfish (Kyriacou et al., 2014). This is supported by the calculated intake of fish in present-day hunter–gatherer societies (Table 4.1).

Insects and other invertebrates are likely to have contributed to the diet of Paleolithic hunter–gatherers (as with other great apes—see Chapter 5). In addition, the insect-derived food, honey, is often neglected food item in the Paleolithic diet. The use of honey and beeswax preceded the Neolithic revolution. There is evidence from rock art that late Paleolithic hunter–gatherers in Spain "harvested" honey and beeswax from wild bees (Crane, 1999; Dams and Dams, 1977). Surprisingly, honey represents about 12% of food brought back to camps by the present-day Hadza people in Tanzania (Pontzer et al., 2012).

4.3.3 Comparison of Hunter–Gatherer and Farming Societies

It is often assumed that farming was, and still is, a superior system than hunting and gathering

TABLE 4.4   Differences in Intake of Specific Minerals Between Paleolithic Hunter–Gatherers[a] and People on a Western Diet

Nutrient	Paleolithic hunter–gatherer diet[a] (mg)	Western diet (mg)
Calcium	653	392
Iron	28.5	4.9
Sodium	768	4000
Potassium	10,500	3050

[a] Assumes consumption of 65% of energy from plants and 35% from animals.
Data from Eaton, S.B., Eaton, S.B.III, Konner, M.J., 1997. Paleolithic nutrition revisited: a twelve-year retrospective on its nature and implications. Eur. J. Clin. Nutr. 51, 207–216.

with the latter dismissed as scavenging. For example, it is argued that improved technologies are much more advantageous to agricultural societies than hunter–gatherer societies (Fig. 4.3). However, there are examples in history where, in a competition between farming and hunter–gatherer societies, the latter prevailed. For example, hunter–gatherer Inuits and Eskimos ultimately prevailed over the Viking or Norse settlements in Greenland (Riede, 2011), and also presumably the Native Americans prevailed over Viking settlers in Vineland, North America. It has been argued that climatic changes were responsible for the demise of the farmer societies in Greenland with catastrophic reductions of crop and livestock production due to an unstable niche, whereas the hunter–gatherer populations thrived (reviewed by Riede, 2011). In contrast, it is argued that the shift from hunter–gatherer societies to farming in Southern Europe and the Fertile Crescent was due to increasing populations and an unstable foraging niche (Rowley-Conwy and Layton, 2011). There is evidence that hunter–gatherers existed

at the same time as farmers in Neolithic Europe (Bollongino et al., 2013), although the farmer societies ultimately prevailed.

4.3.4 Dogs and Paleolithic Hunter–Gatherers

While domestication of animals is associated with the Neolithic revolution, dogs were domesticated at least 15,000 years ago (Pang et al., 2009; Shannon et al., 2015) or even around 30,000 years ago (Germonpré et al., 2009; Skoglund et al., 2015; Druzhkova et al., 2013), both during the Paleolithic era. Analysis of mitochondrial DNA supports the argument that dogs accompanied people migrating across the Bering Strait (Leonard et al., 2002; van Asch et al., 2013) about 13,800 years ago (Waters et al., 2011) or even 15,000–16,500 years ago (Goebel et al., 2008). This is supported by the presence of dog fossils in North America dating from about 9000 years ago (Lupo and Janetski, 1994; Morey and Wiant, 1992). Dogs, therefore, played a role in, at least some, Paleolithic hunter–gatherer societies in aiding hunting and for protection. For example, it has been proposed that dogs played a role in hunting and then herding of reindeer in Scandinavia (Riede, 2011). It has been suggested that the human–dog partnership played a critical role in human's competition with the Neanderthals and the ultimate demise of the Neanderthals (Shipman, 2015). Some present-day hunter–gatherers, such as the Inuit and Eskimos possess dogs with their dogs having no detectable genetics coming from European dogs (van Asch et al., 2013). For more details on the human–animal relationship, see Chapters 10 and 12.

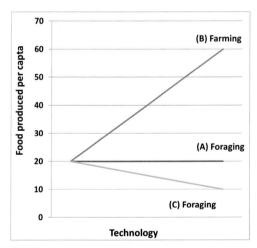

FIGURE 4.3 **Conceptual view of the impact of improving technology (tools, strategies, etc.) on food foraged by hunter–gatherer societies.** (A) Hunter–gatherers *(red line)* when populations of prey species are stable. (C) Hunter–gatherers *(green line)* when there is overhunting of prey species. In contrast, with farming (B), improved technology increases food production per capita *(red line)*.

4.3.5 Nutrient Deficiencies in Paleolithic Societies

It is possible that Paleolithic people had nutrient deficiencies when populations did not have access to specific essential nutrients, such as iodine, because of geography (Eaton and Konner, 1985). For instance, some soils are deficient in iodine leading to plants with low iodine content and in turn livestock with low iodine content (Eaton and Konner, 1985).

4.3.6 Paleolithic Hunter–Gatherers and the Extinction of Megafauna

At the end of the Pleistocene and beginning of the Holocene epochs, there were extensive extinctions of megafauna in Europe, Asia, North and South America, and Australia (see Chapter 21). This was arguably unprecedented and has been attributed to either human activity (hunting) and/or shifts in ecosystems due to climate change (reviewed by Sandom et al., 2014). Migrations of humans brought them into contact with populations of large mammals and birds that were naive to the new carnivores, the hunters. There is evidence that the migrations of hunter–gatherers (humans or *Homo sapiens*) led to the extinction of megafauna in the Americas (Martin, 1973; Surovell et al., 2016) (discussed in Section 5.2.5) and Australia (Miller et al., 2005; Rule et al., 2012). It is therefore not surprising that Africa predominantly was immune to the megafauna extinctions. Sandom et al. (2014) provide strong evidence for the role of humans in the loss of megafauna based on a global comprehensive analysis and more accurate dating of fossil remains. For instance, hunter–gatherers contributed to the extinction of megafauna in Europe and Asia including wild horses and European Steppe bison (Lorenzen et al., 2011). Moreover, populations of mammoth appear to have been greatly decreased by climate change and consequent shifts in ecosystems, making them vulnerable to overhunting (MacDonald et al., 2012; Nogués-Bravo et al., 2008). There were also the effects of rising sea levels separating populations (Mann et al., 2015).

ANIMALS AND PALEOLITHIC ART

There are widespread examples of Paleolithic art showing realistic, and often exclusively, representations of wild animals as paintings or etchings on cave walls and ceilings (Ghosh, 2014). The earliest depiction of an animal dates to 35,400 years ago (reviewed by Marchant, 2016). This was a painting of a babirusa in a cave on the island of Sulawesi, Indonesia (Fig. 4.4A). Other examples of Paleolithic art are found in caves, such as in Lekeitio in Spain (BBC, 2016) and in the Lascaux Caves and Chauvet Cave of Pont d'Arc in France (see Fig. 4.4B–D). These date from 14,500 to 10,000 years ago. Another representation of animals by people prior to the adoption of agriculture is the effigy mounds of present-day Iowa, Wisconsin, Northern Illinois, and Southern Minnesota in the USA. These are low relief (0.5 m high) and were constructed in the shapes of birds, bear (Fig. 4.4E), deer, bison, lynx, turtle, and panthers during the Late Woodland Period (CE/AD 600–1250 or 1400–750 BP) (National Park Service, In Press). Another group of such mounds was constructed in what is present-day New Mexico (Fig. 4.4F). It is questioned whether these depictions of animals as paintings in caves or as mounds was art or religion or a combination

FIGURE 4.4 **Paleolithic art depicting animals.** (A) A painting of a babirusa from a cave on the island of Sulawesi (Indonesia) dating from over 30,000 years ago. (B) Paleolithic cave painting of a bison (from about 16,000 years ago) from Altimera Caves in Spain (editor from Wikipedia). (C) Paleolithic cave painting of wild cattle (from 17,000 years ago) from Lascaux Caves in France (editor from Wikipedia). (D) Etching of wild horses and goats from a cave in Spain dated to over 14,000 years ago. (E) Aerial photograph of a bear effigy mound with the outline more easily shown by the addition of limestone. Courtesy National Parks Service. (F) Marching Bear Group of mounds (also showing bird) from an aerial photograph with the outline of the bears shown by the addition of agricultural lime. *Courtesy National Parks Service.*

4.3.7 Transitions From Hunting Animals to Warfare With Other Humans

There is a debate whether Paleolithic hunter–gatherer societies were peaceful or had battles with others (Bowles, 2009). The prevailing evidence supports the latter. Violent intergroup fighting using tools for hunting occurred in hunter–gatherer societies in Sub-Saharan Africa at least by the beginning of the Holocene, about 10,000 BP and prior to the onset of the Neolithic era. Human remains from 27 people from Nataruk, west of Lake Turkana, provide strong evidence of their violent deaths in a massacre (Mirazón Lahr et al., 2016). The bodies were left unburied in a lagoon. The cause of deaths were determined to be either from blunt force trauma to the skulls by wooden clubs or to wounds in the neck presumed to be from arrows and/or spear points. We are seeing the tools used for hunting being used as weapons of war prior to there being significant stores of food or land to be acquired in warfare. Moreover, there is an archaeological evidence of attacks on settlements in Nubia (present-day southern Egypt and North Sudan) 12,000–14,000 BP (Kelly, 2005) and pictorial representations of battles from Australia (10,000 BP) (Taçon and Chippindale, 1994). Analysis of prehistorical skulls in Southern California shows higher frequency of damage in the Channel Islands (19%) than the mainland (7.5%) (Walker, 1989). The difference may be attributable to greater competition for resources on the Channel Islands (Walker, 1989). These reports suggest that warfare developed much earlier than previously thought. Many of the existing hunter–gatherer societies in Asia, Australia, North and South America, and South Africa participate or have taken part in raiding and warfare (Moreno, 2011). It is argued that if a group's territory was "invaded," then it is easy to suppose that a hunting party might transform to a raiding or war party.

4.4 THE PALEOLITHIC DIET

4.4.1 Basis of the Paleolithic Diet

The ancestors of modern humans were hunter–gatherers for at least 2 million years with a shift to agriculture beginning only 11,300 years ago. It is reasonable to consider that humans were, and hence are, adapted to a hunter–gatherer or Paleolithic diet. It has been proposed that "the diets of modern-day hunter–gatherers may represent a reference standard for modern human nutrition and a model for defense against certain diseases of affluence" (Cordain et al., 2000). Some nutritionists have argued that the Western diet adopted over the last hundred years is responsible for the increases in cardiovascular and other diseases (Cordain et al., 2005; Eaton and Konner, 1985; Eaton et al., 1997). The relationship between animal products and disease is addressed in Chapter 3.

The Paleolithic diet is also called the Stone Age diet, hunter–gatherer diet, or caveman diet. It consists of lean meat, fish, eggs (limited to 2 per day), fruits, vegetables, nuts, and seeds. It excludes dairy products, legumes (e.g., peas and beans), grains (wheat, barley, oats), potatoes (but roots are included), refined sugar together with high fructose corn syrup, alcohol (but wine is allowed), processed oils, and salt (Klonoff, 2009; Mayo Clinic, 2014). "Proponents of the Paleolithic diet believe that modern humans are genetically adapted to a Paleolithic diet and not to the current so-called civilized diet" (Klonoff, 2009) with the 11,300 years since the beginning of the Neolithic agricultural revolution insufficient for evolutionary changes to human genetics. However, people have gained amylase genes for digesting starch (Perry et al., 2007) and a proportion of the human population has lactase persistence allowing milk consumption (Tishkoff et al., 2007) (see Chapter 19) and shifts in the microbiome (Harper and Armelagos, 2013.). The rationale for the diet is also weakened by

Paleolithic hunter–gatherers in fact consuming at least some grain, not limiting their input of eggs and not having wine available. Moreover, the Paleolithic diet is based on the view that the hunter–gatherers' diet was 35% animal- and 65% plant-derived (Eaton and Konner, 1985). This is similar to the situation in some present-day hunter–gatherers, for example, the Hadza (37%) (Pontzer et al., 2012) and !Kung San (Belovsky, 1987) and is within the range for hunter–gatherer diets (Cordain et al., 2000). However, the average split between animal and plant materials in the diet is 60:40 in existing hunter–gatherer societies (Cordain et al., 2000).

4.4.2 Positive Effects of the Paleolithic Diet

In healthy adults on a Paleolithic diet for 3 weeks, there is decreased body weight and waist or abdominal circumference (indicating abdominal fat) (Osterdahl et al., 2008). The effect on body fat is seen in other studies with decreased body fat and/or abdominal circumference in the following: postmenopausal women (Mellberg et al., 2014), obese postmenopausal women (Ryberg et al., 2013), and type 2 diabetics compared to those on a diabetes diet (Jönsson et al., 2009). Compared to those on an isocaloric healthy diet, people with metabolic syndrome on a Paleolithic diet had greater body weight loss and HDL-cholesterol together with lower circulating concentrations of triglyceride (Boers et al., 2014); metabolic syndrome being a predictor of cardiovascular and other diseases, and is based on waist circumference, blood pressure, and circulating concentrations of glucose and lipids. There are other effects on body lipids; for example, obese postmenopausal women on a Paleolithic diet had decreased circulating concentrations of triglyceride and liver concentrations of triglyceride (Ryberg et al., 2013).

Consumption of a Paleolithic diet is followed by reductions in blood pressure in, for instance, healthy adults (systolic) (Osterdahl et al., 2008), obese postmenopausal women (diastolic) (Ryberg et al., 2013), type 2 diabetics (diastolic) compared to a diabetes diet (Jönsson et al., 2009), and people with metabolic syndrome (diastolic and systolic) compared to those on an isocaloric healthy diet (Boers et al., 2014). People with either ischemic heart disease or type 2 diabetes on a Paleolithic diet also experienced improved satiety after eating compared to Mediterranean or diabetes diets, respectively (Jönsson et al., 2010, 2013). It is not surprising therefore that people on a Paleolithic diet consume less energy (Jönsson et al., 2010) as do obese postmenopausal women (Ryberg et al., 2013). The effects of the consumption of a Paleolithic diet has been compared to the effects of the consumption of a diabetes diet that is high in fiber and complex carbohydrates from vegetables, fruits, nuts, legumes, fish that are low in fat, and sources of monounsaturated and polyunsaturated fats, such as avocados, almonds, pecans, walnuts, olives, and canola, olive, and peanut oils. People with type 2 diabetes did better with the Paleolithic diet, there being greater weight loss and decreased circulating concentrations of triglyceride. Compared to a Mediterranean diet, the Paleolithic diet also has effects on glucose metabolism with an improved glucose tolerance in people with ischemic heart disease (Lindeberg et al., 2007) and reduced fasting blood glucose in obese postmenopausal women (Ryberg et al., 2013).

4.4.3 Long-Term Acceptance of the Paleolithic Diet

The long-term acceptance of the Paleolithic diet is questioned with the diet missing favorites, such as bread, candy, cakes, cookies, potatoes (including French fries), rice, ice-cream, lattés, and beer. There have been efforts to modify the Paleolithic diet with the omission of meat and eggs (Bligh et al., 2015) and other refinements (Konner and Boyd-Eaton, 2010). These may be very useful clinically but it is not clear how they can be viewed as Paleolithic, given that the hunter–gatherer diet was so high in animal-derived foods as discussed earlier.

4.5 HUNTER GATHERING CONTINUES IN FISHING (CAPTURE FISH)

Fishing (capture fish) from marine and freshwater areas produces about 91.3 million metric tons per year globally (FAO, 2014) (Table 4.5 and Fig. 4.4). This is a significant contribution to our protein needs and compares to a total meat production of 310 million metric tons per year (FAOStats, 2016). The level of fish harvested from marine waters has not increased over the past 20 years due to nonsustainable commercial fishing (FAO, 2014) (Fig. 4.5). There were 4.72 million fishing vessels in the world in 2012 with an overcapacity of fishing fleets (FAOStats, 2016). Capture fishing from inland waters is disproportionately important in Africa and to a lesser extent in Asia (FAOStats, 2016).

Table 4.6 summarizes the top 10 countries for "capture fish." The top capture fish species are the following: Peruvian anchovy, walleye pollock, skipjack tuna, *Sardinellas nei*, Atlantic herring, chub mackerel, scads, yellowfin tuna, Japanese anchovy, largehead hairtail, Atlantic cod, and sardine (FAO, 2014). In 2012, the amount of shrimp species harvested was 3.4 million tonnes while that of cephalopods was 4 million tonnes (FAO, 2014). The number of capture fishers globally is about 50 million people coming from the following continents:

- Asia 40 million people
- Africa 5.9 million
- South America and Caribbean 2.3 million
- North America 0.3 million
- Europe 0.6 million

Not only has the amount of fish harvested by captured fishing remained constant but also the number of overfished and fully fished species increased over the past 20 years (FAO, 2014). The following is an assessment of fishing globally:

- At biologically unsustainable level and therefore overfished: 28.8% (compared to 10% in 1975)
- Fully fished: 61.3%
- Underfished: 9.9% (compared to 40% in 1975) (FAO, 2014)

As the intensity of capture fish increases, the amount harvested for many fish species will reach a steady level with diminishing returns compared to effort (Fig. 4.6) or threatens the survival of the population or entire species of fish (Fig. 4.6). It is interesting that this has the same pattern as illustrated in Fig. 4.3 for human hunter–gatherer societies with overhunting threatening the existence of some species.

TABLE 4.5 Comparison of the Importance of Fishing (Capture Fish) and Aquaculture Globally (FAO, 2014)

Fishing type	2007	2012
Capture (in million metric tons)	90.8	91.3 (up 0.5%)
Aquaculture (in million metric tons)	49.9	66.6 (up 33.5%)

(A)

(B)

FIGURE 4.5 **Commercial capture fishing.** (A) Large-scale commercial fishing. (B) Fishing in Oman.

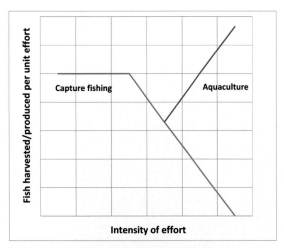

FIGURE 4.6 **Conceptual changes in fish captured per unit effort with intensity of fishing (e.g., number and size of trawlers).** Initially, when fish stocks are adequate, greater effort results in increased fish harvested, but there is a steady level of fish captured per unit effort *(blue line)*. When fishing stocks are declining due to overfishing, there is progressively less amount of fish captured per unit effort *(blue line)*. In contrast, with aquaculture *(red line)*, there is increased fish produced per unit effort.

TABLE 4.6   Top Countries for Fishing (Capture Fish) in 2012 (FAO, 2014)

Ranking	Country	Capture (in million metric tons)
1	China	13.9
2	Indonesia	5.4
3	USA	5.1
4	Peru	4.8
5	Russian Federation	4.1
6	Japan	3.6
7	India	3.4
8	Chile	2.5
9	Vietnam	2.4
10	Myanmar	2.3

4.6 RESIDUAL IMPACT OF HUNTING ON NUTRITION IN WESTERN COUNTRIES

Venison from hunted deer and elk are consumed by hunters in the USA as are wild pigs, pheasants, ducks, wild turkeys, and other birds. It is difficult to get quantitative information on the impact of these on overall nutrition. It is likely to be low even though the proportion of people consuming deer and elk is high with 67% of people surveyed in the USA having consumed venison (Abrams et al., 2011). Meat from hunted animals has contributed to the diet of miners and people building roads and railroads.

Venison is donated to humanitarian organizations in the USA and Canada. Based on organizations responding to surveys, 1012 metric tons of venison per year and other hunted meat were documented as donated to humanitarian organizations (Avery and Watson, 2009). This extrapolates to a total of 2872 metric tons per year adjusting for organizations not responding to the survey. Such donations not only provide meat as part of food assistance but also increased the number of deer harvested where there is overpopulation (Hindreth et al., 2011).

There are risks from consumption of hunted meat, such as toxicity from the use of lead shot. This is particularly apparent in populations that consume large proportions of their diet from hunted animals and birds as was observed in a study of 50 adult men from Nuuk, Greenland (Johansen et al., 2006).

4.7 CONCLUSIONS

Knowledge of present-day hunter–gatherers is providing insights into Paleolithic hunter–gatherers. Moreover, research is revealing much on the Paleolithic hunter–gatherers. The shift from a hunter–gatherer society in the Neolithic revolution is discussed in detail in Chapter 5, which also considers additional comparisons of early farmers and Paleolithic hunter–gatherers. It is interesting to see that there is a parallel shift from hunter–gatherer capture fishing to aquaculture (fish farming) occurring today.

References

Abrams, J.Y., Maddox, R.A., Harvey, A.R., Schonberger, L.B., Belay, E.D., 2011. Travel history, hunting, and venison consumption related to prion disease exposure, 2006–2007 FoodNet Population Survey. J. Am. Diet. Assoc. 111, 858–863.

Avery, D., Watson, R.T., 2009. Distribution of venison to humanitarian organizations. In: Watson, R.T., Fuller, M., Pokras, M., Hunt, W.G. (Eds.), Ingestion of Lead from Spent Ammunition: Implications for Wildlife and Humans. The Peregrine Fund, Boise, Idaho, USA.

BBC News, 2016. Cave Art: Etchings Hailed as 'Iberia's Most Spectacular.' Available from: http://www.bbc.com/news/world-europe-37654544

Belovsky, G.E., 1987. Hunter-gatherer foraging: a linear programming approach. J. Anthropol. Archaeol. 6, 29–76.

Bentley, G.R., 1985. Hunter–gatherer energetics and fertility: a reassessment of the !Kung San. Hum. Ecol. 13, 79–109.

Berbesque, J.C., Marlowe, F.W., Shaw, P., Thompson, P., 2014. Hunter–gatherers have less famine than agriculturalists. Biol. Lett. 10, 20130853.

Bligh, H.F., Godsland, I.F., Frost, G., Hunter, K.J., Murray, P., MacAulay, K., Hyliands, D., Talbot, D.C., Casey, J., Mulder, T.P., Berry, M.J., 2015. Plant-rich mixed meals based on Palaeolithic diet principles have a dramatic impact on incretin, peptide YY and satiety response, but show little effect on glucose and insulin homeostasis: an acute-effects randomised study. Br. J. Nutr. 113, 574–584.

Boers, I., Muskiet, F.A., Berkelaar, E., Schut, E., Penders, R., Hoenderdos, K., Wichers, H.J., Jong, M.C., 2014. Favourable effects of consuming a Palaeolithic-type diet on characteristics of the metabolic syndrome: a randomized controlled pilot-study. Lipids Health Dis. 13, 160–172.

Bollongino, R., Nehlich, O., Richards, M.P., Orschiedt, J., Thomas, M.G., Sell, C., Fajkosová, Z., Powell, A., Burger, J., 2013. 2000 years of parallel societies in Stone Age Central Europe. Science 342, 479–481.

Bowles, S., 2009. Did warfare among ancestral hunter-gatherers affect the evolution of human social behaviors? Science 324, 1293–1298.

Blurton-Jones, N., Hawkes, K., Draper, P., 1994. Foraging returns of !Kung adults and children: why didn't !Kung children forage. J. Anthropol. Res. 50, 217–248.

Cordain, L., Eaton, S.B., Miller, J.B., Mann, N., Hill, K., 2002. The paradoxical nature of hunter-gatherer diets: meat-based, yet non-atherogenic. Eur. J. Clin. Nutr. 56 (Suppl. 1), S42–S52.

Cordain, L., Eaton, S.B., Sebastian, A., Mann, N., Lindeberg, S., Watkins, B.A., O'Keefe, J.H., Brand-Miller, J., 2005. Origins and evolution of the Western diet: health implications for the 21st century. J. Clin. Nutr. 81, 341–354.

Cordain, L., Miller, J.B., Eaton, S.B., Mann, N., Holt, S.H., Speth, J.D., 2000. Plant-animal subsistence ratios and macronutrient energy estimations in worldwide hunter-gatherer diets. Am. J. Clin. Nutr. 71, 682–692.

Cordain, L., Miller, J., Mann, N., 1999. Scant evidence of periodic starvation among hunter-gatherers. Diabetologia 42, 383–384.

Crane, E., 1999. The World History of Beekeeping and Honey Hunting. Routledge, New York.

Dams, M., Dams, L.R., 1977. Spanish art rock depicting honey gathering during the Mesolithic. Nature 268, 228–230.

Diamond, J., 2002. Evolution, consequences and future of plant and animal domestication. Nature 418, 700–707.

Druzhkova, A.S., Thalmann, O., Trifonov, V.A., Leonard, J.A., Vorobieva, N.V., Ovodov, N.D., Graphodatsky, A.S., Wayne, R.K., 2013. Ancient DNA analysis affirms the canid from Altai as a primitive dog. PLoS One 8, e57754.

Eaton, S.B., Eaton, III, S.B., Konner, M.J., 1997. Paleolithic nutrition revisited: a twelve-year retrospective on its nature and implications. Eur. J. Clin. Nutr. 51, 207–216.

Eaton, S.B., Konner, M., 1985. Paleolithic nutrition. A consideration of its nature and current implications. N. Engl. J. Med. 312, 283–289.

FAO, 2014. The State of World Fisheries and Aquaculture. Available from: http://www.fao.org/3/d1eaa9a1-5a71-4e42-86c0-f2111f07de16/i3720e.pdf

FAOStats, 2016. Available from: http://www.fao.org/faostat/en/#home

Frisch, R.E., 1987. Body fat, menarche, fitness and fertility. Hum. Reprod. 2, 521–533.

Geraci, J.R., Smith, T.G., 1979. Vitamin C in the diet of Inuit hunters from Holman, Northwest Territories. Arctic 32, 135–139.

Germonpré, M., Sablin, M.V., Stevens, R.E., Hedges, R.E.M., Hofreiter, M., Stiller, M., Despré, V.R., 2009. Fossil dogs and wolves from Paleolithic sites in Belgium, the Ukraine and Russia: osteometry, ancient DNA and stable isotopes. J. Archaeol. Sci. 36, 473–490.

Ghosh, P., 2014. Cave Paintings Change Ideas About the Origin of Art. BBC News Available from: http://www.bbc.com/news/science-environment-29415716

Gluckman, P.D., Hanson, M.A., 2006. Evolution, development and timing of puberty. Trends Endocrinol. Metab. 17, 7–12.

Goebel, T., Waters, M.R., O'Rourke, D.H., 2008. The Late Pleistocene dispersal of modern humans in the Americas. Science 319, 1497–1502.

Hamilton, M.J., Lobo, J., Rupley, E., Youn, H., West, G.B., 2014. The Ecology and Energetics of Hunter-Gatherer Residential Mobility. Santa Fe Institute Working Paper

2014-09-034. Available from: http://www.santafe.edu/media/workingpapers/14-09-034.pdf

Harper, K.N., Armelagos, G.J., 2013. Genomics, the origins of agriculture, and our changing microbe-scape: time to revisit some old tales and tell some new ones. Am. J. Phys. Anthropol. 152 (Suppl. 57), 135–152.

Hawkes, K., O'Connell, J.F., Blurton, N.G., 1995. Hadza children's foraging: juvenile dependency, social arrangements, and mobility among hunter–gatherers. Curr. Anthropol. 36, 688–700.

Heller, C.A., 1949. The Alaskan Eskimo and the white man's diet. J. Home Econ. 41, 177–178.

Hindreth, A.M., Hygnstrom, S.E., Hams, K.M., VerCauteren, K.C., 2011. The Nebraska deer exchange: a novel program for donating harvested deer. Wildl. Soc. Bull. 35, 195–200.

Howell, N., 1979. Demography of the Dobe !Kung. Academic Press, New York.

Huachuca Illustrated, 2017. Apache Scouts in the Punitive Expedition. http://net.lib.byu.edu/estu/wwi/comment/huachuca/HI1-23.htm.

Johansen, P., Pedersen, H.S., Asmund, G., Riget, F., 2006. Lead shot from hunting as a source of lead in human blood. Environ. Pollut. 42, 93–97.

Jönsson, T., Granfeldt, Y., Ahrén, B., Branell, U.C., Pålsson, G., Hansson, A., Söderström, M., Lindeberg, S., 2009. Beneficial effects of a Paleolithic diet on cardiovascular risk factors in type 2 diabetes: a randomized cross-over pilot study. Cardiovasc. Diabetol. 16, 35.

Jönsson, T., Granfeldt, Y., Erlanson-Albertsson, C., Ahrén, B., Lindeberg, S., 2010. A Paleolithic diet is more satiating per calorie than a Mediterranean-like diet in individuals with ischemic heart disease. Nutr. Metab. 7, 85.

Jönsson, T., Granfeldt, Y., Lindeberg, S., Hallberg, A.C., 2013. Subjective satiety and other experiences of a Paleolithic diet compared to a diabetes diet in patients with type 2 diabetes. Nutr. J. 2, 105.

Kelly, R., 2005. The evolution of lethal inter-group violence. Proc. Natl. Acad. Sci. USA 102, 15288–15294.

Konner, M., 2005. Hunter–gatherer infancy and childhood: the !Kung and others. In: Hewlett, B.S., Lamb, M.E. (Eds.), Hunter–Gatherer Childhoods: Evolutionary, Developmental and Cultural Perspectives. Aldine Transaction, New Brunswick, NJ, pp. 19–64.

Konner, M., 2016. Hunter–gatherer infancy and childhood in the context of human evolution. In: Meehan, C.L., Crittenden, A.N. (Eds.), Childhood: Origins, Evolution, and Implications. School of Advanced Research Press, Santa Fe, NM, pp. 123–154.

Konner, M., Boyd-Eaton, S., 2010. Paleolithic nutrition: twenty-five years later. Nutr. Clin. Pract. 25, 594–602.

Klonoff, D.C., 2009. The beneficial effects of a Paleolithic diet on type 2 diabetes and other risk factors for cardiovascular disease. J. Diabetes Sci. Technol. 3, 1229–1232.

Kyriacou, K., Parkington, J.E., Marais, A.D., Braun, D.R., 2014. Nutrition, modernity and the archaeological record: coastal resources and nutrition among Middle Stone Age hunter–gatherers on the Western Cape coast of South Africa. J. Hum. Evol. 77, 64–73.

Leonard, W.R., Robertson, M.L., 1997. Comparative primate energetics and hominid evolution. Am. J. Phys. Anthropol. 102, 265–281.

Leonard, J.A., Wayne, R.K., Wheeler, J., Valadez, R., Guillén, S., Vilà, C., 2002. Ancient DNA evidence for Old World origin of New World dogs. Science 298, 1613–1616.

London, D.S., Beezhold, B., 2015. A phytochemical-rich diet may explain the absence of age-related decline in visual acuity of Amazonian hunter–gatherers in Ecuador. Nutr. Res. 35, 107–117.

Lindeberg, S., Jönsson, T., Granfeldt, Y., Borgstrand, E., Soffman, J., Sjöström, K., Ahrén, B., 2007. A Palaeolithic diet improves glucose tolerance more than a Mediterranean-like diet in individuals with ischaemic heart disease. Diabetologia 50, 1795–1807.

Lorenzen, E.D., Nogués-Bravo, D., Orlando, L., Weinstock, J., Binladen, J., Marske, K.A., Ugan, A., Borregaard, M.K., Gilbert, M.T., Nielsen, R., Ho, S.Y., Goebel, T., Graf, K.E., Byers, D., Stenderup, J.T., Rasmussen, M., Campos, P.F., Leonard, J.A., Koepfli, K.P., Froese, D., Zazula, G., Stafford, Jr., T.W., Aaris-Sørensen, K., Batra, P., Haywood, A.M., Singarayer, J.S., Valdes, P.J., Boeskorov, G., Burns, J.A., Davydov, S.P., Haile, J., Jenkins, D.L., Kosintsev, P., Kuznetsova, T., Lai, X., Martin, L.D., McDonald, H.G., Mol, D., Meldgaard, M., Munch, K., Stephan, E., Sablin, M., Sommer, R.S., Sipko, T., Scott, E., Suchard, M.A., Tikhonov, A., Willerslev, R., Wayne, R.K., Cooper, A., Hofreiter, M., Sher, A., Shapiro, B., Rahbek, C., Willerslev, E., 2011. Species-specific responses of Late Quaternary megafauna to climate and humans. Nature 479, 359–364.

Lupo, K.D., Janetski, J.C., 1994. Evidence of the domesticated dogs and some related canids in the eastern Great Basin. J. Calif. Gt. Basin Anthropol. 16, 199–220.

Malthus, T., 1798. An Essay on the Principle of Population. Joseph Johnson, London.

Mann, D.H., Groves, P., Reanier, R.E., Gaglioti, B.V., Kunz, M.L., Shapiro, B., 2015. Life and extinction of megafauna in the ice-age Arctic. Proc. Natl. Acad. Sci. USA 112, 14301–14306.

Marchant, J., 2016. A journey to the oldest cave paintings in the world: the discovery in a remote part of Indonesia has scholars rethinking the origins of art—and of humanity. Available from: http://www.smithsonianmag.com/history/journey-oldest-cave-paintings-world-180957685/?no-ist

Martin, P.S., 1973. The discovery of America: the first Americans may have swept the Western Hemisphere and decimated its fauna within 1000 years. Science 179, 969–974.

National Park Service, In Press. Effigy Mound Builders. Available from: https://www.nps.gov/efmo/learn/historyculture/effigy-moundbuilders.htm

Mayo Clinic, 2014. Paleo Diet: What is it and Why is it so Popular? Available from: http://www.mayoclinic.org/healthy-lifestyle/nutrition-and-healthy-eating/in-depth/paleo-diet/art-20111182

MacDonald, G.M., Beilman, D.W., Kuzmin, Y.V., Orlova, L.A., Kremenetski, K.V., Shapiro, B., Wayne, R.K., Van Valkenburgh, B., 2012. Pattern of extinction of the woolly mammoth in Beringia. Nat. Commun. 3, 893.

Mellberg, C., Sandberg, S., Ryberg, M., Eriksson, M., Brage, S., Larsson, C., Olsson, T., Lindahl, B., 2014. Long-term effects of a Palaeolithic-type diet in obese postmenopausal women: a 2-year randomized trial. Eur. J. Clin. Nutr. 68, 350–357.

Miller, G.H., Fogel, M.L., Magee, J.W., Gagan, M.K., Clarke, S.J., Johnson, B.J., 2005. Ecosystem collapse in Pleistocene Australia and a human role in megafaunal extinction. Science 309, 287–290.

Mirazón Lahr, M., Rivera, F., Power, R.K., Mounier, A., Copsey, B., Crivellaro, F., Edung, J.E., Maillo Fernandez, J.M., Kiarie, C., Lawrence, J., Leakey, A., Mbua, E., Miller, H., Muigai, A., Mukhongo, D.M., Van Baelen, A., Wood, R., Schwenninger, J.-L., Grün, R., Achyuthan, H., Wilshaw, A., Foley, R.A., 2016. Inter-group violence among early Holocene hunter–gatherers of West Turkana, Kenya. Nature 52, 394–398.

Moreira, F., 2010. Tracking History. Professional Tracking Services. Available from: http://professionaltrackers.com/tracking-classes/tracking-history/

Moreno, E., 2011. The society of our "out of Africa" ancestors (I): the migrant warriors that colonized the world. Commun. Integr. Biol. 4, 163–170.

Morey, D.F., Wiant, M.D., 1992. Early Holocene domestic dog burials from the North American Midwest. Curr. Anthropol. 33, 224–229.

Nogués-Bravo, D., Rodríguez, J., Hortal, J., Batra, P., Araújo, M.B., 2008. Climate change, humans, and the extinction of the woolly mammoth. PLoS Biol. 6, e79.

Nutrition Canada, 1975. The Eskimo Survey Report, Department of National Health and Welfare, Ottawa, 148 pp.

Osterdahl, M., Kocturk, T., Koochek, A., Wändell, P.E., 2008. Effects of a short-term intervention with a Paleolithic diet in healthy volunteers. Eur. J. Clin. Nutr. 62, 682–685.

Pang, J.F., Kluetsch, C., Zou, X.J., Zhang, A.B., Luo, L.Y., Angleby, H., Ardalan, A., Ekström, C., Sköllermo, A., Lundeberg, J., Matsumura, S., Leitner, T., Zhang, Y.P., Savolainen, P., 2009. mtDNA data indicate a single origin for dogs south of Yangtze River, less than 16,300 years ago, from numerous wolves. Mol. Biol. Evol. 26, 2849–2864.

Pastoors, A., Lenssen-Erz, T., Ciqae, T., Kxunta, U., Thao, T., Bégouën, R., Biesele, M., Clottes, J., 2015. Tracking in caves: experience based reading of Pleistocene human footprints in French caves. Cambridge Archaeol. J. 25, 551–564.

Pastoors, A., Lenssen-Erz, T., Breuckmann, B., Ciqae, T., Kxunta, U., Rieke-Zapp, D., Thao, T., 2016. Experience based reading of Pleistocene human footprints in Pech-Merle. Quat. Int. 430, 156–162.

Patton, G.C., Viner, R., 2007. Pubertal transitions in health. Lancet 369, 1130–1139.

Perry, G.H., Dominy, N.J., Claw, K.G., Lee, A.S., Fiegler, H., Redon, R., Werner, J., Villanea, F.A., Mountain, J.L., Misra, R., Carter, N.P., Lee, C., Stone, A.C., 2007. Diet and the evolution of human amylase gene copy number variation. Nat. Genet. 39, 1256–1260.

Pontzer, H., Raichlen, D.A., Wood, B.M., Mabulla, A.Z., Racette, S.B., Marlowe, F.W., 2012. Hunter–gatherer energetics and human obesity. PLoS One 7, e40503.

Pontzer, H., Raichlen, D.A., Wood, B.M., Emery Thompson, M., Racette, S.B., Mabulla, A.Z., Marlowe, F.W., 2015. Energy expenditure and activity among Hadza hunter–gatherers. Am. J. Hum. Biol. 27, 628–637.

Raubenheimer, D., Rothman, J.M., Pontzer, H., Simpson, S.J., 2014. Macronutrient contributions of insects to the diets of hunter–gatherers: a geometric analysis. J. Hum. Evol. 71, 70–76.

Riede, F., 2011. Adaptation and niche construction in human prehistory: a case study from the southern Scandinavian Late Glacial. Trans. R. Soc. B 366, 793–808.

Rowley-Conwy, P., Layton, R., 2011. Foraging and farming as niche construction: stable and unstable adaptations. Trans. R. Soc. B 366, 849–862.

Rule, S., Brook, B.W., Haberle, S.G., Turney, C.S., Kershaw, A.P., Johnson, C.N., 2012. The aftermath of megafaunal extinction: ecosystem transformation in Pleistocene Australia. Science 335, 1483–1486.

Ryberg, M., Sandberg, S., Mellberg, C., Stegle, O., Lindahl, B., Larsson, C., Hauksson, J., Olsson, T., 2013. A Palaeolithic-type diet causes strong tissue-specific effects on ectopic fat deposition in obese postmenopausal women. J. Intern. Med. 274, 67–76.

Sandom, C., Faurby, S., Sandel, B., Svenning, J.-C., 2014. Global Late Quaternary megafauna extinctions linked to humans, not climate change. Proc. R. Soc. B 281 (1787), 20133254.

Shannon, L.M., Boyko, R.H., Castelhano, M., Corey, E., Hayward, J.J., McLean, C., White, M.E., Abi Said, M., Anita, B.A., Bondjengo, N.I., Calero, J., Galov, A., Hedimbi, M., Imam, B., Khalap, R., Lally, D., Masta, A., Oliveira, K.C., Pérez, L., Randall, J., Tam, N.M., Trujillo-Cornejo, F.J., Valeriano, C., Sutter, N.B., Todhunter, R.J., Bustamante, C.D., Boyko, A.R., 2015. Genetic structure in village dogs reveals a Central Asian domestication origin. Proc. Natl. Acad. Sci. USA 112, 13639–13644.

Shipman, P., 2015. The Invaders: How Humans and Their Dogs Drove Neanderthals to Extinction. Harvard University Press, Cambridge, MA.

Skoglund, P., Ersmark, E., Palkopoulou, E., Dalén, L., 2015. Ancient wolf genome reveals an early divergence of domestic dog ancestors and admixture into high-latitude breeds. Curr. Biol. 25, 1515–1519.

Surovell, T.A., Pelton, S.R., Anderson-Sprecher, R., Myers, A.D., 2016. Test of Martin's overkill hypothesis using radiocarbon dates on extinct megafauna. Proc. Natl. Acad. Sci. USA 113, 886–891.

Taçon, P., Chippindale, C., 1994. Australia's ancient warriors: changing depictions of fighting in the rock art of Arnhem Land. N. T. Camb. Archaeol. J. 4, 211–248.

Tishkoff, S.A., Reed, F.A., Ranciaro, A., Voight, B.F., Babbitt, C.C., Silverman, J.S., Powell, K., Mortensen, H.M., Hirbo, J.B., Osman, M., Ibrahim, M., Omar, S.A., Lema, G., Nyambo, T.B., Ghori, J., Bumpstead, S., Pritchard, J.K., Wray, G.A., Deloukas, P., 2007. Convergent adaptation of human lactase persistence in Africa and Europe. Nat. Genet. 39, 31–40.

US Army Campaigns, 2015. Mexican Expedition Campaign. Available from: http://www.history.army.mil/html/reference/army_flag/mexex.html

US Immigration and Customs Enforcement (ICE), 2007. ICE Shadow Wolves. Available from: https://www.ice.gov/factsheets/shadow-wolves

van Asch, B., Zhang, A.B., Oskarsson, M.C., Klütsch, C.F., Amorim, A., Savolainen, P., 2013. Pre-Columbian origins of Native American dog breeds, with only limited replacement by European dogs, confirmed by mtDNA analysis. Proc. R. Soc. B 280, 20131142.

Walker, P.L., 1989. Cranial injuries as evidence of violence in prehistoric Southern California. Am. J. Phys. Anthropol. 80, 313–323.

Waters, M.R., Stafford, T.W., McDonald, H.G., Gustafson, C., Rasmussen, M., Cappellini, E., Olsen, J.V., Szklarczyk, D., Jensen, L.J., Gilbert, M.T.P., Willerslev, E., 2011. Pre-Clovis mastodon hunting 13,800 years ago at the Manis site, Washington. Science 334, 351–353.

Weisdorf, J.L., 2005. From foraging to farming: explaining the Neolithic revolution. J. Econ. Surv. 19, 561–586.

Yockelson, M., 1997. The United States Armed Forces and the Mexican Punitive Expedition: Part 1. Available from: https://www.archives.gov/publications/prologue/1997/fall/mexican-punitive-expedition-1.html

Further Reading

Jafarishorijeh, S., Bannasch, D.L., Ahrens, K.D., Wu, J.T., Okon, M., Sacks, B.N., 2011. Phylogenetic distinctiveness of Middle Eastern and Southeast Asian village dog Y chromosomes illuminates dog origins. PLoS One 6, e28496.

Animals and Hominid Development

Colin G. Scanes

University of Wisconsin–Milwaukee, Milwaukee, WI, United States

5.1 OVERVIEW

It might be questioned why the evolution of humans should be included in a book on animals and human society. It is contended that the critical event in the evolution of humans was the development of an omnivorous diet with increasingly effective hunting and latter fishing techniques. There are at least 12 concepts of how and/or why hominids evolved (Strauss, 2015). These include the following:

1. Tool-making
2. Killing (Raymond Dart's "killer ape" hypothesis)
3. Sharing food
4. Swimming (with hairlessness improving speed)
5. Throwing
6. Hunting
7. Trading monogamous sex for food (meat)
8. Cooking meat to supply energy and other nutrients for the enlarging brain
9. Cooking tubers for food
10. Walking on two feet
11. Adaptability
12. Cooperative activities

It is argued that the progressive acquisition of an omnivorous diet with meat, organ meats, and marrow encompasses virtually all of these concepts together with the need to communicate, to pass on acquired knowledge, and to minimize conflict with others in the group. The evolution of hominins, their diets, and their anatomy are discussed in this chapter (see quotation in sidebar). It is argued that the driving forces for hominid evolution were to maximize food quality (protein, energy) and quantity (considered in more detail later), to escape predators and to reproduce successfully.

During hominid evolution, altruism for others (see quotation in sidebar) and cooperation within the family or the group societies was balanced by individual initiative and aggression to others of the same and different species. Without man's inventiveness, cooperation, and communication leading to the development of agriculture (see Chapter 6), our natural state would be sad (see quotation in sidebar). The Neolithic revolution was an essential prerequisite for the development of culture encompassing the arts and sciences. With our best characteristics, so much has been achieved and so much is possible.

Quotations Applicable to the Development of Humans

Greatness: applied in this case to early hominid species that were human ancestors or blind alleys in hominid evolution.

"Some are born great, some achieve greatness, and some have greatness thrust upon them" (William Shakespeare) to which should be added "and some just die out."

Altruism:

"Greater love hath no man than this, that a man lay down his life for his friends." John 15:13 (King James Bible).

The natural state of human life:

No arts; no letters; no society; and which is worst of all, continual fear, and danger of violent death: and the life of man, solitary, poor, nasty, brutish and short. *Thomas Hobbes (1651)*

5.2 HOMINID EVOLUTION

5.2.1 Geological Epochs

Hominid evolution occurred principally in the Pliocene and Pleistocene epochs. The geological epochs are shown as follows (International Commission on Stratigraphy, 2015):

Holocene epoch	Present to 11,700 years ago
Pleistocene epoch	11,700–2.6 million years ago
Pliocene epoch	2.6–5.3 million years ago
Miocene epoch	5.3–23 million years ago
Oligocene epoch	23–34 million years
Eocene epoch	34–56 million years

5.2.2 Classification of Primates

The classification of the primates enables a first examination of relatedness between groups (Box 5.1).

5.2.3 Primate Evolution

The split between the Platyrrhini (New World monkeys) and the Catarrhini (Old World monkeys and apes) occurred about 40 million years ago in the Eocene epoch (Bosinger et al., 2011). Old World monkeys (Cercopithecoidea) and the apes (Hominoidea) diverged in the Oligocene. Based on genomic and fossil evidence, the last common ancestor of the Cercopithecoidea and the Hominoidea existed about 25 million years ago (Bosinger et al., 2011; Stevens et al., 2013). The last common ancestor of chimpanzee and humans (*Pan–Homo* divergence) was about 6 million years and the last common ancestor of chimpanzee/ humans and gorillas was about 15 million years ago (Bosinger et al., 2011; Carrigan et al., 2012).

5.2.4 Evolution of the Human Branch or Hominidae (Tribe Hominini)

There are multiple species in the human evolutionary branch (Fig. 5.1; Table 5.1). The evolution of our species, *Homo sapiens*, encompasses transitions from populations of *Australopithecus africanus*, *Australopithecus garhi*, *Homo habilis*, *Homo erectus*, and *Homo heidelburgensis*. Changes in the biology and diet are considered in Sections 5.3 and 5.4, respectively. What is particularly interesting is that different species of hominids existed at the same time. For example, anatomically modern humans, Neanderthals (*Homo neanderthalensis*), *Homo denisova*, and *Homo floresiensis* all lived 30,000–40,000 BP, while about 1.7 million BP, there were at least four species of hominins temporally coexisting.

5.2.5 Evolution Within Atomically Modern Humans (*Homo sapiens*)

H. sapiens have lived from at least 195,000 years ago to the present (Fig. 5.1) with evidence of

BOX 5.1 CLASSIFICATION OF PRIMATES

Order Primates (based on Grove, 2005; Tree of Life, 2015).

Suborder Strepsirhini (lemurs, bush-babies, and lorises)
Suborder Haplorhini
Infraorder Tarsiformes (or Tarsii) (tarsiers)
Infraorder Simiiformes (monkeys and apes)
Parvorder Platyrrhini (New World monkeys)
Parvorder Catarrhini (Old World monkeys and apes)
Superfamily Cercopithecoidea (Old World monkeys)
Superfamily Hominoidea (apes)

The Superfamily Hominoidea consists of two families:

Family Hylobatidae or lesser apes: gibbons
Family Hominidae or great apes consisting of the following existing species: orangutan (genus *Pongo*), gorilla (genus *Gorilla*), chimpanzee (genus *Pan*), bonobo (genus *Pan*), and humans (genus *Homo*).

According to Wood and Richmond (2000), relationships within the Family Hominidae are the following (ITIS, 2016):
Family Hominidae (hominids) (considered by the Integrated Taxonomic Information System (ITIS) to be valid)

Subfamily Ponginae
Genus *Pongo* (pongines)
Subfamily Gorillinae
Genus *Gorilla*

Subfamily Homininae (hominines) (considered by ITIS to be valid but includes Gorilla)
Tribe Panini
Genus *Pan* (panins) (chimpanzees and bonobos)
Tribe Hominini (hominins)
Subtribe Australopithecina (australopiths)
Genus *Ardipithecus*
Genus *Australopithecus*
Genus *Paranthropus*
Subtribe Hominina (hominans)
Genus *Homo* (considered by ITIS to be valid)

Questions on the Classification of Hominids

Groves (2001) argues that chimpanzees should be categorized in the same genus as humans, *Homo*. The corollary to this is that the common ancestor of chimpanzees, bonobos, and humans together with extinct intermediate and other forms, would be considered within a single genus. Instead, there are more genera and many species. This view is supported on genetic/genomic grounds (Curnoe and Thorne, 2003). Based on genomic analyses, humans, chimpanzees, and bonobos should be considered as a single genus (Curnoe and Thorne, 2003). Moreover, the multiple human ancestral "species" including Neanderthals, *H. erectus* and present humans would be all considered as *Homo sapiens* (Curnoe and Thorne, 2003). In this chapter, the traditional species names are employed.

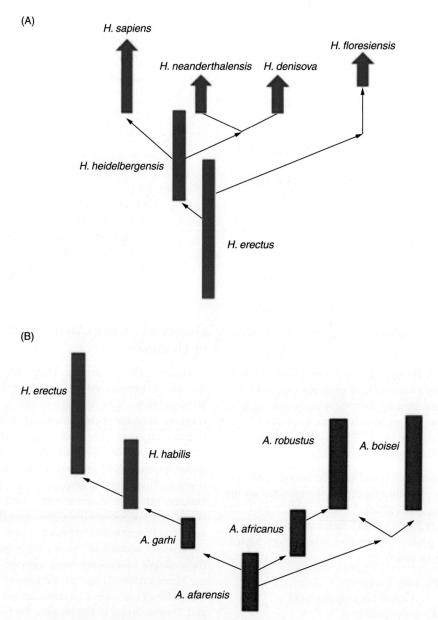

FIGURE 5.1 **Evolution of human-like hominids.** (A) Evolution within the genus *Homo*. (B) Evolution within the early hominid ancestors of humans.

their presence dating from 195,000 year ago in what is Ethiopia today (McDougall et al., 2005). The species developed and lived in Eastern and Southern Africa. The Paleolithic diet was similar to today's hunter–gatherers with meat and plants each representing about half the dietary energy. Between 80,000 and 60,000 BP new technologies were developed including more effective hunting weapons, such as bone-tipped throwing spears and possibly also wooden

TABLE 5.1 Examples of Hominins

	Time in years ago	Location	Strategies to obtain and eat meat
A. afarensis or "Lucy" and "First Family"	2.9–3.8 million	Open forest or extensively wooded areas in Eastern Africa	Gatherer of plant materials, scavengers (taking what large carnivores had left behind facilitated by use of stone tools), and hunter of small animals
A. garhi	~2.5 million	East Africa	Scavenger/hunter–gatherer (butchering large animals with stone tools and eating flesh and bone marrow)
A. africanus (bipedal + climbs trees)	3.3–2.1 million	Southern Africa	Scavenger/hunter–gatherer
H. habilis (handy man)	2.4–1.4 million	Savannah (grasslands) of Eastern, Northern and Southern Africa	Hunter–gatherer
H. erectus	143,000 to 1.9 million	Africa but extending to Asia and probably Europe	Hunter–gatherer
H. heidelbergensis	200,000–700,000	Eastern and Southern Africa, Europe and possibly Asia	Hunter–gatherer (hunting large animals, e.g., elephants, rhinoceroses, and hippopotamus plus deer and horses)
H. neanderthalensis (Neanderthal man)	28,000–200,000	Middle East and Europe	Hunter–gatherer
H. sapiens (modern man)	Present to 200,000	All continents except Antarctica	Hunter–gatherer followed by farming

arrows. In addition, there were novel tools for working skins including end-scrapers and awls (Mellars, 2006). Concomitant enhancements in efficiency of hunting, followed by increases in human population, would be expected. Human movements would follow migrating animals and become more extensive particularly at times of climatic and/or environmental changes (Mellars, 2006).

The climatic changes with the glaciations resulted in major shifts in the ecosystems on which the humans depended. There was a "bottleneck" in human population in Africa about 60,000 years ago with low populations and extinction threatened (Fagundes et al., 2007). There were migrations of hunter–gatherer *H. sapiens* "out of Africa" during the Pleistocene:

- First to the Near East and the Arabian Peninsula by about 50,000–60,000 BP (Henn et al., 2012). The first evidence for *H. sapiens* outside Africa is dated to 93,000 years ago and found in Jebel Qafzeh in present-day Israel (reviewed by Tattersall, 2009).
- Spreading East across Asia via a coastal route.
  - Through present-day Iran to the Indian subcontinent to Southeast Asia in about 50,000 BP (Henn et al., 2012) and then on to Australia.
  - Arriving in present-day Australia about 45,000 years ago (Henn et al., 2012).
- To Europe about 45,000 years ago (Henn et al., 2012).

BOX 5.2 FIRST AMERICANS, ANIMALS, AND EXTINCTION OF MEGAFAUNA

The first Americans were successfully spreading across North America and then South America. The first Americans brought dogs with them based on studies of mitochondrial DNA (van Asch et al., 2013). In what might be described as irony, a projectile tip made of a mastodon bone was found imbedded in a rib of a mastodon (*Mammut americanum*) with radiocarbon dating placing it at 13,800 BP (Waters et al., 2011). In addition to the evidence of mammoths (*Mammuthus primigenius*) at around this time in what is Wisconsin today (reviewed by Waters et al., 2011), early Americans were hunting horses (*Equus*

conversidens) and camel (*Camelops hesternus*) in an ice-free corridor in the rolling prairie about 13,300 BP (Waters et al., 2015). It is argued that overhunting was responsible for the synchronous extinction of the Late Pleistocene megafauna in North America between 13,800 and 11,400 BP (Alroy, 2001; Faith and Surovell, 2009). The extinction of sloths was later in South America than North America consistent with a delay in the arrival of people (Steadman et al., 2005). This view of prehistoric anthropomorphic-caused extinctions is not uniformly endorsed (Grayson and Meltzer, 2002, 2004).

- Spreading east across Asia through Central Asia to East Asia arriving 35,000–40,000 BP (Henn et al., 2012).
- Then on to the Bering land–bridge between Siberia and Alaska, arriving about 20,000–25,000 BP (reviewed by Goebel et al., 2008).
- Then crossing to North America about 15,000 BP and spreading along the coast (Henn et al., 2012) (for impact on animals, see Box 5.2).
- Spreading through South America 12,000–14,000 BP (Henn et al., 2012).

In addition, there were many Neolithic and later migrations. For example, people spread across Polynesia; arriving in Tonga in 2838 BP (Burley et al., 2012), to the Society Islands between AD 1025–1120 and to the remaining islands in East Polynesia between 1190 and 1290 (Wilmshurst et al., 2011). They carried with them pigs and chickens (Thomson et al., 2014).

When early peoples arrived in Europe and Asia, there was some interbreeding with the existing populations of *H. neanderthalensis* and of *H. denisova* before their demises. It is estimated that 3.6% of the Neanderthal genome is in the

genome of most Europeans, while 6%–8% of the genome of Denisovans is in the Melanesian gene pool (Lowery et al., 2013). Based on genomic evidence, interbreeding between Neanderthals and *H. sapiens* occurred between 47,000–65,000 BP probably in the Near East (Hublin, 2009; Kuhlwilm et al., 2016; Sankararaman et al., 2012). Moreover, a skull exhibiting hybrid features of human and Neanderthal characteristics has been reported from a cave in Israel dating to 54,700 BP (Hershkovitz et al., 2015). Based on the greater Neanderthal ancestry in East Asians, Wall et al. (2013) concluded that interbreeding between Neanderthals and modern humans did not "occur at a single time and place."

5.3 CHANGES IN ANATOMY DURING HUMAN EVOLUTION

5.3.1 Brain

Brain size is the hallmark of the development of humans. Increased brain size facilitated improving tools and techniques in hunting and exploitation of marine animals. Table 5.2

TABLE 5.2 Changes in Brain Size During Hominid Evolution[a]

Time in years ago in million years	Species	Brain size (cm³)	Brain size as a percentage of body weight	Number of neurons (billions)
Present	Chimpanzee (*Pan troglodytes*)	320–480	0.5	28
Present	Bonobo (*Pan paniscus*)	300–400	0.5	28
5–7	Common ancestor of chimpanzees and humans[b]	400	0.5	28
3.9–3.0	*A. afarensis*	375–550	1.25	35
3.0–2.4	*A. africanus*	420–500	1.25	31
~2.5	*A. garhi*	400–500	1.25	30-35
2.7–2.3	*P. aethiopicus*	410	1.25	33
1.9–1.4	*Australopithecus robustus*	530	1.25	35
1.9–1.6	*H. habilis*	610	1.8	40
1.8–1.5	Early *H. erectus*	860	1.8	54
0.5–0.3	Late *H. erectus*[c]	980	1.8	62
0.6–0.3	*H. heidelbergensis*	1000	1.8	76
0.4–Present	Modern humans (*H. sapiens*)	1200–1400	1.8	86
0.2–0.03	*H. neanderthalensis*	1450	1.8	85

[a] *Based on Leonard et al. (2010); Herculano-Houzel and Kaas (2011).*
[b] *Assumed same as present chimpanzee.*
[c] H. floresiensis *(95,000–17,000 years ago) Brain size 417 cc.*

summarizes the changes in brain size during human evolution. Clearly, brain size increased 3.6-fold between the common ancestor of chimpanzees and humans and modern humans. The increase in brain size was due to gradual increases and stepwise large increases; these occurred 100,000 1.0 and 1.8 million years ago (Shultz et al., 2012). Changes in brain size were not associated with paleoclimatic changes (Shultz et al., 2012). As the brain became larger during hominid development, the cortex grew disproportionately to achieve 80% of the brain in *H. sapiens* (reviewed by Hofman, 2014).

It has been argued that decreases in the energetically expensive intestine allowed increases in the weight of another energetically expensive tissue, the brain (Aiello, 1997; Aiello and Wheeler, 1995). However, brain size is not correlated with gut mass across multiple mammals (Navarrete et al., 2011). Instead, brain size is

negatively correlated with adipose tissue supporting energy trade-offs (Navarrete et al., 2011).

5.3.2 Gastrointestinal Tract

There are marked differences in the anatomy of the gastrointestinal tract between humans and other great apes (Milton, 2003; Stevens, 1988). The gastrointestinal tract is disproportionately larger across other primates (2.9% of body weight) than in humans (1.7% of body weight) (calculated from Aiello, 1997; Aiello and Wheeler, 1995). Moreover, there are marked differences in the relative sizes of different regions of the gastrointestinal tract between great apes and modern humans (Table 5.3). In particular, the relative size of the small intestine is 2.7-fold greater in humans while the colon is 2.7-fold larger in the great apes. This reflects the need for microbial fermentation in great apes.

TABLE 5.3   Comparison of Relative Size of Regions of the Gastrointestinal Tract (as Percentage of Gut Volume) Data from Milton, K., 2003. The critical role played by animal source foods in human (*Homo*) evolution. J. Nutr. 133 (Suppl. 2), 3886S–3892S

Species	Stomach	Small intestine	Large intestine or colon
Orangutan	17	28	55
Gorilla	24	13	54
Chimpanzee	19	23	53
Mean nonhuman great apes	20	21	54
Human	23	56	20

Dietary energy would provide chimpanzees with 2340 kcal day^{-1} with high fermentation of fiber but only 1706 kcal day^{-1} if there was low fermentation of fiber (Milton and Demment, 1988). Another major difference in human evolution is the shift in gut microbiome (Moeller et al., 2014).

5.3.3 Muscular–Skeletal System

An overall pattern within hominin evolution was an initial decline in stature/height and body weight followed by consistent increases (Table 5.4). In contrast, Grabowski et al. (2015) view that there was no increase in body size between the genera *Australopithecus* to early *Homo* but with *H. erectus* having a larger size. During human evolution, there have been a series of anatomical changes in the muscular–skeletal system including in the shift to bipedalism. Another example is the shift within the shoulders allowing throwing at high velocities; a feature achieved by about 2 million years (Roach and Richmond, 2015; Younga et al., 2015). Similarly, the human hand evolved with a grip again allowing throwing but also the production of tools for hunting (Marzke, 2013; Young, 2003).

Jaw size and size of teeth (Table 5.4) decreased during the development of the hominins, reflecting increased meat consumption and cutting the meat with tools. Molar sizes were markedly smaller in *H. erectus* than in earlier hominids with a series of further reductions in the evolution of *H. sapiens* (Organ et al., 2011).

TABLE 5.4   Changes in Body Size During Hominid Evolution[a]

	Adult height in cm (feet and inches)	Adult weight in kg (lb)	Post canine teeth surface area (mm^2)
Early *Australopithecus*[a]	136 (4'6")	41 (90)	—
A. africanus[a]	121 (4'1")	35.5 (78)	460[c]
H. habilis[a]	118 (3'10")	32 (70)	478[c]
H. erectus[a]	165 (5'5")	54 (119)	383[c]
H. heidelbergensis[a]	166 (5'5")	56.5 (126)	—
H. neanderthalensis[a]	164 (5'5")	59.5 (131)	—
H. sapiens Paleolithic[b]	168 (5'8")	60.5 (133)	334[c]

[a] Based on Human Origins and Nature Knowledge.
[b] Data from Hermanussen (2003) also see Table 6.3 in Chapter 6.
[c] Data from Snodgrass et al. (2009).

The combination of consumption of meat and use of tools has been calculated to reduce the force and number of chews required (Zink and Lieberman, 2016). In addition, cooking would greatly reduce the time and energy required for chewing. A mutation about 2.4 million years ago resulted in the human *MYH16* gene becoming a pseudogene. In great apes, an important protein in masticatory muscles of the jaw is myosin heavy chain 16, translated from the *MYH16* gene transcript (Grus and Zhang, 2013; Stedman et al., 2004). Thus, the pseudogenization of the *MYH16* gene resulted in jaw muscles that were less strong. Presumably this was not disadvantageous. Other pseudogenizations in human evolution reduced the olfactory repertoire (discussed in Chapter 1).

5.3.4 Other Changes

Multiple other changes in biology occurring during human evolution include loss of body hair facilitating prolonged running and hence hunting.

5.4 DIET(S) OF HUMAN ANCESTORS?

5.4.1 Introduction to Hominid Diets

Ungar and Sponheimer (2011) concluded that "diet changes are considered key events in human evolution." The evidence is from tooth size and shape, microwear of teeth, stone tools, butchered bones, and stable isotope ratios.

5.4.2 Primate Diets

5.4.2.1 *Overview*

Is it possible to know what these common ancestors ate? The diet of Hominoids in the Miocene was predominantly fruit and other plant materials (Kay, 1977). It is likely that the diet of the common ancestor of chimpanzees, bonobos, and humans was predominantly soft fruit (Andrews and Martin, 1991).

5.4.2.2 *Gorilla Diets*

Among the great apes (family Hominidae), gorillas are herbivorous exclusively consuming multiple types of vegetative foods (especially bark) (Yamagiwa and Basabose, 2006) with very little animals/meat consumed.

5.4.2.3 *Chimpanzee and Bonobo Diets*

Chimpanzees and bonobos are frugivores with over half of their food being fruit (Table 5.5) or herbivores with over 95% of their diet from plants. Observations on populations of chimpanzees support them being ripe fruit specialists (Watts et al., 2012). Based on δ^{15}N in protein (discussed in Section 5.5.2), the diets of chimpanzees

TABLE 5.5 Contribution of Dietary Items to the Nutrition of Chimpanzees and BonobosData from NRC (National Research Council), 2003. Feeding ecology, digestive strategies, and implications for feeding programs in captivity. In: Nutrient Requirements of Nonhuman Primates: Second Revised Edition. National Academies Press, Washington, DC. Available from: http://www.nap.edu/read/9826/chapter/3#17

Food item	Chimpanzee	Bonobo
Fruit	64 (19–99)	52 (1–100)
Leaves	16 (0–56)	14 (0–28)
Vegetation (pith, stem, and stalks, etc.)	7 (0–27)	24 (0–55)
Seeds	3 (0–30)	3 (0–6)
Flowers	2 (0–18)	2 (0–7)
Bark or root		2 (0–11)
Bark	2 (0–26)	
Root, wood etc.	2 (0–41)	
Prey	4 (0–28)[a,b]	2 (0–3)[a,c]

[a] Includes insects.
[b] Includes hunting monkeys.
[c] Monkeys not hunting.

range from omnivorous with significant consumption of vertebrate animal protein to exclusively herbivorous with marked leguminous plant intake (Schoeninger et al., 1999).

Some male chimpanzees hunt (Goodall, 1986). In at least some populations of chimpanzees, groups of males hunt primates and ungulates with one species of monkey, the Western red colobus (*Procolobus badius*) representing 91% of their prey in one population (Basabose, 2002; Mitani and Watts, 1999). Adult male chimpanzees consume much more meat than females (Boesch and Boesch-Achermann, 2000; Fahy et al., 2013). There is very limited consumption of meat in bonobos. Bonobos also consume some small mammals and eggs (Bermejo et al. 1994). They have been observed to be eating flying squirrels (Ihobe, 1992).

Chimpanzees consume termites, honeybees, driver ants, and insect larvae (Hamad et al., 2014); these comprising only about 2%–4% of daily food intake (Deblauwe, 2009; Deblauwe and Janssens, 2008; Goodall, 1986; World Chimpanzee Foundation, 2015). Similarly, bonobos consume some insect larvae together with earthworms (Bermejo et al. 1994). These levels of consumption might seem to be inconsequential but this intake is important as it provides specific essential amino-acids, vitamins, (e.g., vitamin B_{12}), and minerals, such as iron and manganese (Deblauwe, 2009; Deblauwe and Janssens, 2008).

5.4.3 Diet of the Human Branch of the Hominidae

5.4.3.1 Diet of Australopithecus

There is evidence from Sr/Ca that *Australopithecus* consumed a mix of plant and animal matter but not the form of the animal materials; be it insects/larvae/earthworm, and/or marrow (scavenged) or meat (hunted) from herbivores (Sponheimer et al., 2005). As long as 3.39 million years ago, one group of early hominids, *Australopithecus afarensis*, was using stone tools

to butcher large ungulate animals (McPherron et al., 2010). By about 2.6 million years ago, hominids of the human branch were butchering large animals with stone tools and then eating the calorie-rich flesh and bone marrow (reviewed by Pobiner, 2013). It is unclear whether this was a characteristic of one, two, or all the extant species: *A. africanus*, *A. garhi*, or *Paranthropus aethiopicus*. The most likely is *A. garhi*.

The plants consumed by *Australopithecus* and *Paranthropus* include the following:

- *Ardipithecus ramidus* predominantly C_3 plants (Henry et al., 2012)
- *Australopithecus anamensis* diet predominantly C_3 foods
- *A. afarensis* increased C_4 plants savannah and woodlands (Cerling et al., 2013; Levin et al., 2015; Wynn et al., 2013)
- *Australopithecus sediba* predominantly C_3 plants (Henry et al., 2012)
- *Paranthropus boisei*, approximately 80% low-quality C_4 plants (Cerling et al., 2011; Lee-Thorp et al., 2010)

Ye and Gu (2011) argue that the shift in hominin diet, accompanying the transition from forest to savannah in *Australopithecus*, was the shift to harder grassland seeds and underground tubers.

5.4.3.2 Diet of Homo habilis

Hominids dating to 1.95 Ma, presumably *H. habilis*, were consuming land herbivores along with aquatic animals including fish, such as catfish, turtles, and crocodiles, and hence obtaining critical nutrients for brain growth (Archer et al., 2014; Braun et al., 2010).

There were large decreases in the number of species of large carnivores (>21.5 kg) in East Africa between about 2 and 1.5 million years ago and again between 1.5 million years ago and the present (Werdelin and Lewis, 2005, 2013). This is thought to be due to competition from ancestral hominin hunters (Werdelin and Lewis, 2005, 2013).

5.4.3.3 Diet of Homo erectus

H. erectus had markedly increased energy expenditures made possible by the increase in animal materials (Leonard and Robertson, 1997). The increased consumption of animal products was followed by/caused the anatomical changes associated with meat eating in *H. erectus* with reductions in the size of the teeth, jaws, and intestines together with increases in brain size (reviewed by Pobiner, 2013). With the energy demands of a large brain, hominids are envisioned as hunting for fat (Ben-Dor et al., 2011). The extinction of *H. erectus* has been ascribed to overhunting of elephants and competition from other hominins hunting for smaller but faster animals (Ben-Dor et al., 2011). The body size of *H. erectus* is linked to latitude and hence environment temperature (Antón, 2007).

5.4.3.4 Diet of Neanderthals (Homo neanderthalensis)

Neanderthals (*H. neanderthalensis*) are thought to be functionally omnivores consuming animals and plants in their diet after cooking. Meat would be a major source of protein (Sistiaga et al., 2014). This is supported by the following:

- Analysis of the wear of teeth indicates that the diet of Neanderthals was predominantly from animals but there were some plant materials eaten (Fiorenza, 2015).
- Analysis of $\delta^{15}N$ values with 3%–5% greater than herbivores and similar to carnivores due to enrichment (Richards and Trinkaus, 2009).
- Analysis of Neanderthal feces detected biomarkers for animals and plants; respectively, coprostanol for meat and 5β-stigmastanol for plants (Sistiaga et al., 2014).

The Neanderthal animal diet included meat from large mammals (megafauna) together with smaller animals including ibex, chamois, rabbits, turtles, and marine organisms (Bocherens et al., 2005; Smith, 2015; Yravedra and Cobo-Sánchez, 2015). Megafauna is thought to be the major source of meat from animals including woolly rhinoceros, woolly mammoth, rhinoceros, deer, and wild horses (Smith, 2015).

5.4.3.5 Diet of Homo floresiensis

Skeletal remains of another human-like species have been found on Island of Flores. *H. floresiensis* was short (107 cm or 3'6" in height) with a small brain (Table 5.2) and lived between 18,000 BP and 95,000 BP (reviewed by Baab, 2016). There is evidence for *H. floresiensis* hunting or scavenging for rodents, young stegodonts, Komodo dragons, and other fauna (reviewed by Baab, 2016).

5.4.3.6 Diet of Paleolithic Modern Humans (Homo sapiens)

Attempts have been made to reconstruct what our ancestors ate before the advent of agriculture based on the diets of present-day hunter–gatherers (see Chapter 4). These receive 45%–65% of the energy in their diet from animals from hunting and fishing (Cordain et al., 2000). The percentage of the protein requirements coming from animals was higher. The δN values for early modern Paleolithic humans (*H. sapiens*) are high, somewhat greater than in Neanderthals (Richards and Trinkaus, 2009). This would suggest people were consuming marine/aquatic fauna as a significant part of their diet about 27,000–34,000 years ago (Richards and Trinkaus, 2009).

5.4.3.7 Importance of Meat on Brain Development

In today's humans, the brain uses about a fifth of the calories consumed by the body (Raichle and Gusnard, 2002). The resting metabolic rate has been estimated for females (reviewed by Aiello and Wells, 2002):

- *H. sapiens* 1.64
- *Homo ergaster* 1.53

- *A. afarensis* 1.0 (by definition and used as a reference point)

Meat was critically important to the development and functioning of the brain for its high-energy requirements and specific nutrient requirements (Table 5.2). Moreover, meats contain essential fatty acids, docosahexaenoic and arachidonic together with omega-3 fatty acids that are important for brain development (Crawford, 1992; Innis, 2007). Increased consumption of meat during human evolution would have provided the necessary energy and essential fatty acids.

5.4.3.8 Cooking Meat and Human Evolution

The development of man's ability to use fire enabled cooking of meat. There is a debate as to when in human development fire was first used. Some consider that fire was first used by human ancestors (*H. erectus*) about 1.8 million years ago. The use of fire for cooking greatly reduced the time needed for chewing of meat and improved the digestibility of foods. Wrangham (2009) has argued that fire and hence cooking increased calorie availability. This had profound implications allowing greater brain development in *H. erectus* (Table 5.2). Moreover, there was a reduced size of intestines, teeth, and jaw muscles. Other researchers conclude that cooking meat became common only after 500,000 years ago (Zink and Lieberman, 2016).

5.4.3.9 Conclusions on the Changes in Diet in Human Evolution

Based on the diets of chimpanzees, bonobos, and gorillas, it is reasonable to conclude that the last common ancestor of humans, chimpanzees, and bonobos was consuming at most 5% of their energy from animal tissues. Increases in meat consumption (Table 5.6) played a critical part in human evolution. Consumption of meat and other animal products was essential to brain development. In addition, it allowed the gastrointestinal tract to become smaller and

TABLE 5.6 Contribution of Dietary Energy From Meat in Hominids. Assumed Same as Neanderthal (*H. neanderthalensis*)

	Percentage of dietary energy from meat
Gorilla	0
Chimpanzee	4
H. erectus	40–60[a]
Neanderthal (*H. neanderthalensis*)	40–60
Present human–gatherer	45–65
Neolithic through medieval (estimated)	~15

[a] *Assumed same as Neanderthal* (Homo neanderthalis).

more concentrated in the small intestine. Energy requirements for chewing and digestion were greatly reduced.

5.5 APPROACHES TO UNDERSTANDING THE DIET OF HOMINID ANCESTORS

5.5.1 Morphological Evidence

Tooth size, enamel thickness, microstructure, and microwear are used to provide evidence for diet (Delezene et al., 2013; Ungar, 2011). Tooth size is constant in *Ardipithecus*, *Australopithecus*, and *Paranthropus*, but there is a progressive reduction in tooth size during evolution within the genus *Homo* (Evans et al., 2016; Quam et al., 2009). Microware of the teeth of *A. anamensis* and *A. afarensis* suggest that tough foods were not a major part of their diets (Ungar et al., 2010).

5.5.2 Stable Isotopes

The analysis of the stable isotopes in the collagen of bones and tooth enamel or fatty acids from bone (Colonese et al., 2015) is enabling a fuller understanding of what people ate in the past. There are differences in how plants and animals metabolize different isotopes of the same element. This is seen, for instance, with ^{13}C versus ^{14}C in C_3 plants (found in trees, shrubs, and herbs together with temperate or cool growing grasses including wheat) and C_4 plants (Savannah grasses together with tropical plants, such as maize/corn and millet). The ^{13}C in *Australopithecus* (>4 million years ago) and present-day great apes was consistent with the consumption of C_3 plants, such as fruits, and presumably forest living. By about 3.2 million years ago, there was some shift to the consumption of grasses (or the animals that consumed them) and at least some Savannah living presumably followed by a further shift toward C_4 plants in the genus *Homo* (Cerling et al., 2013).

The ratio of ^{15}N to ^{14}N ($\delta^{15}N$) depends on the trophic level of the diet (differing between plants, herbivorous animals, omnivorous animals, and carnivorous animals). Studies determining the nonradioactive isotopes can be used. The $\delta^{15}N$ values are 3%–5% greater in collagen than protein in the diet. A herbivore will therefore have a $\delta^{15}N$ value about 5% greater than plant protein. A carnivore will have about 9% greater $\delta^{15}N$ value than plant protein (or 4% greater than the protein in herbivore flesh). A carnivore consuming other carnivores will have a $\delta^{15}N$ value of 12%. Some marine animals may have a $\delta^{15}N$ value of 16%–20%. Based on the $\delta^{15}N$ value of collagen in bone it is possible to estimate the diet of an animal or person (method reviewed by Richards and Trinkaus, 2009). Based on the $\delta^{15}N$ values at the level of carnivore species, it was concluded that Neanderthals were predominantly carnivorous (Richards and Trinkaus, 2009). Moreover, $\delta^{15}N$ values in skeletal remains of fossil hominids have been used to estimate the duration of breast-feeding (Waters-Rist et al., 2011).

5.5.3 Other Isotopes

Other isotopes determined in tooth enamel provide evidence of diet including strontium isotope ratios (Balter et al., 2012).

5.6 PREDATION AND HUMAN DEVELOPMENT

A. afarensis lived in woody grassland based on $\delta^{18}O$ and $\delta^{13}C$ of paleosol carbonates (Aronson et al., 2008). They were, therefore, more vulnerable to predation. Assemblages of bones from australopithecines closely resemble the remains of prey from such animals as saber-toothed cats, hyenas, and leopards (Hart and Sussman, 2005). Based on high-resolution X-ray computed tomography analysis of the bones of Lucy (*A. afarensis*) from 3.2 million years ago, it was concluded that "her cause of death was a vertical deceleration event or impact following a fall from considerable height" (Kappelman et al., 2016). This would support arborealism by *A. afarensis* presumably to escape predators. It is likely that the *A. afarensis* and other early hominins nest in trees in a manner identical to today's chimpanzees living in open savannahs (Hernandez-Aguilar et al., 2013).

Children were particularly vulnerable to being prey. The skull and mineralized brain of a 3-year-old *A. africanus*, the "Taung child," from 2 million years ago, were found in 1924 by Raymond Dart. It was initially viewed that the child was killed by large cats but more recent analysis demonstrates that it was killed by an eagle. This was based on adjacent eggshells and bones of small mammals and other small animals together with the close similarity to assemblages for living eagles (Berger and Clarke, 1995, 1996; Berger and McGraw, 2007; reviewed Smithsonian, 2016). The ongoing danger of predation during hominid development is supported by tooth marks, consistent with the bite of a leopard, into the skull of a fossil hominid 1–2 million years ago in Southern Africa (Hart and Sussman, 2005). Fire was likely used first to protect these hominids from predators before being used for cooking.

"It is suggested that predation, initially upon early hominids by powerful carnivores but latterly by humans on their hunted prey, served as a powerful stimulus for the development of intelligence" (Brain, 1995).

Quotations About Predators Eating People

"Lions and tigers, and bears, oh my!" Dorothy in Wizard of Oz by L. Frank Baum (1856–1919), author.

"When a man wants to murder a tiger he calls it sport; when a tiger wants to murder him he calls it ferocity." George Bernard Shaw (1856–1950) an Irish playwright.

"You don't see sick animals in the wild. You don't see lame animals in the wild, and it's all because of the predator: the lion, the tiger, the leopard, all the cats." Tippi Hedren (1930 to date), actress and activist for the conservation of lions and tigers.

"An infallible method of conciliating a tiger is to allow oneself to be devoured." Konrad Adenauer (1876–1967), Chancellor of Federal Republic of Germany (West Germany).

TABLE 5.7 Comparison of the Contribution of Dietary Energy Coming From Carbohydrate, Fat, and Protein in Existing Hominids

	Carbohydrate	Fat	Protein
Gorilla[a]	73	3	24
Chimpanzee[a]	73	6	21
Human hunter–gatherer[a]	~30	~43	~27
Western[a]	53	33	14
Recommended[b]	45–65	20–35	10–35

[a] Based on Leonard et al. (2010).
[b] Institute of Medicine. Dietary Reference Intakes 2002.

likely to be markedly higher in hunter–gatherers. Interestingly, we think of fat in our diet as "bad" but hunter–gatherers consume much more.

The decrease in gastrointestinal tract size during human evolution was accompanied by, and perhaps allowed, the increases in brain size (Table 5.2). Table 5.8 illustrates the "expensive tissue hypothesis" with reductions in gastrointestinal tract weights and as a percentage of

TABLE 5.8 Comparison on the Weight and BMR of Brain and Gastrointestinal Tract Between Modern Humans and Those Expected by Projection of the Averages Across Primates

Parameter	Human	Expected based on pattern in primates/projected from average in primates to human body weight
Weight (kg)		
Brain	1.3	0.45
GIT	1.1	1.88
BMR (W)		
Brain	14.6	5.0
GIT	13.4	22.9

BMR, Basal metabolic rate; GIT, gastrointestinal tract.
Based on Aiello, L.C., Wheeler, P. 1995. The expensive tissue hypothesis: The brain and the digestive system in human and primate evolution. Curr. Anthropol. 36, 199–221; Aiello, L.C., 1997. Brains and guts in human evolution: the expensive tissue hypothesis. Braz. J. Genet. 20, 141–148

5.7 CONCLUSIONS

A shift to a much greater consumption of meat is argued to be a major driving-force for human evolution. Teaford and Ungar (2000) concluded that there was a shift in the diet of early hominids between 4.4 and 2.3 million years ago. A contrary view is that the shift to meat occurred in the genus *Homo* line or specifically *H. erectus*. Snodgrass et al. (2009) point to the increase in brain size (Table 5.2), body size (Table 5.3), and diet all occurring in *H. erectus*.

Changes in macronutrients in human evolution are summarized in Table 5.7. Based on today's hunter–gatherers, the difference between chimpanzees and gorillas of the greatest magnitude is fat consumption as a percentage of energy in the diet. There was only a modest increase in protein, but specific essential amino-acids are body weight declining from 2.9% in primates to 1.7% in humans. This is matched by increases in brain size and neuron number (Table 5.2) (based on Aiello, 1997; Aiello and Wheeler, 1995).

The increase in brain size was pivotal in hominids producing tools of increasing sophistication for hunting. The development of hunting might be envisioned as the following ratchet system:

Development of tools ⟶ Better tools ⟶ Improved hunting techniques

↗ ↘ ↑ ↘ ↑

Brain Some meat ⟶ Larger brain More meat ⟶ Larger brain

Then repeated and repeated.

The existence of tools, albeit primitive, and hunting in social groups were essential prerequisites for the development of the following:

- clothing and shelter
- language

This ultimately led to the panoply of today's technologies. Therefore, it is argued that without meat, there would be no humans!

A Synthesis on the Drivers of Human Evolution

In *H. sapiens*, there are 100 billion neurons in the brain with an estimated storage capacity of 1.25×10^{12} bytes (reviewed by Hofman, 2014). Brain enlargement was essential to the development and furthering of cognition together with social interactions, cooperation, and development of language.

The driving forces for brain enlargement might include the following (Aiello and Wells, 2002; Babbitt et al., 2011; Joffe, 1997):

- Climatic changes leading to a greatly expanded savannah and hence a novel environment. There were also the adoption of bipedal walking and, particularly, running was critical to success in this environment to protect against predators (while retaining at least initially the ability to climb trees, also for protection).
- Increased risks of predation in open habitats leading to enlarged social groups.

- Larger social groups requiring greater social skills including much later acquisition of language.
- Larger brains requiring shifts in energy consumption (e.g., protein and fat from animals) and allocation. There were tradeoffs with increased energy allocation for the brain and less for muscles. Moreover, there is an increased expression of genes controlling aerobic metabolism.
- Larger brains requiring a longer period before puberty—"childhood." Progressively less sexual dimorphism in size.
- Longer juvenile period or "childhood" requiring shifts in parenting and sexual partnering with joint male and female responsibilities for raising juveniles.
- Increasingly monogamous relationships with females making the strategic decision that nurturing males are more attractive (sexual selection).

Sex and Reproduction

All human societies have marriages or long-term monogamous partnership with three characteristics: (1) mutual obligations, (2) expectation that the relationship will last through pregnancy, lactation, and childrearing, and (3) sex (Buss and Schmitt, 1993). In addition, polygyny is practiced in many societies. There is also short-term mating with the sexual relationship lasting from hours to months (Buss and Schmitt, 1993). Shifts in sex and reproductive behaviors have been critical to human evolution (Darwin, 1871; Jones and Ratterman, 2009). Buss and Schmitt (1993) contend that human mating is inherently strategic, either consciously or unconsciously (Buss and Schmitt, 1993). Mating is dependent on the following:

Competition between males → Mating ← Attractiveness of male to female

Production of a quality neonate is dependent on the following:

Genetics of male   Genetics of female

↘   ↙

Mating

↓

Number of spermatozoa   Female at fertile time

↘   ↓   ↙

Fertilization

↓   ←   Nutrition and so on of mother

Neonate

Reproductive success can be assessed by the quality of the offspring at sexual maturity, and is dependent on the following:

Quality neonate   Parental investment   ←   Mate quality

↘   ↙

Quality sexually mature offspring

References

Aiello, L.C., 1997. Brains and guts in human evolution: the expensive tissue hypothesis. Braz. J. Genet. 20, 141–148.

Aiello, L.C., Wheeler, P., 1995. The expensive tissue hypothesis: the brain and the digestive system in human and primate evolution. Curr. Anthropol. 36, 199–221.

Aiello, L.C., Wells, J.C.K., 2002. Energetics and the evolution of the genus Homo. Annu. Rev. Anthropol. 31, 323–338.

Alroy, J., 2001. A multispecies overkill simulation of the end-Pleistocene megafaunal mass extinction. Science 292, 1893–1896.

Andrews, P., Martin, L., 1991. Hominoid dietary evolution. Philos. Trans. R. Soc., B 334, 199–209.

Antón, S.C., 2007. Climatic influences on the evolution of early Homo? Folia Primatol. 78, 365–388.

Archer, W., Braun, D.R., Harris, J.W., McCoy, J.T., Richmond, B.G., 2014. Early Pleistocene aquatic resource use in the Turkana Basin. J. Hum. Evol. 77, 74–87.

Aronson, J.L., Hailemichael, M., Savin, S.M., 2008. Hominid environments at Hadar from paleosol studies in a framework of Ethiopian climate change. J. Hum. Evol. 55, 532–550.

Baab, K., 2016. The place of Homo floresiensis in human evolution. J. Anthropol. Sci. 94, 5–18.

Babbitt, C.C., Warner, L.R., Fedrigo, O., Wall, C.E., Wray, G.A., 2011. Genomic signatures of diet-related shifts during human origins. Proc. R. Soc. B 278, 961–969.

Balter, V., Braga, J., Télouk, P., Thackeray, J.F., 2012. Evidence for dietary change but not landscape use in South African early hominins. Nature 489, 558–560.

Basabose, A.K., 2002. Diet composition of chimpanzees inhabiting the montane forest of Kahuzi, Democratic Republic of Congo. Am. J. Primatol. 58, 1–21.

Ben-Dor, M., Gopher, A., Hershkovitz, I., Barkai, R., 2011. Man the fat hunter: the demise of Homo erectus and the emergence of a new hominin lineage in the Middle Pleistocene (ca. 400 kyr) Levant. PLoS One 6, e28689.

Berger, L.R., Clarke, R.J., 1995. Eagle involvement in accumulation of the Taung child fauna. J. Hum. Evol. 29, 275–299.

Berger, L.R., Clarke, R.J., 1996. The load of the Taung child. Nature 379, 778–779.

Berger, L.R., McGraw, W.S., 2007. Further evidence for eagle predation of, and feeding damage on, the Taung child. S. Afr. J. Sci. 103, 296–298.

Bermejo, J., Illera, G., Sabater Pi, J.S., 1994. Animals and mushrooms consumed by bonobos (Pan paniscus). Int. J. Primatol. 15, 879–898.

Bocherens, H., Drucker, D.G., Billiou, D., Patou-Mathis, M., Vandermeersch, B., 2005. Isotopic evidence for diet and subsistence pattern of the Saint-Césaire I Neanderthal: review and use of a multi-source mixing model. J. Hum. Evol. 49, 71–87.

Boesch, C., Boesch-Achermann, H., 2000. The Chimpanzees of the Taï Forest: Behavioural Ecology and Evolution. Oxford University Press, Oxford, UK.

Bosinger, B.E., Zachary, P., Johnson, Z.P., Guido Silvestri, G., 2011. Primate genomes for biomedicine. Nat. Biotechnol. 29, 983–984.

Brain, C.K., 1995. The importance of predation to the course of human and other animal evolution. S. Afr. Archaeol. Bull. 50, 93–97.

Braun, D.R., Harris, J.W., Levin, N.E., McCoy, J.T., Herries, A.I., Bamford, M.K., Bishop, L.C., Richmond, B.G., Kibunjia, M., 2010. Early hominin diet included diverse terrestrial and aquatic animals 1.95 Ma in East Turkana, Kenya. Proc. Natl. Acad. Sci. USA 107, 10002–10007.

Burley, D., Weisler, M.I., Zhao, J.X., 2012. High precision U/Th dating of first Polynesian settlement. PLoS One 7, e48769.

Buss, D.M., Schmitt, D.P., 1993. Sexual strategies: perspectives on human mating. Psychol. Rev. 100, 204–232.

Carrigan, M.A., Uryasev, O., Davis, R.P., Zhai, L.M., Hurley, T.D., Benner, S.A., 2012. The natural history of class I primate alcohol dehydrogenases includes gene duplication, gene loss, and gene conversion. PLoS One 7, e41175.

Cerling, T.E., Manthi, F.K., Mbua, E.N., Leakey, L.N., Leakey, M.G., Leakey, R.E., Brown, F.H., Grine, F.E., Hart, J.A., Kaleme, P., Roche, H., Uno, K.T., Wood, B.A., 2013. Stable isotope-based diet reconstructions of Turkana Basin hominins. Proc. Natl. Acad. Sci. USA 110, 10501–10506.

Cerling, T.E., Mbua, E., Kirera, F.M., Manthi, F.K., Grine, F.E., Leakey, M.G., Sponheimer, M., Uno, K.T., 2011. Diet of Paranthropus boisei in the early Pleistocene of East Africa. Proc. Natl. Acad. Sci. USA 108, 9337–9341.

Colonese, A.C., Farrell, T., Lucquin, A., Firth, D., Charlton, S., Robson, H.K., Alexander, M., Craig, O.E., 2015. Archaeological bone lipids as palaeodietary markers. Rapid Commun. Mass Spectrom. 29, 611–618.

Cordain, L., Brand-Miller, J., Eaton, S.B., Mann, N., Holt, S.H.A., Speth, J.D., 2000. Plant-animal subsistence ratios and macronutrient energy estimations in worldwide hunter-gatherers. Am. J. Clin. Nutr. 71, 682–692.

Crawford, M.A., 1992. The role of dietary fatty acids in biology: their place in the evolution of the human brain. Nutr. Rev. 50, 3–11.

Curnoe, D., Thorne, A., 2003. Number of ancestral human species: a molecular perspective. Homo 53, 201–224.

Darwin, C., 1871. The descent of man, and selection in relation to sex. Murray, London.

Deblauwe, I., 2009. Temporal variation in insect-eating by chimpanzees and gorillas in Southeast Cameroon: extension of niche differentiation. Int. J. Primatol. 30, 229–252.

Deblauwe, I., Janssens, G.P., 2008. New insights in insect prey choice by chimpanzees and gorillas in southeast Cameroon: the role of nutritional value. Am. J. Phys. Anthropol. 135, 42–55.

Delezene, L.K., Zolnierz, M.S., Teaford, M.F., Kimbel, W.H., Grine, F.E., Ungar, P.S., 2013. Premolar microwear and tooth use in *Australopithecus afarensis*. J. Hum. Evol. 65, 282–293.

Evans, A.R., Daly, E.S., Catlett, K.K., Paul, K.S., King, S.J., Skinner, M.M., Nesse, H.P., Hublin, J.J., Townsend, G.C., Schwartz, G.T., Jernvall, J., 2016. A simple rule governs the evolution and development of hominin tooth size. Nature 530, 477–480.

Fahy, G.E., Richards, M., Riedel, J., Hublin, J.J., Boesch, C., 2013. Stable isotope evidence of meat eating and hunting specialization in adult male chimpanzees. Proc. Natl. Acad. Sci. USA 110, 5829–5833.

Fagundes, N.J., Ray, N., Beaumont, M., Neuenschwander, S., Salzano, F.M., Bonatto, S.L., Excoffier, L., 2007. Statistical evaluation of alternative models of human evolution. Proc. Natl. Acad. Sci. USA 104, 17614–17619.

Faith, J.T., Surovell, T.A., 2009. Synchronous extinction of North America's Pleistocene mammals. Proc. Natl. Acad. Sci. USA 106, 20641–206415.

Fiorenza, L., 2015. Reconstructing diet and behaviour of Neanderthals from Central Italy through dental macrowear analysis. J. Anthropol. Sci. 93, 119–133.

Goebel, T., Waters, M.R., O'Rourke, D.H., 2008. The late Pleistocene dispersal of modern humans in the Americas. Science 319, 1497–1502.

Goodall, J., 1986. Chimpanzees of Gombe. Belknap Press, Cambridge, MA, USA.

Grabowski, M., Hatala, K.G., Jungers, W.L., Richmond, B.G., 2015. Body mass estimates of hominin fossils and the evolution of human body size. J. Hum. Evol. 85, 75–93.

Grayson, D.K., Meltzer, D.J., 2002. Clovis hunting and large mammal extinction: a critical review of the evidence. J. World Prehist. 16, 313–359.

Grayson, D.K., Meltzer, D.J., 2004. North American overkill continued? J. World Prehist. 31, 133–136.

Groves, C., 2001. Primate taxonomy. Smithsonian Institute, Washington, DC.

Grove, C.P., 2005. Wilson and Reeder's Mammal Species of the World, 3rd Edn. Johns Hopkins University Press, Baltimore.

Grus, W.E., Zhang, J., 2013. Human Lineage-Specific Gene Inactivation. In: eLS. John Wiley & Sons, Ltd, Chichester, UK.

Hamad, I., Delaporte, E., Raoult, D., Bittara, F., 2014. Detection of termites and other insects consumed by African great apes using molecular fecal analysis. Sci. Rep. 4, 4478.

Hart, D., Sussman, R.W., 2005. Man the Hunted: Primates, Predators and Human Evolution. Westview Press, Cambridge, MA, USA.

Henn, B.M., Cavalli-Sforza, L.L., Feldman, M.W., 2012. The great human expansion. Proc. Natl. Acad. Sci. USA 109, 17758–17764.

Henry, A.G., Ungar, P.S., Passey, B.H., Sponheimer, M., Rossouw, L., Bamford, M., Sandberg, P., de Ruiter, D.J., Berger, L., 2012. The diet of *Australopithecus sediba*. Nature 487, 90–93.

Herculano-Houzel, S., Kaas, J.H., 2011. Gorilla and orangutan brains conform to the primate cellular scaling rules: implications for human evolution. Brain Behav. Evol. 77, 33–44.

Hernandez-Aguilar, R.A., Moore, J., Stanford, C.B., 2013. Chimpanzee nesting patterns in savanna habitat: environmental influences and preferences. Am. J. Primatol. 75, 979–994.

Hershkovitz, I., Marder, O., Ayalon, A., Bar-Matthews, M., Yasur, G., Boaretto, E., Caracuta, V., Alex, B., Frumkin, A., Goder-Goldberger, M., Gunz, P., Holloway, R.L., Latimer, B., Lavi, R., Matthews, A., Slon, V., Mayer, D.B., Berna, F., Bar-Oz, G., Yeshurun, R., May, H., Hans, M.G., Weber, G.W., Barzilai, O., 2015. Levantine cranium from Manot Cave (Israel) foreshadows the first European modern humans. Nature 520, 216–219.

Hermanussen, M., 2003. Stature of early Europeans. Hormones (Athens) 2, 175–178.

Hobbes, T., 1651. Leviathan or the Matter, Forme, and Power of a Commonwealth, Ecclesiastical and Civil. Available from: https://archive.org/details/leviathanormatt01hobbgoog.

Hofman, M.A., 2014. Evolution of the human brain: when bigger is better. Front. Neuroanat. 8, 15.

Hublin, J.J., 2009. Out of Africa: modern human origins special feature: the origin of Neanderthals. Proc. Natl. Acad. Sci. USA 106, 16022–16027.

Ihobe, H., 1992. Observations of the meat-eating behavior of wild bonobos (*Pan paniscus*) at Wamba, Republic of Zaire. Primates 33, 247–250.

Innis, S.M., 2007. Dietary ($n - 3$) fatty acids and brain development. J. Nutr. 137, 855–859.

International Commission on Stratigraphy, 2015. Available from: http://www.stratigraphy.org/ICSchart/ChronostratChart2015-01.pdf.

ITIS (Integrated Taxonomic Information System), 2016. Available from: http://www.itis.gov/servlet/SingleRpt/SingleRpt?search_topic=TSN&search_value=180091.

Joffe, T.H., 1997. Social pressures have selected for an extended juvenile period in primates. J. Hum. Evol. 32, 593–605.

Jones, A.G., Ratterman, N.L., 2009. Mate choice and sexual selection: what have we learned since Darwin? Proc. Natl. Acad. Sci. USA 106 (Suppl. 1), 10001–10008.

Kappelman, J., Ketcham, R.A., Pearce, S., Todd, L., Akins, W., Colbert, M.W., Feseha, M., Maisano, J.A., Witzel, A., 2016. Perimortem fractures in Lucy suggest mortality from fall out of tall tree. Nature 537 (7621), 503–507.

Kay, R., 1977. Diets of the early Miocene hominoids. Nature 268, 628–630.

Kuhlwilm, M., Gronau, I., Hubisz, M.J., de Filippo, C., Prado-Martinez, J., Kircher, M., Fu, Q., Burbano, H.A., Lalueza-Fox, C., de la Rasilla, M., Rosas, A., Rudan, P., Brajkovic, D., Kucan, Ž., Gušic, I., Marques-Bonet, T., Andrés, A.M., Viola, B., Pääbo, S., Meyer, M., Siepel, A., Castellano, S., 2016. Ancient gene flow from early modern humans into Eastern Neanderthals. Nature 530, 429–433.

Lee-Thorp, J.A., Sponheimer, M., Passey, B.H., de Ruiter, D.J., Cerling, T.E., 2010. Stable isotopes in fossil hominin tooth enamel suggest a fundamental dietary shift in the Pliocene. Philos. Trans. R. Soc. B 365, 3389–3396.

Levin, N.E., Haile-Selassie, Y., Frost, S.R., Saylor, B.Z., 2015. Dietary change among hominins and cercopithecids in Ethiopia during the early Pliocene. Proc. Natl. Acad. Sci. USA 112, 12304–12309.

Leonard, W.R., Robertson, M.L., 1997. Comparative primate energetics and hominid evolution. Am. J. Phys. Anthropol. 102, 265–281.

Leonard, W.R., Snodgrass, J.J., Robertson, M.L., 2010. Evolutionary perspectives on fat ingestion and metabolism in humans. In: Montmayeur, J.-P., le Coutre, J. (Eds.), Fat Detection. Taste, Texture, and Post Ingestive Effects. CRC Press, Boca Raton, pp. 3–19.

Lowery, R.K., Uribe, G., Jimenez, E.B., Weiss, M.A., Herrera, K.J., Regueiro, M., Herrera, R.J., 2013. Neanderthal and Denisova genetic affinities with contemporary humans: introgression versus common ancestral polymorphisms. Gene 530, 83–94.

Marzke, M.W., 2013. Tool making, hand morphology and fossil hominins. Philos. Trans. R. Soc. B 368, 20120414.

McDougall, I., Brown, F.H., Fleagle, J.G., 2005. Stratigraphic placement and age of modern humans from Kibish, Ethiopia. Nature 433, 733–736.

McPherron, S.P., Alemseged, Z., Marean, C.W., Wynn, J.G., Reed, D., Geraads, D., Bobe, R., Béarat, H.A., 2010. Evidence for stone-tool-assisted consumption of animal tissues before 3.39 million years ago at Dikika, Ethiopia. Nature 466, 857–860.

Mellars, P., 2006. Why did modern human populations disperse from Africa ca. 60,000 years ago? A new model. Proc. Natl. Acad. Sci. USA 103, 9381–9386.

Milton, K., 2003. The critical role played by animal source foods in human (Homo) evolution. J. Nutr. 133 (2), 3886S–3892S.

Milton, K., Demment, M.W., 1988. Chimpanzees fed high and low fiber diets and comparison with human data. J. Nutr. 118, 1082–1088.

Mitani, J., Watts, D., 1999. Demographic influences on the hunting behavior of chimpanzees. Am. J. Phys. Anthropol. 109, 439–454.

Moeller, A.H., Li, Y., Mpoudi Ngole, E., Ahuka-Mundeke, S., Lonsdorf, E.V., Pusey, A.E., Peeters, M., Hahn, B.H., Ochman, H., 2014. Rapid changes in the gut microbiome during human evolution. Proc. Natl. Acad. Sci. USA 111, 16431–16435.

Navarrete, A., van Schaik, C.P., Isler, K., 2011. Energetics and the evolution of human brain size. Nature 480, 91–93.

Organ, C., Nunn, C.L., Machanda, Z., Wrangham, R.W., 2011. Phylogenetic rate shifts in feeding time during the evolution of Homo. Proc. Natl. Acad. Sci. USA 108, 14555–14559.

Pobiner, B., 2013. Evidence for meat-eating by early humans. Nat. Educ. Knowl. 4, 1.

Quam, R., Bailey, S., Wood, B., 2009. Evolution of M1 crown size and cusp proportions in the genus Homo. J. Anat. 214, 655–670.

Raichle, M.E., Gusnard, D.A., 2002. Appraising the brain's energy budget. Proc. Natl. Acad. Sci. USA 99, 10237–10239.

Richards, M.P., Trinkaus, E., 2009. Isotopic evidence for the diets of European Neanderthal and early modern humans. Proc. Natl. Acad. Sci. USA 106, 16034–16039.

Roach, N.T., Richmond, B.G., 2015. Clavicle length, throwing performance and the reconstruction of the Homo erectus shoulder. J. Hum. Evol. 80, 107–113.

Sankararaman, S., Patterson, N., Li, H., Pääbo, S., Reich, D., 2012. The date of interbreeding between Neanderthals and modern humans. PLoS Genet. 8, e1002947.

Schoeninger, M.J., Moore, J., Sept, J.M., 1999. Subsistence strategies of two "Savanna" chimpanzee populations: the stable isotope evidence. Am. J. Primatol. 49, 297–314.

Shultz, S., Nelson, E., Dunbar, R.I., 2012. Hominin cognitive evolution: identifying patterns and processes in the fossil and archaeological record. Philos. Trans. R. Soc. Lond. B 367, 2130–2140.

Sistiaga, A., Mallol, C., Galván, B., Summons, R.E., 2014. The Neanderthal meal: a new perspective using faecal biomarkers. PLoS One 9, e101045.

Smith, G.M., 2015. Neanderthal megafaunal exploitation in Western Europe and its dietary implications: a contextual reassessment of La Cotte de St Brelade (Jersey). J. Hum. Evol. 78, 181–201.

Smithsonian, 2016. What does it mean to be human? Available from: http://humanorigins.si.edu/evidence/human-fossils/fossils/taung-child.

Snodgrass, J.J., Leonard, W.R., Robertson, M.L., 2009. The energetics of encephalization in early hominids. In: Hublin, J.-J., Richards, M.P. (Eds.), The Evolution of Hominin Diets: Integrating Approaches to the Study of Palaeolithic Subsistence. Springer, Dordrecht, pp. 15–29.

Sponheimer, M., de Ruiter, D., Lee-Thorp, J., Späth, A., 2005. Sr/Ca and early hominin diets revisited: new data from modern and fossil tooth enamel. J. Hum. Evol. 48, 147–156.

Steadman, D.W., Martin, P.S., MacPhee, R.D., Jull, A.J., McDonald, H.G., Woods, C.A., Iturralde-Vinent, M., Hodgins, G.W., 2005. Asynchronous extinction of Late Quaternary sloths on continents and islands. Proc. Natl. Acad. Sci. USA 102, 11763–11768.

Stedman, H.H., Kozyak, B.W., Nelson, A., Thesier, D.M., Su, L.T., Low, D.W., Bridges, C.R., Shrager, J.B., Minugh-Purvis, N., Mitchell, M.A., 2004. Myosin gene mutation correlates with anatomical changes in the human lineage. Nature 428, 415–418.

Stevens, C.E., 1988. Comparative physiology of the vertebrate digestive system. Cambridge University Press, Cambridge, UK.

Stevens, N.J., Seiffert, E.R., O'Connor, P.M., Roberts, E.M., Schmitz, M.D., Krause, C., Gorscak, E., Ngasala, S.,

Hieronymus, T.L., Temu, J., 2013. Palaeontological evidence for an Oligocene divergence between Old World monkeys and apes. Nature 497, 611–614.

Strauss, M., 2015. 12 Theories of how we became human, and why they're all wrong. National Geographic. Available from: http://news.nationalgeographic.com/2015/09/150911-how-we-became-human-theories-evolution-science/.

Tattersall, I., 2009. Human origins: Out of Africa. Proc. Natl. Acad. Sci. USA 106, 16018–16021.

Teaford, M.F., Ungar, P.S., 2000. Diet and the evolution of the earliest human ancestors. Proc. Natl. Acad. Sci. USA 97, 13506–13511.

Thomson, V.A., Lebrasseur, O., Austin, J.J., Hunt, T.L., Burney, D.A., Denham, T., Rawlence, N.J., Wood, J.R., Gongora, J., Girdland Flink, L., Linderholm, A., Dobney, K., Larson, G., Cooper, A., 2014. Using ancient DNA to study the origins and dispersal of ancestral Polynesian chickens across the Pacific. Proc. Natl. Acad. Sci. USA 111, 4826–4831.

Tree of Life, 2015. Available from: http://tolweb.org/Catarrhini/16293.

Ungar, P.S., 2011. Dental evidence for the diets of Plio-Pleistocene hominins. Am. J. Phys. Anthropol. 146 (Suppl. 53), 47–62.

Ungar, P.S., Sponheimer, M., 2011. The diets of early hominins. Science 334, 190–193.

Ungar, P.S., Scott, R.S., Grine, F.E., Teaford, M.F., 2010. Molar microwear textures and the diets of *Australopithecus anamensis* and *Australopithecus afarensis*. Philos. Trans. R. Soc., B 365, 3345–3354.

van Asch, B., Zhang, A.B., Oskarsson, M.C., Klütsch, C.F., Amorim, A., Savolainen, P., 2013. Pre-Columbian origins of Native American dog breeds, with only limited replacement by European dogs, confirmed by mtDNA analysis. Proc. R. Soc. B 280, 20131142.

Wall, J.D., Yang, M.A., Jay, F., Kim, S.K., Durand, E.Y., Stevison, L.S., Gignoux, C., Woerner, A., Hammer, M.F., Slatkin, M., 2013. Higher levels of Neanderthal ancestry in East Asians than in Europeans. Genetics 194, 199–209.

Waters, M.R., Stafford, Jr., T.W., McDonald, H.G., Gustafson, C., Rasmussen, M., Cappellini, E., Olsen, J.V., Szklarczyk, D., Jensen, L.J., Gilbert, M.T., Willerslev, E., 2011. Pre-Clovis mastodon hunting 13,800 years ago at the Manis site, Washington. Science 334, 351–335.

Waters, M.R., Stafford, Jr., T.W., Kooyman, B., Hills, L.V., 2015. Late Pleistocene horse and camel hunting at the southern margin of the ice-free corridor: reassessing the age of Wally's Beach, Canada. Proc. Natl. Acad. Sci. USA 112, 4263–4267.

Waters-Rist, A.L., Bazaliiskii, V.I., Weber, A.W., Katzenberg, M.A., 2011. Infant and child diet in Neolithic hunter-fisher-gatherers from Cis-Baikal, Siberia: intra-long bone stable nitrogen and carbon isotope ratios. Am. J. Phys. Anthropol. 146, 225–241.

Watts, D.P., Potts, K.B., Lwanga, J.S., Mitani, J.C., 2012. Diet of chimpanzees (*Pan troglodytes schweinfurthii*) at Ngogo,
Kibale National Park, Uganda, 1. Diet composition and diversity. Am. J. Primatol. 74, 114–129.

Werdelin, L., Lewis, M.E., 2005. Plio-Pleistocene Carnivora of eastern Africa: species richness and turnover patterns. Zool. J. Linn. Soc. 144, 121–144.

Werdelin, L., Lewis, M.E., 2013. Temporal change in functional richness and evenness in the Eastern African Plio-Pleistocene carnivoran guild. PLoS One 8, e57944.

Wilmshurst, J.M., Hunt, T.L., Lipo, C.P., Anderson, A.J., 2011. High-precision radiocarbon dating shows recent and rapid initial human colonization of East Polynesia. Proc. Natl. Acad. Sci. USA 108, 1815–1820.

Wood, B., Richmond, B.G., 2000. Human evolution: taxonomy and paleobiology. J. Anat. 197, 19–60.

World Chimpanzee Foundation, 2015. Available from: http://www.wildchimps.org/wcf/english/files/chimp4.htm.

Wrangham, R., 2009. Catching fire—how cooking made us human. Profile, London.

Wynn, J.G., Sponheimer, M., Kimbel, W.H., Alemseged, Z., Reed, K., Bedaso, Z.K., Wilson, J.N., 2013. Diet of *Australopithecus afarensis* from the Pliocene Hadar Formation, Ethiopia. Proc. Natl. Acad. Sci. USA 110, 10495–10500.

Yamagiwa, J., Basabose, A.K., 2006. Diet and seasonal changes in sympatric gorillas and chimpanzees at Kahuzi-Biega National Park. Primates 47, 74–90.

Ye, K., Gu, Z., 2011. Recent advances in understanding the role of nutrition in human genome evolution. Adv. Nutr. 2, 486–496.

Young, R.W., 2003. Evolution of the human hand: the role of throwing and clubbing. J. Anat. 202, 165–174.

Young, N.M., Capellini, T.D., Roach, N.T., Alemsegede, Z., 2015. Fossil hominin shoulders support an African ape-like last common ancestor of humans and chimpanzees. Proc. Natl. Acad. Sci. USA 112, 11829–11834.

Yravedra, J., Cobo-Sánchez, L., 2015. Neanderthal exploitation of ibex and chamois in southwestern Europe. J. Hum. Evol. 78, 12–32.

Zink, K.D., Lieberman, D.E., 2016. Impact of meat and Lower Palaeolithic food processing techniques on chewing in humans. Nature 531 (7595), 500–503.

Further Reading

Boutton, T.W., Lynott, M.J., Bumsted, M.P., 1991. Stable carbon isotopes and the study of prehistoric human diet. Crit. Rev. Food Sci. Nutr. 30, 373–385.

Salesse, K., Dufour, É., Castex, D., Velemínský, P., Santos, F., Kuchařová, H., Jun, L., Brůžek, J., 2013. Life history of the individuals buried in the St. Benedict Cemetery (Prague, 15th–18th centuries): insights from (14)C dating and stable isotope (δ(13)C, δ(15)N, δ(18)O) analysis. Am. J. Phys. Anthropol. 151, 202–214.

Singman, J.L., McLean, W., 1995. Daily Life in Chaucer's England. Greenwood Press, Westport, Connecticut.

6

The Neolithic Revolution, Animal Domestication, and Early Forms of Animal Agriculture

Colin G. Scanes

University of Wisconsin–Milwaukee, Milwaukee, WI, United States

6.1 INTRODUCTION

Two of the most important events in human history were the Neolithic Revolution (i.e., the transition from hunter–gatherer to agricultural societies, initially growing crops followed by raising domesticated livestock) and the Industrial Revolution (i.e., the era of mechanization of manufacturing) (Jones, 2001; Weisdorf, 2005) (Table 6.1). Both of these were accompanied by large increases in economic development, as reflected by per capita income and population growth (Table 6.1). In addition, it is argued that the biomedical/public health revolution is a third such event; again with increased per capita income and in population growth. In contrast, during Paleolithic times, human population increased but only as the land area occupied expanded with groups of people spreading through Eurasia, to Australia, to the Americas, and across Polynesia.

6.2 NEOLITHIC REVOLUTION

6.2.1 Overview

The Neolithic (literally New Stone Age) Revolution with the beginning of agriculture occurred independently in multiple geographical locations. These included the Middle East (the Fertile Crescent), East Asia along the Ganges, South Asia (the Indus Valley), Mesoamerica, the Andes, and West Africa. The Neolithic agricultural revolution can be viewed as a shift from extractive or exploitive to productive food-producing societies. Alternatively, the transition can be viewed as incremental with late Paleolithic hunter–gatherer societies having settlements or at least base-camps, tools for harvesting, and so on. The Neolithic Revolution first encompassed the domestication of cereals and legumes with domestication of animals occurring somewhat later. Corollaries to the Neolithic

TABLE 6.1   Human Population and Growth Rate Over the Past 300,000 Years

Time (years)	Human population	Growth rate (%)
Series A[a]		
300,000 BP	1 million	
10,000 BP (8,000 BC or BCE)[b]	5 million	0.00055[c]
Series B[d]		
12,000 BP (10,000 BC or BCE)	5 million	
5,000 BP	49 million	0.033
2,500 BP	165 million	0.048
CE/AD 1,000 (1,000 BP)	335 million	0.047
CE 1,500 (500 BP)	481 million	0.072
CE 1,900	1.64 billion	0.306
CE 1,950	2.58 billion	0.906
CE 2,000	6.09 billion	1.718

BCE, Years before common era; BP, years before present; CE, common era (equivalent to AD).
[a] Deevey (1960).
[b] Estimated at the beginning of the Holocene/Neolithic.
[c] Estimated over 290,000 years.
[d] Averages reported by Manning (2008).

Revolution were the following: musculoskeletal stress, poorer overall nutrition accompanied by reduced height, the appearance of dental disease, and a huge increase in fertility.

Multiple explanations have been proposed for the shift from a hunter–gatherer society to agriculture. Explanations include overhunting, population pressure, climate change or other environmental shifts, human behavior with leisure to experiment with plants, problems living on marginal lands, and opportunity (unintentional) coupled with a ratchet effect with increasing population demanding more food. The only consensus is that there is no single explanation that is "entirely satisfactory" (Weisdorf, 2005).

The Neolithic Revolution entailed close to all of the postinfant population working as farm labor (Weisdorf, 2005). Farming was "more trouble than it saved" (Weisdorf, 2005). As techniques improved, surpluses of food would be occurring for the first time in human history allowing the development of specialized occupations and leading to hierarchal organizational structures (Diamond, 1997; Weisdorf, 2005).

6.3 TRANSITION FROM HUNTER–GATHERER TO FARMING SOCIETIES IN THE NEAR EAST (EPIPALEOLITHIC TO NEOLITHIC TRANSITION)

6.3.1 The Epipaleolithic to Neolithic Transition

Arguably the first site for the Neolithic Revolution was the Middle East, specifically the Levant and Anatolia (see sidebar and Fig. 6.1).

Definitions

- Anatolia: the Asian parts of the present-day country of Turkey (Fig. 6.1).
- Levant: present-day Israel, Palestine, Syria, Lebanon, and Jordan together with parts of Iraq. The region is bounded by the Mediterranean Sea to the West, the Taurus mountains of southern Anatolia (present-day Turkey) to the North, the Arabian Desert and Mesopotamia to the East, and the Sinai desert to the South (Fig. 6.1).
- Fertile Crescent: where agriculture originated in the Levant (later Canaan), Anatolia, and Mesopotamia (Fig. 6.1).
- Neolithic: New Stone Age when agriculture originated.
- Paleolithic: Old Stone Age with Upper Paleolithic being the later parts of the Old Stone Age and the Epipaleolithic being the final part of the Upper Paleolithic.

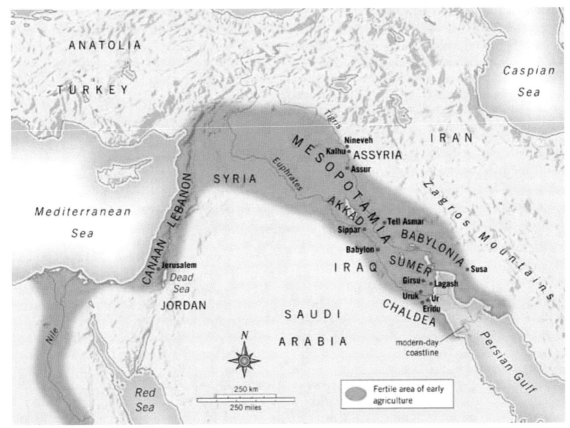

FIGURE 6.1 **A map of part of the Middle East showing the location of the Fertile Crescent.** *Source: Courtesy Ted Mitchell, Coventry Public School.*

6.3.2 The Epipaleolithic Natufian Peoples of the Levant

6.3.2.1 Overview

The Natufian hunter–gatherers lived in settlements in the Levant over 10,000 years ago. There are differences in the details of the timeline for Natufians. Dates get progressively earlier in the more recent research. For example, dates range between about 13,000 and 10,000 years before present (BP) (Karasik et al., 2000), or between 12,500 and 10,300 BP (Eshed et al., 2006) or between 15,000 and 11,700 BP (Eitam et al., 2015). This chapter will employ the timelines of Eitam et al. (2015):

- Early Natufian (15,000–13,500 BP)

- Late Natufian (13,500–11,700 BP)
- Neolithic preceramic (or prepottery) populations (11,700–7,500 BP)

It has been suggested that the shift from hunter–gatherer to farming in the Levant was an adaptive response to abrupt climate change and consequent changes in vegetation. There was a catastrophic breakdown in the system of foraging and societal collapse (deMenocal, 2001; Moore and Hillman, 1992; Weiss and Bradley, 2001). In contrast, Rosen and Rivera-Collazo (2012) considered that the hunter–gatherers were "highly adaptable." This might militate against the catastrophic breakdown scenario.

6.3.2.2 Natufian Animal Connections

There are examples that are strongly indicative of the importance of animals in Natufian societies. For instance, the grave of a presumed Natufian Sharman woman in the Levant contained remains of the following animals: tortoises (50 shells), a wild boar, an eagle, aurochs (wild cattle), a leopard, and martens (2) (Grosman et al., 2008). In addition, there is evidence that Natufians had "feasts," perhaps celebratory or ceremonial, consuming aurochs and turtles (*Testudo graeca*) (Munrom and Grosman, 2010).

6.3.2.3 Dietary Changes in the Hunter–Gatherer to Neolithic Transition

6.3.2.3.1 PLANT-BASED FOODS

The Natufian hunter–gatherers and the subsequent Neolithic farmers of the southern Levant consumed cereals; initially wild and subsequently domesticated. Evidence for this comes from dental wear, which do not exhibit major changes of pattern of wear between the Levant hunter–gatherers and Neolithic farmers (Eshed et al., 2006). Natufian hunter–gatherers harvested wild grains including barley prior to full ripening using flint sickles. Subsequently, Late Natufians produced flour using rock-cut mortars (Eitam et al., 2015). Preserved plant remains indicate that human hunter–gatherers were collecting cereal grains in the same area

as early as 23,000 BP (Weiss et al., 2004). Eitam et al. (2015) proposed that the Early Natufian peoples (15,000–13,500 BP) were consuming meals of hulled grains (groats) and porridge but later they transitioned to unleavened bread from more finely ground grain produced in the Late Natufian (13,500–11,700 BP).

6.3.2.3.2 ANIMAL-BASED FOODS

The Natufians and subsequent Neolithic farmers hunted animals including the following species: mountain gazelles (*Gazella gazella*), goitered gazelles (*Gazella subgutturosa*), Persian fallow deer (*Dama dama mesopotamica*), roe deer (*Capreolus capreolus*), wild cattle/aurochs (*Bos primigenius*), and wild boar (*Sus scrofa*) together with Greek tortoises (*T. graeca*), brown hares (*Lepus capensis*), and Chukar partridges (*Alectoris chukar*) (Bar-Yosef, 1998; Rosen and Rivera-Collazo, 2012).

Table 6.2 summarizes the changes in large- and medium-sized mammals consumed based on bones prior to and after the Neolithic Revolution in the Levant. Clearly, hunter–gatherers consumed principally antelope and deer together with some wild cattle, sheep/goats, pigs, and equids. There was little difference in the animals consumed at the beginning of the Neolithic, during the Prepottery A Neolithic, presumably reflecting ongoing hunting. However, there was a precipitous decrease in consumption

TABLE 6.2   Changes in the Proportion of Bones (% ± SEM) From Medium and Large Mammals in Basecamps/Settlements in the Levant

	Early natufian	Late natufian	Prepottery neolithic A	Prepottery neolithic B
Gazelles	71.8 ± 6.0	79.3 ± 8.3	75.4 ± 8.8	12.0 ± 2.0
Deer	16.7 ± 4.6	11.3 ± 5.6	10.1 ± 4.6	0.8 ± 0.5
Cattle	3.4 ± 1.5	3.2 ± 1.8	5.2 ± 2.1	14.6 ± 2.2
Pigs	5.7 ± 0.7	2.9 ± 1.9	8.8 ± 2.9	14.8 ± 4.3
Equids	0.4 ± 0.4	0 ± 0.0	0.3 ± 0.3	2.3 ± 1.6
Sheep/goats	4.6 ± 1.4	3.5 ± 2.0	2.9 ± 1.1	60.9 ± 7.9

of antelope and deer in the subsequent Prepottery B Neolithic. Instead, consumption switched to primarily domesticated sheep/goats together with some domesticated pigs and cattle.

Levels at Aşıklı Höyük With Similar Systems at Other Sites

Levels in an archaeological site, such as Aşıklı Höyük allow dating of items at the same level to about the same age. The upper levels are the youngest (starting with level 1) and the lower levels are the oldest (level 5).

Similarly, excavations at Aşıklı Höyük in Central Anatolia (present-day Turkey) have demonstrated a shift from hunted animals to captive/domesticated animals. At level 4 (see side bar), hunting provided about 50% of meat for the diet including large game, such as deer, onager, small game such as hare, turtles, hedgehogs, bustards, partridges and ducks, and river fish, predominantly carp (Stiner et al., 2014). Sheep and/or goat bones made less than half of the bones at level 4 (11,000–10,200 BP) rising to more than 85% at level 2 (10,000 and 9,500 cal BP) (Stiner et al., 2014). Hunted animals progressively became less important making up

less than 10% of the bones at the end of level 2 (Stiner et al., 2014).

6.4 IMPACTS OF THE NEOLITHIC REVOLUTION ON HUMAN HEALTH, NUTRITION, POPULATION, AND GROWTH

6.4.1 Impact on Health and Nutrition

Estimates of the decrease in meat consumption following the Neolithic Revolution are as much as an 80% decline (Hermanussen, 2003). It is not surprising, therefore, that there is evidence for nutritional deficiencies, such as anemia in Neolithic populations (Papathanasiou, 2005). Moreover, the shift to agriculture was associated with a reduction in stature indicating a poorer overall nutrition (reviewed by Richards, 2002). There were decreases in adult height, respectively, 4.0% in men and 4.3% in women between Paleolithic and Neolithic (Table 6.3). Skeletal remains provide evidence for musculoskeletal stress "indicative of increased physical activity and heavy workloads" (Papathanasiou, 2005). Along with the decrease in stature, there is shift in diet with much lower intakes of protein and fat in medieval diets than Paleolithic diets (Table 6.4).

TABLE 6.3 Changes in Height in Europe From the Paleolithic to Modern Periods

| | Height in cm (feet and inches) | | Population density (no. of people per km²) |
	Men	Women	
Paleolithic	174 (5′8 1/2″)	162 (5′3 3/4″)	0.1
Neolithic	167 (5′6 1/2″)	155 (5′1″)	4
Classical/Greek/Roman	170 (5′7″)	157 (5′1 3/4″)	31
19th century	166 (5′6 3/8″)	157 (5′1 3/4″)	70[a]
Late 20th century	182.5 (5′11 7/8″)	169 (5′6 1/2″)	

[a] Based on Rothenbacher (2000).

Calculated from Hermanussen, M., 2003. Stature of early Europeans. Hormones (Athens) 2, 175–178.

TABLE 6.4   Comparison of the Contribution of Dietary Energy Coming From Carbohydrate, Fat, and Protein in Existing Hominids

	Carbohydrate	Fat	Protein
Human hunter–gatherer[a]	~30	~43	~27
Neolithic through Medieval[b]	66	18	16
Western[c]	53	33	14

[a] Based on Leonard et al. (2010).
[b] Calculated from Singman and McLean (1995) using the approach of Resor (2014) for prosperous English peasant in 1400 and for Native American in South Dakota in 1400 (Boutton et al., 1991).
[c] Institute of Medicine (2002).

People in the Paleolithic had good teeth with little caries (tooth decay). Following the Neolithic Revolution, there was an increase in caries and teeth were worn down and pitted by the shift to consumption of carbohydrate (reviewed by Richards, 2002). Other studies, however, indicate low caries in the teeth in Neolithic populations (Papathanasiou, 2005). The basis for this difference is not readily apparent.

The Neolithic Revolution influenced the mean age of death of 31–32 years for Natufian hunter–gatherers (Eshed et al., 2004; Karasik et al., 2000)a with small increases in life expectancy at both birth (Natufian hunter–gatherers 24.6 years vs. Neolithic farmers 25.5 years) and mean age of adults mean age at death (Natufian hunter–gatherers 31.2 years old vs. Neolithic farmers 32.1 years old) (Eshed et al., 2004). However, a marked gender difference exists. Neolithic men lived longer than hunter–gatherers. In contrast, Natufian hunter–gatherer women living longer than Neolithic women farmers presumably due to deaths associated with labor and the increased fertility (Eshed et al., 2004). In the North American southwest prereproductive life expectancy at birth decreased but life expectancy at 15 years old increased as agriculture spread and corn became progressively more important to the diet and as people lived in one place (called sedentism) (Kohler and Reese, 2014).

6.4.2 Impact on Population and Fertility

The Neolithic or agricultural revolution resulted in a demographic transition and major increases in population (Table 6.1) and population density (Table 6.3). The population of hunter–gatherers rose at a very low rate constrained by the carrying capacity of the land (see Chapter 4, Fig. 4.1). The increase in Paleolithic global populations parallels the increase in range as humans migrated from Africa to Asia, Europe, the Americas, and Australia. The increase in the growth rate of human populations increased by as much 60-fold with the Neolithic Revolution (Table 6.1). "Population, when unchecked, goes on doubling itself every twenty-five years or increases in a geometrical ratio" (Malthus, 1798). This is seen in the USA with the population rising from 2.5 million in 1776 to 5.3 million in 1800 to 11.1 million to 1825 and 23.2 million in 1850 (US Census, in press); the population growth being unchecked as more land came into cultivation.

The birth rate of the Natufian hunter–gatherers was low and seemingly declining (Fig. 6.2) (Bocquet-Appel, 2008, 2011). With the transition to cereal and legume production in the Neolithic Revolution, there was a marked increase in birth rates leading to a concomitant increase in the proportion of young people (<18 years old) and to growing populations (Fig. 6.2) (Bocquet-Appel, 2008, 2011). This conclusion is

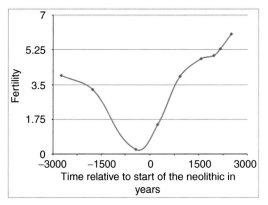

FIGURE 6.2 **Changes in human fertility before and after the Neolithic Revolution.** There was a significant increase (regression: adjusted R^2 0.95, $P < .0001$) in fertility between immediately prior to the Neolithic Revolution and about 3000 years after its beginning (calculated by the author). *Source: Data from Bocquet-Appel, J.-P., 2008. The Neolithic demographic transition, population pressure and cultural change. Comp. Civiliz. Rev. 58, 36–49; Bocquet-Appel, J.-P., 2011. The agricultural demographic transition during and after the agriculture inventions. Curr. Anthropol. 52, S497–S510.*

based on estimates of the proportion of immature skeletal remains, for instance, in the Levant (Bocquet-Appel, 2008) together with analysis of ethnographic data for transitions from hunter–gatherer to farming societies in historical times (Bocquet-Appel, 2011). Similarly, in the North American southwest, as agriculture expanded between 3100 and 1000 BP there were demographic shifts with increases in birth rates (Kohler and Reese, 2014). Contributing to the increase in birth rate was the earlier menarche in Neolithic than hunter–gatherer communities (Hermanussen, 2003).

6.5 TOWARD AN EXPLANATION OF THE NEOLITHIC REVOLUTION

Why farm? Why give up the 20-hour work week and the fun of hunting in order to toil in the sun? Why work harder, for food less nutritious and a supply more capricious? Why invite famine, plague, pestilence and crowded living conditions? Harlan (1992).

It is argued that the shift to farming involves the following:

1. The shift being incremental after settled camps/villages were established with initially food storage, then wild plants being cultivated, then domesticated crops but with ongoing hunting and foraging (progressively contributing a lower proportion of the diet).
2. An increase in the level of control and incentives for work and innovation to the individual or household (Table 6.4). Jones (2001) concluded that "In an economy with a well-functioning system of property rights, inventors are allowed to earn the returns." It is argued that the Neolithic Revolution was such a system with inventors and workers earning returns.
3. A "ratchet effect" with it being only possible to meet the food needs of increased populations by agriculture.
4. With increasing farmer populations, there would be competition with adjacent hunter–gatherers for land and animals to be hunted. This would be particularly intense at times of drought.
5. Development of "belief systems" and cultures that valued hard work and innovation.

Table 6.5 summarizes the differences in incentives for hunter–gatherers and for farmers following the Neolithic Revolution. The advantages of farming include much greater control, some increase in the reliability of food supplies, and greatly decreased constraints on fertility.

For hunter–gatherers, a new activity, such as fishing, could enable a new plateau in population or increased carrying capacity. Alternatively, a new technology increasing the efficiency and/or effectiveness of hunting might result in a temporary increase in food availability. However, this could, with overhunting, be followed by a decline to pretechnology levels or potentially catastrophic extinctions of local prey animals. Initially, any increase in population would be matched by food

TABLE 6.5   A Synthesis of the Impacts of the Neolithic Revolution

	Hunter–gatherer societies	Early agricultural societies
Time working	Increase in food availability but storage of food problematic. With increased population, reduced effectiveness of labor. Therefore, low incentive to work longer than required for needs.	Increase in food availability Therefore, incentive to work longer to provide long-term needs of household.
Impact of new technology	Increase in food availability but could be temporary with local extinctions. Therefore, low incentive to innovate particularly as individual inventor gains little.	Increase in food availability. Therefore, high incentive to innovate particularly as individual inventor or household gains much.

Control of reliable food supply	Group but subject to vagaries of nature
	Ranking
	  Control: low–moderate
	  Reliability: moderate
		Individual or household but somewhat subject to vagaries of weather and pests.
		Ranking
		  Control: moderate–high
		  Reliability: moderate–high

Population	Increased population leads to reduced food availability per person and/or more time required for hunting but with a diminishing return. Therefore, incentive to keep population constant.	Increased food production with per person food availability unchanged. Therefore, little or no incentive to keep population constant.
Nomadic life	Yes Incentive: pursue food resources Disincentives: 1. Storage of food is problematic 2. Energy costs in migrations 3. Reliability/low knowledge of new locale No Incentive: 1. Storage of food possible 2. Knowledge of locale Disincentive: Exhaustion of resources in locale	No Incentives: 1. Ready food storage 2. Allows cultivation 3. Facilitates use of seeds adapted to locale Yes: to pastoralism later

but this is likely to be temporary if the group was operating close to carrying capacity. The solution to this is nomadism over a larger range and/or out-migration of some members of the group.

6.6 DOMESTICATION OF CROPS

6.6.1 Overview

Domestication of crops is included as they were domesticated first in the Neolithic Revolution. It is argued that domestication of plants was an essential prerequisite to the domestication of animals for the following reasons:

1. The development of knowledge of how to domesticate.
2. The availability of crop residues for animal feeds.
3. The needs for animal products as a source for protein, minerals, and vitamins (discussed in Chapter 3).
4. Settlements with increasing populations progressively further from habitats of wild animals suitable for hunting.

The process of domestication can be summarized as follows:

> Unintentional process by which human intervention, selection and replanting (i.e., environmental manipulation) eventually gave rise to strains of plants and animals that depended upon human assistance for their survival and upon which humans in turn depended sustenance (Weisdorf, 2005).

Domesticated grasses are shown in Fig. 6.3.

6.6.2 Near East Crops (Wheat, Barley, and Legumes)

6.6.2.1 Wheat

There are two major forms of wheat today:

1. Common wheat (*Triticum aestivum*)
2. Durum or hard wheat (*Triticum durum*)

Common wheat (Fig. 6.3C) accounts for 95% of global wheat production and is used for producing bread, cakes, crackers, cookies, and noodles. Durum wheat is used to produce modern-day macaroni, pasta, and semolina products. Durum wheat (tetraploid) is also derived from the hybridization of domesticated emmer wheat (*Triticum dicoccum*) with goat grass (*Aegilops speltoides*) (based on Faris, 2014; Peng et al., 2011).

The first domesticated forms of wheat date from about 10,000 BP and were the following:

- Einkorn wheat (*Triticum monococcum* ssp. *aegilopoides*)
- Emmer wheat (*Tr. dicoccum*)

Both were cultivated before domestication (reviewed by Faris, 2014).

FIGURE 6.3 **Domesticated grasses.** (A) Rice; (B) maize (corn); and (C) wheat. *Source: Part A: Courtesy USDA Agricultural Research Service; Photo by Keith Weller; Part B: Shutterstock; Part C: Courtesy USDA Agricultural Research Service; Photo by Scott Bauer.*

Einkorn wheat (*Tr. monococcum* ssp. *aegilopoides*) was domesticated from wild einkorn wheat (*Tr. monococcum* ssp. *monococcum*) about 10,000 BP (reviewed by Faris, 2014). Einkorn wheat was domesticated in the region of the Karacadag mountains in present-day southeastern Turkey (reviewed by Faris, 2014). A critical mutation that allowed and facilitated the domestication of Einkorn wheat is shown as follows (reviewed by Faris, 2014):

Brittle rachis type (Br1^{3A})	→	Nonbrittle rachis type (br1^{3A})
(shattering and scattering grain)		(nonshattering and thus enabling harvesting and threshing)
	Mutation	

Initially, both forms were cultivated but the nonbrittle form became dominant (reviewed by Faris, 2014).

6.6.2.2 Wild Emmer Wheat

Wild emmer (*Triticum turgidum* ssp. *dicoccoides*) was gathered as early as 19,000 BP (reviewed by Faris, 2014). Emmer wheat (*Tr. dicoccum* or *Tr. turgidum* ssp. *dicoccum*) was also domesticated in the region of the Karacadag mountains as early as 9000 BP (reviewed by Faris, 2014). Domestication again involved the use of a mutation to a nonbrittle rachis form (reviewed by Faris, 2014).

Founder Crops From the Fertile Crescent

Emmer wheat and einkorn wheat are viewed as two of the eight founder crops from the Fertile Crescent (Fig. 6.1) domesticated 10,000–11,000 years ago along with the following (Zohary, 1999):

- Barley (*Hordeum vulgare*)
- Lentil (*Lens culinaris*)
- Pea (*Pisum sativum*)

- Chickpea (*Cicer arietinum*)
- Bitter vetch (*Vicia ervilia*)
- Flax (*Linum usitatissimum*)

Wild ancestors of the eight founder crops existed in a core area (Lev-Yadun et al., 2000) where all the ancestral wild species of plants are found (Abbo et al., 2010). Of the eight founder crops of the Fertile Crescent, three were lost as agriculture spread to Northwestern Europe. These were chickpea, bitter vetch, and lentil (Bakels, 2014). In contrast, the opium poppy was added (Bakels, 2014).

6.6.2.3 Other Crops

Barley (*H. vulgare* L.) was domesticated from a wild grass, *Hordeum spontaneum*, in the Fertile Crescent about 10,000 BP (Badr et al., 2000). Chickpea or garbanzo beans are used today in hummus, falafel, or ground to flour in India and Italy. The legume, chickpea (*C. arietinum* L.), was domesticated from its wild progenitor (*Cicer reticulatum* Ladizinsky) in eastern Anatolia (present-day Turkey) (van Oss et al., 2015). It appears that this was a monophyletic domestication (van Oss et al., 2015).

6.6.3 East Asia Crops (Millet and Rice)

Asian rice (*Oryza sativa*) is a major component of human diets (Fig. 6.3A). It was first domesticated from a grass, wild rice (*Oryza rufipogon*), 8000–9000 BP in East Asia (present-day China). There is debate as to whether the initial single domestication event occurred along the lower Yangtze River (Gross and Zhao, 2014; Molina et al., 2011) or along the Pearl River. The former is based on archaeological and genetic evidence while the latter by the similarity of domesticated rice and local wild rice varieties. The key domestication trait of nonshattering was fixed about 7000 BP (Molina et al., 2011).

The proportion of nonshattering rice increased between 6900 and 6600 BP (Fuller et al., 2009). During the same timeline, the proportion of plant remains associated with humans rose from 8% to 24% along the lower Yangtze River (Fuller et al., 2009) indicating a shift in consumption.

There are two major varieties (frequently referred to as subspecies) of domesticated rice; the sticky and short-grained dry high-land japonica or sinica rice (*Oryza sativa japonica*) and long-grained low-land indica (*Oryza sativa indica*) (Khush, 1997). There are alternate views as to whether these represent the results of one or two domestication events. The prevailing view is that there were at least two domestications for rice; japonica being domesticated in present-day southern China approximately 9000 BP and indica on the Indian subcontinent about 5000 BP (He et al., 2011; Londo et al., 2006). This was followed by gene flow predominantly from *japonica* to *indica* (Yang et al., 2012). African domesticated rice, *Oryza glaberrima*, is thought to have been domesticated in the Niger River delta in West Africa (Khush, 1997).

6.6.3.1 Millet

Millet was the first domesticated grain in East Asia; common millet (*Panicum miliaceum*) being domesticated beginning about 10,300 BP (Lu et al., 2009).

6.6.4 American Crops (Corn/Maize, Potato)

6.6.4.1 Maize (Corn)

Maize (*Zea mays* ssp. *mays*) (Fig. 6.3B) was domesticated from its wild progenitor, teosinte (*Zea mays* ssp. *parviglumis*), in the present-day Mexican highlands about 9000 BP (Matsuoka et al., 2002).

6.6.4.2 Potato

It is frequently held that the potato, *Solanum tuberosum*, was domesticated from its wild progenitor, *Solanum brevicaule* complex (Spooner et al., 2005). This single domestication event is supposed to have occurred in the Andes in southern present-day Peru (Spooner et al., 2005). However, there is evidence for hybridization with germplasm from lowland south–central present-day Chile (Ghislain et al., 2009).

6.7 ANIMAL DOMESTICATION

6.7.1 Overview

Table 6.6 summarizes the domestication of mammalian species. The domestication trajectory can be described as follows:

Wild → Captured → Breeding in captivity → Isolated breeding → Selection, etc. → Domesticated

Selection includes intentional and unconscious selection by humans together with genetic drift, genetic bottlenecks, and selective pressure for prospering under captive/domesticated conditions (based on Larson et al., 2014; Marshall et al., 2014). Among the traits selected across animal species were docility, reproductive shifts, and decreased size (Larson et al., 2014). It is likely that, in addition, there was unintentional selection in capture or bias with slower and/ or younger animals more likely to be captured. It has been argued that there was little selection in the management of herds early in domestication (Marshall et al., 2014). This is referred to as *laissez faire* herd management. Intentional selection was very low for females early in domestication (Marshall et al., 2014).

Evidence for domestication includes the following (Evin et al., 2013; reviewed by Zeder, 2008):

- Bones indicating disproportionate killing of young males for meat and hence promoting survival of females for reproduction.
- Reduction of body size as evidenced by bone and tooth size.

TABLE 6.6   Summary of Domestications of Mammals

Domesticated animals	Ancestor	Number of domestications	Date	Location
Alpaca (*Vicugna pacos*) [Valid synonym *Lama pacos*]	Vicuña (*Vicugna vicugna*)	One	6,000–7,000 BP	South America: Andes
Bactrian camel (*Ca. bactrianus*)	Wild camel (*Camelus bactrianus ferus*)	One	5,000 BP	Central Asia to present-day Western China
Cat (*Felis catus*)	Near Eastern wildcat (*Felis silvestris lybica*)	One (1)	9,500 BP	Fertile Crescent
Cattle (*B. taurus taurus*)	Wild ox/wild cattle/auroch of following subspecies	Two (2)	10,500 BP	Southwest Asia
Zebu cattle (*B. taurus indicus*)	(*B. taurus primigenius, B. taurus namadicus*)		8,000–7,500 BP	Indus Valley (Indian subcontinent)
Dog (*Canis lupus familiaris*)	Wolf (*Canis lupus*)	Multiple	15,000–30,000 BP	Eurasia
Donkey (*E. asinus*)	Nubian wild ass (*E. africanus africanus*) and Somalian wild ass (*E. africanus somaliensis*)	Two	5,000–8,000 BP	Northeastern Africa
Dromedary (*Ca. dromedarius*)	Thomas' camel (*Ca. thomasi*)	One	3,200 BP	Arabian peninsula
Goat (*Capra hircus hircus*)	Wild goat or bezoar (*Capra hircus aegagrus*)	One (1)	10,000 BP	Fertile Crescent
Horse (*E. caballus*)	Wild horse (*E. ferus*)		5,500 BP	Steppes of western Central Eurasia
Llama (*Lama glama*)	Guanaco (*Lama glama guanicoe*)		6,000–7,000 BP	Andes
Pig (*S. scrofa*)	Eurasian Wild boar or wild pigs (*S. scrofa*)	Two (2) + potentially more	10,500 BP	1. Near East 2. Central China/ Yellow River
Sheep (*Ovis aries aries*)	Asiatic Mouflon (*Ovis aries orientalis*)		11,000 BP	Southwest Asia
Water buffalo (*Bu. bubalis*)	Wild water buffalo (*Bu. arnee*)	Two (2)	4,000–5,000 BP	1. Indus Valley (Indian subcontinent) 2. China/Indochina border region

Diamond (2002) posed the question as to why certain animals were or were not domesticated. Animals were not domesticated if they have a poor disposition; being "nasty" or "vicious."

Features of animals that promote domestication include the following:

- a suitable diet can be readily provided
- rapid growth rate

- high fecundity and low generation time
- easily bred in captivity
- calmness and low panic response in enclosures
- social hierarchy

Diamond (2002) differentiates between tamed and domesticated animals in that the latter are genetically and phenotypically distinct from their wild ancestors. Examples of tamed but not domesticated animals are African and Asian elephants.

6.7.2 Cattle

Introgression

Gene flow or the transfer of genes from one species to another. This is extended to include gene flow from wild ancestral or related species to domesticated animals or plants.

Taxonomy

Traditionally, European and humped domestic cattle are assigned to two species, respectively, *Bos taurus* and *Bos indicus*. These can interbreed and should not be considered separate species. According to the Integrated Taxonomic Information System (ITIS), they should be classified as *B. taurus taurus* and *B. taurus indicus* (ITIS, 2017). The wild ancestors of domestic cattle were aurochs, which were classified as *Bos primigenius.* The *taurus* species name takes precedence as it was used first going back to Carl Linnaeus from 1758. Alternatively, the International Commission on Zoological Nomenclature does *not* consider the term, *B. primigenius*, to be invalid (emphasis added). This chapter will employ the ITIS system of *B. taurus taurus* and *B. taurus indicus* as employed by Decker et al. (2014).

Ancestral Wild Cattle

According to IUCN (2015) but changing the species name to reflect the ITIS system, there were three subspecies:

- *B. taurus primigenius* from Europe, Middle East, and West Asia. *B. taurus primigenius* was the progenitor of taurine cattle and became extinct when the last was killed in 1627 (Loftus et al., 1994).
- *Bos taurus namadicus* until extinction found on the Indian subcontinent. *B. taurus namadicus* was the ancestor of indicine-type cattle (Chen et al., 2010).
- *Bos taurus mauretanicus* from North Africa. *B. taurus primigenius* contributed genes to African cattle (Decker et al., 2014) (also see next subsection).

Domesticated Cattle

Cattle were domesticated from wild cattle. It is generally thought that cattle was domesticated independently twice in two different geographical locations, respectively, in Anatolia (Fig. 6.1) and the Indus Valley (on the Indian subcontinent) (Loftus et al., 1994). This led to the two major groupings of domestic cattle:

- *B. taurus taurus* or taurine or European-type cattle (humpless) or traditionally found in Europe, West Asia, East Asia, and Africa (except East Africa). These were domesticated in Anatolia (adjacent to the Fertile Crescent, Fig. 6.1) (Loftus et al., 1994). Bones from domesticated cattle have been dated to the Early Prepottery Neolithic (10,800–10,300 BP) in the Fertile Crescent (specifically the middle Euphrates valley and the high Tigris valley) (reviewed by Bollongino et al., 2012).
- *B. taurus indicus* or indicine-type cattle or cattle with humps. These were domesticated in the Indus Valley of the Indian subcontinent between 8000 and 7500 BP with about 80 founder female aurochs (reviewed by Bollongino et al., 2012). The indicine-type

cattle includes the native breeds of the Indian subcontinent together with Zebu cattle of East Africa (Chen et al., 2010). Brahman cattle are derived from such indicine cattle.

Fig. 6.4 illustrates the domestication of cattle. There were introgressions (interbreeding) from other populations of wild cattle with evidence following gene flows:

- North African wild aurochs to taurine cattle in North Africa in African breeds of cattle (Decker et al., 2014)
- European wild aurochs to taurine cattle in Western cattle (Italy, Bonfiglio et al., 2010; Mona et al., 2010; British Isles, Park et al., 2015)
- Indicine cattle from the Indian subcontinent introgressed into East African cattle twice, first between 3000 and 4000 BP and the second after CE 700 (or AD) (Decker et al., 2014).

Fig. 6.4 illustrates the domestication of cattle. Domesticated taurine cattle were initially limited to between the Levant, Central Anatolia, and Western Iran. They reached the following:

- Southeastern Europe by 8800 BP
- Southern Italy by 8500 BP
- Central Europe by 8000 BP
- Northern Europe with cheese being made as early as 7000–8000 BP (Salque et al., 2013)

Dairying was occurring as early as in 6000–7000 BP (Dunne et al., 2012) in the then savannah-like Sahara (Kuper and Kröpelin, 2006). Fig. 6.5 illustrates the historical development of cattle. Water buffalo (*Bubalus bubalis*), yak (*Bos grunniens*), gaur (*Bos gaurus*), and banteng (*Bos javanicus*) are the other bovids that were domesticated.

6.7.3 Pigs

Pig classification (according to ITIS)
Domesticated pig: *Sus scrofa scrofa*

There have been multiple loci for the domestication of pigs (Larson et al., 2005) including the following locations:

- Anatolia in West Asia about 10,500 BP (Zeder, 2008),
- East Asia (present-day China) (Larson et al., 2010),
- Western Europe 5000–6000 BP (either introgression by use of local female wild pigs for breeding for repeated generations or domestication of indigenous wild boar) (Larson et al., 2007),
- Additional possible domestication events in India, in Southeast Asia, and in Taiwan (Larson et al., 2010).

Fig. 6.6 illustrates the domestication of pigs.

6.7.4 Sheep and Goats

Goat classification (according to ITIS)
Domesticated goat: *Capra hircus hircus*
Progenitor bezoar: *Capra hircus aegagrus*
Sheep classification (according to ITIS)
Domesticated sheep: *Ovis aries aries*
Progenitor: *Ovis aries orientalis*

It has been concluded that there were six wild bezoar lineages in domestic goats. This suggests that there was a long process of capturing wild females to supplement the captive herds being domesticated (Naderi et al., 2007).

6.7.5 Horses

Horse classification (according to ITIS)
Domesticated horse: *Equus caballus*
Progenitor wild horse: *Equus ferus*

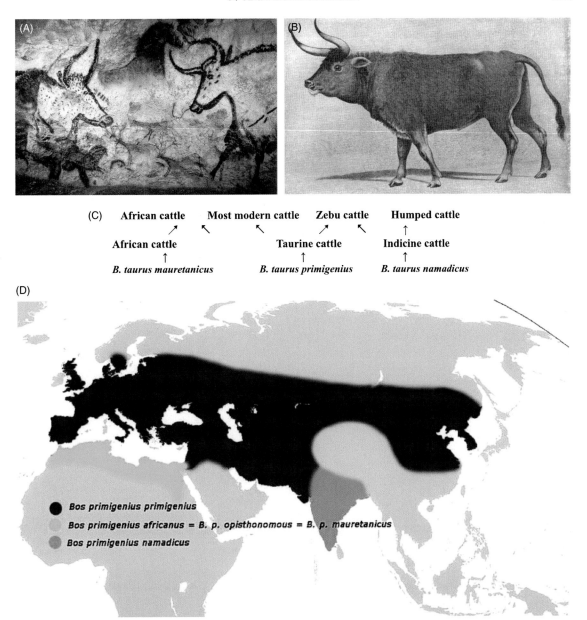

FIGURE 6.4 **Domestication of cattle from wild cattle or aurochs.** (A) Cave painting (Lascaux, France) of an auroch bull. (B) Painting of an auroch bull from about AD/CE 1600. (C) Domestication and development of cattle. At least three domestication events are likely with cattle being domesticated from populations of wild cattle: *B. taurus primigenius* (also known as *B. primigenius primigenius*) in the Fertile Crescent, *B. taurus mauretanicus* (also known as *B. primigenius mauretanicus* or *B. primigenius africanus*) in North Africa, and *B. taurus namadicus* (also known as *B. primigenius namadicus*) on the Indian subcontinent. Present North African cattle are a mixture of African and taurine cattle. Zebu cattle are the result of introgression of indicine cattle genes into taurine cattle. Humped cattle may include introgression of taurine genes. (D) Distribution of wild cattle prior to domestication. *Source: Part A, B, and D: From Wikimedia Commons.*

FIGURE 6.5 **Cattle from history to present.** (A) Egyptian hieroglyph from 1200 BCE (1200 BC) showing farmer plowing with cattle. (B) Cattle and other livestock are readily moveable wealth in overland migrations. (C) *B. taurus indicus* or indicine cattle (also known as *B. indicus*). (D) Modern dairy cattle (*B. taurus taurus*) with greatly increased milk production due predominantly to genetic improvement together with better nutrition and health. *Source: Part A, C: From Wikimedia Commons; Part B: Courtesy New York Public Library; Part D: Courtesy USDA Agricultural Research Service; Photo by Keith Weller.*

FIGURE 6.6 **Domestication of pigs from Eurasian wild pigs.** (A) Wild boar; (B) domesticated sow with piglet. *Source: Part A: From Wikimedia Commons; Part B: Courtesy USDA Agricultural Research Service; Photo by Scott Bauer.*

Horses were domesticated from the wild progenitor, *E. ferus*, on the steppes of western Central Eurasia (modern-day Ukraine, parts of Russia, and northwest Kazakhstan about 5500 BP (Outram et al., 2009). By 5500 BP, horses were bridled (potentially allowing riding). Moreover, there is evidence for the consumption of horse milk and meat 5500 years ago (Outram et al., 2009). Domesticated horse spread across Eurasia along with the transmission of knowledge of horses including about reproduction and by movement of stallions. There appears to have been multiple introgressions with female wild horses (Jansen et al., 2002; Warmuth et al., 2012). Horses were used to draw chariots as early as 4000 BP (reviewed by Jansen et al., 2002).

6.7.6 Other Livestock

6.7.6.1 Domestication of Donkeys or Asses

Domesticated donkeys (*Equus asinus*) originated from Nubian wild ass (*Equus africanus africanus*) and Somalian wild ass (*Equus africanus somaliensis*) in Northeast Africa (Beja-Pereira et al., 2004; Kimura et al., 2011; Rossel et al., 2008). There appears to have been two domestication events in Northeast Africa; the first in present-day southern Egypt, Sudan, and Eritrea and from Nubian wild asses together with a second in the Horn of Africa (present-day Somalia) from Somalian wild asses. Estimates of the time of the domestications range from about 5000 BP, based on archeological evidence of donkey skeletons in Egyptian tombs, and possibly as early as 8500 BP (Beja-Pereira et al., 2004; Kimura et al., 2011; Rossel et al., 2008).

Donkeys were critically important for transportation including in ancient Egypt and along the Silk Road from China to the east. Chinese donkeys again are derived from domesticated donkeys of Nubian wild ass (*E. africanus africanus*) and Somalian wild ass (*E. africanus somaliensis*) descents (Han et al., 2014).

Donkeys are a source of meat, pulling carts, riding, pack animals for carrying loads (beasts of burden), and milk.

6.7.6.2 Domestication of Water Buffalo

There is strong evidence that there were two independent domestications of water buffalo in 4000–5000 BP resulting in the swamp and river water buffalo (*Bu. bubalis*); both being derived from wild water buffalo (*Bubalus arnee*) (Lei et al., 2007; Yindee et al., 2010). The river water buffalo was domesticated in the Indus Valley of the Indian subcontinent (Nagarajan et al., 2015). The swamp buffalo was domesticated in what is today's China/Indochina border region with later introgressions from wild water buffalo along the Mekong River (Zhang et al., 2015). Swamp buffalo were transported to the north along the Yangtze River valley in prehistory (Zhang et al., 2015).

6.7.6.3 Domestication of Camels

The Camelids originated in North America with the ancestors of the South America camelids migrating south and the ancestors of camels migrating to Eurasia (reviewed by Barreta et al., 2013). The Bactrian camel (*Camelus bactrianus*) was domesticated from the wild camel (*Camelus bactrianus ferus* or *Camelus ferus*) (Ji et al., 2009). It is thought that the Bactrian camel originated in Central Asia in Bactria (present-day Afghanistan and Tajikistan) or further east (present-day western China) about 5000 BP (Potts, 2005).

The dromedary (*Camelus dromedarius*) is thought to have been domesticated from the now extinct Thomas' camel (*Camelus thomasi*) (Gautier, 1966) in the Arabian peninsula (Beech et al., 2009). The dating of the domestication of the dromedary range from as early as 5500 BP (Beech et al., 2009) to as late as 3200 BP (Ben-Yosef et al., 2012; Uerpmann and Uerpmann, 2002); the latter supported by the dromedary being first found in the Levant about 2970 BP (Sapir-Hen and Ben Yousef, 2013).

6.7.6.4 Domestication of Alpacas and Llama

Llamas (*Lama glama*) were domesticated from guanaco (*Lama glama guanicoe*) about 6000–7000 BP in the Andes (Kadwell et al., 2001). Alpacas (*Vicugna pacos*) were domesticated from vicuña or vicugna (*Vicugna vicugna*) about 6000–7000 BP in the Andes (Kadwell et al., 2001; Wheeler, 1995).

6.7.7 Dogs

Dogs were domesticated from gray wolves in the Paleolithic era by hunter–gatherers considerably before the Neolithic Revolution. Domestication of dogs is covered in Section 5.6.6.7.

6.7.8 Cats

The domestic cat (*Felis silvestris catus*) was domesticated from members of a subspecies of the wildcat, specifically the Near Eastern wildcat (*Felis silvestris lybica*) (Driscoll et al., 2007). Cats are thought to have been domesticated in the Fertile Crescent around the beginning of the Neolithic Revolution from as few as five founders (Driscoll et al., 2007). The dating of the domestication of cats is not well established. There is archaeological evidence for the presence of domestic cats in present-day Cyprus in 9500 BP (Vigne et al., 2004). Domestic cats were traded with the remains of cats found in Shaanxi in Central China dating back to about 5400 BP (Hu et al., 2014). There is still doubt whether cats should be considered domesticated or semidomesticated or commensal (Montague et al., 2014).

The 40–50 breeds of cats were developed by cat fanciers within the last 150 years. These were produced from seven geographically distinct breeds/populations dating back over 150 years (Driscoll et al., 2007; Montague et al., 2014). Genetically, random-bred populations of cats can be assigned to the following geographical groupings: Europe, Mediterranean, Egypt, Iraq/ Iran, Arabian Sea, India, Southeast Asia, and East Asia (Montague et al., 2014).

6.7.9 Poultry

Table 6.7 provides a summary of the domestications of poultry. Domestication of chickens will be considered first (Fig. 6.7). Charles Darwin (1868) considered that chickens were derived from red junglefowl (*Gallus gallus*) on the Indian subcontinent. There is now strong evidence for multiple domestications of red junglefowl resulting in chickens as early as 10,000 BP in the following: Southeast Asia and two areas in East Asia, respectively southern and northern China (Liu et al., 2006; Miao et al., 2013; Xiang et al., 2014). There was also some contribution from the gray junglefowl (*Gallus sonneratii*) on the Indian subcontinent (Liu et al., 2006; Tixier-Boichard et al., 2011). The *raison d'etre* for domestication included cock-fighting, meat, and eggs. The critical importance of chickens in the past is supported by the colonizing Polynesians bringing chickens in their voyages across the Pacific even as far as South America (Storey et al., 2007). Given the importance of docility and calmness in domesticated animals, it is interesting to note that the stress response of red junglefowl is markedly different from that of domestic chickens (Fallahsharoudi et al., 2015).

The progenitor of Chinese breeds of domesticated ducks was the Mallard duck (*Anas platyrhynchos*), but in a few breeds, there is evidence of gene flow from the spot-billed duck (*Anas poecilorhyncha*) (Li et al., 2010). The Muscovy duck (*Cairina moschata*) was domesticated by 1965 BP (CE/AD 50) in the northern Andes or present-day southern Peru (Stahl, 2003). There have been at least two domestications of geese in, respectively, present-day China and Egypt with, respectively, the swan goose (*Anser cygnoides*) and Greylag goose (*Anser anser*) domesticated (Shi et al., 2006). Domestication is estimated at 3000 BP (FAO, 2005). Turkeys (*Meleagris gallopavo*) were domesticated

TABLE 6.7   Summary of Domestication of Birds

Domesticated animals	Ancestor	Number of domestications	Date	Location
Domestic chicken	Jungle fowl	Multiple	7,000–10,000 BP	East Asia (present-day China), Southeast Asia and India
Domestic duck (*Anas platyrhynchos*)	Mallard duck (*A. platyrhynchos*)	One?	2,500 BP	East Asia (present-day China)
Domestic goose (*Ans. anser*)		Two	3,000 BP	
Chinese (*Ans. cygnoides*)	Swan goose (*Ans. cygnoides*)			East Asia (present-day China)
European	Greylag goose (*Ans. anser*)			Present-day Egypt
Domesticated Muscovy duck (*Cairina moschata*)	Muscovy duck (*Cai. moschata*)	One	1,965 BP (CE/AD 50)	South America, northern Andes present-day southern Peru
Domestic turkey (*M. gallopavo*)	Wild turkeys (*M. gallopavo*)	One	1,900–2,300 BP	Mesoamerica, present-day Mexico

For detailed discussion and citations, see text.

FIGURE 6.7   **Domestication of chickens from red junglefowl.** (A) Red junglefowl; (B) laying chicken; and (C) chicks. *Source: Part A: From Wikimedia Commons; Part B: Courtesy USDA Agricultural Research Service; Photo by Stephen Ausmus; Part C: Courtesy USDA Agricultural Research Service; Photo by Keith Weller.*

from wild turkeys about 2000 years ago in Mesoamerica, specifically present-day Mexico (Thornton et al., 2012).

6.7.10 Aquaculture Species

Fish have been raised in captivity for about 4000 years. For instance, about 2000 years ago, the Romans cultured carp that had been captured on the Danube River (Balon, 1995). Examples of domesticated fish are given in Table 6.8 and Fig. 6.8A–B. There have been large numbers of species of freshwater and marine finfish and shellfish domesticated. This has allowed the large increases in aquaculture (see Chapters 7 and 8) (see aquaculture facility, Fig. 6.8C). Of domesticated aquatic animals, 50% were domesticated in the last 30 years (Duarte et al., 2007). However, only a few of the fish can be viewed as fully domesticated not depending on wild populations for restocking (Teletchea and Fontaine, 2014.).

6.7.11 Conclusions

There were peaks in animal domestications (Tables 6.4 and 6.5). The first was about 10,000 years ago. It might be questioned why the Neolithic Revolution did not occur earlier or later? It is contended that domestication of animals followed an inevitable pathway once animals were held in captivity. For instance, there is evidence that sheep and/or goats were held captive in round houses in Aşıklı Höyük in Central Anatolia before 10,200 BP (Stiner et al., 2014). This is based on the buildup of dung (Stiner et al., 2014).

TABLE 6.8   Examples of the Domestication of Fish

Domesticated animals	Ancestor	Number of domestications	Date	Location
Atlantic salmon (*Salmo salar*)	Atlantic salmon (*S. salar*)	One	Late 1960s	Norway
Channel catfish (*Ictalurus punctatus*)	Channel catfish (*I. punctatus*)	One?	1870s 1967	USA Commercial aquaculture
Coho salmon (*Oncorhynchus kisutch*)	Coho salmon (*O. kisutch*)	One	First steps 1900 Aquaculture 1980s	USA USA
Common carp (*Cyprinus carpio*)	Common carp (*Cy. carpio*)	One or two	2000 BP	China and Western Europe
Nile tilapia (*Oreochromis niloticus*)	Nile tilapia (*Or. niloticus*)	One	4000 BP 1940s and 1970s	Egypt Worldwide
Rainbow trout (*Oncorhynchus mykiss*)	Rainbow trout (*O. mykiss*)	?	1874 Expanded in 1950s	Worldwide
Sea trout (*Salmo trutta*)	Sea trout (*S. trutta*)	Multiple	1739 1841	Germany UK
Silver carp (*Hypophthalmichthys molitrix*)	Silver carp (*H. molitrix*)	One?	800–1000 BP (CE/AD 1000–1200)	China

Data from FAO, 2005–2011. Available from: http://www.fao.org/fishery/culturedspecies/Cyprinus_carpio/en, http://www.fao.org/fishery/culturedspecies/Oreochromis_niloticus/en, and http://www.fao.org/fishery/culturedspecies/Oncorhynchus_mykiss/en.

FIGURE 6.8 **Domestication of fish.** (A) Atlantic salmon. (B) Rainbow trout. (C) Catfish are raised in aquaculture ponds. The colors represent differences in algae. *Source: Part A: Courtesy USDA Agricultural Research Service; Photo by Troutlodge, Inc; Part B: Courtesy USDA Agricultural Research Service; Photo by Stephen Ausmus; Part C: Courtesy USDA Agricultural Research Service; Photo by Scott Bauer.*

6.8 EARLY FORMS OF ANIMAL AGRICULTURE

6.8.1 Animals in the Neolithic

Agriculture allowed greatly increased population density (Table 6.3). Around 10,000 BP, Neolithic farmers carried their technology including seeds on primitive boats to present-day Greece via Cyprus (Fernández et al., 2014). Farming was established in Central Europe by 7500 BP but were the farmers descendants of the hunter–gatherers or immigrants bringing the new technologies? Genomic evidence strongly favors the latter for men but probably not women (Bramanti et al., 2009; Szécsényi-Nagy et al., 2015). Similarly, farming was spread to present-day Italy and the Iberian Peninsula around 7400 BP by seagoing immigrants bringing Neolithic crops, domestic animals, and knowledge (Zilhão, 2001). The Neolithic maritime colonization of Europe is analogous to the movements to the Polynesian islands.

Indo-European Languages and Spread of Agriculture by Neolithic Men

The Indo-European languages consist of most of the languages of Europe together with the languages of Iran, Afghanistan, and the Indian subcontinent. The branches of this language group are the following:

- Celtic including Welsh, Breton, and Gaelic
- Germanic including German, Dutch, Danish, Norwegian, Swedish, and English
- Italic → Latin → Romance languages, such as Catalan, French, Italian, Portuguese, and Spanish
- Baltic: Lithuanian and Latvian
- Slavic
  - South (Bulgarian, Croatian, Serbian, and Slovenian)
  - West (Czech, Polish, and Slovak)
  - East (Belarusian, Russian, and Ukrainian)
- Albanian
- Hellenic Greek
- Armenian
- Indo-Iranian (e.g., Bengali, Hindi, Pashto Persian, and Urdu)

These are all derived from a common ancestral language, Proto-Indo-European. There are two views as to where the Proto-Indo-European language was spoken before spreading out. One is that the language was spread by Kurgan horsemen from an area north of the Black Sea stitching to the Caucasus in the east some 5000–6000 years ago. Alternatively, it is thought that the Proto-Indo-European language was spread from Anatolia with out-migrating Neolithic farmers about 9000 BP (Gray and Atkinson, 2003). The model of spread of language with spread of Neolithic agriculture is supported by the analysis of 87 languages with the initial Indo-European divergence occurring between 7800 and 9800 BP (Bouckaert et al., 2012; Gray and Atkinson, 2003). Moreover, an Anatolian origin is supported (Bouckaert et al., 2012). The Farming/Language Dispersal Model with men bringing agricultural technology, their language, and genes as indicated by the presence of their Y chromosomes is also see in the southwest of present-day USA and Mesoamerica (Kemp et al., 2010).

6.8.2 Pastoralism

Definition of Pastoralism

Pastoralism is a subsistence livelihood with nomadic societies grazing herbivorous livestock on poor range land. The pastoralist management system is can be categorized as the following:

- Nomadic: exclusive pastoralists migrating in an irregular manner to new pastures for grazing.
- Transhumant: exclusive pastoralists with regular back and forth migrations, such as semiannual vertical movements from highlands at the end of the summer to lowlands in the winter and then back.
- Agropastoral: both raising livestock and crops.
- Enclosed pastoralism.

FIGURE 6.9 **Pastoralists in South Sudan selling cattle.**
Source: From Wikimedia Commons.

Pastoralists predominantly live in arid and semiarid rangelands typically in tropical or subtropical regions. Pastoralism continues to be important globally. For instance, it is estimated that about 240 million people in sub-Saharan Africa live in pastoralist societies (Fig. 6.9). The number of pastoralists is declining because of advances in agriculture including irrigation, urbanism, and social stigma for the nomadic peoples (FAO, 2001). However, it is increasingly recognized that pastoralism still contributes substantially to the economies of countries across Africa (African Union, 2010) supplying large numbers of livestock to domestic and international markets and to the nutrition of people (African Union, 2010). Examples of pastoralist societies include the following (FAO, 2001):

- Herders of cattle (Africa) (e.g., Zulu, Masai, and Kikuku)
- Herders of sheep and goats (e.g., Bedouin Arabs of the Middle East)
- Herders of horses, yak, camels, sheep, and goats (Central Asia)
- Herders of llamas, alpacas, and sheep in the Andes of South America

- Herders of reindeer by the Sami or Lapps of northern circumpolar Norway and Finland and the people of Northern Mongolia

Pastoral societies rely on their livestock for food (meat, milk, and blood), leather for clothing and tents, and feces for fuel for cooking with usually no crops grown (FAO, 2001). However, when pastoralists are not nomadic they also produce crops (FAO, 2001). Pastoralists' dogs are frequently critical to managing livestock. Moreover, their chickens feed on insects growing in the livestock feces (FAO, 2001).

There is evidence of the early development of pastoralism from Neolithic village-based livestock production in the Central Zagros mountains in present-day Western Iran (Abdi, 2003). Pastoralism featured in the Old Testament of the Bible with Abraham, Isaac, and Jacob being nomadic pastoralists. The continuing conflict between settled farmers and nomadic pastoralists is illustrated in the fight between the Biblical Cain and Abel (van den Blink et al., 1991). Associated with this are the importance of property rights and fencing for farmers irrespective of whether raising crops or livestock. Alternatively, a nomadic life is crucial for many pastoralists. The conflict extended to the conflict between ranchers and farmers in the American West exemplified by the song "The farmer and the cowman should be friends" written by librettist Oscar Hammerstein and from the musical Oklahoma (1943).

6.9 CONCLUSIONS

Among the cultural consequences of the Neolithic Revolution were the following:

- Development of large villages with meeting spaces.
- Some large villages transitioning and differentiating toward states initially as supralocal chiefdoms.

- The Neolithic Revolution enabled some people to have specialized roles not directly related to food (Diamond, 1997). However, the majority of people remained producing food.
- Development of hierarchies (village chief to chiefs of groups of villages to kings of states or statelets).

These developments were crucial to the development of civilization with written languages, accounting systems, buildings, and technology.

The disadvantages of settlements with a resident population and food storage during the Neolithic Revolution include the following:

- Large increase in rodents (Bocquet-Appel, 2011) and hence the possibility of zoonotic diseases.
- Contamination of drinking water with fecal microorganisms due to the absence of latrines and the buildup in feces (Bocquet-Appel, 2011) and later manure from domesticated animals.
- Increased likelihood of warfare as there were stores of food (and later, other accumulated possessions) to take.
- Depletion of plant nutrients in the soil (Weiss et al., 2004).

The primary consequence of the first two disadvantages would be increased mortality of children after the elevated birthrate (Bocquet-Appel, 2011).

There is evidence for Neolithic warfare in Europe including skull damage and a disproportionate ratio of injuries to adult men rather than women or children. Examples have been reported in Central Europe (Meyer et al., 2015), Balkans 9000 BP to around 7500 BP (Roksandic et al., 2006), Iberian Peninsula (Jiménez-Brobeil et al., 2009), and Scandinavia (5900–3700 BP) (Fibiger et al., 2013).

References

Abbo, S., Lev-Yadun, S., Gopher, A., 2010. Agricultural origins: centres and non-centres: a near eastern reappraisal. Crit. Rev. Plant Sci. 29, 317–328.

Abdi, K., 2003. The early development of pastoralism in the Central Zagros mountains. J. World Prehist. 17, 395–448.

African Union, 2010. Policy framework for pastoralism in Africa: securing, protecting and improving the lives, livelihoods and rights of pastoralist communities. Available from: ea.au.int/en/sites/default/files/Policy%20Framework%20for%20Pastoralism.pdf?q=dp/rea/sites/default/files/Policy%20Framework%20for%20Pastoralism.pdf.

Badr, A., Muller, K., Schafer-Pregl, R., El Rabey, H., Effgen, S., Ibrahim, H.H., Pozzi, C., Rohde, W., Salamini, F., 2000. On the origin and domestication history of barley (Hordeum vulgare). Mol. Biol. Evol. 17, 499–500.

Bakels, C., 2014. The first farmers of the Northwest European plain: some remarks on their crops, crop cultivation and impact on the environment. J. Archaeol. Sci. 51, 94–97.

Balon, E.K., 1995. Origin and domestication of the wild carp, Cyprinus carpio: from Roman gourmets to the swimming flowers. Aquaculture 129, 3–48.

Barreta, J., Gutiérrez-Gil, B., Iñiguez, V., Saavedra, V., Chiri, R., Latorre, E., Arranz, J.J., 2013. Analysis of mitochondrial DNA in Bolivian llama, alpaca and vicuna populations: a contribution to the phylogeny of the South American camelids. Anim. Genet. 44, 158–168.

Bar-Yosef, O., 1998. The Natufian culture in the Levant, threshold to the origins of agriculture. Evol. Anthropol. 6, 159–177.

Beech, M., Mashkour, M., Huels, M., Zazzo, A., 2009. Prehistoric camels in south-eastern Arabia: the discovery of a new site in Abu Dhabi's Western Region, United Arab Emirates. Proc. Semin. Arabian Stud. 39, 17–30.

Beja-Pereira, A., England, P.R., Ferrand, N., Jordan, S., Bakhiet, A.O., Abdalla, M.A., Mashkour, M., Jordana, J., Taberlet, P., Luikart, G., 2004. African origins of the domestic donkey. Science 304, 1781.

Ben-Yosef, E., Shaar, R., Tauxe, L., Ron, H., 2012. A New Chronological Framework for Iron Age Copper Production in Timna (Israel). Bull. Am. Schools Orient. Res. 366, 1–41.

Bocquet-Appel, J.-P., 2008. The Neolithic demographic transition, population pressure and cultural change. Comp. Civiliz. Rev. 58, 36–49.

Bocquet-Appel, J.-P., 2011. The agricultural demographic transition during and after the agriculture inventions. Curr. Anthropol. 52, S497–S510.

Bonfiglio, S., Achilli, A., Olivieri, A., Negrini, R., Colli, L., Liotta, L., Ajmone-Marsan, P., Torroni, A., Ferretti, L., 2010. The enigmatic origin of bovine mtDNA haplogroup R: sporadic interbreeding or an independent event of Bos primigenius domestication in Italy? PLoS One 5, e15760.

Bollongino, R., Burger, J., Powell, A., Mashkour, M., Vigne, J.D., Thomas, M.G., 2012. Modern taurine cattle descended from small number of near-eastern founders. Mol. Biol. Evol. 29, 2101–2104.

Bouckaert, R., Lemey, P., Dunn, M., Greenhill, S.J., Alekseyenko, A.V., Drummond, A.J., Gray, R.D., Suchard, M.A.,

Atkinson, Q.D., 2012. Mapping the origins and expansion of the Indo-European language family. Science 337, 957–960.

Boutton, T.W., Lynott, M.J., Bumsted, M.P., 1991. Stable carbon isotopes and the study of prehistoric human diet. Crit. Rev. Food Sci. Nutr. 30, 373–385.

Bramanti, B., Thomas, M.G., Haak, W., Unterlaender, M., Jores, P., Tambets, K., Antanaitis-Jacobs, I., Haidle, M.N., Jankauskas, R., Kind, C.J., Lute, F., Terberger, T., Hiller, J., Matsumura, S., Forster, P., Burger, J., 2009. Genetic discontinuity between local hunter–gatherers and central Europe's first farmers. Science 326, 137–140.

Chen, S., Lin, B.Z., Baig, M., Mitra, B., Lopes, R.J., Santos, A.M., Magee, D.A., Azevedo, M., Tarroso, P., Sasazaki, S., Ostrowski, S., Mahgoub, O., Chaudhuri, T.K., Zhang, Y.P., Costa, V., Royo, L.J., Goyache, F., Luikart, G., Boivin, N., Fuller, D.Q., Mannen, H., Bradley, D.G., Beja-Pereira, A., 2010. Zebu cattle are an exclusive legacy of the South Asia Neolithic. Mol. Biol. Evol. 27, 1–6.

Darwin, C., 1868. The Variation of Animals and Plants Under Domestication. John Murray, London, UK.

Decker, J.E., McKay, S.D., Rolf, M.M., Kim, J., Molina Alcalá, A., Sonstegard, T.S., Hanotte, O., Götherström, A., Seabury, C.M., Praharani, L., Babar, M.E., Correia de Almeida Regitano, L., Yildiz, M.A., Heaton, M.P., Liu, W.S., Lei, C.Z., Reecy, J.M., Saif-Ur-Rehman, M., Schnabel, R.D., Taylor, J.F., 2014. Worldwide patterns of ancestry, divergence, and admixture in domesticated cattle. PLoS Genet. 10, e1004254.

deMenocal, P.B., 2001. Cultural responses to climate change during the late Holocene. Science 292, 667–673.

Deevey, E.S., 1960. The human population. Sci. Am. 203, 195–204.

Diamond, J., 1997. Guns, Germs, and Steel: The Fates of Human Societies. W.W. Norton & Company, New York, NY.

Diamond, J., 2002. Evolution, consequences and future of plant and animal domestication. Nature 418, 700–707.

Driscoll, C.A., Menotti-Raymond, M., Roca, A.L., Hupe, K., Johnson, W.E., Geffen, E., Harley, E.H., Delibes, M., Pontier, D., Kitchener, A.C., Yamaguchi, N., O'Brien, S.J., Macdonald, D.W., 2007. The Near Eastern origin of cat domestication. Science 317, 519–523.

Duarte, C.M., Marbá, N., Holmer, M., 2007. Rapid domestication of marine species. Science 316, 382–383.

Dunne, J., Evershed, R.P., Salque, M., Cramp, L., Bruni, S., Ryan, K., Biagetti, S., di Lernia, S., 2012. First dairying in green Saharan Africa in the fifth millennium BC. Nature 486, 390–394.

Eitam, D., Kislev, M., Karty, A., Bar-Yosef, O., 2015. Experimental barley flour production in 12,500-year-old rock-cut mortars in Southwestern Asia. PLoS One 10, e0133306.

Eshed, V., Gopher, A., Gage, T.B., Hershkovitz, I., 2004. Has the transition to agriculture reshaped the demographic structure of prehistoric populations? New evidence from the Levant. Am. J. Phys. Anthropol. 124, 315–329.

Eshed, V., Gopher, A., Hershkovitz, I., 2006. Tooth wear and dental pathology at the advent of agriculture: new evidence from the Levant. Am. J. Phys. Anthropol. 130, 145–159.

Evin, A., Cucchi, T., Cardini, A., Strand Vidarsdottir, U., Larson, G., Dobney, K., 2013. The long and winding road: identifying pig domestication through molar size and shape. J. Archaeol. Sci. 40, 735–743.

Fallahsharoudi, A., de Kock, N., Johnsson, M., Ubhayasekera, S.J., Bergquist, J., Wright, D., Jensen, P., 2015. Domestication effects on stress induced steroid secretion and adrenal gene expression in chickens. Sci. Rep. 5, 15345.

FAO, 2001. Pastoralism for the new millennium. Available from: http://www.fao.org/docrep/005/y2647e/y2647e02.htm.

FAO, 2005. Origins and breeds of domestic geese. Available from: http://www.fao.org/docrep/005/y4359e/y4359e03.htm.

Faris, J.D., 2014. Wheat domestications: key to agricultural revolutions past and future. In: Tuberosa, R., Graner, A., Frison, E. (Eds.), Genomics of Plant Genetic Resources. Springer, New York, NY, pp. 439–464.

Fernández, E., Pérez-Pérez, A., Gamba, C., Prats, E., Cuesta, P., Anfruns, J., Molist, M., Arroyo-Pardo, E., Turbón, D., 2014. Ancient DNA analysis of 8000 B.C. near eastern farmers supports an early Neolithic pioneer maritime colonization of Mainland Europe through Cyprus and the Aegean Islands. PLoS Genet. 10, e1004401.

Fibiger, L., Ahlström, T., Bennike, P., Schulting, R.J., 2013. Patterns of violence-related skull trauma in Neolithic Southern Scandinavia. Am. J. Phys. Anthropol. 150, 190–202.

Fuller, D.Q., Qin, L., Zheng, Y., Zhao, Z., Chen, X., Hosoya, L.A., Sun, G.P., 2009. The domestication process and domestication rate in rice: spikelet bases from the Lower Yangtze. Science 323, 1607–1610.

Gautier, A., 1966. Camelus thomasi from the Northern Sudan and its bearing on the relationship C. thomasi: C. bactrianus. J. Paleontol. 40, 1368–1372.

Ghislain, M., Núñez, J., Herrera Mdel, R., Spooner, D.M., 2009. The single Andigenum origin of Neo-Tuberosum potato materials is not supported by microsatellite and plastid marker analyses. Theor. Appl. Genet. 118, 963–969.

Gray, R.D., Atkinson, Q.D., 2003. Language-tree divergence times support the Anatolian theory of Indo-European origin. Nature 426, 435–439.

Gross, B.L., Zhao, Z., 2014. Archaeological and genetic insights into the origins of domesticated rice. Proc. Natl. Acad. Sci. USA 111, 6190–6197.

Grosman, L., Munro, N.D., Belfer-Cohen, A., 2008. A 12,000-year-old Shaman burial from the southern Levant (Israel). Proc. Natl. Acad. Sci. USA 105, 17665–17669.

Han, L., Zhu, S., Ning, C., Cai, D., Wang, K., Chen, Q., Hu, S., Yang, J., Shao, J., Zhu, H., Zhou, H., 2014. Ancient DNA provides new insight into the maternal lineages and domestication of Chinese donkeys. BMC Evol. Biol. 14, 246.

Harlan, J.R., 1992. Crops and Man. American Society of Agronomy, Madison, WI.

He, Z., Zhai, W., Wen, H., Tang, T., Wang, Y., Lu, X., Greenberg, A.J., Hudson, R.R., Wu, C.I., Shi, S., 2011. Two evolutionary histories in the genome of rice: the roles of domestication genes. PLoS Genet. 7, e1002100.

Hermanussen, M., 2003. Stature of early Europeans. Hormones (Athens) 2, 175–178.

Hu, Y., Hu, S., Wang, W., Wu, X., Marshall, F.B., Chen, X., Hou, L., Wang, C., 2014. Earliest evidence for commensal processes of cat domestication. Proc. Natl. Acad. Sci. USA 111, 116–120.

Institute of Medicine, 2002. Dietary reference intakes for energy, carbohydrate, fiber, fat, fatty acids, cholesterol, protein, and amino acids. National Academies Press, Washington, DC, USA.

ITIS (Integrated Taxonomic Information System), 2017. Available from: http://www.itis.gov/servlet/SingleRpt/SingleRpt?search_topic=TSN&search_value=183838#.

IUCN, 2015. Red list of threatened species. Available from: http://www.iucnredlist.org/details/136721/0.

Jansen, T., Forster, P., Levine, M.A., Oelke, H., Hurles, M., Renfrew, C., Weber, J., Olek, K., 2002. Mitochondrial DNA and the origins of the domestic horse. Proc. Natl. Acad. Sci. USA 99, 10905–10910.

Ji, R., Cui, P., Ding, F., Geng, J., Gao, H., Zhang, H., Yu, J., Hu, S., Meng, H., 2009. Monophyletic origin of domestic bactrian camel (Camelus bactrianus) and its evolutionary relationship with the extant wild camel (Camelus bactrianus ferus). Anim. Genet. 40, 377–382.

Jiménez-Brobeil, S.A., du Souich, P., Al Oumaoui, I., 2009. Possible relationship of cranial traumatic injuries with violence in the south-east Iberian Peninsula from the Neolithic to the Bronze Age. Am. J. Phys. Anthropol. 140, 465–475.

Jones, C.I., 2001. Was an industrial revolution inevitable? Economic growth over the very long run. Adv. Macroecon. 1, 1–43.

Kadwell, M., Fernandez, M., Stanley, H.F., Baldi, R., Wheeler, J.C., Rosadio, R., Bruford, M.W., 2001. Genetic analysis reveals the wild ancestors of the llama and the alpaca. Proc. R. Soc. Lond. B. 268, 2575–2584.

Karasik, D., Arensburg, B., Pavlovsky, O.M., 2000. Age assessment of Natufian remains from the land of Israel. Am. J. Phys. Anthropol. 113, 263–270.

Kemp, B.M., González-Oliver, A., Malhi, R.S., Monroe, C., Schroeder, K.B., McDonough, J., Rhett, G., Resendéz, A.,

Peñaloza-Espinosa, R.I., Buentello-Malo, L., Gorodesky, C., Smith, D.G., 2010. Evaluating the farming/language dispersal hypothesis with genetic variation exhibited by populations in the Southwest and Mesoamerica. Proc. Natl. Acad. Sci. USA 107, 6759–6764.

Khush, G.S., 1997. Origin, dispersal, cultivation and variation of rice. Plant Mol. Biol. 35, 25–34.

Kimura, B., Marshall, F.B., Chen, S., Rosenbom, S., Moehlman, P.D., Tuross, N., Sabin, R.C., Peters, J., Barich, B., Yohannes, H., Kebede, F., Teclai, R., Beja-Pereira, A., Mulligan, C.J., 2011. Ancient DNA from Nubian and Somali wild ass provides insights into donkey ancestry and domestication. Proc. R. Soc. B Biol. Sci. 278, 50–57.

Kohler, T.A., Reese, K.M., 2014. Long and spatially variable Neolithic demographic transition in the North American southwest. Proc. Natl. Acad. Sci. USA 111, 10101–10106.

Kuper, R., Kröpelin, S., 2006. Climate-controlled Holocene occupation in the Sahara: motor of Africa's evolution. Science 313, 803–807.

Larson, G., Albarella, U., Dobney, K., Rowley-Conwy, P., Schibler, J., Tresset, A., Vigne, J.D., Edwards, C.J., Schlumbaum, A., Dinu, A., Balaçsescu, A., Dolman, G., Tagliacozzo, A., Manaseryan, N., Miracle, P., Van Wijngaarden-Bakker, L., Masseti, M., Bradley, D.G., Cooper, A., 2007. Ancient DNA, pig domestication, and the spread of the Neolithic into Europe. Proc. Natl. Acad. Sci. USA 104, 15276–15281.

Larson, G., Dobney, K., Albarella, U., Fang, M., Matisoo-Smith, E., Robins, J., Lowden, S., Finlayson, H., Brand, T., Willerslev, E., Rowley-Conwy, P., Andersson, L., Cooper, A., 2005. Worldwide phylogeography of wild boar reveals multiple centers of pig domestication. Science 307, 1618–1621.

Larson, G., Liu, R., Zhao, X., Yuan, J., Fuller, D., Barton, L., Dobney, K., Fan, Q., Gu, Z., Liu, X.H., Luo, Y., Lv, P., Andersson, L., Li, N., 2010. Patterns of East Asian pig domestication, migration, and turnover revealed by modern and ancient DNA. Proc. Natl. Acad. Sci. USA 107, 7686–7691.

Larson, G., Piperno, D.R., Allaby, R.G., Purugganan, M.D., Andersson, L., Arroyo-Kalin, M., Barton, L., Vigueira, C.C., Denham, T., Dobney, K., Doust, A.N., Gepts, P., Gilbert, M.T.P., Gremillion, K.J., Lucas, L., Lukens, L., Marshall, F.B., Olsen, K.M., Pires, J.C., Richerson, P.J., Rubio de Casas, R., Sanjur, O.I., Thomas, M.G., Fuller, D.Q., 2014. Current perspectives and the future of domestication studies. Proc. Natl. Acad. Sci. USA 111, 6139–6146.

Lei, C.Z., Zhang, W., Chen, H., Lu, F., Liu, R.Y., Yang, X.Y., Zhang, H.C., Liu, Z.G., Yao, L.B., Lu, Z.F., Zhao, Z.L., 2007. Independent maternal origin of Chinese swamp buffalo (Bubalus bubalis). Anim. Genet. 38, 97–102.

Leonard, W.R., Snodgrass, J.J., Robertson, M.L., 2010. Evolutionary perspectives on fat ingestion and metabolism in humans. In: Montmayeur, J.-P., le Coutre, J. (Eds.), Fat

Detection. Taste, Texture, and Post Ingestive Effects. CRC Press, Boca Raton, FL, pp. 3–19.

Lev-Yadun, S., Gopher, A., Abbo, S., 2000. The cradle of agriculture. Science 288, 1602–1603.

Li, H.F., Zhu, W.Q., Song, W.T., Shu, J.T., Han, W., Chen, K.W., 2010. Origin and genetic diversity of Chinese domestic ducks. Mol. Phylogenet. Evol. 57, 634–640.

Liu, Y.P., Wu, G.S., Yao, Y.G., Miao, Y.W., Luikart, G., Baig, M., Beja-Pereira, A., Ding, Z.L., Palanichamy, M.G., Zhang, Y.P., 2006. Multiple maternal origins of chickens: out of the Asian jungles. Mol. Phylogenet. Evol. 38, 12–19.

Loftus, R.T., MacHugh, D.E., Bradley, D.G., Sharp, P.M., Cunningham, P., 1994. Evidence for two independent domestications of cattle. Proc. Natl. Acad. Sci. USA 91, 2757–2761.

Londo, J.P., Chiang, Y.C., Hung, K.H., Chiang, T.Y., Schaal, B.A., 2006. Phylogeography of Asian wild rice, *Oryza rufipogon*, reveals multiple independent domestications of cultivated rice, *Oryza sativa*. Proc. Natl. Acad. Sci. USA 103, 9578–9583.

Lu, H., Zhang, J., Liu, K.B., Wu, N., Li, Y., Zhou, K., Ye, M., Zhang, T., Zhang, H., Yang, X., Shen, L., Xu, D., Li, Q., 2009. Earliest domestication of common millet (*Panicum miliaceum*) in East Asia extended to 10,000 years ago. Proc. Natl. Acad. Sci. USA 106, 7367–7372.

Malthus, T., 1798. An Essay on the Principle of Population. J. Johnson, St. Paul's Church-Yard, London.

Manning, S., 2008. Year-by-Year World Population Estimates: 10,000 B.C. to 2007 A.D. Available from: http://www.scottmanning.com/content/year-by-year-world-population-estimates/.

Marshall, F.B., Dobney, K., Denham, T., Capriles, J.M., 2014. Evaluating the roles of directed breeding and gene flow in animal domestication. Proc. Natl. Acad. Sci. USA 111, 6153–6158.

Matsuoka, Y., Vigouroux, Y., Goodman, M.M., Sanchez, G.J., Buckler, E., Doebley, J., 2002. A single domestication for maize shown by multilocus microsatellite genotyping. Proc. Natl. Acad. Sci. USA 99, 6080–6084.

Meyer, C., Lohr, C., Gronenborn, D., Alt, K.W., 2015. The massacre mass grave of Schöneck-Kilianstädten reveals new insights into collective violence in Early Neolithic Central Europe. Proc. Natl. Acad. Sci. USA 112, 11217–11222.

Miao, Y.W., Peng, M.S., Wu, G.S., Ouyang, Y.N., Yang, Z.Y., Yu, N., Liang, J.P., Pianchou, G., Beja-Pereira, A., Mitra, B., Palanichamy, M.G., Baig, M., Chaudhuri, T.K., Shen, Y.Y., Kong, Q.P., Murphy, R.W., Yao, Y.G., Zhang, Y.P., 2013. Chicken domestication: an updated perspective based on mitochondrial genomes. Heredity 110, 277–282.

Molina, J., Sikora, M., Garud, N., Flowers, J.M., Rubinstein, S., Reynolds, A., Huang, P., Jackson, S., Schaal, B.A., Bustamante, C.D., Boyko, A.R., Purugganan, M.D., 2011. Molecular evidence for a single evolutionary origin

of domesticated rice. Proc. Natl. Acad. Sci. USA 108, 8351–8356.

Mona, S., Catalano, G., Lari, M., Larson, G., Boscato, P., Casoli, A., Sineo, L., Di Patti, C., Pecchioli, E., Caramelli, D., Bertorelle, G., 2010. Population dynamic of the extinct European aurochs: genetic evidence of a north-south differentiation pattern and no evidence of post-glacial expansion. BMC Evol. Biol. 10, 83.

Montague, M.J., Li, G., Gandolfi, B., Khan, R., Aken, B.L., Searle, S.M., Minx, P., Hillier, L.W., Koboldt, D.C., Davis, B.W., Driscoll, C.A., Barr, C.S., Blackistone, K., Quilez, J., Lorente-Galdos, B., Marques-Bonet, T., Alkan, C., Thomas, G.W., Hahn, M.W., Menotti-Raymond, M., O'Brien, S.J., Wilson, R.K., Lyons, L.A., Murphy, W.J., Warren, W.C., 2014. Comparative analysis of the domestic cat genome reveals genetic signatures underlying feline biology and domestication. Proc. Natl. Acad. Sci. USA 111, 17230–17235.

Moore, A.M.T., Hillman, G.C., 1992. The Pleistocene to Holocene transition and human economy in Southwest Asia: the impact of the Younger Dryas. Am. Antiq. 57, 482–494.

Munrom, N.D., Grosman, L., 2010. Early evidence (ca. 12,000 B.P.) for feasting at a burial cave in Israel. Proc. Natl. Acad. Sci. USA 07, 15362–15366.

Naderi, S., Rezaei, H.R., Taberlet, P., Zundel, S., Rafat, S.A., Naghash, H.R., el-Barody, M.A., Ertugrul, O., Pompanon, F., Econogene Consortium, 2007. Large scale mitochondrial DNA analysis of the domestic goat reveals six maternal lineages with high haplotype diversity. PLoS One 2, e1012.

Nagarajan, M., Nimisha, K., Kumar, S., 2015. Mitochondrial DNA variability of domestic river buffalo (*Bubalus bubalis*) populations: genetic evidence for domestication of river buffalo in Indian subcontinent. Genome Biol. Evol. 7, 1252–1259.

Outram, A.K., Stear, N.A., Bendrey, R., Olsen, S., Kasparov, A., Zaibert, V., Thorpe, N., Evershed, R.P., 2009. The earliest horse harnessing and milking. Science 323, 1332–1335.

Papathanasiou, A., 2005. Health status of the Neolithic population of Alepotrypa Cave, Greece. Am. J. Phys. Anthropol. 126, 377–390.

Park, S.D., Magee, D.A., McGettigan, P.A., Teasdale, M.D., Edwards, C.J., Lohan, A.J., Murphy, A., Braud, M., Donoghue, M.T., Liu, Y., Chamberlain, A.T., Rue-Albrecht, K., Schroeder, S., Spillane, C., Tai, S., Bradley, D.G., Sonstegard, T.S., Loftus, B.J., MacHugh, D.E., 2015. Genome sequencing of the extinct Eurasian wild aurochs, *Bos primigenius*, illuminates the phylogeography and evolution of cattle. Genome Biol. 16, 234.

Peng, J.H., Sun, D.F., Nevo, E., 2011. Domestication evolution, genetics and genomics in wheat. Mol. Breed. 28, 281–301.

Potts, D.T., 2005. Bactrian camels and bactrian-dromedary hybrids. The Silk Road 3/1: 49–58. Available from:

http://www.silk-road.com/newsletter/vol3num1/7_bactrian.php.

Resor, C.W., 2014. Dietary requirements of a medieval peasant. Available from: http://people.eku.edu/resorc/Medieval_peasant_diet.htm.

Richards, M.P., 2002. A brief review of the archaeological evidence for Palaeolithic and Neolithic subsistence. Eur. J. Clin. Nutr. 56, 1270–1278.

Roksandic, M., Djurić, M., Rakocević, Z., Seguin, K., 2006. Interpersonal violence at Lepenski Vir Mesolithic/Neolithic complex of the Iron Gates Gorge (Serbia–Romania). Am. J. Phys. Anthropol. 129, 339–348.

Rosen, A.M., Rivera-Collazo, I., 2012. Climate change, adaptive cycles, and the persistence of foraging economies during the late Pleistocene/Holocene transition in the Levant. Proc. Natl. Acad. Sci. USA 109, 3640–3645.

Rossel, S., Marshall, F., Peters, J., Pilgram, T., Adams, M.D., O'Connor, D., 2008. Domestication of the donkey: timing, processes, and indicators. Proc. Natl. Acad. Sci. USA 105, 3715–3720.

Rothenbacher, F., 2000. The European population: a historical data handbook for 21 European countries from 1850–1945. EURODATA Newsl. 2000, 1–10.

Salque, M., Bogucki, P.I., Pyzel, J., Sobkowiak-Tabaka, I., Grygiel, R., Szmyt, M., Evershed, R.P., 2013. Earliest evidence for cheese making in the sixth millennium BC in northern Europe. Nature 493, 522–525.

Sapir-Hen, L., Ben Yousef, E., 2013. The introduction of domestic camels to the Southern Levant: evidence from the Aravah valley. Tel Aviv 40, 277–280.

Shi, X.W., Wang, J.W., Zeng, F.T., Qiu, X.P., 2006. Mitochondrial DNA cleavage patterns distinguish independent origin of Chinese domestic geese and Western domestic geese. Biochem. Genet. 44, 237–245.

Singman, J.L., McLean, W., 1995. Daily Life in Chaucer's England. Greenwood Press, Westport, CT.

Spooner, D.M., McLean, K., Ramsay, G., Waugh, R., Bryan, G.J., 2005. A single domestication for potato based on multilocus amplified fragment length polymorphism genotyping. Proc. Natl. Acad. Sci. USA 102, 14694–14699.

Stahl, P., 2003. Pre-Columbian Andean animal domesticates at the edge of empire. World Archaeol. 34, 470–483.

Stiner, M.C., Buitenhuis, H., Duru, G., Kuhn, S.L., Mentzer, S.M., Munro, N.D., Pöllath, N., Quade, J., Tsartsidou, G., Özbaşaran, M., 2014. A forager-herder trade-off, from broad-spectrum hunting to sheep management at Aşıklı Höyük, Turkey. Proc. Natl. Acad. Sci. USA 111, 8404–8409.

Storey, A.A., Ramírez, J.M., Quiroz, D., Burley, D.V., Addison, D.J., Walter, R., Anderson, A.J., Hunt, T.L., Athens, J.S., Huynen, L., Matisoo-Smith, E.A., 2007. Radiocarbon and DNA evidence for a pre-Columbian introduction of Polynesian chickens to Chile. Proc. Natl. Acad. Sci. USA 104, 10335–10339.

Szécsényi-Nagy, A., Brandt, G., Haak, W., Keerl, V., Jakucs, J., Möller-Rieker, S., Köhler, K., Mende, B.G., Oross, K., Marton, T., Osztás, A., Kiss, V., Fecher, M., Pálfi, G., Molnár, E., Sebők, K., Czene, A., Paluch, T., Šlaus, M., Novak, M., Pećina-Šlaus, N., Ősz, B., Voicsek, V., Somogyi, K., Tóth, G., Kromer, B., Bánffy, E., Alt, K.W., 2015. Tracing the genetic origin of Europe's first farmers reveals insights into their social organization. Proc. R. Soc. B Biol. Sci. 282, 20150339.

Teletchea, F., Fontaine, P., 2014. Levels of domestication in fish: implications for the sustainable future of aquaculture. Fish Fisher. 15, 181–195.

Thornton, E.K., Emery, K.F., Steadman, D.W., Speller, C., Matheny, R., Yang, D., 2012. Earliest Mexican Turkeys (Meleagris gallopavo) in the Maya Region: implications for pre-Hispanic animal trade and the timing of turkey domestication. PLoS One 7, e42630.

Tixier-Boichard, M., Bed'hom, B., Rognon, X., 2011. Chicken domestication: from archeology to genomics. C. R. Biol. 334, 197–204.

Uerpmann, H.-P., Uerpmann, M., 2002. The appearance of the domestic camel in South-East Arabia. J. Oman Stud. 12, 235–260.

US Census, in press. Selected Historical Decennial Census Population and Housing Counts. Available from: http://www.census.gov/population/www/censusdata/hiscendata.html.

van den Blink, R., Bromley, D.W., Chavas, J.-P., 1991. The economics of Cain and Abel: Agro-pastoral property rights in the Sahel. USAID working paper 11. Available from: http://pdf.usaid.gov/pdf_docs/PNACB225.pdf.

van Oss, R., Abbo, S., Eshed, R., Sherman, A., Coyne, C.J., Vandemark, G.J., Zhang, H.B., Peleg, Z., 2015. Genetic Relationship in Cicer Sp. expose evidence for geneflow between the cultigen and its wild progenitor. PLoS One 10, e0139789.

Vigne, J.D., Guilaine, J., Debue, K., Haye, L., Gérard, P., 2004. Early taming of the cat in Cyprus. Science 304, 259.

Warmuth, V., Eriksson, A., Bower, M.A., Barker, G., Barrett, E., Hanks, B.K., Li, S., Lomitashvili, D., Ochir-Goryaeva, M., Sizonov, G.V., Soyonov, V., Manica, A., 2012. Reconstructing the origin and spread of horse domestication in the Eurasian steppe. Proc. Natl. Acad. Sci. USA 109, 8202–8206.

Weisdorf, J.L., 2005. From foraging to farming: explaining the Neolithic revolution. J. Econ. Surv. 19, 561–586.

Weiss, H., Bradley, R.S., 2001. Archaeology. What drives societal collapse? Science 291, 609–610.

Weiss, E., Wetterstrom, W., Nadel, D., Bar-Yosef, O., 2004. The broad spectrum revisited: evidence from plant remains. Proc. Natl. Acad. Sci. USA 101, 9551–9555.

Wheeler, J.C., 1995. Evolution and present situation of the South American Camelidae. Biol. J. Linnean Soc. 54, 271–295.

Xiang, H., Gao, J., Yu, B., Zhou, H., Cai, D., Zhang, Y., Chen, X., Wang, X., Hofreiter, M., Zhao, X., 2014. Early Holocene chicken domestication in northern China. Proc. Natl. Acad. Sci. USA 111, 17564–17569.

Yang, C.C., Kawahara, Y., Mizuno, H., Wu, J., Matsumoto, T., Itoh, T., 2012. Independent domestication of Asian rice followed by gene flow from japonica to indica. Mol. Biol. Evol. 29, 1471–1479.

Yindee, M., Vlamings, B.H., Wajjwalku, W., Techakumphu, M., Lohachit, C., Sirivaidyapong, S., Thitaram, C., Amarasinghe, A.A., Alexander, P.A., Colenbrander, B., Lenstra, J.A., 2010. Y-chromosomal variation confirms independent domestications of swamp and river buffalo. Anim. Genet. 41, 433–435.

Zeder, M.A., 2008. Domestication and early agriculture in the Mediterranean Basin: origins, diffusion, and impact. Proc. Natl. Acad. Sci. USA 105, 11597–11604.

Zhang, Y., Lu, Y., Yindee, M., Li, K.Y., Kuo, H.Y., Ju, Y.T., Ye, S., Faruque, M.O., Li, Q., Wang, Y., Cuong, V.C., Pham, L.D., Bouahom, B., Yang, B., Liang, X., Cai, Z., Vankan, D., Manatchaiworakul, W., Kowlim, N., Duangchantrasiri, S., Wajjwalku, W., Colenbrander, B., Zhang, Y., Beerli, P., Lenstra, J.A., Barker, J.S., 2015. Strong and stable geographic differentiation of swamp buffalo maternal and paternal lineages indicates domestication in the China/Indochina border region. Mol. Ecol. 25 (7), 1530–1550.

Zilhão, J., 2001. Radiocarbon evidence for maritime pioneer colonization at the origins of farming in West Mediterranean Europe. Proc. Natl. Acad. Sci. USA 98, 14180–14185.

Zohary, D., 1999. Monophyletic vs. polyphyletic origin of the crops on which agriculture was founded in the Near East. Genet. Resour. Crop Evol. 46, 133–142.

Further Reading

Huang, X., Kurata, N., Wei, X., Wang, Z.X., Wang, A., Zhao, Q., Zhao, Y., Liu, K., Lu, H., Li, W., Guo, Y., Lu, Y., Zhou, C., Fan, D., Weng, Q., Zhu, C., Huang, T., Zhang, L., Wang, Y., Feng, L., Furuumi, H., Kubo, T., Miyabayashi, T., Yuan, X., Xu, Q., Dong, G., Zhan, Q., Li, C., Fujiyama, A., Toyoda, A., Lu, T., Feng, Q., Qian, Q., Li, J., Han, B., 2012. A map of rice genome variation reveals the origin of cultivated rice. Nature 490, 497–501.

7

Animal Agriculture: Livestock, Poultry, and Fish Aquaculture

Colin G. Scanes

University of Wisconsin–Milwaukee, Milwaukee, WI, United States

7.1 INTRODUCTION

Agricultural production has made great strides over the last 50 years based on data from the United Nations Food and Agriculture Organization (FAOStats, 2016). Table 7.1 summarizes the increase in production of cereals and animal products between 1991 and 2013. Production of major cereal crops, namely, corn/maize, rice, and wheat have increased markedly more than the rate of human population growth (Table 7.1). Global production of pork and eggs has increased, respectively, by 60.5 and 87.6% between 1991 and 2013; about twice the rate of human population growth over the same time period (32.8%). Chicken meat production has increased by 157.6%; over fivefold that of human population (Table 7.1). Animal agriculture accounts for over 40% of agriculture globally based on gross value of production (FAO, 2015a). It is argued that the increase in agricultural production is due to the application of scientific principles, such as selective breeding, optimizing nutrition and animal and plant health through vaccination, and pesticides.

Terminology

Just as we avoid terms that are offensive for people, we should also avoid emotive or derogatory terms in discussing agriculture or any other topic for that matter. For instance, are the terms "factory farm" or "industrialized animal agriculture" helpful? They are used, for instance, by groups, such as the Humane Society of the United States that some refer to as activist and others as animal welfare advocates (HSUS, 2010)? Parenthetically, it is recognized that the term industrialized animal agriculture is used by some social scientists. The term "factory farm" is imprecise as it can represent intensive animal production or animals housed in doors in confinement; the latter having a straightforward meaning. The term "industrialized animal agriculture" can be viewed as ambiguous as it encompasses all or only some of the following: large scale, corporate ownership, vertical integration (ownership/control/coordination of production and processing), and confinement (animals in indoor facilities). In this chapter, the terms vertical integration, contracting, and confinement will be employed. Equally, does the term "family farm" add meaning? It is used widely in advertising due to its emotive value. However, it adds little meaning as farms can be both operated by a family and structured as a corporation. Moreover, a family-owned farm can be large or small.

TABLE 7.1 Changes in Production of Agricultural Products in Million Metric Tons in the World From 1961 to 2013

	1961	1991	2013	% Increase
Corn (maize)	205.0	494.5	1016.7	496
Rice	215.6	518.7	745.7	346
Wheat	222.4	546.9	713.2	321
Milk (cow)	313.6	471.7	635.6	203
Pork	24.7	70.4	113.0	457
Chicken	7.55	37.3	96.1	1273
Eggs (chicken)	14.4	36.4	68.3	474
Beef	27.7	53.6	64.0	231
Population (in billions)	3.08	5.36	7.12	231

Data from FAOStats, 2016. United Nations Food and Agriculture Organization Statistics. Available from: http://faostat3.fao.org/browse/ Q//E; http://faostat.fao.org/site/535/default.aspx#ancor.*

There have been large changes in agricultural practices, particularly in the USA, in other Western countries, and in China. In 1900, over 40% of the people employed in the USA were employed in production agriculture producing about 8% of GDP (Table 7.2). In contrast, by 2000, less than 2% of people were employed, full-time or part-time, in agriculture producing 0.7% of the GDP. However, the number of people with "off-farm" employment rose markedly from as little as 30% in 1930 (and presumably much less in 1900) to 93% in 2002 (data from Dimitri et al., 2005). The shift from employment in agriculture provided labor for the burgeoning manufacturing and other economic sectors with the increase in productivity in agriculture surpassing that of US manufacturing (Dimitri et al., 2005). On-farm productivity increased as follows with real farm GDP per person rising (Gardner, 2002):

- From 1900 to 1940, 1.0% per year reflecting mechanization (e.g., tractors and milking machines).
- From 1940 to 1980, 2.8% per year reflecting improved mechanization, automation, and science-based agriculture.
- Since 1980, productivity increases reflect further automation and developments in science-based agriculture.

Science-based agriculture encompasses advances in plant and animal breeding, improved animal and plant nutrition, and control measures for pests and diseases. Corn provides an example of the science-based change with 4.1-fold improvement in yield between 1945 and 2000. While the price of corn in nominal dollars showed a small increase, correcting for inflation the price of corn in 1945 would have a buying power of US$11.77 today. Corn

TABLE 7.2   Changes in Agriculture in the 20th Century in the USA

	1900	1945	2000
Percentage of population employed in production agriculture (%)	41	16	1.9
Number of people engaged in production agriculture in millions	9.8	—	2.7[a]
US population in millions	76.0	139.9	282.1
Agriculture as a percentage of GDP	>8.0	6.8	0.7
Size of farms (ha)	57	77	174
Number of commodities per farm	5.1	4.6	1.1
Indices of agricultural production and productivity			
Corn production (billion bushels)[b]	2.66	2.58	9.97
Corn yield (bushels per acre[b]) (price in nominal $ per bushel)	28.1 (0.35)	33.1 (1.23)	136.9 (1.85)

[a] This is similar to the figures for 2015—2.6 million full- and part-time jobs in production agriculture (farming) (ERS, 2015).
[b] One acre being 0.405 ha; one bushel of corn being 25.4 kg.
Data from Bureau of Labor Statistics, 2001. American Labor in the 20th Century. Available from: http://www.bls.gov/opub/mlr/cwc/american-labor-in-the-20th-century.pdf; Bureau of Labor Statistics, 2016. Charting the labor market: data from the Current Population Survey (CPS). Available from: http://www.bls.gov/web/empsit/cps_charts.pdf; Dimitri, C., Effland, A., Conklin, N., 2005. The 20th Century Transformation of U.S. Agriculture and Farm Policy. Economic Information Bulletin Number 3. USDA Economic Research Service. Available from: http://www.ers.usda.gov/media/259572/eib3_1_.pdf; NASS, 2015a. Crop production historical track records. Available from: http://www.usda.gov/nass/PUBS/TODAYRPT/croptr15.pdf.

is predominantly used to feed livestock and poultry. Increases in agricultural production and productivity have predominantly reduced costs to consumers.

7.2 BIOLOGY UNDERGIRDING USE OF LIVESTOCK AND POULTRY FOR FOOD

7.2.1 Overview

The use of livestock and poultry is undergirded by their unique biology. Pigs and chickens are attractive as providers of meat due to their efficiency as synthesizers of protein, particularly muscle protein, reflected in their rapid growth. In addition, they exhibit tremendous ability to produce large numbers of off-spring. The latter is employed with chickens as a source of eggs. What makes ruminants (cattle, sheep, goats, and water buffalo) particularly attractive as providers of meat and milk is their ability to use plant

feedstocks containing cellulose and hemicellulose that are not digestible by people (see Section 7.2.2). Genetics is an aspect of the biology of these domesticated animals that has allowed the large increase in production (Table 7.1) with the increases in production efficiency covered in Section 7.3.

7.2.2 Metabolism in Nonruminants

The metabolism of nonruminants, such as pigs and chickens is summarized as follows:

Protein → digested in small intestine → amino-acids absorbed → synthesis of new proteins in animal.

Starch (and other carbohydrates) → digested in small intestine → glucose absorbed → employed as energy source by animal or stored as glycogen or used as a substrate to synthesize fatty acids, which in turn is stored as triglyceride in adipose tissue.

FIGURE 7.1   **Cattle on rangeland not suitable for crop production.** The nutrients from range plants include their cellulosic components. These are fermented to VFA together with methane by the microbial population of the rumen. *Source: Courtesy USDA Natural Resource Conservation Service; Photo by Keith Weller.*

Triglyceride → digested in small intestine → fatty absorbed → stored as triglyceride in adipose tissue or used as energy source.

7.2.3 Metabolism in Ruminants

Ruminants, such as cattle (Fig. 7.1), buffalo, sheep (Fig. 7.2), and goats, are herbivorous mammals that are distinguished by having a four chambered rumen. They are foregut anaerobic fermenters with the microbial population producing volatile fatty acids (VFA) from cellulose,

hemicellulose, and starch but not lignin in the feed:

Cellulose, hemicellulose or starch (+fats) → VFA (acetate, propionate, and butyrate) → digested in small intestine → absorbed in rumen

Camels, llamas, and alpacas are not true ruminants, instead having a three-chamber "rumen" producing VFA (reviewed in FAO, 2007). In addition to producing VFAs, methane is produced by ruminants and *Camelidae*. This methane is a significant contributor to greenhouse

FIGURE 7.2 **In mountain rangeland, sheep can graze on plants such as larkspur that is toxic to cattle.** *Source: Courtesy USDA Agricultural Research Service.*

gases and hence to climate change. This is discussed in more detail in Section 18.4. The VFAs are predominantly acetate, propionate, and butyrate (Bergman, 1990):

- Acetate: 40%–75% → Energy and fatty acids (stored as triglyceride).
- Propionate: 15%–40% → Glucose by gluconeogenesis in ruminant liver (this is particularly important for dairy animals as milk sugar, lactose, is a disaccharide synthesized from glucose).

- Butyrate: 10%–20% → Ketone bodies in rumen epithelial cells.

About 55 moles of VFA are produced daily in cattle (Sharp et al., 1982) representing about 70% of the animal's energy requirements (Bergman, 1990).

Proteins are synthesized from plant protein and nonprotein nitrogen in the rumen. This is digested in the small intestine and the amino-acids absorbed:

Protein and nonprotein nitrogen → microbial protein → digested in small intestine → amino-acids absorbed

The rumen lumen contains a complex microbiome of bacteria, archaea, fungi, protozoa, and viruses with bacteria the most numerous organism being present at 10^{10}–10^{11} g^{-1} of content and making up more than 50% of the cell mass (Brulc et al., 2009; Creevey et al., 2014). The microbial population of the rumen also synthesizes vitamins.

7.2.4 Disease Control

There have been two epidemiological transitions in human history (discussed in Chapters 6 and 15) (Perry et al., 2013):

1. The Neolithic Revolution with large increases in human population density and close proximity with livestock and hence the risk of transmission of zoonotic diseases.
2. The science-based epidemiological transitions of the 20th century.

A series of approaches to reduce the impact of disease on livestock and poultry are employed including vaccination, control of vectors, antibiotics, parasiticides, biosecurity, disease resistance, optimizing nutrition, and diagnostic testing. Vaccination is an invaluable tool for disease prevention in livestock and poultry (reviewed by Morton, 2007).

7.3 LIVESTOCK AND POULTRY FOR MEAT PRODUCTION

7.3.1 Overview

> **Word Origins and Animal Agriculture**
>
> Interesting factoid of the origin of words (etymology) related to animal agriculture.
>
> The origins of the words of meats comes from Old French: beef (from Old French boef, cf. Modern French boeuf), pork (Old and Modern French porc), veal (from Old French veel, cf. Modern French for calf, veau), mutton (from Old French moton, cf. Modern French mouton), and poultry (from Old French pouletrie, derived from Old French poulet), while the names of the animals (cow, bull, sow, boar, pig, sheep, ewe, ram, chicken, and deer) come from Anglo-Saxon or Old English, a Germanic language. This is because of the Norman (French) invasion of England in 1066. The nobles who spoke French ate the meat while the peasants who raised the animals spoke Anglo-Saxon/Old English. Today's English language is a melding of Old French and Old English. The words reindeer and slaughter in the English language come from the Viking words "hreindyri" and "slatra" following the Viking invasions of England (793–1066).

The global production of meat is summarized in Table 7.3, with pigs being the top meat species followed by chicken and cattle. These three species provide 88% of the meat produced in the world. Production of meat is now focused in Asia and the Americas (Table 7.4) with China being the top producer of meat followed by the USA and Brazil (Table 7.5).

7.3.2 Cattle

There are over 800 breeds of domestic cattle and these are assigned to two species, which can readily interbreed. These are respectively, *Bos taurus* (non-humped or taurine cattle) and *Bos indicus* (humped or indicine or zebu cattle) (see Section 6.6.2). These are used for milk and meat together with traditional locomotive power (e.g., pulling a plow or cart).

Cattle are raised on rangeland (Fig. 7.1) or on improved pasture in cow calf and then stocker operations. In the USA, most beef cattle are "finished" on a high-grain (corn) diet fattening them prior to slaughter. Alternatively, beef cattle can have their entire lives on pasture with the meat sold at premium prices. The meat from grass-fed beef has less fat and higher levels of some fatty acids (see Sections 3.5.2 and 3.5.3).

In 1961, beef was the top meat globally based on production (Table 7.1). It is recognized that the consumption of beef is not accepted by Hindus. Production of beef has increased in the past 50 years at the same rate as human population growth but at a slower rate than pork and chicken production (Table 7.1). The top countries for beef production are summarized in Table 7.6 with the USA continuing to be the nation producing the most beef. There is considerable scope to increase the efficiency of beef production in developing countries with extensive ranch land with improved genetics and pasture management.

7.3.3 Pigs

> **Bringing Home the Bacon**
>
> This expression for earning a living and taking care of a family is thought to originate from Little Dunmow, Essex (United Kingdom) where a couple received a side of bacon from the local priory for their devotion. This was included in the Wife of Bath's Tale by Geoffrey Chaucer in 1395 (Phrases UK, 2016).

TABLE 7.3   World Production of Meat From Different Species in 2013

Meat type (species)	Production (million metric tons)	Top country for production	Production (million metric tons)
All meat	310.4	China	85.2
Pork (pig meat)	113.0	China	53.8
Chicken	96.1	USA	17.4
Beef (cattle meat)	64.0	USA	11.7
Mutton and lamb (sheep meat)	8.6	China	2.1
Turkey	5.5	USA	2.6
Goat	5.4	China	2.0
Duck	4.3	China	3.0
Water buffalo	3.5	India	1.6
Goose	2.7	China	2.6
Game meat	2.0	Papua New Guinea	0.4
Rabbit	1.8	China	0.7

Data from FAOStats, 2016. United Nations Food and Agriculture Organization Statistics. Available from: http://faostat3.fao.org/browse/Q/*/E; http://faostat.fao.org/site/535/default.aspx#ancor.

TABLE 7.4   Comparison of Global Production of Meat by Continent in 2013

Region	Production (million metric tons)
World	310.4
Africa	17.4
Americas	97.6
Asia	131.2
Europe (European Union)	58.0 (44.4)
Oceania	6.3

Data from FAOStats, 2016. United Nations Food and Agriculture Organization Statistics. Available from: http://faostat3.fao.org/browse/Q/*/E; http://faostat.fao.org/site/535/default.aspx#ancor.

TABLE 7.5   Top Countries for Meat Production in 2013

	Country or grouping	Production (million metric tons)
1	China	85.2
2	USA	42.6
3	Brazil	26.0
4	Germany	8.2
5	India	6.2

Data from FAOStats, 2016. United Nations Food and Agriculture Organization Statistics. Available from: http://faostat3.fao.org/browse/Q/*/E; http://faostat.fao.org/site/535/default.aspx#ancor.

TABLE 7.6   Top Countries for Beef Production in 2013

	Country or grouping	Production (million metric tons)
1	USA	11.7
2	Brazil	9.7
3	China	6.4
4	Argentina	2.8
5	Australia	2.3

Data from FAOStats, 2016. United Nations Food and Agriculture Organization Statistics. Available from: http://faostat3.fao.org/browse/Q/*/E; http://faostat.fao.org/site/535/default.aspx#ancor.

Pork is globally the meat of choice for many people (Table 7.1) but is not consumed by Muslims and Jews (see Chapter 11). The top pork-producing countries are summarized in Table 7.7 with China being by far the top producing country.

Globally, pigs are now raised predominantly in confinement facilities (indoors with controlled environments) (Fig. 7.3). The feed is formulated for optimal growth and animal health measures, such as vaccination and biosecurity, are implemented. Genetic selection has markedly

TABLE 7.7   Top Countries for Pork (Pig Meat) Production in 2013

	Country or grouping	Production (million metric tons)
1	China	53.8
2	USA	10.6
3	Germany	5.5
4	Spain	3.4
5	Brazil	3.3

Data from FAOStats, 2016. United Nations Food and Agriculture Organization Statistics. Available from: http://faostat3.fao.org/browse/ Q/*/E; http://faostat.fao.org/site/535/default.aspx#ancor.

improved the growth rate and the number of offspring.

There are three systems for housing pregnant pigs: (1) gestation stalls for individual pigs [2 m (6.5 ft) long and 0.76 m (2.5 ft) wide], (2) group pens, and (3) free-range systems. According to the AVMA (2015), the advantages of gestation stalls are the greatly reduced risk of injuries from aggressive interactions with other pigs, allowing pigs to stand and lie down, some stereotypical behaviors and individualized diets, together with facilitating veterinary care. In contrast, the disadvantages are decreased activity, increased risk of injury, and appearance to the public resulting in controversy and banning of the system in some places.

7.3.4 Chickens

Chicken as a Sign of Prosperity

A slogan of the 1928 presidential campaign by candidate and elected president Herbert Hoover was "a chicken every pot."

Until the middle of the 20th century, chicken meat was a by-product of egg production with meat harvested from old layers. However, chicken meat production was transformed with the application of science-based production raising chickens on deep litter in ventilated barns. Globally, chicken production rose an astonishing 12.7-fold between 1961 and 2013 (Table 7.1) (FAOStats, 2016). The top chicken-producing countries are summarized in Table 7.8 with the USA continuing to be the top producer.

The basis for the increase in chicken production has been the following:

1. Genetics with meat-type chickens selected for growth rate, feed conversion, and breast meat. The change in growth rate is seen by a comparison of random bred chickens representing chickens in the mid-1950s and then meat birds at the beginning of the 21st century (Table 7.9). Growth rate was improved by 4.62-fold while breast weight increased by 7.97-fold. A corollary of the increased growth was decreased immune functioning (e.g., weights of primary immune organs and antibody response to a challenge; Table 7.9).
2. Optimizing nutrition also increased growth rate with, for instance, a 27.8% increase between 1957 and 2001 (Havenstein et al., 2003).
3. Improvements in animal health with biosecurity, vaccination, parasiticides, and so on.

7.3.5 Sheep and Goats

Globally, sheep and goats are often used on poor rangeland for meat and milk production (Fig. 7.2). The top producers of lamb/mutton (sheep meat) in 2013 were the following (FAOStats, 2016):

1. China 2.08 million metric tons
2. Australia 0.66 million metric tons
3. New Zealand 0.45 million metric tons

FIGURE 7.3 **Examples of concentrated animal feeding operation.** (A) Pigs in a confinement facility. (B) Cattle in a confinement facility. *Source: Courtesy USDA Natural Resources Conservation Service; Photos by Jeff Vanuga.*

TABLE 7.8   Top Countries for Production of Chickens in 2013

Ranking	Country or grouping	Production (million metric tons)
1	USA	17.4
2	China	13.4
3	Brazil	12.4

Data from FAOStats, 2016. United Nations Food and Agriculture Organization Statistics. Available from: http://faostat3.fao.org/browse/Q/*/E; http://faostat.fao.org/site/535/default.aspx#ancor.

TABLE 7.9   Effect of Selection on Growth and Other Characteristics in Chickens

	Nonselected (randombred) chickens	Commercial genetically selected chickens
Growth rate (body weight at day 42)	578	2672
Feed: gain at day 42	2.25	1.62
Breast weight as % of body weight at 43 days old	11.6	20.0
Primary immune organs as % of body weight	0.76	0.53
Antibody response +10 days postprimary immunization with sheep red blood cells	2.10	0.59
Antibody response +10 days postsecondary immunization	2.29	1.43

Calculated from Cheema, M.A., Qureshi, M.A., Havenstein, G.B., 2003. A comparison of the immune response of a 2001 commercial broiler with a 1957 randombred broiler strain when fed representative 1957 and 2001 broiler diets. Poult. Sci. 82, 1519–1529; Havenstein, G.B., Ferket, P.R., Qureshi, M.A. 2003. Growth, livability, and feed conversion of 1957 versus 2001 broilers when fed representative 1957 and 2001 broiler diets. Poult. Sci. 82, 1500–1508; Havenstein, G.B., Ferket, P.R., Qureshi, M.A. 2003. Carcass composition and yield of 1957 versus 2001 broilers when fed representative 1957 and 2001 broiler diets. Poult. Sci. 82, 1509–1518.

The top producers of goat meat in 2013 were the following (FAOStats, 2016):

1. China 2.00 million metric tons
2. India 0.66 million metric tons
3. Pakistan 0.30 million metric tons

7.4 LIVESTOCK FOR MILK PRODUCTION

Mammals are unique in producing milk to nourish their newborn young. This characteristic has long been employed by agriculture. Milk production is dominated by two species: dairy cattle and water buffalo, but there is also significant production from sheep, goat, and camel milk (Table 7.10). Cattle, sheep, and goats for milk production can be maintained on pasture or rangeland. Increasingly, dairy cattle are kept in sheltered facilities where they are milked either 2 or 3 times daily. An optimized feed is brought to cows.

Using the USA as an example of a Western industrialized country, milk production increased progressively, for instance, between 1900 and 1950, 1950 and 2000, and 2000 and 2014 (Table 7.11). Between 1900 and 1950, there were increases in the number of dairy cows (21%) and the average yield (48%) (Table 7.11). Since then cow numbers have declined as milk yield per cow increased (Table 7.11). The changes in yield can be attributed to improvements to predominantly the following:

- genetics (allowed by the extensive use of artificial insemination)
- disease control
- nutrition

The improvements in yield, mechanization, and economies of scale have affected the structure of the industry with the number of dairy farms declining dramatically in the USA and Western countries (Table 7.11). "The shift to larger dairy farms is driven largely by the economics of dairy farming. Average costs of production, per hundredweight of milk produced, are lower in larger herds, and the differences are substantial" (MacDonald and Newton, 2014). In the last 22 years, there has been a large increase in the number of very large dairy farms with, for instance, the number of operations with over 2999 dairy cows increasing from 31 in

TABLE 7.10   World Production of Milk From Different Species in 2013

Milk type (species)	Production (million metric tons)	Top country for production	Production (million metric tons)
All milk	768.6	India	135.6
Cow	635.6	USA	91.3
Water Buffalo	102.0	India	70.0
Goat	18.0	India	5.0
Sheep	10.1	China	1.5
Camel	2.9	Somalia	1.1

Data from FAOStats, 2016. United Nations Food and Agriculture Organization Statistics. Available from: http://faostat3.fao.org/browse/Q//E; http://faostat.fao.org/site/535/default.aspx#ancor.*

TABLE 7.11   Changes in Milk Production in Dairy Cattle and in the Number of Cows in the USA

	1900	1950	2000	2014
Dairy cow numbers (millions)	18.1	22.0	18.2	8.6
Milk produced per cow (lb)	3,600	5,314	9,210	22,258
Milk production (million metric tons)	29.6	53.3	76.3	86.8
Number of farms with dairy cows	>4,000,000	3,681,627	116,874[a]	51,470

[a] *Data for 1997.*

Based on Farrington, E.H., 1910. Development of dairying. Correspondence College of Agriculture; Blayney, D., 2002. The changing landscape of U.S. milk production. Statistical Bulletin No. (SB-978). Available from: http://www.ers.usda.gov/publications/sb-statistical-bulletin/sb978.aspx; ERS, 2007. Changes in the Size and Location of U.S. Dairy Farms. Available from: http://www.ers.usda.gov/media/430528/err47b_1_.pdf; MacDonald, J.M., Newton, D., 2014. Milk Production Continues Shifting to Large-Scale Farms. Available from: http://www.ers.usda.gov/amber-waves/2014-december/milk-production-continues-shifting-to-large-scale-farms.aspx#.Vv1_NmA_N7M; NASS, 2015b. Milk production. Available from: http://www.usda.gov/nass/PUBS/TODAYRPT/mkpr0215.pdf

1992 to 440 in 2014 in the USA (MacDonald and Newton, 2014).

7.5 POULTRY FOR EGG PRODUCTION

Eggs predominantly come from chickens, ducks, and geese with chickens being, by far, the predominant egg-producing species (Table 7.12). China is the top producer of hens, ducks, and geese eggs (Table 7.12). Other major producers of chicken eggs are the USA and Brazil.

TABLE 7.12   World Production of Eggs From Different Species in 2013

Egg type (species)	Production (million metric tons)	Top country for production	Production (million metric tons)
All eggs	73.9	China	21.1
Chicken	68.3	China	24.7
Other (predominantly duck and goose)	5.6	China	4.3

Data from FAOStats, 2016. United Nations Food and Agriculture Organization Statistics. Available from: http://faostat3.fao.org/browse/Q//E; http://faostat.fao.org/site/535/default.aspx#ancor.*

TABLE 7.13   Changes in Egg Production in the USA

	1900	2014
Egg produced per hen per year	100	265
Mortality (%)	40	5

Data from United Egg Producers, in press. American egg farming: how we produce an abundance of affordable, safe food. Available from: http://www.unitedegg.org/information/pdf/American_Egg_Farming.pdf.

The changes in the efficiency of egg production between 1900 and 2014 in the USA are summarized in Table 7.13. Production per hen increased 2.65-fold. This can be attributed to the following:

- Markedly improved genetics.
- Optimization of diet particularly providing sufficient calcium for the production of the egg shell.
- Reduction of infectious disease reflected in part by the decrease in mortality.

In addition, the majority of laying hens were moved to cages or aviaries within indoor facilities and thus eliminating the risks of predation.

Commercial egg production is carried out in conventional cages globally. Conventional cages are constructed of wire and can be multilevel. Feed and water is provided automatically as is removal of eggs and manure. An example of the area allowed in such systems is each cage housing five hens with an area of 560 cm^2 (Tactacan et al., 2009). A European Union ban on conventional layer cages became effective January 1, 2012. Layers now must be housed in either enriched cages or alternate systems. Irrespective of which of these systems is employed, hens must be provided with "a nest, perching space, litter to allow pecking and scratching, and unrestricted access to a feed trough" (European Union, 1999).

7.6 INTENSIVE AND EXTENSIVE PRODUCTION OF LIVESTOCK AND POULTRY ENCOMPASSING ANIMAL WELFARE

7.6.1 Overview

Increasingly livestock and poultry are reared together in large numbers in confinement facilities with controlled environments (Tables 7.14 and 7.15; Fig. 7.3). The global increase in cereal, pig, and poultry production (Table 7.1) is predominantly due to intensive agriculture with its high requirements for inputs (Table 7.14). Inputs for plant agriculture, particularly intensive plant agriculture, include fertilizers, pesticides, and seeds with improved genetics. The latter is achieved by selection often following mutagenesis by chemicals or radiation treatment or genetic modification by gene insertion. Inputs for animal agriculture include feed, vitamin and minerals, antibiotics, ecto- and endoparasite control, vaccines, and improved genetics of livestock or poultry due to selection.

The differences between intensive and extensive agriculture are summarized in Table 7.14. A major difference between intensive and extensive agriculture is the land (Table 7.14). Intensive agriculture uses high-quality, often high-priced, land. If rainfall is insufficient, irrigation will be employed. With livestock and poultry, intensive agriculture will use the following:

- Confinement (indoor) facilities for pigs and poultry (meat-type and layers).
- Cattle or sheep maintained outdoors but brought inside for milking and/or seasonally when weather is inclement.
- Beef or dairy cattle maintained outdoors in pens with shelters (such as a feedlot for beef cattle).

Not only does intensive agriculture use high-priced land but also that animal facilities require significant capital investment and costs for labor. Moreover, there are significant costs

TABLE 7.14   Comparison of Intensive and Extensive Agriculture

	Intensive agriculture	Extensive agriculture
Land	High quality high cost	Poor and low cost
Rainfall	High	Low
Labor and/or capital for machinery, automation, and animal facilities	High	Low
Inputs	High	Low
Irrigation	Maybe	No
Distance to centers of population or transportation	Low	High
Production	High	Low
Productivity	High	Low
Environmental impact: water and land (see Chapter 18)	High	Low
Environmental impact: greenhouse gases (see Chapter 18)	Low	High
Impact on animal welfare		
1. *Freedom from hunger and thirst*	Zero to low risk	Medium risk of lack of energy or specific nutrients. Some risk of lack of water.
2. *Freedom from discomfort*	Low risk in temperature-controlled environments.	Even when shelter is provided—medium risk from extremes of temperature and exposure to rain and snow.
3. *Freedom from pain, injury or disease*	With good management, animal health programs and veterinary care—low to medium risk. With overcrowding and/or poor husbandry, risks increase.	With appropriate animal health programs, risk is low to medium from predators, diseases from wildlife, and potentially toxic plants.
4. *Freedom to express normal behavior*	Low to medium depending on degree of crowding and design of facilities.	Zero with freedom to roam
5. *Freedom from fear and distress*	Low risk when animal welfare considerations are integral to management. With poor management or abuse, risks rise to medium or even high, particularly prior to slaughter.	Low to medium risk with predators and lack of veterinary care. With poor management or abuse, risks rise to medium to high, particularly prior to slaughter.

Based in part on University of Reading, Environmental Changes in Farm Management, http://www.ecifm.rdg.ac.uk/.

TABLE 7.15　US Environmental Protection Agency Definitions of Types of CAFOs (EPA, 2015)

	Small CAFO	Medium CAFO	Large CAFO
Beef cattle or cow/calf	<300	300–999	>999
Dairy cattle	<200	200–699	>699
Pigs above 55 lb (25 kg)	<750	750–2,500	>2,499
Pigs below 55 lb (25 kg)	<3000	3,000–9,999	>9,999
Laying hens or broilers (meat) chickens	<9000	9,000–29,999	>29,999

CAFO, Concentrated animal feeding operation.

for inputs including feeds and electric usage for lighting, ventilation, and possibly temperature control. To more than balance these, there is high efficiency of production and usually profitability. Proximity to both of transportation and processing together with centers of population is advantageous to market the animal products. In contrast to extensive agriculture, there is much less production of greenhouse gases (see Chapter 18). The disadvantages of animal-intensive agriculture include the following:

- Feeds (e.g., corn and soybeans) particularly for pigs and poultry can also be consumed directly as food for people.
- The land used for the production of animal feeds could be used for growing vegetables and other plants directly consumed by people.
- The perception of lower animal welfare (see box on animal welfare).
- Substantial risks of environment contamination from the excreta.

Intensive plant agriculture also has disadvantages. These include loss of soil, depletion of aquifers and contamination of water. There are disadvantages of the use of chemical fertilizers. These include increases in sap feeding insects, such as aphids due to the impact of nitrogen fertilizer on plant-soluble nitrogen (Matson et al., 1997). Moreover, there is leaching of both nitrogen and phosphate into ground water and hence rivers and streams (see Chapter 18).

In a study from the US Midwest, conventional tillage was reported to result in a 47% loss of soil carbon between 1907 and 1950 (Matson et al., 1997). With a shift to reduced tillage between 1970 and 1990, the loss from 1907 was reduced to 39% (Matson et al., 1997). Furthermore, the adoption of no-till farming (zero tillage) has the advantage over conventional tillage being the increased sequestration of carbon in the soil ($57 \text{ g C m}^{-2} \text{ year}^{-1}$) (West and Post, 2002). In conservation tillage, some (>30%) of the crop residue is left on the land. This prevents soil erosion, improves water retention, and reduces the carbon footprint (Fernandez-Cornejo et al., 2013). Conservation tillage has been adopted in the USA being employed by 30% of farmers for soybean production in 1990 rising to 62% in 2006 (Fernandez-Cornejo et al., 2013). Researchers at the United States Department of Agriculture Economic Research Service and the Environmental Protection Agency concluded that the adoption of herbicide-tolerant genetically engineered soybeans "has a positive and highly significant (P value < 0.0001) impact on adoption of conservation tillage among US soybean farmers" (Fernandez-Cornejo et al., 2013). Genetically engineered [GE also known as genetically modified or genetically modified organisms] crops have been widely adopted in the USA with 90% of land planted for corn being GE corn and 93% of land planted for soybean being GE soybeans (Fernandez-Cornejo et al., 2014).

Extensive livestock production uses land that is generally unfit for plant agriculture; there is very low rainfall. Production per unit land area is low but it is questioned what else can be done with the land. The alternative position is that the ranges employed by extensive agriculture could support large numbers of wild animals. An advantage of extensive agriculture is its low impact on the environment. The disadvantages of extensive agriculture include the following:

- low productivity
- distance to loci for consumption of the animal products
- extremes of temperature and the presence of predators adversely influencing animal welfare (see box on animal welfare)

- high production of greenhouse gases

There is considerable potential to advance extensive livestock production including improving the ranch land (Ayuko, 1981). These efforts, however, well-intentioned, can have adverse consequences. For instance, there was a comprehensive effort in Inner Mongolia to improve extensive livestock production with the introduction of superior livestock genetics, construction of shelters, drilling of wells, cultivating grasslands, and planting forage crops (Li and Li, 2012). Unfortunately, this has resulted in overgrazing, disruption of the indigenous ecosystem, and the shift in authority from the local community level to the central government (Li and Li, 2012).

Animal Welfare

The principles of animal welfare, the five freedoms, were developed by the Farm Animal Welfare Council, established in 1979 by the British Government. These five freedoms are the following (Farm Animal Welfare Council, 2012):

1. *Freedom from hunger and thirst* (interpreted as the following: provision of, or access to, water and an optimal nutrition).
2. *Freedom from discomfort* (provision of either a temperature-controlled ventilated environment or adequate shelter).
3. *Freedom from pain, injury, or disease* (provision of an environment where the risk of injury or death is minimized and of veterinary care and other animal health measures).
4. *Freedom to express normal behavior* (provision of sufficient space per animal *to express normal behavior* including "to stand up, lie down, turn around, groom themselves, and stretch their limbs" (Brambell Report, 1965).
5. *Freedom from fear and distress* (provision of an environment to eliminate as far as possible "mental suffering").

These five freedoms were built on the Brambell Report (1965) and are widely supported.

How to determine stress in an animal?

Temple Grandin has been a leader in advancing the welfare of livestock. She reported that the blood levels of the stress hormone, cortisol, are markedly elevated in cattle treated roughly (Grandin, 1997).

Are extensive or intensive animal agricultural practices associated with animal welfare issues per se?

Responsible knowledgeable farmers and ranchers ensure that their animals have access to water and feed to provide optimal nutrition (Freedom 1). Not only does this ensure welfare but also maximizes production.

There are practices that impact animal welfare including dehorning, tail docking, castration, beak trimming, and branding. These are associated with increased release of the stress hormones, cortisol in livestock, corticosterone in poultry, and of specific white blood cells, namely, the neutrophil:lymphocyte ratio (in poultry, heterophil:lymphocyte ratio).

7.6.2 Sustainable Agriculture

The goals of sustainable agriculture are the following: environmental stewardship (including stewardship of soil, water, and wildlife also reducing dependence on fossil energy/decreasing emission of greenhouse gases), "economic profitability, and social and economic equity" (Agricultural Sustainability Institute, 2017) (also see Chapter 14 and essay by Mark Rasmussen). The latter encompasses social responsibility to workers, small farmers, rural communities, and providing safe food for consumers (Agricultural Sustainability Institute, 2017). The US Secretary of Agriculture stated the need to "balance goals of improved production and profitability, stewardship of the natural resource base and ecological systems, and enhancement of the vitality of rural communities" (USDA, 1996) and this describes sustainable agriculture.

Sustainable agriculture also features strongly in the United Nations Sustainable Development Goals. These were developed by an intergovernmental Open Working Group and adopted in 2014 by the General Assembly (United Nations). Goal 2 is to "end hunger, achieve food security and improved nutrition, and promote sustainable agriculture." Among the specific targets are the following:

- "By 2030 double the agricultural productivity and the incomes of small-scale food producers, particularly women, indigenous peoples, family farmers, pastoralists and fishers, including through secure and equal access to land, other productive resources and inputs, knowledge, financial services, markets, and opportunities for value addition and nonfarm employment."
- "By 2030 ensure sustainable food production systems and implement resilient agricultural practices that increase productivity and production, that help maintain ecosystems, that strengthen capacity for adaptation to climate change, extreme weather, drought,

flooding and other disasters, and that progressively improve land and soil quality."

The United Nations Sustainable Development Goals are said to build on the progress made toward the Millennium Development Goals but the reduction in hunger and poverty are also due to the tremendous economic development in China (Council on Foreign Relations, 2015). Moreover, the goals are broad and progress toward them is difficult to quantify (Council on Foreign Relations, 2015). Rather than concentrating on eliminating hunger or extreme poverty, the focus is on so many targets that success is extremely unlikely (Economist, 2015).

Sustainable agriculture requires a balance between production, profitability, protecting the environment, and societal considerations. Given this, it is not always clear the extent to which an agricultural system can be viewed as sustainable. Sustainable agriculture includes but is not limited to organic agriculture (considered in Section 7.7). Intensive conventional agriculture might be considered as sustainable if there are programs of environmental stewardship to protect the soil, such as no-till farming and water quality, such as riparian buffers. Extensive conventional agriculture is considered as sustainable particularly when there is environmental stewardship. This is despite the high production of greenhouse gases per unit meat production compared to intensive production (see Chapter 18).

7.7 SPECIALIZED PRODUCTION SYSTEMS INCLUDING ORGANIC FARMING

7.7.1 Overview

There is increasing consumer demand for animal products from specified production systems, such as organic, natural, and cage-free. These are considered in the next subsections.

Terminology

Organic: The term organic will be used in this volume. It is a widely recognized term originating from Lord Northbourne (1940) (reviewed by Kuepper, 2010). USDA organic regulations define "organic" as follows: "Organic production is a production system that...respond[s] to site-specific conditions by integrating cultural, biological, and mechanical practices that foster cycling of resources, promote ecological balance, and conserve biological diversity."

It should be noted that the specific requirements are defined somewhat differently in different countries (explained later). It is recognized that organic is a philosophical concept rather than based on scientific principles.

Natural: This is defined by various marketing systems but not governmental entities.

7.7.2 Organic Production of Livestock and Poultry

Quotations and Views on Organic Agriculture

It is vitally important that we can continue to say, with absolute conviction, that organic farming delivers the highest quality, best-tasting food, produced without artificial chemicals or genetic modification, and with respect for animal welfare and the environment, while helping to maintain the landscape and rural communities

Charles, Prince of Wales, United Kingdom (1948–).

In contrast, von Borell and Sørensen (2004) concluded that "Practical experience shows that organic livestock production is certainly no guarantee of good animal health and welfare." Based on the differences in production efficiencies, it is estimated that organic agriculture can support 3 to 4 million people globally compared to the present population of 7 billion and the projected population of 9 billion by 2050 (Connor, 2008). On the other hand, there are reports of greater diversity of wildlife birds in organic than conventional farming with, for instance, more bird species (Chamberlain et al., 2010; Smith et al., 2010) and more butterflies (Hodgson et al., 2010).

7.7.2.1 Overview

Organic agriculture has lower yields for plant agriculture and lower efficiency of production than conventional agriculture. This is countered by the reduced costs for inputs and higher prices obtained by producers. There has been a large increase in the consumption of organic food globally. Table 7.16 summarizes the increases in livestock and poultry being reared under organic certification in the USA. Organic livestock and poultry production

TABLE 7.16 Increases in Organic Livestock and Poultry Production in the USA Based on the Number of Certified Animals

	1997	2003	2011
Beef cattle	4,429	27,285	106,181
Dairy cattle	12,897	74,435	254,771
Pigs	482	6,564	12,373
Sheep	705	6,564	5,914
Laying hens	537,826	1,591,181	6,663,283
Meat or broiler chickens	38,285	6,301,014	28,644,354
Turkeys	750	217,353	504,315

Data from ERS (USDA Economic Research Service), in press. Available from: http://www.ers.usda.gov/data-products/organic-production.aspx.

requires organic feeds and/or organic pasture management, specific standards for animal welfare and housing including access to outside, extensive record keeping, USDA certification, and precludes the use of antibiotics performance enhancers, but allows vaccination (Agricultural Marketing Service, 2017; Coffey and Baier, 2012; USDA, 2017). Organic production of seed crops for feeds (Fig. 7.4) for organically produced livestock and poultry excludes the use of synthetic fertilizers, pesticides, and genetically modified organisms (Agricultural Marketing Service, 2017).

FIGURE 7.4　**Harvested organic feed corn in a northeast Ohio farm.** *Source: Courtesy USDA Natural Resource Conservation Service; Photo by Scott Bauer.*

The following are either permitted or prohibited for organic production in the USA (based in part from News Staff, 2010; also Agricultural Marketing Service, 2017):

1. Organic crops
   a. Legumes fixing nitrogen, Yes
   b. Crop rotation, Yes
   c. Fertilizers
      - Organic fertilizer (e.g., manure), Yes
      - Organic fertilizer (sewage sludge), No
      - Unprocessed rock phosphate Yes (and analogous source of potassium)
      - Synthetic nitrogen fertilizer, No
   d. Genetics
      - Genetically improved seed including use of mutagens (chemical or radiation) causing genetic modifications, Yes
      - Genetic modification (by gene transfer in biotechnology), No
   e. Pesticides
      - Synthetic, No
      - Natural, Yes
2. Organic livestock and poultry
   a. Antibiotics, No
   b. Chemical disinfectants (e.g., chlorine and sodium hypochlorite), Yes
   c. Feed
      - From organic crops, Yes
      - From conventional crops, No
      - Trace minerals, Yes
      - Vitamins (natural or, much more likely, synthetic), Yes
   d. Genetics
      - Selection programs including use of candidate genes, Yes
      - Genetic modification (gene transfer by biotechnology), No
   e. Hormones
      - Oxytocin can be used therapeutically, Yes
      - Somatotropin or other performance enhancers, No
   f. Parasite treatments, usually no, but for exceptions, see Section 7.7.2.2.
   g. Vaccination (discussed next)

7.7.2.2 Animal Welfare, Disease Control, and Organic Animal Production

Consumers of organic meat and other animal products expect a higher level of animal welfare and more humane treatment in organic agriculture than in conventional production but is this the case? (reviewed by Hughner et al., 2007). The tenets of organic animal agriculture are that with lower animal densities and concomitant stress and organic feed, the animals will be more resistant to infectious and metabolic diseases and thus veterinary drugs should not be required (Kijlstra and Eijck, 2006). Access to the outdoors also exposes animals to pathogens and parasites (reviewed by Kijlstra and Eijck, 2006). Vaccination is viewed as "may prevent serious outbreak" of disease (reviewed by Kijlstra and Eijck, 2006). Organic standards differ globally with vaccination permitted in the USA but only allowed in the United Kingdom under some circumstances "if there is a known disease risk on the farm, or neighboring land, which cannot be controlled by any other means" (Soil Association, 2017).

The available research evidence does not support the tenet that organically raised animals are per se healthier than conventional livestock. For instance, cattle from organic dairy herds have either the same or higher somatic cell counts (indicative of mastitis) in their milk compared to cattle from conventional farms (reviewed by Sutherland et al., 2013). There are also problem areas, such as parasite control and achieving balanced diets (Hovi et al., 2003). Gastrointestinal nematodes and other parasites are a particular problem for the organic production of cattle, sheep, pigs, and poultry due to restrictions of permitted treatments and the animals' access to the outdoors, for instance, pastures (reviewed by Sutherland et al., 2013). Comparing organic with conventional farms, there are more parasitic helminths in sheep and higher levels of parasitic nematodes (e.g., *Ascaris suum*) and coccidian protozoa (*Eimeria*) in pigs (reviewed by Sutherland et al., 2013).

Few effective veterinary therapies are permitted in organic livestock or poultry production (Kijlstra and Eijck, 2006). Sutherland et al. (2013) concluded that "there is still a lack of scientifically evaluated, organically acceptable therapeutic treatments that organic animal producers can use when current management practices are not sufficient to maintain the health of their animals." However, to assure humane treatment of animals, according to the Agricultural Marketing Service of the USDA, "if approved interventions fail, the animal *must* still be given all appropriate treatment(s)" (emphasis added) including antibiotics but then any products cannot be sold as organic. Synthetic parasiticides are prohibited for animals prior to marketing or for milk production. However, three parasiticides, ivermectin, fenbendazole, and moxidectin, are permitted for organic cattle production in the USA for either external and/or internal parasites when other approaches are ineffective (Agricultural Marketing Service, 2012, 2015).

7.7.2.3 Production Efficiency

Organic plant agriculture has lower yields (34%) compared to conventional systems based on a metaanalysis of multiple studies (Seufert et al., 2012). Organic production of corn was reported with the following yields (Pimental et al., 2005):

- animal organic farming: 3.7 million kcal ha^{-2}
- legume organic farming: 3.5 million kcal ha^{-2}
- conventional agriculture: 5.2 million kcal ha^{-2}

Moreover, there is much higher requirement for labor for organic production (Pimental et al., 2005). The corollary is that the costs to produce organic crops are likely to be higher.

7.7.2.4 Nutritional Quality of Food From Organic Agriculture

There is considerable anecdotal discussion of the nutritional quality of organic foods. However, in a comprehensive review of organic foods, there were no differences detected for calcium, copper, magnesium, potassium, zinc, and fat (Dangour et al., 2009). Based on a series of metaanalyses, there are some differences between organic and conventionally raised plants, meat, and milk. For instance, the levels of pesticides are lower in organic crops while the concentrations of putative antioxidant nutrients is higher (Barański et al., 2014). Średnicka-Tober et al. (2016a) concluded from a metaanalysis that for most nutrients there was no evidence for differences between meats from organic or conventionally raised animals. In contrast, there are elevated concentrations of total polyunsaturated fatty acids (PUFA) and $n - 3$ PUFA in organic meat and milk (Średnicka-Tober et al., 2016a,b). Organic milk is also reported to have higher concentrations of conjugated linoleic acid, α-tocopherol, and iron but lower concentrations of iodine and selenium (Średnicka-Tober et al., 2016b). The effects on PUFA and conjugated linoleic acid are similar to the difference between grass-fed and corn-fed cattle (discussed in Chapter 3).

7.7.2.5 Conclusions

The top three reasons why consumers purchase organic foods are the following (Shaw Hughner et al., 2007):

1. Perceived healthfulness and improved nutritional qualities.
2. Tastes better.
3. Better for the environment.

Is there truth to these?

There are some differences between meat and milk from organic and conventionally raised (see preceding section) but it is not clear what organic practices make the difference. Body composition is likely to be influenced by access to different foods, such as pastures. There is a lack of good evidence for the assertion that organic foods taste better. It seems that taste has not always been a priority to the food industry. A concern for organically raised vegetables is the use of animal manure. There are risks of food-borne diseases from the consumption of uncooked vegetables, such as salad contaminated with pathogens shed by

livestock (discussed in Chapter 18). The issue of whether organic agriculture is better for the environment is open to question. Organic agriculture does not allow the use of chemical pesticides or genetically modified crops. This restricts the use of "no-till" systems. Moreover, the corollary to the lower production efficiency of organic agriculture is that more land is required for the same total production. Parenthetically, it should be noted that there is a clear case for regulations on pesticides and their use and predominantly this is the practice. The issues related to the differences in welfare with organic versus conventional production and the lack of effective therapies is addressed in Section 7.7.2.2. It is concluded that organic food may or may not have better attributes than food raised in a conventional manner.

7.7.3 Natural Beef and Pork

Natural beef or pork is marketed under specific brands and requirements. Generally, the requirements for natural beef or pork are similar to those of organic. However, there are differences:

- government certification is not required
- organic feed grains or pastures are not required

Products can be labeled as hormone-free and natural.

7.7.4 Free-Range or Cage-Free Poultry or Raised With Access to the Outdoors and/or Pasture

There has been an increasing interest by researchers on the different production systems for poultry. For egg-laying hens, there is a similar level of production between conventional and enriched cages but lower in aviary-type cages (Table 7.17). Hen mortality has been reported to be markedly greater in cage-free aviaries (Karcher et al., 2015). This is due to keel injuries in the aviary and may reflect inadequacies in aviary design or genetic selection of laying hens for cages. There were no differences in antibody response or in the

TABLE 7.17 Comparison of Egg Production and Mortality in Conventional, Enriched, and Cage-Free Aviary Systems (Karcher et al., 2015)

	Conventional	Enriched	Aviary
Eggs per hen	352	363	340
Egg weight	58.5	59.1	58.4
Feed conversion	2.02	1.99	2.12
Mortality (%)	4.7	5.1	11.5

two indices of stress, circulating concentrations of corticosterone and heterophil:lymphocyte ratios in hens, in conventional cages or enriched environments (Tactacan et al., 2009).

There are differences in the composition of eggs from free-range chickens to caged layers with increased polyunsaturated fatty acids and $n-3$ fatty acids (Table 7.18). However, there are reports of disadvantages. Organically free-range raised hens had higher mortality and lower egg production compared to those in conventional cages (Leenstra et al., 2012). In a study of 19 flocks of free-range egg-laying hens, all had parasitic nematode eggs in their feces (Sherwin et al., 2013). Predation was reported in 40% of free-range flocks of chickens (Bestman and Wagenaar, 2014).

There are also systems for slow growth meat chickens, such as Label Rouge in France and

TABLE 7.18 Comparison of Composition of Eggs From Free-Range and Conventionally Housed Hens (Anderson, 2011)

	Conventionally housed hens	Free-range hens
Fatty acids (%)	7.88[a]	8.11[b]
Saturated fatty acids (%)	2.55	2.55
Monounsaturated fatty acids (%)	3.67[a]	3.80[b]
Polyunsaturated fatty acids (%)	1.25[a]	1.36[b]
$n-3$ Fatty acids (mg 50 g^{-1})	70.6[a]	84.3[b]
Cholesterol (mg 50 g^{-1})	163	165
Vitamin A (mg 50 g^{-1})	160	156
β-Carotene (mg 50 g^{-1})	2.77[a]	10.5[b]
Vitamin E (mg 50 g^{-1})	1.37	1.19

Different superscript letters indicate difference $P<0.05$.

various heritage breeds in the USA. Slow growth is achieved based principally on the genetics of the birds and nutritional restriction by raising on pasture.

7.8 FINFISH AQUACULTURE

The amount of fish obtained by capture fishing has been essentially flat since 1990 (considered in Section 5.5). In contrast, fish obtained by aquaculture has grown from 4.7 million metric tons in 1980 to 13.1 million metric tons in 1990 reaching 66.6 million metric tons (valued at US$137.7 billion) in 2012 (FAO, 2015b):

- inland aquaculture, 41.9 million metric tons (41.1% increase since 2007)
- mariculture, 24.7 million metric tons (23.5% increase since 2007)

Definition of Aquaculture

Aquaculture (either inland or mariculture) is raising finfish or shell-fish in human-constructed aqueous environments. Inland aquaculture is conducted in large ponds while mariculture employs enclosures. Aquaculture originated at least 2000 years ago but the growth of this approach has occurred predominantly in the last 50 years due to the following:

- New species have been domesticated.
- There are improved feeds and genetics.
- Systems for production, such as water recirculation and disease control using vaccines and antibiotics.

Together these facilitate efficient and profitable production. However, there can be adverse effects on the environment due to waste leakage or run-off.

The top countries for aquaculture in 2012 are summarized in Table 7.19 (FAO, 2015b). FAO (2015b) is "promoting a coherent approach for the

TABLE 7.19   Top Countries for the Production of Finfish by Either Inland Aquaculture or Mariculture

Ranking	Country	Inland aquaculture	Ranking	Country	Mariculture
1	China	23.3	1	Norway	1.3
2	India	3.8	2	China	1.2
3	Vietnam	2.1	3	Chile	0.8
4	Indonesia	2.1	4	Indonesia	0.6

Data from FAO, 2015b. The State of World Fisheries and Aquaculture. Available from: http://www.fao.org/3/a-i3720e.pdf.

sustainable, integrated and socio-economically sensitive management of oceans and wetlands, focusing on capture fisheries, aquaculture."

While China has an inland aquaculture and mariculture sector with a large variety of species including carp (Table 7.20), some countries focus on specific species:

- Norway, Atlantic salmon
- Vietnam, catfish
- USA, channel catfish

Production in the USA is focused in states along the Gulf of Mexico (#1 Mississippi, 2# Alabama, 3# Arkansas, and #4 Texas) with 6668 metric tons or US$361 million in production in 2015 (NASS, 2016). There are 69,900 acres (28,300 ha) of catfish-rearing ponds in the USA as of January 2016 (NASS, 2016).

Among the reasons for the growth of aquaculture are genetic improvement of the aquaculture species, good nutrition, programs for control of diseases, and rearing systems where predation is greatly reduced. Another factor is the production of relatively high-value products. This is particularly important in developing countries where exports of fish from aquaculture are valued at US$34 billion per year (FAO, 2015b).

TABLE 7.20   Top Aquaculture Species Globally by Production in 2012

Ranking	Species	Production (million metric tons)
1	Grass carp (*Ctenopharyngodon idellus*)	5.0
2	Silver carp (*Hypophthalmichthys molitrix*)	4.2
3	Common or pond carp (*Cyprinus carpio*)	3.8
4	Saltwater clam (*Ruditapes philippinarum*)[a]	3.8
6	Whiteleg or Pacific white shrimp (*Penaeus vannamei*)[a]	3.2
7	Bighead Carp (*Hypophthalmichthys nobilis*)	2.9
8	Major or Indian carp (*Catla catla*)	2.8
9	Crucian carp (*Carassius carassius*)	2.5
10	Atlantic salmon (*Salmo salar*)	2.1
11	Rohu or Labeo rohita, a carp-like fish (*Roho labeo*)	1.6
12	Milkfish (*Chanos chanos*)	0.9
13	Rainbow trout (*Oncorhynchus mykiss*)	0.9
14	Giant tiger prawn or Asian tiger shrimp (*Penaeus monodon*)[a]	0.9
15	Chinese razor clam (*Sinonovacula constricta*)[a]	0.7

[a] *Invertebrate aquaculture species covered in Section 10.6.2.2.*
From FAO, 2012. Fishery and aquaculture statistics 2012. Available from: http://www.fao.org/3/a-i3740t.pdf.

7.9 INTERSPECIES AND INTRASPECIES HYBRIDS

7.9.1 Overview

Interspecies and intraspecies hybrids are produced for domestic animals. These are particularly attractive as brood stock for fish for aquaculture.

Interspecies Hybrids

Interspecies hybrids are produced by crossing two distinct but related species. The cross can be of species within the same genus but this is not necessarily the case. Interspecies hybrids are most commonly infertile. Hybrids frequently express hybrid vigor with improved performance.

Intraspecies Hybrids

While *B. taurus* and *B. indices* may appear to be separate species, both were domesticated from a single wild species (see Section 6.6.2). Intraspecies crosses exhibit hybrid vigor whether for corn (see box), cattle, or meat-type chickens.

Hybrid Corn or Why Have Farmers Been Buying Seed From Seed Companies?

The simple answer at least for corn is that purchased seed is superior. The principle of hybrid vigor and its application to corn (maize) was described over a hundred years ago (Shull, 1908, 1909). Beginning in the 1920s, Henry A. Wallace (1880–1965) began selling hybrid corn seed (Crow, 1998). This was the first commercial production of hybrid corn seed. His company was Hi-Bred Corn that

later became Pioneer Hi-Bred, now a division of Dupont. There was rapid adoption of the new technology of hybrid corn by farmers with, for instance, 10% using it in 1935, then increasing to 90% in 1939 (Crow, 1998).

Annual Increase in Corn Production in the USA

- Prior to hybrid corn (open-pollinated) (1860–1930), 0.0012 metric tons ha^{-1}.
- Double cross hybrid corn (1930–60), 0.061 metric tons ha^{-1}.
- Single cross hybrid corn (1960–2000), 0.104 metric tons ha^{-1} (calculated from Crow, 1998).

7.9.2 Mules

Mules are an interspecies hybrid produced by the breeding of a male donkey (jack or jackass) with a female horse (mare). Mules have advantages over donkeys as pack animals (carrying freight) and people, particularly in mountainous areas. Horses have 64 chromosomes while donkeys have 62 chromosomes. Mules are the oldest known hybrid. It is suggested that mules were first produced in Northern Anatolia (present-day Turkey) and used in Ancient Egypt beginning 5000 years ago (Babb, 2015).

7.9.3 Beefalo (Cattlo)

According to the American Beefalo Association, beefalo are interspecies hybrids between beef cattle (*B. taurus*) (5/8) and American bison (*Bison bison*) (3/8). The hybrids are both successfully produced and fertile despite bison and beef cattle being from a different genus. However, the genetic separation between these bovids is only about 5 million years (Verkaar et al., 2004). Beefalo were first produced in about 1880 and are reputed to have greater hardiness, growth rates, and ease of calving (American Beefalo Association, 2016).

7.9.4 Hybrid Poultry

There is one interspecies in poultry, namely, the Moulard (or Moullard or Mulard). This is an infertile hybrid between domesticated muscovy (or muscovy ducks) (*Cairina moschata*) and domesticated Pekin ducks (*Anas platyrhynchos domestica*). Both the parent species have 80 chromosomes at the diploid stage with 14 macrochromosomes, 2 sex chromosomes, and 64 microchromosomes (Islam et al., 2014). Moulards are useful in the production of duck meat, particularly the breast meat steak (*Magret de Moullard*) and *foie gras* (fatty liver used in Pâté). Intraspecies crosses are used in broiler chickens with four grandparent lines that are crossed to produce two parent lines. The system takes advantage of the phenomenon of hybrid vigor.

7.9.5 Hybrid Finfish

Hybrid catfish are produced by crossing female channel catfish (*Ictalurus punctatus*) with the male blue catfish (*Ictalurus furcatus*) (Dunham and Masser, 2012). The hybrid catfish has markedly superior growth rate (>25%) (Dunham and Masser, 2012). Another hybrid catfish is produced by crossing the African (*Clarias gariepinus)* and Thai catfish (*Clarias macrocephalus*) (FAO, 1997).

Other hybrid fish include the following (FAO, 1997):

- Hybrid tilapia (*Oreochromis niloticus* × *Oreochromis aureus*). The offspring are predominantly male.
- Hybrid bass: striped bass (*Morone saxatilis*) × white bass (*Morone chrysops*) (Fig. 7.5).
- Hybrid trout: lake trout (*Salvelinus namaycush*) × brook trout (*Salvelinus fontinalis*).
- Hybrid salmon: Atlantic salmon (*Salmo salar)* × European or brown trout (*Salmo trutta*) with improved growths rates.

FIGURE 7.5 **Hybrid striped bass, also known as sunshine bass, is an interspecies hybrid of striped bass (M. saxatilis) × white bass (M. chrysops).** Production of fish by aquaculture is increasing and is projected to surpass harvested fish. The advantages of aquaculture include optimizing genetics, nutrition, and environment. *Source: Courtesy USDA Agricultural Research Service. Photo by Peggy Greg.*

- Hybrid salmon/trout: rainbow trout (*Oncorhynchus mykiss*) × coho salmon (*Oncorhynchus kisutch*).

7.10 PRODUCTION AND MARKETING SYSTEMS

7.10.1 Overview

The traditional method of marketing livestock and poultry is the spot or cash market

TABLE 7.21   Types of Production Systems

System	Responsibility for production decisions	Method of payment of farm operator
Spot or cash	Farm operator	Price negotiated at time of sale based on market price
Marketing contract	Farm operator but contract may specify quantity and quality to be delivered and time of delivery	Price determined prior to or during production
Production contract	Contractor controls some decisions	Farm operator paid for services
Vertical integration	Single company controls decisions	Farm operator/manager paid for performance

Based on MacDonald, L., Perry, J., Ahearn, M., Banker, D., Chambers, W., Dimitri, C., Key, N., Nelson, K., Southard, L., 2004. Contracts, markets, and prices: organizing the production and use of agricultural commodities. Agricultural Economic Report No. 837, November. Available from: www.ers.usda. gov/Publications/aer837/.

(Table 7.21). The individual farmer makes all the decisions with cash markets and takes all the financial risks. There have been shifts to other methods of linking the farmer with the market (Table 7.21). These include the following:

- Marketing contract is where a farmer and processor agrees to a price, delivery date, and quality terms well before marketing.
- Production contract is where a farmer or grower owns or rents the production facility and performs the day to day care of the livestock or poultry following a management guide developed by the integrator. The integrator usually owns and processes the animals. In addition, the integrator is responsible for some, or all, of the following: the animal genetics, feed, and veterinary care. The integrator may further process the meat, for example, to ready-to-eat meals. The grower is paid based on prearranged metrics, such as animal weight and mortality.
- Vertical integration (Fig. 7.6.) for broiler chicken production. "Vertical integration reduces quantity and/or quality risk, generates efficiencies in moving product through the system, and captures the profits from both levels of the production and marketing process. Integration assures raw materials or customers, and may help

avoid market power that might exist in supply or customer markets" (Hayenga et al., 2000), where animal genetics/breeding, livestock, or poultry production and its management, feeds, packing (slaughter and processing), and possibly further processing are all under a single company or corporate entity.

Production of meat-type chickens is either by production contracts or vertical integration (MacDonald et al., 2004). There was a rapid increase in marketing/production contracts in the 1990s (Table 7.22) with 2% packer-owned pigs in 1993 increasing to 25% in 2000 (MacDonald et al., 2004). These systems have the following effects (Hayenga et al., 2000):

1. The level of independence of, or control by, the individual farmer is greatest with cash markets and lowest with production contracts.

Independence of a farmer = Cash markets > Marketing contracts > Production contracts

2. The financial risk to the individual farmer is greatest with cash markets and lowest with production contracts.

Financial risk = Cash markets > Marketing contracts > Production contracts

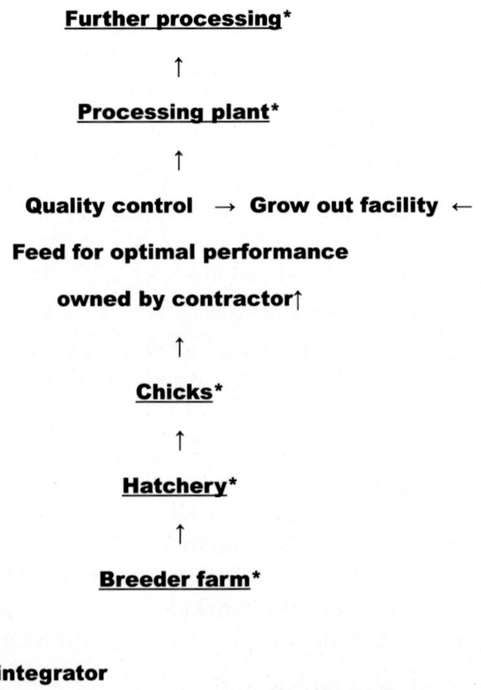

Further processing*

↑

Processing plant*

↑

Quality control   →   **Grow out facility**   ←

Feed for optimal performance

owned by contractor↑

↑

Chicks*

↑

Hatchery*

↑

Breeder farm*

* Owned by integrator

FIGURE 7.6   **A model of vertical integration in the production of meat-type chickens.** *Source: Based on National Chicken Council, in press. Available from: http://www.nationalchickencouncil.org/industry-issues/vertical-integration/.*

TABLE 7.22   Shift to Contracts and Vertical Integration in Pig Production in the USA in the 1990s

Year	Percentage contracts and vertical integration (%)
1993	13
1997	50
2000	75

Based on MacDonald, L., Perry, J., Ahearn, M., Banker, D., Chambers, W., Dimitri, C., Key, N., Nelson, K., Southard, L., 2004. Contracts, markets, and prices: organizing the production and use of agricultural commodities. Agricultural Economic Report No. 837, November. Available from: www.ers.usda.gov/Publications/aer837/.

The corollary is that there are differences in the access to capital.

3. The access to capital for the individual farmer is lowest for those who use cash markets and highest for production contracts.

Access to capital = Production contracts > Marketing contracts > Cash markets

7.11 IMPACTS OF LIVESTOCK AND POULTRY IN POVERTY ALLEVIATION

There is still much that can be done to improve agriculture particularly to raise the living standards of the poorest people on the planet. Possession of livestock provides a "safety net" to some of the poorest people globally, reducing their vulnerability. Organization, such as the Heifer Project and the Gates Foundation are providing cattle, sheep, goats, and chickens to alleviate poverty globally.

Livestock diseases can be devastating to the poorest people in the world. The death of a single cow is catastrophic to a family with one cow. Livestock pandemics can kill whole populations of livestock populations and devastate the economies and peoples of some of the poorest countries with, for instance, contagious bovine pleuropneumonia costing the South African country, Botswana, $300 million (Perry and Grace, 2009). Moreover, there can be closing of markets (domestic and export) to animal products leading to a collapse in prices (Perry and Grace, 2009).

Another example of an animal disease impacting some of the poorest people is Trypanosomosis. This severely impacts cattle production in Africa with reductions in calving or kidding or lambing rates, calf survival, and milk production (FAO, 2005). Another disease impacting producers particularly in developing countries is the viral disease, foot and mouth disease (FMD). There have been efforts to eradicate FMD with success having been attained in North America and most of Europe. However, FMD remains a major problem for livestock producers globally with costs estimated at between US$6.5 and 21 billion (Knight-Jones and Rushton, 2013).

7.12 CONCLUSIONS

Production of meat, milk, and eggs has been increasing tremendously in the last 50 years (Table 7.1) building on the advances in yield of feed crops including corn/maize (Table 7.2). It is argued that there have been two transitions in livestock and poultry production:

- The Neolithic Revolution with the domestication of livestock and poultry and their widespread use (discussed in Chapter 6).
- The science-based expansion of production. The sciences include in order of importance genetics (discussed in an essay in this volume by Paul Siegel), nutrition (animal feeds are discussed in an essay by Lucy Waldron in this volume), and multidisciplinary approaches for disease control.

References

Agricultural Marketing Service, 2012. National Organic Program; Amendments to the National List of Allowed and Prohibited Substances (Livestock). Federal Register 77, No. 94,28472.

Agricultural Marketing Service, 2015. Parasiticides: Fenbendazole, Ivermectin, Moxidectin. Available from: https://www.ams.usda.gov/sites/default/files/media/Fenbendazole%20TR%202015.pdf

Agricultural Marketing Service, 2017. Organic requirements. Available from: https://www.ams.usda.gov/sites/default/files/media/Organic%20Livestock%20Requirements.pdf

Agricultural Sustainability Institute, 2017. Available from: http://asi.ucdavis.edu/programs/sarep/about/what-is-sustainable-agriculture

American Beefalo Association, 2016. Available from: http://americanbeefalo.org

Anderson, K.E., 2011. Comparison of fatty acid, cholesterol, and vitamin A and E composition in eggs from hens housed in conventional cage and range production facilities. Poult. Sci. 90, 1600–1608.

AVMA, 2015. Welfare issues. Available from: https://www.avma.org/KB/Resources/LiteratureReviews/Pages/Welfare-Implications-of-Gestation-Sow-Housing.aspx

Ayuko, L.J., 1981. Assessment of potential for rangeland improvement. Available from: http://www.fao.org/wairdocs/ilri/x5543b/x5543b16.htm

Babb, D., 2015. History of the mule. American Mule Museum. Available from: http://www.mulemuseum.org/history-of-the-mule.html

Barański, M., Srednicka-Tober, D., Volakakis, N., Seal, C., Sanderson, R., Stewart, G.B., Benbrook, C., Biavati, B., Markellou, E., Giotis, C., Gromadzka-Ostrowska, J., Rembiałkowska, E., Skwarło-Sońta, K., Tahvonen, R., Janovská, D., Niggli, U., Nicot, P., Leifert, C., 2014. Higher antioxidant and lower cadmium concentrations and lower incidence of pesticide residues in organically grown crops: a systematic literature review and meta-analyses. Br. J. Nutr. 112, 794–811.

Bergman, E.N., 1990. Energy contributions of volatile fatty acids from the gastrointestinal tract in various species. Physiol. Rev. 70, 567–590.

Bestman, M., Wagenaar, J.P., 2014. Health and welfare in Dutch organic laying hens. Animals (Basel) 4, 374–390.

Brambell Report, 1965. Report of the Technical Committee to Enquire Into the Welfare of Animals Kept Under Intensive Livestock Husbandry Systems. Her Majesty's Stationery Office, London, UK.

Brulc, J.M., Antonopoulos, D.A., Miller, M.E., Wilson, M.K., Yannarell, A.C., Dinsdale, E.A., Edwards, R.E., Frank, E.D., Emerson, J.B., Wacklin, P., Coutinho, P.M., Henrissat, B., Nelson, K.E., White, B.A., 2009. Gene-centric metagenomics of the fiber-adherent bovine rumen micro-biome reveals forage specific glycoside hydrolases. Proc. Natl. Acad. Sci. USA 106, 1948–1953.

Chamberlain, D.E., Joys, A., Johnson, P.J., Norton, L., Feber, R.E., Fuller, R.J., 2010. Does organic farming benefit farmland birds in winter? Biol. Lett. 6, 82–84.

Coffey, L., Baier, A.H., 2012. Guide for organic livestock producers. Available from: https://attra.ncat.org/attra-pub/summaries/summary.php?pub=154

Connor, D., 2008. Organic agriculture cannot feed the world. Field Crops Res. 106, 187–190.

Council on Foreign Relations, 2015. The World Economy in 2016: Watch China. Available from: https://www.cfr.org/expert-brief/world-economy-2016-watch-china

Creevey, C.J., Kelly, W.J., Henderson, G., Leahy, S.C., 2014. Determining the culturability of the rumen bacterial microbiome. Microb. Biotechnol. 7, 467–479.

Crow, J.F., 1998. 90 years ago: the beginning of hybrid maize. Genetics 148, 923–928.

Dangour, A.D., Dodhia, S.K., Hayter, A., Allen, E., Lock, K., Uauy, R., 2009. Nutritional quality of organic foods: a systematic review. Am. J. Clin. Nutr. 90, 680–685.

Dimitri, C., Effland, A., Conklin, N., 2005. The 20th Century Transformation of U.S. Agriculture and Farm Policy. Economic Information Bulletin Number 3. USDA Economic Research Service. Available from: http://www.ers.usda.gov/media/259572/eib3_1_pdf

Dunham, R., Masser, M., 2012. Production of hybrid catfish. Southern Regional Aquaculture Center, USA, publication 190.

Economist, 2015. The 169 commandments: the proposed sustainable development goals would be worse than useless. Available from: http://www.economist.com/news/leaders/21647286-proposed-sustainable-development-goals-would-be-worse-useless-169-commandments

EPA (Environmental Protection Agency), 2015. Animal feeding operations. Available from: http://cfpub.epa.gov/npdes/home.cfm?program_id=7

ERS, 2015. Available from: http://www.ers.usda.gov/data-products/ag-and-food-statistics-charting-the-essentials/ag-and-food-sectors-and-the-economy.aspx

European Union, 1999. Available from: http://ec.europa.eu/food/animals/welfare/practice/farm/laying_hens/index_en.htm

FAO, 1997. Available from: http://www.fao.org/docrep/005/w7611e/w7611e7.htm

FAO, 2005. Socio-economic consequences for poor livestock farmers of animal diseases and VPH problems. Available from: http://www.fao.org/docrep/005/y3542e/y3542e04.htm

FAO, 2007. Camels and camel milk. Feeds and digestions. Available from: http://www.fao.org/docrep/003/x6528e/x6528e07.htm

FAO, 2015a. World Agriculture: Towards 2015/2030. An FAO perspective. Available from: http://www.fao.org/docrep/005/y4252e/y4252e07.htm

FAO, 2015b. The State of World Fisheries and Aquaculture. Available from: http://www.fao.org/3/a-i3720e.pdf

FAOStats, 2016. United Nations Food and Agriculture Organization Statistics. Available from: http://faostat3.fao.org/browse/Q/*/E; http://faostat.fao.org/site/535/default.aspx#ancor

Farm Animal Welfare Council, 2012. Available from: http://webarchive.nationalarchives.gov.uk/20121007104210/http:/www.fawc.org.uk/freedoms.htm

Fernandez-Cornejo, J., Hallahan, C., Nehring, R., Wechsler, S., Grube, A., 2013. Conservation tillage, herbicide use, and genetically engineered crops in the United States: the case of soybeans. AgBioForum 15, 231–241.

Fernandez-Cornejo, J., Wechsler, S., Livingston, M., Mitchell, L., 2014. Genetically Engineered Crops in the United States. Economic Research Report No. 162, 60 pp.

Gardner, B.L., 2002. American Agriculture in the Twentieth Century: How It Flourished and What It Cost. Harvard University Press, Cambridge, MA.

Grandin, T., 1997. Assessment of stress during handling and transport. J. Anim. Sci. 75, 249–257.

Havenstein, G.B., Ferket, P.R., Qureshi, M.A., 2003. Growth, livability, and feed conversion of 1957 versus 2001 broilers when fed representative 1957 and 2001 broiler diets. Poult. Sci. 82, 1500–1508.

Hayenga, M., Schroeder, T., Lawrence, J., Hayes, D., Vukina, T., Ward, C., Purcell, W., 2000. Meat packer vertical integration and contract linkages in the beef and pork industries: an economic perspective. Special Report for the American Meat Institute, May 22. Available from: http://www2.econ.iastate.edu/faculty/hayenga/AMIfullreport.pdf

Hodgson, J.A., Kunin, W.E., Thomas, C.D., Benton, T.G., Gabriel, D., 2010. Comparing organic farming and land sparing: optimizing yield and butterfly populations at a landscape scale. Ecol. Lett. 13, 1358–1367.

Hovi, M., Sundrum, A., Thamsborg, S.M., 2003. Animal health and welfare in organic livestock production in Europe: current state and future challenges. Livest. Prod. Sci. 80, 41–53.

HSUS, 2010. An HSUS report: the impact of industrialized animal agriculture on rural communities. Available from: http://www.humanesociety.org/assets/pdfs/farm/hsus-the-impact-of-industrialized-animal-agriculture-on-rural-communities.pdf

Hughner, R.S., McDonagh, P., Prothero, A., Shultz, II., C.J., Stanton, J., 2007. Who are organic food consumers? A compilation and review of why people purchase organic food. J. Consum. Behav. 6, 94–110.

Islam, F.B., Uno, Y., Nunome, M., Nishimura, O., Tarui, H., Agata, K., Matsuda, Y., 2014. Comparison of the chromosome structures between the chicken and three anserid species, the domestic duck (*Anas platyrhynchos*), Muscovy duck (*Cairina moschata*), and Chinese goose (*Anser cygnoides*), and the delineation of their karyotype evolution by comparative chromosome mapping. J. Poult. Sci. 51, 1–13.

Karcher, D.M., Jones, D.R., Abdo, Z., Zhao, Y., Shepherd, T.A., Xin, H., 2015. Impact of commercial housing systems and nutrient and energy intake on laying hen performance and egg quality parameters. Poult. Sci. 94, 485–501.

Kijlstra, A., Eijck, I.A.J.M., 2006. Animal health in organic livestock production systems: a review. J. R. Neth. Soc. Agric. Sci. 54, 77–94.

Knight-Jones, T.J.D., Rushton, J., 2013. The economic impacts of foot and mouth disease—what are they, how big are they and where do they occur? Prev. Vet. Med. 112, 161–173.

Kuepper, G., 2010. A brief overview of the history and philosophy of organic agriculture. Kerr Center for Sustainable Agriculture. Available from: http://kerrcenter.com/wp-content/uploads/2014/08/organic-philosophy-report.pdf

Leenstra, F., Maurer, V., Bestman, M., van Sambeek, F., Zeltner, E., Reuvekamp, B., Galea, F., van Niekerk, T., 2012. Performance of commercial laying hen genotypes on free range and organic farms in Switzerland, France and The Netherlands. Br. Poult. Sci. 53, 282–290.

Li, W., Li, Y., 2012. Managing rangeland as a complex system: how government interventions decouple social systems from ecological systems. Ecol. Soc. 17, 9.

MacDonald, J.M., Newton, D., 2014. Milk Production Continues Shifting to Large-Scale Farms. Available from: http://www.ers.usda.gov/amber-waves/2014-december/milk-production-continues-shifting-to-large-scale-farms.aspx#.Vv1_NmA_N7M

MacDonald, L., Perry, J., Ahearn, M., Banker, D., Chambers, W., Dimitri, C., Key, N., Nelson, K., Southard, L., 2004. Contracts, markets, and prices: organizing the production and use of agricultural commodities. Agricultural Economic Report No. 837, November. Available from: www.ers.usda.gov/Publications/aer837/

Matson, P.A., Parton, W.J., Power, A.G., Swift, M.J., 1997. Agricultural intensification and ecosystem properties. Science 277, 504–509.

Morton, D.B., 2007. Vaccines and animal welfare. Rev Sci Tech Off Int Epiz 26, 157–163.

NASS (National Agricultural Statistics Service of USDA), 2016. Catfish production. Available from: http://usda.mannlib.cornell.edu/usda/current/CatfProd/CatfProd-02-05-2016.pdf

News Staff, 2010. Organic food—what it means and the list of artificial ingredients allowed. Available from: http://www.science20.com/news_articles/organic_food_what_it_means_and_list_artificial_ingredients_allowed-82014

Northbourne, L, 1940. Look to the Land. J.M. Dent, London.

Perry, B., Grace, D., 2009. The impacts of livestock diseases and their control on growth and development processes that are pro-poor. Philos. Trans. R. Soc. B Biol. Sci. 364, 2643–2655.

Perry, B.D., Grace, D., Sones, K., 2013. Current drivers and future directions of global livestock disease dynamics. Proc. Natl. Acad. Sci. USA 110, 20871–20877.

Phrases UK, 2016. Available from: http://www.phrases.org.uk/meanings/bring-home-the-bacon.html

Pimental, D., Hepperly, P., Hanson, J., Douds, D., Seidel, R., 2005. Environmental, energetic, and economic comparisons of organic and conventional farming systems. Bioscience 55, 573–582.

Seufert, V., Ramankutty, N., Foley, J.A., 2012. Comparing the yields of organic and conventional agriculture. Nature 485, 229–232.

Sharp, W.M., Johnson, R.R., Owens, F.N., 1982. Ruminal VFA production with steers fed whole or ground corn grain. J. Anim. Sci. 55, 1505–1514.

Shaw Hughner, R., McDonagh, P., Prothero, A., Shultz, II, C.J., Stanton, J., 2007. Who are organic food consumers? A compilation and review of why people purchase organic food. J. Consum. Behav. 6, 1–17.

Sherwin, C.M., Nasr, M.A., Gale, E., Petek, M., Stafford, K., Turp, M., Coles, G.C., 2013. Prevalence of nematode infection and faecal egg counts in free-range laying hens: relations to housing and husbandry. Br. Poult. Sci. 54, 12–23.

Shull, G.H., 1908. The composition of a field of maize. Am. Breeders Assoc. Rep. 4, 296–301.

Shull, G.H., 1909. A pure line method of corn breeding. Am. Breeders Assoc. Rep. 5, 51–59.

Smith, H.G., Dänhardt, J., Lindström, A., Rundlöf, M., 2010. Consequences of organic farming and landscape heterogeneity for species richness and abundance of farmland birds. Oecologia 162, 1071–1079.

Soil Association, 2017. Available from: http://www.soilassociation.org/whatisorganic/organicanimals/antibiotics

Średnicka-Tober, D., Barański, M., Seal, C., Sanderson, R., Benbrook, C., Steinshamn, H., Gromadzka-Ostrowska, J., Rembiałkowska, E., Skwarło-Sońta, K., Eyre, M., Cozzi, G., Krogh Larsen, M., Jordon, T., Niggli, U., Sakowski, T., Calder, P.C., Burdge, G.C., Sotiraki, S., Stefanakis, A., Yolcu, H., Stergiadis, S., Chatzidimitriou, E., Butler, G., Stewart, G., Leifert, C., 2016a. Composition differences between organic and conventional meat: a systematic literature review and meta-analysis. Br. J. Nutr. 115, 994–1011.

Średnicka-Tober, D., Barański, M., Seal, C.J., Sanderson, R., Benbrook, C., Steinshamn, H., Gromadzka-Ostrowska, J., Rembiałkowska, E., Skwarło-Sońta, K., Eyre, M., Cozzi, G., Larsen, M.K., Jordon, T., Niggli, U., Sakowski, T.,

Calder, P.C., Burdge, G.C., Sotiraki, S., Stefanakis, A., Stergiadis, S., Yolcu, H., Chatzidimitriou, E., Butler, G., Stewart, G., Leifert, C., 2016b. Higher PUFA and n-3 PUFA, conjugated linoleic acid, α-tocopherol and iron, but lower iodine and selenium concentrations in organic milk: a systematic literature review and meta- and redundancy analyses. Br. J. Nutr. 115, 1043–1060.

Sutherland, M.A., Webster, J., Sutherland, I., 2013. Animal health and welfare issues facing organic production systems. Animals (Basel) 3, 1021–1035.

Tactacan, G.B., Guenter, W., Lewis, N.J., Rodriguez-Lecompte, J.C., House, J.D., 2009. Performance and welfare of laying hens in conventional and enriched cages. Poult. Sci. 88, 698–707.

USDA, 1996. Available from: http://www.usda.gov/oce/sustainable/Council%20Memorandum.pdf

USDA, 2017. Organic Agriculture. Available from: http://www.usda.gov/wps/portal/usda/usdahome?contentidonly=true&contentid=organic-agriculture.html

Verkaar, E.L.C., Nijman, I.J., Beeke, M., Hanekamp, E., Lenstra, J.A., 2004. Maternal and paternal lineages in cross-breeding bovine species. Has wisent a hybrid origin? Mol. Biol. Evol. 21, 1165–1170.

von Borell, E., Sørensen, J.T., 2004. Organic livestock production in Europe: aims, rules and trends with special emphasis on animal health and welfare. Livest. Prod. Sci. 90, 3–9.

West, T.O., Post, W.M., 2002. Soil organic carbon sequestration rates by tillage and crop rotation: a global data analysis. Soil Sci. Soc. Am. J. 66, 1930–1946.

Further Reading

Bureau of Labor Statistics, 2014. Available from: http://www.bls.gov/opub/mlr/2014/article/agriculture-occupational-employment-and-wages.htm

United Nations Department of Economic and Social Affairs, 2017. Available from: https://sustainabledevelopment.un.org/?page=view&nr=164&type=230&menu=2059

THE AVATARS OF ANIMAL AGRICULTURE: THE GLOBAL IMPACT OF LIVESTOCK IN ALLEVIATING HUNGER

Kenneth M. Quinn

The World Food Prize Foundation, Des Moines, IA, United States

The first essential component of social justice is adequate food for all mankind.

–*Dr. Norman E. Borlaug, Founder of the World Food Prize*

Visitors to the World Food Prize Hall of Laureates located in Des Moines, Iowa are treated to a profusion of artwork and exhibits, including murals, mosaics, a 20-foot tall stained glass window, sculptures, paintings, and interactive displays arrayed on the building's 3 levels, all designed to tell the story of human food production through the ages.

As part of this exposition, inscribed in gold leaf in the squinches of the Ruan Laureate Room on the main level of the Hall are the names of 21 diverse individuals judged to have played a significant role throughout the history of food production and agriculture. Among these iconic figures are names like Mendel (genetics), Burbank (horticulture), Pasteur (food safety), and Von Liebig (fertilizer).

Animal husbandry was the one category for which it seemed almost impossible to identify a well-known pioneer. The difficulty in finding a key person in animal agriculture may stem from the fact that the domestication of animals for food began in the Middle East more than 10,000 years ago. As they provided a constant source of protein-rich food, animals contributed to the spread of human civilization across the globe in the following millennia. The horse was domesticated on the steppes of modern-day Ukraine approximately 6000 years ago. Livestock production and animal traction developed into an integral part of farming practices and agricultural economies from ancient times up to our present day, while animal products have been a mainstay in the diet of rich societies and sought after—but not always attained—by poorer ones. It was only in the 1950s that over 50% of the plowing in the world was done by mechanical means.

Eventually, after considerable research, the name Walstan was settled on. Saint Walstan, born in England in the late 10th century, was revered as the patron saint of farms, farmers, ranchers, and husbandrymen. Walstan had dedicated his life and service to farming and especially to the care of farm animals. He died in the pasture tending his cattle, and his devotion was such that as his body was brought back on a cart pulled by oxen, legend has it that a spring of fresh water miraculously appeared on the spot where the animals stopped. Walstan represented all of those countless humans who had contributed to shaping one of the most critical elements of human development—animal agriculture.

There has not, however, been such a problem of identifying modern avatars of animal agriculture. Over the past 30 years, The World Food Prize has honored a number of exceptional laureates for achievements in animal agriculture and animal science—underscoring the critical importance of livestock in providing adequate food and nutrition for human populations around the world.

Among those laureates who have been recognized for breakthrough achievements in increasing the quantity, quality, or availability of animals raised for food are Verghese Kurien of India in 1989; Edward F. Knipling and Raymond C. Bushland of the United States in 1992; Walter Plowright of the United Kingdom in 1999; and Jo Luck of the United States in 2010.

- Dr. Kurien established a program known as "Operation Flood," which organized dairy farmer cooperatives that produced, processed, and marketed milk for rural and urban areas in India. Reaching nearly 250 million people in the 1960s and 1970s, Operation Flood was the largest agricultural development program of its kind in the world. Kurien's role as the "Father of the White Revolution" transformed India from a milk-deficient nation to the world's largest milk producer by the end of the 20th Century. Dairy farming became India's largest self-sustaining industry and provided millions of children and adults with high-quality nutrition.

- Drs. Knipling and Bushland developed the Sterile Insect Technique (SIT), which was first used to combat screwworms, insects that prey on warm-blooded animals, especially cattle herds. Proven effective in controlling outbreaks of a wide range of insect pests throughout the world, SIT has been a breakthrough in protecting animals and agricultural products to feed the world's human population and has eliminated screwworm in North and most of Central America. Significantly, this approach is also the model upon which some efforts to control the Zika virus are being developed.

- As the world's population and demand for food and agricultural resources grew, the eradication of major animal plagues has been crucial. Dr. Plowright developed a practical vaccine against the devastating cattle disease rinderpest, commonly known as cattle plague, which has eradicated it from most regions of the world. This contribution to the world's food supply has been staggering: statistics show that during the 30 years following the initiation of Plowright's rinderpest vaccinations in the 1960s, over 70 million tons of meat and more than 1 billion tons of milk were added to food production totals in the developing world. The increase in healthy cattle—integral throughout Asia and Africa for fertilizing soils, planting and cultivating crops, and carrying loads—also boosted production rates on subsistence farms worldwide.

- Jo Luck spearheaded the effort to build the nongovernmental organization Heifer International into one of the premier hunger-fighting organizations anywhere in the world, bringing food- and income-producing animals to extremely poor families in Africa, Asia, and Latin America. These projects have guided families to self-reliance, and provided the opportunity for improved livelihoods through animal husbandry, technical training, and community training. Her innovative approaches at Heifer included issuing a call to action to grassroots supporters to make individual financial contributions that collectively sponsored more than 30 kinds of livestock and animals—from bees to water buffaloes—along with trees, seeds, and training that are provided to recipients.

- Jo Luck witnessed firsthand the empowerment of smallholder farmers through livestock when she said: "I was in Thailand, and I saw the oldest little woman… She was out there hanging onto this thing behind this great big water buffalo and just working her heart out. She said 'let me tell you something: if I die my family will miss me terribly, but if this water buffalo dies, they lose all their hope, future and success and means of income.' You know, she knew the truth, and she understood the truth. And that's what agriculture and animal agriculture do."

The global impact of livestock in agriculture and development has been a topic of discussion by experts at the World Food Prize Norman

Borlaug Dialogue international symposiums, which take place in Des Moines every mid-October around UN World Food Day. The following are highlighted points from recent conferences.

- Commenting on the nutritional value of consuming protein through meat, milk, and eggs, Dr. Christopher Nelson, CEO of Kemin Industries, a global nutrition ingredients manufacturer, said "Chronic malnutrition oftentimes has to do with lack of protein. A person needs 52 grams of protein daily (including 240 milligrams of lysine and 80 milligrams of methionine), and although vegetable sources can provide many of the protein components, the quantities are oftentimes insufficient to meet requirements for overall human development—which is especially important for growing children."

- Economist Dr. Robert Thompson expressed concern about how the world is going to keep pace with a growing demand for animal protein in the diets of newly developed countries in Asia and Africa: "We have had major productivity growth in animal agriculture. In dairy farming we're producing almost five times as much milk per cow as we did at the end of World War II—with 80 percent less feed. But, we've got to keep raising productivity in animal agriculture, as well as in plant agriculture if we are going to feed the world's larger population better than today at reasonable cost without destroying the environment."

- The managing director of Godrei Agrovet, an animal feed and agribusiness in India, Mr. Balram Yadav, expressed a similar concern with respect to the fastest growing country in the world: "As India becomes richer and its middle class expands, we have seen a rapid growth of animal protein industry. Though our per capita consumption of animal protein is much lower than the world average and

even the recommendations of WHO, it is rising rapidly. From 2008 to 2013, 37 percent of agricultural growth output came from animal protein. Within this category, the output of poultry has risen the fastest."

- A farmer and professor at Zamorano University in Honduras, Dr. Isidro Ochoa, talked about the comparatively quick recovery by the dairy industry in that country after the devastating effects to all agriculture due to Hurricane Mitch in 1998. He further stated that since then, various constraints—water scarcity, high feed costs, diseases, and inadequate genetic research—have made it difficult for livestock producers to deliver better milk and meat products.

- Mr. Pierre Ferrari, President at Heifer International, emphasized that livestock can bridge the gap between "values-based holistic development" that asks "How can smallholder agriculture achieve the necessary scale so as to be able to feed the world and cool the planet" and "market development mechanisms" that ask "How can large-scale land or other investments benefit smallholder farmers and the rural poor." For example, a partnership between the Bill and Melinda Gates Foundation and Heifer International mobilized 180,000 dairy farmers in East Africa. Upon meeting with the board of directors of a 7000-member dairy group in Kenya, Mr. Ferrari was struck by "the sophistication of the chairman of the board, who had been a two-cow farmer four years ago and now was in discussion on balance sheet construction and repayment of debt."

- The topic of zoonotic diseases—animals to humans—was touched upon by Mr. Shouchun Wang, founder and chairman of the Shandong Xiantan Company, the leading broiler processor in Shandong Province: "In 2012, avian influenza impacted the poultry industry both large-scale and small holder producers." "...for our farmers what we must do is to do a better job of preventing disease outbreaks among flocks...and we hope from the efforts of the government, scientists, and also farmers that the problem can be solved in the future."

- Mr. Jeff Simmons, President of the animal health company Elanco, pointed to the rapidly growing demand by developing countries to include more meat in their diets: "There are three billion people on our planet who are striving to cross the chasm from an exclusively plant-based diet to eating meat, milk, and eggs for the first time." He cited a worrisome trend of decreasing egg production (fewer eggs per hen), which, in his view, will make it difficult for poultry and egg production to keep pace with the increase in demand from a growing, more affluent population, unless technology and innovation are applied in the livestock industry worldwide.
- Concerning the role of livestock in environmental stewardship, mediation and agrobiodiversity consultant Ms. Anita Idel stated that "To be protective for the climate, cattle need good grasslands. And good grasslands need cattle. And both need good grazing management... Being ruminants, cattle do what we can't. We would die being fed on grass and hay. But cattle don't only survive this way, but in digesting grass and hay they are producing milk and meat. Fed with grass and hay, cattle are not in competition with human food needs... Grazing is a crucial management tool to enhance the vigor of mature perennial grasses."
- Dr. Kwesi Atta-Krah, former Director General of Bioversity International, emphasized the environmental synergies of crop/livestock integration: "I have seen projects in Kenya where just the insertion of one dairy cow into the farming system of a household brings transformation... the positive link between livestock systems where the manure is actually managed and used, where trees, such as lucina are planted, which provide a fodder source—these elements create a well integrated system."
- The manifold benefits of livestock to smallholder farmers was further affirmed by Mr. Sahr Lebbie, Africa Area Vice President at Heifer International: "we have identified four M's when it comes to livestock's impact—milk, money (because the more of it they can produce, the more they are able to sell some surplus). The third M is manure, which they use to fertilize their lands. And the fourth is muscle, because many of these farmers use oxen to plough their lands."

As our species confronts what we consider the single greatest challenge in human history—can we sustainably and nutritiously feed the close to 10 billion people who will be on our planet by the year 2050?—it seems clear that animal agriculture will need to play an increasingly important role as an important source of proteins and vitamins in the global diet. As science endeavors to improve nutrition through the biofortificaiton of plants and food crops, animals stand as impressive examples of an historic biofortification process dating back thousands of years and continuing to the present time.

This legacy, and the accomplishments of so many World Food Prize laureates, should inspire the next generation of animal scientists and husbandryfolk as they finish the work of mitigating livestock's contributions to climate change, sustainably managing our grasslands and pastures, delivering on the human rights of herders and nomads, and ensuring that our crop and animal resources are justly produced and equitably distributed to a growing and hungry population.

THOUGHTS ON THE EVOLUTION OF THE DOMESTIC CHICKEN AS FOOD FOR HUMANS

Paul Siegel

Virginia Tech, Blacksburg, VA, United States

The Neolithic period of 8,000–10,000 years ago saw the initial development of animal husbandry including the beginning of the domestication of chickens. Initially popular for sport, cultural, and religious reasons, the evolution of the domestic chicken as part of the human diet is a complex process (Lawler, 2014; Smith and Daniel, 1975; Stevens, 1991). Although originating in the jungles of Southeast Asia, today chickens thrive throughout the world (Robinson et al., 2014) under a plethora of husbandry settings demonstrating their plasticity in adaptability.

Promiscuity, social groupings of males and females, maternal behavior, incubation behavior, precocial young, and general dietary habits were traits that initially favored the domestication of the chicken (Hale, 1969). Today, due to human intervention, however, some are no longer relevant. For example, the development of artificial incubation and brooding systems has made incubation and maternal behaviors moot.

Great changes in commercial poultry production during the last half of the 20th century and the beginning of the 21st century have resulted in the development of a highly sophisticated industry. The fixed wing aircraft has allowed for rapid movement of eggs and chicks so that similar, if not the same, egg and meat stocks are produced globally under a range of husbandry settings from low to high intensities. Although it continues to have an important role, the popular dual-purpose chicken of the 19th and early 20th centuries has been superseded by "human-driven artificial selection" of stocks that excel in the production of either eggs or meat. This change from dual-purpose to specialized meat and egg stocks was an outcome based on sound science. Pedigree selection and use of heterosis allowed geneticists to develop populations of chickens that efficiently allocated resources to growth (meat) or reproduction (egg yield). Of course, a balance was needed because a "chicken is still a chicken" and although emphasis was on efficient egg or meat production, biologically egg stocks needed to grow and meat stocks had to reproduce.

Advances in veterinary medicine, nutrition, physiology, and immunology, interacting in concert with breeding programs, facilitated the expression of the genetic potential of chickens bred specifically for maximizing egg or meat production under a range of environments. These synergies facilitated the advances in performance of commercial egg and meat chickens. The past half century has seen an increase in total egg production of one egg per year in egg stocks (Pelletier et al., 2014). During the same time period, broilers attained a market weight of 2.2 kg 1 day earlier each year (Collins et al., 2014; Havenstein et al., 2003; Zuidhof et al., 2014). These consistent incremental changes involving an efficient transfer of plant to animal protein over an extended period of years have had considerable environmental consequences in land utilization and waste. Whether this rate of change can continue in the face of feeding an expanding middle class world population that has a preference for animal protein, is, in my opinion, problematic. This opinion has relevancy in the context that chicken meat and eggs are an animal source of food with little religious or cultural taboos and that concomitant with improved economic conditions, there are increases in consumption of meat. Therefore, while chickens lend themselves to a wide range of production practices making them globally accessible, it is not realistic to assume that they will be the panacea for providing inexpensive animal protein for feeding an expanding demographically diverse world population.

Although the chicken has a long history as a food source, the basis for rapid advances in efficiency of production seen in recent years is built on a sound scientific foundation (Siegel, 2014). There are numerous examples of the advantages of the chicken as a model organism for animal research. That embryogenesis occurs outside the body of the mother has great value for research in developmental biology. Contributions to genetics commenced during the first decade of the 20th century are legendary. Bateson demonstrated Mendelian genetics in animals, and the chicken was the first example of complementary gene action (Bateson and Punnett, 1905–1908). Other examples where the chicken was the basis for scientific inquiry include behavior (the peck-order), biochemistry (folic acid), virology (Rous sarcoma), and immunology (B-cells).

The emergence of the chicken as a source of food coincided with that of an expanding human population. The parallelisms during the past and current centuries have been remarkable. The rediscovery of Mendelian genetics at the turn of the 20th century laid the basis for its application in the 1950s and, in concert with advances in computer and molecular technologies, facilitated the emergence of stocks specifically bred for egg or meat production. Although the role of dual-purpose and indigenous chickens is diminishing, they still remain an important source of meat and eggs where efficient conversion of feedstuffs to either meat or eggs has less relevance.

As we look to the future with the "carbon footprint" and "water footprint" of an increasing human population worldwide coupled with reductions in arable land and water, efficient production of poultry becomes increasingly relevant. Yet, are limits being approached? Namely, will the genetic changes made in the latter half of the last century and those in the beginning of the current century abate? If so, when? The

balance of breeding for meat or eggs is delicate, that is, meat stocks must reproduce and egg stocks must grow. Cloning of individuals, which results in a lack of genetic variation, is not a viable alternative because the reproductive cycle of the chicken while short for farm animals is slow relative to pathogens making clones vulnerable in host–pathogen relationships. Overall, a lack of genetic variation in individuals and populations increases their vulnerability to environmental insults.

In the context of time, the changes seen in the recent past do not necessarily reflect those for the long term. For example, a plethora of Food and Agriculture Organization reports suggest short term (e.g., 10 years) that poultry meat production will grow more than 2% annually. How will this occur? Via numbers of chickens, efficiency of production, genetics…?? This prediction does not imply that conversion of plant to animal protein will improve at this rate in short term, and certainly not long term to alleviate the increased demand on the efficient production of foodstuffs. Not addressed is whether increases in numbers of humans will parallel changes in living standards and increases in meat consumption.

There is also the caveat when viewing percentages, "percentage is the fertile mother of fallacy" (source unknown). Enhancing the efficiency of poultry production is a process and to view it as a destination in providing animal protein to an ever increasing and economically diverse world population is not realistic. During the last 50+ years we have seen increases in egg production of 20%, and improved feed efficiency (egg mass/feed consumed) of 35% in egg stocks. For meat stocks, the time to reach market weight has been reduced by 50% as has feed conversion. It is not realistic to assume that this rate of change will continue for the next 50 years. A further caveat is that these changes have not occurred globally, but on a regional basis where there is an expanding middle class in the developing nations—not necessarily where the greatest reproductive increases are occurring in human populations. Moreover, increased consumption of a product can occur because people prefer it, it is less expensive than competing products, or both. Again, the percentage caveat—namely, consumption per person or total (can increase with an increasing population).

Viewed on a regional basis, caution should be exercised when there are changes in percentages because denominators are not always the same, and reports of percentage changes in percentages are not valid. As we look to the future, the calculus involves quantitative and qualitative availability of land, and perhaps more important, water. Climate-wise, the world is not static and longer-term complexities come into play. With reductions in land due to human population increases and competition for water, will costs of production of poultry meat and eggs increase? Will husbandry systems go vertical to meet competition for available land for plants and humans? With vertical concentration of production, will disease control become a greater challenge?

There will be a competition for resources between an expanding human population and the production of food to feed them. Malthus was probably correct—the difficulty is in predicting when it will occur. Perhaps we are already seeing unsustainable numbers on a regional basis with the mass migrations of human populations from the less to the more developed nations. While poultry is important in feeding an expanding world population, it is only one component in the paradigm.

References

Bateson, W., Punnett, R.C., 1905. Experimental studies in the physiology of heredity. In: Peters, J.A. (Ed.), Classical Papers in Genetics. Prentice-Hall, Englewood Cliffs, NJ, pp. 42–59.

Collins, K.E., Klepper, B.H., Ritz, C.W., McLendon, B.L., Wilson, S.L., 2014. Growth, livability, feed consumption, and carcass composition of the Athens-Canadian Random Bred 1955 meat-type chicken versus the 2012 high-yielding Cobb-500 broiler. Poult. Sci. 93, 2953–2962.

Hale, E.B., 1969. Domestication and the evolution of behavior. In: Hafez, E.S.E. (Ed.), The Behavior of Domestic Animals. Williams and Wilkins, Baltimore, MD, pp. 22–42.

Havenstein, G.B., Ferket, P.R., Qureshi, M.A., 2003. Growth livability and feed conversion of 1957 versus 2001 broilers when fed representative 1957 and 2001 broiler diets. Poult. Sci. 82, 1502–1508.

Lawler, A., 2014. Why Did the Chicken Cross the World? Atria Books–Simon and Schuster, New York, NY.

Pelletier, N., Ibarburu, M., Xin, H., 2014. Comparisons of the environmental footprint of the egg industry in the United States in 1960 and 2010. Poult. Sci. 93, 241–245.

Robinson, T.P., Wint, G.R.W., Conchedda, G., Van Boeckel, T.P., Ercol, V., Palamara, E., Cinardi, G., D'Aietta, L., Hay, S.I., Gilbert, M., 2014. Mapping the global distribution of livestock. PLoS One 9, 1–13.

Siegel, P.B., 2014. Evolution of the modern broiler and feed efficiency. Annu. Rev. Anim. Biosci. 2, 375–385.

Smith, P., Daniel, C., 1975. The Chicken Book. Little and Brown, Boston, MA.

Stevens, L., 1991. Genetics and Evolution of the Domestic Fowl. Cambridge University Press, New York, NY.

Zuidhof, M.J., Schneider, B.L., Carney, V.L., Korver, D.R., Robinson, F.E., 2014. Growth, efficiency, and yield of commercial broilers from 1957, 1978, and 2005. Poult. Sci. 93, 2970–2982.

IMPACT OF ANIMAL FEEDS

Lucy Waldron
LWT Animal Nutrition Ltd, Feilding, New Zealand

Over the decades, the animal feeds industry has influenced, and been influenced, by human consumers on many levels. The modern industry most people are familiar with is a far cry from the small flocks and herds of the 1940s and before, where animals were produced on a minor scale by local farmers, people kept pigs in their back yards, and roasted chicken was a luxury.

After the Second World War, governments and companies focused on producing enough meat, milk, and eggs to feed the populations across the globe, many of which had been subject to rationing for a prolonged period of time. This meant that changes were needed in how animals were managed and fed to generate meats and eggs, which were cheaper for consumers to purchase and in high enough quantities to meet market demand. With this aim in mind, many animal production systems, such as pigs and poultry, underwent intensification to produce cheaper food products. Research into factors affecting yields of meat and eggs and time needed to slaughter or point of lay helped facilitate such developments.

Improvements in genetics and nutrition from research and farm trials combined to generate high-performing, earlier-developing animals, with major improvements in body weight, carcass, milk and egg yields, and feed conversion ratio (FCR). For example, in the 1950s, a 6-week-old broiler weighed just over half a kilogram with an FCR of 2.35. Modern intensive farming produces meat chickens that weigh 2.8 kg with an FCR of less than 1.70 for the same age, and since the 1960s, egg production from layers has increased by 64% with a 20% reduction in feed ingested. These developments occurred alongside the appearance of large farms with housed animals kept under controlled conditions and fed diets specifically to meet their faster growing needs. Changes in management and feed were needed to meet the far higher genetic potential for growth and egg production. Hence, temperature and lighting controls were introduced, alongside optimized diets, which were targeted at certain growth phases of the animal. Cattle became housed in "feed lots" where their diet was brought to them as a total mixed ration, and they had little access, if any, to grazing.

Such changes in systems created increased efficiency and improved return on investment for the farmer, as well as making previously luxurious products, such as chicken breast meat, more available and affordable for consumers. Less manpower on the farm was required to generate meat and eggs, due to automated feeding and watering systems, and the use of cages for layers with machinery-driven egg collection and sorting equipment. Thus, a modern industry concerned with genetics, feed, and ingredient production arose to cater for the demands of meat and egg production. This saw major investment in research into nutrition and feeding management, as well as the education and employment of personnel with specializations to support the industry. Associated industries, such as the production of day-old chicks, came into being to supply the large intensive farms and, at the other end of the process, abattoirs and factories for preparing carcasses and meat products for retailers were built, generating employment for semiskilled and skilled labor.

As meat and eggs have become more abundant and cheaper to buy, consumers have become more discerning regarding how their food is produced, particularly those from middle- and high-income brackets. Since the 1990s, these consumer sectors expressed increasing concern regarding how animals are fed—especially the use of chemicals, inorganic compounds, and pharmaceuticals in the feed. From the research side, a better understanding of certain scientific areas, for example environmental excretion from animals, mycotoxins, and producing meat and eggs containing nutrients that can promote human health, has led to improvements in production systems and the generation of "value added" products.

The Environment

To examine environmental concerns, the background to these issues should be mentioned. In certain parts of the world, such as North America and Northern Europe, in the late 1990s higher levels of minerals and other compounds were found in groundwater supplies that were needed for human drinking water, including nitrates, phosphorus, copper, and zinc. This manifested in several ways—but the most obvious were increasing algal blooms in water systems destined for human consumption via treatment plants. The growth of algae and other toxin-generating microorganisms is promoted by the abundance of substrate sources, such as nitrogenous compounds and minerals, in the water they colonize. These problems resulted in investigations into excretion from intensive farming, which were considered the main source of such pollution, as manure spread on soil results in percolation of indigested nutrients into water courses.

Feed enzymes had already been developed from the 1980s onward, but mainly as a support to increasing nutrient supply and limiting antinutritional factors from various cereals that were deleterious to animal production. The awareness of environmental issues encouraged research into the use of enzymes to control excretion, for example, proteases to reduce the amount of nitrogenous material in excreta from chickens and pigs. Phytase enzymes came to the fore in controlling phosphorus excretion—which was a major factor in water quality for many countries, and, as a result, phytase is now included in the majority of pig and poultry diets in the developed world. Optimizing rumen efficiency by adding live yeasts to diets and controlling protein intake to maximum levels of 15% is now known to be important in preventing high levels of nitrogen being excreted in urine from cattle. The inclusion of "natural" forms of minerals, that is, those chelated to small peptides rather than the sulfate and oxide ores previously used in feed, has been researched and shows more efficient uptake and storage in the animal; hence, lower levels are required in feeds resulting in major reductions in excretion rates. Environmental issues linked with intensive agricultural practices continues to be an important political and research topic around the world, as water becomes a more limited commodity for human consumption, and hence the environmental industry has grown at a pace to match such concerns.

Antibiotics in Feed

In the mid-1990s, residues in meat and eggs from antibiotics used in animal feed to control disease were highlighted in Scandinavian countries as being linked to the increasing ineffectiveness of human therapeutic treatments. There are various arguments for and against this proposition, which lingers to this day. For the animal production industry, it meant major changes in dietary ingredients.

Antibiotics for animal feed use had been identified in the 1960s, whereby those compounds deemed less useful for human medicine were allowed to be used in prophylactic doses to prevent disease in farm animals and promote growth and welfare. However, increasing antibiotic resistance in human populations meant that these less effective compounds were now being revisited as "drugs of last resort" for controlling infections. As a result, from the late 1990s, prophylactic use of antibiotics in feed was increasingly limited or banned. Although some parts of the world still have access to in-feed antibiotics at the time of writing, more are now considering following suit and limiting or banning their use as prophylactics.

To counter the loss of antibiotics in feed and maintain productive performance and animal welfare, specialist feed ingredient companies began research into nonpharmaceutical alternatives to antibiotics, spawning a large industry producing probiotics (gut active beneficial bacteria), prebiotics (substrate for promoting beneficial bacterial growth), specialist fibers, organic acids, and a range of plant- and yeast-derived "natural" alternatives. This has led to a far greater understanding of the synergistic role that bacteria play in the gut of a wide variety of species. These products, either singly or in combination, have been shown to be effective in promoting correct bacterial populations in the gut of animals, making it more difficult for pathogenic organisms to colonize the host and cause disease.

Controlling Feed Ingredient Quality

As nutritional science has advanced, so too our understanding of the compounds and potential problems within feed materials has escalated. This started with examining the suitability and improvement of feedstuffs from the animal's perspective, for example, the application of feed enzymes to prevent the negative activity of antinutritional factors in the gut, such as nonstarch polysaccharides from grain. Better processing of certain feedstuffs, such as heat treatment for soybean meal to negate lectin- and trypsin-inhibitor compounds, which reduce digestion and can promote autoimmune responses, is now applied before feeds are manufactured. Novel essential dietary elements, such as minerals can now be supplied in a form that is akin to those

compounds in nature (bound to small protein units), rather than being from the more inefficient use of inorganic mineral ores. These do not interfere with other minerals on the basis of valency (electrical charge), are selectively and preferentially taken up from the gut and considered to be more effectively stored in tissues.

Naturally occurring toxins from fungal growth are better recognized and analyzed, which has resulted in legislation in many developed countries controlling the import of potentially affected feedstuffs. In some countries, this means that shiploads that exceed the permitted levels can be denied entry at port, or may be blended with other, toxin-free feedstuffs to reduce the levels. Many feed companies routinely add mycotoxins binders to prevent harm to the animals consuming the feed, which can include liver damage, reproductive problems, and impaired immunity. From the consumer's point of view, mycotoxins can accumulate in various edible tissues of the body and in animal products, such as milk. This poses a danger for humans eating these products as such toxins are linked to various diseases and can be, for example, carcinogenic. Mycotoxins can be a particular problem when feedstuffs are grown with restrictions on the use of fungicides due to concerns regarding residues, which themselves can be a problem for consumers, so a balance between cereal quality and the prevention (by binding) of toxins in final feeds and animal products needs to be reached for producers.

Value-Added Meat and Eggs

There is a sizeable sector of the human population that is concerned with eating healthily and requires foods that can be of benefit to them to resolve real or prevent perceived issues. These consumers are typically willing to pay extra for animal products that contribute to improved nutrient intake.

Eggs are a very adaptable vehicle for various nutrients, and there has been a rise in the specialist market for eggs from hens that have been fed, for example, certain minerals or omega oils. The transfer of such nutrients and compounds into eggs has been shown to be efficient, resulting in marketing claims for eggs that are based on heart health, high antioxidant levels, and so on. Cows fed higher mineral diets, such as selenium, produce milk high in this important antioxidant mineral, which has been associated with various protective effects against disease in humans. More recently, algae, which are high in certain omega oils, such as docosahexaenoic acid, associated with immunity, disease prevention, and reproduction in humans, has been used in research in layer hen diets to boost docosahexaenoic acid levels in eggs. Mineral levels in meat can be increased when animals are specifically fed higher levels in feed. Increasing antioxidants in the diet can be passed into the meat, leading to better shelf life, less drip loss, as well as allowing marketing claims for human nutritional benefits.

Cheap Food Production Versus Increased Controls in Feeding Animals

Although changes in feed regulations have been initially driven by the concerns of wealthier consumers, this has filtered down via more enforced legislation to food typically consumed by poorer sectors of society. Issues regarding obesity and malnutrition in the lower quartiles of the human population have led to changes in quality control for cheap meat products, particularly those used in the "fast food" industry. As such, producers for these markets still have to follow the same rules regarding feed quality regarding, for example, mycotoxins limits and antibiotic-free diets, even though the final consumed food is sold at a much cheaper level. This has led to a dilemma between the extra costs the producer has to fund to meet regulations, and the price the lower end consumer is willing to pay. There is a price threshold that consumers need to pay to ensure that producers can meet quality requirements for supplying meat and eggs. The adage "you get what you pay for" is certainly true in food—if you are not willing to pay decent amounts of money for meat and eggs, then you cannot reasonably demand that they be produced at higher cost. Making the purchase of good, traceable food a priority in all consumers, including those on low income, is essential to maintain the best feeding standards and practices that meet modern society's expectations. However, economic comparisons have shown that the percentage spent on average on food is now far lower compared to the 1970s, with proportionally more being spent on rents and mortgages in modern times. Balancing the demands for welfare, food safety, and quality against higher production costs and what consumers are willing and able to pay is becoming an important political issue in many global regions.

SUSTAINABLE LIVESTOCK PRODUCTION

Mark Rasmussen*, Colin Scanes**
*Iowa State University, Ames, IA, United States
**University of Wisconsin–Milwaukee, Milwaukee, WI, United States

Introduction

It is an oft-repeated statement that humans are growing more distant from biology and nature in our technological civilization. Few people today have the opportunity to be raised on livestock farms where experience with farm animals becomes routine. The origins of the relationship between humans and animals has been lost in the mists of time, although we do know that animal domestication began thousands of years ago with small animals, such as goats, sheep, and dogs. In our development of agriculture and the improved nutrition that is derived from it, protein was frequently in short supply until animals could be used to provide that essential nutritional ingredient. Livestock production in its many forms solved this dilemma, as well as other human problems.

Currently, opinion persists in some quarters that humans should rely only on plant and vegetable crops for sustenance and eliminate animal products from their diet. The argument is that this would be more environmentally and socially friendly. This viewpoint suffers from several deficiencies. First, it disregards the long-running interdependent relationship humans have had with animals, a relationship that continues to this day. This opinion also ignores the essential fact that integrating crop and livestock production in agricultural systems is the best means available to achieve sustainability in our agricultural and farming systems. Nature has created intertwined biological cycles that include animals and plants and the science of ecology recognizes the value of those interactions. Many plants cannot reproduce without the involvement of animals and animals certainly cannot exist without plants. This plant/animal interaction also applies to sustainable agriculture where a key principle is to pattern our agricultural systems after nature. Sustainability theory acknowledges that in the long run, agriculture only works when it mimics and patterns itself after nature. Livestock production complements crop production in ways that are difficult to replicate if one is separated from the other.

Sustainable livestock production can be summarized as a system of farming practices that contribute to long-term productivity. It provides livestock products to humans in a way that ensures that the "whole" agricultural system can persist in an equilibrium state with the environment, society, and nature. Furthermore, it also must be compatible with accepted social and religious requirements of the civilization in which it is practiced. In this essay we will discuss several aspects of sustainable livestock production and offer the reader a broad overview of the field.

Feed and Forage

A useful place to start is a consideration of the basic supply chain that provides the most essential input for any livestock enterprise: feed and forage. The production of feed requires a cropping system that can provide feed and the nutrients it contains to a diverse population of livestock at various ages and stages of production. Young animals require different rations than mature animals or animals kept for breeding. Globally, the feed supply encompasses a wide variety of plants including root crops (turnips), coarse cereal grains (maize and barley), protein grains (soybean), harvested forages (alfalfa and silages), and grazed pastures or grasslands of native plants. Further down the supply chain, crop cultivation and grasslands also require inputs. These include water (natural or irrigation), nutrients (synthetic fertilizers, animal manures, and other biological sources), soil, and the tools necessary to sow, cultivate, harvest, and store feed products. Soil should be regarded as the most fundamental resource in any sustainable farming system. No farming system can be considered sustainable if it abuses or degrades soil. A cropping system that maintains or improves soil health is vital to all forms of agriculture. Soil must be managed in ways that avoid organic matter depletion, soil erosion, compaction, and other forms of degradation.

Water for crop production is another essential resource, especially in the wake of climate change. Rainfall amounts vary widely across the earth. Some geographic locations may receive too much rain and be subject to flooding and erosion if the land is tilled excessively or overgrazed. In particularly wet locations, it may be difficult to farm the land using typical tillage practices because of excessively wet conditions. The presence of livestock, especially grazing livestock, can be critical in such regions. Sustainable livestock systems are possible under wet conditions because ruminants can be grazed or fed harvested forages even in areas of high precipitation.

In contrast, other geographic regions suffer from inadequate rainfall and in these areas many forms of agriculture require water supplementation in the form of irrigation. Irrigation, however, is dependent upon adequate, sustainable supplies of water. Some regions that have successfully irrigated crops in very dry regions of the world will be unable to do so in the future due to water shortages. One example is the Kingdom of Saudi Arabia where underground aquifers of fossil water have been used to irrigate maize and alfalfa to support a domestic dairy industry. However, as the supplies of local groundwater are depleted, the Kingdom has resorted to importing forages to support the domestic dairy industry. In time, this too will cease, as logistical costs to sustain the system rise to unacceptable levels. Since the 1950s, the Great Plains states of Texas, Oklahoma, and Kansas have fostered a beef feedlot industry by using water from the Ogallala aquifer. Although this aquifer contains large quantities of water, the southern portion of the aquifer has been overused and water levels have declined dramatically in recent years. Pumping costs and lack of water are already forcing the cattle feeding industry to migrate north to Nebraska and eventually much of this region will be forced to revert to dryland agriculture.

In addition to the primary nutrients required for livestock production, it is also important to consider other feed ingredients, such as minerals and vitamins. Sometimes these nutrients may be available in the major feeds provided to animals, but in many cases they must be added to animal rations as nutritional supplements. The kinds and amounts of supplements needed depend upon the species of livestock, their stage of growth or production, the feeds used, and the geographic region where production occurs. Some examples of vitamin and mineral nutrients includes macrominerals (calcium and phosphorous), trace minerals (zinc and iodine), as well as vitamins, such as A, D, and the B vitamins like niacin. A large body of nutritional knowledge gathered over decades of research provides many specific details on the individual amounts required, the symptoms of deficiency or excess, and the best supplements to use for each of these nutrients. Most mineral nutrients are obtained from extractive mining and purification industries, while most vitamins are obtained from synthetic chemical or biochemical processes. Care must be taken to manage existing resources for the long term so that the necessary supplies are available for future generations of livestock.

Livestock Facilities

Other important considerations in sustainable livestock production are the facilities and physical infrastructure that provide an adequate level of care and welfare for the animals. In nomadic societies, the facilities required may be minimal and call for nothing more than dependable water supplies, open grazing lands, and catch facilities for occasional confinement for inventory counts or capture purposes.

In other societies with a legal system of private property, fence laws, and livestock containment regulations, more elaborate livestock facilities are necessary. Fences become necessary to contain animals on private property and avoid encroachment on another owner's land. Modern livestock production systems have taken containment to its functional limits. In these facilities, commonly known as concentrated animal feeding operations, livestock are housed in very dense environments to minimize production costs. Such facilities can range from open feedlots for beef cattle to hog barns with pens placed over manure storage pits to wire battery cages for laying hens. In such facilities, large numbers of animals are confined to minimize the individual space given to each animal. Considerable controversy has arisen over concentrated animal feeding operations and the welfare of the animals, especially with regard to battery cages for hens and gestation crates for sows.

All facilities need to be managed to minimize risks to animal health and well-being, whether stemming from infectious disease, injury, or predators. Facilities that minimize stress reduce illness and injury. Truly sustainable livestock facilities also must diminish or eliminate the environmental externalities imposed on others or the environment. Prominent examples of such externalities include manure runoff, which can cause fish kills and harm other aquatic life, odors that diminish the quality of life of neighbors, and

practices that release mineral nutrients and damage water quality. The Dead Zone in the Gulf of Mexico and water quality issues in the Chesapeake Bay are examples of environmental damage that have arisen from unsustainable agricultural practices.

Technology

The use of technology has modified many aspects of livestock production. In most cases, technology has increased productivity, reduced the labor requirement per animal, and allowed large populations of animals to be cared for with decreased effort. Automated systems of feeding, water delivery, and waste handling have altered the practices of more traditional "hands on" animal husbandry.

Technology in the form of breeding, hybridization, and genetic selection also has changed livestock production. An understanding of genetics allowed humans to "create" animals with desired characteristics through breeding and selection. More recently, technology has allowed even greater changes through the use of artificial insemination, in vitro fertilization, and cloning. In its most current and controversial form, molecular technology has allowed genetic modification through artificial gene transfer, which can move genes across species boundaries. The long-term sustainability of some of these newer technologies remains a subject of discussion and debate.

Medications and Drugs

Livestock production also relies upon an adequate supply of vaccines, drugs, antibiotics, and growth promotants. The use of these compounds has been beneficial to livestock enterprises and, in some cases, proved to be essential for the animals' health and welfare. These compounds have been used to control diseases, infectious microbes, harmful insects, and parasites, especially in high-level livestock confinement units.

Their use has allowed the proliferation of highly confined systems, reduced death losses, relieved animal suffering, improved growth and performance, and helped provide safe products for human consumption.

These medications are very diverse in their origins and manufacture, although most are produced under strict clinical procedures typically used by drug manufacturers. Most medications are subject to strict regulation and labeling requirements. This class of drugs includes hormones and other medications that alter reproductive and metabolic processes. Some are used to suppress estrous while others (beta agonists) alter the ratio of fat and lean deposition. Most compounds in this class are used to enhance growth and therefore influence some aspects of economics and profitability. Some types of livestock production operations (organic) attempt to minimize or eliminate the use of medications. Their concerns center on the long-term safety aspects and the impacts upon the animal and environment.

Some of the most controversial issues in agriculture arise from the use of these products and the ways they impact consumer and export markets. One category of products, the growth promotants, is particularly contentious in export markets. While the use of many of these products is permitted in the United States, many other countries do not allow their use. As a result, segregated production channels must be established for products destined for export.

Other concerns related to sustainability include availability, price, regulatory oversight, animal and human safety, and loss of effectiveness through resistance. Antimicrobial resistance is an especially important issue with potential effects on livestock and human health. When microbes (bacteria) are exposed to antimicrobial compounds, they can become resistant to a given compound. The antimicrobial may kill a majority of the population, but those surviving microbes may flourish even in the presence of the antimi-

crobial. As a result, the resistant microbes will predominate and can cause disease or injury that the antimicrobial originally was able to prevent.

Sustainable farming practices that use such drugs very carefully help prolong their effectiveness. Unfortunately, such practices can be difficult to apply in real-life situations. A decision to use or not use a drug must balance several issues, including individual animal welfare, whole herd or flock health, and the risk of use and resistance development. The transfer of resistant zoonotic agents, those capable of infecting animals and humans, is one of the most serious concerns with respect to drug-resistant organisms.

Environment

Livestock production, like other agricultural practices, has an impact upon the environment. Ever since humans developed agriculture they have altered environments and affected the existence of other species. Even the most widely dispersed forms of agriculture, such as nomadic and open grazing can alter an environment. As livestock production has consolidated and intensified, its impact upon the environment has become more damaging. As a result, livestock producers have been required to more carefully manage environmental impacts that arise from production operations through the use of advanced production practices and regulation. Large facilities (those above certain animal unit thresholds) are required to file applications and obtain permits, typically from state agencies, before facility construction can begin.

Increasingly, the general public has become more aware of agricultural issues and production practices. Livestock production is no exception to this trend and as a result the industry has had to deal with air, water, and soil quality problems. Regulations vary, but most dictate facility locations, set back requirements, waste handling procedures, carcass disposal, and water runoff control. Some critical aspects of high-density livestock production, such as odor have been left unregulated, much to the displeasure of many rural neighbors who live downwind of such facilities.

Recently, additional food product safety regulations also require facilities to control insect and vermin, undergo periodic inspection, and abide by compliance orders. Small livestock operations tend to be exempt from most regulations based upon the assumption that their environmental impact is smaller and better able to be tolerated. The accuracy of this assumption is subject to debate.

Economics

Sustainable businesses must generate a profit to stay in operation. Livestock production has changed in recent decades to become a highly capital-intensive enterprise. It requires substantial investment in animals, feed supplies, production facilities, and other inputs. For example, with the value of a dairy cow at $1700/head, a modest-sized dairy consisting of 300 cows would require a capital investment of more than $500,000. Other facilities and operating costs easily increase the total to over $1 million. Larger operations call for proportionately greater investments; a large dairy of 5000 cows requires a multimillion dollar investment. In the absence of personal wealth, funds of this magnitude must be obtained from farm partnerships, investor pools, commercial banks, or the farm credit system. These large capital requirements are an important component of sustainability and present a significant hurdle to beginning livestock farmers.

In addition to the investment capital risks, marketing risks also impact agricultural sustainability prospects. Because production is cyclical, agricultural markets are inherently variable and all segments of the industry experience fluctuating revenue and income cycles, either short or long term. For enterprises with frequent production and sales like dairy farming, revenues can be evenly spaced and continuous through the year. However, other livestock enterprises have much longer cycles. Beef cow enterprises may receive revenue only once or twice a year when calves and cull cows are sold. Infrequent income receipts require greater cash balances and longer-term financial management if the operation is to be financially sustainable.

Human Factors

An often overlooked aspect of sustainable livestock production is that related to human capital. Even though many livestock enterprises now use technology to make production more efficient, all enterprises still require people. People are needed to monitor and manage the health and well-being of the animals, as well as to keep records and troubleshoot daily problems. A significant part of the livestock production industry, in spite of technological developments, still relies on the skill and experience of the people. Good livestock skills are an art and knowing how to recognize and care for sick animals is an acquired trait, usually obtained through experience and training. In some cases, it calls for knowledge passed from generation to generation. Human intervention also is important when handling animals to administer treatments or injections. Such interventions require skill in directing the movement of animals in a manner that is consistent with worker and animal safety.

As agriculture has consolidated and expanded, enterprises have become more dependent on external labor beyond the available family members, which had been the historical norm. As a result, many livestock enterprises have struggled to maintain an adequate pool of skilled and experienced laborers who are motivated to care for animals properly. The livestock production industry has received severe criticism over animal welfare issues that often have occurred as a result of a poorly trained or unmotivated work force. In many industry segments, producers have had to rely upon unskilled immigrant labor to fill positions. Additionally, the industry has had problems attracting adequate numbers of livestock professionals such as nutritionists and veterinarians. These professions require many years of education and educational expense.

It is anticipated that labor will continue to be one of the livestock industry's most vulnerable areas with regard to sustainability.

Consumer Factors

No agricultural enterprise can be sustainable if its products lack a market. Consumers provide the demand side of the economic enterprise. Entire industries have been eliminated when consumer concerns were ignored or dismissed. Agricultural products are no exception, and consumers must be considered very important to any livestock sustainability analysis. Livestock products traditionally have been the object of consistently high consumer demand, but in recent decades many factors related to consumer attitudes and purchasing behavior have exerted greater influence on marketing and sales. Concern over the nutritional aspects of livestock products (saturated fat content) have influenced sales and forced changes to comply with consumer preferences. Fluid milk consumption has declined as consumers have altered their purchasing patterns when a wider assortment of beverages appeared on the market. Poultry consumption (especially chicken) has increased as intensive production provided abundant supplies at relatively low prices. A fundamental factor of the food business is the inelastic demand for food; typically when one food enjoys increased sales, another food segment sees a sales decline.

Markets also have changed as consumers have become more aware of farming practices related to nutrition, health, animal welfare, and the environment. Food safety concerns have affected the market significantly when recalls occur. Some consumer segments have been willing to absorb price increases in exchange for products produced using alternative livestock production practices, such as genetically modified organism-free, antibiotic-free, cage-free, and organic. In many cases, the livestock industry has responded sluggishly to these emerging consumer trends. Additional changes can be expected as consumers become more aware of agricultural production issues and continue to adjust their purchasing behavior accordingly. Astute producers will adapt to these market forces to be successful.

The livestock industry will continue to respond to the forces of the marketplace. Many changes will be required to cope with external factors, such as erratic weather, climate change, resource management, and consumer demands. Trends toward improved sustainability will continue as producers are motivated or required to do things differently. Agriculture, like many segments of society, has a long way to go to achieve moderate levels of sustainability, but inevitably change will come to the industry. Those enterprises that cannot or will not change eventually will be replaced by those that will. Activists and opinion leaders will continue to press the industry for improvements as they encourage consumers to vote with their food purchasing dollars. Government oversight also will continue to evolve as food safety and other concerns work their way through the legal, political, and policy process. It is a challenging but exciting time to be associated with the livestock industry as it continues to respond to outside challenges and alter its practices to be more successful and sustainable.

Invertebrates and Their Use by Humans

Colin G. Scanes

University of Wisconsin–Milwaukee, Milwaukee, WI, United States

8.1 INTRODUCTION

There are multiple uses of invertebrate animals. The biology and relationships of the invertebrates is summarized in the textbox.

The Invertebrates

Over 95% of the animals on the planet are invertebrates. The major invertebrate groups include:

- Arthropods (phylum Arthropoda) including insects (class Insecta), arachnids (class Arachnida) (e.g., spiders, mites, and ticks), and crustaceans (subphylum Crustacea) (e.g., shrimp, crabs, and lobsters)
- Annelids (phylum Annelida) (e.g., earthworms)
- Mollusks (phylum Mollusca) (e.g., clams, mussels, oysters, squid, and octopus)
- Nematode worms (phylum Nematoda) (e.g., *Caenorhabditis elegans*, parasitic worms)
- Flatworms (phylum Platyhelminthes) (e.g., liver fluke and tapeworm)
- Echinoderms (phylum Echinodermata) (e.g., starfish and sea urchins)
- Cnidaria (phylum Cnidaria)

The fossil record clearly indicates that the major phyla of animals were present in the Cambrian period (490–540 million years ago) in the "Cambrian explosion" but the fossil record before that is uncertain (reviewed by Budd, 2013). Based on comparisons of genome, proteins, and anatomy, the evolution of multicellular animals (Metazoa) can be placed into two major groups (Hejnol et al., 2009):

- Bilateria (bilateral symmetry)
- Others, for example, sponges (Demospongiae) and Cnidaria (e.g., jellyfish and sea anemones)

The divergence time is estimated as 1298 million years ago (Hedges et al., 2004). In turn the Bilateria (animals with bilateral symmetry or having left and right sides the same) have two major groups (Hejnol et al., 2009):

- "Superphylum" Deuterostomia
- "Superphylum" Protostomia

The divergence time has been estimated as 940 or 976 million years ago (Hedges et al., 2004; Wang et al., 1999). In turn, the deuterostomes include the phyla Chordata (e.g., vertebrates), Hemichordata, and Echinodermata. The protostomes contain multiple phyla:

- Arthropoda and Nematoda together in a grouping called Ecdysozoa
- Annelida and Mollusca together in a grouping called Trochozoa or Lophotrochozoa
- Platyhelminthes

8.2 ARTHROPODS AND THEIR USE BY PEOPLE

8.2.1 Overview

Arthropods are used by humans in the following manner:

- Crustaceans as a source of food; considered in Section 8.4.
- Insects as food; considered in Section 8.5.
- Honeybees as pollinators and producers of honey and wax; considered in Section 8.6.
- Silk worms as producers of silk; considered in Section 8.7.

8.3 CRUSTACEANS AS FOODS

8.3.1 Overview

The anatomy of crustaceans is typically made up of a head, thorax, and abdomen together with branched appendages. There are over 30,000 species of crustaceans with krill being essential to many marine food chains. Crustaceans that are popular to eat include crabs, crawfish,

lobster, prawns, and shrimp. These are harvested predominantly from aquaculture production but capture harvesting is still important.

8.3.2 Capture Harvesting of Crustaceans

Definitions

- Fish are divided into two groupings: finfish (addressed in Chapter 7) and shellfish (crustaceans and mollusks).
- Fish are either harvested by capture harvesting (by commercial fishing) or raised in aquaculture/mariculture.
- Aquaculture is where fish are raised in enclosures (Fig. 6.8C) with feed and other essentials for life provided by humans. Aquaculture employs domesticated fish. It can use seawater (in mariculture) or freshwater.

Capture harvesting is commercial fishing on seas and oceans together with rivers and lakes.

In 2012, capture harvest of shrimp species was 3.4 million metric tons (mt) (FAO, 2014). Among the top 25 marine capture species in 2013 are the following two crustacean species (FAO, 2014):

- Akiami paste shrimp (*Acetes japonicus*) 585,433 mt
- Gazami crab (*Portunus trituberculatus*) 503,885 mt

Capture harvesting is addressed in Chapter 4 on hunter–gatherers.

8.3.3 Aquaculture Production of Crustaceans

There were about 6 million mt of crustaceans produced by aquaculture in 2013. Table 8.1 summarizes the top countries for aquaculture. The

TABLE 8.1 Top Countries for Aquaculture Production of Crustacean Species in 2012 (FAO, 2014)

Ranking	Country	Aquaculture production (million metric tons)
1	China	3.6
2	Thailand	0.6
3	Vietnam	0.5
4	Indonesia	0.4
5	India	0.3
6	Bangladesh	0.1

following are the major crustaceans produced by aquaculture (FAO, 2014):

- Whiteleg shrimp (*Penaeus vannamei*), 3.3 million mt in 2013
- Giant tiger prawn (*Penaeus monodon*), 804,000 mt in 2013
- Chinese river crab (*Eriocheir sinensis*), 730,000 mt in 2013
- Red swamp crawfish (*Procambarus clarkii*), 652,000 mt in 2013
- Giant river prawn (*Macrobrachium rosenbergii*), 203,000 mt in 2013
- Indo-Pacific swamp crab (*Scylla serrata*), 180,000 mt in 2013
- Indian white prawn (*Penaeus indicus*), 23,600 mt in 2013

Crustaceans, such as shrimp, are raised with high-quality feed (Fig. 8.1).

8.4 CONSUMPTION OF INSECTS

Eating Insects

"The use of edible insects has been trivialized" (Ayieko and Oriaro, 2008) and viewed through the biases of Westerners.

"To many people the idea of eating insects evokes only feelings of disgust" (Vane-Wright, 1991).

Insects (entomophagy) can provide a valuable source of high-quality protein, minerals, such as zinc, vitamins, such as vitamin B_{12}, and essential fatty acids (FAO, 2013). There are over 500 insect species that are viewed as edible including aquatic insects, ants (larvae, pupae, and winged), bees, beetles (grubs and adults), caterpillars (e.g., silkworm larvae and the mopane caterpillar), cicadas, grasshoppers, locusts, mealworms, wasps, and winged termites (Fig. 8.2).

Estimates of the number of people consuming insects are as high as 2 billion from Africa,

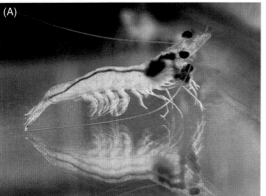

FIGURE 8.1 **Shrimp are a popular invertebrate in the diet.** (A) Shrimp (*Penaeus vannamei*) raised on soy-based feeds. (B) Grilled barbeque-style shrimp. *Source: Part A: Courtesy USDA Agricultural Research Service.*

FIGURE 8.2 **Insects as food.** (A) A woman harvesting grasshoppers in Laos, Southeast Asia. Example of insects as food: (B) mealy worms and (C) grasshoppers. *Source: Part A: Courtesy FAO, 2013. Nutritional value of insects for human consumption. In: Edible Insects Future Prospects for Food and Feed Security, pp. 67–88. Available from: http://www.fao.org/docrep/018/i3253e/ i3253e06.pdf; https://www.pinterest.com/entophile/bugs-as-food/.*

Asia, and Latin America (Barennes et al., 2015). Insects are invaluable particularly to children in Africa such in Nigeria (Banjo et al., 2006). Edible insects are a significant source of protein in rural Thailand (Yhoung-Aree et al., 1997) and India (Shantibala et al., 2014). Similarly, in a study of people in Laos, 97% were regular consumers of insects with the top five insect species eaten there being cicadas, crickets, grasshoppers, short-tailed crickets, and eggs of weaver ants (Barennes et al., 2015). Edible Diptera species including lake flies (*Chaoborus* and *Chironomus*

sp.) are consumed by the Luo communities in Kenya close to the Lake Victoria basin (Ayieko and Oriaro, 2008).

It has been concluded that edible insects are either more healthy or equally healthy than meat products (Payne et al., 2016) and have a lower impact on the environment (Oonincx and De Boer, 2012; Oonincx et al., 2010). There appears to be considerable potential for the growing role of insects in addressing human nutritional needs (FAO, 2013; Vane-Wright, 1991).

Cochineals are insects that are native to South America (Rodriguez and Niemeyer, 2001). These feed on prickly pear cactus. They accumulate a red compound, the cochineal dye, specifically carmine acid. These were cultivated by the Aztecs and Incas. The dye has been used in coloring clothing, such as the redcoats, the British army in the Revolutionary war in which the USA became independent. In addition, cochineal is used as a food coloring, for example, in pink Frappuccinos. Peru is the major producer of cochineal supplying 80% of global carmine (International Development Research Centre, 1998).

8.5 HONEYBEES, HONEY, AND WAX

8.5.1 Overview

Global production of honey and beeswax in 2013 (FAOStats, 2015) was the following:

- Honey production, 1.66 million mt (increased 41% since 1990)
- Beeswax production, 65,000 mt

According to the United States Department of Agriculture (USDA) National Agricultural Statistics Service (NASS), there are 2.74 million bee colonies in the USA (USDA NASS, 2015). Honeybees (*Apis mellifera*) play a critically important role in pollinating wild and domesticated flowering plants, such as clover, fruit trees together with garden and wild plants. Pollination by bees has been valued as contributing $200 billion per year to the global economy (Gallai et al., 2009).

Bees (Fig. 8.3) are social insects that transport pollen grains from the male organ (the anther) of the flower of one plant to the female organ (the stigma) in the flower of another plant allowing fertilization and hence sexual reproduction (combination of the genes from different plants). The flowers contain sweet fluid, nectar. This provides the bees with nutrients for immediate use and for storage. The bees transport the nectar to the hives and transform it into honey in honeycombs. The disaccharide, sucrose, in nectar is the same as cane sugar. The honeybees convert the sugar in nectar to monosaccharides, glucose and fructose, by enzyme action and

FIGURE 8.3 **Honeybees are critically important as pollinators and also provide honey.** (A) A honeybee. (B) Honeycomb with honey accumulated by honeybees. *Source: Part A: Courtesy USDA Agricultural Research Service; Part B: Shutterstock.*

regurgitation. Moreover, much of the water is lost by evaporation.

Bees

    Phylum: Arthropod
    Class: Insects
    Order: Hymenoptera
    Genus/species: Predominantly *A. mellifera* together with *Apis dorsata*

8.5.2 Composition of Honey

There are marked differences in the composition of honeys and in the glycemic index ranging from 40 to 87 compared to glucose. Honey has the following composition (Bee source, http://beesource.com/):

- Monosaccharides
  - Fructose 38% (range 27%–52%)
  - Glucose 31% (range 20%–33%)
- Disaccharides, 9%
- Oligosaccharide, 4%
- Other solids including organic acids and free amino-acids, 1%
- Water, 17%

The color of honey varies from white, through extra-light amber, light amber, pale amber, amber, to dark amber depending on the flowers from which the nectar is obtained (USDA NASS, 2015).

8.5.3 Global Honey Production

According to the Food and Agriculture Organization (FAOStats, 2015), the leading countries for honey production in 2013 were the following:

1. China 466,400 mt
2. Turkey 94,694 mt
3. Argentina 80,000 mt
4. Ukraine 73,713 mt
5. Russian Federation 68,446 mt

6. USA 67,812 mt
7. Mexico 56,907 mt
8. Ethiopia 45,000 mt
9. Iran 44,000 mt
10. Brazil 35,365 mt

8.5.4 Honey Production in the USA

According to the USDA NASS (USDA NASS, 2015), honey production in the USA was 178 million pounds in 2014 valued at $384.8 million. Production of homey in the USA is declining (American Beekeeping Federation, 2012) reflecting the vulnerability of bees (Gallai et al., 2009). The top five states for honey production in 2014 were the following:

1. North Dakota 33.3 million pounds
2. Montana 14.9 million pounds
3. South Dakota 14.8 million pounds
4. Florida 13.4 million pounds
5. California 10.9 million pounds

8.5.5 Food Safety and Honey

Honey can contain the spores of *Clostridium botulinum* (Nevas et al., 2005, 2006) and hence can be a hazard to young children and others who are immune suppressed.

8.5.6 Beeswax

Bees construct combs composed of large numbers of hexagonal cells. These are made of beeswax and store honey. Beeswax is used in gelatin capsules, chewing gum, glazes, and nonfood uses, such as cosmetics, waterproofing, crayons, furniture polish, candles, and waxes (e.g., for skis and bow strings). It consists of esters of long-chain fatty acids and fatty alcohols (~40%), complex wax esters (21%), paraffinic hydrocarbons (14%), fatty acids ~13%, and minor quantities of fatty alcohols (1%) (Kuznesof, 2005). Beeswax is marketed as either white or yellow beeswax.

8.5.7 Domestication of Bees

Based on the unique chemical "fingerprint" of beeswax, bees were in close proximity with early Neolithic communities in the Near East (Larson and Fuller, 2014). It seems that early farmers domesticated bees over 5000 years ago (Roffet-Salque et al., 2015). Today's bee keepers manage domesticated bees. The use of honey and bees wax preceded the Neolithic Revolution (see Chapters 5 and 6).

There is increasing knowledge of the domestication of bees with a lack of the usual domestication "bottlenecks" (reflecting the genetics of small numbers of individual animals) in populations of honeybees (*A. mellifera*) (Wallberg et al., 2014). Moreover, it is not surprising with the extensive use of bees by the Romans 2000 years ago that the Italian honeybee is disproportionately important to the genetics of beekeeping around the world (Wallberg et al., 2014).

8.5.8 Impact of Honey and Other Sources of Fructose in Our Diets

8.5.8.1 *Overview*

Humans get fructose in their diet from the following:

- Honey, which contains significant amounts of fructose.
- Fruits, which contain fructose, the fruit sugar.
- Refined sugar or sucrose (from cane sugar or sugar beet); a disaccharide made up of one glucose molecule linked to one fructose molecule.
- High-fructose corn syrup; this is produced from the polysaccharide starch in corn. The starch is converted to glucose and then some are converted to fructose.

According to the USDA's Economic Research Service, sugar consumption was 90 g day^{-1} in 1970 in the USA. This declined over the next 35 years but recently is increasing due to concerns about high-fructose corn syrup. There has been a switch to high-fructose corn syrup and about a 20% increase in total sugar consumption. The advantages of high-fructose corn syrup are price and that it is sweeter than glucose.

8.5.8.2 *Fructose and Health*

There is a controversy about the effects of high-fructose corn syrup. There are two forms of high-fructose corn syrup: high-fructose corn syrup 55 (fructose 55% and glucose 42%) and high-fructose corn syrup 42 (fructose 42% and glucose 53%). While the Food and Drug Administration considers it as generally recognized as safe, some nutritionists argue that high-fructose corn syrup is a major cause of the obesity epidemic in the USA and globally (Bray et al., 2004). Others disagree (Jacobson, 2004; Klurfeld et al., 2013) considering that the problem with obesity is excess intake of calories and insufficient activity (exercise) to consume those calories.

It is not clear why the concern is over high-fructose syrup and not honey. The discussion on fructose is confounded by peoples' views on natural food (honey is natural but high-fructose corn syrup is not), producers and manufacturers with financial interests, and researchers comparing fructose with glucose (when both sucrose and high-fructose corn syrup contain both glucose and fructose). There are differences between the metabolism of fructose and that of glucose with fructose absorbed in the small intestine by a different mechanism (GLUT 5) than glucose (Arcot and Brand-Miller, 2005).

8.6 SILK WORM AND SILK

8.6.1 Overview

Silk is produced by the mulberry silkworm (*Bombyx mori*), these being insect larvae or domesticated silk moth caterpillars. The silkworms feed on mulberry leaves.

Phylum: Arthropoda
Class: Insecta
Order: Lepidoptera (moths and butterflies)
Genus and species: *B. mori* (silk moth)

The silk is collected from the cocoons of the silkworm larva. According to the United Nations Food and Agriculture Organization (FAOStats, 2015), worldwide production of silk worm cocoons was 574,000 mt for 2013 with China the number one producer with 380,002 mt. This reflected an increase of 70.7% since 2000 (calculated from FAOStats, 2015).

The two major proteins in silk are the following:

- Fibroin (a protein also found in spider's webs; a unique structure of repeated sequence of the following amino-acids: Gly–Ser–Gly–Ala–Gly–Ala)
- Siricin

Because of the very compact structure of the protein, fibroin, silk is very strong. Silk has been produced for at least 5000 years starting in China. Sericulture is raising silkworms.

8.7 MOLLUSKS AND THEIR USE BY PEOPLE

8.7.1 Overview

Mollusks are harvested as an important part of the diet (see Section 8.7.1). In addition, mollusks are the source of pearls and mother of pearl (see Section 8.7.2) and some pigments (see Section 8.7.3).

8.7.2 Molluscan as Food

Among the popular mollusks in foods are mussels, oysters (Fig. 8.4), squid, and scallops.

FIGURE 8.4   **Oysters on the half shell.** *Source: Courtesy Wikipedia.*

Mollusks are harvested from capture harvesting and from aquaculture production.

8.7.2.1 Capture Harvesting of Mollusks

The total catch of cephalopods in 2012 exceeded 4 million mt (FAO, 2014). Among the top 25 marine capture species in 2013 are the following molluscan species:

- Jumbo flying squid (*Dosidicus gigas*) 847,292 mt
- Argentine shortfin squid (*Illex argentinus*) 525,383 mt

8.7.2.2 Aquaculture

Production of mollusks by aquaculture is growing. Table 8.2 summarizes the top countries for aquaculture production of mollusks.

The major species of mollusks produced by aquaculture are the following (FAO, 2014):

TABLE 8.2 Top Countries for Aquaculture Production of Mollusks in 2012 (FAO, 2014)

Ranking	Country	Aquaculture production (million metric tons)
1	China	12.3
2	Vietnam	0.4
3	Republic of Korea	0.4
4	Japan	0.3
5	Chile	0.3

1. Japanese carpet shell (*Ruditapes philippinarum*) 3.9 million mt in 2013
2. Pacific cupped oyster (*Crassostrea gigas*) 556,000 mt in 2013
3. Blue mussel (*Mytilus edulis*) 198,000 mt in 2013
4. Yesso scallop (*Patinopecten yessoensis*) 170,000 mt in 2013
5. Mediterranean mussel (*Mytilus galloprovincialis*) 117,000 mt in 2013
6. American cupped oyster (*Crassostrea virginica*) 108,000 mt in 2013
7. New Zealand mussel (*Perna canaliculus*) 83,600 mt in 2013
8. Northern quahog (*Mercenaria mercenaria*) 24,000 mt in 2013

8.7.3 Mollusks and Cultured Pearls

Pearls are naturally formed in marine oysters (predominantly of two genera of clams: *Pinctada* and *Pteria*) in response to an irritant. Calcium carbonate is laid down in concentric layers around the nucleus. Cultured or farmed pearls are produced in oysters. They are seeded with spherical shell beads. The sources of the beads are the shells of freshwater mussels from the Tennessee and Mississippi rivers. It takes between 3 months to 2 years to produce a 2–7 mm pearl. There are about 40 tons of cultured pearls produced each year. The major producers of cultured pearls include Japan, China, and Mexico. The internal shell of the oysters is called "mother of pearl."

> Phylum: Mollusca
> Class: Bivalve
> Order: Pterioidea
> Gena: *Pinctada* and *Pteria*

8.7.4 Molluscan Animal Dyes

Clothes were dyed with molluscan dyes during historical times. One example is Tyrian or royal purple used to dye clothes, for instance, for Roman and Byzantine emperors (Margariti et al., 2013). This is an organic bromide and comes from blue sea snail. Wearing the color purple was a mark of being an emperor or a member of his family. Another dye from sea snails was used to give clothes a blue color—known as Biblical blue (PhysOrg, 2013).

Many Cephalopods (octopus, squid, etc.) produce a black dye that is released as a "smoke screen" or a defense against predators. The inks contain melanin and free amino-acids including taurine together with dopamine and its precursor L-DOPA that disrupt the sensory mechanisms of the predator (Derby et al., 2007; reviewed by Derby, 2014). Cephalopod ink, usually from cuttlefish, is used to color foods in Italian and Spanish cuisine, such as black pasta, black rice, pasta in octopus ink sauce, or pork and squid in ink soup (reviewed by Derby, 2014).

8.8 ECHINODERMS AND THEIR USE BY HUMANS

Echinoderms are used for some niche foods. The global production of sea urchins and sea cucumber is estimated at 178,000 mt (FAO, 2012).

> Phylum: Echinodermata
> Class: Holothuroidea (sea cucumbers)
> Class: Echinoidea (sea urchins)

8.8.1 Sea Cucumbers

Sea cucumbers live on ocean floors and play an important role in recycling nutrients. They have an exoskeleton and a single gonad. Not only are sea cucumbers harvested but are also grown in aquaculture. Japanese sea cucumber (*Stichopus japonicus*) has been produced in Japan and China since, respectively, the 8th and 16th centuries CE (AD).

8.8.2 Sea Urchins

Sea urchins are spherical sedentary marine animals with moveable spines and an exoskeleton; they consume algae. Sea urchins (red, black, or purple) are heavily harvested to meet consumer demand for gonads particularly in Japan and France (Fig. 8.5). Sea urchins are harvested for their gonads with a market price of over US$100 per kilogram. The gonads can be erroneously called roe. The gonads are usually consumed raw as sushi and called uni, with three grades:

- Grade A uni, gold or bright yellow color
- Grade B uni, less vivid yellow
- Grade C uni, "left overs"

Alternatively, sea urchin gonads are salted, pickled, or made into paste.

There has been a decline in the yield from capture fishing over the last 15 years from 115,000 to about 82,000 mt (Carboni et al., 2012). Along with this has been the increasing interest in production by aquaculture (Carboni et al., 2012; McBride, 2005).

8.9 EDIBLE SEA SQUIRTS (TUNICATES)

> Phylum: Chordata
> Subphylum: Tunicata

Tunicates are animals that attach to rocks and other marine surfaces (Fig. 8.6). Many tunicates are not considered as edible. However, some edible tunicates are captured or produced by aquaculture. Capture sea squirts include the following:

- sea violet (*Microcosmus vulgaris*)
- sea tulip (*Pyura pachydermatina*)
- piure (*Pyura chilensis*) in Chile
- *Halocynthia roretzi*

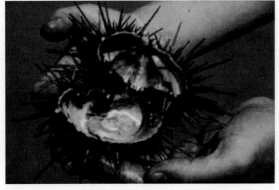

FIGURE 8.5 **Edible sea urchin.** *Source: Courtesy University of California–Davis.*

FIGURE 8.6 **A wild sea squirt living on a rocky shore.** *Source: Courtesy NOAA; http://www.lib.noaa.gov/retiredsites/korea/main_species/sea_squirt.htm.*

Aquaculture species include the following:

- sea pineapple (*H. roretzi*)
- sea peach (*Halocynthia aurantium*)
- *Styela clava*
- *Styela plicata*

The production of sea squirts by aquaculture is estimated at about 10,000 mt globally with Korea and Japan being the major producers (FAO, 2012) (Fig. 8.6). Consumption is mainly in Asia, Chile, and around the Mediterranean Sea (reviewed by Lambert et al., 2016). The advantages of tunicates are that they have a high protein content but low fat. The disadvantages are that they bio-accumulate toxicants in the water (Lambert et al., 2016).

8.10 EDIBLE JELLYFISH

Jellyfish consist of a bell and trailing tentacles; often containing toxins. They float in the oceans and move in ocean currents and due to their own pulsating movements. A few species of jellyfish, such as cannonball jellyfish (*Stomolophus meleagris*) and jelly blubber (*Catostylus mosaicus*) are used for human food. About half a million metric tons of edible jellyfish (phylum Cnidaria) are harvested per year (Kitamura and Omori, 2010; You et al., 2016):

- Southeast Asia, 169,000 mt wet weight
- China capture harvested, 300,000 mt wet weight
- China aquaculture, 60,000 mt wet weight

8.11 BIOMEDICAL USES OF INVERTEBRATES

8.11.1 Invertebrate Animal Models

The genomes of *Drosophila melanogaster* (fruit fly) and *C. elegans* (a nematode) were among the first genomes to be reported. There are multiple other invertebrate genomes sequenced

(The Genome Institute at Washington University, 2016) with many more underway.

8.11.2 Jellyfish (*Aequorea victoria*) Green Fluorescent Protein

Green fluorescent protein from a jellyfish (*Ae. victoria*) is used in biomedical research. For instance it is used to label and follow cell movements during the development of transgenic mice and zebrafish (reviewed by Progatzky et al., 2013). The use of green fluorescent protein from jellyfish was recognized in the Nobel Prize for chemistry (2008).

8.11.3 Therapy Maggots

Maggots are blowfly larvae. They are used in maggot therapy for cleaning wounds of dead tissue. Medical Maggots have been cleared for use by the FDA (Medical Maggots, 2017).

Blowflies Phylum: Arthropoda Class: Insects Order: Diptera

8.12 OTHER USES

There are other uses of invertebrate animal species, such as vermiculture.

8.12.1 Vermiculture

Earthworms (phylum Annelida), usually red wigglers (*Eisenia foetida*) or European night crawlers (*Eisenia hortensis*), are used to compost organic materials, such as pig and cattle manure, agricultural and yard waste, and food waste (e.g., cafeteria, coffee shop, restaurant, and groceries). The worm castings result in organic fertilizer that can be applied to the land. The process of composting is usually odorless. There is a growing group of vermiculture enthusiasts and advocates within the organic food, locally produced food, and urban agricultural movements. Canadian night crawlers (*Lumbricus terrestris*) together with European night crawlers are raised by vermiculture to be sold as bait for fishing.

8.13 CONCLUSIONS

Despite the invertebrate species being 97% of the planet's animal species, their usage is limited to a relatively few species. It is argued that there is tremendous scope for utilization of more invertebrates, their proteins and other constituents, and their associated microbiomes.

References

American Beekeeping Federation, 2012. Available from: http://www.abfnet.org/displaycommon.cfm?an=1&subarticlenbr=183.

Arcot, J., Brand-Miller, J., 2005. A preliminary assessment of the glycemic index of honey. A report for the Rural Industries Research and Development Corporation. RIRDC Publication No. 05/027.

Ayieko, M.A., Oriaro, V., 2008. Consumption, indigenous knowledge and cultural values of the lakefly species within the Lake Victoria region. Afr. J. Environ. Sci. Technol. 2, 282–286.

Banjo, A.D., Lawal, O.A., Songonuga, E.A., 2006. The nutritional value of fourteen species of edible insects in southwestern Nigeria. Afr. J. Biotechnol. 5, 298–301.

Barennes, H., Phimmasane, M., Rajaonarivo, C., 2015. Insect consumption to address undernutrition, a national survey on the prevalence of insect consumption among adults and vendors in Laos. PLoS One 10, e0136458.

Bray, G.A., Nielsen, J.N., Popkin, B.M., 2004. Consumption of high-fructose corn syrup in beverages may play a role in the epidemic of obesity. Am. J. Clin. Nutr. 79, 537–544.

Budd, G.E., 2013. At the origin of animals: the revolutionary Cambrian fossil record. Curr. Genomics 14, 344–354.

Carboni, S., Addis, P., Cau, A., Atack, T., 2012. Aquaculture could enhance Mediterranean Sea urchin fishery, expand supply. Glob. Aquacul. Adv. 15, 44–45.

Derby, C.D., 2014. Cephalopod ink: production, chemistry, functions and applications. Mar. Drugs 12, 2700–2730.

Derby, C.D., Kicklighter, C.E., Johnson, P.M., Zhang, X., 2007. Chemical composition of inks of diverse marine molluscs suggests convergent chemical defenses. J. Chem. Ecol. 33, 1105–1113.

FAO, 2012. Fishery and aquaculture statistics. Available from: ftp://ftp.fao.org/fi/Cdrom/CD_yearbook_2012/root/aquaculture/b74.pdf.

FAO, 2013. Nutritional value of insects for human consumption. In: Edible Insects Future Prospects for Food and Feed Security, pp. 67–88. Available from: http://www.fao.org/docrep/018/i3253e/i3253e06.pdf.

FAO, 2014. The state of world fisheries and aquaculture. Available from: http://www.fao.org/3/d1eaa9a1-5a71-4e42-86c0-f2111f07de16/i3720e.pdf.

FAOStats, 2015. Available from: http://faostat3.fao.org/browse/Q/*/E.

Gallai, N., Salles, J.-M., Settele, J., Vaissière, B.E., 2009. Economic valuation of the vulnerability of world agriculture confronted with pollinator decline. Ecol. Econ. 68, 810–821.

Hedges, S.B., Blair, J.E., Venturi, M.L., Shoe, J.L., 2004. A molecular timescale of eukaryote evolution and the rise of complex multicellular life. BMC Evol. Biol. 4, 2–12.

Hejnol, A., Obst, M., Stamatakis, A., Ott, M., Rouse, G.W., Edgecombe, G.D., Martinez, P., Baguñà, J., Bailly, X., Jondelius, U., Wiens, M., Müller, W.E., Seaver, E., Wheeler, W.C., Martindale, M.Q., Giribet, G., Dunn, C.W., 2009. Assessing the root of bilaterian animals with scalable phylogenomic methods. Proc. R. Soc. B Biol. Sci. 276, 4261–4270.

International Development Research Centre, 1998. Carmine dye extraction process and the cochineal insect. Available from: https://idl-bnc.idrc.ca/dspace/bitstream/10625/21568/1/116551.pdf.

Jacobson, M.F., 2004. High-fructose corn syrup and the obesity epidemic. Am. J. Clin. Nutr. 80, 1081.

Kitamura, M., Omori, M., 2010. Synopsis of edible jellyfishes collected from Southeast Asia, with notes on jellyfish fisheries. Plankton Benthos Res. 5, 106–118.

Klurfeld, D.M., Foreyt, J., Angelopoulos, T.J., Rippe, J.M., 2013. Lack of evidence for high fructose corn syrup as the cause of the obesity epidemic. Int. J. Obes. (Lond.) 37, 771–773.

Kuznesof, P.M., 2005. Beeswax. Chemical and technical assessment. Food and Agriculture Organization. Available from: http://www.fao.org/fileadmin/templates/agns/pdf/jecfa/cta/65/beeswax.pdf.

Lambert, G., Karney, R.C., Rhee, W.Y., Carman, M.R., 2016. Wild and cultured edible tunicates: a wild sea squirt living on the rocky shore. A review. Manage. Biol. Invasions 7, 59–66.

Larson, G., Fuller, D.Q., 2014. The evolution of animal domestication. Annu. Rev. Ecol. Syst. 45, 115–136.

Margariti, C., Protopapas, S., Allen, N., Vishnyakov, V., 2013. Identification of purple dye from molluscs on an excavated textile by non-destructive analytical techniques. Dyes Pigm. 96, 774–780.

McBride, S.C., 2005. Sea urchin aquaculture. Am. Fish. Soc. Symp. 46, 179–208.

Medical Maggots, 2017. Medical Maggots™. Available from: http://www.monarchlabs.com/mdt.

Nevas, M., Lindström, M., Hautamäki, K., Puoskari, S., Korkeala, H., 2005. Prevalence and diversity of *Clostridium botulinum* types A, B, E and F in honey produced in the Nordic countries. Int. J. Food Microbiol. 105, 145–151.

Nevas, M., Lindström, M., Hörman, A., Keto-Timonen, R., Korkeala, H., 2006. Contamination routes of *Clostridium botulinum* in the honey production environment. Environ. Microbiol. 8, 1085–1094.

Oonincx, D.G., De Boer, I.J., 2012. Environmental impact of the production of mealworms as a protein source for humans—a life cycle assessment. PloS One 7, e51145.

Oonincx, D.G., van Itterbeeck, J., Heetkamp, M.J., van den Brand, H., van Loon, J.J., van Huis, A., 2010. An exploration on greenhouse gas and ammonia production by insect species suitable for animal or human consumption. PLoS One 5, e14445.

Payne, C.L.R., Scarborough, P., Rayner, M., Nonaka, K., 2016. Are edible insects more or less 'healthy' than commonly consumed meats? A comparison using two nutrient profiling models developed to combat over- and undernutrition. Eur. J. Clin. Nutr. 70, 285–291.

PhysOrg, 2013. Available from: http://phys.org/news/2013-12-israel-elusive-biblical-blue.html.

Progatzky, F., Dallman, M.J., Lo Celso, C., 2013. From seeing to believing: labelling strategies for *in vivo* cell-tracking experiments. Interface Focus 3, 20130001.

Rodriguez, L.C., Niemeyer, H.M., 2001. Cochineal production: a reviving Precolumbian industry. Athena Rev. 2, 76–78.

Roffet-Salque, M., Regert, M., Evershed, R.P., Outram, A.K., Cramp, L.J., Decavallas, O., Dunne, J., Gerbault, P., Mileto, S., Mirabaud, S., Pääkkönen, M., Smyth, J., Šoberl, L., Whelton, H.L., Alday-Ruiz, A., Asplund, H., Bartkowiak, M., Bayer-Niemeier, E., Belhouchet, L., Bernardini, F., Budja, M., Cooney, G., Cubas, M., Danaher, E.M., Diniz, M., Domboróczki, L., Fabbri, C., González-Urquijo, J.E., Guilaine, J., Hachi, S., Hartwell, B.N., Hofmann, D., Hohle, I., Ibáñez, J.J., Karul, N., Kherbouche, F., Kiely, J., Kotsakis, K., Lueth, F., Mallory, J.P., Manen, C., Marciniak, A., Maurice-Chabard, B., Mc Gonigle, M.A., Mulazzani, S., Özdoğan, M., Peric', O.S., Peric', S.R., Petrasch, J., Pétrequin, A.M., Pétrequin, P., Poensgen, U., Pollard, C.J., Poplin, F., Radi, G., Stadler, P., Stäuble, H., Tasic', N., Urem-Kotsou, D., Vukovic', J.B., Walsh, F., Whittle, A., Wolfram, S., Zapata-Peña, L., Zoughlami, J., 2015. Widespread exploitation of the honeybee by early Neolithic farmers. Nature 527, 226–230.

Shantibala, T., Lokeshwari, R.K., Debaraj, H., 2014. Nutritional and antinutritional composition of the five species of aquatic edible insects consumed in Manipur, India. J. Insect Sci. 14, 14.

The Genome Institute at Washington University, 2016. Available from: http://genome.wustl.edu/genomes/category/invertebrates/.

USDA NASS (United States Department of Agriculture National Agricultural Statistics Service), 2015. Available from: http://www.usda.gov/nass/PUBS/TODAYRPT/hony0315.txt.

Vane-Wright, R.I., 1991. Why not eat insects? Bull. Entomol. Res. 81, 1–4.

Wallberg, A., Han, F., Wellhagen, G., Dahle, B., Kawata, M., Haddad, N., Simões, Z.L., Allsopp, M.H., Kandemir, I., De la Rúa, P., Pirk, C.W., Webster, M.T., 2014. A worldwide survey of genome sequence variation provides insight into the evolutionary history of the honeybee *Apis mellifera*. Nat. Genet. 46, 1081–1088.

Wang, D.Y., Kumar, S., Hedges, S.B., 1999. Divergence time estimates for the early history of animal phyla and the origin of plants, animals and fungi. Proc. R. Soc. B Biol Sci 266, 163–171.

Yhoung-Aree, J., Puwastien, P., Attig, G.A., 1997. Edible insects in Thailand: an unconventional protein source? Ecol. Food Nutr. 36, 133–149.

You, K., Bian, Y., Ma, C., Chi, X., Liu, Z., Zhang, Y., 2016. Study on the carry capacity of edible jellyfish fishery in Liaodong Bay. J. Ocean Univ. China 15, 471–479.

Animals in the Military

Colin Salter

University of Wollongong, Wollongong, NSW, Australia

9.1 INTRODUCTION: OTHER ANIMALS ARE NOT QUITE PRESENT

In April 2016, the Macquarie Mint produced a limited-edition run of a $10 silver coin in Australia as part of commemorations for "100 Years of World War 1." Designed to appear as an official government document, the direct marketing campaign included a letter mailed "to selected households" in Australia and referred to the coin as "Symbolising the determination, courage, self-sacrifice and mateship of the Aussie digger." "Digger" is a slang term for soldiers from the Australian and New Zealand Army Corps (ANZAC), formed during World War I (WWI). The character traits described are considered part of the Australian (and to a lesser extent New Zealand) lexicon. Often evoked by politicians, commentators, and other personalities in Australia and New Zealand, the image of the ANZAC is used to evoke notions of nationhood and identity. These notions are evident in the language used to describe the coin itself:

Honouring every Australian who served during the nation's Baptism of Fire, this new silver coin depicts Australia's most powerful emblem of WWI—the iconic Simpson & his donkey *Sealy, C., 2016, Exchange $10 for a $10 silver coin. Personal communication*

Simpson and his Donkey is broadly considered as the "most prominent symbol of Australian courage and tenacity" for their exploits at Gallipoli during WWI (Australian War Memorial, no date). Simpson landed at Gallipoli (at what is now known as ANZAC cove) on April 25, 1915, provided first aid, and assisted in the transport of wounded soldiers. He survived the horrors of war for 3 weeks, before being killed by gunfire on May 19, 1915. Historian Peter Cochrane (1992) describes Simpson as "the pre-eminent legend of Australian heroism and sacrifice" (p. 1). Notorious for recovering wounded soldiers under fire, their actions have become an iconic feature of Australian wartime history. Simpson is presented as being emblematic of certain desired/promoted Australian character traits: tenacity, ingenuity, courage, and mateship. Simpson and his donkey form part of a process of identity formation—expressly nationalistic in this example. In simple terms, Simpson is an Australian legend. The image of a donkey in his *service* emblematizes and reinforces nation and identity in the act of selfless sacrifice—a notion common across representations of perceptions of Other animals, how societies frame their use, attempts to protect specific species, and concern for select individuals (Berger, 1980).

Other Animals?

How to accurately, appropriately, and consistently refer to members of other species represents an ongoing challenge in the humanities and social sciences. In simple terms, humans are animals yet we routinely see ourselves as separate. Such a distinction is founded in Rene Descarte's (1596–1650) description of (nonhuman) animals as soulless automata. In contrast, humans are positioned as possessing an immaterial soul. Having a soul provides for individual and subjective experiences. Many have sought to challenge the dualistic separation of the human from what is described as an irrational nature central to the Cartesian worldview (Regan, 1983). The concept of speciesism made popular by Peter Singer (1975), identified a shared capacity to suffer as a counter to notions of human superiority.

Other animals is used throughout this chapter to illustrate the arbitrary and anthropocentric basis of such distinctions. "Other" also has a double meaning. Other animals are Other in that they are not human, and they are Other in that they are relationally positioned (i.e., socially constructed) as different, with difference implying deficiency. Exposing the arbitrariness of the distinction also highlights implications—exposing that which "frequently is obscured in order to advance a powerful ideological divide that furthers terrible acts of oppression" (Nibert, 2013, p. 6). Nibert identifies the ideological basis of positioning Other animals as inferior to us and what such positioning enables. Reflecting on Singer's analysis in detail, the implications of speciesism, the othering of Other animals makes them available to us. They become usable for our own ends with little consideration for their own intrinsic value and self-worth.

The examples presented in this chapter are explicitly illustrative of the outcomes of the social construction of Other animals as available and usable. The use of Other animals in the specific context of militarism and war is situated within an overarching set of normative everyday practices across Western societies. We can look to our homes for an everyday experience of many:

> Where peaceable co-existence between humans and animals creates possibilities for friendship, such as with companion animals, this bond is placed in question by the modes of discipline, surveillance, containment and control that attend and are inherent to the practice of 'pet ownership' and 'domestication.' The millions of pets 'euthanised' in animal shelters annually highlight that even examples of seemingly happy cohabitation between humans and animals are framed within an 'adopt, foster, euthanize' context of over-arching, and deadly, violence **Wadiwel, 2015, p. 56**

Socially constructing Other animals as *other* enables widespread and varied forms of their use. It is within such a context that Other animals become tools and weapons for military purposes.

Representations of Simpson and his donkey in accounts of their "service" provide a very telling indictor of the social construction of Other animals. Perhaps most obvious is the use of the reductionist moniker "donkey." During his time at Gallipoli, Simpson conscripted up to five different donkeys into his service and each of them were named *Murphy*, *Queen Elizabeth*, *Abdul*, and two named *Duffy*. The use of the singular moniker *donkey* has a number of implications persistent across the use and representation of Other animals for military purposes and provides an illustrative case study to frame this chapter. What we can draw from these representations include a number of longstanding issues across a history of complex and routinely contradictory attitudes toward Other animals, and processes through which memorialization can act to render the individual and the collective animal absent and effectively invisible.

We can look to Stephen Spielberg's 2011 drama *War Horse* for a similarly inconsistent representation of another individual regarded for their "service" in war. *War Horse* provides a romanticized reimagining of the film's namesake, and exemplifies a number of historical and contemporary contradictions in the perception and use of Other animals. As a device of popular culture, the director seeks to memorialize and paint a happier picture of the use of Other animals in warfare. The romanticized nature of the representation is evidenced in discrepancies with the historical basis of the film:

> There was a real-life model for the war horse in the film. His name was Warrior, and he went to war in 1914 with his owner, a British general named Jack Seely. Both survived, defying even the horrors of the Somme. Warrior was celebrated as a hero and died in 1941, in the middle of another war, a few weeks short of his 33rd birthday. Although revered, he was not buried with the expected pomp and ceremony. Instead his body was donated to the war effort and used for meat **Battersby, 2012**

Warrior's exploits in the service of Jack Seely were deemed heroic and considered of significant (human) value. Alfred Munning's painting *Charge of Flowerdew's Squadron*, reproduced as Fig. 9.1, depicts the 1918 cavalry charge at Moreuil Woods in which Warrior carrying Jack was involved (Kenyon, 2011, p. 194).

The film's focus on Warrior (the memorialization and romanticized reimagining of an individual, as with Simpson and his Donkey) acts to render the military use of other animals invisible. All of these other individuals are effectively disappeared, as is any reflection on or consideration of their use. More directly, the memorialization of Warrior provides an overt distraction from the deaths of the 8 million other horses killed during the War—horses killed at the behest of human ends. Separately, memorializing Warrior's *service* as heroic provides a means through which certain forms of violence, be it Warrior becoming pieces of meat or being used as a weapon of human warfare, are normalized or smoothed in such a way that neither acts are seen as forms of violence (Wadiwel, 2015, p. 13).

The use of Warrior as a "war horse" provides an illustration of how his creative agency and autonomy were directed against his own interests. The intention and outcome of such disciplining is to maximize human use value as a delivery device for soldiers, a form of weaponry. We can even read the directing of Warrior's agency in such a way that he would gallop toward gunfire with Jack on his back, as effectively manipulating him to act to expedite his own death.

Warrior's real-life fate illustrates how his celebrated actions were not enough to ensure anything akin to the *honor* bestowed on (human) soldiers. Differently valuable in death, Warrior becoming meat is an example of the ultimate commodification of life through what Joseph Pugliese (2013) terms the "binary logic of anthropocentrism" (p. 33). As a weapon of war, Warrior had (human, military) value in life. Memorialization of his *sacrifice* was not itself enough to see that value extended beyond his death in comparable terms to his *owner*. One form of commodity in life, another in death. Both are human-centered.

Moving beyond the common reference to specific individualized animals, Warrior's corporeal self is just another part of the milieu of horses and other animals killed for human consumption—absent of any of the subjectivity inferred in his memorialization. Warrior became an absent referent, beyond, that is, his unwitting sacrifice to become "rations" for the war effort.

FIGURE 9.1   **The Charge of Flowerdew's Squadron (public domain).**

Being able to describe Warrior's body parts as rations is epistemic: "our knowledge systems do not allow us to see this as violence" (Wadiwel, 2015, p. 33). Warrior becoming meat can be viewed as contradictory *and* a consistent application of anthropocentrism writ large. Such manifestations can and do make us uncomfortable. One way to counter such discomfort is to alter the narrative, to rewrite history. Such a rewriting is clearly evident in the reimagining of Warrior and his fate.

Stephen Spielberg's *War Horse* presents a narrative account of the intertwined lives of Albert and a horse he names Joey. The film opens with a young Albert watching Joey's birth, forming the basis for a lasting relationship. A consistent theme of the film is how both survive and succeed seemingly against all odds. Success is implicitly framed in anthropocentric terms, with the film consciously and unconsciously hinting at manifestations of class and civility. Class is presented in the actions of Ted Narracott, Albert's father, who belligerently outbids his affluent landlord to purchase Joey at the local horse auction and subsequently in his selling of Joey to a gentlemanly Captain John Nicolls in 1914, for the war effort. Joey is sold so Ted can make a rent payment and not lose the farm.

Nicolls is killed during a *civilized* cavalry charge on the Western Front, with his unit effectively wiped out by German machine guns. Joey survives (against all odds) and is captured and subsequently comes into the possession of Emilie, a young French woman, and her grandfather before being recaptured by German soldiers. He is utilized as an artillery horse, pulling guns through the mud and wastelands of the front. In contrast to the romanticized image of the British Cavalry, the film portrays German attitudes toward horses as less civilized. The expendability/replaceability of Other animals for human ends is evident in reference to Joey as strong and that he "should last a month or two." Joey escapes again, only to become entangled in the mass of barbed wire strewn across *no-man's land* between British and German lines.

In one of the more explicit examples of romantic reimaginings in the film, Peter and Colin (German and British soldiers, respectively) work together to free Joey from the barbed wire. Reminiscent of accounts of football matches played between British and German troops between trenches during the Christmas truce of 1914, such as that at Flanders, the scene represents a form of common kindness toward Warrior. Both Peter and Colin are risking their lives to help this hapless horse. Following frank and friendly conversations about the futility of war they flip a coin to determine who takes possession of Joey. Colin wins.

While all this was happening, Albert enlists in the army with the hope of being reunited with Joey. Deployed to the Western Front as an infantryman, Albert is injured during a gas attack after surviving a *heroic* charge across no-man's land toward a machine gun position. While recovering from his injuries, Albert hears accounts of a miraculous horse. He is led, with bandages covering his wounded eyes, to find an injured and weary Warrior being assessed by the army

doctor. The doctor prescribes a bullet for Warrior's troubles. Albert eventually convinces the doctor that Warrior is Joey . They are given a reprieve and Warrior is added to the list of surplus items to be auctioned.

Albert takes up a collection, though cannot match the figure a butcher offers (double the rate for other horses). An unseen man in the crowd makes a large bid and states that he is willing to sell his farm to purchase Joey. Hope prevails, and the butcher withdraws. We come to see the man is Emilie's grandfather, who traveled for 3 days in search of Joey after hearing news of the auction. Albert pleads with Emilie's grandfather to be given Joey, who refuses. When Joey shows an interest in Albert, Emilie's grandfather acquiesces. Hope prevails once again. Emilie's grandfather refuses to take any money in return: "He belongs to you. That is of course what my little girl would have wanted." Albert and Joey return to the family farm in England to happily live out their days.

The remaking of Warrior into Joey symbolically embodies the image of an individual horse as being too valuable (to humans) to become rations. In many ways both Warrior and Joey are absent referents. Their memorialization is almost exclusively about the humans around them. For Joey, how much he meant to Emilie (and her grandfather), to Albert (and Ted), to Peter and Colin, and to all those who had heard accounts of the *miraculous* horse. Memorialization in accounts of their actions are themselves steeped in anthropomorphic and anthropocentric assumptions and values—even insofar as we attempt to ascribe value in and of themselves in the context of their use for human ends. These are directly circumscribed by a frame of instrumental relations—an ontology of usability. In remembering and memorializing the actions of selected individuals in the service of humans, there is no questioning of "whether we should use them in the first place" (Wadiwel, 2015, p. 22). Accounts, such as those of Warrior and the romanticized pop culture reimagining as Joey in *War Horse* fit *our* narrative, much like that of representations of Simpson and his donkey as embodiments of the ANZAC spirit. We identify value in what we want to see as valuable. Recalling the commemorative $10 coin, "determination, courage, self-sacrifice, and mateship." Focusing on these individuals provides a distraction from the largely untold suffering and deaths of all the other individual horses, donkeys, and others forced into military use.

Ontological Usability

Fundamentally, ontology refers to the study of the nature of being; how we make sense of the world around us. Our ontology helps us to understand and to make decisions about our actions and the actions of others. We attach meaning, define, and differentiate between objects and beings based on our ontology. In simple terms, how we perceive and understand our place in the world and those around us.

How we view and understand the existence of Other animals and the relationships we have with them is directly linked to our ontology. In Western (and a number of other) societies, it is normative to view Other animals as available to us. Be this as companions, as tools, or as food.

> The epistemic violence of producing 'the animal' as an inferior entity, and therefore susceptible to all guises of human utility—reproduced, extinguished made captive, hunted, companionised, tortured and experimented upon—already indicates a monstrous endeavour... *Wadiwel, 2015, p. 34*

This instrumental view of Other animals (dominant in the West) positions them as usable, ontologically.

Warrior's eventual fate is just one of many implications for Other animals irrespective of how they may have been utilized for human ends. They are routinely left behind following military actions. Examples include horses used during the charge of the Australian Light Horse at the Battle of Beersheba, dogs used during the American-led invasion and occupation of Vietnam, and in contemporary military incursions in Iraq and Afghanistan (Allsopp, 2015, pp. 49–50). Such abandonments are at times questioned and challenged. For example, following sustained campaigning by soldiers and animal welfare advocates, the passing of the USA's National Defense Authorization Act 2016 provides avenues for dogs used by the military to be returned to the USA. The American Humane Association (2015) described the passage of the Act as "guaranteeing" their return. One direct implication of what we made read as a positive step is that a focus on the return of these dogs effectively renders their use unquestioned. The violence underpinning the situation in which these dogs are transported to a war zone becomes quite imperceptible; the basic structure of domination itself is left intact (Wadiwel, 2015, p. 22).

Representations of Simpson and his donkey, and the outcomes for Warrior and dogs repatriated under the National Defense Authorization Act 2016 are indicative of the multifaceted implications of military use of Other animals. These implications are situated within normative, exploitive, and self-serving anthropocentric assumptions and justifications. Other animals are effectively positioned as ontologically usable. What this means is that in and off itself there is no need for us to justify our use of them. Other animals are socially constructed as being of a lower natural order, much in the same way that racialized humans have been used for millennia as a feature of colonial and imperial conquests. Pugliese (2013) more directly links these two bases for exploitation:

> Racism is predicated on speciesism. At every turn in the documentary history of racism, the spectre of speciesism has always-already inscribed the categorical naming of the racialised other (p. 41).

In the use of (slaves and) Other animals, questions only emerge where either of the following is considered: (1) such use will impact on the property rights of the "owner" or (2) the pain and suffering may be considered unnecessary *to us*. For (1), Other animals as property is a broad cornerstone of contemporary Western societies (Francione, 2007). For (2), *necessary* is specifically defined self-servingly with respect to human ends (Deckha, 2013). In the context of military use, a reason can be as simple as shifting the risk of injury and death away from humans. In parallel, the attributes of selected species—as with dogs in Vietnam and horses during WWI and World War II (WWII)—biopolitically shape their strategic and military value. Their use in saving human lives, whether direct or indirect, provides an attempt to develop an anthropocentric solution to an anthropocentric problem.

Memorialized accounts of Simpson and his donkey are always/already indicative of attitudes toward, and assumptions about, Other animals. His donkeys were a means, a tool, partially considered in the context of their use value. The donkey's values are seen as being emergent through Simpson's ingenuity, not afforded through any inherent value in and of themselves (i.e., as an individual, a subject, with their life having value in itself). The popularized and celebrated *singular* moniker "donkey," when considered beyond this simple framing is itself illustrative of a number of far-reaching implications.

The regular and widespread use of "Simpson and his donkey" is a reductionist representation. There is a lack of consistency and contradictions across accounts of Simpson and the donkeys he appropriated. These include the number of donkeys, names of each donkey, and general confusion about his exploits more broadly. Allsopp (2015) notes: "Simpson had used other donkeys during the same battle, in fact no one really knows how many, some suspect three, other sources state five. Some of the donkeys have

been recorded as being named Duffy, Abdul, and Queen, but again Simpson was heard to call his donkeys other names, such as Elizabeth, as well as several unpleasant words" (p. 154).

Whereas a number of accounts refer to Murphy, Queen Elizabeth, Abdul, and the two Duffies, actual reference to each of them as individuals is largely absent. This process of not naming acts to reduce each of them to the status of a tool with an anthropocentric purpose. Their presence as independent entities in themselves with their own inherent value is displaced. The process enables epistemic violence. Carol Adams (1990) identifies such process as central to the construction of an absent referent. The process of "thingifying" individuals changes the way they are perceived and valued. Shifting away from inherent worth, their meaning now "derives from [their] application or reference to something else" (Adams, 1990, p. 42). Murphy, Queen Elizabeth, Abdul, and the two Duffies become *Simpson's* donkey. That is the source of their value and what they become known or remembered for.

> The absent referent is both there and not there. It is there through inference, but its meaningfulness reflects only upon what it refers to because the originating, literal, experience that contributes the meaning is not there. We fail to accord this absent referent to its own existence *Adams, 1990, p. 42*

The fate of Murphy, Queen Elizabeth, Abdul, and the two Duffies (their existence) is transformed into the other-than-life memorialization of Simpson. They become a singular donkey in his service. They have value and their lives have meaning insofar as they are attached to him and his romanticized exploits. Extending on the construction and enabling of an absent referent that the nonnaming of Murphy, Queen Elizabeth, Abdul, and the two Duffies facilitates, the moniker of a singular donkey directly renders invisible something that is (or at least should be) simply and directly obvious. All but one of Murphy, Queen Elizabeth, Abdul, and the two Duffies were either killed or injured seriously enough to be deemed of insignificant use value and replaced. They were all exchangeable/replaceable.

Murphy, Queen Elizabeth, Abdul, and the two Duffies' individual actions, mobilized to valorize Simpson (including notions of bravery and courage under fire in saving countless human lives) are consolidated into the depersonalized singular and generic *donkey*. Individually, and collectively, they are effectively disappeared. They became (and were) tools for Simpson, invisiblized within and by the binary logic of anthropocentrism:

> In this fundamentalist species bias, human interests come first, with actions that harm nonhuman animals (and ecosystems) being morally permissible as long as human (self) interests are not harmed. This extends beyond utilitarian principles, where nonhuman animals are only given consideration pertaining to their interests being aligned with human interests **Salter, 2014, p. 9**

What happened to Murphy, Queen Elizabeth, Abdul, and the two Duffies, as Simpson's donkeys (and it is important to note this *continues* to happen) is commonplace and normative in human conflict. As is typical in wars, the impacts on Other animals, be it direct, such as those killed as a result of their use, being targeted to prevent their use (Human Rights Watch, 1990), indirectly killed as "collateral damage" (The War You Don't See, 2010), or as a result of broader impacts including unexploded munitions, habitat loss, and the broader ecological impacts of warfare (Andrzejewski, 2014), are largely underreported or not reported at all. We can look to photographer Steve McCurry's first-hand observations for an atypical account. He noted, in the aftermath of the 1991 USA-led Gulf War, coming "across cattle, camels and horses wandering around like zombies" (Tyler, 2014, p. xiii). Here we can locate some of the more-direct-of-the-indirect impacts of war on Other animals. McCurry also noted other, less direct, impacts through speculating on the futures of these individuals: "I guess most died eventually—all the water holes and vegetation were covered in oil" (Tyler, 2014, p. xv). Central to what McCurry

witnessed is an ontology of nonconsideration. Nonconsideration of the impacts of the USA-led military actions, codenamed Operation Desert Shield, on Other animals is normative and widespread. It is ontological as it is premised on the social construction of a species hierarchy in which we, based on self-serving assumptions, place ourselves at the top. The nonconsideration evident in McCurry's account is perpetuated by, and in multiple ways a precursor of, interlinked ontologies of the availability and usability of Other animals. In other words a normative perception that Other animals are available and useable in whichever way and form *we* choose (i.e., Warrior as an explicit weapon of war and as flesh to differentially sustain the war effort). As Dinesh Wadiwel (2015) notes:

> Once we assume we have a right of dominion, then it would seem that ethics is forced to attend to questions of how we use this dominion; that is, how we use animals, rather than whether we should use them in the first place. In other words, ethics becomes a question of how to manage or regulate the effects of our own self proclaimed dominion (p. 22).

The reimagined and romanticized fate of Warrior in Stephen Spielberg's film *War Horse* (represented in the actions of Albert, his fellow soldiers, and others in saving Joey from *becoming* rations) provides a clear example of managing and regulating the ontological usability of animals, or, in Wadiwel's words, how our self-proclaimed dominion is utilized. In the film, our dominion is used to save one individual (Joey) from becoming rations, thereby regulating the effects of our own self-proclaimed dominion over Other animals largely to make us feel better about ourselves. In this act of saving, Joey is effectively remade into an absent referent. In simple terms, Other animals are not quite present.

9.1.1 Other Animals Are Present for Us

The ongoing legacy and persistence of pervasive assumptions, which underpins and enables nonconsideration of the inherent value and individual lives of Other animals, is widespread. We can look to John M. Kistler's (2011) dedication in the opening pages of *Animals in the Military: From Hannibal's Elephants to the Dolphins of the U.S. Navy* for a clear example:

> Dedicated to all the animals of history that have benefited our lives on Earth, and the men and women who have cared for them.

What Kistler infers here is that Other animals have value, and should be offered consideration, insofar as they afford benefit to/for and enrich human lives. Individual animals are clearly an absent referent, at best an afterthought. Kistler's description reflects the fate of Warrior, is broadly representative, and provides an accurate depiction of normative attitudes toward Other animals and the *why* and *how* of their use in human conflict. They are a means to an (human) end. Such assumptions epitomize the history, current, and potential future use of animals in the military (Singer, 2009).

Representations of Simpson and his donkey hint at a relationship between himself and Murphy, Queen Elizabeth, Abdul, and the two Duffies. Narratives of individual Other animals are often constructed along these lines. Personal relationships are routinely evoked, working as a rhetorical and semantic tool to ascribe value to these other-than-human individuals. Such relationships are necessary for any form of individualized consideration, comprising a process with a long history and one that continues through to the present day. It is important to note here that such emotional connections are often real and can be mutual. There is often an interdependency among humans and Other animals, which requires critical and reflective engagement (Taylor, 2014, p. 110). Such connections and relationships can be somewhat reciprocally based or emergent through and relating from the structures of domination inherent in the binary logic of anthropocentrism.

It is important to note that Other animals have agency here, albeit specifically circumscribed and framed within anthropocentric and anthropomorphic constraints. It is in the context of such relationships that human benefits are mobilized as a plea for saving, remembering, and memorializing individual Other animals. For Simpson's donkeys (again noting they are referred to as *his*), their role in helping him to save human lives is central to memorialization. Similarly, Warrior is recalled in reference to his owner Jack Seely. As with the depiction in *War Horse*, there are numerous examples of soldiers seeking to be reunited with "man's best friend" after having been utilized in various roles across recent conflicts. For example, Lance Cpl. David Pond, who returned to the USA after deployment during the occupation of Afghanistan, sought to be reunited with his "military working dog" Pablo. Pond sought out being reunited with Pablo in the hope it will assist in *his* treatment for posttraumatic stress disorder and a traumatic brain injury: "He is my rock, my foundation" (Gutierrez and Connor, 2016). As Pond's case illustrates, Other animals are present for us.

9.1.2 An Anthropocentric Solution to an Anthropocentric Problem

Underlying all of the examples presented in this chapter is largely the nonconsidered use of Other animals. As a strategy, tactic, and a tool in human-to-human violence, the use of Other animals emerges from "a search for anthropocentric solutions to an anthropocentric problem" (Adams, 1997, p. 28). Before more directly engaging with the intent of this maxim, some unpacking is required: "solution" and "problem" are loaded terms. Peace and conflict studies has provided theoretical and conceptual tools to question and challenge the use of warfare as a potential *solution* to human conflict. Rather than seeing conflict as a *problem*, approaches to responding to conflict provide opportunities to move beyond such a simplistic and constructed binary (Barash and Webel, 2008).

Conflict provides an opportunity to reflect on its bases, its foundational elements. Working with conflict, seeking to transform it, provides for an approach with the potential for lasting change. The blunt instrument of warfare imposes the will of one nation, or coalition of nations, over others. The apparent quick fix is not an effective means of lasting change (Lederach, 2003). It is within such a routinely unquestioned context that a number of the problems with waging warfare emerge and are responded to. The injuries caused to soldiers and their lasting impacts (earlier reference to (human) posttraumatic stress disorder provides one example) are influencing military tactics. There are increasing attempts to remove humans from riskier roles and from battlefields entirely. Other animals are being reconstructed as valuable and usable for military purposes—an option that emerges in contexts of apparent crisis.

Adams (1997) developed her maxim of anthropocentric solutions to anthropocentric problems through identifying common elements among responses to a 1996 "outbreak" of the neurodegenerative disease bovine spongiform encephalopathy in farmed cows in the United Kingdom. Colloquially known as "mad cow disease," the outbreak was a direct outcome of feeding them the nervous tissue of farmed sheep, and leading to the slaughtering of 4.4 million individuals:

> The mad cow crisis highlighted the glaring indifference regarding the lives of other animals who have been ontologized as usable. What was new in the mad cow crisis was not this ontology but the unquestioning adherence to it... *Adams, 1997, p. 28*

What Adams identified as common across accounts is a common goal of maintaining the instrumental use of Other animals. Rather than stopping the use of cows as a source of food, the supply process of producing cow parts required refinements. Such an approach embodies an unquestioning adherence to the ontology of usability, which is similarly evident in and a key feature of military/industrial responses to increasing societal awareness of the ongoing human costs of war. Rather than question war, attempts are being made to reduce the potential for human impacts. One example being a return of Other animals to the battlefield (Salter, 2015). Positioned as usable and expendable, Other animals are biopolitically located as suitable/ideal substitutes for humans. Other animals can be trained, *improved*, and otherwise deployed where the risks to humans are considered too great. For example, Pablo was utilized in bomb detection roles, thereby reducing the risk to Lance Cpl. David Pond and others involved in the military occupation of Afghanistan. Whereas deference to the heightened senses of dogs are mobilized to justify their use, they are expressly used to reduce the potential of injury to humans deployed in such roles. An anthropocentric solution for an anthropocentric problem.

9.2 HISTORICAL ROOTS, EARLY USES

Accounts of Simpson and his donkey and the romanticized/reimagined portrayal presented in the film *War Horse* refer to more-recent uses of animals in the military. Robert E. Lubow's (1977) *The War Animals* locates and partially situates the history of use of Other animals for military purposes:

> The history of the use of animals in combat probably dates back to the time of man's first war. When this occurred is in part dependent on one's definition of war, as well as some clear notion of the origin of man. It is reasonable to assume that war's preceded man's first written accounts of himself and therefore the beginnings of animal-man partnership are probably lost (p. 20).

The gendered language notwithstanding, reference to the use of Other animals for human warfare as part of a *partnership* is a common misnomer. It illustrates the depth of normativity in references to status, use, and how notions of *relationship* are constructed/perceived. What is clear is that the use of Other animals for human warfare go back to a millennia. David Nibert's (2013) comparative historical analysis in *Animal Oppression and Human Violence* argues that the use of Other animals as instruments of warfare was enabled and required by the process of domestication. The amassing of military power was deemed necessary to protect land, to maintain large groups of farmed animals, and for a further amassing of economic benefits from such activities (i.e., capitalist expansion). Nibert situates such institutionalized violence as directly entangled with large-scale human-to-human violence, in particular targeting groups of "devalued" humans. Forms of violence including "invasion, conquest, extermination, displacement, coerced and enslaved servitude, gender subordination and sexual exploitation, and hunger" were all linked to expropriation of land for the farming of Other animals (Nibert, 2013, p. 5). One of the many examples used to support his analysis is that of the invasion of the West Indies:

> As conquest proceeded from island to island and then on to the mainland, settlement and colonization were always secured by a base of *cattle*. As the conquered habitations became stabilized, the surplus from his *herds* spread out to virgin islands and founded *wild* populations which supported future advances **Miessner and Morrison, 1991, p. 81, cited in Nibert, 2013, p. 47**

Alongside Nibert's interlinking of institutionalized violence toward Other animals with processes of colonial expansion, the exploitation of Other animals for human warfare is inherently tied to human-to-human violence in organized warfare. Such direct or indirect harm is a constant and continuing element of human society and evident in all uses of Other animals in military service stretching back for as far as written records exist [and likely beyond as Lubow (1977) and Nibert (2013) refer to].

Colonialism, Speciesism, and Racism

In many ways, it is difficult to overlook inherent links between capitalist accumulation and domestication of Other animals identified by Nibert, and how these are situated with the treatment of Other cultures. The social construction of different, Other, cultures as being closer to nature is well documented in the humanities and social sciences (Plumwood, 1993). We can recall here Pugliese's (2013) argument that racism is predicated on speciesism:

> As a fundamentally colonial formation of power, premised on the pivotal role of racism in governing subjects and assigning them positions on racialized hierarchies of life that spanned the right to genocidal extermination (of Indigenous peoples) and of enslavement (of black Africans), biopolitics is informed by a parallel history of speciesism that extends back to the very establishment of human civil and political society—as premised on animal enslavement (domestication) (p. 38).

Drawing on the work of Jim Mason (2005) and Charles Patterson (2002), Pugliese refers to the establishment of animal agriculture as operating as a license for colonialism and the subjugation of Nibert's "devalued" humans. Speciesism and colonialism are marked by technologies of biopolitics, which determine who will live and who will die (Pugliese, 2013, pp. 39–40).

The situating of racialized humans as closer to nature emerged during the Enlightenment and provided a central justification for the premise of colonial expansion. To state more directly, bringing *civilization* to Other cultures and saving them from themselves was a cornerstone of Imperial attitudes. The legacy of such notions lives on today, and is directly engaged with in intersectional analyses across feminist, critical race, and critical animal studies scholarship (Adams and Gruen, 2014; Gaard, 2001; Kim, 2007). Controversy surrounding the live export of sheep and cows from Australia to countries including Indonesia and Egypt provide a contemporary example, and we can look to Jacqueline Dalziell and Dinesh Wadiwel's (2016) "playful" amending of Gayatri Spivak's (1988) critical provocation on British colonialism in India (white men saving brown women from brown men) for an illustration:

> The sentence 'White people saving white animals from non-white people' establishes a truth that Australians kill 'their own' animals in a civilized way, while non-white others do not *Dalziell and Wadiwel, 2016, p. 80*

Dalziell and Wadiwel's reference to *their own* animals, which emerged through the live export controversy, hints at the use of "our" in the context of Australian concerns over the methods used to kill sheep and cattle exported to former colonized countries. The methods of killing *white* animals used by *brown* people were relationally constructed as uncivilized and barbaric. Alongside the racist and neo-colonial implications (the epistemic violence inherent in framing the West as having a monopoly on "humane" methods of killing) of the relational framing of Other cultures as cruel and backward (i.e., not like *us*), the implication of a more suitable form of killing produces two other outcomes: (1) the killing of sheep and cows for human ends is rendered unquestioned and (2) directly emergent from Spivak's original work, an either/or binary is constructed. White people saving brown animals from brown people (so white people can then kill them) is effectively positioned as the only option. Any possibility for questioning the killing of Other animals itself is obfuscated. The response is limited to the form of domination.

The example of the live export controversy illustrates that the use of Other animals (for military purposes) throughout human history is routinely shaped by the construction of either/or binaries and the interlinking of exploitation across class, culture, and ethnicity. The writings of Edward Hautenville Richardson, founder of the first official British facility for training "war dogs," provides an illustration. Richardson's (1920) *British War Dogs, Their Training and Psychology* presents us with an example of entanglements between humans and others (Other animals and "lesser" humans) in warfare:

> A warrior would be preceded by a slave leading a fierce dog, which would attack at word of command, and while it engaged in close combat with the enemy, the master would dash into the conflict with every chance of success (p. 21).

The language used is intentional and clear. Evident is that both the *slave* and the dog were tools based on their lives being socially constructed as having lower value and being worthy of less consideration than their warrior/master. They were and are useful based on their perceived fierceness; assumed (and actual) by those targeted. These individuals were also expendable/replaceable. Referring back to Nibert (2013), they "required no pay," which also made them appealing (Ross, 1922, p. 257).

WWI provides us another example of the interchangeability of exploitation and domination of Other animals and racialized humans. During the East Africa campaign of 1916–17 more than 34,000 donkeys were deployed to transport equipment. Of these, 1042 survived. Environmental factors led to many deaths, with the most significant cause being disease transmission via tsetse fly bites. To protect these donkeys against being bitten (for the sole purpose of prolonging their ability to carry equipment) donkeys were treated with lethal doses of arsenic. In response to the eventual deaths of the treated donkeys (with similarly high casualty rates among horses and mules) "120,000 Africans had to be brought in to lug the guns, and carry the dead animals' packs" (Cooper, 2000, pp. 199–201). Africans were effectively a poor substitute, though one deemed to be better than using white people.

Whereas the replacement of donkeys with Africans provides little direct inference of comparative value, other sources exist that provide clear illustration of such assumptions and values. Kistler (2011) provides a very telling example of intersections between anthropocentric and colonial value constructions of Other animals and racialized humans:

> Almost overnight, horses went from lovely animal to queen of the battlefield. By 1770 BC[E], a horse was worth 7 bulls, or 10 donkeys, or 30 slaves (p. 101).

The context of Kistler's value representation is the apparently unexpected military success of horses, significantly increasing their perceived military value. In what can be seen as lacking critical reflection, Kistler emphasizes this increase in value through relationally positioning horses to other species based on anthropocentric utility (i.e., being worth 7 bulls or 10 donkeys).

In describing the military value of horses in such a way Kistler recenters their expendability/replaceability and obfuscates any potential for questioning their use. In other words, a focus on perceived military usefulness effectively renders nonconsidered many of the assumptions underlying the ability to socially construct Other animals as usable for human ends. Kistler's illustration, however, goes a step further. The last figure equating the value of a horse to 30 racialized humans evokes a response. Some (i.e., nonwhite) humans are socially constructed as having only a fraction of the value of a horse, donkey, or bull. Whereas these human slaves are positioned as lower in value than Other animals, there is one nonnegotiable claim available to them—they could not be eaten (Pugliese, 2013, p. 45).

The relational construction of certain humans as closer to Other animals was also present during WWII. For example, Japanese soldiers were framed as being animal-like in USA propaganda. The social construction of the Japanese Other created numerous implications. Reference to them as "filthy yellow monkeys" operationalized the social construction of a Japanese race, situated closer to nature than a superior culture of the West. Situating Japanese people in such a way "enables the coding of certain other humans as animals that can be killed, as non-humans are, with impunity" (Pugliese, 2013, p. 210). Positioning some humans as Other animals also makes killing them easier, in the normative context of anthropocentrism.

Providing somewhat of a contradiction, animalizing of the Japanese in USA propaganda had somewhat of a complimentary implication:

> Flawlessly dug-in and boundlessly courageous, the Japanese soldiers also seemed to possess a sense of terrain, a lightness of movement and an ability to merge with the environment that the Americans found impossible to match *Allon and Barrett, 2015, p. 133*

The outcome of such a construction of the Japanese as closer to nature was a *creative* response: the military use of "an equally sentient, if not ultimately superior, animal opponent in the form of the Marine Doberman" (Allon and Barrett, 2015, p. 133). Whereas the Japanese were positioned as lesser humans and required the nonhuman skills of Other animals to defeat them, they were perceived to have a number of sensory abilities and cunning, which American soldiers did not possess.

In both of these examples, there is (more than) a double act. Foregrounding the animalization of some humans, with their value being less than Other animals, forms a direct reinforcing of anthropocentric and imperial/colonial social-value constructions. Kistler's example directly acts to promote and facilitate a nonengagement with racism beyond simple mention via its introduction relationally to and alongside the use of Other animals. Whereas mention of slavery may foster a level of response from the reader, being situated in the context of the use of Other animals, the use of Other animals becomes more acceptable from an anthropocentric perspective. The description included on the rear jacket of Kistler's (2011) text provides an indication: "the book reveals the full scope of heroics and horror committed by—and against—animal warriors… reflecting [their] place in human warfare over time, from insects used as stinging projectiles to message delivering pigeons." In short, Other animals have value and the potential to become *heroes* only when useful and used for human ends.

Whereas slavery is largely frowned upon today, as is the broader status of humans as largely disposable/replaceable and someone else's property, entanglements separately engaged with by Nibert and Pugliese (and introduced and nonengaged with by Richardson and Kistler) locate the ongoing and normative positioning of Other animals as ontologically usable. For Other animals, such positioning continues to be a common practice in warfare (and society at large), one fervently pursued by States and some have argued as comprising part of a surprising resurgence:

> Of all the strange changes happening in war in the 21st century, one of the most odd has to be the return of animals to the battlefield. War is not just about hardware and software, it is also now involving what some researchers call 'wetware' *Singer, 2009*

A key feature of the (actual and potential) return of Other animals to the battlefield is a renewed focus on the specific abilities of other species, and how they can be utilized for human ends (Salter, 2015). Psychologist Burrhus Frederic Skinner (1960) noted decades ago:

> Man has always made use of the sensory capacities of animals, either because they are more acute than his own or more convenient (p. 28).

Dinesh Wadiwel's (2015) *The War Against Animals* draws from elements of Barbara Noske's (1997) foundational work on the animal–industrial complex to provide a critical analysis of such creative utilization (Twine, 2012). Drawing on the influential writings of Foucault, Wadiwel articulates an understanding of the use of Other animals as biopolitics that encompasses the exercise of power to make and foster life (to let live), alongside the power to disallow life to the point of death. Biopolitics is about death *and* life. The use of Other animals in the military not only determines when they die. How their lives are allowed to proceed, and in what forms, is central.

Biopolitics

Biopower, drawing on Michel Foucault's (1998) initial conceptualization, refers to the State's management and control of its populace: "the ancient right [of a Sovereign] to take life or let live was replaced by a power to foster life or disallow it to the point of death" (p. 138). State power is exercised through the regulation of one's life—how one lives one's life. Biopolitics is a politics of the body.

The way we view and use Other animals provides a clear articulation of biopolitics. For many, how they live and how they die are closely managed to meet predetermined human ends. Their lives are heavily regulated. How they live, and when they are allowed to die are governed by us, for us. The management of their lives encapsulates a number of "technologies that include capture, enclosure, harness, enforced labour, controlled breeding, castration, branding, and auctioning" (Pugliese, 2013, p. 40).

Exemplifying the regulation of life (and death) central to biopower, "Our relations with animals appear as biopolitical in an almost archetypal way, in so far as they perfectly and efficiently use violence to locate an exact line between life and death" (Wadiwel, 2015, p. 27).

The exercise of power to control Other animals is central to their use for military purposes.

Changing approaches to the use of Other animals for military purposes across history reflect the changing nature of the exercise of power, shifting from directly coercive power toward more invasive forms of regulating life. The trajectory of examples in this chapter provides clear illustration.

9.2.1 Dogs, Pigeons, Elephants

Recalling Richardson's (1920) albeit uncritical reference, dogs were among the first species to be used as weapons in human warfare. For example, they note, "Plutarch and Pliny both mention war dogs in their writings. The dogs were employed as a means of defense against attack, and also as actual weapons of attack" (Richardson, 1920, p. 21). They featured during the Peloponnesian War of 431 to 404 BCE:

> The Corinthians, too, used them for purposes of defense, and the citadel of Corinth had a guard of fifty places in boxes by the sea-shore. Taking advantage of a dark night, the Greeks with whom they were at war disembarked on the coast. The garrison were asleep after an orgy, the dogs alone kept watch and the fifty pickets fell on the enemy like lions: all but one were causalities. Sorter, sole survivor, retiring from the conflict, fled to the town to give warning and roused the drunken soldiers, who came forth to battle. To him alone were the honors of victory, and the grateful town presented to him a collar with the inscription, 'Sorter, Defender and Savior of Corinth,' and erected a monument with his name and those of the forty-nine heroes who fell *Ross, 1922, p. 258*

Accounts of the 386 BCE siege of Rome, the Siege of Mantenea in 362 BCE, at Sardinia in 231

BCE, and Battle of Versella in 101 BCE also include reference to dogs being used (Lubow, 1977, p. 27; Richardson, 1920, pp. 21–22).

Preceding the use of dogs, pigeons were domesticated as far back as 4500 BCE Iraq (Lubow, 1977, p. 27). Wendell Mitchell Levi's (1963) *The Pigeon*, first published in 1946, described pigeons as "serv[ing] its masters from earliest days," as having "no equal" and having a number of other "utilitarian roles" including being burnt alive as a "sin offering," as a religious sacrifice more broadly, for human consumption (and profit generation in this role), and scientific–medical research (pp. 1–5). Their use as a messenger for the exchange of intelligence during times of war is described as "expedient" (Levi, 1963, p. 5). Implicit to all these descriptions is the (human) utility of pigeons as the source of their value. The first known use of pigeons for message delivery was in 2900 BCE Egypt, and the practice continued well into the 20th century (Fang, 2008, pp. 116–117). Following the Egyptian example, Romans used pigeons to convey messages during the siege of Mutina (44–43 BCE), with Greeks using them even earlier (Leighton, 1969, pp. 141, footnote 12). They were used during the Crusades and through to the World Wars of the 20th century. The National Pigeon Service was started in the United Kingdom in 1938, supplying over 200,000 pigeons for use during WWII. In parallel with Nibert's (2013) critical commentary on dogs requiring no *pay*, Levi (1963) notes pigeons were "the only munition of war which was provided free of charge" (p. 7).

Elephants similarly have a long history of human use, stretching back to as far as the Bronze Age (3300–1300 BCE). Their physical strength has been deemed as useful for human ends in multitude ways. The potentially intimidatory value of their size and sheer presence has seen significant mobilization in warfare. Kistler (2011), consistent with his broadly uncritical and overtly anthropocentric accounts of Other animals in the military, provides the following introduction to their use:

> Of all the creatures used by military forces throughout history, only elephants have been regularly trained to mangle enemies in personal combat. From time to time dogs sprang on the enemy, but pachyderms have crushed foes for thousands of years (p. 55).

By way of contrast, being able to dominate elephants became a symbol of Rome's "ability to conquer, to civilize, and to bend both the natural and political worlds to its will" (Shelton, 2006, p. 3). For Rome, in part, this emerged through the use of elephants as weapons against them—as tools of their adversaries. Dominating, torturing, and killing elephants on battlefields and inside arenas as a spectacle of entertainment reflected a desire to exert their will over elephants. They were viewed "as representatives of a natural world which was wild, alien and hostile to human endeavors, and which therefore deserved to be destroyed" (Shelton, 2006, p. 5). Such perspectives are clear in commemorative coins portraying Alexander the Great following his death in 323 BCE. For example, one coin depicts Alexander wearing an elephant-scalp headdress (Scullard, 1974, pp. 76 and plate VIII). The use of trophies of war is itself a form of weaponization, as a means to intimidate adversaries (Perera, 2014). In this instance, dominating elephants was a clear metaphor of an ability to conquer Other peoples and their militaries.

Encapsulated in the desire to dominate elephants was an inability to wholly subsume elephants into human servitude. Perceived as a wild and (partially) tameable species, elephants were considered unpredictable, and at times responded to threats and physical harm on the battlefield in ways incongruent with the aims and intent of their captors. Polybius' account of the battle of Raphia, near Gaza, in 217 BCE provides an example of such inability to control (perceived as unpredictability) elephants: "Most

of Ptolemy's elephants, however, declined the combat, as is the habit of African elephants" (translated by Charles, 2007, p. 307). Whereas Charles (2007) refers to the smaller African elephant used by Ptolemy as being intimidated, even "cowardly" (p. 309), the unwillingness of the individuals to engage could be seen as a refusal to be controlled, even an act of resistance. Similar refusals to obey, to be uncooperative, and to resist their captors are common across other forms of captivity including zoos and circuses (Hribal, 2010).

An outcome of the unwillingness of certain elephants to engage in battle (be it a form of resistance or otherwise) was the impact it would have on the army seeking to mobilize them for apparent military advantage. Following on with Polybius' account of the battle of Raphia: "when Ptolemy's elephants were thus thrown into confusion and driven back on their own lines, Ptolemy's guard gave way under the pressure of the animals" (translated by Charles, 2007, p. 307). Polybius' concise description understates the outcome. Elephants were specifically deployed to charge into lines of soldiers. The impact is quite similar irrespective of whether they are charging toward adversaries, or back toward the lines of the army who sought to mobilize them. Such an outcome (the latter) was at times preplanned by opposing armies, and specific tactics were adopted in seeking to turn elephants away, redirecting their charge toward those seeking to control them.

Invading Roman forces sought to facilitate such behavior during battles with Pyrrhus in Italy. At the city of Maleventum (now known as Benevento), *war pigs* were used as a weapon against elephants (275 BCE). Kistler's (2011) reference to events at Maleventum, reflective of the dedication for his book *Animals in the Military: From Hannibal's Elephants to the Dolphins of the U.S. Navy* further locates underlying assumptions about Other animals in the adoption of this tactic: "The most famous and clever possibility included another animal, the common pig"

(p. 72). Roman author Claudius Aelianus (c.175–235 CE), also known as Aelian, describes the presence of squealing pigs as discomforting to elephants—a potentially effective weapon. Kistler also refers to the use of fire, including flaming javelins, arrows, and torches as weapons to deter elephants as a lead-in to a combination of the two being used, referring to tactics utilized by the besieged Greek city of Megara (266 BCE). Book 4 of Polyaenus' *Stratagems in War* describes their use:

> At the siege of Megara, Antigonus brought his elephants into the attack; but the Megarians daubed some swine with pitch, set fire to it, and let them loose among the elephants. The pigs grunted and shrieked under the torture of the fire, and sprang forwards as hard as they could among the elephants, who broke their ranks in confusion and fright, and ran off in different directions.

Kistler (2011) defers to an unnamed author to suggest that pigs were set alight by Romans at Maleventum, without any reference to the harm caused them (unlike Polyaenus' description of such use as torture, penned some 1950 years previously): "The elephants hated the pigs' squealing shrieks, and they turned to flee" (p. 72). Such use of pigs, with them being set on fire to intimidate elephants, provides a clear illustration of biopolitics: the power to determine how they live, and when and how they die. The use value was specific, limited, and inherently disposable with broadly no consideration of the pain and suffering such use would cause.

The attitudes of Romans toward elephants provide a different manifestation of biopolitics, contrasting with that mobilized by Carthaginians, who viewed elephants as a symbol of military might. These are most recognized in accounts of the exploits of Hannibal, considered as one of the greatest military strategists in history, during the second Punic War (218–201 BCE). Accounts of what transpired as Hannibal marched his forces (including a host of Other animals) across the Alps provides little detail as to the fate of the elephants. They are effectively an absent referent. Consistent with all use of Other animals, their lives were largely inconsequential.

9.3 HORSES: A PERSISTENCE CHOICE IN WAR

Horses are one of the most persistent species used by humans for military purposes. In parallel with the refinement of wheel building techniques and materials, the military use of horse-drawn war chariots began around the 17th century BCE. In the late Bronze Age (1550–1200 BCE), the number of horse and chariots a king possessed was seen as a statement of military might (Drews, 1995, p. 97). They were soon replaced with directly mounted riders.

The Mongols (united under Genghis Kahn) made extensive use of horses during their invasion of expansion through Western and Central Asia during the 13th century CE. Morris Rossabi (1994) describes the vital role of horses in the forging of the Empire: "Genghis Khan and his descendants could not have conquered and ruled the largest land empire in world history without their diminutive but extremely hardy steeds" (p. 48). Recalling Kistler's reference to the value of horses (i.e., deemed equivalent to "7 bulls, or 10 donkeys, or 30 slaves"), cavalrymen would each own three or four reflective of their deemed importance. They were considered weapons, exemplified by Rossabi referring to them as "the intercontinental ballistic missiles of the thirteenth century." Their anthropocentric value extended beyond the use by the Mongols as weapons. As with the fate of Warrior and untold numbers of other horses during WWI, Mongols also consumed their horses.

With a focus further West and a Eurocentric frame, John Edward Morris (1914) referred to a "need" for light cavalry (as an "elementary fact"), in the context of accounts of 4000 light and 300 heavy horseman at Falkirk during the First Scottish War of Independence in 1298 CE:

> The true light horseman, *scutiferi* or *valetti cum equis discopertis qui decanter hobellarii* or *hobdarii*, came originally from Ireland. I have not found the word used before 1296, when a considerable force was brought over from Ireland to make a diversion against the south-west bank of Scotland, where Edward's main army invaded from Berwick towards Dunbar (p. 98).

Romanticized and popularized in modern film and television and used extensively in warfare through the Middle Ages is the image of the knight charging into battle on the back of an armored horse. The romanticized status of the rider, including English class connotations, continued into the World Wars of the 20th century (as Steven Spielberg's *War Horse* illustrates) (War Horse, 2011). For example, Kiester notes (2010):

> Prior to World War I, cavalrymen had been the darlings of every old-time commander. They conjured up a classically romantic image of war: gallant mounted soldiers sweeping out of the mist, with their sabres flashing, to put a panicky enemy to rout (p. 185).

Alongside romantic, classed, notions of the mounted solider, the military effectiveness of cavalry charges as a weapon and tactic of human warfare was shown to be questionable well in advance of WWI. The ill-fated mounted charge of British cavalry at Balaclava in 1854 during the Crimean War provides a clear example. The unit "charged straight into Russian artillery fire that tore them to shreds" (Sorenson, 2014, p. 25).

The actions at Balaclava are memorialized in Alfred Tennyson's 1854 narrative poem *The Charge of the Light Brigade*. Tennyson's implicit reference to Psalm 23 of the Christian Bible's valley of the shadow of death (i.e., "valley of death") to describe the battlefield is a comment on the brutality of war and the lives lost (albeit, not the horses). In parallel, the poem romantically juxtaposes these (human) lives lost in keeping with accounts on cavalry charges and of military service more broadly in its closing stanza: "Honour the charge they made, Honour the Light Brigade, Noble six hundred." This is earlier reinforced in the poem with reference to a sense of unquestioning duty: "Theirs not to reason why, Theirs but to do and die" (Sorenson, 2014, p. 25; Bethune, 1906). As with Other animals (if any consideration is given at all), the most noble of acts is to unquestioningly and patriotically sacrifice oneself for one's country.

Many nations continued to build cavalry forces in the period before WWI. Horses were utilized as pack animals for supplies and ammunition, for the transport of weaponry, for carrying wounded soldiers, and for transporting men. Their historical role as cavalry persisted into the armed conflict. Historians and military strategists continue to analyze such uses, within an anthropocentric frame of human superiority with Other animals being perceived as usable, ontologically. For example, David Kenyon's (2011) *Horsemen in No Man's Land: British Cavalry & Trench Warfare 1914–1918* seeks to alter the historical record on the overall effectiveness of the cavalry, countering the majority of historical accounts. His "modern analytical investigation of the cavalry" (p. 1) has a central focus on strategies and tactics, analyzing deployment and engagement of cavalry units. Introduced as a "keen horseman" in the book's preface, reflection on the use of horses beyond reference to the men (and at times weapons) carried on their backs is nonexistent. In Kenyon's account, horses are an always-there absent referent.

Such analyses of the use of mounted horses in WWI reflect on the introduction of new military technologies of killing, to which they had previously not been subjected. The adoption and extensive use of barbed wire was an effective means of preventing cavalry charges, and the machine gun was quickly proven to be very effective at dispatching the spectacle of the mounted charge:

> Sent to Egypt, the [Australian Imperial Force] Light Horse did not at first have an easy time. At Gallipoli, two brigades, fighting as dismounted infantry, were ordered to make an early-morning charge across a narrow ridge called the Nek. But the Turks had machine-guns strongly positioned in nine lines of trenches, and the first line was literally cut to pieces. By day's end there had been 372 casualties, including 235 Light Horse killed **Kiester, 2010, p. 189**

Across its usage, Light Horse refers to the mounted soldier, and not the horse on whose back they charged into battle. Edwin Kiester provides no detail as to the number of horses "cut to pieces" in the attack on Turkish trenches as they charged across the Nek.

The use of the Light Horse at Gallipoli preceded action during the 1917 Battle of Beersheba, popularized/romanticized in the Australian film *The Lighthorsemen*. In passing reference, Kenyon (2011) refers to their exploits as legendary (p. 18). Simon Wincer's (1987) historical feature follows the Australian 4th Light Horse Brigade during the Palestinian campaign against Turkish forces (The Lighthorseman, 1987). Of note, as with the mythologizing of Simpson and his donkey—consistent with other films released in the years prior, including *Breaker Morant* (1980) and *Gallipoli* (1981)—characters in the film are portrayed as embodying desired/promoted Australian character traits in the context of nation-building and a coming-of-age: tenacity, ingenuity, courage, and "mateship" (situated alongside a form of idealized larrikinism). These representations are expressly, if not always directly, nationalistic and contribute to identity formation and nation-building. Any memorialization of individual horses is always/only situated in the service of their human masters.

The unexpected military success of the charge of the light horse at Beersheba, "across three kilometres of open ground against entrenched infantry supported by artillery and machine guns," spurred the use of the cavalry charge

in other battles (Allsopp, 2015, p. 49). For example, a cavalry squadron of Lord Strathcona's horse regiment charged German lines during the Battle of Moreuil Woods in France on March 30, 1918. The fateful assault is memorialized in Sir Alfred Munnings' 1918 painting *The Charge of Flowerdew's Squadron*, on display in the Canadian War Museum in Ottawa.

In recalling the Battle of Beersheba, Sorenson (2014) describes what transpired at Moreuil Woods as another *one of the last great cavalry charges*. The second squadron's charge was directed squarely at machine gun positions: "As if in a replay of the 'Charge of the Light Brigade,' the Canadian forces were nearly decimated" (p. 26). Sorenson is referring here to the 1854 cavalry charge at Balaclava, memorialized in Tennyson's *The Charge of the Light Brigade*.

Albert Tennyson's *The Charge of the Light Brigade*

Half a league, half a league,
Half a league onward,
All in the valley of Death
Rode the six hundred.
"Forward, the Light Brigade!
Charge for the guns!" he said.
Into the valley of Death
Rode the six hundred.

"Forward, the Light Brigade!"
Was there a man dismayed?
Not though the soldier knew
Someone had blundered.
Theirs not to make reply,
Theirs not to reason why,
Theirs but to do and die.
Into the valley of Death
Rode the six hundred.

Cannon to right of them,
Cannon to left of them,

Cannon in front of them
Volleyed and thundered;
Stormed at with shot and shell,
Boldly they rode and well,
Into the jaws of Death,
Into the mouth of hell
Rode the six hundred.

Flashed all their sabres bare,
Flashed as they turned in air
Sabring the gunners there,
Charging an army, while
All the world wondered.
Plunged in the battery-smoke
Right through the line they broke;
Cossack and Russian
Reeled from the sabre stroke
Shattered and sundered.
Then they rode back, but not
Not the six hundred.

Cannon to right of them,
Cannon to left of them,
Cannon behind them
Volleyed and thundered;
Stormed at with shot and shell,
While horse and hero fell.
They that had fought so well
Came through the jaws of Death,
Back from the mouth of hell,
All that was left of them,
Left of six hundred.

When can their glory fade?
O the wild charge they made!
All the world wondered.
Honour the charge they made!
Honour the Light Brigade,
Noble six hundred!

As with the charge at the Battle of Beersheba, accounts almost exclusively refer to the bravery and sacrifice of those who mounted and directed horses into battle, alongside reflection on the effectiveness of using (mounted) horses in modern military engagements:

> sources differ on what actually took place, and whether indeed the event was a magnificent charge, or a more disorganised scuffle, with at least part of the mounted combat being the attempt of Flowerdew's men to escape the attentions of their supposed victims *Kenyon, 2011, p. 195*

In the wake of massive losses, following repeated cycles of attack and counter-attack, control of Moreuil Woods remained disputed. These losses are indicated in the citation for posthumous award of the Victoria Cross to Lieutenant Gordon Muriel Flowerdew—about 70%. Flowerdew succumbed to his injuries the following day. The citation, included in a supplement to The London Gazette on April 24, 1918, made no mention of the horse on whose back Flowerdew was mounted.

9.3.1 More-Creative Examples of Ontological Usability

The conscription of Other animals for military purposes reflected on in this chapter are situated alongside an array of divergent, even bizarre, ideas and proposals—a number of which were tested. Some continue to be used. Others are the focus of ongoing state-funded military research (Salter, 2015). These include explosive dogs, incendiary bats, bombs guided by cats and pigeons, and the implanting of microelectrical mechanical systems into Other animals. Each provide an example and in many ways contemporary manifestations of the techniques of biopolitics—the assumed power over Other animals to regulate how they live and how they die.

One of the ways in which dogs were used in WWII is illustrative of the levels on nonconsideration afforded Other animals within the mutually reinforcing contexts of human warfare and the binary logic of anthropocentrism. In *Animals in War*, Jilly Cooper (2000) opens her chapter on the use of dogs by describing them as serving "man more nobly in war" than any other species. The chapter closes with the following:

> To end on a truly poignant note, perhaps the most heartbreaking task in the Second World War was performed by the Russian suicide dogs. When the Panzer tanks were rolling towards Moscow, and nothing seemed likely to stop them, the Russians trained little mongrels to hurtle under the tanks, carrying a primed bomb strapped to their backs, and crouch there until dogs and tank were blown to eternity. It is unbearable to think of that last split second of bewilderment and sense of betrayal that must have flashed through the dogs' mind *Cooper, 2000, p. 95*

As is common with references to the unquestioned ontology of usability of Other animals, there is a distinct lack of critical reflection situated within and alongside anthropomorphic romanticism. Calling these individuals "suicide dogs" implies agency, and their actions being of their choosing. What is missing from Cooper's introduction to this use of dogs is a description of how they were manipulated and *trained* to approach tanks: starved individuals were directed toward food, which was left under armored vehicles. Starve and repeat. As Cooper describes it, the actions of the starving dogs (in seeking out food as they had been trained to do through forced deprivation) creates "the illusion that [they] are helping themselves to die" (Wadiwel, 2015, p. 15). Cooper's framing is further reinforced in referring to what they were doing as a "task" and the subsequent inference of a sense of bewilderment and betrayal. It is implied that these individuals knew, at least in that last "split second," what their role was. Of note, Cooper (2000) mobilizes a similar framing of the use of dolphins to carry explosives designed for the destruction of warships (p. 210).

The use of dogs as unwitting suicide bombers (also known as "tank dogs") in WWII (however inventive) can appear quite banal when compared with "imaginative" ideas that emerged from the Office of Strategic Services (OSS). The forerunner to the Central Intelligence Agency,

the OSS proposed and tested an array of uses of Other animals in military conflict. One proposal was outlined in a letter to Franklin D. Roosevelt in 1942:

> The... lowest form of life is the BAT, associated in history with the underworld and regions of darkness and evil. Until now reasons for its creation have remained unexplained.
>
> As I vision it the millions of bats that have for ages inhabited our belfries, tunnels and caverns were placed there by God to wait this hour to play their part in the scheme of free human existence, and to frustrate any attempt to those who dare desecrate our way of life.
>
> This lowly creature, the bat, is capable of carrying in flight a sufficient quantity of incendiary material to ignite a fire *Couffer, 1992, p. 6*

The letter, penned by Lytle S, Adams, was titled "Proposal for surprise attack 'REMEMBER PEARL HARBOUR.'" The only form of redemption for this "lowest form of life" is in the unwitting service in "the scheme of free human existence." Adams' proposal was tested during WWII: bats with surgically attached timed incendiary munitions were encased in a guided delivery device. One design of the *bat bomb* included up to 250 individuals housed inside a metal enclosure. Dropped from a plane, the enclosure would open at a predetermined altitude and release the bats. It was envisioned that they would seek a dark place to shelter (i.e., inside a building). Fire would ensue (Couffer, 1992). The bats themselves were inconsequential beyond being seen as an effective delivery device for the incendiary munitions—much like tank dogs. Following years of research, testing in New Mexico led to a number of unintended structural fires, including the destruction of a US$2 million air force hangar. After a number of unsuccessful design revisions, "the macabre project was finally scrapped" (Hersh, 1968).

Perhaps less sophisticated, albeit similarly aimed to capitalize on perceived/actual traits, was the *cat bomb*. The notion that cats will do anything they can to avoid water situated alongside a perceived ability to be able to contort and direct their falling body to land on their feet was the basis for this proposal. It was hypothesized that a combination of these two abilities would produce desired military outcomes; a cat dropped from a plane above the ocean would *naturally* guide themselves toward anything that is not water—in an act of desperation. If high explosives were strapped to the said cat (or the said cat was strapped to high explosives), this would produce a form of "smart bomb." Dropped above a warship, it was hypothesized that the cat would guide the bomb toward the deck of a warship. Testing of the cat bomb produced less than desirable military results, the experiences and outcomes for the individual cats notwithstanding:

> even unattached to high explosives ... [the cats were] likely to become unconscious long before a Nazi deck seemed an attractive landing place *Harris and Paxman, 1983, p. 203*

More sophisticated than both the ideas behind the bat and cat bombs was *Project Pelican*, which devised an "organic system to direct ground launched missiles against enemy aircraft" (Lubow, 1977, p. 35). The idea was proposed, designed, and researched by (previously introduced) psychologist Burrhus Frederic Skinner, and included extensive testing over more than a decade before finally being scrapped in the 1950s. The guidance system comprised a number of pigeons encased inside a steel sarcophagus. The pigeons were trained to peck at dots on a screen. The location of contact with the screen would direct the path of the bomb, with the dots representing the path to the target. In providing his account of Project Pelican, Skinner (1960) described it as:

> a crackpot idea, born on the wrong side of the tracks intellectually speaking, but eventually vindicated in a sort of middle class respectability (p. 28).

Implicit are notions of thinking outside of the square at significant personal/professional risk, and reaping the rewards from an otherwise unsuccessful project—an abject failure. There are no hints at the implications of Skinner's ideas on the unwitting pigeons, laboratory trained through food deprivation, their impending death rates no mention. An ontology of usability is explicit and common across these *imaginative* proposals. Those who postulated, researched, and promoted the use of dogs, bats, cats, and pigeons with explosives attached to their bodies directly relied on "animal intersubjectivity, coproductivity, and (limited, directed) agency" (Wadiwel, 2015, p. 15). For dogs, hunger and their memory of previous feeding locations; for bats, their desire to seek out a dark place; for cats, a perceived dislike for water; for pigeons, their tendency to peck at objects in search of food. Each of these ideas used the individual Other animals' bodies against them. Central to each of these ideas, the individual Other animals were largely inconsequential in their replaceability/expendability for human ends, such is the binary logic of anthropocentrism.

Research on the potential uses of Other animals continues today, funded by State agencies including the Defense Advanced Research Projects Agency (DARPA), a branch of the Department of Defence in the USA. For example, on April 1, 2014, DARPA publicly announced the establishment of the "Biological Technologies Office" (BTO) as an extension of previous programs under Defense Sciences and Microsystems Technology Offices. The aim of the BTO is to investigate the further use of biological science and its integration into military technologies as a cornerstone of future warfare. Bioengineering is being positioned as central to "harness[ing] the power of natural systems for national security" (DARPA, 2014).

The formal statement of the aims of DARPA in the establishment of the BTO forms a contemporary articulation of biopolitics. Species are being reimagined and remade, in a number of ways more explicitly, with more planning and increasing levels of invasive weaponized manipulation (Salter, 2015). Framed as mechanical devices, as biological/electromechanical hybrids, Other animals are being experimented on and enhanced/disenhanced for human ends (Bateman et al., 2015; Ferrari, 2015; Salter, 2015, 2016). They are socially constructed as enhanced through the implanting of technology into their bodies, providing increasing levels of human control and associated reductions in autonomy and agency. Disenhancements are imagined as techniques that provide a means for humans to manage individual's ability to resist and otherwise exist outside of human control—all with the aim of producing malleable biological weapons platforms for human ends. Individuals are disappeared within technology as discourse. They are absent referents.

Wadiwel labels our relationships with Other animals (exemplified in these examples) as founded on a war, a war we are constantly waging against them. It is a war we fail to notice through epistemic violence and our ability to position our domination over Other animals as peaceable:

> The genealogy of the war against animals is one of continual adaptation and reworking of systems of domination to most effectively capture the agency, escape and vitality of animals and simultaneously maximise human use value **Wadiwel, 2015, p. 16**

Perhaps most important here is the sentence that follows the preceding excerpt: "The façade this process projects is one of seamlessness; absence of hostility; no friction." A perceived absence of hostility evokes an unquestioning of the use of animals. Such normativity or commonplaceness emerges from and reinforces an ontology of usability. Our socially constructed sovereignty over Other animals *precedes* ethics. Such are the bases for contemporary creative adherence to an unquestioned ontology of usability.

9.4 RECOGNITION AND COMMEMORATION AS OBFUSCATION

Alongside the coin introduced in the opening of this chapter, Simpson and his donkey are memorialized at a number of locations in Australia, including the Shrine of Remembrance in Melbourne and the Australian War Memorial in the National Capital. A larger-than-life memorial depicting Simpson and a seemingly less than life size donkey was installed in South Shields, England, where Simpson was born, in 1988.

The Shrine of Remembrance has the more generic and encompassing *The man with the donkey*, a bronze sculpture designed by Wallace Anderson in 1935 (Fig. 9.2):

> The best known of the bearers was John Simpson Kirkpatrick who commandeered a donkey to aid him as casualties grew and manpower was stretched to its

FIGURE 9.2 **The man with the donkey.** *Source: https:// commons.wikimedia.org/wiki/File:The_Man_with_the_Donkey_ statue_oblique_view.jpg, CC BY-SA 3.0.*

limits. Simpson, as he was known, was at the landing at Gallipoli on 25 April 1915 and was killed on 19 May 1915. His story came to exemplify the tenacity and courage of all Anzacs at Gallipoli **Anderson, no date**

Alongside the nation-building, which is central to most memorials of war (with the explicit example here being the notion of the ANZAC legend) is a partly oblique reference to a singular nameless donkey in an otherwise unwitting supportive role. The Australian National Memorial is more explicit. Peter Corlett's bronze sculpture is titled *Simpson and his donkey, 1915* and was completed in 1988 to coincide with settler/colonial Australia's bicentenary: "The sculpture is warm, accessible and above all, a work about humanity" (Fig. 9.3). This

FIGURE 9.3 **Simpson and his donkey, 1915.** *Source: https://commons.wikimedia.org/wiki/File:Simpson_and_his_Donkey_statue_in_Canberra.jpg, CC BY-SA 4.0.*

work of humanity similarly includes reference to a singular donkey.

> In contrast to the human figure it seems less than life size, and with the awkward positioning of its front legs, and its precarious forward lean, we fear it may topple over **Clarke, no date**

The recombination of Murphy, Queen Elizabeth, Abdul, and the two Duffies as singular (as with the *commandeered* donkey in the memorial at The Shrine of Remembrance) is extended with a further depersonalizing reanimalizing. The shrunken less-than-life-size donkey is represented as an "it." These descriptions largely encapsulate the ephemeral *use* of animals by the military. Devoid of individualism, a present yet absent referent.

There are a number of memorializations that provides partial counters to notions of pure utility in the use of Other animals. For example, Australia Post, the national postal service, released as a collectible in October 2015 titled *Animals in War: A Century of Service*. The collection of five stamps includes photos and descriptions for each: a soldier and laden donkey in Syria in 1942; a soldier and "his explosive ordnance detection dog" in Afghanistan in 2008; two soldiers mounted on horseback at Roseberry Park Camp, NSW, in 1914; a soldier of the 5th Australian Pigeon Section, in Queensland in 1945; and a soldier mounted on a camel in Egypt during WWI (Australia Post, 2015). The back of the collectible card enclosing the five stamps provides context for the collection: "This issue commemorates the valour and sacrifice of countless animals during Australia's involvement in war." In seeking to provide recognition for Other animals used by Australian military forces there is an inference of choice. The recognition provided by the collection is inherently anthropomorphic. The semiotics explicitly reinforcing *service*. The reductionist individual value of Other animals is as a means to human ends.

The first photo in the collection implicitly and explicitly references Simpson and his donkeys: "The most famous are Simpson's donkeys, who carried first aid and wounded soldiers during the Gallipoli campaign in World War I." The reductionist singular moniker is replaced with the plural, albeit nameless, "donkeys." References to dogs are centered around saving human lives. There is an uncommon reference to the lives of horses lost on the Western Front during WWI, listed as "millions." Pigeons receive special mention, referencing the award of the Dickin Medal for Gallantry during WWII.

> The [People's Dispensary for Sick Animals, UK] PDSA Dickin Medal, recognised worldwide as the animals' Victoria Cross, is awarded to animals displaying conspicuous gallantry or devotion to duty while serving or associated with any branch of the Armed Forces or Civil Defence Units. The PDSA Dickin Medal is the highest award any animal can receive whilst serving in military conflict **PDSA, no date**

In 1993 the Australian Royal Society for the Prevention of Cruelty to Animals created the Purple Cross Award to recognize the deeds of animals that have shown outstanding bravery and fortitude in the *service* of humans (emphasis added, RSPCA, no date). In May 1997 the second Purple Cross Award was posthumously awarded to Murphy "and all Simpson's other donkeys" for helping save human lives. A ceremony was held at the Australian War Memorial and "a mascot donkey named Simpson received the award," presented by then Australian Deputy Prime Minister Tim Fischer. The Purple Cross has been awarded a total of 8 times, most recently to Sarbi in 2001. Sarbi was "An Australian Special Forces Explosive Detections dog working in Afghanistan," present during a battle for which Lance Cpl. Mark Donaldson was awarded the Victoria Cross for Australia, the highest award in the honor system (RSPCA, no date).

Alongside the award of medals as a form of commemoration, a number of cities and countries have established memorials to Other animals, including the *Animals in War* memorial in London, England (Fig. 9.4). The monument

FIGURE 9.4 **Animals in War memorial, London.**

commemorates *contributions* "in the cause of human freedom" (rear inscription). The front of the memorial is inscribed with:

> This monument is dedicated to all the animals that served and died alongside British and Allied forces in wars and campaigns throughout time.

The inscription at the London memorial, the Purple Cross, and the Dickin Medal infer agency in the memorialized actions on Other animals. They act to obfuscate the epistemic violence of anthropocentrism's binary logic. What makes the London memorial noteworthy is a second, clearly contradictory, inscription. In a larger font, adjacent to this body of text: "They had no choice."

9.5 CONCLUSIONS: USE, MEMORIALIZATION, AND EPISTEMIC VIOLENCE

For as long as humans have been waging war on each other, they have used Other animals for military purposes. Nibert (2013) has argued that such use is foregrounded and intertwined with the domestication of certain species. Rooted in speciesism, the domestication and farming of other animals required the acquisition of more land, leading to expansionist conflict.

Acquisition of land necessitated the displacement of other societies, a process aided by devaluing their populations—most clearly illustrated in colonial and settler/colonial expansion. In parallel, an amassing of military power was essential to protect acquired lands. A specific outcome was the use of Other animals to supplement and support the strategies and tactics of warfare.

Pugliese (2013) has clearly enunciated a link between the binary logic of anthropocentrism and colonialism—the specter of speciesism has always/already inscribed the categorical naming of the racialized other (p. 41). Recalling Kistler's (2011) attempt to highlight the perceived military usefulness of horses, one horse was considered equivalent to 30 slaves (p. 101). The social construction of the Japanese during WWII embodies similar functional aims and outcomes. Identified as having animal-like sensory/mystical abilities, they were positioned as lesser humans, which could only be effectively matched through the use of a species with similar abilities—dogs.

The history of human warfare is typified by the conscription of Other animals. A range of species has been deemed to be useful for military purposes, including dogs, elephants, horses, pigeons, and pigs. Horses have a long history of use, and dogs are a feature of the ongoing occupations of Iraq and Afghanistan. Following on from their use to locate Japanese soldiers during WWII—and in an ironic twist from their being forced to carry explosives during WWI—dogs sensory abilities have come to be utilized in bomb detection roles. A central feature is a reduction of potential harm to humans—an anthropocentric solution to an anthropocentric problem.

Understanding the use of Other animals in the military requires a deeper analysis of power, moving beyond a directly coercive model. Wadwiel's (2015) reflections and extension of Foucault's (1998) influential writing on biopolitics provide conceptual and theoretical means to interrogate structures of domination. The use of animals in the military, and what Wadiwel identifies as a broader war against animals, is emergent from the power to make and foster life (to let live), alongside the power to disallow life to the point of death. Biopolitics is about death *and* life. It is about use and the epistemic violence that enables their normatively unquestioned use.

Alongside such nonquestioning, the *service* of Other animals for military purposes has received memorialization. As accounts of Simpson and his donkey exemplify, Murphy, Queen Elizabeth, Abdul, and the two Duffies are absent referents. Inscribed in the reductionist and singular moniker of donkey in most records and accounts, their value is limited to their use by Simpson. Murphy, Queen Elizabeth, Abdul, and the two Duffies (alongside Warrior, Pablo, and an untold number of others) are never quite present beyond being present for us. Be that in their actual military use or the process of memorialization, which acts to make us feel better about their use for human ends. In short, the memorialization of individual animals acts to obfuscate the structures of domination that render unquestioned the use of Other animals.

Peter W. Singer's (2009) reflections on the resurgence of the use of animals for military purposes (what some have called *wetware*) locates both an increasing awareness of some of the implications of war on humans, and a contemporary manifestation of biopolitics. The establishment of DARPAs BTO was an extension of the *creative* and *imaginative* ideas of the OSS (i.e., bat, cat, and pigeon bombs) and provided a formalization of more-recent research practices of inserting electromechanical systems into the bodies of Other animals. Reflecting state rhetoric of the war on terror, such techniques are designed to "harness the power of natural systems for national security" (DARPA, 2014). The enhancement/disenhancement of Other animals provides perhaps the most easy-to-see examples of biopolitics—an ability to more directly control Other animals lives. The animals themselves are disappeared.

The establishment of the BTO is epitomic of the social construction of Other animals for military purpose, rooted in the binary logic of anthropocentrism. Expressly and directly for human ends, the use of animals in the military is an example of the search for anthropocentric solutions to anthropocentric problem.

References

Adams, C.J., 1990. The Sexual Politics of Meat: A Feminist-Vegetarian Critical Theory. Continuum, New York, NY.

Adams, C.J., 1997. "Mad cow" disease and the animal industrial complex: an ecofeminist analysis. Organ. Environ. 10 (1), 26–51.

Adams, C.J., Gruen, L. (Eds.), 2014. Ecofeminism: Feminist Intersections With Other Animals and the Earth. Bloomsbury, New York, NY.

Allon, F., Barrett, L., 2015. That dog was Marine! Human–dog assemblages in the Pacific war. Anim. Stud. J. 4 (1), 126–147.

Allsopp, N., 2015. A Centenary of Australian Animals at War. Inhouse Publishing, Underwood, QLD.

American Humane Association, 2015. Victory for Nation's Military Dogs: Senate Passes 2016 National Defense Authorization Act with Provisions to Return Military Dogs to U.S. Soil, Reunite Them with Handlers, June 19. Available from: http://www.americanhumane.org/about-us/newsroom/news-releases/victory-for-nations-military-dogs-senate-passes-2016-national-defense-authorization-act-with-provisions-to-return-military-dogs-to-us-soil-reunite-them-with-handlers.html

Anderson, W., no date. The Man with the Donkey 1935, Shrine of Remembrance, Melbourne. Available from: http://www.shrine.org.au/Exhibitions/The-Shrine-Collection/Man-and-donkey

Andrzejewski, J., 2014. War: animals in the aftermath. In: Salter, C., Nocella, A.J., Bentley, J.K.C. (Eds.), Animals and War: Confronting the Military-Animal Industrial Complex. Lexington Books, Lanham, MD, pp. 55–71.

Australia Post, 2015. A century of service: animals in war. Available from: http://www.australiapostcollectables.com.au/blog/a-century-of-service-animals-in-war/

Australian War Memorial, no date. Simpson and his donkey. Forging the nation: federation—the first 20 years. Available from: https://www.awm.gov.au/exhibitions/forging/australians/simpson.asp

Barash, D.P., Webel, C., 2008. Peace and Conflict Studies. Sage, Thousand Oaks, CA.

Bateman, S., Gayon, J., Allouche, S., Gofette, J., Merzano, M. (Eds.), 2015. Inquiring Into Animal Enhancement: Model or Countermodel of Human Enhancement?. Palgrave McMillan, UK.

Battersby, E., 2012. Eight million dead in a single conflict: 5,000 years of war horses, Irish Times, January. Available from: http://www.irishtimes.com/culture/film/eight-million-dead-in-a-single-conflict-5-000-years-of-war-horses-1.444971

Berger, J., 1980. Why look at animals? In: Kaloff, L., Fitzgerald, A. (Eds.), The Animals Reader: The Essential Classic and Contemporary Writings (2007). Berg Publishers, Oxford.

Bethune, E.C., 1906. The uses of cavalry and mounted infantry in modern warfare. R. United Serv. Inst. J. 50 (339), 619–636.

Breaker Morant, 1980. Breaker Morant, Motion Picture, South Australian Film Corporation. Directed by Bruce Beresford.

Charles, M., 2007. Elephants at Raphia: reinterpreting Polybius 5.84-5. Classical Q. 57, 306–311.

Clarke, F., no date. Simpson and his Donkey, 1915, Australian War Memorial. Available from: https://www.awm.gov.au/collection/ART40993/

Cochrane, P., 1992. Simpson and the Donkey: The Making of a Legend. Melbourne University Press, Carlton, VIC.

Cooper, J., 2000. Animals in War. Corgi, London.

Couffer, J., 1992. Bat Bomb: World War II's Other Secret Weapon. University of Texas Press, Austin, TX.

Dalziell, J., Wadiwel, D., 2016. Live exports, animal advocacy, race and 'animal nationalism'. In: Potts, A. (Ed.), Critical Perspectives on Meat Culture. Brill, Leiden, The Netherlands.

DARPA, 2014. DARPA launches biological technologies office: new technology office will merge biology, engineering, and computer science to harness the power of natural systems for national security. Defense Advanced Research Projects Agency. Available from: http://www.darpa.mil/news-events/2014-04-01

Deckha, M., 2013. Welfarist and Imperial: the contributions of anticruelty laws to civilizational discourse. Am. Q. 63 (3), 515–548.

Drews, R., 1995. The End of the Bronze Age: Changes in Warfare and the Catastrophe ca. 1200 B.C. Princeton University Press, Princeton.

Fang, I.E., 2008. Alphabet to Internet: Mediated Communication in Our Lives. Rada Press, St. Paul, MN.

Ferrari, A., 2015. Animal enhancement: technoviosionary paternalism and the colonisation of nature. In: Bateman, S., Gayon, J., Allouche, S., Gofette, J., Merzano, M. (Eds.), Inquiring into Animal Enhancement: Model or Countermodel of Human Enhancement?. Palgrave Macmillan, UK, pp. 13–33.

Foucault, M., 1998. The Will to Knowledge: The History of Sexuality: 1. Penguin Books, London.

Francione, G.L., 2007. Reflections on animals, property, and the law and rain without thunder. Law Contemp. Probl. 1, 9–57.

Gaard, G., 2001. Tools for a cross-cultural feminist ethics: exploring ethical contexts and contents in the Makah whale hunt. Hypatia 16 (1), 1–26.

Gallipoli, 1981. *Gallipoli*, Motion Picture, Village Roadshow. Directed by Peter Weir.

Gutierrez, G., Connor, T., 2016. He Was My Rock: Veteran With PTSD Reunited With Military Dog. NBC News, January 21. Available from: http://www.nbcnews.com/news/us-news/he-was-my-rock-veteran-ptsd-reunited-military-dog-n481166

Harris, R., Paxman, J., 1983. A Higher Form of Killing: The Secret History of Chemical and Biological Warfare. Random House Trade Paperbacks, New York, NY.

Hersh, S.M., 1968. Chemical and Biological Warfare: America's Hidden Arsenal. MacGibbon & Kee, London.

Hribal, J., 2010. Fear of the Animal Planet: The Hidden History of Animal Resistance. AK Press, Oakland, CA.

Human Rights Watch, 1990. 'Mengistu Has Decided to Burn Us Like Wood' Bombing of Civilians and Civilian Targets by the Air Force. News from Africa Watch. Available from: https://www.hrw.org/reports/pdfs/e/ethiopia/ethiopia907.pdf.

Kenyon, D., 2011. Horsemen in No Man's Land: British Cavalry and Trench Warfare, 1914–1918. Pen & Sword Military, Barnsley, South Yorkshire.

Kiester, E., 2010. Windows on WWI: A Collection of Iconic, Lesser-Known and Curious Events from the First Global Conflict. Murdoch Books, Sydney.

Kim, C.J., 2007. Multiculturalism Goes Imperial. Du Bois Rev. 4 (1), 233–249.

Kistler, J.M., 2011. Animals in the Military: From Hannibal's Elephants to the Dolphins of the U.S. Navy. ABC-CLIO, Santa Barbara, CA.

Lederach, J.P., 2003. The Little Book of Conflict Transformation. Good Books, Intercourse, PA.

Leighton, A.C., 1969. Secret communication among the Greeks and Romans. Technol. Cult. 10 (2), 139–154.

Levi, W.M., 1963. The Pigeon, second ed. Levi Publishing, Sumter, SC, revised, reprinted with minor changes and additions.

Lubow, R.E., 1977. The War Animals. Doubleday, Garden City, NY.

Mason, J., 2005. An Unnatural Order: The Roots of Our Destruction of Nature. Lantern Books, New York, NY.

Miessner, F., Morrison, N., 1991. Seeds of Change: Stories of IDB Innovation in Latin America. Inter-American Development Bank, Washington, D.C.

Morris, J.E., 1914. Mounted infantry in mediaeval warfare. Trans. R. Hist. Soc. 8, 77–102.

Nibert, D.A., 2013. Animal Oppression and Human Violence: Domesecration, Capitalism, and Global Conflict. Columbia University Press, New York, NY.

Noske, B., 1997. Beyond Boundaries: Humans and Animals. Black Rose Books, Montreal.

Patterson, C., 2002. Eternal Treblinka: Our Treatment of Animals and the Holocaust. Lantern Books, New York, NY.

PDSA, no date. PDSA Dickin Medal (archived). People's Dispensary for Sick Animals. Available from: https://web.archive.org/web/20140924092934/http://www.pdsa.org.uk/about-us/animal-bravery-awards/pdsa-dickin-medal

Perera, S., 2014. Dead exposures: trophy bodies and the violent visibilities of the nonhuman. Borderlands e-journal 13, 1.

Plumwood, V., 1993. Feminism and the Mastery of Nature. Routledge, London.

Pugliese, J., 2013. State Violence and the Execution of Law: Biopolitical Caesurae of Torture, Black Sites, Drones. Routledge, Oxon.

Regan, T., 1983. The Case for Animal Rights. Berkeley, CA, University of California.

Richardson, E., 1920. British War Dogs, Their Training and Psychology. Skeffington & Son, London.

Ross, E., 1922. The Book of Noble Dogs. The Century, New York, NY.

Rossabi, M., 1994. All the Khan's horses. Nat. Hist. 103 (10), 48.

RSPCA, no date. Purple Cross Award. Royal Society for the Prevention of Cruelty to Animals. Available from: http://www.rspca.org.au/get-involved/rspca-awards/rspca-purple-cross-award

Salter, C., 2014. Introducing the military-animal industrial complex. In: Salter, C., Nocella, A.J., Bentley, L.K.C. (Eds.), Animals and War: Confronting the Military-Animal Industrial Complex. Lexington Books, Lanham, MD, pp. 1–17.

Salter, C., 2015. Animals and war: anthropocentrism and technoscience. NanoEthics 9 (1), 11–21.

Salter, C., 2016. Inquiring into animal enhancement: model or counter-model of human enhancement? NanoEthics 10 (1), 257–260.

Scullard, H.H., 1974. The Elephant in the Greek and Roman World. Cornell University Press, Ithaca, NY.

Shelton, J., 2006. Elephants as enemies in ancient Rome. Concentric 32 (1), 3–25.

Singer, P., 1975. Animal Liberation: A New Ethics for Our Treatment of Animals. Random House, New York, NY.

Singer, P.W., 2009. Man's best friend? The history and future of animals in warfare. Unpublished book chapter.

Skinner, B.F., 1960. Pigeons in a Pelican. Am. Psychol. 15 (1), 28.

Sorenson, J., 2014. Animals as vehicles of war. In: Salter, C., Nocella, A.J., Bentley, J.K.C. (Eds.), Animals and War: Confronting the Military-Animal Industrial Complex. Lexington Books, Lanham, MD, pp. 17–35.

Spivak, G.C., 1988. Can the Subaltern Speak? In: Nelson, C., Grossberg, L. (Eds.), Marxism and the Interpretation of Culture. University of Illinois Press, Urbana, IL, pp. 271–313.

Taylor, S., 2014. Interdependent animals: a feminist disability ethic-of-care. In: Adams, C.J., Gruen, L. (Eds.), Ecofeminism: Feminist Intersections With Other Animals and the Earth. Bloomsbury, New York, NY, pp. 109–126.

The Lighthorseman, 1987. Motion Picture, Hoyts. Directed by Simon Wincer.

The War You Don't See, 2010. Documentary, Dartmouth Films. Directed by John Pilger.

Twine, R., 2012. Revealing the 'Animal–Industrial Complex'—a concept & method for critical animal studies? J. Crit. Anim. Stud. 10 (1), 12–39.

Tyler, A., 2014. Preface. In: Salter, C., Nocella, A.J., Bentley, J.K.C. (Eds.), Animals and War: Confronting the Military–Animal Industrial Complex. Lexington Books, Lanham, MD, pp. xiii–xiii10.

Wadiwel, D.J., 2015. The War Against Animals. Brill, Leiden, The Netherlands.

War Horse, 2011. Motion Picture, Walt Disney Studios. Directed by Stephen Spielberg.

Further Reading

BBC News, 2010. Dickin Medal awarded to bomb sniffing search dog Treo, February 6. Available from: http://news.bbc.co.uk/2/hi/uk_news/8502127.stm

Singer, P.W., 2009. Wired for War: The Robotics Revolution and Conflict in the Twenty-First Century. Penguin Press, New York, NY.

The Shields Gazette, 2015. South Sheilds pays tribute to 'Man with the Donkey' Gallipoli Hero Kirkpatrick, May 19. Available from: http://www.shieldsgazette.com/news/local-news/watch-south-shields-pays-tribute-to-man-with-the-donkey-gallipoli-hero-kirkpatrick-1-7268273

10

Animals in Entertainment

Colin G. Scanes

University of Wisconsin–Milwaukee, Milwaukee, WI, United States

10.1 INTRODUCTION

Entertainment has increased as the leisure time has increased. Economists at the US Federal Reserve Bank have estimated increases in leisure hours as about 7 h for men and 6 h for women in the USA between 1965 and 2003 (Aguiar and Hurst, 2006). A second requisite for entertainment is people having sufficient disposable income. It is surprising how much of entertainment is related to animals.

Quotations About Animals and Entertainment

"The dog is the god of frolic" the American social reformer, Henry Ward Beecher (1813–87).

"Two things only the people actually desire: bread and circuses" Juvenal (Decimus Junius Juvenalis) Roman poet (CE ~57–127).

"Zoo animals are ambassadors for their cousins in the wild" Jack Hanna zookeeper and Director Emeritus of the Columbus Zoo and Aquarium (1947 to date).

"There are always new places to go fishing. For any fisherman, there's always a new place, always a new horizon" Jack Nicklaus golfer (1940 to date).

This chapter includes consideration of the following:

- Zoos, aquaria, animal parks, and game reserves (Section 10.2).
- Animal shows including dog shows, livestock shows, rodeos, circuses, polo and animals in TV shows and movies (animal actors), animal riding, racing (horse, dog, camel, and pigeon), and pleasure carriage (Section 10.3).
- Animal riding, racing (horse, dog, and pigeon races) and equine competition, and pleasure carriages (Section 10.4).
- Hunting and shooting (Section 10.5).
- Fishing (Section 10.6).
- Historical and/or cultural animal fighting or baiting including bull fighting (Section 10.7).
- Legislation and public opinion on animals in entertainment and sport (Section 10.8).

10.2 ZOOS, AQUARIA, AND ANIMAL PARKS

10.2.1 Overview of Wildlife Tourism

Wildlife tourism is thought to have "secure sustainable economic benefits

FIGURE 10.1 **Wildlife-watching tourism is popular in East and Southern Africa.** (A) A photographic safari in Masai Mara Reserve in Kenya. (B) An example of animals watched is the White Rhinoceros (*Ceratotherium simum*) with a cow and calf at a game reserve, South Africa. *Source: Part A: Courtesy Shutterstock; Part B: Courtesy Wikimedia Commons.*

while supporting wildlife conservation" (Higginbottom, 2004). Wildlife tourism can be either nonconsumptive encompassing eco-tourism, nature adventure, or photographic safaris (Fig. 10.1) (the word safari coming from the Swahili for journey; UN WTO, 2014), or consumptive, such as trophy hunting and fishing (Higginbottom, 2004). Nonconsumptive wildlife tourism or animal watching can be segmented into the following (Higginbottom, 2004) (Fig. 10.2):

- Captive-wildlife tourism (in man-made confinement in zoos, wildlife parks, animal sanctuaries, and aquaria).
- Noncaptive wildlife-watching tourism including photographic tourism in scuba diving, for example, over coral reefs, and in national parks (e.g., Kruger National Park in South Africa, Maasai Mara National Reserve in Kenya, and Kakadu National Park in Australia).

The impact of wildlife tourism is estimated at 12 million trips annually growing at 10% a year (UN WTO, 2014).

FIGURE 10.2 **Wildlife-watching and photography in a wetland area.** *Source: Courtesy USDA Agricultural Research Service; Photo by Scott Bauer.*

Animals and Tourism in Sub-Saharan Africa

Tourism based on wild animals is particularly important to the economies of sub-Saharan Africa:

- One in twenty jobs are from tourism (Christie et al., 2013).
- There were 7.7 million people employed in tourism in Africa in 2010 (World Bank et al., 2010).
- Tourism contributed $9 and $1.3 billion to the economies of South Africa and Tanzania in 2010 (World Bank et al., 2010).
- Directly and indirectly, tourism contributed about 10% of sub-Saharan Africa's GDP ($170 billion a year) in 2013.
- More than 36 million people visited Africa (Economist, 2014).
- South Africa is viewed as having the highest potential for growth (the top tourism-ready country) in sub-Saharan Africa with Namibia, Kenya, Botswana, and Tanzania viewed also as tourism-ready countries (World Economic Forum, 2015).

There is some fragility to African tourism with a marked decline in the outbreak of Ebola despite it being far from the major tourist sites (Economist, 2014).

Nature, predominantly animal-based tourism, generated US$3.2 billion in Southern Africa in 2000–01 (South Africa $2.3 billion, Tanzania $0.3 billion, and Namibia $0.25 billion) (Economists at Large, 2013). Free-ranging-wildlife–watching in Africa focuses on the big five African animals (African elephant, Cape buffalo, leopard, lion, and rhinoceros), and to a lesser extent on the gorilla and other great apes, birds, whales, and dolphins. Lemur tracking is also popular (UN WTO, 2014). Trophy hunting is discussed in Section 10.5.5 and contributes about 7% of nature-based tourism.

10.2.2 History and Economic Impact and Magnitude of Zoos and Aquaria

The development of modern zoos was from menageries that were open to the public—Tiergarten Schönbrunn [Vienna, Austria) in 1752 and the Ménagerie du Jardin des Plantes (Paris, France) in 1793] (National Geographic, 2015). The Zoological Society of London was established in 1826 but opened to the public in the 1850s and known as the London Zoo. The Berlin Zoo opened to the public also about 1850. There were progressively more scientific modern zoos in Europe and North America with the founding of the Smithsonian National Zoo in 1889 by Act of Congress (National Zoo, 2016). Zoos were viewed as localities for science, public education, and arguably national/civic pride. In the case of the latter, zoos contained animals from colonized countries. Zoos increasingly became leaders in the conservation of endangered animals. In addition, zoos stimulate the economic development (discussed later). The first aquarium was at the London Zoo in 1857 (Forteath, 2010). Large aquaria (Table 10.1) have been constructed for public enjoyment, profit, civic/national pride, and as anchors for economic development.

According to the Association of Zoos and Aquariums (AZA), accredited zoos and aquariums had a significant economic impact in the USA in 2010 (Fuller, 2011):

- 165.5 million visitors
- $3.01 billion in operations expenditures (supporting 85,000 jobs)
- $469 million in construction
- with economic multipliers, the total economic impact of zoos and aquaria is $16.0 billion

These figures exclude *concessions* in the zoos and aquariums together with restaurants and hotels catering to visitors of the zoo, aquaria, and animal parks.

TABLE 10.1   Largest Aquariums in the World

Ranking	Aquarium	Location	Size in liters (million)	Size in gallons (million)
1	Chimelong Ocean Kingdom	Hengqin Island, China	49	12.9
2	S.E.A. Aquarium	Singapore	45	11.9
3	Georgia Aquarium	Atlanta, GA	24	6.3
4	Dubai Mall Aquarium	Dubai, UAE	10	2.64
5	Okinawa Churaumi Aquarium	Okinawa, Japan	7.5	1.98
6	L'Oceanografic	Valencia, Spain	7.0	1.85

10.2.3 Aquaria

The largest aquaria in the world are listed in Table 10.1. Other very large aquaria globally are the following:

- Turkuazoo (Istanbul, Turkey)
- Monterey Bay Aquarium (Northern California)
- uShaka Marine World (Durban, South Africa)
- Shanghai Ocean Aquarium (Shanghai, China)
- Aquarium of Genoa (Italy)
- Aquarium of Western Australia (Perth, Australia)

The largest aquaria in the USA based on attendance are listed in Table 10.2. Also see Fig. 10.1.

10.2.4 Zoos

Among the large zoos in the world based on acreage and number of animal species in alphabetical order are the following: Beijing Zoo (China), Berlin Zoological Garden (Germany), Bronx Zoo (New York), Columbus Zoo and Aquarium (Ohio), Henry Doorly Zoo and Aquarium (Nebraska), London Zoo (England), Moscow Zoo (Russian Federation), National Zoological Gardens of South Africa, San Diego Zoo (California), Toronto Zoo (Canada). There is a high attendance at zoos (Fig. 10.3). The largest zoos in the USA are listed in Table 10.3 while the largest animal parks in the USA are shown in Table 10.4.

10.2.5 Wild Animal Watchers

The National Survey of Fishing, Hunting, and Wildlife-Associated Recreation estimated that there were 71.8 million animal watchers in the USA in 2011 (US Fish and Wildlife Service, 2012). Birds are the major class of animals watched (Figs. 10.4 and 10.5). This was calculated to

TABLE 10.2   Largest Aquaria in the USA Based on Attendance

Ranking	Name	Attendance in millions
1	SeaWorld Orlando	5.9
2	SeaWorld San Diego	4.1
3	Georgia Aquarium	2.7
4	SeaWorld San Antonio	2.5
5	Steinhart Aquarium (San Francisco, CA) attendance	2.4

FIGURE 10.3   **Great Grey Heron (*Ardea cinerea*).** *Source: Courtesy USDA Natural Resource Conservation Service; Photo by Lynn Betts.*

TABLE 10.3   Largest Zoos in the USA

Ranking	Name	Location	Attendance in millions
1	San Diego Zoo	San Diego, CA	3
2	Lincoln Park Zoo	Chicago, IL	3
3	Saint Louis Zoological Park	St Louis, MO	3
4	National Zoological Park	Washington, DC	2.7
5	Bronx Zoo	Bronx, NY	2.1

TABLE 10.4   Largest Animal Parks in the USA

Ranking	Name	Location	Attendance in millions
1	Disney's Animal Kingdom Park	Florida	3
2	Busch Gardens Tampa Bay	Chicago	3
3	San Diego Wild Animal Park	San Diego	3

FIGURE 10.4   **Families enjoy outings at zoo.** (A) A family interacting with giraffes. (B) Viewing the flamingos at the San Diego Zoo. *Source: Part A: Courtesy Shutterstock; Part B: Courtesy Wikimedia Commons.*

FIGURE 10.5   **Large aquariums are popular worldwide.** (A) Whale shark at the Georgia Aquarium. (B) Tropical fish on coral reef in large aquarium. *Source: Part A: Courtesy Wikimedia Commons; Part B: Courtesy Shutterstock.*

TABLE 10.5   Expenditures by Animal Watchers, Hunters, and Anglers in the USA

	Expenditures of animal watchers and sportsmen in the USA (in billion US$)
Animal-watching	54.9
Fishing	41.8
Hunting	33.7
Hunting and fishing	14.3

Data from US Fish and Wildlife Service, 2012. Wildlife & Sport Fishing Restoration Program. National Survey of Fishing, Hunting, and Wildlife-Associated Recreation. Available from: http://www.census.gov/prod/2012pubs/fhw11-nat.pdf.

generate US$54.9 billion (US Fish and Wildlife Service, 2012). This is 61% of the expenditures on hunting and fishing in the USA (Table 10.5).

10.3 ANIMAL SHOWS

10.3.1 Livestock Shows

Livestock *shows* are present in many countries. In the USA, there are national and regional shows (Fig. 10.6) including the following:

- American Royal; American Royal Livestock, Horse Show, and Rodeo Kansas City, MO.
- North American International Livestock Exposition, Louisville, KY.
- National Western Stock Show Denver, CO.
- Southwestern Exposition and Livestock Show/Fort Worth Stock Show and Rodeo.
- Houston Livestock Show and Rodeo, Houston, TX, with paid attendance of 2.4 million people in 2015 and a direct economic impact of US$101 million from people outside the metropolitan region (Smith, 2010).

In addition in the USA, state and county fair include livestock, horse, poultry, and rabbit competitions. Among the largest state fairs are the following:

- The State Fair of Texas Dallas, TX (attendance, ~3 million)
- Minnesota State Fair St. Paul, MN (attendance, 1.7 million)

FIGURE 10.6   **Livestock show are held in state, county, and other fairs.** (A) Cowboy with grandson at the Houston Livestock Show. (B) Livestock judging by children in 4H program at county fair. (C) Livestock show at county fair. *Source: Part A–B: Courtesy Wikimedia Commons; Part C: Courtesy Shutterstock.*

- Iowa State Fair, Des Moines, IA (attendance, 1.0 million)
- Wisconsin State Fair Milwaukee/West Allis (attendance, 1.0 million)

While attendance at state and county fairs aggregates at over 20 million attendees, it should be stressed that the fairs include midway rides (carnivals); performers; exhibitions; and fruit, vegetable, and cake competitions along with livestock. They are a piece of Americana. Arguably the largest livestock and equine show in England is the Royal Three Counties Show, Malvern, Worcestershire, United Kingdom. Exhibiting animals can be an outstanding experience for young people.

10.3.2 Dog Shows

Dog shows continue to be popular with the two largest being the following:

- Westminster Kennel Club Dog Show: first show in 1877 and held annually at Madison Square Gardens, New York, NY. Includes 2800 American Kennel Club registered dogs and is televised (Westminster Kennel Club Dog Show, 2016).
- Crufts: inaugurated in 1891 with 125th anniversary in 2016; held at National Exhibition Centre (Birmingham, England); highest attendance is 160,000 in 2008 (Crufts, 2016).

10.3.3 Circuses

Beginning 2500 years ago, the people of Rome watched chariot races at the Circus Maxima. It is estimated that 150,000 people could attend (Encyclopedia Romana, 2016a). The modern circus is thought to have originated in London (England) in 1768 with horsemanship events combined with clowns and vaulting (reviewed by Zanola, 2010). There is a lack of information on attendance at circuses (Zanola, 2010). In Italy, there are about 100 circus units with one location having 132,000 seats available at over 44 days for the circus (Zanola, 2010). In the USA, a corporation, Feld Entertainment, operates the Ringling Bros. and Barnum & Bailey Circus and has sales of $1.0 billion from 5000 shows attracting 30 million (Forbes, 2014). However, the sales and attendance figures also include Disney On Ice, clowns, and monster truck rallies sales (Forbes, 2014).

There is a controversy over the use of wild animals in circuses (PETA, 2015). In the USA, circuses, such as Ringling, are phasing out the use of Asian elephants (National Geographic, 2016). At the time of writing, a proposed ban of wild animals in circuses in England had not been passed. Operators of traveling circuses in England with wild animals are required to have a license under the "Welfare of Wild Animals in Travelling Circuses (England) Regulations 2012" (UK Government, 2016a).

10.3.4 Rodeos and Mexican Rodeos *Charreadas*

Rodeo events use the *riding* and roping skills and speed of cowboys (*vaqueros*). The governing body for rodeos is the Professional Rodeo Cowboys Association (PRCA). Among the events are the following:

- bareback bronc riding
- barrel racing
- bull riding
- calf roping
- saddle bronc riding
- steer wrestling
- team roping (see calf roping)
- tie-down roping

The PRCA sanctions 600 rodeos in the USA and Canada with estimates of attendance varying between 5.4 (Humphreys and Humphreys, 2005) and 30 million people (PRCA, 2016).

10.3.5 Animal Actors

Animal actors are considered in the textbox.

Examples of Animal Actors

Cats (TV and movies)
    The Adventures of Milo (tabby cat) and Otis (pug)
    Lucky in ALF (1986–90)
Dogs (TV)
    Buck (Briard) in Married with Children (1987–97)
    Comet (golden retriever) in Full house (1987–95)
    Eddie (Jack Russell terrier) in Frasier (1993–2004)
    Isis, Pharaoh and Tiaa (yellow Labrador retrievers) in Downton Abby (2010–16)
    Jagger (German shepherd) on Angie Tribeca (2016)
    Lassie (rough collie) in Lassie (1954–74)
Dogs (movies)
    Toto (Brindle Cairn Terrier) Wizard of Oz (1937)
    Max (Belgian Malinois) (2015)
Monkeys (TV and movies)
    Ross's Marcel (white-throated capuchin monkey) in Friends (1994–2004)
    Chimpanzees in the Tarzan movie series
    Capuchin in The Hangover Part II (2011)
Dolphin (TV and movies)
    Flipper (movie 1963 and TV series 1964–67)
Horses
TV
    Mister Ed (1961–66)
    Budweiser Clydesdales: used for promotion (particularly Superbowl and Holiday season ads)
Movies
    Black Beauty (seven movies)
    Seabiscuit (2003)
    Secretariat (2010)
    War horse (2011)

10.3.6 Polo

Polo is a game for teams of riders on horseback (four per team) and horses (referred to as polo ponies but in fact, horses). It is played on a grass field 300 yards (274 m) long and 160 yards (146 m) wide (with side boards). The aim for each team is to hit a wooden or plastic ball with a long mallet between the goal posts (any height). The game originated in Persia (present-day Iran) about 2500 years ago. The United States Polo Association is the official source of rules and tournaments in the USA (US Polo Organization, 2016). While not an Olympic sport, it is played in over 70 countries including the USA, the United Kingdom, and Argentina. The World Polo Championship is played every 3 years. There is also an intercollegiate polo (men's and women's) played between universities in the USA.

10.4 ANIMAL RIDING, RACING, AND PLEASURE CARRIAGES

10.4.1 Horse Riding

Horses, ponies, and donkeys are ridden for pleasure and competition (Figs. 10.7 and 10.8). The economic impact of horses in the USA is considered in an essay elsewhere in this volume by Karyn Malinowski.

10.4.2 Horse-Drawn Carriages

Horse-drawn carriages are found in many tourism destinations. Efforts to end this practice

FIGURE 10.7 **Horse in Hawaii.** *Source: Courtesy Agricultural Research Service, USDA; Photo by Scott Bauer.*

FIGURE 10.8 **An exhibitor and her horse in a competition.** *Source: Courtesy USDA Tennessee Walking Horse; Photo by Steve R. Kendrot.*

in New York's Central Park have not moved forward (WSJ, 2015).

10.4.3 Horse Racing

10.4.3.1 Horse Racing in the USA

Attendance at horse racing in the USA was estimated to be 6.0 million people (Humphreys and Humphreys, 2005). Horse racing in the USA has a US$10.6 billion direct impact based on an economic impact survey in 2005 (American Horse Council, 2005).

Thoroughbred racing is a predominantly flat racing in the USA (see side bar). In addition, Standardbred horses engage in harness races (trotters and pacers) with the major races for trotters being the Hamiltonian and the Kentucky Futurity and for pacers being the Meadowlands Pace and the North America Cup.

Equine Definitions

Steeplechase

The steeplechase race originated in Ireland as a race from "steeple to steeple" or from village to village each with a church steeple.

Triple Crown

The term "Triple Crown" originated in England for the 2000 Guineas, the Epsom Derby, and the St. Leger Stakes.

The US Triple Crown races are 1.25–1.5 miles (2.0–2.5 km) on dirt tracks for 3-year-old fillies, colts, and geldings. The three races are the following:

- The Kentucky Derby held at Churchill Downs (Louisville, KY)
- The Preakness Stakes held at Pimlico Race Course (Baltimore, MD)
- The Belmont Stakes at Belmont Park (Elmont, NY)

Dressage

Dressage is from the French for training equine. This requires the horse and rider to perform intricate patterns of movement.

Driving Events

Driving events have horses harnessed to a horse-drawn vehicle or carriage containing a driver. The horse(s) are controlled by the driver who uses the reins and a whip. In addition, there may also be a second or third person (groom and navigator) to assure balance. The number of horses varies from one (singles), two (pairs or tandems), three (unicorns), and four (four-in-hands).

10.4.3.2 Horse Racing in the United Kingdom

Attendance at race courses in the United Kingdom in 2014 was 5.6 million people at 60 different race courses (Deloitte, 2014). Direct impact of horse racing in the United Kingdom in 2014 was estimated as £1.1 billion, but adding indirect and induced impact brings the impact to £3.34 billion (Deloitte, 2014). In Great Britain, horse races that allow professional jockeys are one of the following:

- Flat: The top races are the 2000 Guineas, the Epsom Derby (first run in 1780), and the St. Leger Stakes.
- National Hunt race with hurdles: the horses jumping over fences.
- National Hunt steeplechase racing with the horses jumping over fence obstacles, ditches, and water jump with the top event steeplechase race being the Grand National (held at Aintree, near Liverpool, England).
- In addition, there are "point-to-point" races that allow amateur jockeys.

10.4.4 Olympic and World Equestrian Games

There are three Olympic equestrian events: (1) show jumping, (2) dressage, (3) 3-day event (show jumping, dressage, and cross-country). The one equestrian Paralympic event is dressage.

The World Equestrian Games is held on even years between Olympics under the auspices of the Fédération Equestre Internationale. The events are show jumping, dressage, 3-day event (show jumping, dressage, and cross-country) plus driving (four horses), endurance riding, and vaulting. Driving is a Para-equestrian event. Accompanying the FEI are national associations and events.

To assure the welfare of horses, the highest standards of safety, health, nutrition, housing, transportation, and veterinary care are required for horses in the World Equestrian and Olympic Games and these standards are rigorously enforced (Atock and Williams, 1994; FEI, 2016). Moreover, the standards are stated to prohibit abuse of the horses during events and training (Atock and Williams, 1994).

10.4.5 Other Horse Events

The USA Equestrian, formerly the American Horse Shows Association, oversees 2700 competitions with 26 breeds and types of completion.

10.4.6 Endurance Races

Recognized endurance races are 50, 75, and 100 miles (80, 121, and 161 km) long (US Equestrian Federation, 2016).

10.4.7 Dog Races

Greyhound racing is still popular in some areas. For instance, in the United Kingdom, there were 2 million spectators at over 5000 events with 7000 people directly employed (Greyhound Board of Great Britain, 2015). In addition, there are sled dog races. The Iditarod Trail Sled Dog Race was first ran to Nome in 1973.

10.4.8 Pigeon Racing

Among the organizations for pigeon fanciers and racers are the following: Australian National Pigeon Association, American Racing Pigeon Union, Canadian Pigeon Fanciers Association, Chinese Racing Pigeon Association, International Federation American Homing Pigeon Fanciers Inc., and Royal Pigeon Racing Association (the United Kingdom).

Using statistics from the Chinese Racing Pigeon Association, less than half returned home in races (715 returned home out of 1591 pigeons released) (Li et al., 2016).

10.5 HUNTING AND SHOOTING

Definitions in the USA

The following definitions are based on the US Fish and Wildlife Service (2012):

- Sportspersons or sportsman: people who fish and/or hunt.
- Hunters: sportspersons who hunt including licensed hunters using rifles and shotguns and those without a license (e.g., bow, muzzleloaders, other primitive firearms, or pistols or handguns)
- Anglers: sportspersons who fish including licensed hook and line anglers together with those without a license use such methods as spears.

Definitions in the United Kingdom

The following definitions are based on the Cambridge Dictionary (2016):

- Sportsman: someone who plays sports, such as soccer (football) and rugby, both well and fairly.
- Huntsman: people who go hunting for foxes on horseback with hounds for pleasure.
- Hunting: activity of people who fox hunt on horseback with hounds for pleasure.
- Shooting: the sport of shooting animals, such as grouse and pheasants; also means any action of shooting a gun.
- Anglers: people who fish.

10.5.1 Overview

In the USA, sportspersons spent a total of $89.8 billion in 2011 on items used for hunting and/or fishing (Table 10.5) (US Fish and Wildlife Service, 2012). Shooting sports contribute £2 billion (~$3 billion) to the UK economy with 2 million hectares (4.9 million acres) actively managed for conservation (BASC, 2014).

10.5.2 Hunting in the USA

According to the National Survey of Fishing, Hunting, and Wildlife-Associated Recreation, in 2011, there were about 13.7 million adult hunters (Fig. 10.9) in the USA; increasing 9.6% since 2006 (Table 10.6). The participation rate in hunting in the USA is about 6% (Table 10.6). Hunting is predominantly an activity of white men with participation rates differing by gender (men 11% women 1%) and ethnicity (white 7%, African American 2%) (US Fish and Wildlife Service, 2012). About a third of hunters hunt on publicly owned lands compared to 84% who hunt on privately owned land (US Fish and Wildlife Service, 2012).

Using definitions of the US Fish and Wildlife Service (2012), most hunters in the USA hunt big games, followed by small games, migratory birds, and then other animals as follows:

- big game (e.g., deer, elk, wild turkey, and bear), 11.6 million people
- small game (e.g., squirrels, rabbits, pheasants, quail, and grouse), 4.5 million people
- migratory birds (waterfowl and doves), 2.6 million people
- other animals (e.g., raccoons, feral pigs, and coyotes), 2.2 million people

10.5.3 Regulations and Hunting

Unregulated overhunting was responsible for huge decreases in populations of bison and white-tailed deer (Fig. 10.10) in the USA. Intensive hunting in the 1800s resulted in a precipitous drop in the population of white-tailed deer from 25–30 to 0.3–0.5 million in 1910 (Table 10.7)

FIGURE 10.9 **Hunting is a popular activity in the USA.** (A) Hunter with 10-point buck. (B) Dog and hunter working together. *Source: Courtesy Wikimedia Commons.*

TABLE 10.6   Number of Hunters[a] and Anglers in the USA in Millions With Participation Rate in Parentheses

	2006	2011
Hunters	12.5	13.7 (6%)
Anglers	30.0	33.1 (14%)

a Using US meaning of the word.
Data from US Fish and Wildlife Service, 2012. Wildlife & Sport Fishing Restoration Program. National Survey of Fishing, Hunting, and Wildlife-Associated Recreation. Available from: http://www.census.gov/prod/2012pubs/fhw11-nat.pdf.

TABLE 10.7   Changes in White-Tailed Deer and Bison Populations in the USA (US Fish and Wildlife Service, 2012; VerCauteren, 2003)

Year	White-tailed deer	Bison
1600	25–30 million	30–60 million
1900	500,000	1,000
1990	25–32 million	250,000

(VerCauteren, 2003). Similarly, the number of bison was decreased by 99.98% between 1600 and 1900 due to massive hunting (Table 10.7) (US Fish and Wildlife Service, 2012). Among multiple reasons, the bison were to provide meat for workers building transcontinental railroads and for their hides (US Fish and Wildlife Service, 2012).

Cessation of hunting was followed by increases in bison numbers to 12,500 in 1919 and to 250,000 in 1990 (Table 10.7) (US Fish and Wildlife Service, 2012). With white-tailed deer, hunting was initially stopped completely. This was followed by tightly regulated hunting (initially from bucks and then allowing hunting for does) (Table 10.7). Populations of deer now meet or exceed those before European settlement. The

FIGURE 10.10   **White-tailed deer.** *Source: Courtesy Agricultural Research Service, USDA; Photo by Scott Bauer.*

complete recovery of white-tailed deer numbers can be attributed to tight and thoughtful regulation of hunting and to the dearth of natural predators. This complete recovery can also overshoot with, at least in some places, an increase in population of deer.

10.5.4 Shooting/Hunting in the United Kingdom

10.5.4.1 Shooting in the United Kingdom

Examples of species where shooting is permitted (BASC, 2015):

- Mammals: red deer, Silka deer, fallow deer, roe deer, Chinese water deer, Muntjac deer, wild boar, hare, and rabbit together with pest species, such as brown rats, mink, foxes, and gray squirrel.
- Birds: pheasants, partridges, grouse, ducks (e.g., mallard, teal, pintail, and tufted duck), woodcock, and geese.

10.5.4.2 Fox Hunting With Hounds

A traditional rural activity in England was fox hunting with hounds (Fig. 10.11). The fox was pursued by the hounds and the members of the hunt on horseback. This was made illegal in 2004. However, simulated hunting was allowed with the hounds followed by the hunters on horseback pursuing a scent trail. If a fox is flushed out by the hounds, it is killed by them. The "fox" hunts attracted 250,000 people to a series of meets on December 26, 2014 (Daily Telegraph, 2015).

10.5.5 Trophy Hunting in Africa

In sub-Saharan Africa, trophy hunting is largely limited to Southern Africa (South Africa largest, well developed in Botswana, Namibia, and Zimbabwe) together with Tanzania (East Africa) (Christie et al., 2013; Lindsey, 2008). Most of the hunters come from Western Europe (particularly Germany and the United Kingdom)

FIGURE 10.11 **Fox hunting with hounds originated in England.** This hunt is in Kentucky. *Source: Courtesy Shutterstock.*

and the USA (Borge et al., 1990). About 1.4 million km^2 is used for trophy hunting with this being larger than the national parks where hunting is not permitted (Lindsey, 2008). The growth of trophy hunting in Southern African is due to game ranching and permitted hunting of dangerous species, such as elephants, buffaloes, lions, and leopards (Lindsey, 2008). In addition, in West Africa, there is a substantial number of tourist–hunters shooting birds (Lindsey et al., 2007). There is a considerable opposition to trophy hunting from such groups as the International Fund for Animal Welfare, the Humane Society of the United States (HSUS), and Born Free USA/Born Free International.

The advantages of trophy hunting include the following:

- The economic impact that trophy hunting in Southern and East Africa generates is estimated at US$201 million per year (Lindsey et al., 2007). This is calculated as 7% of animal-based tourism.
- Trophy hunting is said to play an important role in animal conservation (Lindsey, 2008). It is claimed that trophy hunting has been

critical in the recovery of populations of bontebok, black wildebeest, cape mountain zebra, and white rhino (Lindsey, 2008).

- Hunting operators conduct operations against poaching and illegal wildlife trade (Lindsey, 2008).

In a position paper for the International Fund for Animal Welfare, the HSUS, and Born Free USA/Born Free International, the economic impact of trophy hunting was questioned (Economists at Large, 2013). For instance, a lack of transparency of the data on the economic impacts was noted and that only 3% of trophy hunting expenditures went to "Area and community development" (Economists at Large, 2013). This was interpreted as "only 3 percent of the money actually reaches the rural communities where hunting occurs" (International Fund for Animal Welfare, 2013). This seems to be refuted by the estimates that three-quarters of the revenue from trophy hunting is retained in country (Booth, 2010) with accompanying employment locally and nationally.

The problems of trophy hunting include the following:

- Corruption such that shooting exceeds quotas and government ministers are encouraged to permit specific operations (World Economic Forum, 2015).
- Inadequate regulation of the hunting industry (Lindsey, 2008).
- In the event of a downturn in trophy hunting, poaching increases (Cruise, 2015).
- Perimeter fencing around game reserves. This prevents migrations of animals (Lindsey et al., 2007)
- Small game ranches can be overstocked resulting in habitat degradation (Lindsey et al., 2007)
- Game ranchers kill "nonhuntable" predators (e.g., wild dogs or cheetahs) (Lindsey et al., 2007)

To assure wildlife-based land uses, Lindsey et al. (2012) recommended restricting the number of lions shot to sustainable levels rather than cessation of lion hunting (also supported by Packer et al., 2011 based on quantitative data).

10.6 FISHING

10.6.1 Angling in the USA

Angling is popular in the USA with one in seven people participating. While the participation rate for the US adult population engaging in angling is 14% (Table 10.6), there are differences by ethnic group (16% white, 10% African American, and 5% Hispanic) (US Fish and Wildlife Service, 2012).

According to the National Survey of Fishing, Hunting, and Wildlife-Associated Recreation, most US anglers fish freshwater, followed by salt water, and finally the Great Lakes. The top fish species are the following (US Fish and Wildlife Service, 2012):

- Freshwater excluding Great Lakes: 27.1 million anglers (top fish species: black bass, panfish, trout, catfish, crappie, and other bass) (Fig. 10.12).
- Marine: 8.1 million anglers (striped bass, flatfish, red drum, sea trout, blue fish, and salmon) (Figs. 10.13 and 10.14).
- Great Lakes: 1.7 million anglers (top fish species: walleye, black bass, perch, and salmon).

10.6.2 Fishing in the United Kingdom

10.6.2.1 England

In England and Wales, anglers are required to have a rod license or junior rod license. There are 6.1 million people who participated in freshwater and/or marine fishing. The majority of anglers in England and Wales participate in freshwater fishing but 37.2% took part in sea angling. Freshwater fishing is categorized as either of the following (Brown, 2012):

FIGURE 10.12 **Fly fisherman on the South Fork of the Holston River.** *Source: Courtesy USDA Natural Resource Conservation Service; Photo by Jeff Vanuga.*

FIGURE 10.13 **Marine fishing in the surf.** *Source: Courtesy NOAA; Photo by William B. Folsom.*

- game fishing (salmon, trout, and char)
- coarse fishing (nonsalmonids)

The largest proportion of anglers were coarse fishermen The largest proportion of anglers were coarse fishermen fishing most often in still-waters as opposed to rivers and streams (58%). Anglers also participate in angling competitions. Over 38% of anglers were introduced to fishing by their parent; 19% by another family member and 27% by a friend (Brown, 2012). Concern for the environment especially on water quality (pollution) is a significant issue for anglers (Brown, 2012).

10.6.2.2 Scotland

Recreational fishing in Scotland is a major source of tourism (SCSTG, 2016) with anglers contributing £113 million ($158 million) to the Scottish economy (Radford et al., 2014). Game fish anglers make up to 65% of angling in Scotland. This includes fly fishing. The game fish species are the following salmonids: brown trout (*Salmo trutta*), rainbow trout (*Oncorhynchus mykiss*), grayling (*Thymallus thymallus*), Atlantic salmon (*Salmo salar*), plus migrating sea trout (*S. trutta*). It should be noted that sea and brown trout are the same species but different populations. Among the most important coarse fish are pike. In the economically struggling region of the Scottish Highlands, angling generates 81 full jobs and £24.6 million to the economy (Radford et al., 2014).

FIGURE 10.14 **Anglers after catching a salmon.** *Source: Courtesy NOAA.*

10.7 HISTORICAL AND/OR CULTURAL ANIMAL FIGHTING AND BAITING

10.7.1 Overview

It is illegal in many countries to conduct a number of practices that were common in the past where animals were used as entertainment involving cruelty, such as bear baiting (multiple English bulldogs vs. a black bear chained to a post), cock fighting (e.g., the United Kingdom and the USA), dog fighting (e.g., North America), and fox hunting with dogs (the United

Kingdom). Some of these are culturally important to some groups. It is questioned whether or how much we should be sensitive to different cultures in immigrants from different countries and allow practices we disagree with? Alternatively, are there practices so morally reprehensible that they should be illegal globally irrespective of the culture and traditions of the particular country or region?

10.7.2 Bullfighting

Bullfighting is an entertainment or blood sport found in Spain, Mexico, Venezuela, Peru, Columbia, and Ecuador. Bulls are first wounded by the picadors stabbing the bull. The latter is then encouraged to charge the matador who will end the contest by killing the bull. While bullfighting is no longer legal in one region (Catalonia), it is considered as an important part of Spanish culture and continues. In the variant of bull fighting in both Portugal and Southern France, the bull is not killed in the arena.

10.7.3 Cockfighting

Cockfighting is an ancient activity. It is illegal in many places including the United Kingdom and all 50 States in the USA. It continues legally, for instance, in parts of Latin America and Southeast Asia particularly in Indonesia and the Philippines, and illegally elsewhere. With the accompanying betting, it is said to be a multi-billion dollar activity globally. Under the 2014 Farm Bill, it is a federal crime to attend a cockfight (LA Times, 2014).

10.7.4 Dogfighting

Dogfighting is illegal in many countries including Australia, the United Kingdom, and the USA. However, according to the Huffington Post (2014), there are more than 10,000 organized and illegal dog fights in the USA per year.

10.7.5 Roman Coliseum and Animals

Amphitheaters were constructed throughout the Roman Empire with the largest being the Colosseum or Flavian amphitheater with a capacity of over 65,000 (built in AD/CE 72). The colosseum was used for gladiatorial combats, killing criminals or captured prisoners from wars, and killing or hunting animals, such as bears and bulls, together with exotic African or Asian animals, such as lions, elephants, giraffes, rhinoceroses, tigers, and leopards. Combats between animals and those to be executed focused on spectacle and not evenness with, for instance, the person tied to a post and killed by the animals. Among the people savagely killed in this way is often said to be Christian martyrs with Christians being "thrown to the lions." Doubts have been expressed about this, however (Moss, 2013). The animals used in the amphitheaters were cared for by *Bestiarii* who also encouraged them to fight (Encyclopedia Romana, 2016b).

10.8 LEGISLATION AND PUBLIC OPINION ON ANIMALS IN ENTERTAINMENT AND SPORT

10.8.1 Overview

There have been campaigns against the use of animals in all or some forms of entertainment (League Against Cruel Sports, 2015; PETA, 2015).

For instance, the League Against Cruel Sports (2015) was founded in 1924 in England with a goal to "ban fox hunting, stag hunting, otter hunting, hare hunting, and hare coursing in the UK"; these activities employing dogs. It is first important to consider what is "cruelty" (see Section 10.8.2).

10.8.2 Animal Cruelty

What is animal cruelty? Animal cruelty can be the following:

- Deliberate abuse, such as beating, shooting or stabbing, torturing, tormenting or setting them on fire.

- Neglect or not providing necessary food, water, shelter, or veterinary care.

Veterinary care encompasses disease prevention with prophylactics, such as vaccinations and flea/tick control or disease treatment (e.g., antibiotics), following injury, and humane euthanasia (HSUS, 2016a,b). It is argued that the converse of animal cruelty is animal welfare. The welfare of animals used for entertainment should be paramount and normally is. Are there people who consciously abuse animals? The answer is yes. If there are laws against something, some people break them. Is there a link between cruelty to animals and antisocial/criminal behavior in people? This is discussed next.

10.8.3 Animal Cruelty and Antisocial/Criminal Behavior

Criminologists, psychiatrists, and other social scientists have linked animal cruelty to subsequent violent criminal behavior (reviewed by Arluke and Madfis, 2014). For instance, Kellert and Felthous (1985) concluded that "childhood cruelty toward animals occurred to a significantly greater degree among aggressive criminals than among nonaggressive criminals or noncriminals." The evidence is regarded as "weak or inconsistent" as frequently animal cruelty is not categorized as to species (e.g., companion animals vs. pest rodents), distance of action (personally injuring or killing vs. shooting from a distance) (Arluke and Madfis, 2014), together with cultural acceptability. The need to refine the concept of animal cruelty has been emphasized (Arluke and Madfis, 2014).

Physically torturing animals is a type of animal cruelty that has been shown to be a predictor of sadistic serial killing (Levin and Arluke, 2009). Prior childhood animal abuse was found in 43% of school shooters (Arluke and Madfis, 2014). What is the situation with noncriminals? About 28% of undergraduate students self-reported animal abuse as children

(Levin and Arluke, 2009). However, when animal cruelty was limited to dogs and cats and physical cruelty (e.g., strangulation, bludgeoning, or beating to death) the reported incidence was less than 1% (Levin and Arluke, 2009). In contrast, 40% of school shooters had previously committed acts of cruelty to dogs or cats in close physical proximity (Arluke and Madfis, 2014). This is referred to as "up-close and personal" cruelty (Arluke and Madfis, 2014).

The following were reported in a study of inmates at medium- or maximum-security prisons (Henderson et al., 2011):

- ~80% self-reported hitting animals
- ~33% shot or kicked animals
- ~20% had sex with animals

"The age at which offenders began committing animal cruelty and having sex with animals were predictive of adult interpersonal violence" (Henderson et al., 2011). In a comparison of childhood animal abuse in an incarcerated population, urban people abused cats, dogs, and wild animals while rural people abused cats (Tallichet and Hensley, 2005). About a quarter of children with conduct disorder are cruel to animals (reviewed by Miller, 2001). Studies on the abuse of animals have focused on incarcerated people and/or mass murderers. Studies have not, to the author's knowledge, examined the impacts of benign or at least more benign interactions with animals, such as companion animals, horses, hunting, fishing, equestrian events, horse riding, horse races, attending zoos, circuses, and so on, and whether they increase the development of antisocial and/or criminal behaviors. The intuitive answer is they do not. Let us consider two of the most benign interactions between people and animals: companion animals and therapy animals.

10.8.4 Positive Interactions With Animals

10.8.4.1 Overview

Just as "laughter is pleasure" and experiencing pleasures is important to both physical and mental health (Pattillo and Itano, 2001), the presence of animals adds to the quality of life and hence health.

10.8.4.2 Companion Animals

Companion animals are covered elsewhere in the book. However, it is suggested that companion animals are an important source of entertainment. For instance, walking for leisure is increased in people who have a dog but not cat (Cutt et al., 2008; Yabroff et al., 2008). In the USA, the welfare of companion animals is regulated by state laws. The position of the organization PETA on having companion animals is that "it would have been in the animals' best interests if the institution of 'keeping'—i.e., breeding animals to be kept and regarded as 'pets'—never existed."

10.8.4.3 Animals for Therapy Service

Animals, such as dogs and cats, are increasingly present in therapeutic settings including nursing homes. The intuitive reason is that animals add to the quality of life for the residents in the same way that people enjoy their pets. There is an evidence, albeit limited, that the presence of animals has a therapeutic effect (Filan and Llewellyn-Jones, 2006). For instance, Kamioka et al. (2014) concluded that the randomized controlled trials "conducted have been of relatively low quality" but animal-assisted therapy "may be an effective treatment for mental and behavioral disorders, such as depression, schizophrenia, and alcohol/drug addictions." Moreover, the increased agitation and depression in people with dementia during extended stays in nursing homes is reduced by the presence of animals (Majić et al., 2013). The use of animals as therapeutic aids was arguably initiated by placing cats in each ward of a nursing home as a "mascot" (Brickel, 1979). The presence of a cat was reported by staff as enhancing resident responsiveness (Brickel, 1979).

The usefulness of animals goes beyond dogs and cats. For instance, there was a marked reduction in the global severity index in people with mental health problems after a program of inter-actions with horses called the "Equine-Assisted Experiential Therapy" (Klontz et al., 2007). Bird feeders enhanced the happiness of residents in nursing homes (Banziger and Roush, 1983). Loss of body weight is a problem in people with de-mentia. There was an increase in food consumed in people with Alzheimer's disease when eating in the presence of an aquarium with fish (Ed-wards and Beck, 2002).

10.8.4.4 Ethical Issues, Public Service, and Zoos

Minteer and Collins (2013) concluded that "the practice of keeping animals in zoos and aquariums is one of the more intriguing areas of conflict within the animal ethics–conserva-tion ethics debate." In the position paper Com-mitting to Conservation: The World Zoo and Aquarium Conservation Strategy, zoos and aquaria accept "their responsibility that comes with maintaining and caring for animals" (Bar-ongi et al., 2015). This is coupled with the goals of advancing conservation and public educa-tion. Indeed, the question whether it is ethical to have animals in zoos and aquaria was answered in the American Veterinary Medical Association news by "only if conservation and animal wel-fare are the focus" (Kuehn, 2002).

There are differences in the viewpoints of ani-mal welfare/rights organization. For instance, HSUS considers that "Zoos are a fact of life" but stress the need for humane, professional care and the need to keep animals "in displays re-sembling their natural habitats as closely as pos-sible." In contrast, PETA (2015) completely "op-poses zoos." Another animal rights organization states that "Zoo teach us it is alright to cage animals if we justify it with an excuse (we need them for conservation, research, public educa-tion.…" The Animal Welfare Institute has been strongly advocating to end killer whale shows at SeaWorld (Newsweek, 2015).

In the USA, zoos, marine mammal parks, and circuses are regulated under the Animal Welfare Act of 1966 with the requirement of licensing for those who exhibit animals to the public (USDA NAL, in press). Moreover, there is regular inspec-tion to insure compliance (USDA NAL, in press). In addition, the best zoos are also accredited by the AZA (AZA, 2016; Kuehn, 2002). Standards for accreditation promote animal welfare via "superior animal husbandry." The AZA defines animal welfare "as an animal's collective physi-cal, mental, and emotional states over a period of time." There is a profound need for research aims to identify reliable behavioral indicators of the welfare of animals in zoos (Watters, 2014). An important question is would animal conser-vation be where it is today were it not for the educational role of zoos.

10.9 CONCLUSIONS

Acknowledging that our interactions with animals give pleasure, create employment, and can aid conservation, we are left with the follow-ing questions:

1. Should we take an absolutist view that animals are equal in value to people and there is no place for any entertainment role for them?
2. Can we take the position on cruelty advanced by Supreme Court justice Potter Stewart "I know it when I see it" when considering pornography (*Jacobellis v. Ohio*, 1964)?
3. Can we try to balance pleasure and discomfort in humans and animals?
4. Can we legislate and regulate to assure high standards of animal welfare? Are these standards based on anthropomorphizing or on sound science? Are these enforceable? Some activities are not regulated. For instance, people exhibiting livestock and poultry at county and state fairs in the USA are not regulated by the Animal Welfare Act of 1966 (USDA NAL, in press). However, state laws against abuse of animals still apply.
5. Is rhetorical argument a good substitute for investigating a topic?

6. Can we be tolerant of the views of others and other cultures?

Other questions might be

1. Can the killing of an animal used for entertainment ever be justified? What if it is performed humanely? What if the animal is injured? What if it enables populations of the species to be restored?
2. Can discomfort of an animal used for entertainment be justified?

References

Aguiar, M., Hurst, E., 2006. Measuring trends in leisure: the allocation of time over five decades. Available from: http://www.frbsf.org/economic-research/files/mtl.pdf

American Horse Council, 2005. The economic impact of the horse industry on the United States. National Report.

Arluke, A., Madfis, E., 2014. Animal abuse as a warning sign of school massacres: a critique and refinement. Homicide Stud. 18, 7–22.

Atock, M.A., Williams, R.B., 1994. Welfare of competition horses. Rev. Sci. Tech. 13, 217–232.

AZA (Association of Zoos and Aquariums), 2016. Available from: https://www.aza.org/animal-husbandry-and-welfare/

Banziger, G., Roush, S., 1983. Nursing homes for the birds, a control-relevant intervention with bird feeders. Gerontologist 23, 527–531.

Barongi, R., Fisken, F.A., Parker, M., Gusset, M., 2015. Committing to Conservation: The World Zoo and Aquarium Conservation Strategy. WAZA Executive Office. Available from: http://www.waza.org/files/webcontent/1.public_site/5.conservation/conservation_strategies/committing_to_conservation/WAZA%20Conservation%20Strategy%202015_Portrait.pdf

BASC (The British Association for Shooting and Conservation), 2014. The value of shooting: the economic, environmental and social contribution of shooting sports to the UK. Available from: http://basc.org.uk/wp-content/uploads/downloads/2014/07/The-Value-of-Shooting2014.pdf

BASC, 2015. Available from: http://basc.org.uk/game-and-gamekeeping/quarry-species-shooting-seasons/

Booth, V.R., 2010. Contribution of hunting tourism: how significant is this to national economies. Joint publication of FAO (Food and Agriculture Organization of the United Nations) and CIC (International Council for Game and Wildlife Conservation).

Borge, L., Nelson, W.C., Lietch, J.A., Leistritz, F.L., 1990. Economic Impact of Wildlife-Based Tourism in Northern Botswana. Agricultural Economics Report no. 262. Available from: http://ageconsearch.umn.edu/bitstream/23121/1/aer262.pdf

Brickel, C.M., 1979. The therapeutic roles of cat mascots with a hospital-based geriatric population, a staff survey. Gerontologist 19, 368–372.

Brown, A., 2012. The National Angling Survey 2012. Available from: http://www.resources.anglingresearch.org.uk/sites/resources.anglingresearch.org.uk/files/National_Angling_Survey_Report_2012.pdf

Cambridge Dictionary, 2016. Available from: http://dictionary.cambridge.org/dictionary/english/sportsman

Christie, I., Fernandes, E., Messerli, H., Twining-Ward, L., 2013. Tourism in Africa: harnessing tourism for growth and improved livelihoods. World Bank. Available from: http://www.worldbank.org/content/dam/Worldbank/document/Africa/Report/africa-tourism-report-2013-overview.pdf

Crufts, 2016. Available from: http://www.crufts.org.uk

Cruise, A., 2015. Is trophy hunting helping save African elephants? Fees from trophy hunting of elephants that are supposed to help local communities—and elephants—often don't. Available from: http://news.nationalgeographic.com/2015/11/151715-conservation-trophy-hunting-elephants-tusks-poaching-zimbabwe-namibia/

Cutt, H.E., Knuiman, M.W., Giles-Corti, B., 2008. Does getting a dog increase recreational walking? Int. J. Behav. Nutr. Phys. Act. 5, 17.

Daily Telegraph, 2015. Available from: http://www.telegraph.co.uk/news/earth/countryside/11418998/Ten-years-on-from-the-hunting-ban-has-anything-really-changed.html

Deloitte, 2014. Economic impact of British Racing in 2013. Available from: http://www.britishhorseracing.com/wp-content/uploads/2014/03/EconomicImpact-Study2013.pdf

Economist, 2014. The ignorance epidemic: the virus is claiming new victims—African tourism and football. Available from: http://www.economist.com/news/middle-east-and-africa/21632641-virus-claiming-new-victimsafrican-tourism-and-football-ignorance

Economists at Large, 2013. The $200 million question: how much does trophy hunting really contribute to African communities? A report for the African Lion Coalition, prepared by Economists at Large, Melbourne, Australia. Available from: http://www.ifaw.org/sites/default/files/Ecolarge-2013-200m-question.pdf

Edwards, N.E., Beck, A.M., 2002. Animal-assisted therapy and nutrition in Alzheimer's disease. West. J. Nurs. Res. 24, 697–712.

Encyclopedia Romana, 2016a. University of Chicago. Circus Maxima. Available from: http://penelope.uchicago.edu/~grout/encyclopaedia_romana/circusmaximus/circusmaximus.html

Encyclopedia Romana, 2016b. University of Chicago. Gladiators. Available from: http://penelope.uchicago.edu/~grout/encyclopaedia_romana/gladiators/gladiators.html

FEI (Fédération Equestre Internationale), 2016. Available from: http://www.fei.org

Filan, S.L., Llewellyn-Jones, R.H., 2006. Animal-assisted therapy for dementia: a review of the literature. Int. Psychogeriatr. 18, 597–611.

Forbes, 2014. Available from: http://www.forbes.com/sites/ryanmac/2014/01/28/ringling-bros-owner-not-clowning-around-with-business-cannons-to-billionaire-status/#16805eee3608

Forteath, N., 2010. Mariculture of aquarium fishes. In: Hoagland, P. (Ed.), Marine Policy & Economics: A Derivative of the Encyclopedia of Ocean Sciences. Academic Press/Elsevier, New York, NY, pp. 203–210.

Fuller, S., 2011. The economic impact of spending for operations and construction by Association of Zoos and Aquariums accredited zoos and aquariums. Available from: https://www.aza.org/uploadedFiles/Press_Room/News_Releases/AZA%20Impacts%202011.pdf

Greyhound Board of Great Britain, 2015. Economic impact of British greyhound racing industry in 2014. Available from: http://cagro-greyhounds.co.uk/wp-content/uploads/2014/03/Greyhound-Racing-Economic-Impact-2014.pdf

Henderson, B.B., Hensley, C., Tallichet, S.E., 2011. Childhood animal cruelty methods and their link to adult interpersonal violence. J. Interpers. Violence 26, 2211–2227.

Higginbottom, K., 2004. Wildlife tourism: an introduction. In: Higginbottom, K. (Ed.), Wildlife Tourism: Impacts, Management and Planning. Common Ground Publishing, Altona, VIC, pp. 1–14.

HSUS, 2016a. Abuse. Available from: http://www.humanesociety.org/issues/abuse_neglect/

HSUS, 2016b. Zoos and other exhibitors. Available from: http://www.humanesociety.org/issues/zoos/

Huffington Post, 2014. Available from: http://www.huffingtonpost.com/2014/06/19/dog-fighting_n_5502623.html

Humphreys, B.R., Humphreys, B.R., 2005. The size and scope of the sports industry in the United States. IASE/NAASE Working Paper Series, No. 08-11. Available from: http://web.holycross.edu/RePEc/spe/HumphreysRuseski_SportsIndustry.pdf

International Fund for Animal Welfare, 2013. New report: economics of trophy hunting in Africa are overrated and overstated. Available from: http://www.ifaw.org/united-states/news/new-report-economics-trophy-hunting-africa-are-overrated-and-overstated

Kamioka, H., Okada, S., Tsutani, K., Park, H., Okuizumi, H., Handa, S., Oshio, T., Park, S.J., Kitayuguchi, J., Abe, T., Honda, T., Mutoh, Y., 2014. Effectiveness of animal-assisted therapy: a systematic review of randomized controlled trials. Complement. Ther. Med. 22, 371–390.

Kellert, S.R., Felthous, A.R., 1985. Childhood cruelty toward animals among criminals and noncriminals. Hum. Relat. 38, 1113–1129.

Klontz, B.T., Bivens, A., Leinart, D., Klontz, T., 2007. The effectiveness of equine-assisted experiential therapy: results of an open clinical trial. Soc. Anim. 15, 257–267.

Kuehn, B.M., 2002. Is it ethical to keep animals in zoos? AVMA News. Available from: https://www.avma.org/news/javmanews/pages/021201d.aspx

LA Times (Los Angeles Times), 2014. Available from: http://www.latimes.com/nation/politics/politicsnow/la-pn-farm-bill-catfish-eggs-cockfighting-20140204-story.html

League Against Cruel Sports, 2015. Available from: http://www.league.org.uk/

Levin, J., Arluke, A., 2009. Refining the link between animal abuse and subsequent violence. In: Linzey, A. (Ed.), The Link Between Animal Abuse and Violence. Sussex Academic Press, Eastbourne, UK, pp. 163–171.

Li, Z., Courchamp, F., Blumstein, D.T., 2016. Pigeons home faster through polluted air. Sci. Rep. 6, 18989.

Lindsey, P.A., 2008. Trophy hunting in sub-Saharan Africa: economic scale and conservation significance. Best Practices in Sustainable Hunting 41–47. Available from: ftp://ftp.fao.org/docrep/fao/010/aj114e/aj114e09.pdf

Lindsey, P.A., Balme, G.A., Booth, V.R., Midlane, N., 2012. The significance of African lions for the financial viability of trophy hunting and the maintenance of wild land. PLoS One 7 (1), e29332.

Lindsey, P.A., Rouletb, P.A., Romanacha, S.S., 2007. Economic and conservation significance of the trophy hunting industry in sub-Saharan Africa. Biol. Conserv. 134, 455–469.

Majić, T., Gutzmann, H., Heinz, A., Lang, U.E., Rapp, M.A., 2013. Animal-assisted therapy and agitation and depression in nursing home residents with dementia: a matched case-control trial. Am. J. Geriatr. Psychiatry 21, 1052–1059.

Miller, C., 2001. Childhood animal cruelty and interpersonal violence. Clin. Psychol. Rev. 21, 735–749.

Minteer, B.A., Collins, J.P., 2013. Ecological ethics in captivity: balancing values and responsibilities in zoo and aquarium research under rapid global change. ILAR J. 54, 41–51.

Moss, C., 2013. The Myth of Persecution: How Early Christians Invented a Story of Martyrdom. HarperCollins, New York, NY.

National Geographic, 2016. Available from: http://news.nationalgeographic.com/2016/01/160111-ringling-elephants-retire/

National Geographic, 2015. Zoos. Available from: http://education.nationalgeographic.org/encyclopedia/zoo/

National Zoo, 2016. Available from: https://nationalzoo.si.edu/AboutUs/History/

Newsweek, 2015. Animal rights groups 'skeptical'of Sea-World's plan to end killer whale shows. Available from: http://www.newsweek.com/critics-skeptical-seaworld-end-killer-whale-shows-392612

Packer, C., Brink, H., Kissui, B.M., Maliti, H., Kushnir, H., Caro, T., 2011. Effects of trophy hunting on lion and leopard populations in Tanzania. Conserv. Biol. 25, 142–153.

Pattillo, C.G.S., Itano, J., 2001. Laughter is the best medicine: and it's a great adjunct in the treatment of patients with cancer. Am. J. Nurs. 101, 40–43.

PETA (People for the Ethical Treatment of Animals), 2015. Available from: http://www.peta.org/issues/animals-in-entertainment/animals-used-entertainment-factsheets/

PRCA (Professional Rodeo Cowboys Association), 2016. Available from: https://www.prorodeo.org/Portal/Home/Leadership/AboutPrca.aspx

Radford, A., Riddington, G., Anderson, J., 2014. The economic impact of game and coarse angling in Scotland. Available from: http://www.gov.scot/resource/doc/47171/0014600.pdf

SCSTG (Scottish Country Sports Tourism Group), 2016. Available from: http://www.countrysportscotland.com/fishing-scotland/

Smith, B., 2010. The economic impact of Houston Livestock Show and Rodeo™. Available from: http://www.rodeo-houston.com/Portals/0/Downloads/AboutUs/hlsr_econ_impact_2010.pdf

Tallichet, S.E., Hensley, C., 2005. Rural and urban differences in the commission of animal cruelty. Int. J. Offender Ther. Comp. Criminol. 49, 711–726.

UK Government, 2016a. Available from: https://www.gov.uk/government/publications/2010-to-2015-government-policy-animal-welfare/2010-to-2015-government-policy-animal-welfare

UN WTO (World Tourism Organization), 2014. Towards measuring the economic value of animal watching tourism in Africa. Available from: http://apta.biz/wp-content/uploads/2014/10/UNWTO-Wildlife-Study_Report.pdf

USDA NAL (National Agricultural Library), in press. Animal Welfare Act. Available from: https://awic.nal.usda.gov/zoo-circus-and-marine-animals

US Equestrian Federation, 2016. Available from: https://www.usef.org

US Fish and Wildlife Service, 2012. Wildlife & Sport Fishing Restoration Program. National Survey of Fishing, Hunting, and Wildlife-Associated Recreation. Available from: http://www.census.gov/prod/2012pubs/fhw11-nat.pdf

US Polo Organization, 2016. Available from: http://www.us-polo.org/

VerCauteren, K., 2003. The deer boom: discussions on population and range expansion of the white-tailed deer. In: Hisey, G., Hisey, K. (Eds.), Bowhunting Records of North American White-Tailed Deer. second ed. Pope and Young Club, Chatfield, MN, pp. 15–20.

Watters, J.V., 2014. Searching for behavioral indicators of welfare in zoos: uncovering anticipatory behavior. Zoo Biol. 33, 251–256.

Westminster Kennel Club Dog Show, 2016. Available from: http://www.westminsterkennelclub.org

World Bank, Africa House, and ATA (Africa Travel Association), 2010. The State of Tourism in Africa. Joint publication of Africa House at New York University (NYU), the Africa Travel Association (ATA), and the World Bank. Available from: http://www.africatravelassociation.org/userfiles/StateofTourismII_FINAL.pdf

World Economic Forum, 2015. Top 10 tourism-ready countries in Africa. Available from: http://www.weforum.org/agenda/2015/06/top-10-tourism-ready-countries-in-africa

WSJ (Wall Street Journal), 2015. Proposed ban on horse-drawn carriages gains little traction. Available from: http://www.wsj.com/articles/proposed-ban-on-horse-drawn-carriages-gains-little-traction-1436749053

Yabroff, K.R., Troiano, R.P., Berrigan, D., 2008. Walking the dog: is pet ownership associated with physical activity in California? J. Phys. Act. Health 5, 216–228.

Zanola, R., 2010. Major influences on circus attendance. Empirical Econ. 38, 159–170.

Further Reading

Animal Ethics, in press. Available from: http://www.animalethics.org.uk/zoos.html

PETA, 2016a. Animal rights uncompromised: 'pets.' Available from: http://www.peta.org/about-peta/why-peta/pets/

PETA, 2016b. Animal rights uncompromised: zoos. Available from: http://www.peta.org/about-peta/why-peta/zoos/

The Australian, 2015. Available from: http://www.theaustralian.com.au/sport/cruelty-allegations-threaten-future-of-dog-racing/news-story/4867e761675e90a6d0be1d03dfc1cd81

UK Government, 2016b. Available from: http://www.legislation.gov.uk/ukpga/1970/30/contents Conservation of Seals Act 1970

US Fish and Wildlife Service, in press. Available from: https://www.fws.gov/bisonrange/timeline.htm

THE HORSE–HUMAN BOND

Karyn Malinowski

Rutgers–The State University of New Jersey, New Brunswick, NJ, United States

Horses and humans have a long and varied history dating back thousands of years. The earliest evidence indicating the importance of horses to humans comes from faunal remains and cave paintings in the south of France and Spain around 15,000 years ago where they were skinned and butchered for food (Clutton-Brock, 1992). Horses and humans have forged intimate interspecies relationships for centuries as coworkers in battle, for agricultural pursuits, and for transportation, resulting in the modernization of society as we know it (Garcia, 2010). Horses have always been seen as a status symbol of wealth and power by those fortunate enough to own them. While horses initially may have been a food source for humans, and became the impetus for American industrialization and urbanization, the horse's role in society has changed from that of a work animal to one being used for sport, recreation, racing, equine-assisted activities and therapies (EAAT), and as companions and contributors to the pastoral landscape (Hausberger et al., 2008; Keaveny, 2008). In his review of two books on the role of horses in the industrialization and urbanization of America, Rawson (2011) found that neither urbanization nor industrialization would have been possible without the assistance from horses, which historians usually treat as a relic of the preurban and preindustrial past. Nineteenth century America was portrayed as a society composed of symbiotic species, rather than a society of humans who just happened to use horses. An example used to depict the depth of interaction between horses and humans was the intimate relationship between horse and driver where the well-trained delivery horse knew the route, as well as the driver and stopped in the correct locations by itself. In this scenario the horse and driver were really like coworkers with the driver playing the lead character.

Cultural differences in the approach to the horse–human relationship, which continue to exist today, have been evident from ancient and classical history. Social interactions and contact between horses and humans reflect differences in the two main approaches: (1) a cooperative approach based upon understanding the behavior of the horse and (2) an approach based on human dominance and horse submission (Van Dierendonck and Goodwin, 2005).

Equestrian sport has been linked with social status throughout history and an interesting dichotomy exists between genders and their interactions with horses. Recreational and amateur riders are predominately female, while most professional riders and trainers are male (Robinson, 1999). This dichotomy extends to the racing world where most caretakers of the animals are female and most of the trainers, jockeys, and drivers are male (Van Dierendonck and Goodwin, 2005).

The animal–human bond has become a focus of scientific investigation in recent years and horses are included in this body of literature, as well as traditional companion animals, such as dogs and cats. An article by Keaveny (2008) reported results of an interpretive phenomenological investigation of the relationship between humans and their horses using participant observations, in-depth interviews, and written open-ended questions including the critical incident technique (CIT). The study found that horse insiders do not come to this passion by choice and that they are "born" to be a "horse person." The horse person loves and reveres horses, needs to be around them, and makes large sacrifices in terms of time, money, and energy to be with them often at the dismay of other family members.

Throughout history, anecdotal evidence has supported the idea that animals can be "healing." As a whole, there has been an increased interest in the broader field of animal-assisted therapy (AAT), using a wide variety of therapeutic animals, including horses, over the past several decades. In a comprehensive metaanalysis, Nimer and Lundahl (2007) found that "Overall, AAT was associated with moderate effect sizes in improving outcomes in four areas: Autism-spectrum symptoms, medical difficulties, behavioral problems, and emotional well-being." Dimitrijevic (2009) also highlights AAT for use in the treatment of "Psychiatric patients afflicted with depression, schizophrenia, phobias and addiction problems" and notes that, "loneliness is easier to endure in the company of animals." There has also been success in the use of AAT for treating numerous physical ailments, such as AIDS, rheumatoid arthritis, and cardiovascular disease. Patients who participate in AAT report a more expedient reduction of symptoms, when compared with patients who were not exposed to therapeutic animals (Dimitrijevic, 2009). Finally, from a neurobiological perspective, research has shown that interacting with animals stimulates the production of circulating oxytocin, which is known to have calming effects on the body and to aid in relaxation (VanFleet and Faa-Thompson, 2010).

The Professional Association of Therapeutic Horsemanship International (PATH Intl.) is a federally registered 501(c3) nonprofit, formed in 1969 as the North American Riding for the Handicapped Association to promote EAAT for individuals with special needs. There are currently more than 4,600 certified instructors and equine specialists and 866 member centers around the globe to help nearly 62,000 children and adults, including nearly 4,000 veterans and active-duty military personnel, with physical, cognitive, and emotional challenges find strength and independence through the power of horses each year. The mission of the Professional Association of Therapeutic Horsemanship International is to promote safety and optimal outcomes in EAAT for individuals with special needs (PATH International, 2016).

Emergent practices in equine-facilitated learning are exploring increasingly creative forms of horse–human expression and learning. These practices offer goal-oriented activities related to horse care and management, such as grooming, lunging, activities on the ground when unmounted on the horse, and actual riding. These activities can be conducted individually or in a group setting. These activities assist in assessing intra- and interpersonal relationship styles, leadership approaches, and a range of emotional and behavioral issues. Results of an unpublished pilot qualitative study suggest that there may be an esthetic component of horse–human interaction leading to individual emotional and spiritual well-being. Another important benefit is that of learning to shift from dominance to collaboration in relationships impacting people's personal and professional lives. Participating in equine-facilitated activities may also influence one's conditioned values and beliefs and result in a positive reframing of self-image and relational skills (Garcia, 2010).

A 2008 article by Keaveny reported results of an interpretive phenomenological investigation of the relationship between humans and their horses using participant observations, in-depth interviews, and written open-ended questions including the critical incident technique. Two phenomena that help explain human attraction to small companion animals like dogs and cats are present to a lesser degree with horses. Anthropomorphism, the projection of human characteristics onto animals, and neoteny, the "cuteness" factor, are much less in horses because anthropomorphism is more common among predators and while horses are majestic and powerful, they are not usually considered "cute." Horses think (being a prey animal) and socialize differently than humans, spending most of their time with other horses and their vision (almost 360-degree field of view) and sense of hearing and smell are much more acute than humans. As a result, though horses and humans share physical time together, they do not perceive their environment in the same manner.

Data collected during a 1-year time frame utilized the CIT, a well-established research method in the services-marketing literature to explore the human–horse bond. The CIT encourages informants to describe an incident identified as the defining moment in the development of the bond with the horse. Data analysis revealed multiple themes of friendship and emotional support. While dogs and cats give unconditional love to their owners, horses receive unconditional love from humans.

A goal of the study was to introduce the concept that animal companionship is not limited to household pets. Several themes emerged from this research. First, riding a horse adds a level of physicality, intimacy, and intensity not experienced with household companion animals. A second theme is that the horse–human relationship is a working partnership. Mutual trust and respect are two core values embodied in this partnership. A third theme is that horse–rider bonds develop in the aftermath of getting through a tough situation together or when the horse is in trouble and is rescued by a human. Horses and riders also bond after accomplishing a difficult task together (Keaveny, 2008).

People have always considered "horse people" to be "different." A recent study by Wolframm et al. (2015) investigated the dispositional tendencies of rider cohorts, in relation to age, nationality, riding discipline, competitive level, and level of risk-seeking behaviors. Results showed that riders in their late teens and early twenties considered themselves less agreeable and conscientious, and more emotionally unstable than older riders. Competitive riders were more extroverted and conscientious than recreational riders and riders competing in high-risk disciplines, such as racing, polo, and eventing considered

themselves less agreeable and conscientious. The fact that older riders were more agreeable, conscientious, and emotionally balanced might suggest a role for them as coaches and mentors to younger riders.

While much of the research conducted investigating the horse–human bond has centered on the positive aspects of interactions with horses on humans, very little has been conducted looking at the impact of humans on horses. Research is needed to assess the physiological effects on horses after interactions with humans besides measurements of heart rate and behavior. There currently exist minimally invasive techniques to measure stress and well-being, including heart rate variability and various hormones, such as cortisol.

Measurement of heart rate variability is a noninvasive measure that can be used to study the autonomic nervous system, especially the balance between sympathetic and vagal activity. While used successfully in human medicine in research and clinical studies for years, it has become increasingly popular in animal research to investigate changes in sympathovagal balance related to psychological and environmental stress. It also has the potential to help assess the underlying neurophysiological processes of the stress response and to the state of well-being in animals (Von Borell et al., 2007). In an issue of NAHRHA's *Strides* magazine, recent studies in horses used in EAAT using heart rate variability were described. It was demonstrated that the introduction of an unknown person to a horse did not alter the horse's heart rate variability. The rhythm of the person's heart rate variability was more important than whether or not the horse had an association with the person. These researchers found that a relationship exists between horses and humans that can be measured by heart rate variability to potentially assess stress and well-being in both the horse and human during their interaction. Heart rate variability frequency cycles from horses influence the human's cycle but not vice versa. Initial findings support the fact that the autonomic state of the horse has greater influence on the human response than the other way around. Findings also indicated that it may be the role of the human to initiate bonding with the horse. Human heart rate variability before and after interaction with horses can be used to assess the ability of the horse to transfer positive emotions to humans exhibiting stress (Gehrke and Kaye, 2010).

From an animal welfare standpoint, it is time to conduct sound, scientific studies that objectively look at physiological stress and the possible positive effects that interactions with humans have on horses (Beck and Katcher, 2003). Only then can we begin to fully understand what we know as the horse–human bond.

References

Beck, A.M., Katcher, A.H., 2003. Future directions in human-animal bond research. Am. Behav. Sci. 47, 79–93.

Clutton-Brock, J., 1992. Horsepower: The History of the Horse and Donkey in Human Society. Harvard University Press, Cambridge, MA.

Dimitrijevic, I., 2009. Animal-assisted therapy—a new trend in the treatment of children and adults. Psychiatr. Danub. 21, 236–241.

Garcia, D.M., 2010. Of equines and humans: toward a new ecology. Ecopsychology 2, 85–89.

Gehrke, E.K., Kaye, E., 2010. Strides (spring issue). North American Horseback Riding for the Handicapped Association, Denver, CO.

Hausberger, M., Roche, H., Henry, S., Visser, E.K., 2008. A review of the human–horse relationship. Appl. Anim. Behav. Sci. 109, 1–24.

Keaveny, S.M., 2008. Equines and their human companions. J. Bus. Res. 61, 444–454.

Nimer, J., Lundahl, B., 2007. Animal-assisted therapy: a meta-analysis. Anthrozoos 20, 225–238.

PATH International, 2016. Available from: http://www.pathintl.org/

Rawson, M., 2011. A horse is a horse, of course, but also much more: recovering the animal contribution to the urbanization and industrialization of America. J. Urban Hist. 37, 614–618.

Robinson, I.H., 1999. The human-horse relationship: how much do we know? Equine Vet. J. Suppl. 28, 42–45.

Van Dierendonck, Goodwin, D., 2005. Social contact in horses: implications for human-horse interactions. In: de Jorge, F.H., van den Bos, R. (Eds.), The Human–Animal Relationship. Royal van Gorcum, Assen, The Netherlands, pp. 65–81.

VanFleet, R., Faa-Thompson, T., 2010. The case for using animal assisted play therapy. Br. J. Play Ther. 6, 4–18.

Von Borell, E., Langbein, J., Despres, G., Hansen, S., Leterrier, C., Marchant-Forde, J., Marchant-Forde, R., Minero, M., Mohr, E., Prunier, A., Valance, D., Veissier, I., 2007. Heart rate variability as a measure of authonomic regulation of cardiac activity for assessing stress and welfare in farm animals—a review. Physiol. Behav. 92, 293–316.

Wolframm, I.A., Williams, J., Marlin, D., 2015. The role of personality in equestrian sports: an investigation. Comp. Exerc. Physiol. 11, 133–144.

THE ECONOMIC IMPACT OF HORSES IN THE UNITED STATES

Karyn Malinowski

Rutgers–The State University of New Jersey, New Brunswick, NJ, United States

Horses and the equine industry have made a difference in human lives for hundreds of years and millions of people worldwide participate in equine-related activities through employment, recreation, and sport. Besides the economic impact, horses and the farms on which they are housed provide a bucolic landscape; critically important in densely populated areas of the country where agriculture is a shrinking segment of society.

The US horse industry is extremely diverse and global, with segments ranging from horse racing and pari-mutuel wagering, sport horse competition, recreational riding, to equine-assisted therapy and activities for the physically and mentally challenged. In 2014 global trade of horses was approximately $2.4 billion and the value of horses exported by the United States totaled $456 million, making the United States the world's leader in the export of horses by value. The value of horses exported from the United States exceeds the value of live cattle, swine, sheep, and goats combined and is twice the value of all exported live poultry (Carolan, 2016).

The total economic impact of the US equine industry is $102 billion (American Horse Council Foundation, 2005) and the value of horses in our society is "priceless." A survey of horse owners and users, conducted by the Rutgers Equine Science Center in 2016 (Rutgers Equine Science Center, 2016), showed that equine enthusiasts care deeply about the well-being of their animals and rated the level of importance of equine health and issues, such as lameness, nutrition, disease, ethical treatment of horses, and care of the older horses as extremely important.

Economic Impact

Of the total economic impact ($102 billion) of the US equine industry reported by the American Horse Council Foundation in 2005, approximately $32 billion is generated from the recreational segment, $28.8 billion from horse shows, $26.1 billion from racing, and $14.7 billion from other industry segments, such as sport competition. The industry also provides 1.4 million full-time equivalent jobs. Of the 9.2 million horses in the United States 3.9 million are used for recreation, 2.7 million for showing, 845,000 for racing, and 1.75 million for other activities, such as sport horse competition, driving, therapeutic riding programs, and so on (Table 1). The 9.2 million

TABLE 1   Size of the US Horse Industry

Activity	No. of horses	Economic impact ($ billions)	Full-time jobs
Racing	844,531	$26.124	383,826
Showing	2,718,954	$28.788	380,416
Recreation	3,906,923	$31.975	435,082
Other	1,752,439	$14.651	212,010
Total	9,222,847	$102 Billion	1,411,333

American Horse Council Foundation, Deloitte, 2005. The Economic Impact of the Horse Industry on the United States. Deloitte, Washington, DC.

horses in the United States are owned by 1.96 million people, with an additional 2 million non-horse-owning people affiliated with the industry through family or as volunteers. The 1.96 million people owning horses make up 41.9% of the total participation in the equine industry, while the approximately 700,000 employees make up 15%.

Despite the obvious impact of the US horse industry on the gross domestic product, there has been concern expressed by leaders of all breeds and disciplines regarding the continued decline in the number of association members, foals born, and horses registered in recent years. In response to this concern, the American Horse Council at its 2014 annual meeting held its National Issues Forum, "Where Have All the Horse Gone"?

Evidence was presented that the industry has experienced several drops in horse numbers and prices in modern history, especially after the 1986 Tax Reform Act, which resulted in a curtailment of industry investment and an across breeds production decline, which resulted in an actual horse shortage. While the economy may have been cited as the single most impactful factor on the recent reduction in US horse numbers, other factors, such as increased cost of horse ownership and participation in equestrian activities, concerns about animal welfare, and increased competition for leisure and gambling dollars were cited as playing a role as well (Capps, 2014).

Horse Racing

For decades, the predominant types of horse racing (Thoroughbred and Standardbred) have been the economic driving engine of the US equine industry and at one time were some of the most popular spectator sports in America and had a majority of the gambling market 40 years ago. Over the past 2 decades, horse racing's popularity has declined as evidenced by a reduction in number of race days and races and pari-mutuel handle. Purses have actually increased, however,

due to the influx of over $8 billion into the US horse racing industry in the past 20 years by the casino industry in states where slot machines and other types of casino gambling have been built at racetracks called racinos, where a portion of casino revenue is used in support of horse racing. In 2013, tracks in 14 states used money from casino gaming to help fund horse racing (American Gaming Association, 2014).

Displayed in Table 2 are current trends in Thoroughbred racing. The number of races declined from 52,257 in 2005 to 41,277 in 2014. Total purses rose from $1.085 billion in 2005 to $1.1 billion in 2014 due to purse enhancements from casino gaming revenue. Pari-mutuel handle decreased from $14.6 billion in 2005 to $10.55 billion. Regarding harness racing, the number of races decreased from 53,028 in 2005 to 44,692 in 2015. Total purses rose from $334.3 million in 2005 to $424.55 million in 2015. Pari-mutuel handle decreased from $2.2 billion in 2005 to $1.5 billion in 2015 (Table 3). Attendance has also declined as evidenced by empty grandstands that once held tens of thousands of people. Attendance has dropped so significantly that many racetracks no longer count numbers of attendees.

The breeding segment of the horse racing industry usually follows the trend of purses and breeder incentives offered. However, the Thoroughbred and Standardbred breeding industries have seen a drop in the annual foal crop, even in states where lucrative breeding incentive programs exist (Table 4).

There are many reasons for the decline in the popularity of horse racing in the United States. The most obvious reason is increased competition for the gambling dollar. In its heyday horse racing was the only legalized form of gambling in the country. Now people have many more opportunities to gamble, many of which require little skill (lotteries and slot machines) with a bigger payout, more entertaining, more convenient, and less dishonest.

TABLE 2   Trends in US Thoroughbred Racing

Year	No. of races	Gross purses ($ billions)	Pari-mutuel handle ($ millions)
2005	52,257	1.085	14,561
2006	51,668	1.120	14,785
2007	51,304	1.181	14,725
2008	50,120	1.165	13,662
2009	49,368	1.098	12,315
2010	46,379	1.031	11,419
2011	45,417	1.061	10,770
2012	45,086	1.128	10,882
2013	43,139	1.127	10,877
2014	41,277	1.112	10,552

From The American Jockey Club. Fact Book, 2015. Available from: http://www.jockeyclub.com/default.asp?section=Resources&area=1

TABLE 3   Trends in US Harness Racing

Year	No. of races	Gross purses ($ millions)	Pari-mutuel handle ($ millions)
2005	53,028	334.34	2,236
2006	53,547	354.71	2,050
2007	56,071	425.63	2,005
2008	54,259	425.10	1,900
2009	52,453	450.125	1,740
2010	48,989	431.42	1,630
2011	48,346	422.98	1,577
2012	46,848	406.62	1,512
2013	46,913	422.41	1,604
2014	44,373	408.98	1,497
2015	44,692	424.56	1,512

From the United States Trotting Association, 2016. Personal communication. Columbus, OH.

TABLE 4   Trends in Total US Standardbred and Thoroughbred Foal Crop

Year	Standardbred foals born in the United States	Thoroughbred foals born in the United States
2005	10,526	35,050
2006	10,009	34,902
2007	9,420	34,356
2008	8,864	32,329
2009	8,448	29,605
2010	7,977	25,932
2011	7,593	22,610
2012	7,388	21,725
2013	6,999	21,275
2014	6,545	20,300

From The American Jockey Club. Fact Book, 2015. Available from: http://www.jockeyclub.com/default.asp?section=Resources&area=1; United States Trotting Association, 2016. Personal communication. Columbus, OH.

The perception by the public that horse racing is corrupt and that drugs are used to chemically enhance the performance of horses is another reason why the sport's popularity has declined. Also, society is now very conscious of animal welfare and well-being and many find racing horses not to be in the animal's best interest. The industry has taken steps to address these issues. In 2015 the Association of Racing Commissioners International formed a new Scientific Advisory Board, which is charged with developing recommendations for which antidoping programs are conducted in the international racing community. In New York where in 2011–12 there was an unusually high break down rate of horses racing during the winter months at Aqueduct Racetrack, a task force was formed to investigate race horse health and safety. The New York Racing Association worked with the New York State Gaming Commission and concerned horsemen and women to implement a set of policies, procedures, and best management practices to improve horse and jockey safety and well-being (New York Racing Association, 2012). Racetracks around the country are also involved in the placement of race horses into second careers at the end of their racing career, which is increasing in popularity nationwide.

Nonracing Sport Competition and Recreation

A majority of the over 9 million horses in the United States are used for nonracing activities, such as showing, sport competition, and recreation totaling over 6.6 million animals involved in these segments of the industry with a combined economic impact of $60.8 billion and employment of over 800,000 people (Table 1). The 2016 Needs Assessment Survey of over 5000 equine enthusiasts found that the majority of respondents (46%) were recreational riders and that 33% were involved in showing and sport horse competition.

References

American Gaming Association, 2014. State of the States. Available from: http://www.americangaming.org/industry-resources/research/state-states

American Horse Council Foundation, Deloitte, 2005. The Economic Impact of the Horse Industry on the United States. Deloitte, Washington, DC.

Capps, T.T., 2014. Where Have All the Horses Gone? American Horse Council National Issues Forum, Washington, DC.

Carolan, R.O., 2016. Equine Disease Forum. National Institute for Animal Agriculture, Denver, CO.

New York Racing Association, 2012. New York Taskforce on Racehorse Health and Safety.

Rutgers Equine Science Center, 2016. Needs Assessment Survey. Rutgers Equine Science Center, New Brunswick, NJ.

Further Reading

Rutgers Equine Science Center, 2007. The New Jersey Equine Industry 2007: Economic Impact. Rutgers Equine Science Center, New Brunswick, NJ.

University of Kentucky, 2012. 2012 Kentucky Equine Survey. University of Kentucky, Lexington, KY.

Animals and Religion, Belief Systems, Symbolism and Myth

Colin G. Scanes*, Pu Chengzhong**

*University of Wisconsin–Milwaukee, Milwaukee, WI, United States
**Centre for the Study of Humanistic Buddhism, Chinese University of Hong Kong, Sha Tin, Hong Kong

11.1 INTRODUCTION

This chapter discusses animals and religions, animals and belief systems, and animals and symbolism. Section 11.2 include animals and religions. This encompasses animals and Christianity, Islam, Judaism, Hinduism, Zoroastrianism, and Buddhism including food laws, attitudes to dogs and cats, and medical drugs and supplies of animal origin. In addition, the section discusses animal sacrifice. There is greater coverage from Judean–Christian–Islamic and Buddhist religious perspectives. Section 11.3 covers animals and belief systems focusing on evolution and "creation science." Sections 11.4 and 11.5 cover, respectively, animal symbolism and animals in myth or legend.

11.2 ANIMALS AND RELIGIONS

11.2.1 An Overview

The rankings of world religions by the number of adherents are summarized in Table 11.1.

It is readily apparent that the top three religions ranked in this manner are Christianity, Islam, and Hinduism. The geographical distribution is by no means uniform with predominantly Islamic countries or those with a high number of Muslims being located in one region stretching from North Africa, through the Middle East and Asia Minor through Iran and the "Stans" (countries with "stan" at their end: Afghanistan, Kazakhstan, Kyrgyzstan, Pakistan, Tajikistan, Turkmenistan, and Uzbekistan), the Indian subcontinent, Malaysia, and Indonesia. In contrast, the regions that have or had a preponderance of Christians are Europe, North and South America, sub-Saharan Africa, and Australasia. This is illustrated by the distribution of adherents to various religions in the USA with 82% of the population identifying as Christians of various denominations (Table 11.2).

Hinduism is focused in India with about 828 million Hindus in India in 2011 (Census India, 2001). Hinduism is derived from one of the oldest major religions in the world,

TABLE 11.1   Worlds Religions by Number of Adherents in 2010 (Pew Forum, 2015a)

Adherents to religion	2010	Projected for 2050
Christians	2.17 billion	2.92 billion
Muslims	1.60 billion	2.76 billion
Unaffiliated	1.13 billion	1.23 billion
Hindus	1.03 billion	1.38 billion
Buddhists	0.49 billion	0.49 billion
Folk religions	0.40 billion	0.45 billion
Other religions	58 million	61 million
Jews	14 million	16 million

TABLE 11.2   Religions in the USA in 2014 (Pew Forum, 2015b)

Religion	Adherents (%)
Evangelical Protestant	38
Roman Catholic	21
Mainline Protestant	17
Historically black Protestants	2
Other Christians	4
Non-Christian faiths (Judaism, Muslim, Buddhist, and Hindu)	3
Unaffiliated	14

Brahminism. It has affinities with other religions originating in India, namely, Jainism, Buddhism, and Sikhism.

The traditional Judeo-Christian view of animals was that God had given humans dominion over animals. This was based on the following: "Then God said, 'Let us make man in our image, after our likeness, and let them have dominion over the fish of the sea, and over the birds of the air, and over the livestock, and over all the earth, and over every creeping thing that creeps on the earth'" (Genesis 1:26 Modern English Version). The early belief that humans had "dominion over animals" was consistent with one of the oldest legal codes, the Code of Hammurabi (1728–1686 BCE, Susa, Iraq) from the cradle of civilization (Szűcs et al., 2012). Moreover, it was consistent with the utilitarian or anthropocentric view of animals of the ancient Greeks and specifically, Aristotle (384–322 BCE), who used them in his extensive biomedical experimentation (Dunn, 2006; Hajar, 2011). "An earlier Judeo-Christian interpretation of the Bible that dominion over animals meant that any degree of exploitation was acceptable has changed for most people to mean that each person has responsibility for animal welfare" (Szűcs et al., 2012). This initially stemmed from the views advanced by Francis of Assisi (~1182–1226) and Thomas Aquinas (1225–74) (Szűcs et al., 2012). The viewpoint of treating animals with kindness has shifted to animal rights, such as views of the Australian moral philosopher, Peter Singer (1946 to date) who described speciesism as "a prejudice or bias in favor of the interests of members of one's own species and against those of members of other species" (Singer, 1975) and others arguing for animal theology in the Christian tradition (Wade, 2004). What is not clear is the extent that all animals (i.e., pests, vectors, and parasites, such as rats, mosquitoes, tapeworm, and plasmodium) are to be considered at the same level. To quote George Orwell (1903–50) from Animal Farm but from a different context, "All animals are equal, but some are more equal than others."

Alternative Viewpoint

Judeo-Christian and Vegetarianism/Veganism

The history of vegetarianism in Europe stems from the Ancient Greeks with the mathematician Pythagoras (~570 to ~495 BCE) who avoided meat based on his belief in reincarnation (Leitzmann, 2014). The development of modern European vegetarianism began about 200 years ago with a number of prominent thinkers (Leitzmann, 2014): "Animals are my friends, and I don't eat my friends," George Bernard Shaw (playwright 1856–1950).

Christian monks and nuns in the Middle Ages in Europe consumed fish, eggs, dairy products, and usually some meat. There were some ascetics, such as St. Francis, who adopted a vegetarian diet or greatly restricted their intake of meat. Some in modern vegetarianism and veganism claim a Judeo-Christian basis with meat not consumed in a prehistoric "paradise," Eden (Genesis 2:9) and in a posthistoric paradise as the following: "The cow will feed with the bear, their young will lie down together, and the lion will eat straw like the ox" (Isaiah 11:7 New International Version).

Meat has been linked to impurity with hedonism occurring in "fleshpots." There are multiple opportunities for Christians to fast or have meatless days as part of their beliefs. For instance, Roman Catholics have meatless Fridays in Lent, particularly Good Friday, and other Holy Days (Godoy, 2012) while people following Greek Orthodox tenets have multiple meatless days.

great personal effort. Verily, your Lord is Compassionate and Merciful; He created horses, mules and donkeys for you to ride and ornament. And He created what you do not know" (Qur'an Sura 16, 5–8). Hinduism, Jainism, and Buddhism are religions that incorporate nonviolence including toward animals in accordance with the principle of Ahimsa. The rise of these religions in India led to the abandonment of animal sacrifice and shifts away from meat consumption (Szűcs et al., 2012).

Animals were associated with ancient Greek gods in Hellenistic culture/religion in the countries around the eastern Mediterranean Sea between the death of Alexander the Great (323 BCE or BC) and the conquest of Greek-ruled Egypt (30 BCE or BC). Examples include the following (Frazier, 1922):

- Dionysus (god of the grape harvest and wine) can be represented as a goat and bull.
- Pan (god of sheep and goat herds together with shepherds) is portrayed with horns, beard, and legs. Interestingly, the Devil is also portrayed with horns, a beard, and cloven hooves like Pan.
- Satyr (demigod) is shown as a man with horse ears, tail, and phallus.
- Demeter (god of harvest) is associated with pigs.
- Attis (demigod) was killed by a boar.

The rest of this section of the chapter will address animals and Christianity, Buddhism, Islam, and Judaism. The latter is included not because of the number of adherents (Table 11.1) but because of its role as the "root" from which Christianity and Islam stemmed. In addition, some attention will be focused on Zoroastrianism (arguably the first monotheistic religion), Hinduism, and animal sacrifice.

Islam also employs the same human dominion over animals system based on the following: "Allah created cattle for you and you find in them warmth, useful services and food, sense of beauty when you bring them home when you take them to pasture. They bear your heavy loads to lands you could not reach except with

11.2.2 Animals and Christianity

11.2.2.1 *Overview*

Animals feature heavily in the Bible, the holy book for Christians (see textbox for examples).

Examples of Animals in the Torah/Old Testament

The creation story:

"And God said, 'Let the water teem with living creatures, and let birds fly above the earth across the vault of the sky'" (Genesis 1:20) and "And God said, 'Let the land produce living creatures according to their kinds: the livestock, the creatures that move along the ground, and the wild animals, each according to its kind. And it was so'" (Genesis 1:24 New International Version).

The snake

"So the LORD God said to the serpent," because you have done this, "Cursed are you above all livestock and all wild animals! You will crawl on your belly and you will eat dust all the days of your life" (Genesis 3:14 New International Version).

Noah's Ark

"You are to bring into the ark two of all living creatures, male and female, to keep them alive with you. Two of every kind of bird, of every kind of animal, and of every kind of creature that moves along the ground will come to you to be kept alive" (Genesis 6:19 New International Version).

Abraham sacrificing a ram

"Abraham looked up and there in a thicket he saw a ram caught by its horns. He went over and took the ram and sacrificed it as a burnt offering instead of his son" (Genesis 22:13 New International Version).

Moses and the Exodus

1. "So I have come down to rescue them from the hand of the Egyptians and to bring them up out of that land into a good and spacious land, a land flowing with milk and honey—the home of the Canaanites, Hittites, Amorites, Perizzites, Hivites, and Jebusites" (emphasis added) (Exodus 3:8 New International Version).

2. "So all the people took off their earrings and brought them to Aaron. He took what they handed him and made it into an idol cast in the shape of a calf, fashioning it with a tool" (Exodus 32:3–4 New International Version).

Animals and prophesy

"The wolf will live with the lamb, the leopard will lie down with the goat, the calf and the lion and the yearling together; and a little child will lead them" (Isaiah 11:6 New International Version).

Courage and faith

"So the king gave the order, and they brought Daniel and threw him into the lions' den. The king said to Daniel, 'May your God, whom you serve continually, rescue you!'" (Daniel 6:16 New International Version).

Examples of Animals in the New Testament

Jesus and fish

Afterward Jesus appeared again to his disciples, by the Sea of Galilee. It happened this way: Simon Peter, Thomas (also known as Didymus), Nathanael from Cana in Galilee, the sons of Zebedee, and two other disciples were together. "I'm going out to fish," Simon Peter told them, and they said, "We'll go with you." So they went out and got into the boat, but that night they caught nothing. Early in the morning, Jesus stood on the shore, but the disciples did not realize that it was Jesus. He called out to them, "Friends, haven't you any fish?"

"No," they answered. He said, "Throw your net on the right side of the boat and you will find some." When they did, they were unable to haul the net in because of the large number of fish. Then the disciple whom Jesus loved said to Peter, "It is the Lord!" As soon as Simon Peter heard him say, "It is the Lord," he wrapped his outer garment around him (for he had taken it off) and jumped into the water. The other disciples followed in the boat, towing the net full of fish, for they were not far from shore, about a hundred yards. When they landed, they saw a fire of burning coals there with fish on it, and some bread. Jesus said to them, "Bring some of the fish you have just caught." So Simon Peter climbed back into the boat and dragged the net ashore. It was full of large fish, but even with so many the net was not torn (John 21:1–11 New International Version).

John the Baptist, camels, and insects

"John's clothes were made of camel's hair, and he had a leather belt around his waist. His food was locusts and wild honey" (Matthew 3:4 New International Version).

Jesus and pigs

"The demons begged Jesus, "If you drive us out, send us into the herd of pigs" (Matthew 8:31 New International Version).

Jesus and an analogy with sheep and goats

He will put the sheep on his right and the goats on his left (Matthew 25:33 New International Version).

Jesus and cattle

Jesus's parable of the prodigal son "Bring the fattened calf and kill it. Let's have a feast and celebrate" (Luke 15:23 New International Version).

Jesus and donkeys

Jesus entering Jerusalem: "As they were untying the colt, its owners asked them, 'Why are you untying the colt?' They replied, 'The Lord needs it.' They brought it to Jesus, threw their cloaks on the colt and put Jesus on it" (Luke 19:33–35 New International Version).

Jesus in the temple expelling animals and money changers

"When it was almost time for the Jewish Passover, Jesus went up to Jerusalem. In the temple courts he found people selling cattle, sheep and doves, and others sitting at tables exchanging money. So he made a whip out of cords, and drove all from the temple courts, both sheep and cattle; he scattered the coins of the money changers and overturned their tables. To those who sold doves he said, "Get these out of here! Stop turning my Father's house into a market!" (John 2:13–16 New International Version) (Fig. 11.1).

Jesus's birth and livestock

Nativity scenes show animals, such as sheep, cattle, and camels around the time of Jesus's birth. According to Ratzinger (2012) (Pope Benedict XVI), this reflects a myth as their presence is not mentioned directly in the Gospels. However,

FIGURE 11.1 **Part of a painting: "Jesus clearing the Temple of moneylenders" by Giovanni Benedetto Castiglione (1609–64) showing imagery of animals that were sacrificed in the second temple in Jerusalem.**

Luke 2:7 states the following: "And she brought forth her firstborn son, and wrapped him in swaddling clothes, and laid him in a manger; because there was no room for them in the inn." A manger is for feeding livestock so at least their presence is implied. Moreover, it is a very attractive part of the traditions.

Animal symbolism and Christianity

The most famous allusion in Christianity to animals is Jesus Christ as the lamb sacrificed for humankind for the forgiveness of sin. Undoubtedly early Christians in Jerusalem attended the temple until its destruction in CE(AD) 70 and presumably participated in animal sacrifices (Vermes, 2012). An early rite in the Church(s) was communion or Eucharist with bread and wine either representing or transubstantiated into the body and blood of Jesus Christ and his sacrifice by death. With the destruction of the temple, the increasing proportion of non-Jews, and the profound symbolism of the Eucharist to early Christians, there was a complete shift from animal sacrifice (also see Section 11.2.8).

11.2.2.2 Animals, Food Laws, and Christians

It is suggested that conversion to Christianity in the Roman Empire in the regions around the Mediterranean Sea was facilitated by the absence of bans on foods, such as pork. Moreover, eating pork set Christians apart from the Jews; Christianity originally being a sect of Judaism (Vermes, 2012). Some Christians do not consume pork including Ethiopian Catholics and Seventh-day Adventists.

11.2.2.3 Cats and Christianity

In the Middle Ages, cats were considered by Christian authorities to be linked to evil and witchcraft. In response to a purported rise in witchcraft and heresy in Germany, Innocent VII (Pope 1484–92) issued a papal bull on witchcraft. This is thought to be associated with politics within the Catholic Church together with misogyny with a fear of women and their cats. Pope Innocent VIII is said to have stated in 1484 that "the cat was the devil's favorite animal and idol of all witches" (Walker-Meikle, 2011).

11.2.2.4 Snake Handling and a Few Christian Sects

There are snake handling Pentecostal Christian sects in Southern Appalachia including the US States of Alabama, Georgia, Kentucky, and Tennessee (Hood, 1998). This is based on literally the following verse: "They shall take up serpents; and if they drink any deadly thing, it shall not hurt them; they shall lay hands on the sick, and they shall recover" (Mark 16:18, King James Bible translation and Modern English Version). A 30-year preacher, George Went Hensley, began snake handling in 1909 with large crowds attracted to open-air spectacles (Ball, 2015). Despite the "Free Exercise Clause of the First Amendment" to the Constitution, various States have attempted to suppress snake handling based on public safety (Ball, 2015).

11.2.3 Animals and Islam

11.2.3.1 Animals and Islamic Food Laws

Consumption of pork is also expressly forbidden for Muslims. "Forbidden to you (for food) are: dead meat, blood, the flesh of swine, and that on which hath been invoked the name of other than Allah" (Qur'an 5:3). Muslim scholars have argued that the prohibition of consuming pork can also be justified on health grounds, such as parasites (see Chapter 17) and zoonotic diseases (see Chapter 18). Muslims can eat shellfish. "Lawful to you is water-game and its use for food—for the benefit of yourselves and those who travel…" (Qur'an al-Maa'idah 5:96). Frogs and crocodiles are, however, forbidden (al-Munajjid, 2016). Compliance with the food laws is high. For instance, few Muslims in the USA (9%) reported that they consume pork compared to 90% not consuming pork in a 2014 survey (Pew Forum, 2015b).

11.2.3.2 Cats and the Prophet Muhammad

The Prophet Muhammad (CE 570–632) is thought to have liked cats. His cat, Abu Hruyrah, is reputed to have saved him from a snake (Nizamoglu, 2016). The cat is considered as a clean animal and the Prophet is said to have stated that cats should be treated as members of the family (Gulevich, 2004; Nizamoglu, 2016).

11.2.3.3 Dogs and Muslims

Dogs are considered as unclean by Muslims (Gulevich, 2004). The Prophet Muhammad is said to have stated that "Angels do not enter a house in which there is a dog or a picture" in Bukhari 4:448. Imam al-Bukhari, a Sunni Muslim, collected hadiths or sayings of the Prophet Muhammad (Gulevich, 2004).

11.2.4 Animals and Judaism (Animals and Jewish Food Laws)

11.2.4.1 Overview

The Torah puts animals into four groups: land animals, marine and freshwater animals, "creeping things" (insects, reptiles, and rodents)," and "flying things" (birds and bats). Among land animals, only some are considered as clean: "You may eat any animal that has a divided hoof and that chews the cud" (Deuteronomy 11:3 New International Version). Any land mammal that does not have both of these qualities is forbidden. The Torah specifies that the following are considered unclean: camels (even though they

chew the cud), rock badgers, rabbits or hares, and pigs. Cattle, sheep, goats, deer, and bison are considered as kosher. Only some marine and freshwater animals are permitted: "Of all the creatures living in the water of the seas and the streams you may eat any that have fins and scales" (Leviticus 11:9). Thus, shellfish, such as lobsters, oysters, shrimp, clams, and crabs, are all forbidden (Kosher Certification Organisation, 2017). Fish, such as tuna, carp, salmon, and herring, are all permitted. Consumption of creeping creatures is prohibited except for locust, cricket, and grasshopper (Leviticus 11:22). In some Jewish communities, kosher birds include chickens, ducks, geese, doves, and quail, among others, while nonkosher birds include owls, pelicans, eagles, ostriches, vultures, and more, together with bats. "These shall be detestable to you from among birds…" (Leviticus 11:13–19; Deuteronomy 14:11–18).

Kashrut is the Jewish food laws. In 1885, Reform Judaism declared the kashrut food laws to be outmoded (Fishkoff, 2009). However, 80% of reform synagogues do not permit pork or shellfish in their buildings (Fishkoff, 2009). In a recent survey, only 16% of Israeli Jews state that they consume pork with 82% not consuming bacon or other pork products (Pew Forum, 2016). Many Jewish (57%) people in the United States reported in a 2014 survey that they consume pork compared to 40% not consuming pork (Pew Forum, 2015b).

11.2.4.2 Development of Food Laws Including Prohibition of Pork

Leviticus written about 450 BCE also prohibits Jews from consuming pork, camel meat, rabbit meat, horse meat, and shellfish together with eating dogs and cats (Harris, 1987). In contrast, Muslims can eat camel meat (Harris, 1987). Levite priests' codification of food laws is thought to have had a basis in preexisting practice (Harris, 1987). It has been argued that by prohibiting pork consumption, Israelites set themselves apart from their neighbors. However, there were shifts away from pork consumption in Mesopotamia and Egypt; for instance, Herodotus (classical Greek 484–425 BCE) indicating Egyptians at that time viewed pigs as unclean (Harris, 1987). Pigs scavenge on dung leading to some religious leaders, such as Rabbi Maimonides (1138–1204), court physician to Saladin in Egypt, to view pigs as "unwholesome," "dirty," and "filthy" (Harris, 1987). Thus, public health might have justified the prohibition on the consumption of pork. However, this is not the case with modern production and processing methods.

Before settling as farmers, the Early Israelites are thought to have been pastoralists (see Chapter 6) with sheep, goats, and cattle (McNutt, 1999). Pigs are said to be unsuitable for herding by pastoralists particularly in arid areas. The pastoralists would not be consuming pork. An interesting argument suggests that the religious prohibition on pork consumption followed a tradition of nonpork consumption coupled with decreased wooded areas suitable for pig production (Harris, 1987).

Evidence That Early Israelites Were Pastoralists

"So Abraham brought sheep and cattle and gave them to Abimelek, and the two men made a treaty" (Genesis 21:27 New International Version).

11.2.5 Animals and Hindus

11.2.5.1 *Overview*

Hinduism is generally considered to be a polytheistic religion but some adherents view that the gods are various manifestations of a single god or universal spirit, Brahma, the creator of the universe. Other important gods are Vishnu, the preserver, and Shiva, the destroyer. It was established about 4000 years ago in what

is today India but there was no single founder or founding event. Instead, it evolved building on local philosophies and beliefs. The name Hindu comes from the Sanskrit for the Indus River.

Hindus believe in a cycle of rebirth with reincarnation of the soul. A central theme of Hinduism is respect for life with care for animals being an integral aspect.

Mahatma Gandhi and Animals

Mahatma Gandhi (1869–1948) was a prominent leader in Indian independence movement who used nonviolence or "Ahimsa" and a noncooperation approach. His philosophy was based on his Hindu religion with inputs of Jainism from his mother. He was a vegetarian. His views on animals are summarized by his quote: "The greatness of a nation and its moral progress can be judged by the way its animals are treated."

11.2.5.2 *Hinduism and Cows*

Hinduism envisions cows or mother cows as sacred and hence the expression "holy cow" (The Atlantic, 2015). India is one of the top two milk-producing nations with water buffalo (see Chapter 7) and dairy cows. Despite this, the slaughter of cows is prohibited in many Indian states. Such bans on beef are detrimental to Muslims together with some non-Hindu tribal units and communities of dalit, also known as untouchables (Biswas, 2015). After their productive lives, some Indian cattles are cared for in shelters or *Gaushalas*. Many are exported for beef with income from this estimated as $4.3 billion (The Atlantic, 2015). In a 2014 survey, 67% of Hindus in the United States reported that they do not consume beef with only 29% consuming beef (Pew Forum, 2015b).

It has been argued that beef consumption occurred in the Vedic period (1000–5000 BCE) with the cow not acquiring its sacred status until about CE 500 (Jha, 2002).

11.2.5.3 *Hinduism and Other Animals*

In Hinduism, gods or goddesses can take the shape of various animals. Karni Mata was a mystic about 600 years ago. She is believed to have had an agreement with the god of death, Yama, that her clan would be reincarnated as rats. There are about 20,000 rats in the Karni Mata Temple in Rajasthan, India (Guynup and Ruggia, 2004).

11.2.6 Animals in Buddhism

Siddhartha Gautama (566–480 BCE) was the founder of Buddhism. He became known as the Buddha or Enlightened One. In his long ministry career of 45 years, animals had always been featured in his preaching. These were initially divided into discourses (*Sūtra*) and disciplinary code (*Vinaya*) with treatises of doctrinal analysis (*Abhidharma*) added later. Animals normally appear in the past-life stories (*Jātaka*) of the Buddha, fables, great events (*apadana*), and similes (Rotman, 2008). In fact, there is a separate genre of past-life stories of the Buddha, and a collection of 557 so-called birth stories existing in the Pali language (a middle Indo-Aryan dialect spoken in northeastern India during the 6th–3rd century BCE) (Chalmers, 1895; Cowell, 1907; Francis, 1905; Francis and Neil, 1897; Rouse, 1895, 1901). Some of the stories also reappeared in later Buddhist traditions in Sanskrit (Speyer, 1895) and Chinese. The Chinese Buddhist canon has preserved many stories of different types dating to different periods and belonging to different Indian Buddhist traditions (Chinese Buddhist Electronic Texts Association, numbers 152, 153, 154, 155).

The contexts in which the animals are used in the Buddhist traditions can be roughly classified into the following categories:

1. Those demonstrating noble spirits of the Buddha when he was still going through different lives in the past in search of the final goal of "Buddhahood." He took the forms of different animals ranging from lions, monkeys, elephants, sparrow, rabbit, and so on. They all convey different values and ethical upholding of the previous lives of the Buddha, which are conductive to his final supreme Enlightenment as a Buddha.
2. Those representing the good character of the Buddha and the bad character of his rival, Devadatta.
3. Those teaching moral lessons: this is represented by the majority of the fables. "Buddhism takes into full account the animal's latent capacity for affection, heroism, and self-sacrifice…" (Story, 1964).
4. Those symbolizing good qualities: Some animals indeed, such as lion, the elephant, the horse, and the serpent, are used as personifications of great qualities. The lion is one of the few animals often used in Buddhism to symbolize the qualities of the Buddha: his nobility, dignity, wisdom, courage, and power (McGinnis, 2004). Some Buddhist narratives convey that the lion is used to glorify the royal lineage of the Buddha's secular family as the title of the Buddha's grandfather was no other term than lion, indicating the nobility of the Buddha's birth. A few aspects of the Buddha's ministry career are also linked with lion. After his enlightenment, the Buddha's is likened to be a lion king, his wisdom the power of the lion (Swearer, 2004). The qualities and strength of a lion not found in other animals are likened to be those virtuous qualities of the Buddha. The fast speed of the lion's running is

likened to the highest degree of the Buddha's meditation. The lion's roar that makes other animals tremble symbolizes the Buddha's preaching destroying the teachings of other religious teachers.

There has long been a wide use of the lion image in Asian Buddhist traditions to symbolize this or that supreme quality. For instance, in Chinese Buddhism, the seat of a chief monk is called the lion seat as that seat symbolizes the dharma preaching of the Buddha. In many Mahayana scriptures, we see quite a number of Buddha's names containing the word lion (*siṃha/siṃgha*). In the Jātaka stories, in a few births the Buddha was born as a lion king. Whereas in Chinese Buddhism, as no lion has been found in Chinese wilderness, they have tiger, thus the animal was an often-seen symbol in the Chinese Buddhist culture, without removing the application of the lion symbol brought from India. Also in Chinese Buddhism, each of the four great celestial Bodhisattvas (Buddha to be) has an animal as their companion-cum-carrying vehicle. Apart from being likened to the lion of mankind, other animals, such as the elephant, bull, buffalo, ad great horse, are also used to symbolically address the Buddha.

As a class of living species, animals are considered in Buddhism as sentient beings together with other living creatures, which are the different forms that a being takes in the cycle of birth and death; animals and human beings are the same in that they both are results of their own past karma conjoined with ignorance. This underlines the idea that humans could be reborn as animals and vice versa, and that animals and humans are interconnected. In cosmological terms, the animal world is different not by space but by state of mind and behavior. As Buddhism advocates that sentient beings are equal in terms that they all possess a Buddha nature (potentiality for becoming a Buddha), so animals can too.

The Buddha is thought to have established positions of meat-consumption permitting meat consumption by his disciples provided the following:

- They did not kill the animals.
- The animals were not killed solely to be eaten by the monks.
- Eating meat from certain animals was prohibited including elephants, horses, snakes, and dogs.

In ancient China, the major dish was stew or geng; containing vegetables, cereals, and meat. The meats included beef, mutton (mature sheep meat), pork, venison, duck, pheasant, turtle, or dog with the poor consuming vegetable stews (Sterckx, 2005). Early Buddhists were not vegetarian at all, and they had their own reasons for not being one (Pu, 2014). However, in later scriptures, consumption of meat was discouraged by the Buddha as with a series of reincarnations, we may be eating our relatives (Kieschnick, 2005). According to Kieschnick (2005), "meat in ancient India was considered a delicacy of great sensual appeal." Buddhists may be vegetarian or meat-eaters (Kieschnick, 2005). It was in China that a permanent and still existing vegetarian diet for monastics developed (Pu, 2014). The practice appeared in the Buddhist community under the influence of the Chinese vegetarian practice observed before the coming of Buddhism. This was then legitimized as a law for monastics at the order of Emperor Wu of the Liang Dynasty in the early 6th century and immediately fixed as a monastic rule in an ordination manual applied to all monastics. With Buddhism preaching to people in neighboring countries, Buddhist monks and nuns were also vegetarian in Vietnam, Korea, and Japan. However, the Japanese monastics abandoned this practice across Buddhist schools in the early years of the 19th century. Moreover, monastics in Cambodia, Japan, Myanmar (Burma), Sri Lanka, Thailand, and Tibet are not vegetarian (Kieschnick, 2005).

Experimentation on animals are also rejected by Buddhism as it is considered to be "never kind" and justified to subject animals to sufferings, and it also creates bad karma for people who do it.

11.2.7 Animals and Zoroastrianism

Zoroastrianism was founded in the region that is now Iran by its prophet, Zarathushtra, also known as Zoroaster. It is arguably the first monotheistic religion. The ethos of Zoroastrianism embraces care for domestic animals particularly dogs and cattle (Foltz and Saadinejad, 2007). In addition, Zoroastrianism endorses killing evil animals, such as snakes (Moazami, 2005). Zoroastrianism was persecuted by the Muslims inspired in Persia between CE 600 and 800 during the 7th–8th centuries). Parsis or Parsee are Zoroastrians who fled to India from persecution. In India, the Parsis seem to have abandoned animal sacrifice arguably due to the influence of their Hindu and Jaina neighbors.

11.2.8 Animal Sacrifice

11.2.8.1 Overview

Some definitions of animal sacrifice are discussed in the textbox.

Meaning of the Word "Sacrifice"

Sacrifice means killing an animal or person as an offering to God or the gods for various purposes including requests for blessings or favors, such as rainfall for crops or good health or victory in battle. Animal sacrifice can also be appeasement or for "purification" after "sins" or acts deemed not pleasing to the deity(ies).

Etymology of the word "sacrifice"

The word, sacrifice, comes from "sacrificial" (Latin). The latter is a compound word comprising "sacer" (Latin for sacred) and "facere" (Latin for to make).

Killing or sacrificing of laboratory animals

Scientists routinely refer to sacrificing laboratory animals as euphemism for killing (Cartwright, 2015) and this is strongly criticized (Rollin, 1981). The use of the term has been justified with the analogy of science to the sacred (Lynch, 1988).

11.2.8.2 Animal Sacrifice and the Mediterranean Region Including Biblical Israel

Animal sacrifice was integral to the religion of ancient Greeks (Fig. 11.2) with this continuing at least into the 2nd century (reviewed by Petropoulou, 2008; Ullucci, 2011). Animal sacrifice was common around the Mediterranean Sea and globally for many early human communities. Ancient Egyptians sacrificed pigs to Osiris, while early in the development of ancient Rome a horse was sacrificed on the field of Mars by the Tiber River (Frazier, 1922). Sacrificing animals was important in Biblical Israel, such as "Then sacrifice one male goat for a sin offering and two lambs, each a year old, for a fellowship offering" (Genesis 23:19 New International Version). Another example of the requirements in biblical Israel of sacrificing is a woman after she has given birth. "When the days of her purification for a son or daughter are over, she is to bring to the priest at the entrance to the tent of meeting (or later the temple) a year-old lamb for a burnt offering and a young pigeon or a dove for a sin offering. He shall offer them before the Lord to make atonement for her, and then she will be ceremonially clean from her flow of blood. These are the regulations for the woman who gives birth to a boy or a girl. But if she cannot afford a lamb, she is to bring two doves or two young pigeons, one for a burnt offering and the other for a sin offering. In this way the priest will make atonement for her, and she will be clean" (Leviticus 12:6–8 New International Version). It is

FIGURE 11.2 **Animal sacrifice in classical Greece and Rome.** (A) A scene showing a bull, ram, and boar about to be sacrificed on the imperial altar in Rome. (B) Sacrifice of a young boar in classical Greece from a cup produced between 510 and 500 BCE. *Source: Part A–B: Courtesy Wikipedia Commons.*

argued that sacrificing animals produced meat for festive meals, with both contributing to a sense of racial identity for the ancient Israelis (Altmann, 2011).

The temple was critically important to the economy of Jerusalem attaching many pilgrims coming to the city before its destruction in CE 70. Animals (sheep, goats, and doves) were sacrificed, burnt as offerings and the remnants consumed (reviewed by Petropoulou, 2008; Reed, 2013) (Fig. 11.1). This is supported by the Bible, Josephus, and archaeological studies. For instance, large numbers of sheep and goat bones were found in Jerusalem's city dump from the early Roman period (Hartman et al., 2013). Analysis of isotopes (^{15}N and ^{13}C) supports a mix of locally produced animals and those being brought from remote sites including at least 37% from desert regions (Hartman et al., 2013).

Jewish Temple in Jerusalem

First temple 957–586 BCE (destroyed by Babylonians).

Second temple 538 BCE to CE 70 (rebuilt 20 BCE and destroyed by Romans).

11.2.8.3 Animal Sacrifice and Christianity

Early Christian teaching was against pagan animal sacrifice and slightly later, Jewish animal sacrifice (reviewed by Petropoulou, 2008; Ullucci, 2011). St. Paul opposed the consumption of meat from animals sacrificed to idols (1 Corinthians 8:4–13 written about CE 55). However, animal sacrifices were occurring at the tomb of St. Felix in Nola near Naples, Italy, who died about CE(AD) 250. These were encouraged by (Saint) Paulinus of Nola (354–431) as part of a program to evangelize the rural area (Trout, 1995). Animal sacrifice continued into the 5th and 6th centuries in Byzantium as evidenced by the sequence of laws prohibiting it; Byzantium being the remnants of the Roman Empire after the fall of Rome (Harl, 1990).

11.2.8.4 Animal Sacrifice and Buddhism

The Buddha together with other non-Brahmin religious leaders has always rejected animal sacrifice. He considers the ritual to be a senseless cruelty and causes harm to rather than advantage for those who perform it (Pu, 2014).

11.2.8.5 Animal Sacrifice and Afro-Caribbean religions

Animal sacrifice continues to be part of Afro-Caribbean religions, such as Cuban Santeria or Yoruba, Palo, and Haitian Voodoo. This has been the subject of confrontations with local and state governments supported by animal activists (Lammoglia, 2008).

11.2.9 Religion and Drugs, Dressings and Implants of Animal Origin

Multiple drugs, dressings, and implants are of animal origin. Table 11.3 summarizes a survey of clergy from different religions on the acceptability of drugs, dressings, and implants of animal origin. Physicians "often overlook these … and may violate patients' religious beliefs. Hindus do not allow bovine products and Muslims avoid porcine products" (Eipe and Oduro-Dominah, 2005). Eriksson et al. (2013) concluded that each religion viewed that "if there are no alternatives, and if the treatment is life-saving, then all religions approved of all treatment modalities regardless of origin." On the other hand, physicians increasingly recognize that these products should be labeled (Shiwani, 2011) and that religious views and personal beliefs (e.g., veganism and vegetarianism) should be taken into account in the design of treatments (Hoesli and Smith, 2011; Sarkar, 2005).

11.3 ANIMALS AND BELIEF SYSTEMS

11.3.1 Overview

This section will concentrate on the belief systems related to evolution and creationism.

It is recognized that animals are vital components of other beliefs/belief systems including animal welfare, animal spirits, animal worship, antivivisection, cats and witches, reincarnation, vegetarianism, and veganism,

11.3.2 Evolution

Arguably the critical first step in the theory of evolution was the publication of Darwin's "Origin of Species" in 1859 (Darwin, 1859). While this was based on the available research from 157 years ago, it has become the unifying theory underlying all of biology. When "Origin of Species" was published, research on the branches of biology and geology, such as paleontology, were in their infancy. Moreover, there was no understanding of genetics and, let alone, genomics.

Up until about 1830, most serious thinkers accepted a young Earth envisioning a fully formed environment with animals, plants, and people in 4004 BCE (BC) (reviewed by Roberts, 2007). "… but by the year 1860, no educated person doubted that the earth was millions of years old" (Roberts, 2007). This was due in large measure to the development of the science of geology and the establishment of the times needed for the deposition of sedimentary rocks. However, as early as 1785, some theologians were accepting an old Earth model with the 6 days of Genesis being interpreted as six expanses of time (reviewed by Roberts, 2007).

TABLE 11.3 Acceptability of Animal Products by Clergy of Different Global Religions

	Bovine-derived drugs, dressings, and implants	Porcine-derived drugs, dressings, and implants	Human-derived implants
Buddhism (Theravada)	Yes	Yes	Yes
Christian	Yes	Yes	Yes
Hindu (Vaishnavism)	No	No	No
Islam (Sunni and Shiite)	Yes	No	Yes
Judaism	Yes	Yes	Yes
Sikhism	No	No	Yes

Based on Eriksson, A., Burcharth, J., Rosenberg, J., 2013. Animal derived products may conflict with religious patients' beliefs. BMC Med. Ethics 14, 48.

11.3.3 Creationism

There are substantial efforts to promote creationism also known as "creation science" or "intelligent design" (Forrest and Gross, 2005). Creationists include individual Young Earth creationists, such as Ken Ham and Kent Hovind, together with groups, such as Answers in Genesis and the Discovery Institute. They oppose modern science, science education, and secularism; with the latter being viewed as synonymous with atheism (Forrest and Gross, 2005). "We come from creation, by God, not from unguided nature, and people who wish to be rational must recognize the effect …. False ideas have to be rooted out for faith to recover" (Dembski and Richards, 2001). Ken Hamm has established the Creation Museum in Kentucky close to Cincinnati Airport. There is a replica of Noah's Ark (called Ark Encounter) associated with the museum. This opened in 2016 and was funded in part by municipal bonds and a state sales tax rebate. The Ark replica includes dinosaurs coexisting with people despite the extinction of all dinosaurs 65 million years ago.

The argument is advanced that in the interests of fairness, creationism and intelligent design should receive equal time/space as evolution (Forrest and Gross, 2005). There is some, albeit very limited, intellectual support for creation (Meyer, 2009). However, the vast majority of scientists accept evolution (Rennie, 2002).

Fallacies Employed by Creationists

Among the fallacies employed by creationists are the following (based on fallacies from Rennie, 2002):

- *Evolution is only a theory.* The word theory is used by scientists for the well-developed body of science with substantial supporting evidence and providing conceptual framework for an entire field.
- *Evolution is not scientific as it cannot be tested.* However, as new discoveries are made in, for instance, genomics and paleontology, they are always consistent with the overall theory or concepts of evolution.
- *There are disagreements between different evolutionary biologists.* While this is true, they are often distorted from disagreements on specific details to refuting the entire theory.
- *If humans descended from monkeys or apes, why are there still monkeys and apes?* This is a misrepresentation as evolution envisions humans, apes, and monkeys having common ancestors.
- *Mutations can only eliminate traits. They cannot produce new features.* Mutations can change a protein or the amount produced. An example is the loss of olfactory genes in human evolution (see Chapter 1) allowing higher brain functioning. Moreover, genes can be duplicated and this coupled with mutations and natural selection leads to entirely different functions.
- *Natural selection might explain microevolution, but it cannot explain the origin of new species and higher orders of life.* There is abundant evidence for speciation (new species) in the present and in fossil history.
- *Evolutionists cannot point to any transitional fossils—creatures that are half reptile and half bird, for instance.* The fossil record is replete with many intermediate forms, such as reptiles with feathers, birds with teeth and claws on their wings, and the development of hominids (Fig. 11.3).

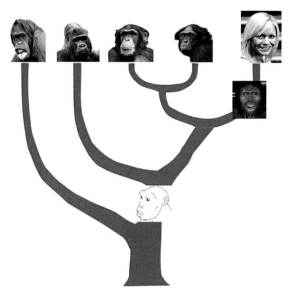

FIGURE 11.3   **A conceptualization of human evolution.**
Source: Courtesy Wikipedia Commons.

Leading creationists have strong debating skills. In contrast, science is based on evidence not clever, but misleading, points. Moreover, creationists have adopted *ad hominem* tactics for over 100 years (Fig. 11.4).

Implicit to accepting creationism is also a belief in Noah's flood. This can be seen as an allegory. However, fundamentalist Christians believe in a world-wide flood with Noah rescuing all the animals on a large boat, the ark (Genesis 6:9–9:28; International Council on Biblical Inerrancy, in press). Based on the bible, this was estimated to have occurred in about 2349 BCE (BC). There are analogous stories from other ancient civilizations in the Middle East (Heidel, 1949; Lorey). These include the Babylonian epic of Gilgamesh found in the Chaldean Flood Tablets in the city of Ur and the Sumerian story of Ziusudra (Lorey). There are 20 points of similarity between the accounts of the Genesis flood and that of Gilgamesh supporting a linkage between them. Thus, the Gilgamesh flood could have been copied from earlier Genesis flood narrative or vice versa or both were derived from a common source (Ontario Consultants for Religious

FIGURE 11.4   **Caricature of Charles Darwin in a misleading and demeaning manner as a monkey from the 1800s.** In a debate on evolution, Samuel Wilberforce (1805–73), Bishop of Oxford (Anglican), asked whether Thomas Henry Huxley "traced his descent from an ape on his mother's or his father's side." An apt response would be "Ridicule is the tribute paid to the genius by the mediocrities" Oscar Wilde (1854–1900). Huxley's response was to dismiss Wilberforce as a man "…plunges into scientific questions with which he has no real acquaintance, only to obscure them by an aimless rhetoric, and distract the attention of his hearers from the real point at issue by eloquent digressions and skilled appeals to religious prejudice." *Source: Courtesy Wikipedia Commons.*

Tolerance, 2016). Arguments for a world-wide flood as opposed to large local floods stretch credulity, such as the "presence of dinosaurs on the ark" and the space and feed for so many animals.

11.3.4 Public Opinion, Evolution Versus Creationism

In the USA, a plurality of people believe in creation and a young Earth (Gallop, 2014):

- Human beings have developed over millions of years from less advanced forms of life, but God guided the process (31%).
- Human beings have developed over millions of years from less advanced forms of life, but God had no part in the process (19%).
- God created human beings pretty much in their present form at one time in the last 10,000 years (42%).

11.4 ANIMALS AS SYMBOLS

11.4.1 Overview

There is abundant evidence for the use of animal symbolism. This section only catches the surface of this. Examples of this animal symbolism discussed are the following: alphabet and symbolism (Section 11.4.2); Christian animal symbols (Section 11.4.3); animals and places and place names (Section 11.4.4); animals, Chinese years, and Zodiacs (Section 11.4.5); animals and emblems (Section 11.4.6); animals and personal names (Section 11.4.7); animal nicknames (Section 11.4.8); and other animal symbolism (Section 11.4.9).

11.4.2 English and Other Western European Alphabet and Animal Symbolism

On the surface, there is no relationship between alphabets and animals. Our alphabet is essentially the same as that used by the Romans but without W, J (I is used for both), and V (used for both U and V). The Roman alphabet is derived from the Greek alphabet via the Etruscans. In turn, the Greek alphabet is derived from that of the Phoenicians and before that Semitic people in what is today, Egypt. The original phonetic letters were based on simplified pictures of words beginning with the specific sound. For instance, *aleph* is Phoenician for the ox and the first letter was a rendition of an ox's head. This became to Greek *alpha* (A α) (Fig. 11.5) and the Roman's **A**. The letter **B** in today's and

FIGURE 11.5   **The A, B, Cs of the Phoenician alphabet.**

the Roman's alphabet was based on the Greek *beta* (B β) and Phoenician *Beth* (as in Bethlehem), the house. The Phoenician *gimel* or *gamal*, the camel, became the Greek's *gamma* (Γ γ). This was used by the Etruscans for the "k" sound and from them to the Romans and to us. Interestingly the etymology of our word, camel, comes from the *camelus* (Latin), *kamelos* (Greek), and *gimel* or *gamal* (Phoenician and Hebrew).

> Another animal-based letter is the Phoenician Nu for snake. ⌐

11.4.3 Christian Animal Symbols

Christian animal symbols through history include fish, peacocks, pigeons, and dolphins (Thomas, 1981). The fish is a very frequently used symbol for Jesus Christ. It is appropriate as Christianity encompasses baptism in water. Moreover, the symbol is based on word play and acronyms and as the Greek word *ichthūs* is from the Latin *Iesous Christos Theou Uios Soter* or Jesus Christ, son of God, savior (Thomas, 1981).

Another example of animal symbols in Christianity are the bees in the Vatican. The great sculptor Gian Lorenzo Bernini (1598–1680) incorporated bees into the canopy in St. Peter's

Basilica. This was because the patron was Urban VIII (Pope from 1623 to 1644). Before his election to the papacy, Urban VIII was Cardinal Maffeo Barberini, and three bees were on the coat of arms of the Barberini family.

11.4.4 Animals and Places and Place Names

There are examples of places being named after animals, such as Oxford (a crossing across the river Thames for cattle) in England, Buffalo in New York State, and Beaverton in Oregon (USA), together with Westbury or Bratton White Horse and the Uffington White Horse (Fig. 11.6).

11.4.5 Animals, Chinese Years, and Zodiacs

Chinese years are named after animals. Livestock and poultry: ox, horse, goat, rooster, and pig; domesticated animals: dog; wild animals: rat, tiger, rabbit, snake, and monkey; and mythical animals: dragon. Similarly, Zodiac animals include Aries (ram), Taurus (bull), Cancer (crab), Leo (lion), Scorpio (scorpion), Capricorn (goat), and Pisces (fish), together with Sagittarius (a centaur).

FIGURE 11.6 **The Uffington White Horse from Oxfordshire (southeast England, northwest of London).** This is 110 m long and first carved into the chalk about 3000 years ago (1000 BCE). *Source: Courtesy Wikipedia Commons.*

11.4.6 Animals and Emblems

Animals are used as the symbols or emblems or names of sports teams. The standard for the Roman legion was the eagle. Another Roman military standard was the dragon (*draco*). The animal names that were used as emblems for Roman legions are shown in Table 11.4.

TABLE 11.4 Symbols of Roman Legions From 100 BCE to CE 450

Symbol	Legion
Animal	
Boar	Legio I Italica, Legio XX Valeria Victrix
Bull	Legio I Germanica (or Augusta), Legio III Gallica, Legio IV Macedonica, Legio VII Claudia Pia Fidelis, Legio VI Victrix, Legio VII Claudia Pia Fidelis, Legio IX Hispana, Legio X Gemina
Lion	Legio IV Flavia Felix, Legio XIII Gemina, Legio XVI Flavia Firma, Legio XVI Gallica
Stork	Legio III Italica Concors (Harmonius)
Wolf (female)[a]	Legio II Italica (Pia)
Mythical animal	
Centaur (half human and half horse)	Legio II Parthica
Pegasus (winged horse)	Legio III Augusta

a With the supposed founders of Rome; the brothers Remus and Romulus (Fig. 11.8).

American professional sports teams are often, but not exclusively, animals: football (Ravens, Panthers, Lions, Bears, Bronchos, Bengals, Jaguars, Colts, Dolphins, Falcons, Rams, Eagles, and Seahawks), baseball (Blue Jays, Cardinals, Cubs, Diamondbacks, Marlins, Orioles, Rays, and Tigers), and ice hockey (Coyotes, Ducks, Panthers, Penguins, and Sharks). Similarly, college and high school teams frequently have an animal name.

11.4.7 Animals and Personal Names

Lion names include Leonard, meaning brave lion. Other lion names include Leonardo, Leon, Lionel, and Leo. One possible derivation of the name Arthur is from "bear." Horse names include Phillip, Phillippa, Pip, Pippa, Coltrane, Colton (colt), Rhiannon/Rhianna, Rosalind (horse and snake), Ross (mighty horse), and Marshal. The etymology of Phillip is from the Greek *Philippos*: *philos* "loving" + *hippos* "horse." Examples of historical figures with names related to horses are the following:

- Philip II (King of Macedonia 359–336 BCE) and father of Alexander the Great.
- Hengist and Horsa were both named after horses. They were the legendary Saxon brothers who were founders of England invited by Vortigern (Anglo-Saxon Chronicle).
- Crazy Horse (American Indian Lakota leader1840–77).

Dragon names are also found. Uther Pendragon was the supposed father of "King" Arthur (Geoffrey of Monmouth, 1136). The name "Pendragon" can be translated as head dragon.

11.4.8 Animal Nicknames

There are a number of cases where people are known for animal nicknames:

- George Crook (US Civil War General, American 1830–90) called the "Grey fox"

- Karl Dönitz (World War II German admiral 1891–1980) called the "Lion"
- Haile Selassie (Ethiopian Emperor 1892–1975) called the "Lion of Judah"
- George Jones (country music singer 1931–2013) called the "The possum"
- Charlie Rich (country music singer 1932–95) called the "Silver fox"
- Richard I King of England (1157–99) called the "Lion-heart"
- Erwin Rommel (World War II German field marshal commanding the Afrika Korps 1891–1944) called the "Desert fox"
- Yamashita Tomoyuki (World War II Japanese general 1885–1946) called the "Tiger of Malaya" by Japanese and the "Beast of Bataan" by Americans

Moreover, there are people who were known for animal names:

- Buffalo Bill (showman 1846–1917)
- Howlin' Wolf (blues singer 1910–76)
- Wolfman Jack (disc jockey 1938–95)

11.4.9 Other Animal Symbolism

There are multiple examples of animals in Aesop's fables (fox and hare) and Grimm's fairy tales (e.g., Little Red Riding Hood and the Big Bad Wolf). There is also considerable symbolism in the White Rabbit and the Cheshire cat in "Alice in Wonderland." Other animal symbolism includes the following:

- Animals on flags (Table 11.5; Fig. 11.7).
- The ironic Cowardly Lion from the "Wizard of Oz."
- Heracles (Hercules) killing a lion.
- The tribe of Judah as a lion: "You are a lion's cub, Judah" (Genesis 49:9 New International Version).
- The lion in the Chronicles of Narnia (C.S. Lewis).
- Romulus and Remus founders of Rome being reared by a female wolf (see Fig. 11.8 for the brothers feeding from female wolf).

TABLE 11.5   Animal and Flags

Animal	Country, state, or county
Animals	
Bear	California and Missouri
Bison	Wyoming
Black swan	Western Australia
Eagle	Prussia (black eagle), Austria, Mexico (white eagle), Serbia (double white eagle), Russian Imperial Flag (double black eagle) (1699–1917), President of the USA, Iowa, Illinois, Michigan, North Dakota, and Utah
Elephant	Variant flag of Thailand
Elk	Michigan
Horse	Pennsylvania, English county of Kent,
Moose	Michigan
Pelican	Louisiana
Rattlesnake	Historical American flag (Gadsden) with "DONT TREAD ON ME"
Mythical animals	
Dragon	Wales (Cymru), Wessex, English county of Somerset

11.5 ANIMALS AND MYTH OR LEGEND

11.5.1 Mythological Animals

In 1941, it was a suggested that dinosaur tracks were the basis of the mythical slaying of a dragon by Siegfried (discussed by Mayor and Sarjeant, 2001). A view has been advanced by Adrienne Mayor that fossilized bones were interpreted as mythical creatures and giants by the ancient Greeks and Romans (Mayor, 2000). For instance, Mayor and her coworkers suggested the following:

- Cyclops, a mythological giant with one eye in the middle of their foreheads was based on skulls of dwarf elephants (Mayor, 2000).
- Confluence of the locations of ancient reports of giants and monsters and deposits of mastodon, mammoth, and other large vertebrate fossils in the Aegean (Mayor, 2000).

- Fossilized mastodon bones from the Island of Samos in the Aegean Sea. These were thought by the ancient Greeks to be from the war elephants of Dionysus from a mythical battle with the Amazons (Solounias and Mayor, 2005).
- The mythical animal, the griffin, is based on fossils of the dinosaur, Protoceratops, from the Gobi desert; the griffin being a composite of the wings, beak, and talons of front feet of an eagle and the rest of a lion (Mayor, 2000).

Moreover, Mayor and her coworkers proposed the following:

- Footprints from three-toed dinosaurs in North Africa are proposed as the basis of the myth of the giant bird of prey, the roc or roc-bird or rukh of Arabian legend (Mayor and Sarjeant, 2001).
- A legend of the Lakota (Sioux, Native American Indians) envisions a battle between giant water monsters or Unktehi

FIGURE 11.7 **Examples of animals on flags and national symbols.** (A) Examples of animals on US flags. (*Top row*): Gadsden flag and flag of the President of the USA. (*Middle row*): Missouri, Michigan, and Wyoming. (*Bottom row*): California, Iowa, and Louisiana. (B) Examples of animals on Australia and UK flags: Western Australia, English county of Kent, nation of Wales (Cymru), and Wessex. (C) Other flags. Eagles on national flags or seals. (*Top row*): Flags of Imperial Russia (double black eagle), Serbia (double white eagle), and Poland (single white eagle). (*Middle row*): Flag of Prussia (black eagle), Great Seal of the USA (eagle), imperial coat-of-arms of Austria (double eagle), and flag of Mexico. (*Bottom row*): Variant flag of Thailand (elephant) and flag of the Grand Duchy of Lithuania. *Source: Courtesy Wikipedia Commons.*

FIGURE 11.8 **Romulus and Remus, the brothers who are the legendary founders of ancient Rome, with the female wolf.** In mythology, the wolf cared for the orphaned brothers. *Source: Courtesy Wikipedia Commons.*

and thunderbirds or Wakinyan with the bones of these still seen in South Dakota. In the same area, geologists today find fossil mosasaurs (marine reptiles) and pterosaurs (flying reptiles) from the Cretaceous period (Mayor, 2005).

Similarly, the dragon myth is thought to be based on one or more of the following: dinosaur fossils, the Nile crocodile, Goanna (monitor lizards), whales or a composite of predators (Stromberg, 2012). Other mythical animals include werewolves and vampires in addition to the kraken, unicorn, the Loch Ness monster, yeti, and bigfoot. Werewolves feature in European folklore. They are usually a man but occasionally a woman trapped in the body of a wolf. Vampires are usually associated with the history or legends of the South Slavs as the "undead" returning from the grave to kill. For instance, an Austrian official reported on the case of Peter Plogojowitz, in Kisilova in northeast Serbia (close to the border with Rumania) (Barber, 1988). Following a series of murders, the Serbian peasant was exhumed 10 weeks after his burial in 1725. The official noted the lack of decomposition, the absence of odor, and the presence of fresh blood in his mouth some 10 weeks after his death and burial (Barber, 1988). A shake was used to "kill" the body of Peter Plogojowitz (Barber, 1988). In fictional vampire stories, the vampires transform into animals, such as a dog or bat in Bram Stoker's Dracula (1897). In this, Dracula was a count from Transylvania and a descendant of Vlad III Dracula or Vlad the Impaler (1431–77), the Prince of Wallachia (part of present-day Romania).

11.6 COMMENTARY

The authors believe that it is important to apply tolerance to the beliefs of others coupled with thoughtful consideration of all related issues. The propensity to try to silence or ridicule the views of others may be understandable but is wrong and frequently counter-productive.

References

al-Munajjid, S.M.S., 2016. Islam Question and Answer. Available from: https://islamqa.info/en/1919

Altmann, P., 2011. Festive Meals in Ancient Israel: Deuteronomy's Identity Politics in Their Ancient Near Eastern Context. De Gruyter, Berlin, Germany.

Anglo-Saxon Chronicle, Available from: http://d.lib.rochester.edu/camelot/text/sutton-vortigern-in-the-anglo-saxon-chronicle

Ball, M.M., 2015. Targeting religion: analyzing Appalachian proscriptions on religious snake handling. Boston University Law Review 1425–1450.

Barber, P., 1988. Vampires, Death, and Burial. Yale University Press, New Haven, CT.

Biswas, S., 2015. Why the humble cow is India's most polarising animal. Available from: http://www.bbc.com/news/world-asia-india-34513185

Cartwright, M., 2015. Why Do Secular Scientists Keep Talking About Animal *Sacrifice*? Slate. Available from: http://www.slate.com/blogs/lexicon_valley/2015/07/09/the_surprising_history_of_scientific_researchers_using_the_word_sacrifice.html

Census India, 2001. Distribution of population by religion. Available from: http://censusindia.gov.in/Census_And_You/religion.aspx

Chalmers, R., 1895. The Jataka (Trans.), vol. I. Cambridge University Press, Cambridge, UK.

Chinese Buddhist Electronic Texts Association (CBETA). Numbers 152–155.

Cowell, E.B., 1907. The Jataka (Trans.), vol. VI. Cambridge University Press, Cambridge, UK.

Darwin, C., 1859. On the Origin of Species. John Murray, London, England.

Dembski, W.A., Richards, J.W., 2001. Unapologetic Apologetics: Meeting the Challenges of Theological Studies. Intervarsity Press, Downers Grove, IL.

Dunn, P.M., 2006. Aristotle (384–322 BC): philosopher and scientist of ancient Greece. Arch. Dis. Child. Fetal Neonatal. Ed. 91, F75–F77.

Eipe, N., Oduro-Dominah, A., 2005. Colloids for vegetarians. Anaesthesia 60, 520.

Eriksson, A., Burcharth, J., Rosenberg, J., 2013. Animal derived products may conflict with religious patients' beliefs. BMC Med. Ethics 14, 48.

Fishkoff, S., 2009. 'Ethical' food, not kosher, is priority for US Reform. The Jewish Chronicle. Available from: http://www.thejc.com/news/world-news/22405/ethical-food-not-kosher-priority-us-reform

Foltz, R., Saadi-nejad, M., 2007. Is Zoroastrianism an ecological religion? J. Study Religion Nat. Cult. 1, 413–430.

Forrest, B., Gross, P.R., 2005. The wedge of intelligent design: retrograde science, schooling, and society. In: Koretta, N. (Ed.), Scientific Values and Civic Virtues. Oxford University Press, Oxford, England, p. 191.

Francis, W.H.T. 1905. The Jataka (Trans.), vol. V. Pali Text Society, Melksham, Wilts, UK.

Francis, H.T., Neil, R.A. 1897. The Jataka (Trans.), vol. III. Pali Text Society, Melksham, Wilts, UK.

Frazier, J.G., 1922. The Golden Bough: A Study in Comparative Religion. MacMillan, London, England.

Gallop, 2014. Available from: http://www.gallup.com/poll/170822/believe-creationist-view-human-origins.aspx

Geoffrey of Monmouth, 1136. The History of the Kings of Britain (Historia Regum Britanniae).

Godoy, M., 2012. Lust, lies and empire: the fishy tale behind eating fish on Friday. PBS. Available from: http://www.npr.org/sections/thesalt/2012/04/05/150061991/lust-lies-and-empire-the-fishy-tale-behind-eating-fish-on-friday

Gulevich, T., 2004. Understanding Islam and Muslim Traditions. Omnigraphics, Detroit, MI.

Guynup, S., Ruggia, N., 2004. Rats rule at Indian temple. National Geographic. Available from: http://news.nationalgeographic.com/news/2004/06/0628_040628_tvrats.html

Hajar, R., 2011. Animal testing and medicine. Heart Views 12, 42.

Harl, K.W., 1990. Sacrifice and pagan belief in fifth- and sixth-century Byzantium. Past Present 128, 7–27.

Harris, M., 1987. The abominable pig. Etnologija. Available from: http://etnologija.etnoinfolab.org/dokumenti/82/2/2009/harris_1521.pdf

Hartman, G., Bar-Oz, G., Bouchnick, R., Reich, R., 2013. The pilgrimage economy of Early Roman Jerusalem (1st century BCE–70 CE) reconstructed from the $\delta^{15}N$ and $\delta^{13}C$ values of goat and sheep remains. J. Archaeol. Sci. 40, 4369–4376.

Heidel, A., 1949. The Gilgamesh Epic and Old Testament Parallels. University of Chicago, Chicago, IL.

Hoesli, T.M., Smith, K.M., 2011. Effects of religious and personal beliefs on medication regimen design. Orthopedics 34, 292–295.

Hood, R.W., 1998. When the spirit maims and kills: social psychological considerations of the history of serpent handling sects and the narrative of handlers. Int. J. Psychol. Religion 8, 71–96.

International Council on Biblical Inerrancy, in press. Available from: http://www.alliancenet.org/international-council-on-biblical-inerrancy

Jha, D.N., 2002. The Myth of the Holy Cow. Verso, London.

Kieschnick, J., 2005. Buddhist vegetarianism in China. In: Roel Sterckx (Ed.), Of Tripod and Palate: Food, Politics and Religion in Traditional China. Palgrave MacMillan, New York, NY, pp. 186–212.

Kosher Certification Organisation, 2017. Available from: http://www.koshercertification.org.uk/whatdoe.html

Lammoglia, J.A., 2008. Legal aspects of animal sacrifice within the context of Afro-Caribbean religions. St. Thomas Law Rev. 20, 710.

Leitzmann, C., 2014. Vegetarian nutrition: past, present, future. Am. J. Clin. Nutr. 1 (100 Suppl.), 496S–502S.

Lorey, F., The Flood of Noah and the Flood of Gilgamesh. Available from: http://www.icr.org/article/noah-flood-gilgamesh/

Lynch, M.E., 1988. Sacrifice and the transformation of the animal body into a scientific object: laboratory culture and ritual practice in the neurosciences. Soc. Stud. Sci. 18, 265–289.

Mayor, A., 2000. The First Fossil Hunters: Dinosaurs, Mammoths, and Myth in Greek and Roman Times. Princeton University Press, Princeton, NJ.

Mayor, A., 2005. Fossil Legends of the First Americans. Princeton University Press, Princeton, NJ.

Mayor, A., Sarjeant, W.A.S., 2001. The folklore of footprints in stone: from classical antiquity to the present. Ichnos 8, 143–163.

McGinnis, M.W., 2004. Buddhist Animal Wisdom Stories. Weatherhill, Boulder, CO.

McNutt, P.M., 1999. Reconstructing the Society of Ancient Israel. SPCK, London, England.

Meyer, S.C., 2009. Signature in the Cell: DNA and the Evidence for Intelligent Design. HarperCollins, New York, NY.

Moazami, M., 2005. Evil animals in the Zoroastrian religion. Hist. Religions 44, 30–317.

Nizamoglu, C., 2016. Cats in Islamic culture. Muslim heritage: discover the golden age of Muslim civilization. Available from: http://muslimheritage.com/article/cats-islamic-culture

Ontario Consultants for Religious Tolerance, 2016. Noah's ark and the flood: points of similarity between the Babylonian and Noachian flood stories. Available from: http://www.religioustolerance.org/noah_com.htm

Petropoulou, M.-Z., 2008. Animal Sacrifice in Ancient Greek Religion, Judaism, and Christianity, 100 BC to AD 200. Oxford University Press, Oxford, England.

Pew Forum, 2015a. The future of world religions: population growth projections, 2010–2050. Available from: http://www.pewforum.org/2015/04/02/religious-projections-2010-2050/

Pew Forum, 2015b. US becoming less religious. Available from: http://www.pewforum.org/files/2015/11/201.11.03_RLS_II_full_report.pdf

Pew Forum, 2016. Jewish beliefs and practices. Available from: http://www.pewforum.org/2016/03/08/jewish-beliefs-and-practices/

Pu, C.Z., 2014. Ethical Treatment of Animals in Early Chinese Buddhism: Concepts and Practices. Cambridge Scholars Publishing, Newcastle, UK.

Ratzinger, J., 2012. Jesus of Nazareth—The Infancy Narratives (Pope Benedict XVI). Random House, New York, NY.

Reed, A.Y., 2013. From sacrifice to slaughterhouse: ancient and modern approaches to meat, animals and civilization. Method and Theory in the Study of Religion. Available from: https://www.academia.edu/4534761/_From_Sacrifice_to_the_Slaughterhouse_Ancient_and_Modern_Approaches_to_Meat_Animals_and_Civilization_

Rennie, J., 2002. 15 Answers to creationist nonsense: opponents of evolution want to make a place for creationism by tearing down real science, but their arguments don't hold up. Available from: http://www.scientificamerican.com/article/15-answers-to-creationist/

Roberts, M.B., 2007. Genesis chapter 1 and geological time from Hugo Grotius and Marin Mersenne to William Conybeare and Thomas Chalmers (1620–1820). In: Piccardi, L., Masse, W.B. (Eds.), Myth and Geology. Geological Society Special Publication, London, pp. 39–49.

Rollin, B.E., 1981. Animal Rights and Human Morality. Prometheus Books, Buffalo, NY.

Rotman, A., 2008. Divine Stories: Divyine Stories Part 1 (Trans.). Wisdom Publications, Boston, MA.

Rouse, W.H.D., 1895. The Jataka (Trans.), vol. II. Cambridge University Press, Cambridge, UK

Rouse, W.H.D., 1901. The Jataka (Trans.), vol. IV. Pali Text Society, Melksham, Wilts, UK

Sarkar, S., 2005. Use of animal products in vegetarians and others. Anaesthesia 60, 519–520.

Shiwani, M.H., 2011. Surgical meshes containing animal products should be labeled. Br. Med. J. 343, 4625.

Singer, P., 1975. Animal Liberation. Harper Collins, New York, NY.

Solounias, N., Mayor, A., 2005. Ancient references to fossils in the land of Pythagoras. Earth Sci. Hist. 23, 207–210.

Speyer, J.S., 1895. Jātakamālā or Garland of Birth-Stories. Oxford University Press, London.

Sterckx, R., 2005. Food and philosophy in early China. In: Roel Sterckx (Ed.), Of Tripod and Palate: Food, Politics and Religion in Traditional China. Palgrave MacMillan, New York, NY, pp. 34–61.

Stromberg, J., 2012. Where did dragons come from? Smithsonian Magazine. Available from: http://www.smithsonianmag.com/science-nature/where-did-dragons-come-from-23969126/?no-ist

Story, F., 1964. The Place of Animals in Buddhism. Buddhist Publication Society, Kandy, Sri Lanka.

Swearer, D.K., 2004. Becoming the Buddha: The Ritual of Image Conservation in Thailand. Princeton University Press, Princeton, NJ.

Szűcs, E., Geers, R., Jezierski, T., Sossidou, E.N., Broom, D.M., 2012. Animal welfare in different human cultures, traditions and religious faiths. Asian-Australas. J. Anim. Sci. 25, 1499–1506.

The Atlantic, 2015. Selling the sacred cow: India's contentious beef industry. Available from: http://www.theatlantic.com/business/archive/2015/02/selling-the-sacred-cow-indias-contentious-beef-industry/385359/

Thomas, C., 1981. Christianity in Roman Britain to AD 500. B.T. Batsford, London, England.

Trout, D., 1995. Christianizing the Nolan countryside: animal sacrifice at the tomb of St. Felix. J. Early Christian Stud. 3, 281–298.

Ullucci, D.C., 2011. The Christian Rejection of Animal Sacrifice. Oxford University Press, Oxford, England.

Vermes, G., 2012. From Jewish to Gentile: How the Jesus movement became Christianity. Biblical Archaeol. Rev. 38, 6.

Wade, R., 2004. Animal theology and ethical concerns. Aust. eJournal Theol. 2, 1–12.

Walker-Meikle, K., 2011. Medieval Cats. British Library Publishing, London, England.

Further Reading

Answers in Genesis, 2016. Available from: https://answersingenesis.org/answers/

Chen, Huiyu 陳懷宇，in press. Lion and the Buddha: the Ornament and Symbolism of Animals in Chinese Buddhist Literature, 〈獅子與佛陀：早期漢譯佛教文獻中的動物裝飾與象徵〉，《政大中文學報》，2010 年，頁55–84.

Creation Museum, 2017. Available from: http://creationmuseum.org

Discovery Institute Center for Science and Culture, in press. Available from: http://www.discovery.org/id/

Foltz, R., 2010. Zoroastrian attitudes toward animals. Soc. Anim. 18, 367–378.

Grey, L., 2002. Animal stories in Buddhist Jātakas. His Lai J. Humanistic Buddhism (西來人間佛教學報) 3, 159–194.

Institute for Creation Research, Available from: http://www.icr.org

Kang Senghui, trnsl. (3rd cent.), Scripture of the Collection of the Six Perfection Practices 六度集經, CBETA, No. 152.

Ken Ham, 2017. Available from: https://answersingenesis.org/answers/

Kent Hovind, 2016. Available from: https://2peter3.com

Liang, Liling 梁麗玲, 2010. Studies in animals stories in Chinese Buddhist translations 《漢譯佛典動物故事之研究》, 文津出版社.

Quran, Available from: http://www.whyislam.org/faqs/restrictions-in-islam/why-do-muslims-abstain-from-pork/

Stewart, J.J., 2010. The question of vegetarianism and diet in Pāli Buddhism'. J. Buddhist Ethics 17, 100–140.

Unknown Translator (prior 5th cent.), Scripture of the Bodhisattva's Original Actions, 菩薩本行經, CBETA, No. 155.

Zhi Qian, trnsl. (3rd cent.), Scripture of the Bodhisattva's Actions, 菩薩本緣經, CBETA, No. 153.

Zhu Fahu, trnsl. (3rd–4th cent.), Scripture of Birth Stories, 生經, CBETA, No. 154.

12

Animals as Companions

Kate Mornement

Pets Behaving Badly, Melbourne, VIC, Australia

12.1 THE HISTORY OF COMPANION ANIMAL (PET) KEEPING

12.1.1 Pet Keeping in Premodern Societies

For the majority of human history people have lived as hunter–gatherers and were heavily reliant on animals for their survival. Initially, animals provided humans with valued resources including food, tools, and pelts; however, the relationship people shared with certain species (e.g., cats and dogs) later evolved to become more mutually beneficial. Whereas the value we place on most animals is based on economic and practical considerations, the importance we place on "companion" animals is a result of the special relationship we continue to share with them. Despite investigations into the historical aspect of the human–companion animal relationship being constrained by the limited amount and uneven distribution of fossilized remains, available evidence from the earliest canine fossils comes from human burial sites indicating a special relationship between early man and dog (Miklósi, 2014). Similarly, evidence from a number of dog burial sites suggest that dogs were at least members of the group, or family, and were also entitled to burial suggesting that the relationship between people and companion animals was an emotional one.

Interestingly, the archaeological record also sheds light on the other extreme of the human–dog relationship. Broken bones and bones with cut and gnaw marks provide strong evidence for butchery, and dogs formed part of the diet for humans living in prehistoric central Europe until the Bronze Age (Bartosiewicz, 1994), in the Maya culture of Mexico (Clutton-Brock and Hammond, 1994), and among the aborigines of New Zealand (Clark, 1997) and Australia where tribes held diverse relationships with dingoes (Meggitt, 1961). Here, dingoes were eaten, kept as pets, used for hunting, and for warmth when sleeping in the cooler months (Fig. 12.1).

In medieval Europe, pet keeping was actively discouraged by religious authorities because it was argued that pets diverted their owner's attention away from God and resources away from humans who were struggling to survive. Pet ownership was considered dirty, unholy, and dehumanizing and was often cited as "evidence" in the trial of witches at the time (Serpell, 2002).

FIGURE 12.1 **Aboriginal cave painting depicting dingoes.** *Source: https://www.flickr.com/photos/82134796@N03/18359881212/in/photostream/.*

Interestingly, throughout the same period pet keeping was an accepted practice of social elites and represented status, wealth, and privilege (Hurn, 2012).

12.1.2 Pet Keeping in Modern Society

In today's society, cats, dogs, and a number of other species we consider companion animals including rabbits, rodents, reptiles, fish, birds, and horses continue to play an important role (Amiot et al., 2016). Companion animals remain popular in most modern Western societies with up to 90% of owners considering their pet as a family member (Carlisle-Frank and Frank, 2006). Current estimates of pet ownership around the world vary from 15% in Turkey (FEDIAF The European Pet Food Industry, 2014) to 40% in the United Kingdom (Pet Food Manufacturing Association, 2016) and up to 63 and 65% in Australia and the United States, respectively (American Pet Products Association, 2016; Animal Health Alliance Australia, 2013, p. 5). Present-day cultures reflect three main aspects of the human–companion animal relationships: their use for food and pelts (which continues in parts of Asia today), their use as working animals, and their use as emotional and social support companions (Miklósi, 2014).

In the Western world, dogs are primarily kept as companions. Until a little over a hundred years ago, most dogs performed work for their owners. Each breed or type had been developed for many years to perform specific tasks, including hunting, herding, and guarding, to which they were well suited. Nowadays, the majority of dogs are not required to perform the working role their breed was originally developed for (Bradshaw, 2011). Although many breeds, in general, and individuals have adapted to this new life as our companions, others remain poorly suited to this role. The same challenges when it comes to adapting to living within our homes are faced by other species commonly kept as pets, especially cats and parrots.

As human society continues to evolve and our population exponentially grows the popularity of dogs as pets appears to have peaked as the pressures we, and the urban environment, place on them increase. We expect dogs to be better behaved than the average young child and yet, as self-reliant as a human adult (Bradshaw, 2011). Behavior once considered normal, indeed desirable, is now perceived as problematic and undesirable. For example, the Border Collie that herds sheep is an asset to a farmer whereas the one that tries to herd children and chases people on bicycles and skateboards is a liability to an owner (Bradshaw, 2011).

These changing expectations we have placed on our canine companions are exacerbated by the way in which we rigidly control their reproduction. For much of our relationship with them we have bred dogs to suit roles that we assigned to them. In performing these roles, it was their ability and function that took precedence rather than their type or appearance. This began to change in the 19th century when breed societies began dividing dogs into breeds, reproductively isolating them and assigning each breed with a single "ideal" appearance, or "breed standard." Most breeders today, in breeding to their breed's standard, strive to breed the perfect-looking dog that will succeed in the show ring rather than a

FIGURE 12.2 **English Bulldog (demonstrating exaggerated physical features including large head and extremely short muzzle).** *Source: Courtesy SheltieBoy, Flickr: AKC Great Falls June 2011, CC BY 2.0, https://commons.wikimedia.org/w/index.php?curid=31509799*

dog well-suited to life as a human companion. Successful show dogs typically make a disproportionately large genetic contribution to the next generation of their breed. Over the years this has resulted in significant inbreeding causing genetically based deformities, diseases, and other functional disadvantages compromising the welfare of many pure-bred dogs (Bradshaw, 2011) (Fig. 12.2).

Although there is a growing rift between dog breeders and those concerned with animal welfare, including scientists, activists, and members of the general public, significant changes to breeding practices are yet to be implemented. This is despite a wealth of new science supporting the need for a shift in societal attitudes and breeding practices (King et al., 2012).

12.1.3 Mechanisms Underlying Human–Animal Relationships

According to Wilson's *Biophilia Hypothesis* (1984), the human desire and inclination to pay attention to and connect with animals is a

biological tendency (Wilson, 1984). The concept of Biophilia refers to a hypothetical human affinity for the living world, as well as the tendency of humans to interact and form emotional attachments with other life forms. According to this theory, Biophilia is considered to be innate to all humans (Borgi and Cirulli, 2016). A number of studies support this hypothesis demonstrating a general proneness toward animals, as well as an increase in social behaviors in the presence of animals (O'Haire et al., 2013), which emerges from early childhood onward. This effect is even observed in children who are challenged socially, such as those with autism spectrum disorders (Muszkat et al., 2015).

The question of why humans are so drawn to animals is yet to be completely clarified. Animals seem to attract the attention of people more so than objects and it has been hypothesized that paying attention to animals conferred a fitness and survival benefit from an evolutionary perspective (Mormann et al., 2011; New et al., 2007).

The study of attitudes toward animals is complex involving evolutionary, psychological, and cultural aspects (Serpell, 2004). Known as the "similarity principle," a growing body of literature has demonstrated that animals that are phylogenetically and/or physically, behaviorally, or cognitively similar to humans tend to be preferred. These animals induce more positive effect and concern for their welfare and conservation (Batt, 2009; Knight, 2008; Tisdell et al., 2006). Perhaps unsurprisingly, humans tend to show negative attitudes toward animals considered phylogenetically distant to humans (e.g., invertebrates, fish, and reptiles). Similarity to people is just one attribute known that explains the enormous variation in people's attitudes toward animals. Physical appearance and anthropomorphic features, such as large eyes and other juvenile traits, have also been shown to influence preferences and attitudes (Borgi and Cirulli, 2013; Serpell, 2004).

Many scholars argue that humans demonstrate a preference for animals they perceive as aesthetically pleasing or "cute" (Archer and Monton, 2011; Herzog, 2001; Knight, 2008). Cuteness is a common measure of attractiveness to a stimulus, especially one associated with infancy and youth (Borgi and Cirulli, 2016). Kindchenschema (or baby schema) was first described by ethologist Konrad Lorenz as a set of facial features (i.e., large eyes and forehead, round face, small nose, and mouth) able to evoke a caregiving and affective response toward infants (Lorenz, 1942). Recent empirical studies support this concept, showing that these traits are often perceived as cute and attractive and thus preferred to those with less infantile faces (Little, 2012; Luo et al., 2011). Moreover, this concept has since been demonstrated by several studies that showed the generalization of the cute response to live animals, cartoon animal characters, and toy animals (Archer and Monton, 2011; Little, 2012). The emergence of the cute response has been shown to occur early during development, present in 3–6 year olds (Borgi et al., 2014) (Fig. 12.3).

Until recently, the neurophysiological basis of attachment and caregiving and how this influences human–animal relationships was an area that received very little scholarly attention (Lenzi et al., 2015). However, researchers have now conducted experiments using functional MRI (fMRI) that show that regions of the brain implicated in emotion, social cognition, and face processing are activated when participants viewed a photograph of either their child or their dog (Stoeckel et al., 2014). Oxytocin, implicated in the neuroendocrine regulation of maternal behavior, has been found to be crucial in the formation of social bonds. Researchers found an association between a dog's gaze and urinary oxytocin levels in their owners following affiliative interactions. Oxytocin levels increased in the dogs as well. These findings suggest that dogs were domesticated by co-opting social cognitive systems in humans involved in social attachment. This phenomenon may contribute to the establishment of the human–animal bond, which demonstrates similarities with the mother–infant relationship (Borgi and Cirulli, 2016).

12.2 MODERN-DAY HUMAN–COMPANION ANIMAL RELATIONSHIPS

12.2.1 The Role of Modern-Day Companion Animals

According to the Oxford English Dictionary, "pets" are defined as tame or hand-reared animals, animals kept for pleasure or companionship, or individuals singled out for special treatment. In the United States, a pet is defined as follows: "For purposes of Housing programs: a domesticated animal, such as a dog, cat, bird, rodent (including a rabbit), fish, or turtle, that is traditionally kept in the home for pleasure rather than for commercial purposes. Common household pet does not include reptiles (except turtles)" (24 CFR 5.306, Title 24—Housing and Urban Development). It is important to acknowledge that culture can determine whether an animal is considered a companion or food.

FIGURE 12.3 **Scottish Fold kitten showing baby schema large eyes and forehead.** *Source: https://commons.wikimedia.org/wiki/File:3mo_lilac_Scottish_Fold_Fanel.jpg*

For example, dogs are kept as pets in the United States, Australia, and Japan but they are eaten in South Korea.

Notably, these definitions do not, however, address the fact that pets are defined in legal terms as the property of their human owners. Pets may be legally bought and sold, in response to consumer trends, and constitute commodities of sorts. In recent years in many municipalities the United States, there has been a trend toward replacement of the term "ownership" to "guardianship" in animal laws, which may have significant implications for the legal rights and protection of companion animals.. Presently, however, the sociocultural (and legal) definition of animals identifies pets as property to their human owners who have a responsibility toward the animal but also the power over their life and death (Hurn, 2012).

Many researchers have concluded that the human–pet relationship is purely utilitarian (defined by the benefits derived by humans, such as through work, protection, and status). Hurn (2012) argues that this does not explain why people expend resources (e.g., time, money, and freedom) on animals kept as pets that do not appear to fulfill any obvious utilitarian role. In addition, many working animals fulfill utilitarian and nonutilitarian (i.e., companionship) roles. For example, sheep dogs who herd for their owners during the day and provide companionship, as pets, at night. In any case, human interactions with pets are fluid and complex (Hurn, 2012). Pets are also employed to perform educational roles in many cultures, such as by (indirectly) teaching children about compassion and empathy (Fifield and Forsyth, 1999). For example, among the Amazonian Guajá, baby monkeys are given to young girls to teach them to provide practical experience caring for an infant. The baby monkeys cling to their human caregiver and require a high level of care and attention, similar to that of a newborn human. Children also learn lessons about life and death through keeping pets; however, experiences through pet keeping differ between individuals and so too will children's attitudes and perceptions toward them (Hurn, 2012).

It is well known that cats and dogs, the most common species kept as pets, exhibit lifelong morphological and behavioral infantile traits. As noted earlier, infantile features in companion animals are thought to explain, in large part, the basis of our attraction to them and may contribute to our motivation for pet keeping (Archer, 1997). The retention of youthful traits into adulthood, known as *neoteny*, is related to domestication and selective breeding over numerous generations for tameness and docility (Belyaev, 1979; Hare et al., 2005). In comparison with their ancestors, domestic dogs are smaller in size with shortened muzzles and smaller skulls, and exhibit life-long behaviors typically seen in young wolves including barking, whining, and attention-seeking. Many of the features humans have selected for aesthetic reasons in certain breeds of dogs have negative consequences for the individuals' welfare (King et al., 2012).

In modern times, pets are found fulfilling roles as substitute humans acting as surrogate children, friends, and protectors. The similarity of the bond between owners and their companion animals and that shared between human parents and their children is of particular interest to researchers (Borgi and Cirulli, 2016). These similarities have been examined within the framework of human attachment theory. Adult–child and human–animal relationships both feature dependency, proximity-seeking, caregiving, affection, and ensure security, comfort, protection, and survival (Ainsworth and Bowlby, 1991; Payne et al., 2015). Research has shown that the language used to communicate with companion animals mimics that used by others to talk to their babies, also known as "motherese" (Burnham et al., 2002). Notably, particularly strong human–animal bonds may not be considered culturally or socially acceptable (Franklin, 1999; Serpell, 1991b; Thomas, 1983) and may have harmful consequences.

FIGURE 12.4   **Morphological variation of the dog.** *Source: Courtesy Mary Bloom, American Kennel Club, http://journals.plos.org/plosbiology/article?id=10.1371/journal.pbio.1000310; CC BY-SA 4.0, https://commons.wikimedia.org/w/index.php?curid=51751579.*

For example, many homeless people prefer to stay on the street than relinquish their pet dog (Hurn, 2012). Similarly, elderly pet owners may be reluctant to seek medical treatment if they believe it might lead to their being hospitalized or placed in aged care and unable to care for their pet (McNicholas et al., 2005b) (Fig. 12.4).

12.2.2 Dogs as Companions

12.2.2.1 Evolution and Domestication

There was a point in human history when dogs, as we know them, did not exist. Instead, there were wolves, jackals, and coyotes. There is some uncertainty as to how modern dogs evolved, such as by artificial selection by humans (where wolf pups were taken from the wild and tamed) or by a process of natural selection (where a changing environment necessitated survival of the fittest and required adaptation to a new niche) (Coppinger and Coppinger, 2002). Regardless of their evolutionary path, dogs are now present in almost every human society around the world; their roles and their use by humans vary significantly (Miklósi, 2014).

It has long been assumed that the wolf was the first species to be domesticated by humans

(Fox, 1978) and that, over time, humans shaped wolves into dogs through artificial selection. However, for over a century, scholars have argued, and continue to do so, over which of the wild canids the dog descended. Moreover, precisely where and when canine domestication took place largely remains under debate. Research suggests the coevolution of dogs and humans may have started as long as 32,000 years ago. Discoveries of patterns of genomic variation indicate that the domestication of dogs began at least 10,000 years ago in southern East Asia (Pang et al., 2009) or the Middle East (von Holdt et al., 2010) and may have involved several source populations or even back-crossing with wolves (Axelsson et al., 2013). Speculation that humans captured wolf puppies and selectively bred them for use in hunting or guarding is now being replaced by theories suggesting that wolves proactively played a role in the domestication process. As humans transitioned from a nomadic to pastoralist lifestyle with the advent of agriculture, it is more probable that wolves scavenged on their waste (Coppinger and Coppinger, 2001). Wolves living close to human camps likely brought with them significant advantages for humans. Sanitation may have improved as food scraps, human faeces, and other waste was scavenged by the wolves. With their acute hearing, wolves may have alerted humans to the approach of a predator or stranger and assisted with hunting, utilizing their highly sensitive sense of smell to track prey.

As humans began selectively breeding dogs they increasingly relied on them for their survival. Indeed, it has been speculated that the human–dog relationship was so beneficial to early humans that it may have given them an advantage over their competitors, the Neanderthals, who did not keep dogs (Taçon and Pardoe, 2002). Archaeological remains of canids indicate several structural changes associated with domestication including a smaller skeleton, a short and compact muzzle, crowded and proportionately smaller teeth, eyes set more toward the front, reduced cranial capacity, widening of the cranium, and a sharply rising stop (Coppinger and Coppinger, 2001; Morey, 1992; Pang et al., 2009). Domestication in dogs has resulted in two different skull morphologies: short and broad and long and narrow. This shortening and widening of the skull is demonstrated in the boxer and similar brachycephalic breeds, whereas the Greyhound and similar breeds exhibit a skull that has lengthened and narrowed. Interestingly, this variation in skull morphology across breeds has been linked to differences in the expression of behavior. Georgevsky et al. found that cephalic index (the ratio of skull width to skull length), body weight, and height covary with behavior.

In addition to changes in morphology, humans have artificially selected for the behavioral traits they desired. These include reduced aggressiveness and increased social cognition (Hare et al., 2012), tameness, hunting ability, and guarding ability. It is estimated that distinctive breed types appeared about 3500 years ago (Clutton-Brock, 1995), with continued selective breeding leading to the development of modern-day dogs, of which there are over 600 types (Lindsay, 2000). The most noted behavioral changes that have occurred in dogs as a result of domestication include the retention of juvenile behaviors, known as "paedomorphism," including barking, tail wagging, and playing. Interestingly, these behaviors, particularly barking, are sometimes perceived as problematic by modern dog owners. For example, dogs that bark incessantly are often the cause of neighborhood disputes (Kobelt et al., 2003) and complaints to local councils.

The success of dogs as a species can be attributed to the ability of their ancestors to adapt to life working in partnership with humans. Population estimates are somewhere in the region of 400 million, although the exact figure is unknown, potentially making dogs the most populous carnivore on the planet today (Coppinger and Coppinger, 2002). The development of dog breeds has resulted in a plethora of dogs that differ significantly morphologically and behaviorally. Although dogs were traditionally bred for their working ability, the majority of dogs living in society today fulfill the role of companion (King et al., 2012); a role in which many fail to adapt.

12.2.2.2 Companion Dogs in Contemporary Society

Present in almost every human society, dogs remain one of the most popular pets in Western cultures (Archer, 1997), although according to Coppinger and Coppinger (2002), the majority of dogs still live as wild animals and scavengers around villages. Available statistics report household ownership rates of 36.5% in the United States (American Veterinary Medical Association, 2012), 24% in the United Kingdom (Pet Food Manufacturers Assocation, 2014), and 39% in Australia (Animal Health Alliance Australia, 2013, p. 11), one of the highest pet ownership rates in the world. Total pet dog populations are estimated to be around 4.2 million in Australia and 70 million in the United States (Animal Health Alliance Australia, 2013, p. 5; Pet Food Manufacturing Association, 2016). As discussed previously, dogs were originally bred to perform specific roles, such as guarding, hunting, and herding and many retain the behavioral traits specific to these roles (King et al., 2012). However, modern dogs are most commonly kept as companions, rarely, if ever, undertaking the roles their ancestors were bred to perform. Despite this, most breeders continue to follow breed standards and a code of ethics, which encourage them to produce dogs possessing temperaments and behaviors suited to their historical role, rather than their modern role as companions (King et al., 2012).

The owner–dog bond has attracted an increase in academic research in recent years, leading to discoveries of several benefits of dogs to people. Owners report many benefits of companion dog ownership, including companionship, friendship, affection, pleasure, and protection (Endenburg et al., 1994; Horn and Meer, 1984; Zasloff and Kidd, 1995), all of which are thought to be important for positive mental health (McNicholas et al., 2005a). Dog ownership is also associated with a range of benefits in children including companionship, comfort and support, improved self-esteem, and learning how to care for another (Triebenbacher, 2000). In addition, exposure to dogs during the perinatal period may reduce the development of allergies in children (Lodge et al., 2012) and pet ownership in childhood may lead to more positive attitudes toward animals in young adults (Paul and Serpell, 1993).

Several health benefits have also been associated with dog ownership (Serpell, 1991a). For example, dog owners have less frequent visits to the doctor and are less likely to be on medication for heart problems (Anderson et al., 1992) and sleeping difficulties (Heady, 1999). Dog ownership is also associated with reduced physiological stress, such as high blood pressure (Allen, 2003; Ramírez and Hernández, 2014) and reduced national health expenditure in Australia (Heady, 1999). In an overview of research in the field, Wells reported that dog ownership is associated with lower incidence of illness and improved recovery. However, not all studies support the health benefits of dog ownership (Bauman et al., 2001; Stallones et al., 1990), and there are limitations in the quality of the science in this area of research. As Herzog (2011) noted, if participants are not randomly assigned to groups and self-select pet ownership, differences between dog owners and nonowners may occur because individuals who perceive themselves to be healthier also choose to have a pet.

In summary, canids have had a long association with humans and this relationship has shaped the behaviors of both species. Living with dogs changed the nature of the human–animal relationship (Taçon and Pardoe, 2002). Despite our long and mutually beneficial association with canines, the human–companion dog bond is not always successful. Many companion dogs are surrendered to animal shelters because of behavioral problems, careless or uneducated owners, or due to unfortunate circumstances, such as their owner falling ill or moving into accommodation that does not allow pets.

12.2.2.3 When the Owner–Companion Dog Bond Breaks Down

Not all human–companion dog relationships are successful and a breakdown in the relationship can occur for a number of different reasons including owner and dog factors. The development of canine behavior problems can be associated with characteristics of a dog's owner and household (Stephen and Ledger, 2006). This was demonstrated in a study by Jagoe and Serpell (1996) in which owners who trained their dogs reported a lower incidence of undesirable behaviors. The same study also found a higher incidence of problem behavior including aggression and separation anxiety in dogs that were allowed to sleep within close proximity to their owners. In addition, it was found that dogs belonging to first-time owners were more likely to develop undesired behavior including aggression, separation anxiety, overexcitability, and phobia of loud noises. These associations suggest that the same dog may behave differently in a household that diverged from these characteristics (Jagoe and Serpell, 1996).

Other studies have also shown that the personality and attitudes of the owner may be associated with dog behaviors (O'Farrell, 1995). For example, Podberscek and Serpell (1997) reported that owners of highly aggressive English Cocker Spaniels were more likely to be shy and tense than owners of dogs with low levels of aggression. Furthermore, Dodman et al. (1997) found that confident and independent dog owners were less likely to have dogs with behavior problems, while neurotic owners were more likely to have dogs that are destructive when left home alone and that display excessive mounting behavior (O'Farrell, 1995).

Unrealistic expectations about dog behavior and management can also contribute to a breakdown in the dog–owner relationship (Scarlett et al., 1999). It is important that potential dog owners understand basic dog behavior and management, including their physiological and psychological needs, in order for the owner–dog relationship to be successful. Behavioral problems commonly arise due to a lack of companionship, training consistency, exercise, and obedience training, and as has previously been established, are contributing factors to the relinquishment of dogs to animal shelters (Jagoe and Serpell, 1996; Marston and Bennett, 2003; Wells and Hepper, 2000).

12.2.2.3.1 REASONS FOR RELINQUISHMENT TO ANIMAL SHELTERS

There remains uncertainty with regard to the reasons why owners relinquish their dogs to animal shelters. An Australian study by Marston et al. (2004) analyzed data on dog admissions for a period of 12 months from three Australian animal shelters. The results showed that one-third of the reasons for surrendering a dog could be characterized as "owner-related" factors. These included accommodation restrictions and moving house, too much work/effort owning the dog, and the owner experiencing health problems. Dog behavior reasons were given for approximately 11% of surrenders. The three most common behavioral problems cited by owners were escaping, boisterousness/hyperactivity, and barking. A remaining one-third of owners surrendering their dog declined to comment as to the reasons why (Marston et al., 2004); it is possible that relinquishers were concerned that such comments may reduce the likelihood of their dog being adopted.

In another animal shelter study, Salmon et al. identified 71 different reasons for the relinquishment of companion animals. The most common reasons were accommodation issues, dog behavior, and owner's financial and lifestyle issues. Further analysis of these data by New et al. found that although moving house was cited as the main reason for relinquishment, behavioral issues were commonly reported as occurring in the month prior to relinquishment. Furthermore, a study conducted in the United States, in which 38 relinquishers were interviewed, found the main reasons given for surrendering their pet were dog behavior problems and the owner experiencing health and/or accommodation difficulties. Behavior problems, such as hyperactivity, inappropriate chewing, elimination, and vocalizations, were cited frequently (Miller et al., 1996). Similarly, studies in the United Kingdom have found that 25.6% to over 30% of owners cited dog behavioral problems as the reason for relinquishing their dog (Wells, 1996). The most frequent types of problems were boisterousness (10%), aggression toward other dogs (9%), and aggression toward people (7.7%) (Ledger and Baxter, 1997). These findings are similar to those reported in a Netherlands study that found the most common reasons for relinquishment to be behavioral; aggression and separation anxiety were most common (van der Borg et al., 1991).

12.2.3 Dogs in Working Roles

Dogs were originally domesticated, in part, for their usefulness to humans in working roles, such as hunting and guarding. Since ancient times, working dogs have served as highly accurate and trainable extensions of human abilities and senses. Despite exponential technological advances, modern machines remain unable to match the effectiveness of trained dogs in several tasks, such as explosives and narcotics detection (Fjellanger et al., 2002; Gazit and Terkel, 2003).

According to Cobb et al. (2015) a working dog is defined as "any domestic dog that is operational in a private industry, government, assistance or sporting context, independently of whether it also performs a role as human companion." In contemporary times, domestic dogs are represented in a wide range of working contexts as guardians, stock herders, detectors, guides, assistants, and racing participants in sporting activities. In recent years, dogs have been trained to perform complex working tasks including search and rescue and as integral components of conservation projects. Used by researchers, managers, and conservationists, a conservation dog team locates biological targets of interest, assisting professionals obtain information about target species. Such teams are gaining world-wide recognition for their efforts. The term "conservation dog" encompasses dogs trained to find faeces (scat detection); match biological-based scents (scent-matching); find live animals, insects, or plants; and sort samples in a lab (discrimination). Many dogs perform dual roles, such as working during the week and as companions on weekends (Cobb et al., 2015).

Working roles are typically undertaken by dogs for reasons of economy, because humans or machines are unable to perform the task or because it is cheaper or easier for a dog. Although research investigating the economic contribution of working dogs is lacking, a recent estimation of the median value of an Australian livestock herding dog (over its lifetime) was calculated to be AUD\$40,000 (Arnott et al., 2014), providing about five times the return on investment (Cobb et al., 2015). Similarly, the investments of resources to breed and train a guide dog for placement with a vision-impaired person have been estimated to be US\$50,000. Although the economic value to the handler has not been assessed, the positive psychosocial impact is significant (Sanders, 2000; Wirth and Rein, 2008).

The United States Department of Defense maintains a population of around 1700 military working dogs (MWD) to provide force-protection support in security and contraband detection of explosives and narcotics. The MWD population is composed primarily of adult large-breed dogs of herding or working breeds (predominately Belgian Shepherd Dogs and German Shepherd Dogs), trained for security and patrol work. Most dogs are dual-trained in detection and security work, creating a capable dog that may be worked in a variety of situations. Sporting and hound breeds are also utilized, albeit less commonly, and are only used for narcotic or explosive detection. Dogs are typically purchased between 12 and 36 months of age, although the US military does not require registration papers, proof of age, or lineage for any dog (Moore et al., 2001).

Due, in part, to their initial cost, extensive training, and valuable role, MWD receive a high standard of husbandry and veterinary care to optimize their capacity to provide the highest quality service for the longest period of time. Research indicates many MWD live to an age equivalent to that of companion dogs of the same breed, if not longer. Death or euthanasia, in the majority of cases, resulted from disease associated with old age (Moore et al., 2001).

People have long recognized the positive effects animals can have on human well-being. For example, in the 19th century, Florence Nightingale suggested a companion bird may be the only source of pleasure for people confined to the same room due to illness (Nimer and Lundahl, 2007). Animal-assisted therapy (AAT), the deliberate use of animals in a therapeutic setting for a specific treatment plan, formalizes this sentiment and has been applied to a wide variety of clinical problems including autistic spectrum symptoms, medical conditions, compromised mental functioning, emotional difficulties, undesirable behaviors, and physical problems. In general, although there are limitations in terms of the quality of the science, AAT appears to offer benefits for people with a wide range of physical and psychological problems (Nimer and Lundahl, 2007). The effects of AAT as a therapeutic agent and the welfare implications for animals used in this context are discussed further in Chapter 13.

As our understanding of canine intelligence and affective experience increases and our training methodologies improve, new roles for working dogs are discovered, such as in the detection of cancer in humans (Pickel et al., 2004).

Despite the widespread use of dogs in diverse working roles, available data suggest that success rates typically average 50% across the working dog industry sectors (Arnott et al., 2014; Branson et al., 2010) meaning half of all dogs bred for working roles fail to become operational. This wastage is problematic for the financial sustainability of the industry and in terms of public perceptions (Spedding, 1995). According to Broom (2010), animal production systems that are not sustainable will not persist into the future. Inefficient systems or poor animal welfare is likely to be unsustainable because it fails to meet the expectations of the general public. Growing awareness of animal use and its implications have led to higher public expectations and lower tolerance for poor welfare. Indeed, shifts in public attitudes demonstrate a change in perceptions of acceptable animal use. For example, setting animals (including dogs) to fight one another is outlawed and the display of exotic animals for entertainment in zoos and circuses is increasingly under scrutiny. This trend will likely lead to further examination of dogs in working roles (Cobb et al., 2015).

12.2.4 Cats as Companions and in Working Roles

12.2.4.1 Evolution and Domestication

The domestication of cats followed a very different pattern to that of dogs and aspects of the history of our relationship with cats remain unclear (Amiot et al., 2016). Wildcats are atypical candidates for domestication because they are obligate carnivores having evolved to live on a diet of meat. Cats are typically solitary and highly territorial becoming very attached to their home range. Furthermore, cats do not perform directed tasks like dogs do and, consequently, their actual usefulness to humans, even as mousers, is debatable (Driscoll et al., 2009b). Hence, there is little reason to believe early agricultural-based communities would have actively sought out and selected the wildcat as a house pet. More likely, wild living cats exploiting human societies were tolerated by people and, over time, gradually diverged from their "wild" relatives (Driscoll et al., 2009a). Thus, whereas adaptation in farmed animals and dogs to human dominion was largely driven by artificial selection, the original domestic cat was a product of natural selection; that is, friendliness and sociability toward humans likely granted cats a survival advantage.

The depiction of pet cats in tomb paintings of the Egyptian New Kingdom suggests that cat domestication dates to at least 3600 BP, but may date back to around 9500 BP in Crete. The genetic evidence for the domestication of cats appears to be most consistent with a single prolonged domestication episode occurring over the broad Near East (Driscoll et al., 2009a).

Cats are the second most popular companion animal and are estimated to reside in 26% of European homes, 29% of Australian homes, and 30.4% of homes in the USA. The total pet cat population is estimated to range from about 3.3 million in Australia, to around 74.1 million in the USA up to 99.2 million in Europe (American Veterinary Medical Association, 2012; Animal Health Alliance Australia, 2013, p. 5; FEDIAF

The European Pet Food Industry, 2014). The American Cat Fanciers Association currently recognizes 41 distinct pedigree breeds for showing; however, by far the most common type of companion cat is the domestic short hair, also known as the "moggie." Rather than a distinct breed, this type of cat is characterized by mixed ancestry, which is thought to confer a number of health benefits as a result of its genetic diversity. The domestic short hair is thought to make up around 90% of the total cat population in the USA and elsewhere.

12.2.4.2 Reasons for Relinquishment to Animal Shelters

Many thousands of cats are relinquished to shelters each year, most of which are euthanized. For example, over 50,000 cats were admitted in 1993 across 16 "open admission" shelters in Massachusetts, USA; of these almost 41,000 (73%) were euthanized. It has been estimated that approximately 1.4 million cats are euthanized in the USA annually (ASPCA, 2016); in the 1990s, euthanasia was reportedly the most common cause of cat deaths (Olson et al., 1991) and may still be the case today (Young, 2013).

The reasons for cat relinquishment are similar to those of other companion animals, with "problem" behaviors, such as "house soiling, incompatibility with other pets, aggressiveness, destructiveness, biting, disobedience, fearful behavior, activeness, and excessive attention seeking" (Wassink-van der Schot et al., 2016) among the most frequently reported. In their survey of Australian cat owners, Wassink-van der Schot et al. (2016) identified several risk factors for problem behaviors in cats, including the breed, age and sex, and socioeconomic factors related to the environment in which they were kept. However, in their study of over 195,000 cat admissions to Australian shelters, Alberthsen et al. (2016) found that most cats were relinquished for owner-related reasons, such as pet restrictions in accommodation, rather than problems associated with cat behavior.

12.2.4.3 Cats in Working Roles

The most prominent working role cats have been utilized for historically, and continue to be used today, is in the control of rodent populations. Conservationists and environmentalists have, however, raised concerns about the effect of cat hunting on nontarget species, such as native animals; mandatory containment in ecologically sensitive areas is increasingly common (Toukhsati et al., 2012).

While not used to the same extent as dogs, cats are also used in a variety of animal-assisted therapy contexts, and are also adopted as "resident" pets in community living arrangements, such as aged care; there is presently a lack of scientific literature to provide an evidence-base for their use in these contexts.

12.2.5 Birds as Companions and in Working Roles

The most commonly kept companion bird species is the parrot and there are estimated to be up to 60 million companion parrots in the United States. The practice of captive parrot breeding, especially in the United States, has seen an exponential increase over the past several decades. The consequences of the rapid growth of the parrot industry have had a number of detrimental effects on wild-living and captive birds (Tweti, 2008). Species within the parrot family range significantly in size and longevity from the relatively small and shorter-lived budgerigars, cockatiels, and lovebirds, to medium-sized conures, amazons, and African grays, to the larger and long-lived cockatoo and macaw species (Engebretson, 2006). The importation of birds into the United States largely began in the 1920s when a young German by the name of Max Stern brought singing canaries from the Hartz Mountains in Germany to sell in New York City. It is said that he singlehandedly popularized bird keeping in the United States. Ten years later, around 800 people in a dozen countries were diagnosed with psittacosis,

commonly referred to as "parrot fever." The disease was transmitted to people by a large importation of birds to the United States from Argentina for the Christmas season. As a result, quarantine laws were tightened (Tweti, 2008).

Following the end of World War II, when air travel became commercialized, exotic birds became available once again and their popularity grew exponentially. In the 1950s bird-admiring hobbyists, typically retirees, began breeding exotic birds in their backyards to supplement their income. Parakeets and finches were popular choices because they were small, easy to handle, colorful, inexpensive to care for and purchase. Over the next 20 years breeders produced millions of birds supplying the marketplace through national chain pet stores. In the early 1970s the United States experienced another significant avian disease outbreak. Velogenic viscerotropic Newcastle disease, with a 95% mortality rate, together with a depopulation strategy by the USA government to control the disease, wiped out approximately 12 million birds in California alone (Tweti, 2008). In 1992 the US government enacted the Wild Bird Conservation Act, closing its borders to the almost half a million wild birds imported annually. The purpose of the Act was to attempt to protect endangered wild bird species populations from depletion due to demand from the pet trade.

12.2.5.1 When the Owner–Companion Bird Relationship Breaks Down

Over the years animal rescue organizations have increased acceptance of animals, beyond that of the typical pet. Nowadays, in addition to cats and dogs, they take in potbelly pigs, big cats (in countries where keeping them is permitted), reptiles, and birds. The most commonly rescued bird is the parrot and parrot rescue organizations continue to be established. The most significant contributing factor to the surrender of parrots is that they are nondomesticated (Engebretson, 2006). Unlike cats and dogs, parrots have not undergone thousands of years of captive breeding for the development of a companion animal with domesticated qualities. The parrot remains a wild animal (Hoppes and Gray, 2010).

Domestic animals have been selectively bred for centuries to enable them to adapt to living in close proximity to humans in captive environments. In parrots, selective breeding to date has focused predominantly on color mutations, not on their suitability as companions (Hoppes and Gray, 2010). In addition, many bird species produced for the pet trade are only a few generations removed from the wild and thus retain most, if not all, of their wild instincts and behaviors (Engebretson, 2006). They are prey animals instinctively observant for predators and changes in their environment. When kept as pets most are caged and their wings are clipped, removing their ability to instinctually fly away when frightened. Many also live with cats and dogs, the very predators they would need to escape in the wild. Parrots are also extremely social flock animals who spend their time foraging for food, building nests, flying, raising young, interacting with flock mates, and keeping watch for predators. The captive parrot shares the desire to perform these same innate behaviors as their wild counterparts; however, their ability to do is extremely limited within the constraints of captivity. Many companion parrots are housed individually, they are caged and cannot fly, they bond to their human caregiver often forming inappropriate pair bond relationships in the absence of conspecifics. In addition, they are noisy, messy, and extremely intelligent, making them relatively high-maintenance pets. These qualities often contrast with the bird owner's expectations of acceptable behavior. Many want a bird that sits quietly, does not bite, and is not messy (Hoppes and Gray, 2010). For these reasons and other ethical considerations, such as depriving parrots their freedom and the ability to perform natural behaviors such as flight, many argue they are not suitable pets. Whereas others disagree referring to the benefits of companion parrots to people (Kidd and Kidd, 1998). However, ethical objections to keeping parrots as companions arise when the benefits to the owner are gained to the detriment of the bird (Engebretson, 2006) (Fig. 12.5).

FIGURE 12.5 **Green Cheeked Conure on owner's shoulder (demonstrating human–bird bond).** *Source: https://commons.wikimedia.org/wiki/File:Pyrrhura_molinae_-pet_on_shoulder-8a.jpg.*

Behavior problems are thought to be common in companion parrots; however, reliable information on this topic is severely lacking in the scientific literature. The most frequent behavior problems include aggression toward the owner, excessive vocalization (screaming), excessive fear, destructive chewing, and feather-picking disorder (Luescher, 2006). Despite the apparent public interest in the fate of unwanted companion birds, involvement from the scientific community to investigate the issue is scarce (Hoppes and Gray, 2010). The issue of captive parrot welfare is ethically significant as the pet parrot industry has grown over the past 2 decades and parrots are now the third most popular pet in the United States (Meehan and Mench, 2006). As their numbers increase, it is becoming evident that ensuring good welfare for companion parrots is a difficult task.

Abnormal and problem behavior is common in companion parrots whose behavioral and environmental needs are not met in captivity. These can include stereotypies, feather picking, bar chewing, pacing, screaming, aggression, fears, and phobias. Research has shown that the development of such problems can be related to insufficient space, lack of environmental complexity, opportunities to forage, and a lack of social interaction with conspecifics (Meehan et al., 2003, 2004). Although the studies suggest that providing a larger cage, environmental enrichment, and socialization can improve the welfare of companion parrots, these changes require an owner with the knowledge, resources, and motivation to make much changes and sustain them throughout their bird's life (Engebretson, 2006). Considering the potential lifespan of common species kept as companion ranges from 15 years (e.g., budgerigar) up to 70+ years for many of the large species (e.g., macaws and cockatoos), problems such as these are often difficult for owners to deal with leading to the bird being neglected, abused, given away, released, or relinquished (Meehan and Mench, 2006) (Figs. 12.6 and 12.7).

12.2.5.2 Birds in Working Roles

The most common uses for birds, other than as a food source, include entertainment, sport (e.g., cock fighting and homing pigeons), as messengers particularly during war (carrier pigeons), and as hunting partners, a practice known as falconry, the hunting of birds or mammals with trained birds of prey (raptors).

The two main objectives to falconry are the supply of food and the sport of watching the prey attempt to escape and the bird of prey chasing it (Prummel, 1997). The bird never becomes tame or domestic, rather it remains a wild animal. A variety of raptor species are utilized in this practice including the goshawk (*Accipiter gentilis*) and sparrowhawk (*Accipiter nisus*), peregrine (*Falco peregrinus*), gyrfalcon (*Falco rusticolus*), and merlin (*Falco columbarius*). Larger birds of prey, such as eagles, are generally considered too heavy to be carried on the hand although they are used by the Kazakh and Kyrgyz people in contemporary Kazakhstan and Kyrgyzstan, as well as diasporas in Bayan-Ölgii, Mongolia and Xinjiang, China (Mayor, 2016).

FIGURE 12.6 **African gray parrot with feather-picking disorder.** *Source: https://commons.wikimedia.org/wiki/File:Feather_picking_african_grey.jpg; https://commons.wikimedia.org/wiki/File:Psittacus_erithacus_-feather_plucking_-pet-6.jpg.*

FIGURE 12.7 **African gray parrot with feather-picking disorder.** *Source: https://commons.wikimedia.org/wiki/File:Feather_picking_african_grey.jpg; https://commons.wikimedia.org/wiki/File:Psittacus_erithacus_-feather_plucking_-pet-6.jpg.*

12.2.6 Other Species Kept as Companions and Used in Working Roles

A number of other mammal species are commonly kept as pets around the world and include domesticated and nondomesticated varieties. These include mice, rats, rabbits, guinea pigs, hamsters, ferrets, hedgehogs, sugar gliders, and other less common pets, often referred to as "exotics."

Although the use of these smaller animals in working roles is uncommon, some, such as the ferret, were originally domesticated because they proved useful to humans. Ferrets were originally kept for their ability to hunt and flush out rabbits from their burrows. More recently, rats have been trained to perform a number of useful tasks including land mine detection in African war zones (Poling et al., 2011).

12.3 COSTS AND BENEFITS OF HUMAN–COMPANION ANIMAL RELATIONSHIPS

Research has identified a number of social, physical, and psychological benefits of human–companion animal relationships. In addition to the benefits experienced by individuals, there are broad societal benefits.

12.3.1 Benefits of Pet Keeping for Humans

12.3.1.1 Economic Benefits of Pets

From an economic perspective, the practice of keeping pets contributes significantly to the world-wide economy every year. It is estimated that the pet industry contributes between US$8 and US$55 billion to the economy in Western countries, including Australia and the United States, annually (Amiot et al., 2016).

12.3.1.2 Health Benefits of Pets

In recent years the human–companion animal bond has attracted much academic research,

leading to discoveries of certain benefits of pets to the psychological and physical well-being of people. Numerous studies have shown that people believe the main advantage of pet ownership is the companionship, friendship, affection, pleasure, and protection they receive (Endenburg et al., 1994; Horn and Meer, 1984; Zasloff and Kidd, 1995), all thought to be important for positive mental health (McNicholas et al., 2005a). Children receive similar psychological benefits from dog ownership including companionship, comfort and support, improved self-esteem, and learning how to care for another (Triebenbacher, 2000). In addition, exposure to dogs during the perinatal period may reduce the development of allergic disease in children (Lodge et al., 2012) and pet ownership in childhood may lead to more positive attitudes toward animals in young adults (Paul and Serpell, 1993).

Researchers have reported an association between companion animal ownership and health benefits. For example, dog owners make fewer annual trips to the doctor, are less likely to be on medication for heart problems (Anderson et al., 1992) and sleeping difficulties (Heady, 1999), and experience a reduction in physiological measures of stress, such as high blood pressure (Allen, 2003; Ramírez and Hernández, 2014). Although the findings are not causal (i.e., healthy people may be more likely to choose to own a pet), the association between general positive health effects and dog ownership has been reported (Serpell, 1991a), along with a potential for pet ownership to reduce national health expenditure, at least in Australia (Heady, 1999). Furthermore, among elderly women and college students living alone, those with a companion animal report less instances of loneliness and depression and are more interested in planning for the future (Goldmeier, 1986; Zasloff and Kidd, 1994). Pets have a calming effect on people with Alzheimer's disease, provide social support for people with limited human social interaction, and are associated with better average indicators of cardiovascular health than nonpet owners (Hart et al., 2006). Some of the most notable research in this field came from two epidemiological studies that found an increased likelihood of surviving a heart attack for people who had a companion dog (Friedmann and Thomas, 1995; Friedmann et al., 1980). The effect was said to match that gained by social support from people.

Pets, especially dogs, also facilitate social interaction with family and friends, as well as facilitating conversations with people outside of the circle of immediate family and friends. A study of people walking their dogs in a park showed that these people received more social approaches and greetings than people without dogs, even those with small children (Messent, 1984). Other studies have found that assistance dogs normalize social interactions for people confined to wheelchairs and for the hearing impaired (Eddy et al., 1988; Hart et al., 1996). The comfort that companion animals provide is the most universally acknowledged benefit of our relationships with them (Hart et al., 2006).

The obesity epidemic is a world-wide health problem for people in modern society placing an immense strain on the healthcare system. This is due, in part, to our increasingly sedentary lifestyle and inadequate exercise. Dogs help combat this problem by motivating many owners to go outdoors and walk their dogs more regularly, with the added benefit of meeting neighbors and friends in the process. Pets are also known to motivate some people to engage in altruistic behaviors, such as visiting the elderly or disabled in nursing homes or partaking in AAT programs in hospitals (Hart et al., 2006). It is important to note, however, that these varied benefits are typically experienced in highly compatible human–animal relationships. Those people who do not enjoy the company of animals, or whose companion animal has severe behavioral problems, are unlikely to experience such benefits.

Despite the number and breadth of such studies, this area of research has itself come under scrutiny and critical analysis (McNicholas et al., 2005b). As Herzog (2011) noted, if participants are not randomly assigned to groups, and, instead, self-select pet ownership, differences between dog owners and nonowners may occur because individuals who perceive themselves to be healthier also choose to have a pet. Despite such concerns, the general consensus that pets are good for certain people remains.

12.3.2 Costs of Pet Keeping for People

Not all studies support the health benefits of pet ownership (Bauman et al., 2001; Stallones et al., 1990). Some of the costs of pet ownership include physical, psychological, and monetary costs, and most pet owners will experience at least one such cost during the course of their pets' lives. For example, pet owners may experience problem behaviors in their pets increasing the likelihood of their pets being surrendered to welfare shelters (Miller et al., 1996; Patronek et al., 1996) and subsequently euthanized. Aggressive behavior, for example, is used as a method of communication and is a normal canine behavior. However, the level of aggression in an individual may be excessive and unacceptable in the dog's immediate environment or dangerous to the general community (Netto and Planta, 1997). Worldwide, dog attacks are a serious public safety issue and can lead to infection, permanent scarring and even death (Blackshaw, 1999).

12.3.2.1 Dog Bite Injury

Perhaps the most significant cost of our close relationship with dogs, at least from a societal point of view, is dog bites. It is estimated that 4.5 million people are bitten by dogs each year, with 19% of these bites being serious enough to require medical attention (Gilchrist et al., 2008). Dog bite injuries often result in pain, infection, emotional distress, and disfiguration, as well as lead to costly health care, such as emergency department visits and hospitalizations (Rhea et al., 2014).

Death due to a dog bite in developed countries is very rare. Available statistics report one death in NZ from 1988 to 2001 (Marsh et al., 2004), an average of 31 deaths per year from 2007 to 2012 in the USA (Weaver, 2014). In the state of New South Wales in Australia, during 2011–12, 5650 attacks by dogs on people and other animals were reported, a 10% increase on the number of reported attacks for the previous year (Cabinet, 2013). Despite these statistics the number of dog-related deaths remains low. In a 2009 study of human fatalities from dog attacks from 1979 through to 2005, the average occurrence of a dog-related death was less than a 1 in 10 million chance of dying (Langley, 2009). Although incidents of dogs attacking people remain relatively infrequent, media around the world focus on news stories involving dog attacks, especially on children. This has created a false impression that dog-related injuries to people have become more frequent and severe than in the past when, in fact, people are more likely to be injured by steps, kitchen utensils, and balloons (Bradley, 2005). Nonetheless, even if dog-related injuries occur infrequently, they remain problematic for the people who are affected and for those dogs involved.

12.3.2.2 Bereavement After the Loss of a Pet

The close relationship many pet owners share with their companion animal can result in significant bereavement experienced when the pet dies. Research has shown that the loss of a much loved companion animal can parallel bereavement for another human (Archer and Winchester, 1994). So strong is some people's attachment to their pets that they experience depression and a significant disruption to their normal functioning following the pet's passing (Sharkin and Knox, 2003). However, the severity of grief is directly correlated to the level of the attachment to the deceased pet (Wrobel and Dye, 2003).

12.3.3 Costs and Benefits of Pet Keeping for the Animals

12.3.3.1 Benefits of Being a Pet

Empirical evidence suggests that humans are not the only ones to benefit from the human–animal relationship and that health benefits may be experienced by companion animals also (Amiot et al., 2016). Through having an owner, pets typically receive food, water, shelter, companionship, and veterinary care when needed. Much loved companion animals owned by financially secure caregivers often receive high-quality food, regular veterinary care, training, and exercise. Physical interactions with owners, such as being petted, have been found to reduce the owner's blood pressure and heart rate, having a calming effect on the animal too (Lynch and McCarthy, 1969; McGreevy et al., 2005). Pets that experience a high level of welfare often live into old age. However, not all pets have it this good.

12.3.3.2 Costs of Being a Pet

Some pets may experience compromised welfare if they are kept in small kennels or cages, mass produced for experimentation (Coppinger and Coppinger, 2002), and more recently, in "puppy farms" to meet the high demand of the pet market. Cats and dogs kept for breeding often live impoverished lives devoid of experience, play, exercise, and socialization. In addition, companion animals that display undesirable behaviors, including aggression toward people, are more likely to be surrendered to welfare shelters (Miller et al., 1996; Patronek et al., 1996) and euthanized as a result.

Most pet animals are kept in isolation or confined spaces, are reliant on their human caregivers for the satisfaction of all of their needs, and must rely on humans, rather than conspecifics, for companionship. This is less problematic for some species, such as dogs who readily accept humans as social partners, than it is for other species, such as parrots. Companion parrots are sensitive, highly social, and active animals whose complex needs are rarely met in a domestic home environment (Luescher, 2006). It is estimated that around 50% of companion parrots are kept in cramped and inadequate conditions. This, together with deprived social interaction for prolonged periods, often result in the development significant stress-related behavior problems.

12.3.4 Successful Human–Companion Animal Relationships

The most successful human–companion animal relationships are based on a respect and understanding of the pet's individual needs and the ability to meet these needs. Successful relationships require a well-informed choice of pet, appropriate match in terms of their pet's personality, exercise and training requirements, and the ability to meet the nutritional, medical, behavioral, and social needs of the individual. Cats and dogs probably account for the majority of successful human–companion animal relationships due to being domesticated and bred, for many generations, to live with people. Our understanding of the needs of these species has also improved over the years thanks to an increase in research being conducted, specifically on their health and behavioral needs. Unfortunately, our understanding of other species, such as companion parrots, lags behind.

When pet owners are able to meet these needs, instances of behavior problems that may place pressure on the relationship are reduced and the pet is more likely to spend its entire life with the original owner.

12.4 THE FUTURE OF HUMAN–COMPANION ANIMAL RELATIONSHIPS

12.4.1 What Will Human–Companion Animal Relationships Look Like in the Future?

As our societies continue to grow, develop, and evolve, our relationships with companion animals will inevitably change. Looking back over the past century provides evidence of this fact and, as such, it is important to think about how our relationships with companion animals is likely to change in the future. Increased urbanization and loss of open spaces in cities throughout the world is likely to have a direct effect on the types of pets we keep and the way

we interact with them. For example, people residing in Australian cities are increasingly preferring smaller dogs in favor of larger breeds, presumably because they are easier to care for in higher density living.

Despite the success of the majority of human–companion animal relationships, there are those that inevitably fail. A breakdown in the bond between owner and pet often results in relinquishment of the pet to a shelter, rescue organization, or veterinarian for euthanasia. The surrender of pets is a complex societal issue resulting in millions of homeless pets world-wide.

12.4.2 Successfully Integrating Companion Animals Into Our Future Societies

Helping modern-day pets thrive within human society is key to securing a healthy and lasting human–companion animal bond into the future. With this in mind, it is critical that governments, policy-makers, urban planners, pet breeders, and pet owners work toward making this a reality. Achieving this goal requires the provision of ongoing public education programs regarding pet husbandry, welfare, behavior, and training as well as ready access to professional help and information to help work through issues that may pose a risk for relinquishment. Ensuring the needs of pets are considered in urban planning and housing design, such as indoor and outdoor spaces for exercise, will help pet owners meet their pets' complex physical and psychological needs, thus reducing or avoiding problem behavior.

It is critical that we work on our understanding of companion animals using the most up to date science so that we can continue to share our lives with them. Pets provide us with many benefits, fulfilling working and companionship roles. As the future unfolds no doubt we will find additional important roles for companion animals to fill. Scientists have made some headway in the last hundred years in our understanding of companion animals; however, there is much more to learn. Now, a new opportunity presents itself to integrate the concepts of wolf and dog behavior, animal welfare science, and new knowledge about cognition and learning allowing us to update our understanding of dogs and other companion animal species and how best to care for them. As new scientific discoveries become more widely disseminated through academic and mainstream media it will be incorporated into the common knowledge of pet keeping (Bradshaw, 2011).

References

Ainsworth, M.D.S., Bowlby, J., 1991. An ethological approach to personality development. Am. Psychol. 46, 333–341.

Alberthsen, C., Rand, J., Morton, J., Bennett, P., Paterson, M., Vankan, D., 2016. Numbers and characteristics of cats admitted to Royal Society for the Prevention of Cruelty to Animals (RSPCA) shelters in Australia and reasons for surrender. Animals 6, 23.

Allen, K., 2003. Are pets a healthy pleasure? The influence of pets on blood pressure. Curr. Dir. Psychol. Sci. 12, 236–239.

American Pet Products Association, 2016. APPA National Pet Owners Survey Statistics: pet ownership and annual expenses. 2015–2016 Survey. Available from: https://www.americanpetproducts.org/pubs_survey.asp.

American Veterinary Medical Association, 2012. U.S. Pet Ownership & Demographics Sourcebook. American Veterinary Medical Association (AVMA), Shaumburg, IL, p. 6.

Amiot, C., Bastian, B., Martens, P., 2016. People and companion animals: it takes two to tango. BioScience 66, 552–560.

Anderson, W., Reid, P., Jennings, G.L., 1992. Pet ownership and risk factors for cardiovascular disease. Med. J. Aust. 157, 298–301.

Animal Health Alliance Australia, 2013. Pet Ownership in Australia. Animal Health Alliance Australia, Canberra.

Archer, J., 1997. Why do people love their pets? Evol. Hum. Behav. 18, 237–259.

Archer, J., Monton, S., 2011. Preferences for infant facial features in pet dogs and cats. Ethology 117, 217–226.

Archer, J., Winchester, G., 1994. Bereavement following death of a pet. Br. J. Psychol. 85, 259–271.

Arnott, E.R., Early, J.B., Wade, C.M., McGreevy, P.D., 2014. Estimating the economic value of Australian stock herding dogs. Anim. Welfare 23, 189–197.

ASPCA, 2016. Shelter Intake and Surrender. Available from: http://www.aspca.org/animal-homelessness/shelter-intake-and-surrender/pet-statistics.

Axelsson, E., Ratnakumar, A., Arendt, M.-L., Maqbool, K., Webster, M.T., Perloski, M., Liberg, O., Arnemo, J., Hedhammar, A., Lindblad Toh, K., 2013. The genomic signature of dog domestication reveals adaptation to a starch-rich diet. Nature 495, 360–365.

Bartosiewicz, L., 1994. Late Neolithic dog exploration: chronology and function. Acta Archeol. Acad. Sci. Hungaricae 46, 59–71.

Batt, S., 2009. Human attitudes towards animals in relation to species similarity to humans: a multivariate approach. Biosci. Horiz. 2, 180–190.

Bauman, A., Russell, S., Furber, S., Dobson, A., 2001. The epidemiology of dog walking: an unmet need for human and canine health. Med. J. Aust. 175, 632–634.

Belyaev, D.K., 1979. Destabilizing selection as a factor in domestication. J. Hered. 70, 301–308.

Blackshaw, J.K., 1999. Meaningful temperament assessment for aggression in dogs—can it be done? In: Proceedings of the Eighth National Conference on Urban Animal Management in Australia Australian Veterinary Association, Gold Coast, Australia, pp. 103–106.

Borgi, M., Cirulli, F., 2013. Children's preferences for infantile features in dogs and cats. Hum. Anim. Interact. Bull. 1, 1–15.

Borgi, M., Cirulli, F., 2016. Pet face: mechanisms underlying human–animal relationships. Front. Psychol. 7, 298.

Borgi, M., Cogliati-Dezza, I., Brelsford, V., Meints, K., Cirulli, F., 2014. Baby schema in human and animal faces induces cuteness perception and gaze allocation in children. Front. Psychol. 5, 414.

Bradley, J., 2005. Dogs Bite But Baloons and Slippers Are More Dangerous. James and Kenneth, Berkley, CA.

Bradshaw, J., 2011. In Defence of Dogs. Penguin, UK.

Branson, N., Cobb, M., McGreevy, P., 2010. Australian Working Dog Survey Report 2009. Australian Government Department of Agriculture, Fisheries and Forestry. Available from: http://www.australiananimalwelfare.com.au/app/webroot/js/kcfinder/upload/files/AW-DIAP_2012_BransonCobbMcGreevy.pdf.

Broom, D.M., 2010. Animal welfare: an aspect of care, sustainability, and food quality required by the public. J. Vet. Med. Educ. 37, 83–88.

Burnham, D., Kitamura, C., Vollmer-Conna, U., 2002. What's new, pussycat? On talking babies and animals. Science 296, 1435.

Cabinet, 2013. Council Reports of Dog Attacks in NSW 2011/2012, p. 17. New South Wales Department of Premier and Cabinet. Available from: http://www.olg.nsw.gov.au/sites/default/files/Annual-reports-of-dog-attacks-in-NSW-2011-12.pdf.

Carlisle-Frank, P., Frank, J.M., 2006. Owners, guardians, and owner–guardians: differing relationships with pets. Anthrozoös 19, 225–242.

Clark, G., 1997. Osteology of the Kuri Maori: the Prehistoric Dog of New Zealand. J. Archaeol. Sci. 24, 113–126.

Clutton-Brock, J., 1995. Origins of the dog: domestication and early history. In: Serpell, J.A. (Ed.), The Domestic Dog: It's Evolution, Behaviour and Interactions with People. Cambridge University Press, Cambridge, pp. 7–20.

Clutton-Brock, J., Hammond, N., 1994. Hot dogs: comestible canids in Preclassic Maya culture at Cuello, Belize. J. Archaeol. Sci. 21, 819–826.

Cobb, M., Branson, N., McGreevy, P., Lill, A., Bennett, P., 2015. The advent of canine performance science: offering a sustainable future for working dogs. Behav. Processes 110, 96–104.

Coppinger, R., Coppinger, L., 2001. Dogs: A Startling New Understanding of Canine Origin, Behaviour and Evolution. Scribner, New York, NY, USA.

Coppinger, R., Coppinger, L., 2002. Dogs: A New Understanding of Canine Origin, Behavior and Evolution. University of Chicago Press, Chicago, IL, USA.

Dodman, N.H., Patronek, G.J., Dodman, V.J., Zelin, M.L., Cottman, N., 1997. Comparison of personality inventories of owners of dogs with and without behavioural problems. Int. J. Appl. Res. Vet. Med. 2, 205–213.

Driscoll, C.A., Clutton-Brock, J., Kitchener, A.C., O'Brien, S.J., 2009a. The taming of the cat. Sci. Am. 300, 68–75.

Driscoll, C.A., Macdonald, D.W., O'Brien, S.J., 2009b. From wild animals to domestic pets, an evolutionary view of domestication. Proc. Natl. Acad. Sci. 106, 9971–9978.

Eddy, J., Hart, L.A., Boltz, R.P., 1988. The effects of service dogs on social acknowledgments of people in wheelchairs. J. Psychol. 122, 39–45.

Endenburg, N., Hart, H., Bouw, J., 1994. Motives for acquiring companion animals. J. Econ. Psychol. 15, 191–206.

Engebretson, M., 2006. The welfare and suitability of parrots as companion animals: a review. Anim. Welfare 15, 263–276.

FEDIAF The European Pet Food Industry, 2014. Fact and Figures 2014. FEDIAF The European Pet Food Industry, Brussels.

Fifield, S.J., Forsyth, D.K., 1999. A pet for the children: factors related to family pet ownership. Anthrozoös 12, 24–32.

Fjellanger, R., Andersen, E., McLean, I., 2002. A training program for filter-search mine detection dogs. Int. J. Comp. Psychol. 15, 277–286.

Fox, M.W., 1978. The Dog: Its Domestication and Behavior. Garland STPM Press, New York, NY.

Franklin, A., 1999. Animals and Modern Cultures: A Sociology of Human–Animal Relations in Modernity. Sage, London.

Friedmann, E., Katcher, A.H., Lynch, J.J., Thomas, S.A., 1980. Animal companions and one-year survival of patients after discharge from a coronary care unit. Public Health Rep. 95, 307.

Friedmann, E., Thomas, S.A., 1995. Pet ownership, social support, and one-year survival after acute myocardial infarction in the Cardiac Arrhythmia Suppression Trial (CAST). Am. J. Cardiol. 76, 1213–1217.

Gazit, I., Terkel, J., 2003. Explosives detection by sniffer dogs following strenuous physical activity. Appl. Anim. Behav. Sci. 81, 149–161.

Gilchrist, J., Sacks, J.J., White, D., Kresnow, M.J., 2008. Dog bites: still a problem? Inj. Prev. 14, 296–301.

Goldmeier, J., 1986. Pets or people: another research note. Gerontologist 26, 203–206.

Hare, B., Plyusnina, I., Ignacio, N., Schepina, O., Stepika, A., Wrangham, R., Trut, L., 2005. Social cognitive evolution in captive foxes is a correlated by-product of experimental domestication. Curr. Biol. 15, 226–230.

Hare, B., Wobber, V., Wrangham, R., 2012. The self-domestication hypothesis: evolution of bonobo psychology is due to selection against aggression. Anim. Behav. 83, 573–585.

Hart, B.L., Hart, L.A., Bain, M.J., 2006. Canine and Feline Behavioral Therapy. Blackwell, Oxford, UK.

Hart, L.A., Zasloff, R.L., Benfatto, A.M., 1996. The socializing role of hearing dogs. Appl. Anim. Behav. Sci. 47, 7–15.

Heady, B., 1999. Health benefits and health cost savings due to pets: preliminary estimates from an Australian national survey. Soc. Res. Indic. 47, 233–243.

Herzog, H., 2001. Some We Love, Some We Hate, Some We Eat: Why It's So Hard to Think Straight About Animals. Harper Perennial, New York, NY.

Herzog, H., 2011. The impact of pets on human health and psychological well-being: fact, fiction, or hypothesis? Curr. Dir. Psychol. Sci. 20, 236–239.

Hoppes, S., Gray, P., 2010. Parrot rescue organizations and sanctuaries: a growing presence in 2010. J. Exot. Pet Med. 19, 133–139.

Horn, J., Meer, J., 1984. The pleasure of their company. Psychol. Today 18, 52–58.

Hurn, S., 2012. Humans and Other Animals: Cross Cultural Perspectives on Human–Animal Interactions. Pluto Press, London.

Jagoe, A., Serpell, J., 1996. Owner characteristics and interactions and the prevalence of canine behaviour problems. Appl. Anim. Behav. Sci. 47, 31–42.

Kidd, A.H., Kidd, R.M., 1998. Problems and benefits of bird ownership. Psychol. Rep. 83, 131–138.

King, T., Marston, L.C., Bennett, P.C., 2012. Breeding dogs for beauty and behaviour: why scientists need to do more to develop valid and reliable behaviour assessments for dogs kept as companions. Appl. Anim. Behav. Sci. 137, 1–12.

Knight, A., 2008. Bats, snakes and spiders, Oh my! How aesthetic and negativistic attitudes, and other concepts predict support for species protection. J. Environ. Psychol. 28, 94–103.

Kobelt, A.J., Hemsworth, P.H., Barnett, J.L., Coleman, G.J., 2003. A survey of dog ownership in suburban Australia—conditions and behaviour problems. Appl. Anim. Behav. Sci. 82, 137–148.

Langley, R.L., 2009. Human fatalities resulting from dog attacks in the United States, 1979–2005. Wilderness Environ. Med. 20, 19–25.

Ledger, R.A., Baxter, M.R., 1997. The development of a validated test to assess the temperament of dogs in a rescue shelter. In: Mills, D.S., Heath, S.E., Harrington, L.J. (Eds.), In: Proceedings of the First International Conference on Veterinary Behavioural Medicine. Universities Federation for Animal Welfare, Birmingham, UK, pp. 87–92.

Lenzi, D., Trentini, C., Tambelli, R., Pantano, P., 2015. Neural basis of attachment-caregiving systems interaction: insights from neuroimaging studies. Front. Psychol. 6, 1246.

Lindsay, S.R., 2000. Handbook of Applied Dog Behavior and Training, Vol. 1: Adaptation and Learning. Iowa State Press, Ames, IA.

Little, A., 2012. Manipulation of infant-like traits affects perceived cuteness of infant, adult and cat faces. Ethology 118, 775–782.

Lodge, C.J., Lowe, A.J., Gurrin, L.C., Matheson, M.C., Balloch, A., Axelrad, C., Hill, D.J., Hosking, C.S., Rodrigues, S., Svanes, C., Abramson, M.J., Allen, K.J., Dharmage, S.C., 2012. Pets at birth do not increase allergic disease in at-risk children. Clin. Exp. Allergy 42, 1377–1385.

Lorenz, K., 1942. Die angeborenen Formen menschlicher Erfahrung. Z. Tierpsychol. 5, 43.

Luescher, A.U., 2006. Manual of Parrot Behavior. Wiley-Blackwell, Ames, Iowa, USA.

Luo, L.Z., Li, H., Lee, K., 2011. Are childrens faces really more appealing than those of adults? Testing the baby schema hypothesis beyond infancy. J. Exp. Child. Psychol. 110, 115–124.

Lynch, J.J., McCarthy, J.F., 1969. Social responding in dogs: heart rate changes to a person. Psychophysiology 5, 389–393.

Marsh, L., Langley, J., Gauld, R., 2004. Dog bite injuries. N. Z. Med. J. 117, U1043.

Marston, L.C., Bennett, P.C., 2003. Reforging the bond—towards successful canine adoption. Appl. Anim. Behav. Sci. 83, 227–245.

Marston, L.C., Bennett, P.C., Coleman, G.J., 2004. What happens to shelter dogs? An analysis of data for 1 year from three Australian shelters. J. Appl. Anim. Welfare Sci. 7, 27–47.

Mayor, A., 2016. The Eagle Huntress Ancient Traditions and New Generations, p. 4. Available from: https://web.stanford.edu/dept/HPS/EagleHuntress2016long.pdf.

McGreevy, P.D., Righetti, J., Thompson, P., 2005. The reinforcing value of physical contact and the effect on canine heart rate of grooming on different anatomical areas. Anthrozoos 18, 236–244.

McNicholas, J., Gilbey, A., Rennie, A., Ahmedzai, A., Dono, J., Ormerod, E., 2005a. Pet ownership and human health: a brief review of evidence and issues. Br. Med. J. 331, 1252–1254.

McNicholas, J., Gilbey, A., Rennie, A., Ahmedzai, S., Dono, J.-A., Ormerod, E., 2005b. Pet ownership and human health: a brief review of evidence and issues. Br. Med. J. 7527, 1252–1253.

Meehan, C., Garner, J., Mench, J., 2004. Environmental enrichment and development of cage stereotypy in Orange-winged Amazon parrots (Amazona amazonica). Dev. Psychobiol. 44, 209–218.

Meehan, C., Mench, J., 2006. Captive parrot welfare. In: Luescher, A.U. (Ed.), Manual of Parrot Behaviour. Blackwell, Ames, IA, pp. 301–318.

Meehan, C., Millam, J., Mench, J., 2003. Foraging opportunity and increased physical complexity both prevent and reduce psychogenic feather picking by young Amazon parrots. Appl. Anim. Behav. Sci. 80, 71–85.

Meggitt, M.J., 1961. The Association Between Australian Aborigines and Dingoes. University of Sydney, Sydney.

Messent, P., 1984. Correlates and effects of pet ownership. In: Anderson, R.K., Hart, B.L., Hart, L.A. (Eds.), The Pet Connection: Its Influence on Our Health and Quality of Life. Center to Study Human-Animal Relationships and Environments, University of Minnesota, Minneapolis, MN, pp. 331–340.

Miklósi, Á., 2014. Dog Behaviour, Evolution, and Cognition. OUP, Oxford.

Miller, D.D., Staats, S.R., Partlo, C., Rada, K., 1996. Factors associated with the decision to surrender a pet to an animal shelter. J. Am. Vet. Assoc. 209, 738–742.

Moore, G.E., Burkman, K.D., Carter, M.N., Peterson, M.R., 2001. Causes of death or reasons for euthanasia in military working dogs: 927 cases (1993–1996). J. Am. Vet. Med. Assoc. 219, 209–214.

Morey, D.F., 1992. Size, shape and development in the evolution of the domestic dog. J. Archaeol. Sci. 19, 181–204.

Mormann, F., Dubois, J., Kornblith, S., Milosavljevic, M., Cerf, M., Ison, M., Fried, I., 2011. A category-specific response to animals in the right human amygdala. Nat. Neurosci. 14, 1247–1249.

Muszkat, M., de Mello, C.B., Muñoz, P.D.O.L., Lucci, T.K., David, V.F., de Oliveira Siqueira, J., Otta, E., 2015. Face scanning in autism spectrum disorder and attention deficit/hyperactivity disorder: human versus dog face scanning. Front. Psychiatry 6, 150.

Netto, W.J., Planta, D.J., 1997. Behavioural testing for aggression in the domestic dog. Appl. Anim. Behav. Sci. 52, 243–263.

New, J., Cosmides, L., Tooby, J., 2007. Category-specific attention for animals reflects ancestral priorities, not expertise. Proc. Natl. Acad. Sci. 104, 16598–16603.

Nimer, J., Lundahl, B., 2007. Animal-assisted therapy: a meta-analysis. Anthrozoös 20, 225–238.

O'Farrell, V., 1995. Effects of owner personality and attitudes on dog behaviour. In: Serpell, J.A. (Ed.), The Domestic Dog: Its Evolution, Behaviour and Interactions with People. Cambridge University Press, Cambridge, pp. 153–160.

O'Haire, M.E., McKenzie, S.J., Beck, A.M., Slaughter, V., 2013. Social behaviors increase in children with autism in the presence of animals compared to toys. PLoS One 8, e57010.

Olson, R.N., Moulton, C., Nett, T.M., Salman, M.D., 1991. Pet overpopulation: a challenge for companion animal veterinarians in the 1990s. J. Am. Vet. Med. Assoc. 198, 1151–1152.

Pang, J., Kluetsch, C., Zou, X.-J., Zhang, A., Luo, L.-Y., Angleby, H., Ardalan, A., Ekström, C., Sköllermo, A., Lundeberg, J., Matsumura, S., Leitner, T., Zhang, Y.-P., Savolainen, P., 2009. mtDNA data indicate a single origin for dogs south of Yangtze River, less than 16,300 years ago, from numerous wolves. Mol. Biol. Evol. 26, 2849–2864.

Patronek, G.J., Glickman, L.T., Beck, A.M., McCabe, G.P., Ecker, C., 1996. Risk factors for relinquishment of dogs to an animal shelter. J. Am. Vet. Med. Assoc. 209, 572–581.

Paul, E.S., Serpell, J.A., 1993. Childhood pet keeping and humane attitudes in young adulthood. Anim. Welfare 2, 321–337.

Payne, E., Bennett, P.C., McGreevy, P.D., 2015. Current perspectives on attachment and bonding in the dog-human dyad. J. Psychol. Res. Behav. Manage. 8, 71–79.

Pet Food Manufacturers Assocation, 2014. Annual Report 2014, TNS, United Kingdom.

Pet Food Manufacturing Association, 2016. Annual Report 2016, Pet Food Manufacturing Association, UK.

Pickel, D., Manucy, G.P., Walker, D.B., Hall, S.B., Walker, J.C., 2004. Evidence for canine olfactory detection of melanoma. Appl. Anim. Behav. Sci. 89, 107–116.

Podberscek, A.L., Serpell, J.A., 1997. Aggressive behaviour in English cocker spaniels and the personality of their owners. Vet. Rec. 141, 73–76.

Poling, A., Weetjens, B., Cox, C., Beyene, N.W., Bach, H., Sully, A., 2011. Using trained pouched rats to detect land mines: another victory for operant conditioning. J. Appl. Behav. Anal. 44, 351–355.

Prummel, W., 1997. Evidence of hawking (falconry) from bird and mammal bones. Int. J. Osteoarchaeol. 7, 333–338.

Ramírez, M.T.G., Hernández, R.L., 2014. Benefits of dog ownership: comparative study of equivalent samples. J. Vet. Behav. 9, 311–315.

Rhea, S., Weber, D.J., Poole, C., Cairns, C., 2014. Risk factors for hospitalization after dog bite injury: a case-cohort study of emergency department visits. Acad. Emerg. Med. 21, 196–203.

Sanders, C.R., 2000. The impact of guide dogs on the identity of people with visual impairments. Anthrozoös 13, 131–139.

Scarlett, J.M., Salmon, M.D., New, J.G.J., Kass, P.H., 1999. Reasons for relinquishment of companion animals in U.S. animal shelters: selected health and personal issues. J. Appl. Anim. Welfare Sci. 2, 41–57.

Serpell, J., 1991a. Beneficial effects of pet ownership on some aspects of human health and behaviour. J. R. Soc. Med. 84, 717–720.

Serpell, J.A., 1991b. Beneficial effects of pet ownership on some aspects of human health and behaviour. J. R. Soc. Med. 84, 717–720.

Serpell, J.A., 2002. Guardian spirits or demonic pets: the concept of the witch's familiar in early modern England, 1530–1712. In: Creager, A.N.H., Jordan, W.C. (Eds.), The Animal/Human Boundary: Historical Perspectives. University of Rochester Press, Rochester, NY, pp. 157–190.

Serpell, J.A., 2004. Factors influencing human attitudes to animals and their welfare. Anim. Welfare 13, 145–151.

Sharkin, B.S., Knox, D., 2003. Pet loss: Issues and implications for the psychologist. Prof. Psychol. Res. Pr. 34, 414.

Spedding, C., 1995. Sustainability in animal production systems. Anim. Sci. 61, 1–8.

Stallones, L., Marx, M.B., Garrity, T.F., Johnson, T.P., 1990. Pet ownership and attachment in relation to the health of U.S. adults, 21 to 64 years of age. Anthrozoos 2, 118–124.

Stephen, J., Ledger, R., 2006. Relinquishing dog owners' ability to predict behavioural problems in shelter dogs post adoption. Appl. Anim. Behav. Sci. 107, 88–99.

Stoeckel, L.E., Palley, L.S., Gollub, R.L., Niemi, S.M., Evins, A.E., 2014. Patterns of brain activation when mothers view their own child and dog: an fMRI study. PLoS One 9, e107205.

Taçon, P.S., Pardoe, C., 2002. Dogs Make Us Human. Nature Australia.

Thomas, K., 1983. Man and the Natural World. Pantheon, London.

Tisdell, C., Wilson, C., Swarna Nantha, H., 2006. Public choice of species for the 'Ark': phylogenetic similarity and preferred wildlife species for survival. J. Nat. Conserv. 14, 97–105.

Toukhsati, S.R., Young, E., Bennett, P.C., Coleman, G.J., 2012. Wandering cats: attitudes and behaviors towards cat containment in Australia. Anthrozoös 25, 61–74.

Triebenbacher, S.L., 2000. The companion animal within the family system: the manner in which animals enhance life within the home. In: Fine, A. (Ed.), Handbook on Animal-Assisted Therapy: Theoretical Foundations and Guidelines for Practice. Academic Press, San Diego, CA.

Tweti, M., 2008. Of Parrots and People: The Sometimes Funny, Always Fascinating, and Often Catastrophic Collision of Two Intelligent Species. Penguin, New York, NY.

van der Borg, J.A.M., Netto, W.J., Planta, D.J.U., 1991. Behavioural testing of dogs in animal shelters to predict problem behaviour. Appl. Anim. Behav. Sci. 32, 237–251.

von Holdt, B.M., Pollinger, J.P., Lohmueller, K.E., Han, E., Parker, H.G., Quignon, P., Degenhardt, J.D., Boyko, A.R., Earl, D.A., Auton, A., Reynolds, A., Bryc, K., Brisbin, A., Knowles, J.C., Mosher, D.S., Spady, T.C., Elkahloun, A., Geffen, E., Pilot, M., Jedrzejewski, W., Greco, C., Randi, E., Bannasch, D., Wilton, A., Shearman, J., Musiani, M., Cargill, M., Jones, P.G., Qian, Z., Huang, W., Ding, Z.-L., Zhang, Y., Bustamante, C.D., Ostrander, E.A., Novembre, J., Wayne, R.K., 2010. Genome-wide SNP and haplotype analyses reveal a rich history underlying dog domestication. Nature 464, 898–902.

Wassink-van der Schot, A.A., Day, C., Morton, J.M., Rand, J., Phillips, C.J., 2016. Risk factors for behavior problems in cats presented to an Australian companion animal behavior clinic. J. Vet. Behav. Clin. Appl. Res. 14, 34–40.

Weaver, S., 2014. Dangerous dog encounters: best practices for police officers, threat assessment, and use of force. J. Law Enforcement 3, ISSN 2161-0231.

Wells, D.L., 1996. The welfare of dogs in an animal rescue shelter. PhD thesis, School of Psychology, The Queens University of Belfast, UK.

Wells, D.L., Hepper, P.G., 2000. Prevalence of behaviour problems reported by owners of dogs purchased from an animal rescue shelter. Appl. Anim. Behav. Sci. 69, 55–65.

Wilson, E.O., 1984. Biophilia. Harvard University Press, Chicago, IL.

Wirth, K.E., Rein, D.B., 2008. The economic costs and benefits of dog guides for the blind. Ophthal. Epidemiol. 15, 92–98.

Wrobel, T.A., Dye, A.L., 2003. Grieving pet death: normative, gender, and attachment issues. OMEGA J. Death Dying 47, 385–393.

Young, J., 2013. The controversy is over: prepubertal neutering is the surgery of choice. In: International Seminar for the Solution of the Overpopulation of Dogs and Cats, October 19, 2013, Portugal.

Zasloff, R.L., Kidd, A.H., 1994. Loneliness and pet ownership among single women. Psychol. Rep. 75, 747–752.

Zasloff, R.L., Kidd, A., 1995. Views of pets in the general population. Psychol. Rep. 76, 1166.

Animals in Medicine and Research

Tiffani J. Howell
La Trobe University, Bendigo, VIC, Australia

13.1 INTRODUCTION: WHY USE ANIMALS AT ALL?

The phrase "animal testing" can conjure strong feelings in people opposed to or in support of such practices. Those in opposition, such as animal rights organizations, point to video footage of laboratory animals in appalling conditions to convince others that all animal testing is bad (PETA Australia, n.d.). While many animals have undoubtedly suffered in the name of scientific or medical advances for the benefit of humans, those in support of animal testing argue that modern-day regulations protect the welfare of animals to ensure that their use meets community standards for ethical treatment.

These regulations are based on laws such as the Animal Welfare Act (1966) in the United States; the Animal Research Act 1985 in New South Wales, Australia; and the Animals (Scientific Procedures) Act 1986 in the United Kingdom. In Western societies, these laws are based on the Judeo–Christian principle that animal use for the benefit of humans is moral and ethical.

Animal welfare regulations in other parts of the world may vary according to religious and cultural practices. This chapter will focus primarily on laws and regulations that govern animal use in science in Western countries.

Animal welfare codes maintain stringent requirements aimed to ensure that animals do not experience any negative welfare state beyond that which is considered absolutely necessary to the research (Home Office, 2014; National Research Council, 2011; NHMRC, 2013). This applies not only to the experiment itself, but also to housing in laboratory conditions, so that all laboratory animals must experience a positive overall welfare state even when they are not being used for a study. Furthermore, some methods used to study animals for scientific knowledge involve only observations of the animals in their natural habitat. These methods are also governed by animal ethics codes if there is any possibility that the research design or the presence of the human experimenter(s) could negatively impact the animal's welfare.

Any researcher who intends to use animal subjects in a study must request approval from their relevant animal ethics committee, such as their university or research institute, before commencing the research. This committee relies on the existing animal welfare laws to determine whether a proposed project will protect the animal's welfare to the fullest extent possible. Membership of the committee comprises not only fellow scientists, but also members of the lay community. This helps ensure that the ethics approval process is not too insular, but rather that it reflects the values and perspectives of society at large.

13.2 HISTORY OF THE USE OF ANIMALS IN SCIENCE AND MEDICINE

Animals have been used in science and medicine for many centuries. In ancient Greece, animals were used in anatomical studies designed to explain the human body (Baumans, 2004). Then there was a long pause in animal studies due to lack of experimental research in general, but animal research began again in the 1600s. During this time, using animals for these purposes was considered morally acceptable because experimenters worked under a Cartesian philosophical framework, which posited that animals could not think, and therefore could not feel pain. However, in the late 1700s, a philosopher named Jeremy Bentham developed a new view, suggesting that animals could suffer, and that this suffering should be an important consideration in deciding whether they are to be used to benefit humans (Baumans, 2004). This approach, called *utilitarianism*, is the basis for many of the laws that govern the use of animals in research, as it focuses on weighing the potential benefits of their use against the suffering caused to the animal subjects (Bentham, 1977; Burns, 2005).

Utilitarianism Versus Rights-Based Approaches to the Use of Animals

There are two dominant ethical perspectives underpinning opinions and attitudes toward the use of animals: utilitarianism and rights (see also Chapter 30). Individuals who take a utilitarian approach tend to agree that the use of animals to benefit humans is moral and ethical in certain circumstances, such as research, medicine, and food production. However, this approval is contingent upon the animal not being subjected to undue suffering, and the level of suffering that is accepted depends on the aim of the research and the species being studied. For instance, people with a utilitarian perspective have a higher acceptance for animal use to help find treatments for human diseases, but less for product testing (Driscoll, 1992). The species also plays a role, with people expecting a higher standard of welfare for animals that are believed to be more "like us" (e.g., primates) than animals such as rats and worms (Davis and Cheeke, 1998; Eddy et al., 1993; Rasmussen et al., 1993). Furthermore, these perceptions may change over time, and in the future perhaps welfare standards will be higher, or lower, than they are currently.

Rights-based perspectives, on the other hand, disapprove of the use of animals for any purpose whatsoever. The belief underlying this approach is that all lives are equally valuable, so harming or killing an animal, even if it could ultimately save a human life, is not acceptable or justifiable (Regan, 1987).

Animal advocacy organizations tend to fall into one of these two camps. For instance, the Royal Society for the Prevention of Cruelty to Animals (RSPCA, or ASPCA in the United States) and World Animal Protection are utilitarian. These organizations accept that animals may be used for human benefit, but they work to ensure that animals in human care experience a welfare state that is as positive as possible (RSPCA Australia, n.d.; World Animal Protection, n.d.). People for the Ethical Treatment of Animals (PETA), on the other hand, is more rights-based. This organization disagrees with using animals for human benefit; they encourage adoption of a vegan diet and disapprove of animals in research (PETA, n.d.).

Animal research continued to increase in popularity after Darwin published his famous work, *On the Origin of Species* (Darwin, 1859), as he successfully illustrated that humans and other animals were biologically similar, making them well suited to research attempting to understand human biological processes (Baumans, 2004). In the late 1800s, this increase continued with the advent of microbiology, which infected healthy animals with pathogens to study how they might potentially affect humans, and continued into the 1980s with the growth of pharmacology, immunology, and toxicology. For example, in the 1930s, there were fewer than 1 million studies published using animals, which increased to nearly 6 million by 1970, and then dropped by nearly half, to just over 3 million, in 1990 (Baumans, 2004), perhaps out of concern about the ethics and welfare of animals in research, which began in the 1970s (Singer, 1977).

More recently, the emergence of genetically modified animals has seen a resurgence in animals used for research purposes. In 2004, there were between 75 million and 100 million vertebrates used in experimental studies (Baumans, 2004). This does not include invertebrate animals used in laboratories (with the exception of cephalopods, such as octopuses and squid), as they are generally not defined as "animals" under the legal definition and are not, therefore, protected by animal welfare regulations. Due to the lack of reporting requirements for work with invertebrate species, it is not possible to estimate, for example, how many fruit flies or nematodes are used annually in laboratory studies. Even though current laws mandate that the welfare of experimental animals be a primary consideration, the ubiquity of animals in research contexts suggests that a large number of scientists believe that their use is justifiable.

13.2.1 Potential Benefits for Humans and Animals

The most obvious benefit of animal research to humans is in advances to health management. This is because animals have traditionally served as models of human biology, meaning that researchers study animal behavior and biology to better understand how the same processes might occur in humans (Fields and Johnston, 2005). Vaccines, drugs, and new medical technologies are first trialed on animal models before tests begin on human subjects. Findings from these studies have advanced knowledge about cell structures and life cycles, and entire genomes for several species have been sequenced (Fields and Johnston, 2005).

Veterinary research with animals has also resulted in huge advances that improve the quality of life of pet animals and livestock animals. Another benefit of research with animals is in environmental conservation, with studies estimating the relative abundance of a particular species in an area helping to inform policy on whether that area should be protected from human development. While there is an ethical and philosophical debate about whether sacrificing an individual animal's welfare is justifiable based on the potential benefit to a large number of other animals (Fraser, 1999), the results of research studies with animals can offer benefits to animals, as well as to humans.

13.2.2 Species Commonly Used in Research

Of the estimated 75–100 million vertebrate animals in research laboratories around the world, the vast majority are rats and mice, which make up over three-quarters of all laboratory animals (mice, 44%; rats, 33%), with birds making up another 10% (Baumans, 2004). Fish are less popular, at 7%, and guinea pigs and rabbits are still less common, at 2% and 1%, respectively.

The wide use of small mammals, such as rats and mice, is based on a number of factors that make them ideal laboratory animals. For instance, they all have a short estrus cycle and produce large litters, enabling researchers to access large numbers of these animals. They also have a short life span. This makes it possible for researchers to study the effects of specific treatments or genetic alterations over the animal's entire life in just a few years, a process that takes many decades in humans. Finally, they are able to cope with life in captivity, reproducing as they would in the wild.

An animal that was unable to survive and reproduce in captivity would not be a suitable laboratory animal. However, the concept of coping well in captivity, at least in previous years, really just meant that animals were capable of breeding and typically did not become so stressed that they died (Barnett and Hemsworth, 1990). This is arguably a low bar for meeting an animal's welfare needs. In fact, even animals that do not die due to the stressors inherent in laboratories are often stressed by routine, noninvasive experiences, such as handling by humans working in the laboratory, or having their cage cleaned (Balcombe et al., 2004). Contemporary minimum animal welfare standards incorporate behavioral measures of stress, such as stereotypies and other abnormal behaviors (Barnett and Hemsworth, 1990).

Observational studies of animals in their natural environment can incorporate much more species diversity than laboratory studies for two reasons. First, there are many animal species that would not be well-suited for life in a laboratory. Large predators, such as bears, could endanger

Phylogenetic Similarity to Humans and Welfare Standards

In addition to small mammals like rodents, other animals are also kept in laboratories, such as cats, dogs, and primates. These animal types present a unique challenge because public license to use animals depends on establishing a higher welfare standard for animals that are more phylogenetically similar to humans (Hagelin et al., 2003). Therefore, in Western countries, the public expectation is that a dog should be treated better than a laboratory rat—or, at least, the community generally accepts a lesser welfare experience for rats than dogs, even if we might prefer that all animals be treated well—and welfare standards for chimpanzees are the highest of any laboratory animal (Gluck, 2014). In fact, the National Institutes of Health in the United States has recently ceased using any chimpanzees in the research studies that it funds (Reardon, 2015), which, over time, will probably reduce the number of chimpanzees used in medical studies anywhere to nil. This outcome reflects the changing perspective within science of the need for these types of animals, with the US Institute of Medicine arguing in 2011 that chimpanzees should only be used if the research could not be done with any other animal (Altevogt et al., 2011).

the safety of the researchers. Endangered species should not be used because every attempt should be made to conserve the remaining individuals of those species, ideally in their natural habitat. Indeed, captive chimpanzees were recently declared an endangered species, and updated regulations now require that researchers obtain a permit from the US Fish and Wildlife Service to research endangered species (Kaiser, 2015). To obtain such a permit, the researchers would need to demonstrate that their research will benefit chimpanzees in the wild (Kaiser, 2015). There are also animals that do not cope well in captivity, becoming distressed or dying; notwithstanding the ethical issues of keeping a stressed animal in captivity, the results of research studies that use such animals may be confounded by fear or pain (Balcombe et al., 2004).

Second, observational studies permit zoologists and biologists to examine the interplay between one species, that species' environment, and other species inhabiting that species' niche (Beier and Noss, 1998). For instance, a zoologist studying tiger behavior will need to observe the behavior of the tigers during interactions with other tigers and other types of animals, as well as how the tiger's environment may affect its behaviors. Therefore, even though the target species may be the tiger, it may be necessary to consider its prey species, and competing predators, living in the same habitat.

13.2.3 Summary

Animals have been used for research purposes for many centuries, with the levels of their use rising and falling based on the research interests and methods of the day. There are many potential benefits to the use of animals in science, with advances in human health being the most obvious advantage. Most animals used in research studies are rodents, but birds, fish, and other mammals are also used sometimes.

13.3 ADVANTAGES AND DISADVANTAGES OF USING LABORATORY ANIMALS

Most members of the general public approve of the use of animals in research, although this level of support often depends on the purpose of the study and the amount of pain inflicted on the animal (Lund et al., 2014). However, the benefits conferred to humans (and, sometimes, to other animals) may come at the cost of the individual animal's welfare, and possibly even its life.

Therefore, the utilitarian framework that underpins laboratory animal welfare regulations requires that the potential benefits must outweigh the negative consequences. This section will discuss some of the advantages and disadvantages of using laboratory animals in scientific studies, and how researchers and policy makers work to try to reduce those disadvantages while enhancing the advantages.

13.3.1 Advantages

The biggest advantage of using laboratory animals in research is the high level of scientific control that they afford researchers. For example, it is possible for research teams to completely control the environment of the animals they work with in a way that would be inconceivable in any other context. This permits researchers to systematically test the effects of an intervention one variable at a time, while ruling out the possibility that factors related to variations in the subject's environment may explain any observed change during the study. Subject to the requirements of animal ethics committees, research scientists are able to decide the size and shape of each individual animal's cage; they can decide to let the animals live alone, or with one other animal of the same or a different species, or with a whole group of animals. They also have control over the ambient temperature, humidity level, and lighting. This can permit researchers to either maintain an ideal environment for the animals, or to manipulate aspects of the environment to determine whether they produce any effects on the animals (Benefiel and Greenough, 1998).

Scientific Control

The extent to which all potentially confounding factors, such as environmental factors, are held constant in a scientific experiment, enable causal conclusions to be drawn about the effect of the independent variable (e.g., an intervention or treatment).

For example, if rats are housed in a variety of uncontrolled settings (e.g., with inconsistent access to food and environmental enrichment), it may not be possible to conclude that changes in the dependent variable (e.g., improvement in blood pressure) are attributable to the independent variable (e.g., blood pressure medication) or due to enhancing aspects of their housing. High-quality science seeks to minimize the effects of confounding factors to increase certainty that any changes observed in the dependent variable following an experiment are actually a consequence of the independent variable under examination.

FIGURE 13.1 **The Zucker fatty rat has a genetic predisposition toward obesity, thus permitting researchers to obtain insights into obesity's effects on health on a shorter timescale than would be possible if they were researching humans alone.** *Source: Photo courtesy Joanna Servaes, licensed under the Creative Commons Attribution-Share Alike 3.0 Unported license.*

Researchers decide what to feed each animal, and how much to feed them, enabling studies into the effects of certain types of foods, or food additives, on the health of the animal (Li et al., 1993). They can provide ample environmental enrichment in the form of toys and other objects to interact with, or they can keep the cage completely barren. Finally, they can control exposure to stimuli. In one study, kittens raised in an environment with moving stripes that were not contingent on their body movements experienced a permanent impairment in motion perception (Held and Hein, 1963). This level of environmental control would simply not be possible outside of a laboratory setting, making it more difficult to ensure that environmental variation did not impact the results of the study.

The second major advantage of laboratory animal use is control of their genetics. It is possible to manipulate the genes of laboratory animals to be more inclined toward development of many health afflictions common in humans. Some laboratory rats rapidly develop tumors (Barth, 1998), Zucker fatty rats are genetically prone to obesity (Kurtz et al., 1989) as shown in Fig. 13.1, and Wistar rats develop diabetes

(Nakhooda et al., 1977). This is important because most laboratory animals, such as rats and mice, have a much shorter life span than humans. Therefore, researchers are able to produce animals that are likely to develop the illness that they are studying, and to measure the effects of certain treatments or lifestyles on that illness, in a matter of months or a few years. It is not ethical to manipulate human genes or behavior to encourage disease development; researchers are reliant on correlational research to determine the effect of genes, behavior, or lifestyle factors on the development of diseases.

Correlational Studies

Correlational research identifies relationships between variables. It does not actually manipulate variables experimentally, which is generally considered requisite for determining cause and effect. This means that it is possible to determine that there is a relationship between, for example, smoking and lung cancer; however, it generally makes no claim as to whether smoking actually *causes* lung cancer. This is because there may be alternate explanations, such as socioeconomic status and lifestyle factors, that could explain this relationship. Long-term, epidemiological studies that take into account many contributing factors can enable greater certainty, but it is useful to remember the old adage that "correlation does not equal causation." Well-controlled, experimental studies that manipulate one variable at a time are used to establish causal relationships. In this way, laboratory studies exposing one group of rats to cigarette smoke, and measuring the incidence of lung cancer in those rats compared to a group that was not exposed to cigarette smoke, allows causal conclusions to be reached.

Phenotype Versus Genotype

An organism's "genotype" refers to its genetic composition. In contrast, an organisms "phenotype" refers to directly observable characteristics, such as physical features and behaviors that arise as a consequence of the interaction between the genotype and the environment.

The third major advantage of research with laboratory animals is that it is possible to do *invasive* studies that would not be ethical with humans. For example, single-neuron electro-encephalography (EEG) recordings in animals, using electrodes implanted inside the brain itself and in a controlled environment, have increased our understanding of how the brain processes incoming stimulus information (Girman et al., 1999). Conversely, noninvasive electrodes that record from the scalp are commonly used in human research (American

Electroencephalographic Society, 1986), but do not provide the level of detail about individual neurons that a single-cell recording can offer.

Another example of a research study that could only be done on a laboratory animal, for ethical reasons, is the famous rat upon whose back grew a human ear (Cao et al., 1997). This was one of the earliest examples of laboratory-grown organs, which could potentially be useful for humans who need transplants in the future.

13.3.2 Disadvantages

The advantages of laboratory animal use make them attractive candidates for many types of research studies; however, there are disadvantages to their use as well. The main disadvantage is possible compromise of the animal's welfare, and this is the issue from which virtually all other disadvantages stem. For example, the precise environmental control that laboratory studies permit may negatively impact on an animal's welfare, such as if it means that their enclosure is devoid of any environmental enrichment or other stimulation. As such controls are strictly governed by animal welfare regulations, which are enforced by animal ethics committees to ensure that all aspects of the study are well justified by the researcher and satisfy the "Three Rs" (Section 13.3.3).

Another disadvantage of using laboratory animals housed in highly controlled environments is that the outcomes may not be "generalizable" to the "real world," meaning that what is observed in a laboratory may not be observed in less controlled conditions. The "reliability" of outcomes, that is, the extent to which a result can be repeated, is related to the concept of generalizability, that is, whether the same outcome applies in a different population of animals and in a different place. However, even small changes in the environment in which laboratory animals are held can sometimes affect the outcome of studies. A result produced in one laboratory may not be reproduced in another laboratory, even if its environment is very similar to the first (Richter et al., 2009).

If the results obtained from a study in a particular laboratory cannot be reproduced in another laboratory that follows the same method, then whether the effects will generalize to the outside world is questionable. Therefore, one of the two most important advantages of using laboratory animals in research studies (i.e., environmental control combined with the ability to manipulate one variable at a time) may also be one of its key disadvantages (i.e., generalizability). The generalizability of an outcome established in a fully controlled environment can be examined by systematically testing the effects of environmental variations. Matched samples, in which two similar animals (e.g., same species, age, height, weight, and sex) are housed in two different environments but all other treatments are kept constant, would be one way to achieve this.

Another major disadvantage of using laboratory animals is end-point management of the animals. Do the animals die as part of the study? If not, what happens to them when the study ends? Under which circumstances would it be acceptable to euthanize an animal? Is it acceptable just because the study is complete, or should the research team try to find homes for the animals to live as pets, or continue to house them in the laboratory at their expense? What if the animals are unlikely to recover, physically or psychologically, from the experimental treatment undertaken? Researchers are required to justify their end-point management to the satisfaction of animal ethics committees. These issues not only affect the animals themselves, but also the researchers responsible for ending their lives. The psychological impacts of euthanizing animals can be profound, and represent a welfare risk for the research team (Close et al., 1996).

13.3.3 The Three Rs: Guidelines for the Use of Animals in Science

There are attempts being made to improve the welfare of animals used in science, both at the policy level and at the level of individual researchers. The key guidelines for accomplishing this goal are known as the Three Rs: Replacement, Reduction, and Refinement.

13.3.3.1 Replacement

Whenever possible, sentient animals (i.e., animals with the capacity to perceive pain), should be replaced with nonsentient alternatives. For example, instead of growing human organs on top of a rat's back, perhaps a 3D printer could be used to generate the organs (Mironov et al., 2003). Other nonsentient alternatives include plants, tissue culture, and microorganisms (Russell and Burch, 1992). For example, as vitamin B is an amino acid required by microorganisms and vertebrates alike, instead of using laboratory rats to measure the effects of vitamin B deficiency, a microorganism could be used for the same purpose. Growing parasites ex vivo in a petri dish would similarly be preferable to growing one inside a living animal (Russell and Burch, 1992).

In Vivo and Ex Vivo Research With Animals

Research with animals may be undertaken in vivo (i.e., inside the organism) or ex vivo (outside the organism). For instance, injecting a live rat with a virus to observe its effects, for example, on the rat's liver, is an example of an in vivo study. In an ex vivo study, tissue, cells, or organs are extracted directly from an organism to test it externally and may be returned to that organism. In vitro studies are closely related, but refer to the study of tissue/cells from a cell line or repository. Both use petri dishes or test tubes to experiment on those cells.

In practice, the principle of replacement does not mean that animals are never used at all, of course. Proposals for new experiments are weighed against the current science of the day. If it is possible for researchers to replace living animals with nonsentient alternatives, this will be required. Researchers must demonstrate that there is no nonsentient alternative to animal use to justify their use to the animal ethics committee in a proposed study.

The principle of replacement is often interpreted as referring to the replacement of vertebrate animals, as opposed to all living animals, which are sometimes replaced by "less sentient" invertebrate species. However, while octopuses are invertebrates, they possess problem-solving abilities and an awareness of their bodies, which would suggest a high level of cognitive functioning (Ikeda, 2009). *Portia* spiders, which prey on other spiders, are able to mimic the vibrations made by the insects that their prey spiders eat to trick them into approaching them (Wilcox and Jackson, 1998). Bees are able to solve the "traveling salesman's dilemma," understanding the most efficient route between three, four, or more different places and expressing that route to other bees through a "waggle dance" (Wong et al., 2008). These animals possess certain cognitive abilities that exceed those of many vertebrate species, making them more "like us" than they appear to be; after all, many humans struggle to calculate the quickest route between several destinations! Currently, many invertebrates, such as spiders and bees, are not considered "animals" for ethics purposes and are therefore not subject to the animal welfare requirements imposed on researchers managing rodents or other vertebrates.

13.3.3.2 *Reduction*

At the present time, complete replacement of sentient animals in research is unrealistic for many studies. For example, if a researcher is trying to understand the cognitive abilities of a honeybee to determine whether they should be considered sentient and therefore subject to animal ethics requirements, it is simply not possible to avoid using honeybees. In the future, it may be possible to model honeybee cognition using neural networks or other nonsentient alternatives; if this is ever the case, the use of honeybees in cognition research will no longer be necessary.

In the meantime, in cases such as this, reducing the number of animals necessary to produce reliable results is the goal (Russell and Burch, 1992). Researchers are therefore required to undertake statistical power analyses to demonstrate how many subjects are needed to observe the expected effect. This draws on past research that has reported the strength of the effect of the experimental manipulation to project the required sample size. For instance, one minimally invasive EEG study with six dogs showed a strong effect size, and produced results that should theoretically be repeatable in another population of dogs, at another time, in another laboratory (Howell et al., 2012). If the effect had been smaller and/or the variability larger, more dogs would have been required. However, given this large effect, it is likely that future studies using the same technique with dogs could achieve similar results, so a small sample size would be sufficient. In a situation where there is not enough available evidence in existing research studies to justify the sample size, pilot studies can help determine effect sizes of a particular treatment on the animal subjects. If the study is expected to require 24 animals, but pilot testing shows that an effect is likely to be observed with only 12 animals, this would reduce the number of animals needed for the study.

Statistical Power

Statistical power is the likelihood of detecting an effect and correctly rejecting the null hypothesis. The stronger the effect, the more likely it will be detected. The larger the sample size, the more sensitive the test will be to detect an effect.

13.3.3.3 *Refinement*

Refinement refers to the development of experimental protocols in such a way as to minimize pain and distress for the animals used (Russell and Burch, 1992). This may mean using a less-invasive technique if one is available, such as using scalp electrodes for an EEG study rather than electrodes implanted directly into the brain. It could also mean that the animals are handled only when necessary, or that their cage is cleaned at the same time that they are already out of the enclosure. As handling is a known stressor for laboratory animals (Balcombe et al., 2004), reducing the number of times that they are handled would be a desired refinement.

These changes could be considered reactive, even if they are implemented before the experiment starts, because they manage animals that are already stressed by laboratory experiences, or because they are changing an established technique to be used in a study. There are proactive refinements that could be incorporated into study design, however. For example, there is evidence that gentling rats (softly handling them) while speaking to them during their first weeks of life makes them tamer; they show fewer behaviors indicating stress, and this effect is long term (Maurer et al., 2008). Therefore, introducing gentle interactions with the rats early on could improve their welfare state during their lives, if it results in them experiencing less stress during routine handling experiences.

Just as pilot studies can help reduce the number of animals needed, they also permit refinement of the technique itself, in addition to the sample size. For instance, if, instead of using minimally invasive subdermal needle electrodes for an EEG study with dogs (Howell et al., 2011), the research team wanted to develop a completely noninvasive technique using scalp electrodes (Kis et al., 2014), pilot studies could help the researchers understand whether this is feasible. They would have the opportunity to use several different electrode placements to find out what is most effective, and also be made aware of the challenges specific to scalp electrode use in dogs, which may not be obvious prior to testing.

Replacement, reduction, and refinement should be used in tandem, not one or the other, to justify the use of animals and, where justified, to optimize their welfare. Wherever live, sentient animals can be replaced with nonsentient material, this should be done. When live, sentient animals are absolutely necessary to the study, it is important to use as few as possible, and the technique should be refined to minimize stress and pain on the animal to the absolute minimum required to obtain reliable data. These techniques, when used together, can help ensure that the welfare state of laboratory animals is as good as possible. Scientists are required to justify their proposed studies on the basis of the Three Rs. Work is presently underway by the National Centre for the Replacement, Refinement and Reduction of Animals in Research (NC3Rs) to determine the impact of the 3Rs.

13.4 NONINVASIVE AND FIELD-BASED RESEARCH IN ANIMAL STUDIES

It is typical to think of "animal research" as invasive laboratory-based studies; however, there has been a lot of knowledge gained through noninvasive studies of captive animals living in artificial environments other than laboratories, and through observational studies of animals in their native habitat. For example, noninvasive marine mammal research with captive dolphins has helped researchers understand more about their cognitive capabilities (Herman et al., 1999; Tschudin, 2001; Tschudin et al., 2001). However, the ethics of housing marine mammals has recently come under scrutiny, especially after the release of a controversial film called "Blackfish" (Brammer, 2015). A clearer benefit of observational research is the outcome of studies estimating the number of endangered species in a particular habitat, which can provide evidence in support of conservation practices (JNCC and Defra, 2012).

While the connection between "conservation" and "animals in research" may not seem immediately obvious, this relationship becomes clearer when considering animals that are observed in field-based studies. It is often estimates of certain species in their natural habitat that provide us with information about human impacts on ecology, or the impact of invasive species on native species (Department of the Environment, n.d.; JNCC and Defra, 2012). For example, the tiger quoll (*Dasyurus maculatus*), the largest marsupial carnivore native to mainland Australia, is critically endangered in the Otway Ranges National Park in the State of Victoria, due to habitat destruction and the introduction of nonnative predators, such as foxes and cats. It is an apex predator, so its conservation is important to the ecology as a whole, and the Conservation Ecology Centre has been set up to help conserve and increase their dwindling numbers; they even have a program using scent detection dogs to find evidence of the quolls (Conservation Ecology Centre, 2013).

The results of research examining the impact of human activity on native animals can also impact policy. For instance, habitat corridors have been established to link national parks, permitting animals to move freely between those areas (Beier and Noss, 1998). In the United States, the Endangered Species Act of 1973 requires the protection of critical habitats for more than 2000 species listed as threatened or endangered (US Fish and Wildlife Service, 1973). In Australia, the Environment Protection and Biodiversity Conservation Act of 1999 lists over 1700 species that are threatened or endangered, and the UK Post-2010 Diversity Framework contains strategies to conserve and value biodiversity throughout the United Kingdom and its overseas territories (JNCC and Defra, 2012). These policies and strategies are all based on research highlighting the importance of environmental conservation, many of which include studies of animals in their natural habitats (Department of the Environment, n.d.; JNCC and Defra, 2012). Without basic research estimating the number of a certain species in an area, it would be difficult to take policy decisions aimed at conserving animals, as we would have no evidence that any species are impacted by habitat loss or other human activities.

Research Methods in Observational Studies

There are several ways in which a researcher may observe animals in their natural habitat. For instance, a researcher may follow an animal's tracks, or count the number of tracks it finds in a particular area. The researcher may collect scat (excrement) to differentiate individual animals and/or observe feeding habits, or take video and/or audio recordings of animals over a period of time. Approval to conduct research that involves handling or trapping of live animals, disruption to their natural habitat (including studies in which the researcher plays audio recordings of, for example, the animal's predators, prey, or other individuals of the same species, to measure how an animal responds to the noises) is required from an animal ethics committee.

13.4.1 Pet Animals

For decades, there have been studies using species of laboratory animals that are often kept as pets, such as cats and dogs (Overall and Dyer, 2005). However, a relatively recent phenomenon is the use of *actual* pets in research studies. That is, pet owners are recruited to bring their pets in for research studies, often focused on better understanding their cognitive abilities (Bensky et al., 2013; Miklósi, 2008)

FIGURE 13.2 **A pet Weimaraner observes her owner holding her favorite toy in a research study.** The owner was behind her in an adjoining room with a large window, but she was able to see him in the mirror. This and other the noninvasive study methods are used by pet animal cognition researchers in attempts to understand animal cognitive abilities. *Source: Photo by Tiffani Howell.*

or their relationship with their owner (Kubinyi et al., 2009; Ley et al., 2009). As owners in Western societies typically consider their pets to be members of the family (Kubinyi et al., 2009), it is important that any study using pet animals be minimally invasive and not likely to cause distress, such as one study examining whether dogs can use a mirror to determine the location of their owner (see Fig. 13.2 for a photo of a dog watching its owner in the mirror). Therefore, the use of pets in scientific research applies a higher standard of the Three Rs; this raises the question as to whether the welfare standards applied to pets should apply to all animals, especially when the same species is treated differently as a function of their pet or nonpet status.

13.4.1.1 Considerations for Using Pet Animals in Research

There are several issues to consider when using pet animals in research. First, the standard of welfare is arguably higher for pet animals compared to laboratory animals. Laboratory animals may spend much of their lives in a cage, either individually or in group housing depending on the needs of the species and the study requirements. The levels of environmental enrichment provided in these cages may vary from completely barren to highly stimulating (Wolfer et al., 2004). Laboratory animals also spend a high proportion of their lives in research studies compared to pet animals who spend most of their time engaging in other activities.

Pet animals live in human homes, where, at least in principle, they receive adequate environmental enrichment and regular contact with their owners and other pets in the home. Precisely because they are considered to be members of the family, they are assumed to experience better welfare overall than laboratory animals. Whether this is actually true is difficult to confirm, as there are no objective tools available to measure pet animal welfare due to the large variation in the ways they are managed by their owners. This makes management practices difficult to standardize relative to, for example, livestock. There is also very little behavioral research to even provide guidance about what the average pet needs to experience optimal welfare (Howell et al., 2016a,b).

In fact, the assumption that pet animals experience good welfare may be incorrect. A survey examining pet management practices among a representative sample of dog, cat, bird, and rabbit owners in Victoria, Australia suggests that, while many owners are appropriately meeting their pet's environmental, diet and exercise, behavioral, social, and health needs, there is room for improvement among some owners. For instance, children are not always supervised during interactions with their pets, some pets regularly exhibit behavioral problems that may be indicative of psychological or physical distress, and not all pets are vaccinated as often as recommended by veterinarians (Howell et al., 2015, 2016a,b).

Another consideration when using pet animals instead of laboratory animals is that pet animals may be considered more like animals in the "real world" than laboratory animals, even if the laboratory animals are pet species, such as dogs (Miklósi, 2014). This may be relevant when using animals as models of human conditions because humans also live in the "real world" rather than a well-controlled laboratory. Research with pet animals requires an acceptance that it is not possible to control the environment, and there will also typically be less genetic control than is possible in laboratory animal studies. However, there are ways around these issues. For example, a researcher may choose to recruit pets that all live in the same home, and/or animals that are all related. In a study of dog behavioral responses to an auditory stimulus, the researcher could contact a breeder to work with the dogs living in his or her home, which all come from the same genetic lines. This would provide a higher level of genetic and environmental control than would be possible if participant recruitment was aimed at any pet dog.

However, just as one disadvantage of using laboratory animals with highly controlled genetics and environment is a potential lack of ability to generalize to other animals of that species, this would also apply with dogs from the same breeder. Perhaps a different breed, or even a different line from the same breed, would respond differently to the same auditory stimuli. In such situations, it is important to consider the aim of the research. In a study using laboratory animals to develop a new vaccine for use in humans, within-species differences may not be as important as in a study aimed at understanding typical behavior for that species.

13.5 MEDICAL ADVANCES DUE TO ANIMAL RESEARCH

A considerable number of advances in human and veterinary medicine have come from animal research. Vaccines and disease treatments for humans and animals alike have resulted from research studies with animals.

13.5.1 Human Vaccines and Treatment for Diseases

As animals have been used as models of human disease, several vaccines come from animal-based studies, which have advantages and disadvantages. For instance, influenza vaccines are grown in eggs (Kaiser, 2006). The benefit of this is, of course, that it is possible to vaccinate against the flu. The disadvantage is that the flu vaccine must be administered very carefully to individuals who are allergic to eggs (James et al., 1998). Mice, rabbits, guinea pigs, and monkeys have served as animal models for tuberculosis vaccine development (Orme et al., 2001), but all have their disadvantages. For instance, monkeys may become aggressive and they are very expensive to house, and in mice it is difficult to measure a hypersensitivity to the vaccine using a skin prick test (Orme et al., 2001). More recently, monkeys have been used as a model in developing an AIDS vaccine, thus far proving elusive due to specific properties of HIV, which make it capable of rapid mutation and adaptation (Walker and Burton, 2008).

In addition to vaccines, medications for existing illnesses have also been developed using animals. For example, ACE inhibitors, drugs that lower blood pressure, are made from snake venom (Costa-Neto, 2005). Likewise, stroke research often uses animal models, although a limitation of these studies is that they typically use healthy, inbred animals with none of the comorbid illnesses common in people who have had a stroke (Macrae, 2011). As mentioned in Section 13.2, a genetically modified rat has become a model for cancer research (Barth, 1998).

Not all animal models for diseases need to be laboratory animals. For instance because of their extensive inbreeding to select for particular traits, dog breeds have a large degree of genetic homogeneity (Irion et al., 2003). Therefore, dogs have been used as models for naturally occurring human cancers that are commonly observed in dogs. Using dogs as a model permits researchers to potentially pinpoint the gene, or set of genes, which is responsible for cancer development (Rowell et al., 2011).

While it would appear that animal-based research studies should ultimately result in translational studies with humans, this is not always the case. Basic knowledge about human biology and genetics has increased tremendously due to research using animal models (Fields and Johnston, 2005). Unfortunately, however, only around one-third of research studies with animals, which were subsequently highly cited by other researchers, went on to be adapted for studies with human subjects (Hackam and Redelmeier, 2006). There are substantial barriers to the translation of basic research into clinical studies; these include structural issues, such as limited research funding and lack of interest by pharmaceutical companies to undertake translational research that does not have an obvious financial benefit (Hörig et al., 2005). However, there are cultural problems as well. For instance, physicians are not always aware of the need for collaboration with scientists to undertake translational studies, and some scientists may struggle to accept the lack of control that is necessary in research with human subjects (Hörig et al., 2005).

13.5.2 Veterinary Vaccines and Treatment for Diseases

Veterinary science proceeds in large part through animal research. Thanks to studies with animals, veterinarians are now able to vaccinate against, or diagnose and treat, illnesses that would otherwise kill our beloved pets, or negatively impact their quality of life. Dogs may have hip replacement (Olmstead, 1987) or knee replacement (Shino et al., 1984) surgeries, and they are vaccinated against the deadly and painful parvovirus (Spibey et al., 2008). Cats may have chemotherapy (Zwahlen et al., 1998) and receive vaccinations against feline leukemia (Tartaglia et al., 1993). Furthermore, research

into the management of livestock production animals by Temple Grandin and others has been used to improve farming and slaughter practices in intensive systems (Barnett and Glatz, 2004; Grandin, 1998, 2000, 2005), which may improve the welfare of billions of animals. This is in addition to the extensive scientific research that has aimed at treating common illnesses in livestock animals (Radostits et al., 2006).

13.5.3 Summary

There have been many medical advances due to research with animals. Whether animals are truly needed in medical research is debated on ethical grounds, but disease treatments and vaccines for humans and animals have been developed based on studies with animal models. Humans are able to be vaccinated against the flu and tuberculosis, and treated for high blood pressure. More is now known about how the human body works in general, including processes such as learning and memory (Angelucci et al., 1999), the process and development of disease (Götz and Ittner, 2008), and the therapies that may treat disease (Arap et al., 1998), all through research with animals. Pet animals no longer have to die painful deaths from parvovirus, but can be vaccinated against it. Despite differences in opinion as to whether the use of animals in research justifies the results, it is difficult to deny that many important medical advances have come from studies using animals.

13.6 SERVICE AND THERAPY ANIMALS

In addition to being subjects of research studies aimed at improving health or improving conservation practices, some animals have been employed as service or therapy animals (see Fig. 13.3 for an example of a service dog). There are many types of service animal (for a review, see Howell et al., 2016c), such as guide

FIGURE 13.3 **US Air Force wounded veteran, August O'Neill, being kissed by his service dog, Kai.** Service dogs perform a wide variety of tasks for their owners, which are intended to directly mitigate the effects of the owner's disability. *Source: Photo by Senior Airman Jette Carr, in the public domain of the United States, licensed under the terms of the United States Government Work.*

dogs (Batt et al., 2008) and diabetes alert dogs (Rooney et al., 2013; Wells et al., 2008). There are also dogs that alert epileptic owners of an impending seizure (Dalziel et al., 2003), and that help improve cognitive and social outcomes for children with autism spectrum disorder (Smyth and Slevin, 2010). This last category is becoming increasingly popular in recent years, and they are believed to work by providing the child with something to focus on to relax and bond with others (Smyth and Slevin, 2010).

Therapy animals, on the other hand, are part of a therapist-led and/or -designed program intended to convey some therapeutic benefit to individuals (Jegatheesan et al., 2015). They are often not owned by the people for whom they are providing the therapeutic outcomes. There is some evidence that therapy animals may help children feel comfortable when learning to read (Jalongo, 2005), to improve mental health outcomes for dementia patients (Filan and Llewellyn-Jones, 2006), to treat depression (Souter and Miller, 2007), and even to help preschoolers improve motor skills (Gee et al., 2007) and object recognition (Gee et al., 2012). The mechanism by which these animals are effective

is unclear. To develop a stronger evidence base around the use of therapy animals, rigorously designed randomized controlled trials are needed. In such a randomized controlled trial, participants would be randomly allocated to either an experimental group (in which they receive exposure to a therapy dog) or a control group (which may involve a comparison treatment, such as toys or other people). Outcome measures would be analyzed before and after participating in the study. It would then be possible to measure a change in any relevant outcomes over time, and determine whether the exposure to a therapy animal made a greater difference than exposure to another type of treatment.

Service Animal Training Process

The training process for service animals is often lengthy and expensive, costing $25,000 AUD (Assistance Dogs Australia, 2016) or more, and with as few as 50% of trainee animals successfully becoming operational (Weiss, 2002). It begins at selection of the animals who will join the training organization. These animals may be chosen from shelters, bought from a specific breeder, or bred by the organization itself. Some organizations even work with pet dogs purchased by owners who request assistance in training the pet to become a service animal.

The training process itself typically lasts up to 2 years, and consists of two components: (1) training for public access rights and (2) impairment-specific training. Training for public access is necessary because service animals are legally permitted to go virtually anywhere the owner goes, and a high level of obedience and good behavior in public is necessary. For impairment-specific training, Assistance Dogs International, a service dog accrediting body, requires that the animal be trained to perform at least three tasks that directly mitigate the impact of the impairment for the owner (Assistance Dogs International, 2014). For a psychiatric service dog, this may include getting medicine for the owner, dialing the emergency line, or "acting up" to give the owner an excuse to leave a distressing situation (Froling, 2009).

13.6.1 Advantages and Disadvantages for Humans

The potential benefits of service and therapy animals for humans are numerous. Many owners of service animals see a tangible improvement in their independence level or the symptoms of their condition. For example, epilepsy alert dogs can reduce the number of seizures experienced by owners (Strong et al., 2002), and diabetes alert dogs reduce paramedic callouts (Rooney et al., 2013). These results are based on self-reports, which may reflect a perception on the part of the owner rather than the reality. However, owners of service dogs often report an improvement in self-reports of quality of life and subjective well-being (Collins et al., 2006). Even if this is a placebo effect in some cases, it is a powerful one.

The disadvantages for humans relate mainly to the expense of obtaining and keeping an animal. At a cost of over $25,000 per animal, this resource is not available to everyone who could potentially benefit from it, although some organizations absorb these costs through donations (Assistance Dogs Australia, 2016; Guide Dogs WA, n.d.).

13.6.2 Advantages and Disadvantages for Animals

The advantages for animals are less clear. Animals in service work may experience their own animal welfare issues (Burrows et al., 2008). For instance, it is common for people with diabetes to have comorbidity with other disorders. If

there is a cognitive impairment, the owner may inadvertently harm the animal through neglect or poor management. Also, children with autism spectrum disorder may unintentionally mistreat their service animal (Burrows et al., 2008). Finally, some animals may be overworked and not get enough play time (Burrows et al., 2008). It is important, therefore, to ensure that animals get a break from the presence of their owner for at least part of the day. Likewise, if an owner is showing signs that he or she is cognitively incapable of coping with the responsibilities of owning an animal, it may be in the animal's best interest to remove it from that situation. Fortunately, according to a recent survey of assistance animal provider organisations, animal welfare is a high priority for providers (Howell et al., 2016c).

13.6.3 Summary

To summarize, service animals work with their owner to help them manage a chronic condition or disability, while therapy animals typically work with more than one person in an environment, such as a school or a nursing home, to provide some sort of therapeutic outcome. Training is extensive, particularly for service animals, and also very expensive. While the effects of these animals on the humans they assist can be profound based on anecdotal reports by owners, the scientific evidence base is not strong enough to draw clear conclusions about their effectiveness. Furthermore, the benefits for the animals themselves are unclear.

13.7 CONCLUSIONS

To conclude, there are many ways in which animals in research have benefited humans and other animals, such as medical advances and increased knowledge about the world around us. However, there is an ethical debate around the use of animals for human benefit, with rights-based arguments against their use, and utilitarian arguments generally in favor provided the suffering to the animal does not outweigh the benefit to humans. Disadvantages of using animals in laboratory-based research are also numerous, and include expense, end-point management, and the ethics of housing captive animals. Attempts are being made to manage these disadvantages through the use of the Three Rs (replacement, reduction, and refinement), but there is still much work to be done to improve well-being in laboratory animals.

Pet animals are a recent source for research studies, which reduces the cost, and may positively impact the welfare, of animals in research. However, the lack of environmental and genetic control can potentially confound the results, which may affect the feasibility of continued medical advancements that have come out of laboratory animal research.

Service and therapy animals may provide a benefit for the people they work with. However, more research is needed with therapy animals, with better quality controls, to better understand how effective they really are, and what may be the mechanism by which they work.

References

American Electroencephalographic Society, 1986. Guidelines in EEG. J. Clin. Neurophysiol. 3, 131–168.

Angelucci, M.E., Vital, M.A., Cesário, C., Zadusky, C.R., Rosalen, P.L., Da Cunha, C., 1999. The effect of caffeine in animal models of learning and memory. Eur. J. Pharmacol. 373, 135–140.

Altevogt, B.M., Pankevich, D.E., Shelton-Davenport, M.K., Kahn, J.P., 2011. Chimpanzees in Biomedical and Behavioral Research: Assessing the Necessity. National Academies Press, Washington, DC.

Arap, W., Pasqualini, R., Ruoslahti, E., 1998. Cancer treatment by targeted drug delivery to tumor vasculature in a mouse model. Science 279, 377–380.

Assistance Dogs Australia, 2016. FAQs. Assistance Dogs Australia, Engadine, NSW, Australia.

Assistance Dogs International, 2014. ADI Minimum Standards and Ethics. Assistance Dogs International, Brownsburg, IN, USA.

Balcombe, J.P., Barnard, N.D., Sandusky, C., 2004. Laboratory routines cause animal stress. J. Am. Assoc. Lab. Anim. Sci. 43, 42–51.

Barnett, J.L., Glatz, P.C., 2004. Developing and implementing a welfare audit in the Australian chicken meat industry. In: Weeks, C., Butterworth, A. (Eds.), Measuring and Auditing Broiler Welfare. CAB International, Wallingford, pp. 231–240.

Barnett, J., Hemsworth, P., 1990. The validity of physiological and behavioural measures of animal welfare. Appl. Anim. Behav. Sci. 25, 177–187.

Barth, R.F., 1998. Rat brain tumor models in experimental neuro-oncology: the 9L, C6, T9, F98, RG2 (D74), RT-2 and CNS-1 gliomas. J. Neurooncol. 36, 91–102.

Batt, L.S., et al., 2008. Factors associated with success in guide dog training. J. Vet. Behav. 3 (4), 143–151.

Baumans, V., 2004. Use of animals in experimental research: an ethical dilemma? Gene Ther. 11, S64–S66.

Beier, P., Noss, R.F., 1998. Do habitat corridors provide connectivity? Conserv. Biol. 12, 1241–1252.

Benefiel, A.C., Greenough, W.T., 1998. Effects of experience and environment on the developing and mature brain: implications for laboratory animal housing. ILAR J. 39, 5–11.

Bensky, M.K., Gosling, S.D., Sinn, D.L., 2013. The world from a dog's point of view: a review and synthesis of dog cognition research. Adv. Study Behav. 45, 209–406.

Bentham, J., 1977. A comment on the commentaries and a fragment on government. In: Burns, J.H., Hart, H.L.A. (Eds.), The Collected Works of Jeremy Bentham. Oxford University Press, London.

Brammer, R., 2015. Activism and antagonism: the 'Blackfish' effect. Screen Educ. 76, 72–79.

Burns, J.H., 2005. Happiness and utility: Jeremy Bentham's equation. Utilitas 17, 46–61.

Burrows, K.E., Adams, C.L., Millman, S.T., 2008. Factors affecting behavior and welfare of service dogs for children with autism spectrum disorder. J. Appl. Anim. Welfare Sci. 11, 42–62.

Cao, Y., Vacanti, J.P., Paige, K.T., Upton, J., Vacanti, C.A., 1997. Transplantation of chondrocytes utilizing a polymer-cell construct to produce tissue-engineered cartilage in the shape of a human ear. Plast. Reconstr. Surg. 100, 297–302.

Close, B., Banister, K., Baumans, V., Bernoth, E.-M., Bromage, N., Bunyan, J., Erhardt, W., Flecknell, P., Gregory, N., Hackbarth, H., 1996. Recommendations for euthanasia of experimental animals: part 1. Lab. Anim. 30, 293–316.

Collins, D.M., Fitzgerald, S.G., Sachs-Ericsson, N., Scherer, M., Cooper, R.A., Boninger, M.L., 2006. Psychosocial well-being and community participation of service dog partners. Disability Rehab. Assist. Technol. 1, 41–48.

Conservation Ecology Centre, 2013. Conservation Dogs. Conservation Ecology Centre, Cape Otway, VIC, Australia.

Costa-Neto, E.M., 2005. Animal-based medicines: biological prospection and the sustainable use of zootherapeutic resources. An. Acad. Bras. Cienc. 77, 33–43.

Dalziel, D.J., et al., 2003. Seizure-alert dogs: a review and preliminary study. Seizure 12 (2), 115–120.

Darwin, C., 1859. On the Origin of Species by Means of Natural Selection: Or, the Preservation of Favored Races in the Struggle for Life. J. Murray, London.

Davis, S.L., Cheeke, P.R., 1998. Do domestic animals have minds and the ability to think? A provisional sample of opinions on the question. J. Anim. Sci. 76, 2072–2079.

Department of the Environment, n.d. Species Expert Assessment Plan. Commonwealth of Australia, Canberra, ACT, Australia.

Driscoll, J.W., 1992. Attitudes toward animal use. Anthrozoos 5, 32–39.

Eddy, T.J., Gallup, G.G., Povinelli, D.J., 1993. Attribution of cognitive states to animals: anthropomorphism in comparative perspective. J. Soc. Issues 49, 87–101.

Fields, S., Johnston, M., 2005. Whither model organism research? Science 307, 1885–1886.

Filan, S.L., Llewellyn-Jones, R.H., 2006. Animal-assisted therapy for dementia: a review of the literature. Int. Psychogeriatr. 18, 597–611.

Fraser, D., 1999. Animal ethics and animal welfare science: bridging the two cultures. Appl. Anim. Behav. Sci. 65, 171–189.

Froling, J., 2009. Service Dog Tasks for Psychiatric Disabilities. International Association of Assistance Dog Partners and Sterling Service Dogs. Sterling Heights, MI.

Gee, N.R., Belcher, J.M., Grabski, J.L., Dejesus, M., Riley, W., 2012. The presence of a therapy dog results in improved object recognition performance in preschool children. Anthrozoos 25, 289–300.

Gee, N.R., Harris, S.L., Johnson, K.L., 2007. The role of therapy dogs in speed and accuracy to complete motor skills tasks for preschool children. Anthrozoos 20, 375–386.

Girman, S.V., Sauvé, Y., Lund, R.D., 1999. Receptive field properties of single neurons in rat primary visual cortex. J. Neurophysiol. 82, 301–311.

Götz, J., Ittner, L.M., 2008. Animal models of Alzheimer's disease and frontotemporal dementia. Nat. Rev. Neurosci. 9, 532–544.

Gluck, J.P., 2014. Moving beyond the welfare standard of psychological well-being for nonhuman primates: the case of chimpanzees. Theor. Med. Bioeth. 35 (2), 105–116.

Grandin, T., 1998. Objective scoring of animal handling and stunning practices at slaughter plants. J. Am. Vet. Med. Assoc. 212, 36–40.

Grandin, T., 2000. Effect of animal welfare audits of slaughter plants by a major fast food company on cattle handling and stunning practices. J. Am. Vet. Med. Assoc. 216 (6), 848–851.

Grandin, T., 2005. Maintenance of good animal welfare standards in beef slaughter plants by use of auditing programs. J. Am. Vet. Med. Assoc. 226 (3), 370–373.

Guide Dogs WA, n.d. Guide Dog FAQs. Guide Dogs WA, Victoria Park, WA, Australia.

Hackam, D.G., Redelmeier, D.A., 2006. Translation of research evidence from animals to humans. J. Am. Med. Assoc. 296, 1727–1732.

Hagelin, J., Carlsson, H.-E., Hau, J., 2003. An overview of surveys on how people view animal experimentation: some factors that may influence the outcome. Public Underst. Sci. 12 (1), 67–81.

Held, R., Hein, A., 1963. Movement-produced stimulation in the development of visually guided behavior. J. Comp. Physiol. Psychol. 56, 872.

Herman, L.M., Abichandani, S.L., Elhajj, A.N., Herman, E.Y.K., Sanchez, J.L., Pack, A.A., 1999. Dolphins (*Tursiops truncatus*) comprehend the referential nature of the human pointing gesture. J. Comp. Psychol. 113, 347–364.

Home Office (Ed.), 2014. Guidance on the Operation of the Animals (Scientific Procedures) Act 1986. Her Majesty's Stationery Office, London.

Hörig, H., Marincola, E., Marincola, F.M., 2005. Obstacles and opportunities in translational research. Nat. Med. 11, 705–708.

Howell, T., Conduit, R., Toukhsati, S., Bennett, P., 2011. Development of a minimally-invasive protocol for recording mismatch negativity (MMN) in the dog (*Canis familiaris*) using electroencephalography (EEG). J. Neurosci. Methods 201, 377–380.

Howell, T.J., Conduit, R., Toukhsati, S., Bennett, P., 2012. Auditory stimulus discrimination recorded in dogs, as indicated by mismatch negativity (MMN). Behav. Processes 89, 8–13.

Howell, T.J., Mornement, K., Bennett, P.C., 2015. Companion rabbit and companion bird management practices among a representative sample of guardians in Victoria, Australia. J. Appl. Anim. Welfare Sci., 1–16.

Howell, T.J., Mornement, K., Bennett, P.C., 2016a. Pet cat management practices among a representative sample of owners in Victoria, Australia. J. Vet. Behav. 11, 42–49.

Howell, T.J., Mornement, K., Bennett, P.C., 2016b. Pet dog management practices among a representative sample of owners in Victoria, Australia. J. Vet. Behav. 12, 4–12.

Howell, T., Bennett, P, and Shiell, A, 2016c. Reviewing assistance animal effectiveness: Literature review, provider survey, assistance animal owner interviews, health economics analysis and recommendations. Report to the National Disability Insurance Agency. La Trobe University, Bendigo. Available from: https://www.ndis.gov.au/medias/documents/hf5/hc0/8799673090078/Assistance-Animals-PDF-1-MB-.pdf

Ikeda, Y., 2009. A perspective on the study of cognition and sociality of cephalopod mollusks, a group of intelligent marine invertebrates. Jpn. Psychol. Res. 51, 146–153.

Irion, D., Schaffer, A., Famula, T., Eggleston, M., Hughes, S., Pedersen, N., 2003. Analysis of genetic variation in 28 dog breed populations with 100 microsatellite markers. J. Hered. 94, 81–87.

Jalongo, M.R., 2005. What are all these dogs doing at school? Using therapy dogs to promote children's reading practice. Child. Educ. 81, 152–158.

James, J.M., Zeiger, R.S., Lester, M.R., Fasano, M.B., Gern, J.E., Mansfield, L.E., Schwartz, H.J., Sampson, H.A., Windom, H.H., Machtinger, S.B., 1998. Safe administration of influenza vaccine to patients with egg allergy. J. Pediatr. 133, 624–628.

Jegatheesan, B., Beetz, A., Choi, G., Dudzig, C., Fine, A., Garcia, R.M., Johnson, R., Ormerod, E., Winkle, M., Yamazaki, K., 2015. The IAHAIO Definitions for Animal Assisted Intervention and Animal Assisted Activity and Guidelines for Wellness of Animals Involved. International Association of Human-Animal Interaction Organizations (IAHAIO), Amsterdam.

Joint Nature Conservation Committee and Defra (Eds.), 2012. UK Post-2010 Biodiversity Framework. Her Majesty's Government, Peterborough, United Kingdom (on behalf of the Four Countries' Biodiversity Group).

Kaiser, J., 2006. A one-size-fits-all flu vaccine? Science 312, 380–382.

Kaiser, J., 2015. An end to US chimp research. Science 350, 1013–11013.

Kis, A., Szakadát, S., Kovács, E., Gácsi, M., Simor, P., Gombos, F., Topál, J., Miklósi, Á., Bódizs, R., 2014. Development of a non-invasive polysomnography technique for dogs (Canis familiaris). Physiol. Behav. 130, 149–156.

Kubinyi, E., Turcsan, B., Miklósi, A., 2009. Dog and owner demographic characteristics and dog personality trait associations. Behav. Processes 81, 392–401.

Kurtz, T.W., Morris, R.C., Pershadsingh, H.A., 1989. The Zucker fatty rat as a genetic model of obesity and hypertension. Hypertension 13, 896–901.

Ley, J.M., Bennett, P.C., Coleman, G.J., 2009. A refinement and validation of the Monash Canine Personality Questionnaire (MCPQ). Appl. Anim. Behav. Sci. 116, 220–227.

Li, H., Cybulsky, M.I., Gimbrone, M., Libby, P., 1993. An atherogenic diet rapidly induces VCAM-1, a cytokine-regulatable mononuclear leukocyte adhesion molecule, in rabbit aortic endothelium. Arterioscler. Thromb. Vasc. Biol. 13, 197–204.

Lund, T.B., Mørkbak, M.R., Lassen, J., Sandøe, P., 2014. Painful dilemmas: a study of the way the public's assessment of animal research balances costs to animals against human benefits. Public Understanding Sci. 23, 428–444.

Macrae, I., 2011. Preclinical stroke research–advantages and disadvantages of the most common rodent models of focal ischaemia. Br. J. Pharmacol. 164, 1062–1078.

Maurer, B.M., Döring, D., Scheipl, F., Küchenhoff, H., Erhard, M.H., 2008. Effects of a gentling programme on the behaviour of laboratory rats towards humans. Appl. Anim. Behav. Sci. 114, 554–571.

Miklósi, A., 2008. Dog Behaviour, Evolution, and Cognition, first ed. Oxford University Press, New York, NY.

Miklósi, Á., 2014. Dog Behaviour, Evolution, and Cognition. Oxford University Press, Oxford.

Mironov, V., Boland, T., Trusk, T., Forgacs, G., Markwald, R.R., 2003. Organ printing: computer-aided jet-based 3D tissue engineering. Trends Biotechnol. 21, 157–161.

Nakhooda, A., Like, A., Chappel, C., Murray, F., Marliss, E., 1977. The spontaneously diabetic Wistar rat: metabolic and morphologic studies. Diabetes 26, 100–112.

National Health and Medical Research Council, 2013. Australian Code for the Care and use of Animals for Scientific Purposes, eighth ed. National Health and Medical Research Council, Canberra, ACT, Australia.

National Research Council, 2011. Guide for the Care and Use of Laboratory Animals, eighth ed. National Research Council, Washington, DC.

Olmstead, M.L., 1987. Total hip replacement. Vet. Clin. North Am. Small Anim. Pract. 17 (4), 943–955.

Orme, I.M., McMurray, D.N., Belisle, J.T., 2001. Tuberculosis vaccine development: recent progress. Trends Microbiol. 9, 115–118.

Overall, K.L., Dyer, D., 2005. Enrichment strategies for laboratory animals from the viewpoint of clinical veterinary behavioral medicine: emphasis on cats and dogs. ILAR J. 46, 202–216.

People for the Ethical Treatment of Animals, n.d. Home. PETA, Norfolk, VA.

People for the Ethical Treatment of Animals Australia, n.d. Animals Used for Experimentation: Product and Animal Testing. People for the Ethical Treatment of Animals Australia, Sydney, NSW, Australia.

Radostits, O.M., Gay, C.C., Hinchcliff, K.W., Constable, P.D., 2006. Veterinary Medicine: A Textbook of the Diseases of Cattle, Horses, Sheep, Pigs and Goats. Elsevier Health Sciences, Amsterdam, Netherlands.

Rasmussen, J.L., Rajecki, D.W., Craft, H.D., 1993. Humans' perceptions of animal mentality: ascriptions of thinking. J. Comp. Psychol. 107, 283–290.

Reardon, S., 2015. NIH to retire all research chimpanzees. Sci. Am. Available from: http://www.scientificamerican.com/article/nih-to-retire-all-research-chimpanzees.

Regan, T., 1987. The Case for Animal Rights, Advances in Animal Welfare Science 1986/87. Springer, New York City, NY, USA, (pp. 179–189).

Richter, S.H., Garner, J.P., Würbel, H., 2009. Environmental standardization: cure or cause of poor reproducibility in animal experiments? Nat. Methods 6, 257–261.

Rooney, N.J., Morant, S., Guest, C., 2013. Investigation into the value of trained glycaemia alert dogs to clients with Type I diabetes. PLoS One 8, e69921.

Rowell, J.L., McCarthy, D.O., Alvarez, C.E., 2011. Dog models of naturally occurring cancer. Trends Mol. Med. 17, 380–388.

Royal Society for the Prevention of Cruelty to Animals, n.d. What We Do. RSPCA Australia.

Russell, W.M.S., Burch, R.L., 1992. The Principles of Humane Experimental Technique. Universities Federation for Animal Welfare, Potters Bar, United Kingdom.

Shino, K., et al., 1984. Replacement of the anterior cruciate ligament by an allogeneic tendon graft. An experimental study in the dog. Bone Joint J. 66 (5), 672–681.

Singer, P., 1977. Animal Liberation. Towards an End to Man's Inhumanity to Animals. Granada Publishing Ltd, London, UK.

Smyth, C., Slevin, E., 2010. Experiences of family life with an autism assistance dog: placing specially trained dogs in families that have a child with autism can bring many benefits. Claire Smyth and Eamonn Slevin evaluated parents' views from a study in Ireland. Learn. Disabil. Pract. 13 (4), 12–17.

Souter, M.A., Miller, M.D., 2007. Do animal-assisted activities effectively treat depression? A meta-analysis. Anthrozoos 20, 167–180.

Spibey, N., et al., 2008. Canine parvovirus type 2 vaccine protects against virulent challenge with type 2c virus. Vet. Microbiol. 128 (1), 48–55.

Strong, V., Brown, S., Huyton, M., Coyle, H., 2002. Effect of trained Seizure Alert Dogs® on frequency of tonic–clonic seizures. Seizure 11, 402–405.

Tartaglia, J., et al., 1993. Protection of cats against feline leukemia virus by vaccination with a canarypox virus recombinant, ALVAC-FL. J. Virol. 67 (4), 2370–2375.

Tschudin, A., 2001. 'Mindreading' mammals? Attribution of belief tasks with dolphins. Anim. Welfare 10, S119–S127.

Tschudin, A., Call, J., Dunbar, R.I.M., Harris, G., van der Elst, C., 2001. Comprehension of signs by dolphins (Tursiops truncatus). J. Comp. Psychol. 115, 100–105.

Walker, B.D., Burton, D.R., 2008. Toward an AIDS vaccine. Science 320, 760–764.

US Fish and Wildlife Service, 1973. Endangered Species Act of 1973. 16 U.S.C. 1531 et seq. US Department of the Interior, Washington, DC.

Weiss, E., 2002. Selecting shelter dogs for service dog training. J. Appl. Anim. Welfare Sci. 5, 43–62.

Wells, D.L., Lawson, S.W., Siriwardena, A.N., 2008. Canine responses to hypoglycemia in patients with type 1 diabetes. J. Altern. Complement. Med. 14 (10), 1235–1241.

Wilcox, R.S., Jackson, R.R., 1998. Cognitive abilities of araneophagic jumping spiders. In: Balda, R.P., Pepperberg, I.M., Kamil, A.C. (Eds.), Animal Cognition in Nature: The Convergence of Psychology and Biology in Laboratory and Field. Academic Press, San Diego, CA, pp. 411–434.

Wolfer, D.P., Litvin, O., Morf, S., Nitsch, R.M., Lipp, H.-P., Würbel, H., 2004. Laboratory animal welfare: cage enrichment and mouse behaviour. Nature 432, 821–822.

Wong, L.-P., Low, M.Y.H., Chong, C.S., 2008. A bee colony optimization algorithm for traveling salesman problem. Second Asia International Conference on Modelling and Simulation. IEEE. pp. 818–823.

World Animal Protection, n.d. Our work. World Animal Protection, London.

Zwahlen, C.H., et al., 1998. Results of chemotherapy for cats with alimentary malignant lymphoma: 21 cases (1993–1997). J. Am. Vet. Med. Assoc. 213 (8), 1144–1149.

ANIMAL EXPERIMENTATION

Larry S. Katz
Rutgers University, New Brunswick, NJ, United States

Introduction

During the past half-century or more, dramatic advances in the diagnosis, prevention, and treatment of human and animal diseases have been realized. Much of this progress depended upon carefully designed and ethically conducted experimentation with many different animal species. The evidence is overwhelming to conclude that virtually every major biomedical advance can be traced back to critical studies using animals.

Animal research has contributed to our understanding and advancement in immunology, contributing to the development of vaccines for polio, diphtheria, small pox, tetanus, whooping cough, mumps, measles, influenza, human papillomavirus, and more. Development and testing of drugs, including antibiotics has saved untold millions of lives. Before antibiotics, even minor injuries often led to amputation or death from gangrene. In the past, permanent disabilities were common from now-treatable conditions, such as pneumonia, strep throat, bronchitis, and ear or dental infections. Surgical techniques that we now take for granted, such as open-heart surgery, spinal cord repair, organ transplantation, and so many more, were first developed and tested using laboratory animals. Thanks to animal research, people who suffer from diabetes, high blood pressure, arthritis, and so many other maladies have excellent quality and length of life.

For continued advances in human and animal health it is critically important that the public understand why animal research remains a vital component in the toolkit of biomedical research advancement and how animal research is regulated. The benefits of animals in this public health endeavor are clear to many, but most are not aware of the ethical, legal, and regulatory environment in which research is conducted by scientists in academia, in the pharmaceutical and healthcare industry, as well as in product safety testing.

Regulations

The oversight of animal use in research has multiple inputs including federal law, public health service (PHS) guidelines, the US Environmental Protection Agency and the Food and Drug Administration (FDA) good laboratory practices, and institutional animal care and use committee work.

Health Research Extension Act

The Public Health Service Policy on the Humane Care and Use of Lab Animals derives authority from the Health Research Extension Act of 1985 'Animals in Research' administered through the National Institutes of Health (NIH), Office of Laboratory Animal Welfare (OLAW). NIH is committed to the highest standards of animal care and use in the research it supports, and upholds federal regulations and policies that ensure ethical use of animal models. NIH encourages the use of the most appropriate models for scientific research including animal models. The knowledge gained from this research is used to develop life-saving treatments for diseases and conditions affecting the health of Americans and the world community. This knowledge ultimately saves lives and improves the quality of life for individuals, their families, and all of society. (NIH, in press)

All institutions that receive funds through the PHS) must comply with provisions of the Health Research Extension Act as administered by the

NIH, OLAW. No activity involving animals may be conducted or supported by the PHS until the institution conducting the activity has provided a written assurance acceptable to the PHS, setting forth compliance with this Policy.

Key elements of the PHS policy include institutional assurances, institutional animal care and use committees (IACUCs), and thorough record keeping. The PHS policy addresses the humane use of animals and the entire institutional program of animal care and use. All programs are expected to include the following: designation of an Institutional Official, appointment of an IACUC, administrative support for the IACUC, standard IACUC procedures, arrangements for a veterinarian with authority and responsibility for animals, adequate veterinary care, formal or on-the-job training for personnel that care for or use animals, an occupational health and safety program for those who have animal contact, maintenance of animal facilities, and provisions for animal care.

The PHS policy is based on a concept of enforced self-regulation. The concept of enforced self-regulation encompasses the following: institutional commitment through an Assurance, the designation of an Institutional Official authorized to assume the obligations imposed by the PHS policy, regular monitoring of the program for animal care and use by an IACUC, IACUC review of protocols that involve the use of laboratory animals, institutional identification and correction of deficiencies, institutional reporting to OLAW, performance standards wherever possible, and use of professional judgment.

The critical step in meeting the expectations of the PHS policy is the effective functioning of the IACUC. The functions of an IACUC are explicitly mandated by the OLAW as follows:

- Review at least once every 6 months the institution's program for the humane care and use of animals.

- Inspect at least once every 6 months all of the institution's animal facilities.
- Prepare reports of the IACUC evaluations.
- Review concerns involving the care and use of animals at the institution.
- Make recommendations to the Institutional Official regarding any aspect of the institution's animal program, facilities, or personnel training.
- Review and approve, require modifications in (to secure approval), or withhold approval of those components of PHS-conducted or PHS-supported activities related to the care and use of animals.
- Review and approve, require modifications in (to secure approval), or withhold approval of proposed significant changes regarding the use of animals in ongoing activities.
- Be authorized to suspend an activity involving animals in accordance with the Policy.

A major responsibility of the IACUC includes the careful evaluation of all proposed work involving vertebrate animals. The governing principles for this evaluation are as follows:

- Procedures with animals will avoid or minimize discomfort, distress, and pain to the animals, consistent with sound research design.
- Procedures that may cause more than momentary or slight pain or distress to the animals will be performed with appropriate sedation, analgesia, or anesthesia, unless the procedure is justified for scientific reasons in writing by the investigator.
- Animals that would otherwise experience severe or chronic pain or distress that cannot be relieved will be painlessly killed at the end of the procedure or, if appropriate, during the procedure.
- The living conditions of animals will be appropriate for their species and contribute

to their health and comfort. The housing, feeding, and nonmedical care of the animals will be directed by a veterinarian or other scientist trained and experienced in the proper care, handling, and use of the species being maintained or studied.

- Medical care for animals will be available and provided as necessary by a qualified veterinarian.
- Personnel conducting procedures on the species being maintained or studied will be appropriately qualified and trained in those procedures.
- Methods of euthanasia used will be consistent with the recommendations of the American Veterinary Medical Association Panel on Euthanasia, unless a deviation is justified for scientific reasons in writing by the investigator.

More detailed information governing OLAW's regulatory role is available (NIH, in press). Explicit recommendations on the proper implementation of a humane animal care program are detailed in the NIH Guide for the Care and Use of Laboratory Animals authored by the Institute for Laboratory Animal Research; Division on Earth and Life Studies; National Research Council. The "Guide" is made available through the National Academies Press (in press).

Animal Welfare Act

Detailed information on the Federal Animal Welfare Act is available through the US Department of Agriculture's (USDA's) Animal and Plant Health Inspection Service (APHIS, in press). This federal law and its associated regulations set the standards for humane care and treatment that must be provided for certain animals that are exhibited to the public, bred for commercial sale, used in medical research, or transported commercially. Facilities using regulated animals for regulated purposes must provide their ani-

mals with adequate housing, sanitation, nutrition, water, and veterinary care, and they must protect their animals from extreme weather and temperatures. Highly trained USDA inspectors, located throughout the United States, conduct routine, unannounced inspections of all facilities licensed/registered under the Animal Welfare Act to make sure these facilities are adhering to the standards set forth in the federal regulations.

Inspectors use the Animal Welfare Act standards and regulations as the baseline by which they determine the level of care being provided. If a facility is meeting the federal standards, the USDA knows that the animals there are receiving humane care and treatment. Conversely, when inspectors cite facilities for items that are not in compliance with the federal standards, USDA Animal Care holds those facilities responsible for properly addressing/correcting those items within a set timeframe. If these noncompliances are not corrected, or if they are of a serious enough nature, USDA has the option to pursue appropriate enforcement actions.

The Animal Welfare Act, which became law in 1966, does not cover every type of animal used in every type of activity. The following are not covered: farm animals used for food or fiber (fur, hide, etc.), cold-blooded species (amphibians and reptiles), horses not used for research purposes, fish, invertebrates (crustaceans, insects, etc.), or rats of the genus *Rattus* and mice of the genus *Mus* that are bred for use in research. Birds are covered under the Animal Welfare Act but the regulatory standards have not yet been established.

Good Laboratory Practices

Not all institutions conducting research or testing on animals are governed by the PHS policy or the Animal Welfare Act based upon the nature and funding of the work or the species involved, respectively. For example, the FDA may review data from private entities doing drug development research with laboratory rodents. To assure that humane care principles and practices are widely adopted, the FDA, the USDA, and the NIH have developed a memorandum of understanding concerning laboratory animal welfare. The participating agencies share a common concern for the care and welfare of laboratory animals used in research and testing.

The FDA is also involved in ensuring proper procedures for the care and use of laboratory animals. The source statute is the Federal Food, Drug, and Cosmetic Act as implemented by the Good Laboratory Practice Regulations. These regulations establish standards for the proper conduct of nonclinical laboratory studies that include animals. Compliance is assessed through an active program of periodic inspections carried out by trained field inspectors. Serious noncompliance is dealt with by procedures ranging from study rejection to laboratory disqualification.

In addition to the legal and public policy framework that regulates the use of animals in science, research, and education, there are professional organizations established to improve the conditions under which animal research is conducted. There is near universal agreement that the better the animal care environment, accounting for the physical and psychological welfare of the animals, the better and more reliable will be the results of scientific animal experimentation. The Association for Assessment and Accreditation of Laboratory Animal Care (in press) (AAALAC, International) and the American Association of Laboratory Animal Science (in press) (AALAS) are two such professional organizations dedicated to improving the conditions for research animals and the societal benefits from animal experimentation.

The AAALAC, International is a not-for-profit organization comprised of experts in animal welfare, animal research, and veterinary medicine dedicated to the improvement of animal care

programs at research and teaching institutions. Evaluation of one's animal care program by AAALAC, International is strictly voluntary. Organizations that submit their animal care program to this extensive review process are committed to providing the best possible care for the animals held under their stewardship. To be assessed for accreditation, applicants thoroughly document the animal care and use program including a description of the management and oversight of animal care and veterinary medicine and a detailed description of the animal environment including the physical description of facilities, the housing options, and the animal management program. In addition to a comprehensive review of an applicant institution's program documentation, a site visit by members of the Council on Accreditation is conducted. Institutions receiving accreditation and who maintain that accreditation through an annual reporting mechanism and periodic "drop-in" visits take seriously their ethical responsibility to provide excellent animal care.

The AALAS is a professional association of laboratory animal scientists who devote their energies to improve the care of laboratory animals used in research to promote the health and well-being of humans and animals. The core values of AALAS hold that the humane and responsible care of laboratory animals is vital to quality research. AALAS members are dedicated to building and disseminating a knowledge base in laboratory animal science for the education and training of those who work in this field.

Summary and Conclusions

Whenever animal use in experimentation may involve pain and/or suffering, we are morally obligated to justify that use. This moral justification does not require the allocation of rights to animals, but rather is compelled by the moral or ethical evaluation of our own actions on another being. In the determination of the "rightness" of action we must weigh the relative costs and benefits to humans and animals, but not accept the mistaken notion of moral equality of all animal species. Thus, while endeavoring to continue to improve the health and welfare of humans and animals we must continue to evolve our research protocols to provide the best type of animal care, keep clear and complete records, and refrain from practices that exploit animals or the environment. One approach to this goal includes the incorporation of the three "Rs" in research protocols as has been adapted from the work of Russell and Burch (1957). We should strive to *reduce* the number of animals in an experiment to an optimal level that provides statistical rigor but does not unnecessarily use too many, we should aim to *refine* protocols to minimize pain and suffering, and we should investigate efforts to *replace* higher vertebrate animals with lower animals or nonanimal alternatives when the goals of the research can be met in this manner.

References

American Association of Laboratory Animal Science (AALAS). Available from: https://www.aalas.org/

Animal and Plant Health Inspection Service, in press. Available from: www.aphis.usda.gov/aphis/ourfocus/animalwelfare/sa_awa/ct_awa_program_information

Association for Assessment and Accreditation of Laboratory Animal Care, International (AAALAC, International). Available from: http://www.aaalac.org/

National Academies Press, in press. Available from: http://www.nap.edu/catalog/12910/guide-for-the-care-and-use-of-laboratory-animals-eighth

National Institutes of Health, in press. Available from: http://grants.nih.gov/grants/olaw/olaw.htm

Russell, W.M.S., Burch, R.L., 1957. The Principles of Humane Experimental Technique. Methuen and Co. Ltd, London, Available from: http://altweb.jhsph.edu/pubs/books/humane_exp/het-toc.

Animals and Human Disease: Zoonosis, Vectors, Food-Borne Diseases, and Allergies

Colin G. Scanes

University of Wisconsin–Milwaukee, Milwaukee, WI, United States

14.1 INTRODUCTION

Animals have a major impact on human health globally with over 5 million deaths attributed to animals due to vector-borne diseases, major zoonotic disease, and snake bites (Table 14.1), and about 18 deaths per 100,000 people in Western countries, such as the USA (Table 14.2). Based on mortality figures in Western countries, such as the USA, people have a much greater effect on the death rate of humans due to diseases (based on our own lifestyle), murder, and suicide than due to animals (Table 14.1).

14.2 ZOONOTIC DISEASES

14.2.1 Overview

Many of the most serious infectious diseases are zoonotic diseases with an estimate of 4 million deaths per year globally based on data from the World Health Organization (WHO, 2015)

for human immunodeficiency virus/acquired immune-deficiency syndrome (HIV/AIDS) and the International Livestock Research Institute (ILRI, 2012) for other zoonotic diseases. Examples of zoonotic diseases are provided in Table 14.3 characterized by disease-causing agents covering first protein prion-caused diseases followed by viral, bacterial, protozoan, and fungal pathogens. Discussion will be focused on a few of the major zoonotic diseases.

Definitions

1. Zoonotic diseases are diseases in which the infectious agents pass from animal to human and, in some cases, vice versa and usually the animal population serves as a reservoir for the infectious agent.

2. Emerging diseases or new emerging diseases are infectious diseases that are

---

newly identified as having human hosts (Morens and Fauci, 2013). These can be the result of changes in exposure to potential pathogens due to economic development and travel. In addition, mutations and recombinations in pathogens may extend the host range. Examples of emerging diseases include Ebola, West Nile virus, and Zika virus diseases.

3. Reemerging diseases are diseases that historically infected humans but are either appearing in new locales or new drug-resistant forms and/or are reappearing after being under control such as tuberculosis (Morens and Fauci, 2013).

TABLE 14.1   Impact of Animals on Human Health Globally: Vector-Borne Diseases, Major Zoonotic Disease, and Snake Bites

	Deaths per year	People affected per year
Vector-borne diseases	1.0 million	1 billion
Malaria[a]	0.6 million	500 million
Dengue[b]	20,000	50–200 million
Zoonotic disease[c]		
HIV	1.5 million	35 million
Gastrointestinal (zoonotic)	1.5 million	2.33 billion
Leptospirosis	123,000	1.7 million
Tuberculosis (zoonotic)	100,000	0.55 million
Rabies (predominantly dog bites)	70,000	70,000
Snake bites[d]	20,000–110,000	0.4–2.4 million
Workers		
Commercial fishing	24,000	

[a] WHO (2016).
[b] Murray et al. (2013).
[c] Data for HIV from WHO (2015) and for other zoonotic diseases from ILRI (2012).
[d] Data from Kasturiratne et al. (2008) and WHO (2013) (Table 14.7).

TABLE 14.2   Comparison of Animal Versus Human Caused Deaths (Age-Adjusted Death Rates) in the USA in 2013

	Age adjusted death rates (per 100,000 people)
Animal caused deaths	
Zoonotic diseases	
Influenza and pneumonia	15.9
HIV	2.1
Animal bites[a]	0.07
Comparisons to human caused deaths	
Heart disease (related to life style)	169.8
Suicide	12.6
Homicide	5.2

[a] Calculated from data by Forrester et al. (2012).
Data from CDC, 2015a. Age-adjusted death rates for selected causes of death, by sex, race, and Hispanic origin: United States, selected years 1950–2013. Available from: http://www.cdc.gov/nchs/data/hus/2014/018.pdf except animal bites.

14.3 PRION ZOONOSIS

14.3.1 Bovine Spongiform Encephalopathy

Bovine spongiform encephalopathy (BSE) or "mad-cow" was first recognized in the United Kingdom in 1986. This is a progressive degenerative neural disease caused by the propagation of a misfolded abnormal protein, the BSE prion (Fig. 14.1). It is estimated that 1–3 million cattle in the United Kingdom were infected with BSE or "mad-cow" disease (Fig. 14.2) (Smith and Bradley, 2003). The disease spread to cattle, for instance, to Europe and Canada via contaminated feed. The disease spread to people via consumption of beef leading to very serious consequences in variant Creutzfeldt–Jakob disease (Smith and Bradley, 2003). In the United Kingdom, 176 people were diagnosed with variant Creutzfeldt–Jakob disease, all of whom are now dead with a median age of death of 28 years old (Andrews, 2012).

TABLE 14.3 Examples of Zoonotic Diseases

Disease	Disease agent	Route of transmission	Source (primary host or reservoir)	Location
Prion zoonotic diseases				
Variant Creutzfeldt–Jakob disease	BSE	Ingestion	Cattle	Most cases UK and European Union
Viral diseases				
AIDS	HIV-1 and HIV-2	Body fluids	HIV-1: Chimpanzees (*Pan troglodytes*) HIV-2: sooty mangabeys (*Cercocebus atys*)	Central and Western sub-Saharan Africa
Ebola virus disease	Ebola virus	Human body fluids (+contact with certain fruit bats)	Fruit bats	Central and Western sub-Saharan Africa
Influenza	Influenza virus	Airborne and contact	Birds and pigs	Global
Yellow fever	Yellow fever virus	Vector: mosquito	Nonhuman primates	South America and Africa
Bacterial diseases				
Anthrax	*Bacillus anthracis*	Contact	Herbivore mammals	Africa, Asia, South America, and parts of Europe
Lyme disease	*Borrelia burgdorferi*	Vector: blacklegged tick (*Ixodes scapularis*)	Deer and wild rodents	Eastern North America
Plague	*Yersinia pestis*	Vector: fleas or by contact	Rats and other rodents	North America, Asia, and so on
Protozoan zoonotic diseases				
Chagas disease	*Trypanosoma cruzi*	Vector: kissing bugs (*Triamoma infestans* and *Rhodnius prolixus*)		Central and South America
Leishmaniases	*Leishmania* (20 species)	Vector: female phlebotomine sandflies		Africa, Asia, and Latin America
Malaria	*Plasmodium falciparum* and *P. vivax*, together with *P. knowlesi, P. malarae*, and *P. ovale*	Vector: *Anopheles* mosquito	*Plasmodium knowlese* in macaques	Africa (together with Asia)
Trypanosomiasis or sleeping sickness	*Trypanosoma brucei*	Vector—Tsetse fly (genus: *Glossina*)		Sub-Saharan Africa
Fungal zoonotic diseases				
Aspergillosis	*Aspergillus* species	Contact	Birds and mammals	Global

AIDS, Acquired immune-deficiency syndrome; BSE, bovine spongiform encephalopathy; HIV-1/2, human immunodeficiency viruses type 1 and 2.

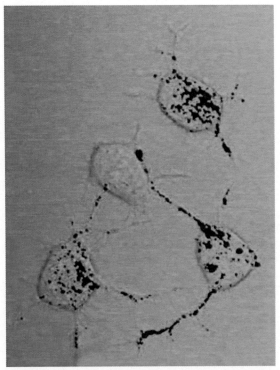

FIGURE 14.1   **The prion disease, bovine spongiform encephalopathy (BSE), causes progressive nervous system degeneration of cattle and people.** Image shows neural tissue from mice infected with another prion disease, scrapie. Prion protein stained in *red* is trafficked between neurons. *Source: Courtesy CDC.*

FIGURE 14.2   **Cattle infected with prions, BSE, exhibit progressive degeneration of the nervous system with nervousness, aggression, abnormal posture, and lack of coordination.** *Source: Courtesy CDC.*

The importance of prions as infective agents was recognized in 1997 with the award of the Nobel Prize in Physiology or Medicine for 1997 to Stanley B. Prusiner for his discovery of "Prions—a new biological principle of infection." The economic cost of BSE in the United Kingdom was estimated as £4 billion (~US$6 billion) (Smith and Bradley, 2003).

Other examples of transmissible spongiform encephalopathies caused by prions include the following (reviewed by Frost and Diamond, 2010):

- chronic wasting disease in deer
- scrapie in sheep and goats
- kuru in humans

14.4 VIRAL ZOONOSES

14.4.1 Overview

The following viral diseases will be considered in this section: HIV/AIDS, influenza, rabies, and Ebola. In addition, dengue will be discussed under vector-transmitted diseases (in Section 14.7.2).

14.4.2 HIV/AIDS

14.4.2.1 Disease

HIV/AIDS is one of the most serious infectious diseases with 1.5 million deaths per year globally (Tables 14.1 and 14.2, Fig. 14.3). HIV infects the immune system reducing its effectiveness (Fig. 14.4). Infection with HIV is

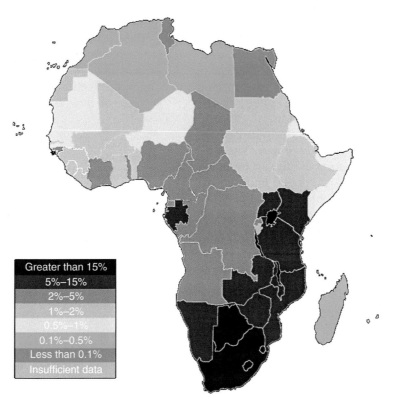

FIGURE 14.3 **Prevalence of HIV infection in Africa.** *Source: From WHO.*

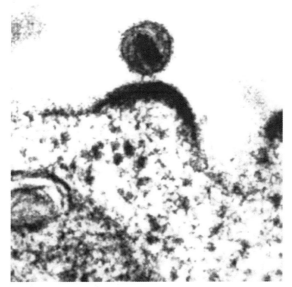

FIGURE 14.4 **HIV is responsible for about 1.5 million deaths per year.** Image shows digitally colorized transmission electron micrograph of an HIV budding from an immune cell. *Source: Courtesy CDC.*

first diagnosed by the presence of antibodies to HIV and the person is called HIV+. People with HIV/AIDS are immunocompromised and prone to opportunistic infections such as *Pneumocystis jiroveci* pneumonia (the most common opportunistic infection causing pneumonia), Kaposi's sarcoma (a virally induced cancer), and Thrush in the throat and mouth caused by another of the fungi, *Candida fungus*. The disease is at present not curable but can be managed. Longevity can be prolonged and quality of life improved for people with HIV/AIDS with treatment with antiretroviral therapy or antiretroviral drugs.

14.4.2.2 Global Distribution and Magnitude

Globally about 75 million people in the world have been infected by HIV with over 34 million deaths (WHO, 2015). Deaths from AIDS continue at about 1.5 million per year with 2.0 million new infections per year. In adults, the virus is transmitted by sexual contact (vaginal, anal, and oral sex) or from blood (e.g., transfusions or shared needles or medical personnel accidentally pricked by improperly disposed hypodermic needles). At the time of writing, there are about 220,000 children who are infected receiving the virus from their mothers with transmission either from milk from nursing mothers or across the placenta.

Number of HIV+ cases by region in 2014 (WHO, 2015):

1. Sub-Saharan Africa, 25.8 million
2. South/SE Asia, 4.1 million
3. Americas, 3.0 million
4. Europe, 2.3 million

In sub-Saharan Africa, about 5% of the adult population is HIV+. There are countries where the incidence exceeds 15% of the adult population (Fig. 14.3). HIV+ people are predominantly in heterosexual men and women (51%–49% of women to men) with infection due to unprotected sex.

According to the Centers for Disease Control and Prevention (CDC), there are about 1.2 million people in the United States infected with HIV. The disease is disproportionately affecting men who have sex with men and African Americans as is illustrated by the demographic data on new infections.

Number of new cases in the USA in 2010 (CDC, 2014):

- 11,200 White MSM (MSM are men who have sex with men)
- 10,600 Black MSM (4.3-fold higher risk than White MSM)
- 6,700 Hispanic MSM (2.1-fold higher risk than White MSM)
- 5,300 Black heterosexual women
- 2,700 Black heterosexual men
- 1,300 White heterosexual women
- 1,200 Hispanic heterosexual women
- 1,100 Black men intravenous drug users
- 850 Black women intravenous drug users

14.4.2.3 Infectious Agent(s)

At least two viruses are the causative agents responsible for the global pandemic, AIDS:

- Human immunodeficiency viruses type 1 (HIV-1), the major causative agent for AIDS
- Human immunodeficiency viruses type 2 (HIV-2).

HIVs are RNA viruses, lentiviruses, and are closely related to simian immunodeficiency viruses (SIV) (reviewed by Sharp and Hahn, 2011). Different primates in sub-Saharan Africa are persistently infected with various SIVs. There are close similarities between HIVs and SIVs (reviewed by Sharp and Hahn, 2011):

- HIV-1 and SIVcpz from chimpanzees (*Pan troglodytes*)
- HIV-2 and an SIV from sooty mangabeys (*Cercocebus atys*)

14.4.2.4 Animal Origins of HIV and the AIDS Pandemic

AIDS was first recognized in 1981. It is thought that SIV was transferred to humans from primates (reviewed by Sharp and Hahn, 2011). One form of SIVcpz is thought to have been transferred from chimpanzees in an area of Cameroon, West Africa about 95 years ago. The first reservoir of HIV-1 was around Kinshasa (in present-day Democratic Republic of the Congo). HIV-2 was an SIV transferred from sooty mangabeys in West Africa (reviewed by Sharp and Hahn, 2011).

14.4.3 Influenza

Influenza is an example of a reemerging zoonotic disease, a series of pandemics in the 20th century and an ongoing problem (Table 14.2). The host and reservoir for influenza viruses (Fig. 14.5) are wild water fowl. The viruses can also infect chickens and some influenza viruses can infect humans and/or pigs. Mutations and/or reassortment can occur in the reservoir populations, poultry, people, or pigs and can influence virulence (reviewed by Taubenberger et al., 2001).

Influenza pandemics in the past hundred years are the following:

- 1918–19 "Spanish" influenza (H1N1 subtype strain)
- 1957–58 "Asian" influenza (H2N2 subtype strain)
- 1968–69 "Hong Kong" influenza (H3N2 subtype strain)
- 2009–10 "Swine-origin" H1N1 influenza (H1N1 subtype strain)

(reviewed by Taubenberger et al., 2001). In addition, there is evidence for influenza pandemics over the preceding one, or even two, thousand years (Lina, 2008).

The 1918–19 pandemic killed about 40 million people globally with 675,000 dying in the USA (US Department of Commerce, 1976; reviewed by Taubenberger et al., 2001). Death rates were particularly high in young adults (reviewed by Taubenberger et al., 2001). Moreover, it is estimated that 28% of the world's population, 140 million people, were clinically infected (reviewed by Taubenberger et al., 2001). The virus responsible for the 1918–19 influenza pandemic has been characterized from tissue in histological slides or frozen in a grave in the permafrost (Taubenberger and Kash, 2011; Taubenberger et al., 2005).

14.4.4 Rabies

Rabies is a viral zoonotic disease of many mammalian species but also bats, cats, foxes,

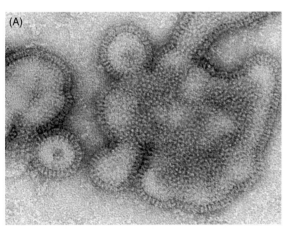
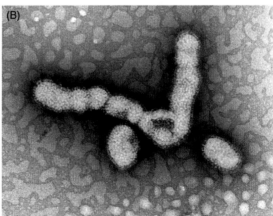

FIGURE 14.5 **Pandemics of influenza have resulted in many deaths with the 1918–19 pandemic killing about 40 million people globally.** (A) Transmission electron micrograph of influenza virions. (B) Digitally colorized transmission electron micrograph of influenza viruses. *Source: Courtesy CDC.*

raccoons, skunks, and people. It is spread chiefly by animal bites particularly from dogs. Unless treated, rabies is almost always fatal in people and most animals. Rabies vaccination is required legally for dogs and cats in all States in the USA. People who have a high risk of animal bites, such as veterinarians, receive prophylactic vaccination against rabies. If someone is bitten by an animal suspected of having rabies (being rabid), a series of vaccinations are used. Orally active vaccines at baiting stations are employed to control rabies in wildlife.

14.4.5 Ebola

14.4.5.1 Disease

The Ebola virus disease (also called Ebola or Ebola hemorrhagic fever) is a deadly zoonotic disease with mortalities up to 90%. The animal hosts of the virus are fruit bats (family: Pteropodidae) in Central and West Africa. Once transferred to people, the virus can be transmitted to other people in body fluids. The disease was first identified in 1976 with two outbreaks in that year, one in South Sudan and one in the Democratic Republic of Congo; the latter near the Ebola River. The largest outbreak of Ebola was in West Africa (with confirmed cases in Liberia, Sierra Leone, Guinea, and Nigeria) beginning in 2014 (CDC, 2016a). There have been over 11,300 deaths. In addition in 2014, there was an unrelated outbreak of Ebola in the Democratic Republic of Congo.

14.5 BACTERIAL ZOONOSES

14.5.1 Overview

Bacterial zoonotic diseases include the following:

- Tuberculosis caused by *Mycobacterium tuberculosis* (Table 14.1)
- Leptospirosis caused by bacteria of the genus *Leptospira* (Table 14.1)
- Anthrax caused by *Bacillus anthracis* (Table 14.3)

Anthrax has been weaponized as a biological weapon of mass destruction. Two other zoonotic bacterial diseases will be considered, namely, plague and Lyme disease (Table 14.3).

14.5.2 Plague

Plague is one of the zoonotic diseases that has devastating effects on people causing major pandemics. There are three forms of plague:

- Bubonic plague: infected lymph nodes. It is vector-transmitted by fleas of rats particularly black rats (*Rattus rattus*) but also other rodents.
- Pneumonic plague: infected lungs (death rate 90%). Transmission by fleas and by aerosol from coughing (person to person).
- Septicemic plague: infection of bloods.

All forms of plague are caused by the bacterium, *Yersinia pestis* (Fig. 14.6). Plague is found in Africa, Asia, South America, and the USA, albeit

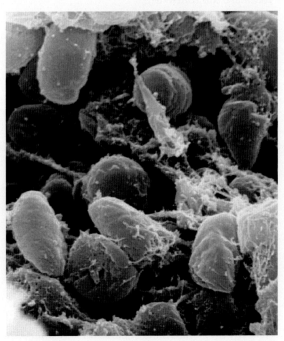

FIGURE 14.6 **Plague is caused by the bacterium, *Yersinia pestis*.** The vector is rat fleas. *Source: Courtesy CDC.*

rarely (Arizona, California, Colorado, and New Mexico). They can be treated with antibiotics such as streptomycin, gentamicin, and ciprofloxacin.

14.5.2.1 *Plague in History*

There have been at least three pandemics caused by the bacterium *Y. pestis*:

- Plague of Justinian AD/CE 541–543 but persisting for 200 years in the Byzantine Empire and other parts of Europe with disease organism identified as *Y. pestis* (Harbeck et al., 2013).
- Black Death (pandemic) occurred in Europe in 1347–51 (Benedictow, 2004) coming from Central Asia (disease organism identified by Haensch et al., 2010). Rats, the host of the vector, were transported via ships. Millions of people died with the death rate estimated as about 20% although there are some estimates of up to 65% (Benedictow, 2004). The pandemic persisted for 400 years with peaks between 1650 and 1700:
  - Great Plague of Seville (Spain)
  - Great Plague of Marseilles (France)
  - Great Plague of Vienna (Austria)
  - Great Plague of London (England) (1665–66): According to Daniel Defoe (1660–1731), the author of Robinson Crusoe, the city authorities ordered the slaughter of 40,000 dogs and 200,000 cats on the mistaken belief that these animals were responsible for spreading plague (Defoe, 1665). This made the situation worse as it removed the predators of rats!
- Third pandemic originating in Yunnan Province of China (mid-1850s) and spreading globally including the Great Plague of San Francisco (1900–04). Researchers in Hong Kong isolated the bacterium responsible for plague, *Y. pestis* (previously known as *Pasteurella pestis*) (Yersin, 1894). During and after World War II, weaponized plague was developed.

These pandemics were among the most devastating in human history.

14.5.3 Lyme Disease

Lyme disease is caused by the spirocete, *Borrelia burgdorferi* (Table 14.3). It is a vector-borne and a zoonotic disease. It is transmitted by the blacklegged ticks (or deer tick, *Ixodes scapularis*) (Fig. 14.7). Other hosts for blacklegged ticks include deer, foxes, coyotes, other mammals, birds, and amphibians. These ticks are small with the nymph stage being less than 2 mm across. The tick normally needs to be attached 36 h for transmission of the bacteria, *Bo. burgdorferi*, so that if the tick is removed, the chances of transmission are greatly reduced. According to the CDC, symptoms of Lyme disease include

FIGURE 14.7 **Adult deer tick (*Ixodes scapularis*) is a vector for Lyme disease.** *Source: Courtesy USDA Agricultural Research Service; Photo by Scott Bauer.*

the following: fever, headache, fatigue, and a skin rash.

There are very different estimates of the number of people who are infected with *Bo. burgdorferi* (Hubálek, 2009):

- North America, 16,500
- Asia, 3,500
- North Africa, 10

In contrast, the CDC estimated 25,000 confirmed cases of Lyme disease in the USA together with another 8,000 probable cases (CDC, 2015b). With less strict criteria for Lyme disease, estimates in the USA rise to over 300,000 annually (Stricker and Johnson, 2014). Irrespective of the differences, the CDC (2015a) tabulates Lyme disease as the fifth most common Nationally Notifiable Disease.

14.6 PROTOZOAN ZOONOSES

14.6.1 Overview

Protozoan diseases have a severe impact globally where the effects are predominantly in developing countries with over 600,000 deaths annually (Table 14.1). Protozoan diseases include Chagas disease, Leishmaniases, and Trypanosomiasis or sleeping sickness (Table 14.3). Another protozoan zoonotic disease is malaria and this is discussed in Section 14.7.2 under vector-transmitted diseases). Examples of protozoan diseases in Western countries include Cryptosporidiosis. These are discussed in Chapter 18 in Section 18.3.6.4.

14.7 VECTOR-BORNE DISEASES

A second way that animals impact human health is that they can be vectors passing disease organisms from person to person or from animal to person. Vector-transmitted diseases are responsible for about 1 million deaths per year globally (Table 14.1). Examples of zoonotic diseases are provided in Table 14.3 with the vectors detailed in Table 14.4. Two of the most serious vector-borne diseases globally are dengue and malaria and these are considered in, respectively, Sections 14.7.2 and 14.7.3 (see also Chapter 16). Other vector-borne diseases are discussed under zoonosis, namely, plague (Section 14.5.2) and Lyme disease (Section 14.5.3) (Table 14.5).

Definition of Vector-Borne Diseases

Vector-borne diseases are transmitted from person to person or animal to animal or animal to person by a biting or blood-sucking animal such as mosquitoes (Fig. 14.8), fleas, ticks and lice. Mosquitoes are the vectors for arboviruses (arthropod-borne viruses) with *Aedes aegypti* mosquitoes being the vectors for yellow fever, Zika virus, dengue, and chikungunya (CDC, 2016b).

14.7.1 Dengue

Dengue or dengue fever is the leading cause of disease and death in the tropics. The WHO estimates that there are between 50 and 200 million infections of dengue per year globally (Table 14.1), predominantly in Asia (Southeast Asia including Thailand together with the Philippines) and Latin America. About 40% of the world's population in over 100 countries is at risk of being infected by dengue viruses. Dengue is endemic in Puerto Rico and cases have been confirmed in Florida. There have been multiple epidemics of severe dengue (or dengue hemorrhagic fever). Dengue viruses are transmitted by the vector, *A. aegypti* mosquitoes (Fig. 14.8). There is presently no vaccine for dengue.

TABLE 14.4   Examples of Vector-Borne Diseases

Disease	Disease agent	Immediate vector	Location
Viral diseases			
Dengue	DENV-1, DENV-2, DENV-3, and DENV-4	Mosquitos	
Yellow fever	YFV	Mosquitos	South America and Africa
Zika virus disease	ZIKV	*Anopheles* mosquito	South America and Africa
Bacterial diseases			
Epidemic or louse-borne relapsing fever	*Borrelia recurrentis*	Body lice (*Pediculus humanus corporis*)	
Epidemic typhus#	*Rickettsia prowazekii*	Body lice (*P. humanus corporis*)	
Lyme disease	*Borrelia burgdorferi*	Blacklegged tick (*Ixodes scapularis*)	Eastern North America
Plague	*Yersinia pestis*	Rat fleas (*Xenopsylla cheopis*)	
Trench fever	*Bartonella quintana*	Body lice (*P. humanus corporis*)	
Tularemia	*Francisella tularensis* (infective dose 10–50 bacteria)	Ticks and deer flies	North America
Protozoan diseases			
Chagas disease	*Trypanosoma cruzi*	Kissing bugs (*Triamoma infestans* and *Rhodnius prolixus*)	Central and South America
Leishmaniases	*Leishmania* (20 species)	Female phlebotomine sandflies	
Malaria	*Plasmodium falciparum* and *P. vivax*, together with *P. knowlese, P. malarae*, and *P. ovale*	*Anopheles* mosquito	Africa (together with Asia)
Trypanosomiasis or sleeping sickness	*Trypanosoma brucei*	Tsetse fly (genus: *Glossina*)	Sub-Saharan Africa

DENV, Dengue virus; YFV, yellow fever virus; ZIKV, Zika virus.

TABLE 14.5   Vectors of Diseases considered under the vector

Bug-borne disease (insects of order *Hemiptera*)
• Chagas (caused by protozoan *Trypanosoma cruzi*)
There is evidence for the presence of human pathogens in bedbugs (*Cimex lectularius* and *Cimex hemipterus*). There is ongoing research on whether bedbugs are vectors for the transmission of human diseases (Delaunay et al., 2011).
Flea-borne diseases (insects of order *Siphonaptera*)
• Plague (caused by bacterium *Yersinia pestis*)
Fly-borne diseases (insects of order *Diptera*)
• Mosquito-borne diseases
  ◦ Chikungunya caused by the chikungunya virus.
  ◦ Dengue caused by the dengue virus (400 million people infected per year globally).
  ◦ Dog heart worm, the parasitic nematode *Dirofilaria immitis* infecting dogs, cats, other carnivores, and also humans.
  ◦ Eastern equine encephalitis is caused by the Eastern equine encephalitis virus and infecting horses and humans.
  ◦ Malaria caused by the parasitic protozoan *Plasmodium*.
  ◦ West Nile disease is caused by the West Nile virus.
  ◦ Western equine encephalitis caused by the Western equine encephalitis virus, infecting horses and people.
  ◦ Yellow fever caused by the yellow fever virus.
• Tsetse fly-borne diseases
  ◦ Trypanosomiasis (humans and cattle) caused by protozoan Trypanosomes
• Horse flies-borne diseases
  ◦ Equine infectious anemia or swamp fever in horses (caused by equine infectious anemia virus—an RNA virus)
  ◦ Trypanosomiasis caused by protozoan Trypanosomes
Lice-borne disease (flightless insects of order Phthiraptera)
• Epidemic relapsing fever (caused by bacterium *Borrelia recurrentis*)
• Epidemic typhus (caused by bacterium *Rickettsia prowazekii*)
• Trench fever (caused by bacterium *Bartonella quintana*)
Mite-borne disease
• Scrub typhus caused by bacterium *Orientia tsutsugamushi*
• Rickettsialpox caused by bacterium *Rickettsia akari*
Tick-borne diseases (arachnids of order *Ascari*)
• Lyme disease caused by bacterium *Borrelia burgdorferi*
• Rocky mountain spotted fever caused by bacterium *Rickettsia rickettsii*
• Tularemia caused by bacterium *Francisella tularensis*.

FIGURE 14.8   **Mosquitos are vectors of multiple diseases.** (A) Female *Anopheles* mosquito with needle-like labrum being inserted through the skin to draw a blood meal. (B) Female *Anopheles* mosquito ingesting blood meal. *Source: Courtesy CDC; Photo by James Gathany.*

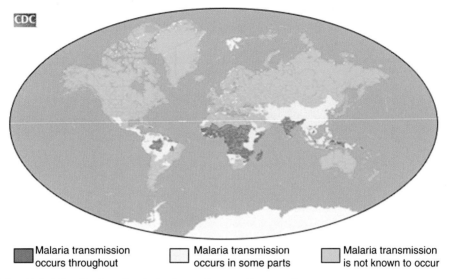

Malaria transmission occurs throughout

Malaria transmission occurs in some parts

Malaria transmission is not known to occur

FIGURE 14.9 **Global distribution of malaria.** *Source: Courtesy Shutterstock.*

14.7.2 Malaria

According to the WHO there are about 214 million cases of malaria per year with 438,000 deaths (mostly children) in 2015. The global distribution of malaria is shown in Fig. 14.9. Malaria is caused by the protozoa *Plasmodium falciparum* and *Plasmodium vivax,* together with *Plasmodium knowlese, Plasmodium malarae,* and *Plasmodium ovale* (Fig. 14.10). These multiply in the liver and then infect the erythrocytes. If not treated, malaria can be life threatening particularly in children. The vector for malaria is the *Anopheles* mosquito.

14.7.3 Other Vector-Borne Diseases

Other vector-borne diseases include Lyme disease, plague, and tick-borne relapsing fever (Fig. 14.11).

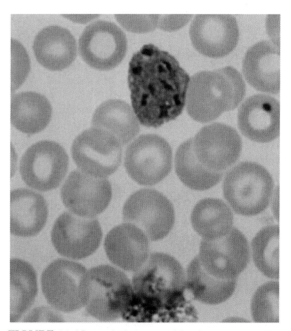

FIGURE 14.10 **Malaria is caused by the protozoa, principally *Plasmodium falciparum* and *P. vivax.*** There are about 214 million people infected with malaria per year. Image shows *Plasmodium* shown in Giemsa-stained photomicrograph of blood. *Source: Courtesy CDC.*

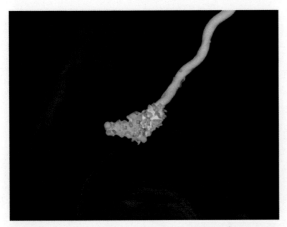

FIGURE 14.11 **Ticks and lice are the vectors for the spread of the spirochete, *Borrelia hermsii*, causing tick-borne relapsing fever.** *Source: Courtesy CDC.*

14.8 ANIMAL PRODUCTS, FOOD SAFETY, AND FOOD-BORNE DISEASES

14.8.1 Overview: What Are the Causes of Food-Borne Disease?

Another way that animals contribute to human disease is through food where infectious agents can be spread in animal products such as meat, milk, eggs, fish, and vegetables grown in fields contaminated with livestock excreta. Based on epidemiological studies of the USA, the major causes of food-borne diseases are the following:

- bacterial, 62%
- viral, 27%
- protozoan, 9%

The CDC has estimated that the proportion of food-borne disease that result in hospitalization come from the following food types (Painter et al., 2013):

- meat and dairy, 45.5%
  - dairy, 16.2%
  - poultry, 11.5%
  - eggs, 7.1%

- beef, 5.4%
- pork, 5.1%
- plants (fruits, vegetables, etc.), 40.9%
- aquatic animals (fish, seafood, etc.), 5.6%

14.8.2 Viral Diseases From Animal Products

The most common food-borne viral diseases are caused by the following (Atreya, 2004):

- noroviruses (Norwalk virus) with vomiting, nausea, and diarrhea
- hepatitis A and E viruses with inflammation of the liver
- rotaviruses causing gastroenteritis particularly in children
- astroviruses causing gastroenteritis

Mollusks, such as oysters and mussels, are filter feeders. If the waters are contaminated with human sewage, any pathogens will be found in the mollusks. Oysters are frequently consumed raw and there is a risk of food-borne diseases. Most (52%) of the shellfish-borne diseases in the USA between 1991 and 1998 were due to Norwalk and Norwalk-like viruses (Shieh et al., 2000). It is thought that sewage contamination was due to not following handling guidelines (Shieh et al., 2000). Noroviruses have also been detected in oysters in Europe (Boxman et al., 2006). After a major storm, norovirus was detected in oysters due to sewage contamination (Grodzki et al., 2012). Oysters are also a source of hepatitis A with it being estimated that 13% of shellfish-borne diseases between 1894 and 1990 in the USA were due to hepatitis A (Shieh et al., 2000).

Rotaviruses are usually considered as species-specific (Cook et al., 2004). Strains of rotaviruses from cattle and pigs exhibit large variability and these animals are suggested to be a reservoir for human infection (Midgley et al., 2012). Moreover, people particularly in rural areas are likely to be ingesting rotaviruses from livestock (Cook et al., 2004).

14.8.3 Bacterial Diseases From Animal Products

There is still a problem with bacteria in foods and food preparation causing disease globally. According to the CDC, the following summarizes the incidence of food-borne bacterial diseases in the USA (CDC, 2012):

- *Salmonella*: 1,000,000 people with the disease per year (380 deaths per year) (Fig. 14.12)
- *Campylobacter* spp.: 850,000 people with the disease per year (76 deaths per year)
- *Escherichia coli*: 203,000 people with the disease per year (21 deaths per year)
- Listeriosis (*Listeria monocytogenes*): 1,600 people with the disease per year (250 deaths per year)
- *Vibrio*: 53,000 people with the disease per year (48 deaths per year)

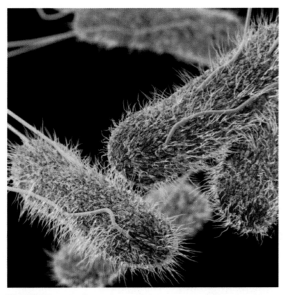

FIGURE 14.12 *Salmonella* **is responsible for about a million cases of food-borne illness in the USA.** Figure shows computer-generated image of *Salmonella* bacteria based on scanning electron microscopy. There are multiple short fimbriae and flagella allowing the bacteria to move.

FIGURE 14.13 **Ground beef can be a source of food-borne disease.** Cooking to 160°F (71°C) is effective in destroying pathogenic bacteria. *Source: Courtesy USDA Agricultural Research Service; Photo by Stephen Ausmus.*

In the USA, the CDC monitors outbreaks and attempts to determine the causative factor. For example, salmonellosis can be due to contamination of ground beef (Fig. 14.13) or pork but can also be from nonfood-related epidemiology such as contact with backyard poultry or pet turtles, geckos, or hedgehogs. The discussion on *E. coli* will be omitted as the environment contaminated with *E. coli* and the potential for contaminated vegetables to be consumed by people are discussed in Chapter 18 (Section 18.3.6.2). Another bacteria responsible for food-borne disease are of the genus *Vibrio*. *Vibrio parahemolyticus* and *Vibrio vulnificus* were responsible for, respectively, 29 and 9% of shellfish-borne diseases in the USA between 1991 and 1998 (Shieh et al., 2000).

14.8.4 Protozoal Diseases From Animal Products

Protozoa causing food-borne diseases include the following: *Toxoplasma, Cryptosporidium, Cyclospora* (linked to imported food), and *Giardia intestinalis* (could be transmitted from contaminated water when camping). *Toxoplasma* and *Cryptosporidium* are considered in more detail in Chapter 18.

14.9 ANIMALS AND ALLERGIES

14.9.1 Overview

Based on studies in Europe, 5%–20% of children have allergies (WHO, 2007).

Definitions

Allergic Reaction

An allergic reaction is an immunoglobulinE (IgE)-mediated response to an allergen.

Allergies

IgE-mediated allergic responses are seen in rhinoconjunctivitis ("hay fever"), asthma, and eczema or dermatitis.

Tests for Allergies

Skin tests are widely used to confirm IgE-mediated allergic reactions.

Oral Food Challenge

Double-blind placebo-controlled food challenge or oral food challenge (OFC). Statistics on the prevalence of food allergies is confounded by only 10% of countries reporting data based on OFC-proven IgE-mediated food allergy.

Anaphylaxis

Anaphylaxis is a life-threatening allergic response with a rash, swollen throat and tongue, swelling around the eyes and mouth, rapid heart rate, constriction of airways, and potentially also diarrhea and/or vomiting.

14.9.2 Allergies to Animal Allergens in Household Dust

Household dust consists of particulate manner from the exterior and organic matter; the latter includes dead skin from people and dander (dead skin) from pets (Layton and Beamer, 2009). Allergies can be to pet dander or to house dust mites or to their feces. House dust mites live on humans together with animal dander (reviewed by Nadchatram, 2005). People are continuously shedding dead skin cells at a rate of about 150 mg nitrogen (crude protein) per day (Calloway et al., 1971). The majority of human dead skin is lost to the inner layers of our clothing and bedding or lost when showering/bathing. The remainder becomes part of household dust. Similarly, pets are continuously shedding skin. As they do not have clothes or regular showers, pet dander adds to household dust.

Household allergens (cat, dog, and mite) are associated with morbidity in children with asthmatics (Gent et al., 2012). In households with asthmatic children, there are significant concentrations of cat, dog, cockroach, and house mite allergens (Leaderer et al., 2002):

- cat allergen >1 μg g^{-1}, 42% of households
- dog allergen >0.5 μg g^{-1}, 61% of households
- cockroach allergen >0.6 U g^{-1}, 28% of households
- mite allergen >0.5 μg g^{-1}, 68% of households

14.9.3 Allergies and Companion Animals

The concentration of cat antigens in mattress dust is as follows (Heinrich et al., 2006):
- current cat owners, 61.4 μg g^{-1}
- past cat owner, 1.4 μg g^{-1}
- never owned cat, 0.4 μg g^{-1}

There are allergic responses to other companion animals such as horses (Gawlik et al., 2009).

14.9.4 Allergies, Insects, and Arachnids

14.9.4.1 House Dust Mites

Dust mites (arachnids) are found throughout the world. Examples include the European house dust mite (*Dermatophagoides pteronyssinus*) and the American house dust mite (*Dermatophagoides farinae*). Allergens from house dust mites cause rhinitis and asthma. Globally, between 65 and 130 million people (1%–2%)

exhibit sensitization to house dust mite allergens. However, this rises to 50% in asthmatic patients (Calderón et al., 2015; de Vries et al., 2005).

The number of house dust mites (*Dermatophagoides* spp.) in household dust is as follows (Korsgaard, 1998):

- floor dust, 12.2 mites g^{-1}
- mattress dust, 100-fold greater

14.9.4.2 Bees, Wasps, and Hornets (Insects of the Order Hymenoptera)

Rates of incidence of anaphylaxis were estimated as 3.2 per 100,000 person years, from a study in Rochester, Minnesota in 1993 (Klein and Yocum, 1995) and 0.9 episodes of anaphylaxis per 1000 emergency department visits in Madrid (Moro Moro et al., 2011). Venom from bees, wasps, and hornets was the trigger for 29% of anaphylaxis in people going to emergency rooms in Florida and where the triggers were identified (Harduar-Morano et al., 2011). In contrast, Hymenoptera venom (bees, wasps, and hornets) were the triggers for 3.3% of episodes in Spain (Moro Moro et al., 2011). Estimates of hospital admission from insect stings in the United Kingdom rose from 0.1 per 100,000 population per year in 1998 to 0.5 per 100,000 population per year in 2012 (Turner et al., 2015). The rate of fatalities from insect stings was 0.009 per 100,000 population per year in the United Kingdom with mean age for fatalities being 59 years old (Turner et al., 2015).

14.9.5 Allergies and Foods

The prevalence of food allergies varies by country irrespective of whether these allergies are determined by and hence reported by OFC or parental reporting as can be seen from the following:

OFC-proven food allergy prevalence in preschool (<5 years old) children

- Australia
  - Australia, 10.0%

- United Kingdom, 5.0%
- Iceland, 1.9%
- Thailand, 1.0%
- Food allergy prevalence in preschool (<5 years old) children based on parental reporting (Prescott et al., 2013)
  - Mozambique, 16.2%
  - USA, 5.9%
  - Hong Kong, 5.3%

Prevalence of food allergies in children irrespective of age rises to 14.8% in the United Kingdom with the most common foods being cow's milk, eggs, peanuts, and seafood (Prescott et al., 2013). Based on the prevalence of food allergies, the WHO and Food and Agriculture Organization recommend that the presence of the following potential allergens (including animal products) should always be listed in packaged foods: shellfish, fish, eggs, and cow's milk together with peanuts, tree nuts, sulfites, and gluten (Hadley, 2006).

While allergic reaction occurs relatively rapidly (within 30 min), this is not the case in the syndrome, delayed anaphylaxis or hypersensitivity to red meat, where the response occurs in 3–6 h (Commins and Platts-Mills, 2013). The allergic response is to an oligosaccharide in beef, pork, veal, and lamb with the IgE directed to galactose-alpha-1,3-galactose (Saleh et al., 2012). There is a link between the anaphylactic reactions to oligosaccharides in red meat and tick bites (Saleh et al., 2012). Moreover, the presence of alpha-gal sIgE is associated with tick bites and also owning a cat (Gonzalez-Quintela et al., 2014).

14.9.6 Animal-Related Occupations and Allergies

Pet shop workers exhibit increased sensitization to dogs and cats (Liccardi et al., 2014). About 2.3% of workers with laboratory animals exhibit laboratory animal allergy symptoms; these being most commonly rhinitis rather than asthma (Elliott et al., 2005).

14.10 LACTOSE INTOLERANCE

The milk sugar, lactose, is a disaccharide composed of galactose and glucose. The cleavage of lactose to its constituent monosaccharides, galactose and glucose, is catalyzed in the small intestine by the enzyme, lactase, or lactase-phlorizin hydrolase.

Lactase

Lactose (galactose~glucose) → Galactose + Glucose → Absorption

The presence of lactase is critically important for the digestion of milk in infants, irrespective of whether the milk is mother's or cow's milk. The levels of lactase decline in midchildhood in most people resulting in lactose intolerance (reviewed by Lee and Krasinski, 1998). Thus, any lactose in the diet is not digested and hence not absorbed in the small intestine. Instead, the lactose passes to the large intestine where it is fermented by intestinal bacteria resulting in "bloating, flatulence, cramps, and nausea" (Itan et al., 2010). Lactose intolerance or the propensity for it is found in about 65% of the population (Gerbault et al., 2011; Itan et al., 2010). There are marked differences in lactose intolerance or the reciprocal, lactase persistence, in different regions of the world with lactase persistence varying between countries and regions (Gerbault et al., 2011):

- United Kingdom and Scandinavia, 92%
- Central Europe, 74%
- Eastern and Southern Europe, 35%
- Indian subcontinent North, 63%
- Indian subcontinent South, 23%
- East and Southeast Asia, <10%

In human history, there have been independent mutations that result in lactase persistence in several populations (Gerbault et al., 2011; Itan et al., 2010). These mutations have facilitated consumption of milk from dairy cattle, goats, sheep, and other animals with consequent increased availability of energy, protein, calcium, and vitamin D initially to Neolithic farmers and pastoralists (reviewed by Gerbault et al., 2011; Lee and Krasinski, 1998). One such mutation is responsible for the lactase persistence in the United Kingdom, Scandinavia, Germany, and Northern France and the people whose ancestors came from this area (Gerbault et al., 2011; Itan et al., 2010). Other mutations that resulted in lactase persistence were in northwest Africa and in Saudi Arabia (Gerbault et al., 2011; Itan et al., 2010).

14.11 ANIMAL BITES, STINGS, AND VENOM

14.11.1 Overview

Animals can cause death to people directly (see also Chapter 16). Fatalities from animals in the USA are as follows (Forrester et al., 2012):

- Other mammals (principally livestock): 112 fatalities per year and 36.4% of all animal-caused deaths.
- Hornets, wasps, and bees: 79 fatalities per year and 28.2% of all animal-caused deaths.
- Dogs: 28 fatalities per year and 13.9% of all animal-caused deaths

Deaths from other mammals are usually accidental deaths, while those from hornets, wasps, and bees are allergic reactions (discussed in Section 14.9.4.2) and dog-caused deaths are due to bites.

Table 14.6 summarizes animal bites in New York City again showing that dog bites are by far the most common (see Section 14.11.3) (Bregman and Slavinski, 2012). Globally, monkey bites are important in some places such as India (WHO, 2013).

14.11.2 Snake Bites

Snake bites with envenomation (release of venom into the bite) results in swelling, hemorrhage, and tissue necrosis leading potentially

TABLE 14.6 Animal Bites in New York City (Bregman and Slavinski, 2012)

Bite source	Percentage of ER visits due to animal bites (%)	ER visits per 100,000 population	Infections from bites (%)
Dog	73.6	56.2	3–18
Cat	13.3	10.2	28–80
Rodent (predominantly rats)	6.6	5.1	2
Other	6.5		

ER, Emergency room of a hospital (also known as accident and emergency or emergency departments).

TABLE 14.7 Incidence and Impact of Snake Bites Globally

	Incidence A	Incidence B
Snake bites	1.2 million	5 million
Envenomations[a]	421,000	2.4 million
Deaths	20,000	110,000
Amputations	—	400,000
Source	Based on Kasturiratne et al. (2008)	Based on WHO (2013)

[a] *Envenomations being snake bites where venom is released into the bite.*

to death or limb amputation (Gutiérrez et al., 2006). High death rates from venomous snakes is noted, for instance, 162 per 100,000 people per year in parts of Nepal (Gutiérrez et al., 2006). Estimates of the impact of snake bites globally is summarized in Table 14.7. The highest number of bites from venomous snakes are in sub-Saharan Africa, Southeast Asia, and India (Kasturiratne et al., 2008). Venomous snake bites in the USA are about 8000 year^{-1} (Consroe et al., 1995).

Gutiérrez and colleagues considered that "public health authorities, nationally and internationally, have given little attention to this problem, relegating snake bite envenoming to the category of a major neglected disease of the 21st century." The only treatment for venomous snake or spider bites is antivenoms produced in horse or sheep (Gutiérrez et al., 2006) (Fig. 14.14). Examples of venomous animals in the USA are snakes such as copperheads (Fig. 14.15),

rattlesnakes, coral snakes, and cotton mouths together with black widow spiders (Fig. 14.16). This is discussed in Section 14.12.

14.11.3 Dog Bites

According to WHO (2013), there are "no global estimates of dog bite incidence" but estimates 85% of animal bites are from dogs with an "estimated 55,000 people dying annually from rabies." In contrast, estimates for dog bites in the USA are the following (Overall and Love, 2001):

- Number, 3.5–4.7 million year^{-1}
- Medical attention sought, 17%–18%
- Hospitalization required, 1%–2%
- Cost, US$102.4 million for emergency services

Globally, dog bites can lead to the transmission of rabies.

14.11.4 Rat Bites

It is estimated that there are 20,000 rat bites per year in the USA with 200 cases of rat bite fever (Elliott, 2007). Rat bites have been associated with poverty (Childs et al., 1998). Rat bites are thought to be predominantly a problem for young children (<6 years) with bites occurring at night (Hirschhorn and Hodge, 1999). Increasingly, pet rats bite children in the household (Elliott, 2007). In a study in New York City, it was calculated that 2.4% of rat bites were sustained by laboratory animal technicians; there also being a higher risk with poverty (Childs et al., 1998).

Capture of wild snakes

↓

Place in snake farm quarantine, administer broad-spectrum antiparasitic
drugs, and perform veterinary surveillance to assure pathogens not
transmissible

↓

Collection of venoms (milking)

↓

Preparation of immunizing doses of venoms

↓

Administer immunizing doses of venoms together with adjuvants to animals
(horses or sheep) for antivenom production

↓

Collection of plasma or sera

↓

Fractionation of plasma immunoglobulins

↓

Quality control of antivenom immunoglobulins

↓

Production of immunoglobulins to snake venoms

FIGURE 14.14   **Production of immunoglobulins to snake venoms or spider venom.** These are used following envenomations.

FIGURE 14.15   **Juvenile copperhead snake (Agkistrodon contortrix).** *Source: Courtesy CDC; Photo James Gathany.*

14.12 ANIMALS AS A SOURCE OF ANTIVENOM

The first horse antivenom was employed
in 1895 (reviewed in WHO, 2010). Fig. 14.14
summarizes how antivenoms are produced
(WHO, 2010):

1. Monospecific antivenoms are used in
   countries or regions with one species of
   venomous snake such as the European adder
   (*Vipera berus*) in the United Kingdom and
   Scandinavia.
2. Polyspecific antivenoms are used where
   there are multiple species of venomous
   snakes. Horses or sheep are immunized
   with a mixture of venoms from the species
   of snakes in the region. For instance there is

FIGURE 14.16 **Female black widow spider (*Latrodectus mactans*) is a source of venom.** *Source: Courtesy CDC; Photo James Gathany.*

a polyvalent antivenom produced in horses for US snakes of the family Crotalidae (North American Crotalidae pit viper: Western diamondback rattlesnake (*Crotalus atrox*), Eastern diamondback rattlesnake (*Crotalus adamanteus*), Mojave rattlesnake (*Crotalus scutulatus*), and copperhead snake (Fig. 14.15) or water moccasin (*Agkistrodon contortrix*)).

3. Antisera can be purified to generate IgG with two binding sites (polyvalent) and further processed using the protease, papain, to Fab fragments with a single binding site (monovalent). This reduces side-effects of the therapy. An example is a Fab fragment antivenom produced in sheep (Consroe et al., 1995).

4. No antivenoms are currently available for multiple species of snakes particularly from Africa and the Indian subcontinent.

14.13 CONCLUSIONS

Animals have profound effects on human health with over 4 billion people affected (Table 14.1). Efforts to address these animal-related diseases are among the most pressing issues to improve human well-being and reduce early death.

References

Andrews, N.J., 2012. Incidence of variant Creutzfeldt-Jakob disease diagnoses and deaths in the UK, January 1994–December 2011. Available from: http://www.cjd.ed.ac.uk/documents/cjdq72.pdf

Atreya, C.D., 2004. Major foodborne illness causing viruses and current status of vaccines against the diseases. Foodborne Pathog. Dis. 1, 89–96.

Benedictow, O.J., 2004. The Black Death 1346–1353: The Complete History. Boydell Press, Woodbridge, UK.

Boxman, I.L., Tilburg, J.J., Te Loeke, N.A., Vennema, H., Jonker, K., de Boer, E., Koopmans, M., 2006. Detection of noroviruses in shellfish in the Netherlands. Int. J. Food Microbiol. 108, 391–396.

Bregman, B., Slavinski, S., 2012. Using emergency department data to conduct dog and animal bite surveillance in New York City, 2003–2006. Public Health Rep. 127 (2), 195–201.

Calderón, M.A., Linneberg, A., Kleine-Tebbe, J., De Blay, F., Hernandez Fernandez de Rojas, D., Vircho, C., Demoly, P., 2015. Respiratory allergy caused by house dust mites: what do we really know? J. Allergy Clin. Immunol. 136, 38–48.

Calloway, D.H., Odell, A.C., Margen, S., 1971. Sweat and miscellaneous nitrogen losses in human balance studies. J. Nutr. 101, 775–786.

CDC, 2012. Available from: http://www.cdc.gov/foodborneburden/PDFs/pathogens-complete-list-01-12.pdf

CDC, 2014. HIV in the United States: at a glance. Available from: http://www.cdc.gov/hiv/statistics/overview/ataglance.html

CDC, 2015a. Age-adjusted death rates for selected causes of death, by sex, race, and Hispanic origin: United States, selected years 1950–2013. Available from: http://www.cdc.gov/nchs/data/hus/2014/018.pdf

CDC, 2015b. Available from: http://www.cdc.gov/lyme/stats/graphs.html

CDC, 2016a. Ebola outbreak in West Africa—Case Counts. Available from: http://www.cdc.gov/vhf/ebola/outbreaks/2014-west-africa/case-counts.html

CDC, 2016b. Surveillance and Control of Aedes aegypti and Aedes albopictus in the United States. Available from: http://www.cdc.gov/chikungunya/resources/vector-control.html

Childs, J.E., McLafferty, S.L., Sadek, R., Miller, G.L., Khan, A.S., DuPree, E.R., Advani, R., Mills, J.N., Glass, G.E., 1998. Epidemiology of rodent bites and prediction of rat infestation in New York City. Am. J. Epidemiol. 148, 78–87.

Commins, S.P., Platts-Mills, T.A., 2013. Delayed anaphylaxis to red meat in patients with IgE specific for galactose alpha-1,3-galactose (alpha-gal). Curr. Allergy Asthma Rep. 13, 72–77.

Consroe, P., Egen, N.B., Russell, F.E., Gerrish, K., Smith, D.C., Sidki, A., Landon, J.T., 1995. Comparison of a new ovine antigen binding fragment (Fab) antivenin for United States Crotalidae with the commercial antivenin for protection against venom-induced lethality in mice. Am. J. Trop. Med. Hyg. 53, 507–510.

Cook, N., Bridger, J., Kendall, K., Iturriza Gomara, M., El-Attar, L., Gray, J., 2004. The zoonotic potential of rotavirus. J. Infect. 48, 289–302.

de Vries, M.P., van den Bemt, L., van der Mooren, F.M., Muris, J.W.M., van Schayck, C.P., 2005. The prevalence of house dust mite (HDM) allergy and the use of HDM-impermeable bed covers in a primary care population of patients with persistent asthma in the Netherlands. Primary Care Respir. J. 14, 210–214.

Defoe, D., 1665. A Journal of the plague year. Available from: http://www.gutenberg.org/files/376/376-h/376-h.htm

Delaunay, P., Blanc, V., Del Giudice, P., Levy-Bencheton, A., Chosidow, O., Marty, P., Brouqui, P., 2011. Bedbugs and infectious diseases. Clin. Infect. Dis. 52, 200–210.

Elliott, S.P., 2007. Rat bite fever and Streptobacillus moniliformis. Clin. Microbiol. Rev. 20, 13–22.

Elliott, L., Heederik, D., Marshall, S., Peden, D., Loomis, D., 2005. Incidence of allergy and allergy symptoms among workers exposed to laboratory animals. Occup. Environ. Med. 62, 766–771.

Forrester, J.A., Holstege, C.P., Forrester, J.D., 2012. Fatalities from venomous and nonvenomous animals in the United States (1999–2007). Wilderness Environ. Med. 23, 146–152.

Frost, B., Diamond, M.I., 2010. Prion-like mechanisms in neurodegenerative diseases. Nat. Rev. Neurosci. 11, 155–159.

Gawlik, R., Pitsch, T., DuBuske, L., 2009. Anaphylaxis as a manifestation of horse allergy. World Allergy Organ. J. 2, 185–189.

Gent, J.F., Kezik, J.M., Hill, M.E., Tsai, E., Li, D.W., Leaderer, B.P., 2012. Household mold and dust allergens: exposure, sensitization and childhood asthma morbidity. Environ. Res. 118, 86–93.

Gerbault, P., Liebert, A., Itan, Y., Powell, A., Currat, M., Burger, J., Swallow, D.M., Thomas, M.G., 2011. Evolution of lactase persistence: an example of human niche construction. Philos. Trans. R. Soc. B Biol. Sci. 366, 863–877.

Gonzalez-Quintela, A., Dam Laursen, A.S., Vidal, C., Skaaby, T., Gude, F., Linneberg, A., 2014. IgE antibodies to alpha-gal in the general adult population: relationship with tick bites, atopy, and cat ownership. Clin. Exp. Allergy 44, 1061–1068.

Grodzki, M., Ollivier, J., Le Saux, J.-C., Piquet, J.-C., Noyer, M., Le Guyader, F.S., 2012. Impact of Xynthia tempest on viral contamination of shellfish. Appl. Environ. Microbiol. 78, 3508–3511.

Gutiérrez, J.M., Theakston, R.D., Warrell, D.A., 2006. Confronting the neglected problem of snake bite envenoming: the need for a global partnership. PLoS Med. 3, e150.

Hadley, C., 2006. Food allergies on the rise? Determining the prevalence of food allergies, and how quickly it is increasing, is the first step in tackling the problem. EMBO Rep. 7 (11), 1080–1083.

Haensch, S., Bianucci, R., Signoli, M., Rajerison, M., Schultz, M., Kacki, S., Vermunt, M., Weston, D.A., Hurst, D., Achtman, M., Carniel, E., Bramanti, B., 2010. Distinct clones of Yersinia pestis caused the Black Death. PLoS Pathog. 6, e1001134.

Harbeck, M., Seifert, L., Hänsch, S., Wagner, D.M., Birdsell, D., Parise, K.L., Wiechmann, I., Grupe, G., Thomas, A., Keim, P., Zöller, L., Bramanti, B., Riehm, J.M., Scholz, H.C., 2013. Yersinia pestis DNA from skeletal remains from the 6(th) century AD reveals insights into Justinianic Plague. PLoS Pathog. 9, e1003349.

Harduar-Morano, L., Simon, M.R., Watkins, S., Blackmore, C., 2011. A population-based epidemiologic study of emergency department visits for anaphylaxis in Florida. J. Allergy Clin. Immunol. 128, 594–600.

Heinrich, J., Bedada, G.B., Zock, J.P., Chinn, S., Norbäck, D., Olivieri, M., Svanes, C., Ponzio, M., Verlato, G., Villani, S., Jarvis, D., Luczynska, C., Indoor Working Group of the European Community Respiratory Health Survey II., 2006. Cat allergen level: its determinants and relationship to specific IgE to cat across European centers. J. Allergy Clin. Immunol. 118, 674–681.

Hirschhorn, R.B., Hodge, R.R., 1999. Identification of risk factors in rat bite incidents involving humans. Pediatrics 104, e35.

Hubálek, Z., 2009. Epidemiology of Lyme borreliosis. Curr. Probl. Dermatol. 37, 31–50.

ILRI (International Livestock Research Institute), 2012. Mapping of poverty and likely zoonoses hotspots. Available from: https://cgspace.cgiar.org/bitstream/handle/10568/21161/ZooMap_July2012_final.pdf

Itan, Y., Jones, B.L., Ingram, C.J., Swallow, D.M., Thomas, M.G., 2010. A worldwide correlation of lactase persistence phenotype and genotypes. BMC Evol. Biol. 10, 36.

Kasturiratne, A., Wickremasinghe, A.R., de Silva, N., Gunawardena, N.K., Pathmeswaran, A., Premaratna, R., Savioli, L., Lalloo, D.G., de Silva, H.J., 2008. The global burden of snakebite: a literature analysis and modelling based on regional estimates of envenoming and deaths. PLoS Med. 5, e218.

Klein, J.S., Yocum, M.W., 1995. Underreporting of anaphylaxis in a community emergency room. J. Allergy Clin. Immunol. 95, 637–638.

Korsgaard, J., 1998. Epidemiology of house-dust mites. Allergy 53 (Suppl. 48), 36–40.

Layton, D.W., Beamer, P.I., 2009. Migration of contaminated soil and airborne particulates to indoor dust. Environ. Sci. Technol. 43, 8199–8205.

Leaderer, B.P., Belanger, K., Trite, E., Holford, T., Gold, D.R., Kim, Y., Jankun, T., Ren, P., McSharry Je, J.E., Platts-Mills, T.A., Chapman, M.D., Bracken, M.B., 2002. Dust mite, cockroach, cat, and dog allergen concentrations in homes of asthmatic children in the northeastern United States: impact of socioeconomic factors and population density. Environ. Health Perspect. 110, 419–425.

Lee, M.-F., Krasinski, S.D., 1998. Human adult-onset lactase decline: an update. Nutr. Rev. 56, 1–8.

Liccardi, G., Steinhilber, G., Meriggi, A., Sapio, C., D'Amato, G., 2014. Sensitization to pets in pet shop workers. Occup. Med. 64, 470–471.

Lina, B., 2008. History of influenza pandemics. In: Raoult, D., Drancourt, M. (Eds.), Paleomicrobiology: Past Human Infections. Springer, New York, NY, pp. 199–211.

Midgley, S.E., Bányai, K., Buesa, J., Halaihel, N., Hjulsager, C.K., Jakab, F., Kaplon, J., Larsen, L.E., Monini, M., Poljšak-Prijatelj, M., Pothier, P., Ruggeri, F.M., Steyer, A., Koopmans, M., Böttiger, B., 2012. Diversity and zoonotic potential of rotaviruses in swine and cattle across Europe. Vet. Microbiol. 156, 238–245.

Morens, D.M., Fauci, A.S., 2013. Emerging infectious diseases: threats to human health and global stability. PLoS Pathog. 9, e1003467.

Moro Moro, M., Tejedor Alonso, M.A., Esteban Hernández, J., Múgica García, M.V., Rosado Ingelmo, A., Vila Albelda, C., 2011. Incidence of anaphylaxis and subtypes of anaphylaxis in a general hospital emergency department. J. Invest. Allergol. Clin. Immunol. 21, 142–149.

Murray, N.E., Quam, M.B., Wilder-Smith, A., 2013. Epidemiology of dengue: past, present and future prospects. Clin. Epidemiol. 5, 299–309.

Nadchatram, M., 2005. House dust mites, our intimate associates. Trop. Biomed. 22, 23–37.

Overall, K.L., Love, M., 2001. Dog bites to humans—demography, epidemiology, injury, and risk. J. Am. Vet. Med. Assoc. 218, 1923–1934.

Painter, J.A., Hoekstra, R.M., Ayers, T., Tauxe, R.V., Braden, C.R., Angulo, F.J., Griffin, P.M., 2013. Attribution of foodborne illnesses, hospitalizations, and deaths to food commodities by using outbreak data, United States, 1998–2008. Emerg. Infect. Dis. 19, 407–415.

Prescott, S.L., Pawankar, R., Allen, K.J., Campbell, D.E., Sinn, J.K.H., Fiocchi, A., Ebisawa, M., Sampson, H.A., Beyer, K., Lee, B.-W., 2013. A global survey of changing patterns of food allergy burden in children. World Allergy Organ. J. 6, 21.

Saleh, H., Embry, S., Nauli, A., Atyia, S., Krishnaswamy, G., 2012. Anaphylactic reactions to oligosaccharides in red meat: a syndrome in evolution. Clin. Mol. Allergy 10, 5.

Sharp, P.M., Hahn, B.H., 2011. Origins of HIV and the AIDS pandemic. Cold Spring Harb. Perspect. Med. 1, a006841.

Shieh, Y., Monroe, S.S., Fankhauser, R.L., Langlois, G.W., Burkhardt, 3rd., W, Baric, R.S., 2000. Detection of Norwalk-like virus in shellfish implicated in illness. J. Infect. Dis. 2 (181 Suppl.), S360–S366.

Smith, P.G., Bradley, R., 2003. Bovine spongiform encephalopathy (BSE) and its epidemiology. Br. Med. J. 66, 185–198.

Stricker, R.B., Johnson, L., 2014. Lyme disease: call for a "Manhattan Project" to combat the epidemic. PLoS Pathog. 10, e1003796.

Taubenberger, J.K., Kash, J.C., 2011. Insights on influenza pathogenesis from the grave. Virus Res. 162, 2–7.

Taubenberger, J.K., Reid, A.H., Janczewski, T.A., Fanning, T.G., 2001. Integrating historical, clinical and molecular genetic data in order to explain the origin and virulence of the 1918 Spanish influenza virus. Philos. Trans. R. Soc. B Biol. Sci. 356, 1829–1839.

Taubenberger, J.K., Reid, A.H., Lourens, R.M., Wang, R., Jin, G., Fanning, T.G., 2005. Characterization of the 1918 influenza virus polymerase genes. Nature 437, 889–893.

Turner, P.J., Gowland, M.H., Sharma, V., Ierodiakonou, D., Harper, N., Garcez, T., Pumphrey, R., Boyle, R.J., 2015. Increase in anaphylaxis-related hospitalizations but no increase in fatalities: an analysis of United Kingdom national anaphylaxis data, 1992–2012. J. Allergy Clin. Immunol. 135, 956–963.

US (United States) Department of Commerce, 1976. Historical Statistics of the United States: Colonial Times to 1970. Government Printing Office, Washington, DC.

WHO (United Nations World Health Organization), 2007. Prevalence of asthma and allergies in children. Available from: http://www.euro.who.int/__data/assets/pdf_file/0012/96996/3.1.pdf

WHO, 2010. WHO guidelines for the production, control and regulation of snake antivenom immunoglobulins. Available from: http://www.who.int/bloodproducts/snake_antivenoms/snakeantivenomguideline.pdf?ua=1

WHO, 2013. Animal bites. Available from: http://www.who.int/mediacentre/factsheets/fs373/en/

WHO, 2015. HIV/AIDS. Available from: http://www.who.int/gho/hiv/en/

WHO, 2016. Malaria. Available from: http://www.who.int/mediacentre/factsheets/fs387/en/

Yersin, A., 1894. La peste bubonique à Hong Kong. Ann. Inst. Pasteur (Paris) 8, 662–667.

Further Reading

D'Amato, G., Liccardi, G., Russo, M., Barber, D., D'Amato, M., Carreira, J., 1997. Clothing is a carrier of cat allergens. J. Allergy Clin. Immunol. 99, 577–578.

ILO, 1999. Report on safety and health in the fishing industry. International Labour Organization, Geneva, May 1999.

Isbister, G.K., Brown, S.G., MacDonald, E., White, J., Currie, B.J., 2008. Current use of Australian snake antivenoms and frequency of immediate-type hypersensitivity reactions and anaphylaxis. Med. J. Aust. 188, 473–476.

White, J., 2000. Bites and stings from venomous animals: a global overview. Ther. Drug Monit. 22, 65–68.

Pest Animals

Samia R. Toukhsati*, Colin G. Scanes**

*Honorary Fellow, The University of Melbourne, Parkville, VIC, Australia
**University of Wisconsin–Milwaukee, Milwaukee, WI, United States

15.1 OVERVIEW OF PESTS

The word, pest, can signify a nuisance, such as ants at a picnic or mosquitoes irritating humans sitting on a veranda on a warm evening. Such nuisances become serious when the nuisance mosquito is a vector transmitting diseases. Alternatively, a pest can be an impediment to human life and/or well-being by damaging housing, spreading diseases, destroying crops, and harming livestock. These are considered here covering such pests as rats and mice (Section 15.2); birds such as pigeons, starlings, mynas, and house sparrows (Section 15.3); biting and stinging pests including snakes, spiders, scorpions, and insects (Section 15.4); and nematodes as pests in crop production (Section 15.5).

15.2 RATS AND MICE

Rats (*Rattus*) and mice (*Mus*) are a large and diverse species of mammals that belong to the superfamily Muroidea. Black and brown rats (also known as "common" rats) and mice are highly adaptable, generalist omnivores that frequently live alongside humans in urban environments. They are known to inhabit many varieties of landscapes, from the tropics to the deserts. Of course, not all rats and mice are pests; some are favorite pets or very useful medical/research subjects. However, when in abundance, rats and mice can have a devastating impact on human interests, such as through competition for food and as vectors of disease (Golden et al., 2015).

While humans are very familiar with two species of rats, namely the black rat (*Rattus rattus*) and the brown rat (*Rattus norvegicus*), there are 61 species in the genus *Rattus*, some of which are endangered. Black rats (also known as the roof rat, ship rat, and house rat among others) and brown rats (also known as the sewer rat, Norway rat, and wharf rat among others) are thought to have evolved in Asia and dispersed worldwide, with the exception of Antarctica (Aplin et al., 2003). Among the most well-known species of mice is the house mouse (*Mus domesticus*). Brown rats can be twice the size of black rats, both of which are many times larger than mice.

Common rats and mice are described as pests due the deleterious impact they have on crops (and thus, competition with humans and livestock for food), other animals (such as through predation of eggs and bird hatchlings), and as vectors (e.g., the Black Death) or reservoirs of disease organisms (e.g., hantavirus). The impact of Pacific rats is discussed in Chapter 18. It has been suggested that rats and mice may be responsible for more human suffering than any other nonhuman vertebrate (Aplin et al., 2003).

Classification of Rats and Mice

- Phylum: Chordata
- Class: Mammalia
- Order: Rodentia
- Superfamily: Muroidea
- Family: Muridae
- Subfamily: Murinae
- Genus: *Rattus* (rats)
- Genus: *Mus* (mice)

15.2.1 Contamination and Disease

Rats and mice are well-known disease vectors (i.e., carriers of disease) and compete with human interests by eating and contaminating their food and the food of their livestock with their feces and urine. Fair weather conditions that favor the growth and availability of food correspondingly favor an increase in populations of Rodentia; in turn, this increases the devastation of crops and the risk of exposure to viral diseases in humans (Golden et al., 2015).

15.2.1.1 *Hantavirus*

Hantavirus, belonging to the Bunyaviridae family, is a zoonotic virus carried by rodents and is typically transmitted to humans via exposure to aerosolized secreta (urine or saliva) or excreta of infected rats or mice (Hartline et al., 2013; Pitts et al., 2013). The Hantavirus causes two febrile diseases that are sometimes fatal in humans, namely, Hantavirus pulmonary syndrome (HPS) and hemorrhagic fever with renal syndrome (Golden et al., 2015). These viruses typically begin with a broad, nonspecific symptom profile that includes fever, muscle aches, and headaches and can graduate to fluid in the lungs and concomitant shortness of breath (in HPS) or low blood pressure and acute renal failure (in hemorrhagic fever with renal syndrome) (Golden et al., 2015). Management is primarily through supportive therapy, such as fluid and electrolyte management, and more intensive therapies, such as kidney dialysis, in the setting of fluid overload (Hartline et al., 2013). These viruses continue to present a significant threat to human health, such as in southwest USA. For instance, in the recent outbreak of HPS in Yosemite National Park (in the Sierra Nevada mountains, California, USA), of the 10 confirmed cases, 3 were fatal (CDC, 2012a). The outbreak was attributed to infected deer mice living alongside tourists (Hartline et al., 2013). There is no known cure or vaccine for Hantavirus; rodent control is the primary strategy to avoid infection.

15.2.1.2 *Lymphocytic Choriomeningitis*

Lymphocytic choriomeningitis, belonging to the Arenaviridae family, is a viral infection with a similar mode of transmission to the Hantavirus in that it is transmitted to humans via aerosolized exposure, or ingestion of, the urine, feces, saliva, or nesting materials from infected rodents, usually the common house mouse (*M. domesticus*) (Vela, 2012), pet hamsters, or guinea pigs (Souders et al., 2015). Early symptoms of infection in humans typically comprise nonspecific febrile symptoms, such as fever, headaches, muscle aches, and gastrointestinal disturbance, which may graduate to very serious illnesses such as acute hydrocephalus (fluid in the brain), myelitis (inflammation of the spinal cord), meningitis (inflammation of the meninges that cover the brain and spinal cord), encephalitis (inflammation of the brain), or meningoencephalitis (inflammation of the meninges and the brain) (Vela, 2012). Treatment via supportive therapy is recommended and while fatalities are rare (<1%), neurological damage may occur at the secondary stage of *Lymphocytic choriomeningitis* (Souders et al., 2015).

15.2.1.3 Leptospirosis

Leptospirosis is a widespread bacterial disease endemic in tropical countries that is transmitted to humans via direct contact with the urine of infected animals or indirectly via water, soil, or food contaminated by the urine of infected animals (Calderón et al., 2014; Mwachui et al., 2015; Shah, 2012). Rodents are among the many species of wild and domesticated mammals that can be carriers of Leptospirosis and are thought to be the main reservoir for transmission of the disease to humans, although occupational exposure from livestock or recreational exposure from contaminated water sources are also prominent sources of disease transmission (Mwachui et al., 2015; Shah, 2012). Symptoms of infection mimic other viral illnesses with several nonspecific symptoms including fever, headaches, muscle aches, and gastrointestinal disturbance, which may graduate to life-threatening illnesses, such as kidney or liver failure or meningitis. As a bacterial infection, Leptospirosis can usually be effectively treated with antibiotics and supportive care of renal, hepatic, and circulatory functions (Shah, 2012).

15.2.1.4 Bubonic plague

Perhaps the most well-known zoonotic bacterium, the bubonic plague is one of the most devastating pandemics known to human history and is responsible for the deaths of between 75 and 200 million people. Rats have long been thought to be reservoirs and vectors of the bubonic plague, or black death, which is caused by the zoonotic *Yersinia pestis* bacteria; transmission to humans has been thought to be primarily attributable to bites from infected fleas from infested black (*R. rattus*) or brown (*R. norvegicus*) rats. Transmission can also occur via direct contact with infected animal tissue via skin abrasions,

exposure to aerosolized bacteria, or contaminated foods (Butler, 2009; Raoult et al., 2013). Recent work that has examined current plague outbreaks in Central Asia, Kazakhstan, has identified gerbils (*Rhombomys opimus*) as the primary host of infected fleas (Kausrud et al., 2007). Moreover, work by these authors that links plague outbreaks to climate conditions suggests that the *Y. pestis* bacteria was continuously reintroduced to Europe, via maritime trade routes, and was unlikely to have otherwise persisted in local disease reservoirs (Schmid et al., 2015). Nonetheless, the *Y. pestis* bacteria can remain dormant in soil for up to several years and outbreaks of bubonic plague are ongoing worldwide, primarily in Africa (Raoult et al., 2013). Common symptoms of infection include high fever, hypotension (low blood pressure), chills, fatigue, and "buboes" (i.e., swollen lymph nodes) typically presenting in the groin and axilla (i.e., armpit). Early and aggressive treatment with antibiotics is essential for recovery (Raoult et al., 2013).

15.2.2 Rodent Pest Impact and Management

Rodents (*R. rattus*, *R. norvegicus*, and *Mus musculus*) continue to rank among the highest known pests to human agriculture, although fewer than 10% of rodent species are responsible for agricultural losses (Singleton et al., 2007). The last comprehensive global estimate of damage from rats and mice was in 1982 with 42 million tons of food destroyed and losses of US$30 billion (Pimentel et al., 2000). Although population statistics are uncertain, it has been estimated that there may be approximately 1.25 billion rats in the USA, primarily concentrated in farms and in urban areas (Pimentel et al., 2000). Using population modeling, it has been estimated that this equates to approximately US$27 billion in annual loses to crop consumption/destruction; this does not take into account the cost of property damage, contamination of crops, disease, and pest control (Pimentel, 2007). Rodent damage to urban businesses in New York City includes "chewed clothing, contaminated (with feces and/or urine) food products, damage to exterior packaging of electronics equipment, and gnawed books" together with gnawed walls, doors, and floors (Almeida et al., 2013). Modeling in Europe has also identified rats as among the highest-ranked invasive species for rates of economic and environmental impact (Nentwig et al., 2010). Rats and mice cause US$1.9 billion in losses per year to agriculture in Southeast Asia (Nghiem et al., 2013). In Australia, mice (*M. domesticus*) are responsible for significant agricultural losses and costs, such as by digging and consuming seeds and seedlings (Brown and Singleton, 1999, 2002) (Fig. 15.1). Plague populations (>1000 p ha^{-1}) are common, occurring on average every 4 years (Brown and Singleton, 1999, 2002).

Historically, rodent pest control management has focused on strategies designed to reduce populations of rodents, such as through the use of chemical rodenticides, such as anticoagulants. Anticoagulants prevent the blood from clotting; death is caused by internal bleeding. Substantial efforts have been made to reduce rat and mouse populations (CDC, 2012b). Notwithstanding the ethicality or otherwise of these approaches to rodent management, three of the main challenges to the efficacy of these methods are the following: (1) rodent neophobia (extreme caution of anything new), (2) the capacity for rodents to learn to associate baited foods with gastric consequences, and (3) a rapidly evolved physiological resistance to rodenticides such as warfarin (Macdonald et al., 1999).

Concerns regarding the specificity, efficacy, and humanness of poison baiting strategies have encouraged new methods, with a shift from pest control toward greater emphasis on pest management (Singleton et al., 2010). "Ecologically Based Rodent Management" provides a framework for evaluating pest management strategies in terms of the impact on nontarget species and the environment, the economic impact and the sustainability of the approach (Singleton et al., 1999). While chemical rodenticides remain widely used, other strategies, such as fertility control measures, are gaining traction but require research investment into the behavior, biology, and ecology of the target species (Singleton et al., 1999). At the present time, poisons remain the most commonly used method of rodent control (primarily anticoagulants), followed by the use of traps and habitat management (Capizzi et al., 2014).

15.3 BIRDS

Many species of birds are considered pests due to their negative impact on agricultural crops, property defacement, noise, displacement of native species, and disease transmission, among other problems. Among the species considered most problematic to human interests are pigeons, mynas, starlings, and sparrows, although this is by no means an exhaustive list.

15.3.1 Pigeon (*Columba livia*)

Rock pigeons (*C. livia*) are a highly adaptable species that have been admired and reviled for thousands of years (Blechman, 2007), with evidence of human–pigeon interactions dating back

FIGURE 15.1   **Mice plague, Australia.**

FIGURE 15.2   **Pigeon plague, Maine, USA.**

FIGURE 15.3   **Starling murmuration.**

10,000 years. The common domestic pigeon was bred from the Rock pigeon, which is also the ancestor of feral pigeons. While the history of domestic pigeons includes medals of honor for the role of homing pigeons as messengers during times of war, more recently, their feral cousins tend to be referred to as the "rats of the sky" and are reviled for their nuisance behaviors, such as damage to buildings due to the corrosive nature of their feces, and concerns regarding the spread of disease (Fig. 15.2).

Classification of Pigeons

Phylum: Chordata
Class: Aves
Order: Columbiformes
Family: Columbidae
Genus: *Columba*

15.3.2 English or Common Starling (*Sturnus vulgaris*)

The English or common starling (*S. vulgaris*) is a highly adaptable and invasive species with established large populations worldwide (Campbell et al., 2012a). Starlings are omnivores with a diverse diet that includes scavenging refuse in times of food scarcity; their choices are limited only by the morphology of their bill and intestinal system. Starlings prefer to forage in cleared areas, such as lawns, and may assist farmers by eating insects that would otherwise destroy crops. However, feeding flocks of up to 20,000 birds can cause substantial losses across a range of fruit and vegetable crops (including cherries, grapes, blueberries, and olives in particular) and cereal industries; losses attributed to starlings are estimated to cost up to US$800 million per year in the US alone (Campbell et al., 2012b) (Fig. 15.3). Starlings are a cavity-nesting species that displace native species through competition for shelter; exploitation of this behavioral trait has been used successfully to trap and control their populations (Campbell et al., 2012a).

Classification of Starlings

Phylum: Chordata
Class: Aves
Order: Passeriformes
Family: Sturnidae
Genus: *Sternus*

15.3.3 Indian or Common Myna (*Acridotheres tristis*)

The Indian or common myna (*A. tristis*) (Fig. 15.4), belonging to the family Sturnidae and a relative of the Starling, is a territorial omnivore that was named among the worst 100 invasive species that present a threat to biodiversity, agriculture, and human interests (Lowe et al., 2000). A native to India and southern Asia, mynas are aggressive and, as an invasive species in many countries throughout the world, compete with native birds for nest cavities in which to breed; the active exclusion of natives to these breeding sites has negatively impacted their reproductive success. Nicknamed the "garbage bird" (Old et al., 2013), recent research suggests that mynas are able to capitalize on habitat modification, which also threatens native species; acting as a "passenger" of habitat change (Grarock et al., 2014). This means that the presence of mynas may restrict the population of native birds in urban environments, but not in natural and rural habitats; accordingly, their impact on native flora and fauna remains uncertain (Old et al., 2013).

FIGURE 15.4 Indian or common myna (*Acridotheres tristis*).

FIGURE 15.5 House sparrow.

Classification of Mynas

Phylum: Chordata
Class: Aves
Order: Passeriformes
Family: Sturnidae
Genus: *Acridotheres*

15.3.4 House Sparrow (*Passer domesticus*)

The common house sparrow (*P. domesticus*) is a human/urban commensal species (i.e., benefits from living alongside humans), native to northern Africa and Eurasia, which is found in most continents throughout the world (Magudu and Downs, 2015) (Fig. 15.5). Sparrows were introduced into the USA in the 1850s–60s and, since then, they have invaded much of North America and have expanded their range to southern Canada, Mexico, and Central and South America (Clergeau et al., 2004; Liebl et al., 2015). Sparrows are opportunistic omnivores that feed on insects, plants, and human food waste; it is possible that they may displace native species through competition for food and nesting sites (Clergeau et al., 2004); however, their negative impact on native biodiversity is contentious (Magudu and Downs, 2015). Nonetheless, sparrows are considered to be a significant agricultural pest in many places due to the damage they cause to fruit, vegetable, and cereal crops (McGregor and McGregor, 2008); their preference for urbanized environments is thought by some to limit this impact (Magudu and Downs, 2015). Notwithstanding its success as an invasive species, some populations of sparrows in their native range, such as the United Kingdom, have seen substantial reductions and are now considered a conservation concern (Magudu and Downs, 2015).

Classification of Sparrows

Phylum: Chordata
Class: Aves
Order: Passeriformes
Family: Passeridae
Genus: *Passer*

15.3.5 Contamination and Disease

Birds are reservoirs of many zoonotic diseases; wild migratory birds can disperse dangerous pathogens around the world (Fuller et al., 2012). Commensal "pest" bird species can also spread and transmit diseases directly to native birds, poultry, humans, livestock, and other human interests via their secreta (e.g., saliva) and excreta. Disease can also be spread via indirect routes, such as via the contamination of food and crops by bird droppings, feathers, ectoparasites, nest debris, and dead birds. This section will examine some of the most problematic avian zoonotic diseases that currently threaten humans and their interests.

15.3.5.1 *Zoonotic Fungi: Cryptococcus neoformans, Histoplasmosis*

C. neoformans and histoplasmosis are examples of potentially fatal fungal diseases that are found worldwide in soil that has been contaminated by bird or bat droppings and are spread via airborne fungi spores that are inhaled following disruption of soil or droppings (Cafarchia et al., 2006; CDC, 2013). Endemic in many regions of the world, these diseases exert their most significant effects in immunocompromised hosts (such as HIV/AIDs-infected patients) and can cause meningitis (acute inflammation around the brain and spinal cord) and lung infections (Malik et al., 2003; Soltani et al., 2013). These opportunistic fungal infections are also seen in companion animals (cats and dogs), livestock (cattle, sheep, and goats), and wild animals (Malik et al., 2003). Infection mimics bacterial pneumonia and is accompanied by respiratory symptoms such as dyspnea (difficulty breathing), fever, chills, cough, and chest pain (Wheat et al., 2016). Most people recover from exposure to these fungal diseases via their natural immunity responses; however, if the infection is severe, chronic, or if it has spread throughout the body, treatment with antifungal medications is indicated (Wheat et al., 2016).

15.3.5.2 *Zoonotic Bacterium: Salmonella, Escherichia coli*

Zoonotic bacterium, such as *Salmonella* (*Salmonella enterica*) and *E. coli*, can live in the intestinal tract of birds (Callaway et al., 2014) and cause several diseases in most vertebrates, most commonly by ingestion of fecal contaminated food (Cheney, 2016). Initially, the genus *Salmonella*, named after the American veterinary pathologist and bacteriologist Daniel E. Salmon, was classified according to the diseases it causes and the species it infects; however, on the basis of more recent work, these have been combined into a single species, *Sa. enterica*, with five subspecies and over 2000 serovars (strains); of these fewer than 100 are known to cause diseases in humans (Cheney, 2016). *Salmonella* is one of the most common causes of gastroenteritis, symptoms of which include diarrhea, nausea, vomiting and stomach cramps, fever, headache, and dehydration (Rothrock et al., 2015). In most humans, these symptoms will resolve within a week; however, it can be lethal in children, the elderly, and in immunocompromised people.

15.3.5.3 Zoonotic Viruses: Highly Pathogenic Avian Influenza, Newcastle Disease

Influenza viruses are common in humans and animals. In particular, birds are a common reservoir of influenza type A, an important zoonotic pathogen, which has a high human morbidity and mortality rate (Fuller et al., 2015). Avian influenza is found in wild birds (e.g., water birds) and poultry (e.g., turkeys and chickens) species and is known to cause severe diseases such as encephalitis (inflammation of the brain) (Bröjer et al., 2012). In recent years, transmission of avian influenza strain (H5N1, H1N1, and H7N9) to humans and livestock animals has highlighted that while control measures are improving, there remains global vulnerability to the significant costs associated with avian pandemics (Mei et al., 2013).

Transmission to humans and other animals is known to have occurred via ingestion, inhalation, or direct transmission to the mucous membranes (Siembieda et al., 2008). Transmission to humans may be attributable to contact with infected birds or other infected mammals (Reperant et al., 2012). Symptoms of infection are broad, ranging from conjunctivitis to severe respiratory illness to multiorgan disease; it is lethal in birds and can be lethal in humans (CDC, 2015), raising concerns of the possibility of a human pandemic (Siembieda et al., 2008). During the period January 2014 to April 2015, a total of 191 human cases of H5N1 from 5 countries were reported to the World Health Organization (WHO) (Hammond et al., 2015). All cases were attributed to infection from poultry and most were hospitalized (98%) with a 32% fatality rate (62/191) (Hammond et al., 2015). Early intervention with antiviral medications, such as oseltamivir, is indicated for H5N1, although cases of antiviral resistance have been recorded (de Jong et al., 2005). Recent outbreaks of avian influenza in China in 2013 can produce significant public health concerns in the community (Fig. 15.6).

FIGURE 15.6 **Bird flu and swine flu masks.**

15.3.6 Avian Pest Control and Management

Birds as pests are associated with substantial health, environmental, and economic burdens. For example, damage to buildings as a consequence of nesting and fecal matter from birds present significant costs to industry and governments. The displacement of native species, distribution of invasive plants via seeds in fecal matter, and noise pollution, are also very costly environmental problems. Threats to human and animal health from avian zoonotic diseases, such as avian influenza, raise significant food safety and pandemic concerns.

In an attempt to counter these problems, many lethal and nonlethal strategies have been adopted for avian pest control (Campbell et al., 2012b). Some nonlethal strategies have been successful and are widely adopted (e.g., the use of nets or spikes to discourage nesting; electrification of surfaces to discourage roosting). The use of lethal control methods, such as shooting, trapping, and baiting, are more controversial and can divide communities, particularly if such methods are perceived as being unethical (i.e., leading to animal suffering). The success or otherwise of control methods is highly variable, with greater impact observed in geographically isolated regions, such as small islands (Campbell et al., 2012b). Historically, there are many examples of pest control efforts gone awry; in one example in Perth, Australia, the narcotic agent used in treated grain to induce stupor to make birds more easy to catch and kill was dosed incorrectly and intoxicated birds were observed crashing into cars and buses, and shop windows, placing everyone in substantial risk of injury (Australian Pest Control Association, 2016).

15.4 BITING AND STINGING PESTS

There are many species of animals, such as snakes, spiders, and scorpions, that present a significant threat to native ecosystems, humans, and livestock. Injuries to humans and other animals from biting and stinging animals can disrupt entire ecosystems and/or present a significant health risk when encountered. It is estimated that approximately 10 million antivenom vials are needed worldwide to manage snake bites and scorpion stings alone (WHO, 2007). There are approximately 20,000 cases of bites or stings treated each year in the USA (Langley, 2008); this section will explore a small selection of the most dangerous species of animals, defined as pests in this context, to humans.

15.4.1 Snakes, Spiders, and Scorpions

There are approximately 3000 species of snakes globally; of these, approximately 10% are considered venomous; however, few are harmful to humans or livestock. Nonetheless, the WHO estimates that approximately 5 million people worldwide are bitten each year by snakes, resulting in over 100,000 fatalities and many more injuries and amputations (WHO, 2015a).

Members of the pit viper family, American rattlesnakes, such as the Eastern diamondback

FIGURE 15.7 **Eastern diamondback snake.**

(*Crotalus adamanteus*), are among the largest venomous species in Northern America. These ectothermic (cold-blooded), large-bodied snakes with triangular heads are known for their characteristic "rattle" at the tip of their tail; this is shaken by the snake to warn off predators (New Wildlife Federation, 2017) (Fig. 15.7). If bitten, *hemotoxic* venom is injected into the bloodstream, which causes hemolysis (destruction of red blood cells), coagulopathy (blood clotting), and organ and tissue damage; bites are potentially fatal and require urgent medical treatment with antivenin. Other snakes, such as the death adder (*Acanthophis antarcticus*) found in Australia and New Guinea, use a *neurotoxic* venom that causes paralysis and death from respiratory failure (Wickramaratna and Hodgson, 2001).

Belonging to the family Araneidae, there are over 40,000 species of spiders worldwide. As a predatory species, all spiders possess fangs and have the capacity to inject venom into their prey to kill or paralyze them. The variety of venoms is extremely diverse; however, few spiders are lethal, or even painful, to humans and large animals; localized cutaneous irritation is common whereas systemic envenomation, while serious, is rare (Isbister and Fan, 2011). Of the species of spiders known to be dangerous to humans, most are confined to a specific region or continent and only two are found worldwide;

FIGURE 15.8   **Female black widow spider.**

Latrodectus (e.g., black widow; Fig. 15.8) and *Loxosceles* (e.g., brown recluse; Fig. 15.9) spiders (Isbister and Fan, 2011). A bite from a *Latrodectus* spider (i.e., latrodectism) typically results in localized and radiating pain from the bite site, headaches, nausea, vomiting, and fatigue (Isbister and Fan, 2011); it is rarely fatal in humans and pain is usually managed effectively with analgesics (Langley, 2008). Bites from *Loxosceles* spiders, such as the brown recluse, usually cause an acute inflammatory response and erythema (i.e., skin irritation) localized to the bite site, which often graduates to skin necrosis (skin cell death) and ulceration (Buch et al., 2015; Isbister

FIGURE 15.9   **Brown recluse spider compared to a USA penny.**

and Fan, 2011); it is rarely systemic and a broad range of treatments are typically administered (e.g., antivenom, antibiotics, and steroids); however, treatment efficacy remains uncertain. While fatalities are rare (<1%), it is estimated that approximately 8000 people are bitten annually by *Loxosceles* spiders, most of which are in South America, and is thus considered a public health problem (Gremski et al., 2014).

Classification of Snakes

Phylum: Chordata
Subphylum: Vertebrata
Class: Reptilia
Order: Squamata
Suborder: Serpentes

Scorpions are predatory arthropods belonging to the same class, Arachnida, as spiders. They are estimated to sting approximately 1 million people worldwide each year (Kassiri et al., 2004) of which approximately 5000 cases are fatal (Aziz, 2015). Thus, scorpion bites are an important public health problem facing many regions of the world, particularly low- to middle-income countries (Skolnik and Ewald, 2013). There are approximately 1500 known species of scorpions worldwide, for which envenomation from approximately 50 species of neurotoxic scorpions is clinically relevant for humans (WHO, 2007); the few deadly species of scorpions all belong to the genus *Centruroides* from the family Buthidae. Neurotoxic venom is injected via a barb at the tip of the tail, which causes symptoms ranging in severity from mild localized effects to lethal systemic effects, including respiratory paralysis, convulsions, and cardiac arrest (Boyer et al., 2009). In the USA, the Arizona bark scorpion (*Centruroides sculpturatus*) is the only scorpion considered dangerous to humans, particularly children; treatment of stings by supportive care is highly effective (Fig. 15.10).

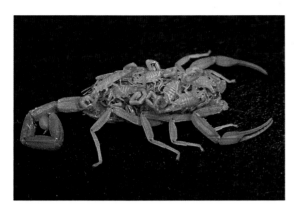

FIGURE 15.10 **Striped bark scorpion female with young on her back.**

> **Classification of Scorpions**
>
> Phylum: Arthropoda
> Subphylum: Cheliceriformes
> Class: Chelicerata
> Class: Arachnida
> Order: Scorpiones

15.5 INSECTS AS PESTS

15.5.1 Overview

Less than 0.5% insect species are considered pests (Sallam, 2013), this ranging from nuisance such as wasps, ants, and mosquitoes; insects acting as vectors; destruction of agricultural crops and forest products; damaging livestock; and impacting housing.

15.5.2 Stinging and Blood-Sucking Insects (Mosquitoes, Ants, Wasps)

Belonging to the family *Culicidae*, there are over 3500 species of mosquito; among these, some species are vectors of serious diseases (such as yellow fever, malaria, and dengue) (discussed in Chapter 14). Female mosquitoes, such as the

Anopheles gambiae, require blood protein to develop their eggs and transmit blood parasites to humans and other animal hosts when they feed (Fig. 15.11). Mosquitoes are drawn to their hosts via the detection of carbon dioxide and other host odor plumes (Spitzen et al., 2016).

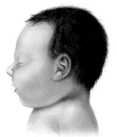

Baby with Typical Head size

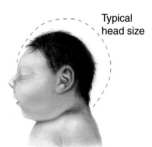

Typical head size

Baby with Microcephaly

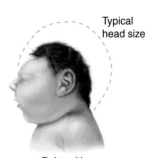

Typical head size

Baby with
Severe Microcephaly

FIGURE 15.11 **Microcephaly.** *Source: From Centers for Disease Control and Prevention, National Center on Birth Defects and Developmental Disabilities.*

The mosquito is the most dangerous pest to humans worldwide; up to 700 million people are infected with mosquito-borne illnesses annually, with more than 1 million deaths (Caraballo and King, 2014). Despite recent control measures (such as diverting mosquitoes away from humans to nearby livestock hosts), blood parasites transmitted by mosquitoes continue to present a major global health problem.

FIGURE 15.12 **Bulldog ant (*Myrmecia pyriformis*).** *Source: From Narendra, A., Reid, S.F., Raderschall, C.A., 2013. Navigational efficiency of nocturnal* Myrmecia *ants suffers at low light levels. PLoS One 8 (3), e58801. CC BY 2.5, https://commons.wikimedia.org/w/index.php?curid=25453303.*

Classification of Mosquitoes

Phylum: Arthropoda
Subphylum: Insecta
Order: Diptera
Suborder: Nematocera
Family: Culicidae

Mosquito-borne diseases can be caused by bacterium, viruses, or parasites. For example, the blood parasite *Plasmodium falciparum* that causes malaria is transmitted by female *An. gambiae* mosquitoes; in 2015, the WHO recorded over 200 million cases of malaria, with almost 450,000 deaths (WHO, 2015b). Serious viral infections, such as dengue and yellow fever, are spread by female vector *Aedes aegypti* mosquitoes, which can be identified by the characteristic white markings on its legs. It is estimated that there are approximately 50 million dengue infections each year, for which there are approximately 20,000 deaths (Whitehorn and Farrar, 2011); it has been estimated that the annual costs of dengue in the Americas range between US$1–4 billion (Shepard et al., 2011). The Zika virus, spread by infected *Ae. aegypti* and *Aedes albopictus* mosquitoes and can cause significant birth defects in the babies of women bitten by infected mosquitoes while pregnant (such as microcephaly; a developmental defect resulting in a smaller brain and head size), has received global coverage; aggressive control and management strategies are underway to combat the spread of this virus (CDC, 2016; Fig. 15.11).

There are over 12,000 species of ants worldwide, belonging to the family Formicidae. The success of ants is largely attributable to their instinct to cooperate for the good of the community (Rhoades et al., 2001). Ant infestations can cause significant problems for humans and their interests, such as by rendering land unusable; however, few species are dangerous to humans. Nonetheless, physiological reactions to ant envenomation continue to present global health and economic problems (Rhoades et al., 2001). Perhaps the most medically important among the small group of venomous ants is the aggressive bulldog ant, *Myrmecia pyriformis* (Fig. 15.12), which is native and virtually endemic to Australia; up to 50,000 people are stung each year and while death is rare, it can be very swift with cardiac arrest known to have occurred in adult male humans within 20 min (McGain and Winkel, 2002).

Classification of Ants

Phylum: Arthropoda
Subphylum: Uniramia
Class: Insecta
Order: Hymenoptera
Suborder: Apocrita
Family: Formicidae

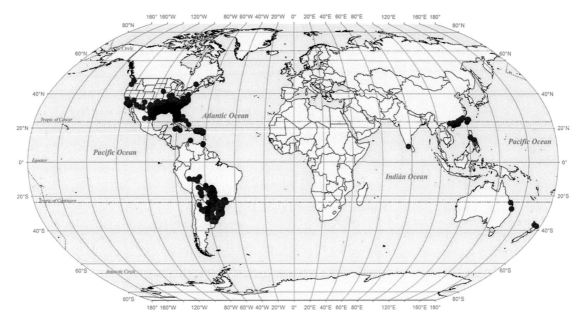

FIGURE 15.13 **Worldwide distribution records of *Solenopsis invicta*.** *Source: From James Wetterer, Sociobiology, CC BY 3.0, https://commons.wikimedia.org/w/index.php?curid=48142753.*

The omnivorous imported fire ant *Solenopsis invicta* is the somewhat less dangerous USA equivalent to the Australian bulldog ant and is very common in southern USA states, the venom of which is lethal to other insects and small animals (Rhoades et al., 2001) (Fig. 15.13). In humans, venom injected by a stinger "burns like fire" and results in characteristic sterile pustules; while their venom is nontoxic to humans, deaths arising from severe allergic reactions (such as anaphylaxis) in sensitized people and secondary infections have been reported (Rhoades et al., 2001). Costs of the fire ant in the USA are estimated at US$1 billion per year the costs including both damage and control (Pimentel et al., 2000).

Wasps, such as hornets and yellow jackets, belonging to the family Vespidae, are yet another species of stinging insect that can become a significant nuisance to humans. There are approximately 5000 species of wasps, the majority of which are solitary with approximately 1000 species living socially in aerial or underground colonies (Pickett and Carpenter, 2010). Nonnative wasps are thought to have been largely introduced accidentally via international trade, such as when wasp queens are present in horticultural produce (Beggs et al., 2011). In particular, *Vespula vulgaris* and *Vespula germanica* are highly successful as invasive, widespread pests, perhaps due to their behavioral and cognitive flexibility (Beggs et al., 2011). Invasive wasps can have substantial negative impact on native ecology through competitive displacement for important food resources, such as honeydew (Beggs et al., 2011) and predation of native species, such as honeybees (Couto et al., 2014). Wasp stings are painful to humans, but rarely life threatening; nonetheless, the risk of stings and possibility of allergic reactions disrupts outdoor activities if wasps are disturbed (Beggs et al., 2011) (Fig. 15.14).

FIGURE 15.14 **Common wasp (*Vespula vulgaris*).** *Source: From M. van Ree. V. vulgaris, CC BY-SA 2.0, https:// commons.wikimedia.org/w/index.php?curid=42072503.*

Classification of Wasps

Phylum: Arthropoda
Subphylum: Uniramia
Class: Insecta
Order: Hymenoptera
Suborder: Apocrita
Family: Vespidae

15.5.3 Insect Pests and Crop Production

Globally, insects are responsible for losses of 35% of preharvest yields with a further 12% of the crop lost during harvesting and later (Popp et al., 2013). Examples of preharvest losses due to insects include the following:

- Colorado potato beetle causing defoliation and thereby reducing yield by 64% (Hare, 1980).
- Insect damage to cassava in sub-Saharan Africa, ~50% lost (Sallam, 2013).
- Larger grain borer (*Prostephanus truncatus*) (invasive species from Central America) damage to maize, for example, in Tanzania, 35% lost (Sallam, 2013).

- Seedpod weevil (*Ceutorhynchus obstrictus*) depresses production of canola in Canada with 20% reductions in yield and Can$5 million in losses per year (Colautti et al., 2006).
- Spotted stem borer (*Chilo partellus*) (invasive species from Asia) damage to maize and sorghum in Eastern and Southern Africa, 30% lost (Sallam, 2013).

Brown Planthoppers and Rice Production

Classification of brown planthopper

Class: Insects
Order: Hemiptera
Species: *Nilaparvata lugens*

The brown planthopper (*N. lugens*) is a major insect pest (reviewed by Sogawa, 1982). Brown planthoppers are phloem sap feeders. They not only damage growing rice but also act as a vector of viruses, including rice grassy stunt tenuivirus and rice ragged stunt oryzavirus, which also damage the rice plants (reviewed by Sogawa, 1982). Another impact of brown planthoppers is the various costs of insecticides (Sarao et al., 2016).

The damage caused by arthropod pests, predominantly insects, on crop production has been estimated in different countries or regions with the following losses reported (Nghiem et al., 2013; Pimentel et al., 2000, 2001):

- Australia, US$1 billion
- Brazil, US$8.5 billion
- India, US$16.8 billion
- South Africa, US$1 billion
- Southeast Asia, US$21.6 billion (impact includes that of weeds and pathogens)
- United Kingdom, US$1 billion
- USA, US$15.1 billion

Pimentel et al. (1991a) reported, based on analysis of the literature and government reports, that insect damage to crops ranged from 7% to 13% on a dollar basis in the USA throughout the 20th century.

In addition to the damage they cause to growing crops, insects cause major losses after harvesting (Sallam, 2013). Among the species responsible for the most damage are the following: weevils [rice weevil (*Sitophilus oryzae*), maize weevil (*Sitophilus zeamais*), and granary weevil (*Sitophilus granarius*)], darkling beetles, meal worms, grain borers, grain beetles, Mediterranean flour moth (*Ephestia kuehniella*), the Angoumois grain moth (*Sitotroga cerealella*), and other moths (Sallam, 2013). The Angoumois grain moth is alone responsible for a 12% loss of stored maize (reviewed by Sallam, 2013). Other arthropods that damage grain include the flour mite (*Acarus siro*).

15.5.4 Impact of Insects on Livestock Production

15.5.4.1 Tsetse Flies

The protozoa trypanosomes cause trypanosomiasis or sleeping sickness in animals and humans in sub-Saharan Africa. It is spread by its vector, the tsetse flies, to wildlife, humans, and livestock, predominantly cattle. Trypanosomiasis has a major impact on cattle production with losses in cattle alone estimated between US$6 and US$12 billion per year in the 1990s (Hursey and Slingenbergh, 1996). The distribution of tsetse flies, and hence trypanosomiasis, impacts the ability to raise cattle and use them for plowing or transportation with greatly elevated morbidity and mortality (reviewed by Alsan, 2015). The presence of tsetse flies has been linked to low population densities and the lack of development of towns in many parts of precolonial sub-Saharan Africa (Alsan, 2015). In contrast, there were civilizations in the past "supported by advanced agricultural systems" in Great Zimbabwe and the Ethiopian highlands where tsetse flies do not survive (Alsan, 2015).

15.5.4.2 Other Insects Impacting Livestock Production

There are multiple insects that impact livestock production including stable flies, bots, and screwworm flies. Losses to cattle production in the USA due to stable flies have been reported as US$2.2 billion per year (Taylor et al., 2012). Among the insects that cause losses to Canadian livestock production is the horn fly (*Haematobia irritans*) impacting cattle with Can$69 million in losses per year and the stable fly (*Stomoxys calcitrans*) affecting dairy cattle with Can$26.8 million losses per year (Colautti et al., 2006). Insects that cause losses in cattle in Brazil are the horn fly (*H. irritans*), US$2.56 billion; cattle grub (*Dermatobia hominis*), US$380 million; New World screwworm fly (*Cochliomyia hominivorax*), US$340 million; and stable fly (*St. calcitrans*), US$340 million (Grisi et al., 2014). Bot flies lay their eggs with the larvae developing under the skin of the host livestock with the principal damage being caused as the larvae emerges.

15.5.5 Impact of Insects on Forestry

There has been increasing concern on the impact of insects on forests. Examples of such impacts include the following:

- Sirex woodwasp (*Sirex noctilio*) damaging pine plantations (>2 million ha) in Argentina, Southern Brazil, and Uruguay with an economic impact of US$7 million per year. A biological control program with the nematode *Deladenus siricidicola* has been implemented (Allard et al., 2003).
- Canadian forestry losses due to Asian longhorn beetle (*Anoplophora glabripennis*) with tree mortality and Can$16 million lost per year; the gypsy moth (*Lymantria dispar*), also tree mortality with Can$121

million lost per year; the beetle emerald ash borer (*Agrilus planipennis*), with Can$15 million lost per year; and the brown spruce longhorn beetle (*Tetropium fuscum*), with Can$175 million lost per year (Colautti et al., 2006).

- Forestry losses in the USA are estimated at US$2.1 billion per year (Pimentel et al., 2001).

15.5.6 Impact of Insects on Recreational Facilities Including Parks

Insects have marked impacts on recreational housing and parks and parkways under municipal and state governments. The impact of damage from boring insects has been estimated as US$1.7 billion in costs per year to local governments. Arthropod pests (predominantly insects) in lawns including golf courses cause considerable damages. This is estimated to cost US$1.5 billion in the USA alone (Pimentel et al., 2000). Insecticide use in the USA is 25% to households and 11% to industrial and government lands (Pimentel et al., 1991b).

15.5.6.1 *Overview*

The impact of damage from boring insects has been estimated as US$1.7 billion in costs per year to local governments and US$830 million lost in residential property values in the USA; with most damaging species being the emerald ash borer (*Ag. planipennis*), hemlock woolly adelgid (*Adelges tsugae*), and gypsy moth (*L. dispar*) (Aukema et al., 2011) with the latter costing US$11 million (Pimentel et al., 2000). Arthropod pests (predominantly insects) in lawns including golf courses cause considerable damages estimated at US$1.5 billion in the USA alone (Pimentel et al., 2000).

In the USA, the most damaging wood-destroying insects are subterranean termites; damages to buildings in the USA being more than US$1 billion per year (Lewis, 2006) while treatment and prevention costing US$2 billion per year (Suiter et al., 2012).

15.5.6.2 *Cockroaches*

Cockroaches are common pests in homes and businesses and in drains and sewers (WHO, 2010). Among the species of cockroaches that have the most impact globally are the following (WHO, 2010):

- American cockroach (*Periplaneta americana*)
- German cockroach (*Blattella germanica*)
- Oriental cockroach (*Blatta orientalis*)

Cockroaches can be carriers for intestinal diseases such as diarrhea, typhoid fever, and cholera together with intestinal parasites and can trigger allergic responses (WHO, 2010). For instance, pathogenic bacteria found in cockroaches in India include the following (Wannigama et al., 2014):

- *E. coli* (intestinal bacteria in humans and other animals).
- *Klebsiella pneumonia*; according the CDC (2012a), infections can lead to pneumonia, bloodstream infections, and meningitis.
- *Pseudomonas aeruginosa*; according the CDC (2012b), infections can lead to pneumonia and bloodstream infections.
- *Enterobacter aerogenes*; viewed as an increasingly important human pathogen/opportunistic (Sanders and Sanders, 1997).

Parasites identified in cockroaches in Africa include hookworms, *Entameba histolytica*, *Trichuris trichuira*, and round worms (*Ascaris lumbricoides*) (Etim et al., 2013). Cockroaches are important triggers in the development of asthma with the following (Shurdut and Peterson, 1999):

- 23%–60% of urban residents in the USA with asthma exhibit sensitivity to cockroach allergens.
- 37% of inner-city children in the USA are sensitive to cockroach allergens.
- 25%–30% of school children in Malaysia, Hong Kong, and southern China are sensitive to cockroach allergens.

15.5.6.3 Termites

> ## Pentagram
>
> One pentagram (Pg) = 10^{15} g or 1 quadrillion grams
> One pentagram = 10^{12} kg or 1 trillion kilograms
> One pentagram = 10^9 t or 1 billion metric tons (US) or 1 billion tonnes (UK)

Forests represent a very important carbon sink with a net 2.4 Pg of carbon added per year (Pan et al., 2011). It is estimated that 72 Pg of carbon is stored in dead wood (Cornwell et al., 2009) including lumber. Termites play a critically important role in the carbon cycle as wood-destroying organisms (Cornwell et al., 2009) and have been reported to represent about 10% of animal biomass in tropical soils and over 90% of soil insect biomass (Donovan et al., 2007). The rates of breakdown of wood by termites are the following (calculated from Maynard et al., 2015):

- Formosan subterranean termite (*Coptotermes formosanus*) 63 mg g^{-1} day^{-1}
- The eastern subterranean termite (*Reticulitermes flavipes*) 42 mg g^{-1} day^{-1}

Termites markedly affect the amount of carbon stored and the rates of production of the greenhouse gas, methane, with rates of methane production from breakdown of wood by termites being the following (calculated from Maynard et al., 2015):

- Formosan subterranean termite (*Co. formosanus*) 14 mg kg^{-1} day^{-1}
- The eastern subterranean termite (*Re. flavipes*) 55 mg kg^{-1} day^{-1}

The termite is a major insect pest. Termites damage unprotected cellulosic materials such as "timber, underground cables, earthen dams, irrigation ditches, and farming equipment" (Ghaly and Edwards, 2011). Annual losses from termites are high with these being in the millions of dollars (reviewed by Ghaly and Edwards, 2011):

- Australia, US$100 million
- China, US$482 million
- India, US$35 million
- Japan, US$800 million
- Malaysia, US$9 million
- USA, US$1 billion

In the USA, the most damaging wood-destroying insects are subterranean termites; damages to buildings including housing being more than US$1 billion per year (Lewis, 2006). Treatment and prevention costs $2 billion per year (Suiter et al., 2012). However, termites are not only responsible for damages to property. There is also increasing evidence that ants and termites can increase wheat production in dryland agriculture (Evans et al., 2011).

15.5.6.4 Boring Insects

It is estimated that the impact of damage from boring insects is US$830 million; this being due to losses in residential property values in the USA due to damage to trees (Pimentel et al., 2000). The most damaging species are the emerald ash borer (*Ag. planipennis*), hemlock woolly adelgid (*Ad. tsugae*), and the gypsy moth (*L. dispar*) (Aukema et al., 2011).

15.5.7 Control of Insects

15.5.7.1 Overview

Protection of crops, livestock, housing, and people from insects can be achieved by pesticides or genetically engineered crops expressing toxins to groups of insects. This can also be done by integrated pest management (IPM) (reviewed by Popp et al., 2013).

TABLE 15.1   Temporal Changes in the Use of Insecticide in Corn and Cotton Production in the USA

| | Pesticide usage (million kg active ingredient) | | | |
| | Insecticide | | Herbicide | |
Year	Usage for corn production	Usage for cotton production	Total usage for 21 crops	Total usage for 21 crops
1960	5.0	30.2	52	16
1980	16.0	8.2	48	213
2000	5.4	18.6	32	161
2008	1.8	2.4	13	179

Comparisons with total usage of insecticides and herbicide in the USA for selected crops: corn, cotton, soybeans, potatoes, wheat, sorghum, peanuts, rice, tomatoes, grapes, apples, barley, grapefruit, lemons, lettuce, peaches, pears, pecans, sugarcane, and sweetcorn.
Data from Fernandez-Cornejo, J., Nehring, R., Osteen, C., Wechsler, S., Martin, A., Vialou, A., 2014. Pesticide use in U.S. agriculture: 21 selected crops, 1960–2008, EIB-124, U.S. Department of Agriculture, Economic Research Service, May 2014.

15.5.7.2 *Insecticides*

Global insecticide use in 2005 was US$7.8 billion; representing 25% of all 4.6 million tons of chemical pesticides (Zhang et al., 2011). Agriculture in the USA in 2005 used insecticides costing US$1.3 billion (Zhang et al., 2011). The usage of insecticides and herbicides in the USA for 21 crops is summarized in Table 15.1.

In the USA, pesticides are evaluated by the US Environmental Protection Agency (EPA) under the Federal Insecticide, Fungicide, and Rodenticide Act. This is to assure that the pesticide does not damage the environment or human health, when used following the label instructions (reviewed by Alavanja, 2009).

The book, Silent Spring, published in 1962 and written by Rachel Carson, led to the banning of the insecticide, DDT (dichlorodiphenyltrichloroethane). The book provided evidence for severe environmental consequences of the use of DDT, which had been used in agriculture from the 1940s until it was banned in the USA in 1972 by the EPA.

15.5.7.2.1 ADVANTAGES OF INSECTICIDES

In the USA, 62% of insecticides are applied to agriculture (Pimentel et al., 1991b). The dollar returns on farmers' use of insecticides are estimated at 1:3–4 (Pimentel et al., 1991b). Popp (2011) concluded that "chemical pesticides will continue to play a role in pest management for the future…there are no practical alternatives" (Popp, 2011) and "without pesticides, food production would drop and food prices would soar" (Popp et al., 2013).

15.5.7.2.2 PROBLEMS WITH INSECTICIDES

There are risks to human health and to the environment by the use of pesticides. Between 1997 and 2003, it is estimated that there were 7238 deaths in China due to pesticides (Zhang et al., 2011) and 45 accidental fatalities per year in the USA (Pimentel, 2007). Animal poisoning from pesticides occurs with, for instance, 120 cattle poisoned with 9 dying and 59 dogs poisoned with 4 fatalities in the USA (Pimentel, 2005). There are links between the incidence of cancer in pesticide applicators using some but not all pesticides (Weichenthal et al., 2010). High exposure to pesticides in pesticide applicators can be reduced by behaviors and training (Payne et al., 2012).

There are estimates that about 50 million people in the USA consume drinking water from groundwater "potentially contaminated by pesticides and other agricultural chemicals" (reviewed by Alavanja, 2009). Moreover, pesticide residues are found in bodies of water in China (Zhang et al., 2011). Crop losses due to pesticide resistance are estimated to be about US$1.5 billion per year (Pimentel, 2005). The Los Angeles Unified School District considers that under the Precautionary Principle, "no pesticide product is free from risk or threat to human health" but that full implementation of this "is not possible at this time and may not be for decades" (Rachel, 2000).

Information to Guide Public Policy: What Is a Poison?

"All things are poison and nothing is without poison; only the dose makes a thing not a poison" Paracelsus (Philippus Aureolus Theophrastus Bombastus von Hohenheim; 1493–1541) physician, scientist, and founder of toxicology.

Natural Safe, Unnatural Unsafe?

Many people believe that if a chemical is natural, it is safe (Ventola, 2010). However, there can be toxicity. Among the toxic natural chemicals include Batrachotoxin produced by the poison dart frog, botulinum toxins, digoxin, morphine, nicotine, snake venom, and strychnine.

"Public policy in the twentieth century was about protecting and expanding the social compact, based on recognition that effective government at the federal level provides rules and services and safety measures that contribute to a better society" Carl Bernstein (1944 to date), journalist.

However, regulations are not necessarily consistent. For instance, the Food and Drug Administration regulates drugs in the USA requiring demonstration of the following: efficacy (do they work?) and safety (are they safe?) together with any therapeutic claims (are they as advertised?). On the other hand, supplements are regulated in the USA by the Food and Drug Administration under the Dietary Supplement Health and Education Act (1994) as foods. Demonstration of efficacy and safety is not required. Statements about therapeutic actions are prohibited but statements about effects on body functioning are allowed; an example of the latter being "glucosamine supplement can state that it improves joint function" (Ventola, 2010). There is no burden for producers of supplements to submit evidence that their products work or are safe (Ventola, 2010). An alternative approach is the "Precautionary Principle." According to the statement from Wingspread (1998), precautionary principle encompasses "the proponent of an activity, rather than the public, should bear the burden of proof" of demonstrating safety. Kriebel et al. (2001) had a series of conclusions supporting the application of the precautionary principle. It sounds reasonable. For instance, "the precautionary principle, then, is meant to ensure that the public good is represented in all decisions made under scientific uncertainty. When there is substantial scientific uncertainty about the risks and benefits of a proposed activity, policy decisions should be made in a way that errs on the side of caution with respect to the environment and the health of the public" (Kriebel et al., 2001). What is not clear is how science-based approval (or reapproval) processes can be developed under this. It is reasonable to have government regulations for the safety of chemicals released into the environment including establishing the toxicity in a range of species, any teratogenic effects, and possible endocrine disruptor (affecting multiple generations) effects.

Pertinent quotations in the discussion and/or development of issues including public policy are the following:
1. "You are entitled to your opinion. But you are not entitled to your own facts." Daniel Patrick Moynihan (1927–2003), US senator for New York for 18 years.

2. In contrast, the antiintellectual view is "people in this country have had enough of experts." Michael Grove (1967 to date), British politician.

3. The Economist (2016) observed that in discussing politics that we are moving to a "post-truth" era. "…That truth is not falsified, or contested, but of secondary importance. …Feelings, not facts, are what matter." Unfortunately, this can be extended to all of public policy.

15.5.7.3 Plants Genetically Engineered to Produce Insecticide Proteins

15.5.7.3.1 OVERVIEW

Plants such a corn (maize) or cotton have been genetically engineered (GE) to produce crystal (Cry) proteins. These Cry proteins are normally produced by a specific bacteria in the soil, namely, *Bacillus thuringiensis* (Bt). These Bt toxins are natural insecticides and hence can also be used in organic agriculture.

Genetically Engineered Plants

Plants can be modified by the process of genetic engineering with the addition of one or more genes from a different organism(s). The resulting plant is referred to as genetically modified or a genetically modified organism. Traditional plant breeding employs selection by plant breeders for specific characteristics. To increase genetic differences, there can be efforts to increase the number of mutations in the genes by using either specific chemicals or radiation to cause mutations. Genetic engineering differs from plant crossing in hybrids when genes can be introgressed from the same or closely related species.

The National Academies' Committee on Identifying and Assessing Unintended Effects of Genetically Engineered Foods on Human Health considers that genetic engineering "is not inherently hazardous" but there can be unintended consequences to the composition of the crop (NRC, 2004).

15.5.7.3.2 ADVANTAGES OF BT CORN AND BT COTTON

There is a series of advantages of Bt corn and Bt cotton. Bt corn and Bt cotton have been available in the USA since 1996 and have been rapidly adopted by farmers with the percentage of the crop rising as follows (USDA Economic Research Service, 2016):

- Bt corn increasing from 8% in 1997 to 79% in 2016. Initially, Bt corn targeted the European corn borer (Hellmich and Hellmich, 2012). New Bt varieties also target the corn rootworm and the corn earworm.
- Bt cotton increasing from 15% in 1997 to 84% in 2016 (Bt target insects pests: tobacco budworm, bollworm, and pink bollworm).

Bt corn has benefited farmers in the USA (Hutchison et al., 2010) and hence the high adoption rate.

Bt cotton was first available in India in 2002. It was rapidly adopted with 7 million farmers using Bt cotton on 10.4 million hectares (90% of cotton-producing land) in 2011. Bt cotton has been opposed by some activists but the available refereed data demonstrate a large increase in yield and profit (Kathage and Qaim, 2012) (Table 15.2).

Prior to the development of Bt corn hybrids, damage due to European corn borer was US$1 billion per year and to the Western corn rootworm (*Diabrotica virgifera virgifera*) was US$1 billion per year (reviewed by Bohnenblust et al., 2014; Hutchison et al., 2010) and 25% of insecticides in the USA was used in the production of corn and cotton (Pimentel et al., 1991b). The adoption of genetically modified Bt corn and

TABLE 15.2 Impact of Bt Cotton on Production Parameters in India (Kathage and Qaim, 2012)

	Conventional cotton	Bt cotton
Yield (kg ha^{-1})	1387	1917 (up 38%)
Seed cost (1000 R ha^{-1})	1.22	3.14
Pesticide cost (1000 R ha^{-1})	4.17	3.12
Profit (1000 R ha^{-1})	11.1	20.6 (up 86%)[a]

R, Rupees (in 2016, 1 R = US$0.015).
[a] *Increase in profit US$142.5 ha^{-1}.*

cotton in the mid-1990s greatly influenced insecticide use (Fernandez-Cornejo et al., 2014) with an 89% decrease in insecticide usage for corn production (Table 15.1). The decline in insecticides for cotton production in 1980 (Table 15.1) was due to the introduction of synthetic pyrethroids, which are based on pyrethrum, a natural compound from plants (chrysanthemum). These were applied at lower rates than older insecticides (Fernandez-Cornejo et al., 2014).

There are lower levels of mycotoxins (toxic chemicals produced by fungi) in Bt corn (Munkvold et al., 1999; multiple studies reviewed by Ostry et al., 2010). This is explained by the lower damage by insects.

15.5.7.3.3 DISADVANTAGES OF BT CORN AND BT COTTON

There is potential for insects to build up resistance to Bt protein. To slow the development of resistance to Bt by the pest species, the US EPA requires refuge with non-Bt corn (Hutchison et al., 2010).

15.5.7.4 Integrated pest management

IPM is an alternative approach to the use of pesticides. IPM approaches include the introduction of predators and specific parasitoids (species of wasps that are internal parasites of other insects). There have been studies reporting that there is no significant effect on yield or profitability for those using IPM compared to those who do not adopt IPM in the fresh tomato market (Fernandez-Cornejo, 1996). In contrast,

income was increased for mango farmers who used a specific IPM approach—spot-sprayed food bait (Muriithi et al., 2016). In a metaanalysis of data of 145 studies, the effectiveness of either parasitoids or predators was demonstrated to be good in reducing the number of pests and increasing pest death (Stiling and Cornelissen, 2005).

15.6 NEMATODES AS PESTS IN CROP PRODUCTION

Various species of nematodes are parasitic in animals and humans (see Chapter 16), in plants, and free-living (nonparasitic) (Schmitt, 1994). Nematodes reduce the production of crops globally by 12% resulting in losses; 15% in developing countries and 9% in industrialized countries (Barker et al., 2010). Losses due to nematodes are estimated at US$78 billion per year (Barker et al., 2010). Reduction of yield due to nematodes in the USA is estimated at US$8 billion per year (Barker et al., 2010). Options to overcome the effects of these parasitic nematodes include the use of nematocides or the development of nematode-resistant crop plants:

- Maize can have over 60 different species of nematodes associated. The evidence of the impact of nematodes on maize production comes from the impact of nematicides with increases in production: doubling in Brazil, 32% increase in the USA, and 80% in South Africa (reviewed by Nicol et al., 2011).

- Nematodes have been reported to cause large decreases in rice yield with losses of up to 60% with *Aphelenchoides besseyi*, about 13% with *Heterodera elachista*, and up to 50% with *Heterodera sacchari* (reviewed by Nicol et al., 2011).
- Similarly, nematodes can cause large decreases in wheat yield with losses due to *Heterodera avenae* of over 20% in Australia and the USA and about 50% in developing countries (reviewed by Nicol et al., 2011). The financial losses in Australia due to *H. avenae* are estimated at A$72 million (reviewed by Nicol et al., 2011). The nematode *Pratylenchus thornei* can have even larger negative consequences (reviewed by Nicol et al., 2011).
- The cyst nematodes *Globodera rostochiensis* and *Globodera pallida* reduce potato production in the UK by 9% resulting in losses of about US$70 million per year (reviewed by Nicol et al., 2011).
- Losses in soybean production in the USA due to soybean cyst nematode (*Heterodera glycines*) were estimated to be 3.47 million metric tons per year between 2006 and 2009 (Koenning and Wrather, 2010); this represents 31% of all losses in soybean production due to plant diseases.

Global sales of nematicides were over US$1 billion in 2015 (Marketsandmarkets, 2016). Gastrointestinal nematodes have a large economic impact on livestock and poultry production with nematodes causing losses of US$7.11 billion in Brazilian cattle production (Grisi et al., 2014).

15.7 CONCLUSIONS

It is argued that some species of animals, including insects, are defined as pests due to their interference with, and destruction of, human interests. Indeed, pests are so destructive that they are responsible for so much human misery.

Losses of crops or damage to livestock can be due to insects, rodents, birds, or nematodes (see Section 15.5). For a subsistence farmer, damage caused by pests can result in insufficient food and dire consequences. Various control measures, such as rodenticides, pesticides, and insecticides, are presently in use; however, these have environmental and health consequences, and also increased production costs. Greater usage of alternative methods of control, such as IPM and genetic modification, may reduce the negative side effects of pest control.

References

Alavanja, M.C.R., 2009. Pesticides use and exposure extensive worldwide. Rev. Environ. Health 24, 303–309.

Allard, G.B., Fortuna, S., Lee Su See, Novotny, J., Baldini, A., Courtinho, T., 2003. Global information on outbreaks and impact of major forest insects pests and diseases. FAO. Available from: http://www.fao.org/docrep/article/wfc/xii/1019-b3.htm.

Almeida, A., Corrigan, R., Sarno, R., 2013. The economic impact of commensal rodents on small businesses in Manhattan's Chinatown: trends and possible causes. Suburban Sustainability 1, 1–11.

Alsan, M., 2015. The effect of the tsetse fly on African development. Am. Econ. Rev. 105, 382–410.

Aplin, K.P., Chesser, T., Have, J.T., 2003. Evolutionary biology of the genus Rattus: profile of an archetypal rodent pest. In: Singleton, G.R., Hinds, L.A., Krebs, C.J., Spratt, D.M. (Eds.), Rats, Mice and People: Rodent Biology and Management. ACIAR Monograph No. 96. P487–P498.

Aukema, J.E., Leung, B., Kovacs, K., Chivers, C., Britton, K.O., Englin, J., Frankel, S.J., Haight, R.G., Holmes, T.P., Liebhold, A.M., McCullough, D.G., Von Holle, B., 2011. Economic impacts of non-native forest insects in the continental United States. PLoS One 6, e24587.

Australian Pest Control Association, 2016. Available from: http://www.pestcontrol.org.au/birds.html.

Aziz, H., 2015. The current concepts in management of animal (dog, cat, snake, scorpion) and human bite wounds. J. Trauma Acute Care Surg. 78, 641–648.

Barker, K.R., Hussey, R.S., Krusberg, L.R., Bird, G.W., Dunn, R.A., Ferris, H., Ferris, V.R., Freckman, D.W., Gabriel, C.J., Grewal, P.S., MacGuidwin, A.E., Riddle, D.L., Roberts, P.A., Koenning, S.R., Wrather, J.A., 2010. Suppression of soybean yield potential in the Continental United States by plant diseases from 2006 to 2009. Plant Management Network. Available from: http://www.plantmanagementnetwork.org/pub/php/research/2010/yield/.

Beggs, J.R., Brockerhoff, E.G., Corley, J.C., Kenis, M., Masciocchi, M., Muller, F., 2011. Ecological effects and management of invasive alien Vespidae. Biocontrol 56, 505–526.

Blechman, A., 2007. Pigeons—The Fascinating Saga of the World's Most Revered and Reviled Bird. University of Queensland Press, St Lucia, Queensland.

Bohnenblust, E.W., Breining, J.A., Shaffer, J.A., Fleischer, S.J., Roth, G.W., Tooker, J.F., 2014. Current European corn borer, *Ostrinia nubilalis*, injury levels in the northeastern United States and the value of Bt field corn. Pest Manage. Sci. 70, 1711–1719.

Boyer, L.V., Theodorou, A.A., Berg, R.A., Mallie, J., Chávez-Méndez, A., García-Ubbelohde, W., Hardiman, S., Alagón, A., 2009. Antivenom for critically ill children with neurotoxicity from scorpion stings. N. Engl. J. Med. 360, 2090–2098.

Bröjer, C., Agren, E.O., Uhlhorn, H., Bernodt, K., Jansson, D.S., Gavier-Widén, D., 2012. Characterization of encephalitis in wild birds naturally infected by highly pathogenic avian influenza H5N1. Avian Dis. 56, 144–152.

Brown, P.R., Singleton, G.R., 1999. Rate of increase as a function of rainfall for house mouse *Mus domesticus* populations in a cereal-growing region in southern Australia. J. Appl. Ecol. 36, 484.

Brown, P.R., Singleton, G.R., 2002. Impacts of house mice on crops in Australia—costs and damage. In: Clark, L., Hone, J., Shivik, J.A., Watkins, R.A., VerCauteren, K.C., Yoder, J.K. (Eds.), Human Conflicts with Wildlife: Economic Considerations. National Wildlife Research Center, Fort Collins, Colorado, pp. 48–58.

Buch, D.R., Souza, F.N., Meissner, G.O., Morgon, A.M., Gremski, L.H., Ferrer, V.P., Trevisan-Silva, D., Matsubara, F.H., Boia-Ferreira, M., Sade, Y.B., Chaves-Moreira, D., Gremski, W., Veiga, S.S., Chaim, O.M., Senff-Ribeiro, A., 2015. Brown spider (*Loxosceles* genus) venom toxins: evaluation of biological conservation by immune cross-reactivity. Toxicon 108, 154–166.

Butler, T., 2009. Plague into the 21st century. Clin. Infect. Dis. 49, 736–742.

Cafarchia, C., Romito, D., Iatta, R., Camarda, A., Montagna, M.T., Otranto, D., 2006. Role of birds of prey as carriers and spreaders of *Cryptococcus neoformans* and other zoonotic yeasts. Med. Mycol. 44, 485–492.

Calderón, A., Rodríguez, V., Máttar, S., Arrieta, G., 2014. Leptospirosis in pigs, dogs, rodents, humans, and water in an area of the Colombian tropics. Trop. Anim. Health Prod. 46, 427–432.

Callaway, T.R., Edrington, T.S., Nisbet, D.J., 2014. Isolation of *Escherichia coli* O157:H7 and *Salmonella* from migratory brown-headed cowbirds (*Molothrus ater*), common Grackles (*Quiscalus quiscula*), and cattle egrets (*Bubulcus ibis*). Foodborne Pathog. Dis. 11, 791–794.

Campbell, S., Cook, S., Mortimer, L., Palmer, G., Sinclair, R., Woolnough, A.P., 2012b. To catch a starling: testing the effectiveness of different trap and lure types. Wildlife Res. 39, 183–191.

Campbell, S., Powell, C., Parr, R., Rose, K., Martin, G., Woolnough, A., 2012a. Can artificial nest-cavities be used as a management tool to assist the control of Common Starlings (*Sterns vulgaris*)? Emu 112, 255–260.

Capizzi, D., Bertolino, S., Mortelliti, A., 2014. Rating the rat: global patterns and research priorities in impacts and management of rodent pests. Mammal Rev. 44, 148–162.

Caraballo, H., King, K., 2014. Emergency department management of mosquito-borne illness: malaria, dengue, and West Nile virus. Emerg. Med. Pract. 16, 1–23.

CDC, 2012a. Outbreak of Hantavirus infection in Yosemite National Park. Available from: http://www.cdc.gov/hantavirus/outbreaks/yosemite-national-park-2012.html.

CDC, 2012b. Evaluation of a neighborhood rat-management program—New York City, December 2007–August 2009. Morbidity and Mortality Weekly Report (MMWR). Available from: http://www.cdc.gov/mmwr/preview/mmwrhtml/mm6137a1.htm.

CDC, 2013. Histoplasmosis outbreak associated with the renovation of an old house—Quebec, Canada, 2013. MMWR 62.

CDC, 2015. Avian influenza A virus infections in humans. Available from: http://www.cdc.gov/flu/avianflu/avian-in-humans.htm.

CDC, 2016. Zika virus. Available from: https://www.cdc.gov/zika/.

Cheney, R.W., 2016. Salmonella. Salem Press Encyclopaedia of Health, Ipswitch, MA.

Clergeau, P., Levesque, A., Lorvelec, O., 2004. The precautionary principle and biological invasion: the case of the House Sparrow on the Lesser Antilles. Int. J. Pest Manage. 50 (2), 82–89.

Colautti, R.I., Bailey, S.A., van Overdijk, C.D.A., Amundsen, K., MacIsaac, H.J., 2006. Characterised and projected costs of nonindigenous species in Canada. Biol. Invasions 8, 45–59.

Cornwell, W.K., Cornelissen, J.H.C., Allison, S.D., Bauhus, J., Eggleton, P., Preston, C.M., Scarff, F., Weedon, J.T., Wirth, C., Zanne, A.E., 2009. Plant traits and wood fates across the globe: rotted, burned, or consumed? Global Change Biol. 15, 2431–2449.

Couto, A., Monceau, K., Bonnard, O., Thiery, D., Sandoz, J.-C., 2014. Olfactory attraction of the hornet *Vespa velutina* to honeybee colony odors and pheromones. PLoS One 9, e115943.

de Jong, M.D., Tran, T.T., Truong, H.K., Vo, M.H., Smith, G.J., Nguyen, V.C., Bach, V.C., Phan, T.Q., Do, Q.H., Guan, Y., Peiris, J.S., Tran, T.H., Farrar, J., 2005. Oseltamivir resistance during treatment of influenza A (H5N1) infection. N. Engl. J. Med. 353, 2667–2672.

Donovan, S.E., Griffiths, G.J.K., Homathevi, R., Winder, L., 2007. The spatial pattern of soil-dwelling termites in primary and logged forest in Sabah, Malaysia. Ecol. Entomol. 32, 1–10.

Etim, S.E., Okon, O.E., Akpan, P.A., Ukpong, G.I., Oku, E.E., 2013. Prevalence of cockroaches (*Periplanata americana*) in households in Calabar: public health implications. J. Public Health Epidemiol. 5, 149–152.

Evans, T.A., Dawes, T.Z., Ward, P.R., Lo, N., 2011. Ants and termites increase crop yield in a dry climate. Nat. Commun. 2, 262.

Fernandez-Cornejo, J., 1996. The microeconomic impact of IPM adoption: theory and application. Agric. Resour. Econ. Rev. 25, 149–160.

Fernandez-Cornejo, J., Nehring, R., Osteen, C., Wechsler, S., Martin, A., Vialou, A., 2014. Pesticide use in U.S. agriculture: 21 selected crops, 1960–2008, EIB-124, U.S. Department of Agriculture, Economic Research Service, May 2014.

Fuller, T., Bensch, S., Muller, I., Novembre, J., Perez-Tris, J., Ricklefs, R.E., 2012. The ecology of emerging infectious diseases in migratory birds: an assessment of the role of climate change and priorities for future research. EcoHealth 9, 80–88.

Fuller, T., Ducatez, M.F., Njabo, K.Y., Couacy-Hymann, E., Chasar, A., Aplogan, G.L., Lao, S., Awoume, F., Téhou, A., Langeois, Q., Krauss, S., Smith, T.B., 2015. Avian influenza surveillance in Central and West Africa 2010–2014. Epidemiol. Infect. 143, 2205–2212.

Ghaly, A., Edwards, S., 2011. Termite damage to buildings: nature of attacks and preventive construction methods. Am. J. Eng. Appl. Sci. 4, 187–200.

Golden, J.W., Hammerbeck, C.D., Mucker, E.M., Brocato, R.L., 2015. Animal models for the study of rodent-borne hemorrhagic fever viruses: Arenaviruses and Hantaviruses. Biomed. Res. Int. 2015, 793257.

Grarock, K., Tidemann, C.R., Wood, J.T., Lindenmayer, D.B., 2014. Are invasive species drivers of native species decline or passengers of habitat modification? A case study of the impact of the common myna (*Acridotheres tristis*) on Australian bird species. Aust. Ecol. 39, 106–114.

Gremski, L.H., Trevisan-Silva, D., Ferrer, V.P., Matsubara, F.H., Meissner, G.O., Wille, A.C., Vuitika, L., Dias-Lopes, C., Ullah, A., de Moraes, F.R., Chávez-Olórtegui, C., Barbaro, K.C., Murakami, M.T., Arni, R.K., Senff-Ribeiro, A., Chaim, O.M., Veiga, S.S., 2014. Recent advances in the understanding of brown spider venoms: from the biology of spiders to the molecular mechanisms of toxins. Toxicon 83, 91–120.

Grisi, L., Leite, R.C., de Souza Martins, J.R., Medeiros de Barros, A.T., Andreotti, R., Duarte Cançado, P.H., Pérez de León, A.A.P., Pereira, J.B., Silva Villela, H., 2014. Reassessment of the potential economic impact of cattle parasites in Brazil. Braz. J. Vet. Parasitol. 23, 150–156.

Hammond, A., Vandemaele, K., Fitzner, J., 2015. Human cases of influenza at the human-animal interface, January 2014–April 2015. Weekly Epidemiol. Rec. 90, 349–362.

Hare, H.D., 1980. Impact of defoliation by the Colorado potato beetle on potato yields. J. Econ. Entomol. 73, 369–373.

Hartline, J., Mierek, C., Knutson, T., Kang, C., 2013. Hantavirus infection in North America: a clinical review. Am. J. Emerg. Med. 31, 978–982.

Hellmich, R.L., Hellmich, K.A., 2012. Use and impact of Bt maize. Nat. Educ. Knowledge 3, 4.

Hursey, B.S., Slingenbergh, J., 1996. The tsetse fly and its effects on agriculture in sub-Saharan Africa. FAO. Available from: http://www.fao.org/docrep/v8180t/v8180T0s.htm.

Hutchison, W.D., Burkness, E.C., Mitchell, P.D., Moon, R.D., Leslie, T.W., Fleischer, S.J., Abrahamson, M., Hamilton, K.L., Steffey, K.L., Gray, M.E., Hellmich, R.L., Kaster, L.V., Hunt, T.E., Wright, R.J., Pecinovsky, K., Rabaey, T.L., Flood, B.R., Raun, E.S., 2010. Areawide suppression of European corn borer with Bt maize reaps savings to non-Bt maize growers. Science 330, 222–225.

Isbister, G.K., Fan, H.W., 2011. Spider bite. Lancet 378, 2039–2047.

Kassiri, H., Feizhaddad, M.H., Abdehpanah, M., 2004. Morbidity, surveillance and epidemiology of scorpion sting, cutaneous Leishmaniasis and Pediculosis capitis in Bandar-Mahshahr County, Southwestern Iran. J. Acute Dis. 3 (3), 194–200.

Kathage, J., Qaim, M., 2012. Economic impacts and impact dynamics of Bt (*Bacillus thuringiensis*) cotton in India. Proc. Natl. Acad. Sci. USA 109, 11652–11656.

Kausrud, K.L., Viljugrein, H., Frigessi, A., Begon, M., Davis, S., Leirs, H., Dubyanskiy, V., Stenseth, N.C., 2007. Climatically driven synchrony of gerbil populations allows large-scale plague outbreaks. Proc. R. Soc. B Biol. Sci. 274, 1963–1969.

Koenning, S.R., Wrather, J.A., 2010. Suppression of soybean yield potential in the Continental United States by plant diseases from 2006 to 2009. Plant Management Network. Available from: http://www.plantmanagementnetwork.org/pub/php/research/2010/yield/.

Kriebel, D., Tickner, J., Epstein, P., Lemons, J., Levins, R., Loechler, E.L., Quinn, M., Rudel, R., Schettler, T., Stoto, M., 2001. The precautionary principle in environmental science. Environ. Health Perspect. 109, 871–876.

Langley, R.L., 2008. Animal bites and stings reported by United States poison control centers, 2001–2005. Wilderness Environ. Med. 19, 7–14.

Lewis, V., 2006. Termite damage and detection: an American perspective. Available from: https://nature.berkeley.edu/upmc/documents/Lewis_2006.pdf.

Liebl, A.L., Schrey, A.W., Andrews, S.C., Sheldon, E.L., Griffith, S.C., 2015. Invasion genetics: lessons from a ubiquitous bird, the house sparrow *Passer domesticus*. Curr. Zool. 61, 465–476.

Lowe, S., Browne, M., Boudjelas, S., De Poorter, M., 2000. 100 of the World's Worst Invasive Alien Species: A Selection from the Global Invasive Species Database. The Invasive

Species Specialist Group (ISSG), a specialist group of the Species Survival Commission (SSC) of the World Conservation Union (IUCN), Auckland, New Zealand.

Macdonald, D.W., Mathews, F., Berdoy, M., 1999. The behaviour and ecology of Rattus norvegicus: from opportunism to kamikaze tendencies. In: Singleton, C.R., Hinds, L.A., Leirs, H., Zhang, Z. (Eds.), Ecologically-Based Management of Rodent. ACIAR Monograph No. 59, pp. 49–80.

Magudu, K., Downs, C.T., 2015. The relative abundance of invasive House Sparrows (Passer domesticus) in an urban environment in South Africa is determined by land use. Afr. J. Wildlife Res. 45, 354–359.

Malik, R., Krockenberger, M.B., Cross, G., Doneley, R., Madill, D.N., Black, D., McWhirter, P., Rozenwax, A., Rose, K., Alley, M., Forshaw, D., Russell-Brown, I., Johnstone, A.C., Martin, P., O'Brien, C.R., Love, D.N., 2003. Avian cryptococcosis. Med. Mycol. 41, 115–124.

Marketsandmarkets, 2016. Available from: http://www.marketsandmarkets.com/PressReleases/nematicides.asp.

Maynard, D.S., Crowther, T.W., King, J.R., Warren, R.J., Bradford, M.A., 2015. Temperate forest termites: ecology, biogeography, and ecosystem impacts. Ecol. Entomol. 40, 1–12.

McGain, F., Winkel, K.D., 2002. Ant sting mortality in Australia. Toxicon 40, 1095–1100.

McGregor, D.B., McGregor, B.A., 2008. Eliminating an avian pest (House Sparrow 'Passer domesticus') population: the role of trapping at a homestead scale. Victorian Nat. 125, 4–10.

Mei, L., Song, P., Tang, Q., Shan, K., Tobe, R.G., Selotlegeng, L., Ali, A.H., Cheng, Y., Xu, L., 2013. Changes in and shortcomings of control strategies, drug stockpiles, and vaccine development during outbreaks of avian influenza A H5N1, H1N1, and H7N9 among humans. Biosci. Trends 7, 64–76.

Munkvold, G.P., Hellmich, R.L., Rice, L.G., 1999. Comparison of fumonisin concentrations in kernels of transgenic Bt maize hybrids and nontransgenic hybrids. Plant Dis. 83, 130–138.

Muriithi, B.W., Affognon, H.D., Diiro, G.M., Kingori, S.W., Tanga, C.M., Nderitu, P.W., Mohamed, S.A., Ekesi, S., 2016. Impact assessment of Integrated Pest Management (IPM) strategy for suppression of mango-infesting fruit flies in Kenya. Crop Prot. 81, 20–29.

Mwachui, M.A., Crump, L., Hartskeerl, R., Zinsstag, J., Hattendorf, J., 2015. Environmental and behavioural determinants of leptospirosis transmission: a systematic review. PLoS Negl. Trop. Dis. 9, e0003843.

Nentwig, W., Kühnel, E., Bacher, S., 2010. A generic impact-scoring system applied to alien mammals in Europe. Conserv. Biol. 24, 302–311.

New Wildlife Federation, 2016. Rattlesnakes. Available from: http://www.nwf.org/wildlife/wildlife-library/amphibians-reptiles-and-fish/rattlesnakes.aspx.

Nghiem, L.T.P., Soliman, T., Yeo, D.C.J., Tan, H.T.W., Evans, T.A., Mumford, J.D., Keller, R.P., Baker, R.H.A., Corlett, R.T., Carrasco, L.R., 2013. Economic and environmental impacts of harmful non-indigenous species in Southeast Asia. PLoS One 8, e71255.

Nicol, J.M., Turner, S.J., Coyne, D.L., den Nijs, L., Hockland, S., Tahna Maafi, Z., 2011. Current nematode threats to world agriculture. In: Jones, J., Gheysen, G., Fenoll, C. (Eds.), Genomics and Molecular Genetics of Plant-Nematode Interactions. Springer, New York, NY, pp. 21–43.

NRC, 2004. Safety of Genetically Engineered Foods: Approaches to Assessing Unintended Health Effects. National Academies Press, Washington, DC.

Old, J.M., Spencer, R.J., Wolfenden, J., 2013. The Common Myna (Sturnus tristis) in urban, rural and semi-rural areas in Greater Sydney and its surrounds. Emu 114, 241–248.

Ostry, V., Ovesna, J., Skarkova, J., Pouchova, V., Ruprich, J., 2010. A review on comparative data concerning Fusarium mycotoxins in Bt maize and non-Bt isogenic maize. Mycotoxin Res. 26, 141–145.

Pan, Y., Birdsey, R.A., Fang, J., Houghton, R., Kauppi, P.E., Kurz, W.A., Phillips, O.L., Shvidenko, A., Lewis, S.L., Canadell, J.G., Ciais, P., Jackson, R.B., Pacala, S.W., McGuire, A.D., Piao, S., Rautiainen, A., Sitch, S., Hayes, D., 2011. A large and persistent carbon sink in the world's forests. Science 333, 988–993.

Payne, K., Andreotti, G., Bell, E., Blair, A., Coble, J., Alavanja, M., 2012. Determinants of high pesticide exposure events in the agricultural health cohort study from enrollment (1993–1997) through phase II (1999–2003). J. Agric. Saf. Health 18, 167–179.

Pickett, K.M., Carpenter, J.M., 2010. Simultaneous analysis and the origin of eusociality in the Vespidae (Insecta: Hymenoptera). Arthropod Syst. Phylogeny 68, 3–33.

Pimentel, D., 2005. Environmental and economic costs of the application of pesticides primarily in the United States. Environ. Dev. Sustain. 7, 229.

Pimentel, D., 2007. Environmental and economic costs of vertebrate species invasions into the United States. In: Managing Vertebrate Invasive Species: Proceedings of an International Symposium. Paper 38. Fort Collins, CO.

Pimentel, D., Lach, L., Zuniga, R., Morrison, D., 2000. Environmental and economic costs of nonindigenous species in the United States. BioScience 50, 53–65.

Pimentel, D., McLaughlin, L., Zepp, A., Lakitan, B., Kraus, T., Kleinman, P., Vancini, F., Roach, W.J., Graap, E., Keeton, W.S., Selig, G., 1991a. Environmental and economic impacts of reducing U.S. agricultural pesticide use. Pimentel, D. (Ed.), Handbook of Pest Management in Agriculture, I, second ed. CRC Press, Boca Raton, FL, pp. 679–718.

Pimentel, D., McLaughton, L., Zepp, A., Lakitan, B., Kraus, T., Kleinman, P., Vancini, F., Roach, W.J., Graap, E.,

Keeton, W.S., Selig, G., 1991b. Environmental and economic impacts of reducing US agricultural pesticide use. BioScience 41, 402–409.

Pimentel, D., McNair, S., Janecka, J., Wightman, J., Simmonds, C., O'Connell, C., Wong, E., Russel, L., Zern, J., Aquino, T., Tsomondo, T., 2001. Economic and environmental threats of alien plant, animal, and microbe invasions. Agric. Ecosyst. Environ. 84, 1–20.

Pitts, R.M., Mauldin, M.R., Thompson, C.W., Choate, J.R., 2013. Evidence of Hantavirus exposure in rodents from North Texas. Western North Am. Nat. 73 (3), 386–391.

Popp, J., 2011. Cost-benefit analysis of crop protection measures. J. Consum. Prot. Food Saf. 6 (Supplement 1), 105–112.

Popp, J., Pető, K., Nagy, J., 2013. Pesticide productivity and food security: a review. Agron. Sustain. Dev. 33, 243–255.

Rachel, 2000. Preferring the Least Harmful Way. Rachel's Environment and Health Weekly 684. Available from: http://www.rachel.org/files/rachel/Rachels_Environment_Health_News_1687.pdf.

Raoult, D., Mouffok, N., Bitam, I., Piarroux, R., Drancourt, M., 2013. Plague: history and contemporary analysis. J. Infect. 66, 18–26.

Reperant, L.A., Kuiken, T., Osterhaus, A.D.M.E., 2012. Influenza viruses. Hum. Vaccin. Immunother. 8, 7–16.

Rhoades, R., 2001. Stinging ants. Curr. Opin. Allergy Clin. Immunol. 1, 343–348.

Rothrock, M.J., Ingram, K.D., Gamble, J., Guard, J., Cicconi-Hogan, K.M., Hinton, Jr., A., Hiett, K.L., 2015. The characterization of Salmonella enterica serotypes isolated from the scalder tank water of a commercial poultry processing plant: recovery of a multidrug-resistant Heidelberg strain. Poultry Sci. 94, 467–472.

Sallam, M.N., 2013. Insect damage: post-harvest operations. International Centre of Insect Physiology and Ecology. FAO. Available from: http://www.fao.org/3/a-av013e.pdf.

Sanders, W.E., Sanders, C.S., 1997. Enterobacter spp.: pathogens poised to flourish at the turn of the century. Clin. Microbiol. Rev. 10, 220–241.

Sarao, P.S., Sahi, G.K., Neelam, K., Mangat, G.S., Patra, B.C., Singh, K., 2016. Donors for resistance to brown planthopper Nilaparvata lugens (Stål) from wild rice species. Rice Sci. 23, 219–224.

Schmid, B.V., Büntgen, U., Easterday, W.R., Ginzler, C., Walløe, L., Bramanti, B., Stenseth, N.C., 2015. Climate-driven introduction of the Black Death and successive plague reintroductions into Europe. Proc. Natl. Acad. Sci. 112, 3020–3025.

Schmitt, D.P., 1994. Plant and soil nematodes: societal impact and focus for the future. J. Nematol. 26, 127–137.

Shah, I., 2012. Leptospirosis. Pediatr. Infect. Dis. 4, 4–8.

Shepard, D.S., Coudeville, L., Halasa, Y.A., Zambrano, B., Dayan, G.H., 2011. Economic impact of dengue illness in the Americas. Am. J. Trop. Med. Hyg. 84, 200–207.

Shurdut, B.A., Peterson, R.K.D., 1999. Human health risks from cockroaches and cockroach management: a risk analysis. Am. Entomol. 45, 142–148.

Siembieda, J., Johnson, C.K., Boyce, W., Sandrock, C., Cardona, C., 2008. Risk for avian influenza virus exposure at human–wildlife interface. Emerg. Infect. Dis. 14, 1151–1153.

Singleton, G.R., Belmain, S.R., Brown, P.R., Hardy, B. (Eds.), 2010. Rodent Outbreaks: Ecology and Impacts. International Rice Research Institute, Los Baños (Philippines).

Singleton, G.R., Brown, P.R., Jacob, J., Aplin, K.P., 2007. Unwanted and unintended effects of culling: a case for ecologically-based rodent management. Integr. Zool. 2, 247–259.

Singleton, C.R., Hinds, L.A., Leirs, H., Zhang, Z. (Eds.), 1999. Ecologically-Based Management of Rodent. ACIAR Monograph No. 59, 494p.

Skolnik, A.B., Ewald, M.B., 2013. Pediatric scorpion envenomation in the United States: morbidity, mortality, and therapeutic innovations. Pediatr. Emerg. Care 29, 98–103.

Sogawa, K., 1982. The rice brown planthopper: feeding physiology and host plant interactions. Annu. Rev. Entomol. 27, 49–73.

Soltani, M., Bayat, M., Hashemi, S.J., Zia, M., Pestechian, N., 2013. Isolation of Cryptococcus neoformans and other opportunistic fungi from pigeon droppings. J. Res. Med. Sci. 18, 56–60.

Souders, H.T., Byler, D., Marupudi, N., Patel, R., McSherry, G., 2015. Protracted symptoms in lymphocytic choriomeningitis: a case report. J. Child Neurol. 30, 644–647.

Spitzen, J., Koelewijn, T., Mukabana, W.R., Takken, W., 2016. Visualization of house-entry behaviour of malaria mosquitoes. Malaria J. 15, 233.

Stiling, P., Cornelissen, T., 2005. What makes a successful biocontrol agent? A meta-analysis of biological control agent performance. Biol. Control 34, 236–246.

Suiter, D.R., Jones, S.C., Forschler, B.T., 2012. Biology of Subterranean Termites in the Eastern United States. Bulletin 1209. University of Georgia, Athens, GA.

Taylor, D.B., Moon, R.D., Mark, D.R., 2012. Economic impact of stable flies (Diptera: Muscidae) on dairy and beef cattle production. J. Med. Entomol. 49, 198–209.

The Economist, 2016. Art of the lie. Available from: http://www.economist.com/news/leaders/21706525-politicians-have-always-lied-does-it-matter-if-they-leave-truth-behind-entirely-art.

USDA Economic Research Service, 2016. Adoption of genetically engineered crops in the United States 1996–2016. Available from: http://www.ers.usda.gov/data-products/adoption-of-genetically-engineered-crops-in-the-us/recent-trends-in-ge-adoption.aspx.

Vela, E., 2012. Animal models, prophylaxis, and therapeutics for Arenavirus infections. Viruses 4, 1802–1829.

Ventola, C.L., 2010. Current issues regarding complementary and alternative medicine (CAM) in the United States. Part 2: Regulatory and safety concerns and proposed governmental policy changes with respect to dietary supplements. Pharm. Ther. 35, 514–522.

Wannigama, D.L., Dwivedi, R., Zahraei-Ramazani, A., 2014. Prevalence and antibiotic resistance of Gram-negative pathogenic bacteria species isolated from *Periplaneta americana* and *Blattella germanica* in Varanasi, India. J. Arthropod Borne Dis. 8, 10–20.

Weichenthal, S., Moase, C., Chan, P., 2010. A review of pesticide exposure and cancer incidence in the Agricultural Health Study cohort. Environ. Health Perspect. 118, 1117–1125.

Wheat, L.J., Azar, M.M., Bahr, N.C., Spec, A., Relich, R.F., Hage, C., 2016. Histoplasmosis. Infect. Dis. Clin. North Am. 30, 207–227.

Whitehorn, J., Farrar, J., 2011. Dengue. Clin. Med. (Lond.) 11, 483–487.

WHO, 2007. WHO plans to increase treatment access for victims of rabies and snakebites. Available from: www.who.int/mediacentre/news/notes/2007/np01/en.

WHO, 2010. Cockroaches: Unhygienic scavengers in human settlements. Available from: http://www.who.int/water_sanitation_health/resources/vector288to301.pdf.

WHO, 2015a. Snake antivenoms. Available from: http://www.who.int/mediacentre/factsheets/fs337/en/.

WHO, 2015b. 10 facts on malaria. Available from: http://www.who.int/features/factfiles/malaria/en/.

Wickramaratna, J.C., Hodgson, W.C., 2001. A pharmacological examination of venoms from three species of death adder (*Acanthophis antarcticus*, *Acanthophis praelongus* and *Acanthophis pyrrhus*). Toxicon 39, 209–216.

Wingspread, 1998. Available from: http://www.sehn.org/wing.html.

Zhang, W.-J., Jiang, F.B., Ou, J.F., 2011. Global pesticide consumption and pollution: with China as a focus. Proc. Int. Acad. Ecol. Environ. Sci. 1, 125–144.

Further Reading

Global Invasive Species Database, 2016. Species profile: *Rattus norvegicus*. Available from: http://www.iucngisd.org/gisd/species.php?sc=159.

National Geographic, 2016. Eastern Diamondback Rattlesnake—*Crotalus adamanteus*. Available from: http://animals.nationalgeographic.com.au/animals/reptiles/eastern-diamondback-rattlesnake/.

The Atlantic, 2015. Bedbugs: a nightmare for the hotel industry. Available from: http://www.theatlantic.com/business/archive/2015/08/bedbugs-scary-worst-hotels-reviews/401036/.

Zhang, X., Meltzer, M.I., Peña, C.A., Hopkins, A.B., Wroth, L., Fix, A.D., 2006. Economic impact of Lyme disease. Emerg. Infect. Dis. 12, 653–660.

Parasites

Colin G. Scanes*, Samia R. Toukhsati**

*University of Wisconsin–Milwaukee, Milwaukee, WI, United States
**Honorary Fellow, The University of Melbourne, Parkville, VIC, Australia

16.1 OVERVIEW OF HUMAN PARASITES

Parasitic infections are caused by the following:

- protozoa (also see Chapter 15)
- nematodes (round worms)
- cestodes (tape worms)
- trematodes (flukes)

Parasitic diseases impact a great many people in the world with the US Centers for Disease Control and Prevention (CDC) estimating that there are over a billion people infected at any one time (CDC, 2017). Moreover, over 85% of people in the world will be infected with one protozoan species, *Toxoplasma gondii*, during their lifetime (Table 16.1).

Parasites are responsible for much human misery and are associated with human morbidity and mortality. Estimates of the global impact of certain parasites on human health are summarized in Table 16.1. Table 16.1 includes estimates of the disability-adjusted life years (DALY) for some parasitic infections. Despite, the wide use of DALY to determine public health priorities, it has been argued that the metric is "…unsuited to parasitic infections. In particular, the current DALY framework fails to acknowledge the non-linear pathologies of infection, the community level dynamics of epidemiology and the co-morbidities of polyparasitism" (Payne et al., 2009).

Tables 16.2–16.4, respectively, summarize the major gastrointestinal parasites, the major parasitic helminths, and those helminth infections that are transmitted via fish. Parasites infecting the gastrointestinal tract (Table 16.2) result in diarrhea and malabsorption of essential nutrients. There is consequent reduced physical and mental development in children. Moreover, in adults intestinal parasites decrease productivity and quality of life. Helminth intestinal parasites vary tremendously in size from very small but visible pinworms to the enormous tapeworms (Table 16.3). Multiple helminth infections (infections caused by worms) are transmitted via raw or uncooked fish (Table 16.4). This is becoming more important in developed countries with increased consumption of raw fish, such as in

Animals and Human Society
http://dx.doi.org/10.1016/B978-0-12-805247-1.00023-X

TABLE 16.1   Impact of Parasitic Diseases Globally

Pathogen	People sick (millions)	Deaths (thousands)	DALY (millions)
Protozoa			
Intestinal protozoa	357	34 or >>100	2.9
E. histolytica	40–50	100	NA
P. falciparum + *Plasmodium vivax, Plasmodium ovale, Plasmodium malariae* (malaria)	214	438	NA
T. gondii (toxoplasmosis)	Up to 6 billion people infected during their lives		
Try. brucei (sleeping sickness)	0.003	NA	NA
Try. cruzi (Chagas disease or American sleeping sickness)	8	10,000	NA
Nematodes (round worm)			
Intestinal nematodes	26.8	2.2	1.3
Lymphatic filariasis	120		
Mansonellosis	114		
Cestodes (tapeworm)	0.6	48	3.7
Trematodes			
Food-borne trematodes	0.2	7.5	2.0
Schistosoma (schistosomiasis)	258	NA	NA

DALY, Disability-adjusted life years; NA, information not available.

Data from Harhay, M.O., Horton, J., Olliaro, P.L., 2010. Epidemiology and control of human gastrointestinal parasites in children. Expert Rev. Anti Infect. Ther. 8, 219–234; Furtado, J.M., Smith, J.R., Belfort, R., Gattey, D., Kevin, L., Winthrop, K.L., 2011. Toxoplasmosis: a global threat. J. Global Infect. Dis. 3, 281–284; Torgerson, P.R., de Silva, N.R., Fèvre, E.M., Kasuga, F., Rokni, M.B., Zhou, X.-N., Sripa, B., Gargouri, N., Willingham, A.V., Stein, C., 2014. The global burden of food borne parasitic diseases: an update. Trends Parasitol. 30, 20–26; Torgerson, P.R., Devleesschauwer, B., Praet, N., Speybroeck, N., Willingham, A.L., Kasuga, F., Rokni, M.B., Zhou, X.N., Fèvre, E.M., Sripa, B., Gargouri, N., Fürst, T., Budke, C.M., Carabin, H., Kirk, M.D., Angulo, F.J., Havelaar, A., de Silva, N., 2015. World Health Organization estimates of the global and regional disease burden of 11 foodborne parasitic diseases, 2010: a data synthesis. PLoS Med. 12, e1001920; Mourembou, G., Fenollar, F., Lekana-Douki, J.B., Ndjoyi Mbiguino, A., Maghendji Nzondo, S., Matsiegui, P.B., Manego, R.Z., Ehounoud, C.H.B., Bittar, F., Raoult, D., Mediannikov, O., 2015. Mansonella, including a Potential New Species, as Common Parasites in Children in Gabon. PLoS Negl. Trop. Dis. 9, e0004155; WHO, 2015. WHO estimates of the global burden of foodborne diseases. Available from: http://apps.who.int/iris/ bitstream/10665/199350/1/9789241565165_eng.pdf?ua=1; WHO, 2015. 10 facts on malaria. Available from: http://www.who.int/features/factfiles/ malaria/en/; WHO, 2015. Chagas disease or American sleeping sickness. Available from: http://www.who.int/chagas/epidemiology/en/; CDC, 2017. Parasites. Available from: https://www.cdc.gov/parasites/ including Flukes http://www.cdc.gov/parasites/fasciola/; Parasites—Angiostrongyliasas (also known as angiostrongyliasis infection) http://www.cdc.gov/parasites/angiostrongylus/; Parasites—Ascariasis http://www.cdc.gov/parasites/ ascariasis/; Parasites—Dracunculiasas (also known as guinea worm disease) http://www.cdc.gov/parasites/guineaworm/; Parasites—Hookworm http://www.cdc.gov/parasites/hookworm/; Parasites: Neglected parasitic infections https://www.cdc.gov/parasites/npi/; Parasites—Schistosomiasis http://www.cdc.gov/parasites/schistosomiasis/epi.html; Taeniasis http://www.cdc.gov/parasites/taeniasis/biology.html; Parasites—Toxocariasis (also known as roundworm infection) (http://www.cdc.gov/parasites/toxocariasis/gen_info/faqs.html); Parasites—Trichuriasis (also known as whipworm infection) http://www.cdc.gov/parasites/whipworm/biology.html.

TABLE 16.2 Effects, Reservoirs, and Transmission of the Major Gastrointestinal Parasites in the World

	Reservoir	Transmission	Effects
Protozoa			
E. histolytica	Humans	Fecal (oral)	Bloody diarrhea
G. intestinalis	Humans and other mammals	Fecal (oral)	Watery diarrhea
Cryptosporidium parvum	Humans and specific mammals (e.g., cattle)	Fecal (oral)	Watery diarrhea
Cyclospora cayetanensis	Unknown	Food and/or water	Watery diarrhea
Nematodes			
The intestinal round worm, *A. lumbricoides*	Humans	Fecal (oral)	Obstruction of intestine or bile duct
Hookworms (*Anc. duodenale*)	Humans	Fecal (oral or pass through the skin of the feet)	Iron deficiency anemia
Hookworms (*N. americanus*)	Humans	Fecal (oral or pass through the skin of the feet)	Iron deficiency anemia
Trichu. trichiura	Humans	Fecal (oral)	Damaged intestinal mucosa
Cestodes			
Pork tapeworm (*Ta. solium*)	Pigs	Undercooked meat	Malabsorption and vitamin/mineral deficiencies; hydatid cysts and epilepsy
Beef tapeworm (*Ta. saginata*)	Cattle	Undercooked meat	
Trematodes		Contaminated water	Obstruction and/or ulceration and/or hemorrhage in intestine

Based on CDC. Available from: https://www.cdc.gov/parasites/index.html.

Japanese cuisine (sushi and sashimi), in Peruvian cuisine (ceviche or raw fish in lemon or lime juice), in Italian cuisine (carpaccio with either salmon or tuna or *pesce crudo*), in Jewish cuisine (lox or raw salmon cured in salt), in American cuisine (tuna tartare), and in Scandinavian cuisine (gravlax or raw salmon cured in salt).

The prevalence of intestinal parasites varies between different countries and populations (Table 16.5). In the Republic of Korea, a country that has achieved status as a high-income nation (World Bank, 2016), the prevalence of intestinal parasites is estimated as 2.6% (Korea Centers for Disease Control and Prevention, 2013). This compares to 24% of children under 5 years of age with intestinal parasites in Ethiopia (G/hiwot et al., 2014), rated by the World Bank (2016) as one of the world's poorest countries. Similarly, a quarter of children (25%) in Malaysia, an upper-middle income economy according to the World Bank (2016), is infected with parasites (Sinniah et al., 2014).

Parasites not only directly impact people throughout the world but also reduce food availability due to impacts on growing crops and on livestock.

TABLE 16.3 Comparison of Various Helminth Parasites (Parasitic Worms)

Species		Disease	Length (metric)	Length (Imperial)
Cestodes				
Beef tapeworm	*Ta. saginata*	Taeniasis eggs cause cysticercosis	up to 25 m	82 ft
Nematodes				
Ascaris parasitic round worm	*A. lumbricoides*	Ascariasis	250 mm	~10.0 in
Pinworm	*En. vermicularis*		10.5 mm in females; 3.5 mm in males	0.375 in for females and 0.125 in for males
Whipworm	*Trichu. trichiura*	Trichuriasis	40 mm	1.6 in
Trematodes				
Liver fluke	*F. hepatica*	Fascioliasis	30 mm	1.2 in

Based on CDC, 2017. Parasites. Available from: https://www.cdc.gov/parasites/ including Flukes http://www.cdc.gov/parasites/fasciola/; Parasites— Angiostrongyliasas (also known as angiostrongyliasis infection) http://www.cdc.gov/parasites/angiostrongylus/; Parasites—Ascariasis http://www.cdc.gov/ parasites/ascariasis/; Parasites—Dracunculiasas (also known as guinea worm disease) http://www.cdc.gov/parasites/guineaworm/; Parasites—Hookworm http://www.cdc.gov/parasites/hookworm/; Parasites: Neglected parasitic infections https://www.cdc.gov/parasites/npi/; Parasites—Schistosomiasis http:// www.cdc.gov/parasites/schistosomiasis/epi.html; Taeniasis http://www.cdc.gov/parasites/taeniasis/biology.html; Parasites—Toxocariasis (also known as roundworm infection) (http://www.cdc.gov/parasites/toxocariasis/gen_info/faqs.html); Parasites—Trichuriasis (also known as whipworm infection) http:// www.cdc.gov/parasites/whipworm/biology.html.

TABLE 16.4 Parasites From the Consumption of Raw Fish

Parasite	Source	Symptoms
Protozoa		
G. lamblia	Home-canned salmon and Chinese fish soup	Nausea, fever, pain, diarrhea
Cestoda		
Di. latum	Raw, undercooked, or marinated fish	Pain, diarrhea, eosinophilia, possibly B$_{12}$ deficiency
Nematoda		
Anisakis simplex	Raw, undercooked, or marinated fish or squid	Acute: Epigastric distress Chronic: diarrhea and "occasional coughing up of larvae"
Trematoda		
Liver fluke		
C. sinensis	Raw or undercooked fish in Southeast and East Asia	Fever, tender enlarged liver (hepatomegaly)
Intestinal flukes		
Echinostoma	Raw or undercooked fish in Southeast and East Asia	Gastroenteritis, anemia, headaches, dizziness, stomach pain, diarrhea, anorexia, eosinophilia
Heterophyes spp.	Middle East, Asia; raw, marinated, or undercooked fish	Abdominal pain, diarrhea
Metagonimus spp.	Middle East, Asia; raw, marinated, or undercooked fish	Abdominal pain, diarrhea
Nanophyetus salmincola	North America; raw, marinated, or undercooked fish	Abdominal pain, diarrhea

Based on Craig, N., 2012. Fish tapeworm and sushi. Can. Fam. Physician 58, 654–658.

TABLE 16.5 Examples of the Prevalence of Helminth Parasites

Country and people	Nematodes	Cestodes	Trematodes	References
Korea	1.7%	0.4%	0.13%	Korea Centers for Disease Control and Prevention (2013)
		H. nana	*Sc. mansoni*	
Ethiopia (children under 5 years of age)	11% (5% *A. lumbricoides*)	10%	9%	G/hiwot et al. (2014)
Ethiopia (children 10–12 years old)	6% *A. lumbricoides*	14	NR	Gelaw et al. (2013)
Coastal Kenya (preschool children)	29% Hookworm, 20% *A. lumbricoides*, 25% *Trichu. trichina*	NR	NR	Brooker et al. (1999)
Nigeria (patients of a hospital in Benin City)	2% *A. lumbricoides*	NR	NR	Akinbo et al. (2011)
Rwanda (tertiary education students)	10% *A. lumbricoides*	NR	NR	Niyizurugero et al. (2013)
Malaysia (children)	20% *Trichu. trichiura*, 10% *A. lumbricoides*, 7% hookworm	NR	NR	Sinniah et al. (2014)
India (children)	*A. lumbricoides* (1% rural, 21% urban), *Trichu. trichiura* (8% urban, 1% Old World) hookworm *Anc. duodenale*	1%	NR	Rayan et al. (2010)

NR, Not reported.

16.2 SINGLE-CELLED PARASITES (PROTOZOA OR PROTISTA)

16.2.1 Overview

Single-celled parasites can infect the blood and multiple organs including the gastrointestinal tract (see Section 16.2.2) and the urogenital tract (see Section 16.2.2). Infections of blood cells from specific protozoan species result in debilitating diseases, such as malaria (discussed in Chapter 15), sleeping sickness, and Chagas disease (Table 16.1).

Classification of Protozoa Including Parasitic Protozoa

It is recognized that traditionally protozoa were placed in a single phylum within the Kingdom Animalia. However, this does not reflect current thinking. There is broad agreement by taxonomists placing all living organisms into two superkingdoms, respectively, the Prokaryota and the Eukaryota, with the following kingdoms (Ruggiero et al., 2015):

- Superkingdom: Prokaryota (comprising the kingdoms Archaebacteria or Archaea and Eubacteria or Bacteria).
- Superkingdom: Eukaryota (comprising the kingdoms Animalia or Metazoa, Protozoa, Plantae, Fungi, and Chromista).

Protozoa and Chromista are both single-celled eukaryotes or protists.

The Kingdom Protozoa is a paraphyletic taxon (Cavalier-Smith, 2013; Tree of Life, in press). The Kingdom Chromista includes algae with chlorophyll *c* together with some parasites frequently thought of as protozoa (Ruggiero et al., 2015).

The Kingdom Protozoa comprises the following phyla: (1) Amoebozoa including parasites, such as *Entamoeba histolytica*; (2) Choanozoa; (3) Euglenozoa including parasites, such as trypanosomes and *Leishmania*; (4) *Loukozoa*; (5) Metamonada including parasites, such as *Trichomonas vaginalis* and *Giardia lamblia*; and (6) Percolozoa (Ruggiero et al., 2015).

The taxon Apicomplexa (including parasitic protozoa, such as *Eimeria*, *Plasmodium*, *Toxoplasma*, and *Cryptosporidium*) was viewed by some as a parasitic phylum of protozoa (Yaeger, 1996).

The infraphylum Apicomplexa is now considered as a member of the superphylum Alveolata and phylum Miozoa (Ruggiero et al., 2015; Tree of Life, in press). Interestingly, these are placed in the Kingdom Chromista not Protozoa (Ruggiero et al., 2015).

Relationships between taxa in the kingdoms Protozoa, Chromista, and other kingdoms are still to be established. Comparative genomics offers considerable promise in elucidating the phylogenic relationships following the sequencing of protozoan and chromist genomes [*Plasmodium falciparum* (reviewed by Berry et al., 2004); *Trypanosoma brucei*, *Try. cruzi*, and *Leishmania major* (El-Sayed et al., 2005); *Leishmania infantum* and *Leishmania braziliensis* (Peacock et al., 2007); *T. gondii* and *Neospora caninum* (Reid et al., 2012); and *Eimeria tenella* (Reid et al., 2014)].

16.2.2 Gastrointestinal Single-Celled Parasites

Infections with pathogenic single-celled parasites, such as *E. histolytica*, result in severe diarrhea and are spread in food, water, or hands contaminated with feces particularly in developing countries. In Rwandan tertiary education students, the prevalence single-celled parasites includes *E. histolytica* (27%), *Tri. intestinalis* (10%), and *Giardia duodenalis* (2%) (Niyizurugero et al., 2013). In children (10–12 years old) in Ethiopia, the prevalence of *Entamoeba histolytica/dispar* was 9% (Gelaw et al., 2013). Infections with single-celled parasites in Malaysian children are the following: *Entamoeba coli*, 3%; *Giardia intestinalis*, 2%; *E. histolytica*, 2%; and *Blastocystis hominis*, 1% (Sinniah et al., 2014).

16.2.3 Toxoplasmosis

The single-celled parasite *T. gondii* causes toxoplasmosis with 60 million people chronically infected in the USA (Furtado et al., 2011). While it can be a zoonotic and a food-borne disease, it can also be congenital (Furtado et al., 2011). Congenital toxoplasmosis can lead to visual impairment (Furtado et al., 2011).

16.2.4 Sexually Transmitted Single-Celled Parasitic Diseases

Trichomoniasis is caused by *Tri. vaginalis*. This is a flagellate single-celled parasite that multiplies in the human urogenital tract and is sexually transmitted. Infection with *Tri. vaginalis* can cause a vaginal discharge and is associated with reproductive tract infections, increased susceptibility to the human immunodeficiency virus, premature births, low birth weight births, and infertility (Ambrozio et al., 2016). The magnitude of this infection is gleaned by the following (reviewed by Ambrozio et al., 2016; CDC, 2017; Field et al., 2016; Harp and Chowdhury, 2011):

- World: Over 160 million people per year infected.
- USA: 1.1 million people per year infected.
- United Kingdom: 0.3% prevalence in sexually experienced women 16–44 years old.
- Southern Brazil: 9% prevalence in women.

> Classification of helminth endoparasites
> Phylum: Nematoda (round worms) (within Superphylum Ecdysozoa as are arthropods)
> Phylum: Platyhelminthes
>
> Class: Cestoda (Tapeworms)
> Class: Trematoda (Flukes)

16.3 NEMATODES

16.3.1 Overview

Nematode infection can come from contaminated food or water or in some cases are soil-transmitted or vector-transmitted (Fig. 16.1). The soil-transmitted helminths (*Ascaris lumbricoides*, hookworm, and whipworm *Trichuris trichiura*) are the most prevalent, infecting an estimated one-sixth of the global population (Table 16.5). Infection rates are highest in children living in sub-Saharan Africa, followed by Asia and then Latin America and the Caribbean. There can be marked urban rural differences in the prevalence of parasites:

- Hookworm in children in Malaysia (urban 0% and rural 12%) (Sinniah et al., 2014)

- *A. lumbricoides* in Indian school-aged children (urban, 21% and rural, 1%) (Rayan et al., 2010).

Nematodes can be spread by vectors as in loiasis (Fig. 16.1) and lymphatic filariasis (see Section 16.3.6).

16.3.2 Angiostrongyliasis

Angiostrongyliasis is caused by either the rat lungworm (*Angiostrongylus cantonensis*) or by *Angiostrongylus costaricensis*. The rat lungworm is found in Southeast Asia, Pacific islands, and Egypt

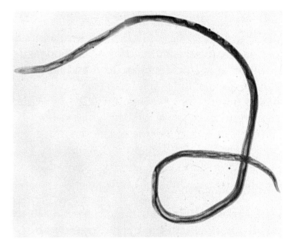

FIGURE 16.1 **Example of the nematode parasitic infection loiasis in West and Central Africa.** It is caused by the African eye worm *Loa loa* and transmitted by deer flies. *Source: Courtesy CDC, 2017. Parasites. Available from: https://www.cdc.gov/ parasites/ including Flukes http://www.cdc.gov/parasites/fasciola/; Parasites—Angiostrongyliasas (also known as angiostrongyliasis infection) http://www.cdc.gov/parasites/angiostrongylus/; Parasites—Ascariasis http://www.cdc.gov/parasites/ascariasis/; Parasites—Dracunculiasas (also known as guinea worm disease) http://www.cdc.gov/parasites/guineaworm/; Parasites—Hookworm http://www.cdc.gov/parasites/hookworm/; Parasites: Neglected parasitic infections https://www.cdc.gov/parasites/npi/; Parasites—Schistosomiasis http://www.cdc.gov/parasites/schistosomiasis/epi.html; Taeniasis http://www.cdc.gov/parasites/taeniasis/biology.html; Parasites—Toxocariasis (also known as roundworm infection) (http://www.cdc.gov/parasites/toxocariasis/gen_info/faqs.html); Parasites—Trichuriasis (also known as whipworm infection) http://www.cdc.gov/parasites/whipworm/biology.html.*

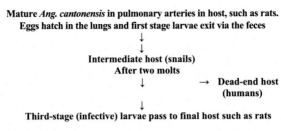

Mature *Ang. cantonensis* in pulmonary arteries in host, such as rats.
Eggs hatch in the lungs and first stage larvae exit via the feces
↓

Intermediate host (snails)
After two molts
↓   →   Dead-end host
(humans)
↓
Third-stage (infective) larvae pass to final host such as rats

FIGURE 16.2   Life cycle of the rat lungworm (*Ang. cantonensis*).

(Ibrahim, 2007) and is associated with eosinophilic meningitis (acute white blood cell inflammatory response to parasitic infection in the cerebrospinal fluid). *Ang. costaricensis* is found in Latin America and the Caribbean and is associated with eosinophilic gastroenteritis (white blood cell inflammatory response to parasitic infection in the esophagus).

Humans are incidental or dead-end hosts consuming larvae, for example, in snails or slugs or water or vegetables contaminated with third-stage infective larvae of *Ang. cantonensis*. The life cycle of *Ang. cantonensis* is shown in Fig. 16.2 (based on CDC, 2017; Ibrahim, 2007).

16.3.3 Ascariasis

Ascariasis is a very common infection in the small intestine caused by the roundworm *A. lumbricoides*. Indeed, it is estimated that to reduce the incidence of ascariasis globally, over a billion children would need to be treated with anthelminth regularly (1–3 times per year) (Horton, 2003; reviewed by Harhay et al., 2010). Ascariasis is usually asymptomatic; however, large infestations can cause severe abdominal pain, weight loss, and respiratory symptoms. Fig. 16.3 summarizes the life cycle of *A. lumbricoides*.

16.3.4 Dracunculiasis or Guinea Worm Disease

Guinea worm disease is caused by the nematode *Dracunculus medinensis*. Most people are

asymptomatic for approximately 1 year, following which the worm will attempt to exit the host (usually via the leg or foot) and there may be associated pain and swelling and secondary infection at the exit site. The life cycle

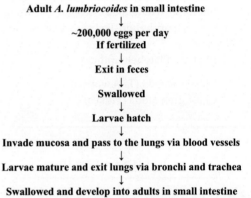

Adult *A. lumbriocoides* in small intestine
↓
~200,000 eggs per day
If fertilized
↓
Exit in feces
↓
Swallowed
↓
Larvae hatch
↓
Invade mucosa and pass to the lungs via blood vessels
↓
Larvae mature and exit lungs via bronchi and trachea
↓
Swallowed and develop into adults in small intestine

FIGURE 16.3   Life cycle of *A. lumbriocoides*. *Source: Based on CDC, 2017. Parasites. Available from: https://www.cdc.gov/parasites/ including Flukes http://www.cdc.gov/parasites/fasciola/; Parasites—Angiostrongyliasas (also known as angiostrongyliasis infection) http://www.cdc.gov/parasites/angiostrongylus/; Parasites—Ascariasis http://www.cdc.gov/parasites/ascariasis/; Parasites—Dracunculiasas (also known as guinea worm disease) http://www.cdc.gov/parasites/guineaworm/; Parasites—Hookworm http://www.cdc.gov/parasites/hookworm/; Parasites: Neglected parasitic infections https://www.cdc.gov/parasites/npi/; Parasites—Schistosomiasis http://www.cdc.gov/parasites/schistosomiasis/epi.html; Taeniasis http://www.cdc.gov/parasites/taeniasis/biology.html; Parasites—Toxocariasis (also known as roundworm infection) (http://www.cdc.gov/parasites/toxocariasis/gen_info/faqs.html); Parasites—Trichuriasis (also known as whipworm infection) http://www.cdc.gov/parasites/whipworm/biology.html). http://www.cdc.gov/parasites/ascariasis/.*

L3 larvae injected into person in water contaminated with secondary host, copepds
↓
Larvae released from dead copepods
↓
Larvae migrate through stomach and small intestine wall
↓
Larvae mature to sexually mature male and female adults
↓
Following fertilization, gravid adult female migrates to skin and gradually emerges releasing L1 larvae
↓
Larvae consumed by copepods
↓
Larvae develop and molt from L1 to L3 stages in copepods
↓
Consumed by person

FIGURE 16.4   Life cycle of the guinea worm.

of the *D. medinensis* is summarized in Fig. 16.4. The eradication program has reduced the incidence from 3.5 million cases per year (1980s) to just 22 cases in 2014. Complete global eradication of Guinea worm disease is within reach (CDC,2017; WHO, 2017b).

16.3.5 Hookworm

It is estimated that between 576 and 740 million people are infected globally with hookworm (CDC, 2017) (Fig. 16.5). Hookworm (*Ancylostoma duodenale* and *Necator americanus*) is transmitted from soil; *Anc. duodenale* from North Africa, Middle East, and India, and *N. americanus* from the Americas, sub-Saharan Africa, Southeast Asia, and China. Hookworms live in the intestinal tract and cause gastrointestinal symptoms,

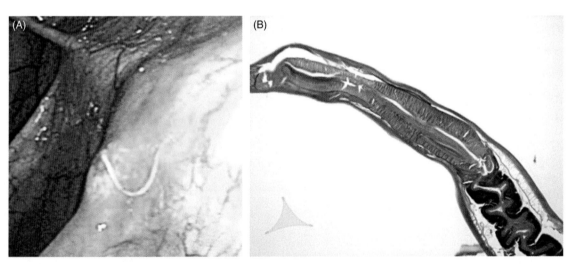

FIGURE 16.5   **Hookworm.** (A) Hookworm shown attached to the mucosa of the colon during colostomy. (B) Hookworm shown as sectioned and stained with hematoxylin and eosin (100× magnification). *Source: Part A–B: Courtesy CDC, 2017. Parasites. Available from: https://www.cdc.gov/parasites/ including Flukes http://www.cdc.gov/parasites/fasciola/; Parasites—Angiostrongyliasas (also known as angiostrongyliasis infection) http://www.cdc.gov/parasites/angiostrongylus/; Parasites—Ascariasis http://www.cdc.gov/parasites/ascariasis/; Parasites—Dracunculiasas (also known as guinea worm disease) http://www.cdc.gov/parasites/guineaworm/; Parasites—Hookworm http://www.cdc.gov/parasites/hookworm/; Parasites: Neglected parasitic infections https://www.cdc.gov/parasites/npi/; Parasites—Schistosomiasis http://www.cdc.gov/parasites/schistosomiasis/epi.html; Taeniasis http://www.cdc.gov/parasites/taeniasis/biology.html; Parasites—Toxocariasis (also known as roundworm infection) (http://www.cdc.gov/parasites/toxocariasis/gen_info/faqs.html); Parasites—Trichuriasis (also known as whipworm infection) http://www.cdc.gov/parasites/whipworm/biology.html.*

such as abdominal pain, diarrhea, and nausea, and can lead to anemia.

16.3.6 Lymphatic Filariasis or Elephantiasis

Lymphatic filariasis or elephantiasis is another disease caused by the nematode with the parasite being members of the family Filarioidea: 90% *Wuchereria bancrofti*, 10% *Brugia* spp. including *Brugia malayi* and *Brugia timori*) (WHO, 2017c). The disease was first described about 600 BCE (BC) by Persian and Hindu physicians (Otsuji, 2011). The adult worms are found in the lymphatic vessels depressing lymph flow and thereby causing edema (WHO, 2017c). The adult females release larvae or microfilaria and these migrate into the blood vessels. The larvae are transmitted by the mosquito vectors of the genera *Culex*, *Anopheles*, and *Aedes* (WHO, 2017c).

In 1996, it was estimated that there were 119 million people with lymphatic filariasis mainly in adults and the elderly and concentrated in sub-Saharan Africa and India (Michael et al., 1996) with 40% in India (Pani et al., 2005). There were 40 million people afflicted with the disfiguring signs of lymphatic filariasis (scrotal hydrocele in 25 million men and 15 million people with elephantiasis) in 2009 and 120 million people infected (WHO, 2017c). Mass drug administration is being employed as a successful strategy to eliminate lymphatic filariasis (WHO, 2017c). The goal is to reduce infection to a level where transmission is nonsustainable (WHO, 2017c). This has been successful with 18 countries including the Republic of Korea and China being free of lymphatic filariasis (WHO, 2017c).

16.3.7 Mansonellosis

Mansonellosis is caused by the filarial nematode *Mansonella perstans* in sub-Saharan Africa and by *Mansonella ozzardi* in South and Central America and the Caribbean. Fig. 16.6 shows *M. perstans* microfilariae. The vectors for *M. perstans* are midges (Mourembou et al., 2015). It is estimated that 114 million people are infected with mansonellosis (Mourembou et al., 2015). The prevalence of *M. ozzardi* in one area in the Amazon was at 41% (Kozek et al., 1982). Mansonellosis is often asymptomatic, but can be associated with inflammation, fever, pain, and skin conditions.

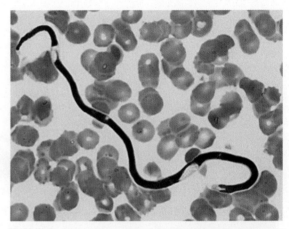

FIGURE 16.6 *M. perstans* microfilariae (190–200 μm) in Wright–Giemsa stained peripheral blood smear (500× magnification with oil). *M. perstans* causes mansonellosis. *Source: Courtesy CDC, 2017. Parasites. Available from: https:// www.cdc.gov/parasites/ including Flukes http://www.cdc.gov/ parasites/fasciola/; Parasites—Angiostrongyliasas (also known as angiostrongyliasis infection) http://www.cdc.gov/parasites/ angiostrongylus/; Parasites—Ascariasis http://www.cdc.gov/ parasites/ascariasis/; Parasites—Dracunculiasas (also known as guinea worm disease) http://www.cdc.gov/parasites/guineaworm/; Parasites—Hookworm http://www.cdc.gov/parasites/hookworm/; Parasites: Neglected parasitic infections https://www.cdc.gov/ parasites/npi/; Parasites—Schistosomiasis http://www.cdc.gov/ parasites/schistosomiasis/epi.html; Taeniasis http://www.cdc.gov/ parasites/taeniasis/biology.html; Parasites—Toxocariasis (also known as roundworm infection) (http://www.cdc.gov/parasites/ toxocariasis/gen_info/faqs.html); Parasites—Trichuriasis (also known as whipworm infection) http://www.cdc.gov/parasites/ whipworm/biology.html.*

16.3.8 Pinworms (*Enterobius vermicularis*)

Pinworms, also known as threadworms or seat worms (*En. vermicularis*) (Fig. 16.7), are probably the most common nematodes infecting people. It was estimated that about between 200 million and 1 billion people are infected with pinworms globally with the highest incidence in children between 5 and 10 years old particularly living in overcrowded conditions (Cook, 1994; Kucik et al., 2004). High rates of infection are

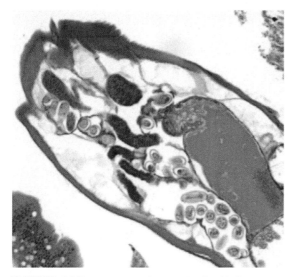

FIGURE 16.7 *En. vermicularis* **stained with hematoxylin and eosin.** This pinworm causes enterobiasis. *Source: Courtesy CDC, 2017. Parasites. Available from: https://www.cdc. gov/parasites/ including Flukes http://www.cdc.gov/parasites/fasciola/; Parasites—Angiostrongyliasas (also known as angiostrongyliasis infection) http://www.cdc.gov/parasites/angiostrongylus/; Parasites—Ascariasis http://www.cdc.gov/parasites/ascariasis/; Parasites—Dracunculiasas (also known as guinea worm disease) http://www.cdc.gov/parasites/guineaworm/; Parasites—Hookworm http://www.cdc.gov/parasites/hookworm/; Parasites: Neglected parasitic infections https://www.cdc.gov/parasites/npi/; Parasites— Schistosomiasis http://www.cdc.gov/parasites/schistosomiasis/epi. html; Taeniasis http://www.cdc.gov/parasites/taeniasis/biology. html; Parasites—Toxocariasis (also known as roundworm infection) (http://www.cdc.gov/parasites/toxocariasis/gen_info/faqs.html); Parasites—Trichuriasis (also known as whipworm infection) http:// www.cdc.gov/parasites/whipworm/biology.html.*

reported in people who are institutionalized; for instance, the incidence being 24.8% of people with developmental disabilities in a center in New Jersey but dropping to less than 1% with a prophylactic program for contacts of those infected (Lohiya et al., 2000). Infection is usually asymptomatic.

The life cycle of *En. vermicularis* encompasses oral ingestion; eggs hatching in the stomach and upper small intestine; larva then migrating to the ileum, cecum, and appendix molting twice; and then metamorphosing into adults. The gravid females (about 10 mm long), each containing about 11,000 fertilized eggs, migrate to the colon and then through the anus (reviewed by Cook, 1994). Pinworms are transmitted by the following (reviewed by Cook, 1994):

- Autoinfection from fingers/fingernails touching infected peri-anal areas.
- Infection from bed sheets with *En. vermicularis* embryonated eggs.
- Infection from dust contaminated with *En. vermicularis* embryonated eggs.
- Retroinfection of larva from the perianal areas.

16.3.9 Strongyloidiasis and *Strongyloides stercoralis*

S. stercoralis (Fig. 16.8) is a soil-transmitted helminth (passed to people from the soil). *S. stercoralis* causes strongyloidiasis with an estimated 30–100 million people infected globally (WHO, 2017a). Intestinal symptoms can include diarrhea, abdominal pain, and weight loss.

16.3.10 Trichinellosis and *Trichinella*

Parasitic nematodes of the genus *Trichinella* cause trichinellosis (trichinosis). Consumption of undercooked meat containing *Trichinella* larvae leads to trichinellosis. The larvae are released in the intestine. This can be asymptomatic with

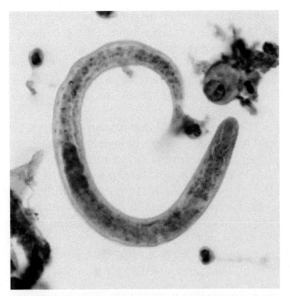

FIGURE 16.8 **Strongyloidiasis caused by *S. stercoralis*.** *Source: Courtesy CDC, 2017. Parasites. Available from: https:// www.cdc.gov/parasites/ including Flukes http://www.cdc.gov/ parasites/fasciola/; Parasites—Angiostrongyliasas (also known as angiostrongyliasis infection) http://www.cdc.gov/parasites/ angiostrongylus/; Parasites—Ascariasis http://www.cdc.gov/ parasites/ascariasis/; Parasites—Dracunculiasas (also known as guinea worm disease) http://www.cdc.gov/parasites/guineaworm/; Parasites—Hookworm http://www.cdc.gov/parasites/hookworm/; Parasites: Neglected parasitic infections https://www.cdc.gov/ parasites/npi/; Parasites—Schistosomiasis http://www.cdc.gov/ parasites/schistosomiasis/epi.html; Taeniasis http://www.cdc.gov/ parasites/taeniasis/biology.html; Parasites—Toxocariasis (also known as roundworm infection) (http://www.cdc.gov/parasites/ toxocariasis/gen_info/faqs.html); Parasites—Trichuriasis (also known as whipworm infection) http://www.cdc.gov/parasites/ whipworm/biology.html.*

Asia (Dupouy-Camet, 2000). Globally *Trichinella* infestation is due to consumption of infected pork but horse meat and wild game (wild pigs and bears) are also significant sources (Dupouy-Camet, 2000). *Trichinella* is also an issue in Western countries. In the USA, between 2008 and 2012, there were 84 confirmed cases of trichinellosis, or less than 0.1 cases per 1 million population per year (Wilson et al., 2015). Where known, the major sources of the *Trichinella* in the USA were bear meat (48.8%) (Fig. 16.9) and

FIGURE 16.9 **Larvae of *Trichinella* sp. encapsulated in the bear muscle tissue.** Trichinellosis is caused by *Trichinella spiralis*. *Source: Courtesy CDC, 2017. Parasites. Available from: https://www.cdc.gov/parasites/ including Flukes http://www.cdc. gov/parasites/fasciola/; Parasites—Angiostrongyliasas (also known as angiostrongyliasis infection) http://www.cdc.gov/parasites/angi-ostrongylus/; Parasites—Ascariasis http://www.cdc.gov/parasites/ ascariasis/; Parasites—Dracunculiasas (also known as guinea worm disease) http://www.cdc.gov/parasites/guineaworm/; Parasites— Hookworm http://www.cdc.gov/parasites/hookworm/; Parasites: Neglected parasitic infections https://www.cdc.gov/parasites/npi/; Parasites—Schistosomiasis http://www.cdc.gov/parasites/schisto-somiasis/epi.html; Taeniasis http://www.cdc.gov/parasites/taeniasis/ biology.html; Parasites—Toxocariasis (also known as roundworm infection) (http://www.cdc.gov/parasites/toxocariasis/gen_info/faqs. html); Parasites—Trichuriasis (also known as whipworm infection) http://www.cdc.gov/parasites/whipworm/biology.html.*

less than 70 larvae or with more larvae causing diarrhea and abdominal pain (reviewed by Wilson et al., 2015). The larvae mature to adults that pass through the intestinal mucosa and migrate to skeletal muscles including the diaphragm via the blood vessels (reviewed by Wilson et al., 2015). This leads to significant health issues or death.

It is estimated that about 11 million people in the world are infected with *Trichinella* particularly in rural areas of South America and pork (26.1%) (Wilson et al., 2015). The rate of trichinellosis has declined from 400 cases per year and 10–15 deaths per year between 1947 and 1951 (Schantz, 1983) but autopsy data from the time indicate *Trichinella* infection in about one-sixth of the population (Stoll, 1947). Trichinellosis is also found in Europe with cases in France, Germany, Italy, and Spain (Dupouy-Camet, 2000). Trichinellosis is rarely found in practicing Muslim countries or in Jewish populations because of the dietary laws forbidding pork consumption (Dupouy-Camet, 2000).

16.3.11 Trichuriasis

Trichuriasis is caused by *Trichuris trichiura* (the human whipworm infection). It is estimated that 604–795 million people are infected with *Trichu. trichiura* (CDC, 2017). These are considered as soil-transmitted helminths. Each whipworm produces between 3,000 and 20,000 eggs per day (CDC, 2017). Mild infestations with few worms are usually asymptomatic; however, severe infestation can be associated with diarrhea, abdominal pain, fatigue, and anemia.

16.3.12 Toxocariasis

Toxocariasis is caused by parasitic roundworms of dogs (*Toxocara canis*) and cats (*Toxocara cati*). The CDC concluded that 70 people, predominantly children, in the USA are rendered blind by toxocariasis annually (CDC, 2017). In addition, 14% of people in the USA have antibodies to *Toxocara*; this being consistent with large-scale exposure (CDC, 2017).

16.4 CESTODES (TAPEWORMS)

16.4.1 Overview

Tapeworms are parasites that inhabit the intestines. They consist of a head (scolex) imbedded into the mucosa, a neck, and a body consisting of a series of segments or proglottids (Fig. 16.10). These contain male and female gonads.

Tapeworms that infect people include the following (WHO, 2015):

- Genus *Taenia* causing Taeniasis and cysticercosis
  - *Taenia solium* (pork tapeworm)
  - *Taenia saginata* (beef tapeworm)
  - *Taenia asiatica* (Asian tapeworm)
  - *Taenia multiceps*
- Genus *Hymenolepis* causing hymenolepiasis
  - Rat tapeworm (*Hymenolepis diminuta*)
  - Dwarf tapeworm (*Hymenolepis nana*)
- Other tapeworms
  - *Dipylidium caninum*
  - *Diphyllobothrium* sp. (fish tapeworm)
  - *Echinococcus granulosus* (dog tapeworm)
  - *Spirometra*

16.4.2 *Taenia*

Ta. solium (pork tapeworm) results in 2.8 million DALY (WHO, 2016). Larval *Ta. solium* infection in the brain (neurocysticercosis) is a major cause of epilepsy in developing countries with it estimated that *Ta. solium* is the cause of 30% of epilepsy cases in many endemic areas where people and roaming pigs live in close proximity (Del Brutto et al., 1992; Harhay et al., 2010; Ndimubanzi et al., 2010). Ingestion of *Ta. solium* eggs can lead to neurocysticercosis. The CDC estimates that there are 1000 new hospitalizations in the USA per year due to neurocysticercosis; this being a major cause of infectious seizures (CDC, 2017). The life cycle of *Ta. solium* and *Ta. saginata* are shown in Fig. 16.11 (WHO, 2016).

16.4.3 *Diphyllobothrium* sp.

There is a growing incidence of the infection diphyllobothriosis with fish tapeworms including *Diphyllobothrium latum*, *Diphyllobothrium*

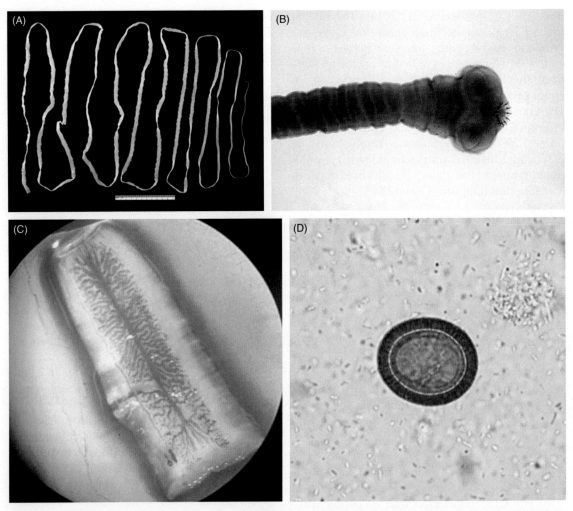

FIGURE 16.10　**Cestodes.** (A) Adult beef tapeworm (*Ta. saginata*). (B) Scolex (head) of pork tapeworm (*Ta. solium*) show-ing hooks and suckers for attachment to the intestine. (C) Mature proglottids from beef tapeworm (*Ta. saginata*) after inject-ing with lactophenol cotton blue. (D) Cestode eggs (diameter 30–34 μm) from *Taenia* sp. *Source: Part A–D: Courtesy CDC, 2017. Parasites. Available from: https://www.cdc.gov/parasites/ including Flukes http://www.cdc.gov/parasites/fasciola/; Parasites—Angiostrongyliasas (also known as angiostrongyliasis infection) http://www.cdc.gov/parasites/angiostrongylus/; Parasites—Ascariasis http://www.cdc.gov/parasites/ascariasis/; Parasites—Dracunculiasas (also known as guinea worm disease) http://www.cdc.gov/para-sites/guineaworm/; Parasites—Hookworm http://www.cdc.gov/parasites/hookworm/; Parasites: Neglected parasitic infections https:// www.cdc.gov/parasites/npi/; Parasites—Schistosomiasis http://www.cdc.gov/parasites/schistosomiasis/epi.html; Taeniasis http://www. cdc.gov/parasites/taeniasis/biology.html; Parasites—Toxocariasis (also known as roundworm infection) (http://www.cdc.gov/parasites/ toxocariasis/gen_info/faqs.html); Parasites—Trichuriasis (also known as whipworm infection) http://www.cdc.gov/parasites/whip-worm/biology.html.*

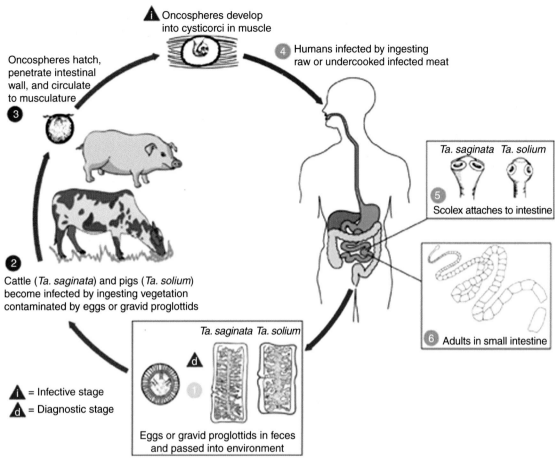

FIGURE 16.11 **Life cycle of beef and pork tapeworms.** *Source: Courtesy CDC, 2017. Parasites. Available from: https://www. cdc.gov/parasites/ including Flukes http://www.cdc.gov/parasites/fasciola/; Parasites—Angiostrongyliasas (also known as angiostrongyliasis infection) http://www.cdc.gov/parasites/angiostrongylus/; Parasites—Ascariasis http://www.cdc.gov/parasites/ascariasis/; Parasites—Dracunculiasas (also known as guinea worm disease) http://www.cdc.gov/parasites/guineaworm/; Parasites—Hookworm http://www.cdc.gov/parasites/hookworm/; Parasites: Neglected parasitic infections https://www.cdc.gov/parasites/npi/; Parasites—Schistosomiasis http://www.cdc.gov/parasites/schistosomiasis/epi.html; Taeniasis http://www.cdc.gov/parasites/taeniasis/biology.html; Parasites—Toxocariasis (also known as roundworm infection) (http://www.cdc.gov/parasites/toxocariasis/gen_info/faqs.html); Parasites—Trichuriasis (also known as whipworm infection) http://www.cdc.gov/parasites/whipworm/biology.html.*

dendriticum, *Diphyllobothrium nihonkaiense,* and *Adenocephalus pacificus* (syn. *Diphyllobothrium pacificum*) (Chen et al., 2014; Kuchta et al., 2013, 2015). The increase in infections is due to consumption of raw or insufficiently cooked fish (Kuchta et al., 2013). Diphyllobothriosis infection is usually asymptomatic, but can be associated with abdominal pain, diarrhea, nausea, and weight loss.

16.5 TREMATODES (FLUKES)

16.5.1 Overview

Trematode infections are either food-borne or Schistosomiasis (snail-borne) (Table 16.6). The major food-borne trematodes infecting humans are the following (Keiser and Utzinger, 2009):

TABLE 16.6   Food-Borne Trematode Infections: Their Site of Infection, Their Intermediate Host and Prevalence

	Intermediate host	Number of infections (million)
Liver fluke		
C. sinensis	Freshwater fish	35
O. felineus	Freshwater fish	1.2
O. viverrini	Freshwater fish	10
F. hepatica and F. gigantica	Freshwater snails	2.4–17.0
Lung fluke		
Paragonimus spp.	Freshwater crabs, crayfish	20.7
Intestinal fluke		
F. buski	Freshwater snails	1.3

Based on Keiser, J., Utzinger, J., 2009. Food-borne trematodiases. Clin. Microbiol. Rev. 22, 466–483.

- Liver flukes (*Clonorchis sinensis*, *Fasciola gigantica*, *Fasciola hepatica* (Table 16.6), *Opisthorchis felineus*, and *Opisthorchis viverrini*).
- Lung flukes (*Paragonimus* spp.).
- Intestinal flukes (e.g., *Echinostoma* spp., such as *Echinostoma ilocanum*, *Fasciolopsis buski*, and the heterophyids, e.g., *Heterophyes heterophyes*).

These are transmitted to people after a certain period in a secondary host. Effects of infection range from no noticeable effect to loss of tissues and inability to work (or even get up) (Fig. 16.12) (Harhay et al., 2010).

16.5.2 Schistosomiasis

Schistosomiasis (bilharzia) is a serious disease with 61.6 million people treated for it globally in 2014. It is also being estimated that 258 million people require preventive treatment for schistosomiasis (WHO, 2016). Schistosomiasis is associated with pain, diarrhea, lower work or school performance and cognition, and infertility (King, 2011). Intestinal schistosomiasis is caused by different species of the genus *Schistosoma* (WHO, 2016):

- *Schistosoma mansoni* (Africa, Middle East, South America, and the Caribbean)

- *Schistosoma guineensis* (sub-Saharan African rainforest)
- *Schistosoma intercalatum* (sub-Saharan African rainforest)
- *Schistosoma japonicas* (East Asia and Southeast Asia particularly Indonesia and the Philippines)
- *Schistosoma mekongi* (Southeast Asia)

Urogenital schistosomiasis is caused by *Schistosoma haematobium* (Africa, Middle East, and Corsica) (CDC, 2017; WHO, 2016).

16.5.2.1 Life Cycle

Larval *Sc. mansoni* (stage called *Cercariae*) are released from snails into water. They penetrate human skin. They are then located in the circulation ending up in the hepatic portal veins where they mature into adults. Paired male and female migrate to the mesenteric venules (veins within folds of tissue that anchor the gastrointestinal organs). The fertilized eggs are shed with the host feces after ulcerating the walls of the colon and rectum. The eggs hatch releasing miracidia and these infect new snails (CDC, 2017). In contrast, with *Sc. haematobium*, the adults are located in the venous plexus of the bladder and the eggs pass through the bladder mucosa and are released into the urine (CDC, 2017; WHO, 2017d).

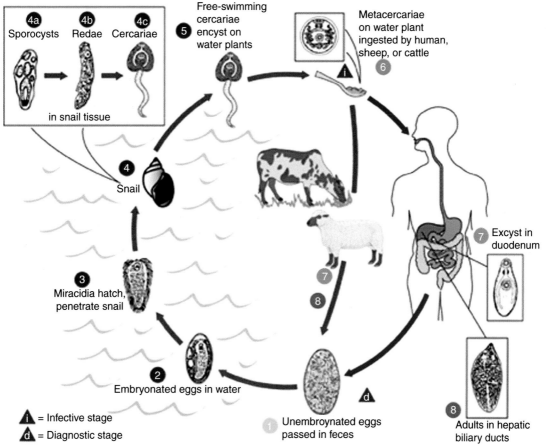

FIGURE 16.12 **Life cycle of liver fluke.** *Source: Courtesy CDC, 2017. Parasites. Available from: https://www.cdc.gov/parasites/ including Flukes http://www.cdc.gov/parasites/fasciola/; Parasites—Angiostrongyliasas (also known as angiostrongyliasis infection) http://www.cdc.gov/parasites/angiostrongylus/; Parasites—Ascariasis http://www.cdc.gov/parasites/ascariasis/; Parasites—Dracunculiasas (also known as guinea worm disease) http://www.cdc.gov/parasites/guineaworm/; Parasites—Hookworm http://www.cdc.gov/parasites/hookworm/; Parasites: Neglected parasitic infections https://www.cdc.gov/parasites/npi/; Parasites—Schistosomiasis http://www.cdc.gov/parasites/schistosomiasis/epi.html; Taeniasis http://www.cdc.gov/parasites/taeniasis/biology.html; Parasites—Toxocariasis (also known as roundworm infection) (http://www.cdc.gov/parasites/toxocariasis/gen_info/faqs.html); Parasites—Trichuriasis (also known as whipworm infection) http://www.cdc.gov/parasites/whipworm/biology.html.*

16.6 ECTOPARASITES

16.6.1 Overview

Ectoparasites that influence humans and livestock include the following arthropods (Hopla et al., 1994):

- Insects (Class Insecta)
  - Bugs (Order Hemiptera), such as bed bugs and kissing bugs
  - Fleas (Order Siphonaptera)
  - Flies (Order Diptera) including blow flies, mosquitoes, and tsetse flies
  - Lice (Order Phthiraptera)
- Arachnids (Class Arachnida)
  - Mites (Order Sarcoptiformes)

- Ticks (Order Parasitiformes; superfamily Ixodoidea)

16.6.2 Bed Bugs

Bed bugs (*Cimex lectularius*) are bloodsucking ectoparasites (external parasites) that have negative impacts on physical health, such as allergies to bites and secondary infections, and on mental health with anxiety, insomnia, and systemic reactions (EPA, 2007). Bed bugs are an increasing problem in hospitals, hotel rooms, and apartments (Bloomberg, 2007; Vaidyanathan and Feldlaufer, 2013). There is limited information on the economic costs and impacts of bed bugs. However, it was reported that "bed bugs reports lowered the value of a hotel room by $21 for leisure travelers and $38 for business travelers."

16.6.3 Fleas

Fleas are predominantly ectoparasites feeding on the blood of mammals (Whitinga et al., 2008). The "so called" human flea, *Pulex irritans* (Table 16.7; Fig. 16.13), is also an ectoparasite of multiple species including guinea pigs, domestic dogs, cats, rats, goats (Bitam et al., 2010), swift foxes (*Vulpes velox*), and burrowing owls (*Athene cunicularia*) (Pence et al., 2004). The "human

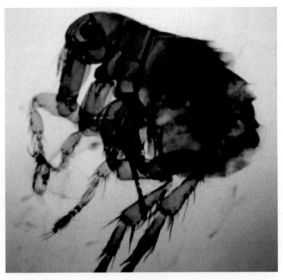

FIGURE 16.13 **Human flea, *Pu. irritans*.** *Source: Courtesy CDC, 2017. Parasites. Available from: https://www.cdc.gov/parasites/ including Flukes http://www.cdc.gov/parasites/fasciola/; Parasites—Angiostrongyliasas (also known as angiostrongyliasis infection) http://www.cdc.gov/parasites/angiostrongylus/; Parasites—Ascariasis http://www.cdc.gov/parasites/ascariasis/; Parasites—Dracunculiasas (also known as guinea worm disease) http://www.cdc.gov/parasites/guineaworm/; Parasites—Hookworm http://www.cdc.gov/parasites/hookworm/; Parasites: Neglected parasitic infections https://www.cdc.gov/parasites/npi/; Parasites—Schistosomiasis http://www.cdc.gov/parasites/schistosomiasis/epi.html; Taeniasis http://www.cdc.gov/parasites/taeniasis/biology.html; Parasites—Toxocariasis (also known as roundworm infection) (http://www.cdc.gov/parasites/toxocariasis/gen_info/faqs.html); Parasites—Trichuriasis (also known as whipworm infection) http://www.cdc.gov/parasites/whipworm/biology.html.*

TABLE 16.7 Arthropod Ectoparasites of Humans

Species		Length	References
Human flea[a]	*Pu. irritans*	~3 mm	CDC (2017)
Head lice	*Pe. humanus capitis*	2–3 mm	CDC (2017)
Body lice	*Pe. humanus corporis*	2.3–3.6 mm	CDC (2017)
Pubic lice	*Pt. pubis*	1.1–1.8 mm	CDC (2017)
Mites	*De. folliculorum*	285 μm	Wesolowska et al. (2014)
Mites	*De. brevis*	185 μm	Wesolowska et al. (2014)
Itch mite	*Sa. scabiei*	Females 375 μm; males 300 μm	CDC (2017)

a Fleas from other animals including dogs (Ct. canis), cats (Ct. felis), rodents (e.g., X. cheopis), and birds can bite people spreading disease.

flea" is not an effective vector but is an intermediate host for Cestodes (Whitinga et al., 2008).

The "human flea" is viewed as a cosmopolitan flea species but related species of fleas are parasites in rodents of the family Muridae (Whitinga et al., 2008). The generally accepted view is that the genus *Pulex* originated in South America then dispersed through North America, crossing the Bering Strait, and finally to Asia and then Europe and Africa. The "human flea" has been identified in pre-Columbian animal mummies in present-day South Peru (Dittmar et al., 2003). There is evidence for the presence of *Pu. irritans* in Western Europe between 1000 and 2000 years ago. This led to the suggestion that fleas came to Western Europe with the migration of Neolithic farmers.

Other species of fleas can feed on humans. These include the following (CDC, 2017):

- Cat flea (*Ctenocephalides felis*)
- Dog flea (*Ctenocephalides canis*)
- Oriental rat flea (*Xenopsylla cheopis*), the primary vector for *Yersinia pestis* (plague)
- Sticktight flea (*Echidnophaga gallinacea*)

16.6.4 Lice

16.6.4.1 Overview

Lice are insect ectoparasites. Their bites are very itchy to the hosts leading to scratching and inflammation. They are obligate parasites that perish if away from the skin of their hosts. Lice show profound species specificity with usually one per mammalian or avian species. Humans are unusual having three genetic types of lice (Table 16.7) living on them:

- Head lice (*Pediculus humanus capitis*) living in the hair and feeding on blood on the scalp.
- Body lice (or clothing lice) (*Pediculus humanus corporis*) living in the clothes and feeding on blood on the body.
- Pubic lice (or colloquially "crabs") (*Pthirus pubis*) living in the coarser pubic hair

Human head and body lice show marked similarities to the chimpanzee louse. Using the molecular clock, it has been estimated that the body lice of chimpanzees and humans diverged 5–6 million years ago (Reed et al., 2004, 2007). This estimate is very close to the last common ancestor of humans and chimpanzees (5–7 million years ago) with the lines diverging (Varki and Altheide, 2005). In contrast, the pubic louse (*Pt. pubis*) has close genomic similarities with the gorilla louse (*Pthirus gorilla*); the last common ancestor of these lice being estimated at 3–4 million years ago (Reed et al., 2004, 2007). This is much later than the divergence between the ancestors of man and gorillas between 8.8 and 18.9 million years ago (Langergraber et al., 2012). It has been suggested that there was interaction between archaic hominids and the ancestors of gorillas resulting in a host switch of the lice. This could be our ancestors killing and eating the ancestors of gorillas although other explanations are possible.

16.6.4.2 Transmission

Lice are transmitted from person to person via shared clothes, bedding or hair brushes, and direct contact including sexual contact; the latter being particularly important for pubic lice.

16.6.4.3 Incidence

16.6.4.3.1 HEAD LICE

It is estimated that there are 6–12 million children between the age of 5 and 11 infested with head lice per year in the USA. Table 16.8 outlines the prevalence of head lice in various countries. The eggs are referred to as nits.

16.6.4.3.2 BODY LICE

While the mean rate of infestation in adults in Brazil is 5.4%, the rate is much higher in slum dwellers at 40.8% (Falagas et al., 2008). Similarly, rates are high in child laborers in slum areas in India (48%), a low socioeconomic village in Turkey (31.1%), and in a socially deprived urban area in Belgium (21.9%) (Falagas et al., 2008). Rates of nits and lice are also high in homeless people in France (22%) (Falagas et al., 2008).

16.6.4.3.3 PUBIC LICE

The incidence of pubic lice is about 2% globally (Dholakia et al., 2014).

There is a series of words in the English language related to lice:

- Cooties: The word is said to be a term from World War I when body/clothing lice was widespread in the soldiers.
- Louse: Can be used to talk about a bad person, as in "he's a louse!"
- Lousy: Louse has given us the word lousy. This means covered in lice or very bad or disgusting or nasty.
- Nitpicking: The difficult process of removing nits gave rise to the word, which means to be very fussy or pedantic.
- To louse up: to mess up.

16.6.5 Lice-Borne Disease

Lice are the vectors for several diseases including the following (Fournier et al., 2002):

- epidemic relapsing fever (*Borrelia recurrentis*)
- epidemic typhus (*Rickettsia prowazekii*)
- trench fever (*Bartonella quintana*)

16.6.6 Mites

16.6.6.1 *Overview*

Mites are very small arachnids (Table 16.7) that cannot be easily seen except with a magnifying glass.

TABLE 16.8 Incidence of Head Lice or Nits in Children in Different Countries

Country	Rate of lice/nit infestation (%)
Australia	13
Brazil	24.1
Egypt	43.6
England	2.4
France	3.3
India	16.6
Poland	1.0
USA	3.6

Calculated from Falagas, M.E., Matthaiou, D.K., Rafailidis, P.I., Panos, G., Pappas, G., 2008. Worldwide prevalence of head lice. Emerg. Infect. Dis. 14, 1493–1494.

Examples of mites that people come across include the following:

- ectoparasitic mites of people, companion animals, poultry, and livestock
- household mites
- mites contaminating stored flour, grains, other foods, and livestock feed

---

There is one word in the English language related to having mange caused by mites—mangy. This is used colloquially to denote shabby.

16.6.6.2 *Ectoparasitic mites*

16.6.6.2.1 SCABIES

Scabies is caused by the itch mite (also known as the scabies mite and mange) (*Sarcoptes scabiei*) (Fig. 16.14) (reviewed by Bandi and Saikumar, 2013). These mites live in the skin and feed on body fluids including the blood and lymph fluid. The female lays about 200 eggs and the hatched larvae burrow into the skin. Distinct subspecies of the scabies mite infect different host

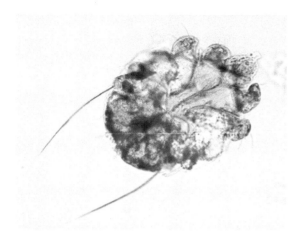

FIGURE 16.14 **The human itch mite, *Sa. scabiei*.** *Source: Courtesy CDC, 2017. Parasites. Available from: https://www.cdc. gov/parasites/ including Flukes http://www.cdc.gov/parasites/fasciola/; Parasites—Angiostrongyliasas (also known as angiostrongyliasis infection) http://www.cdc.gov/parasites/angiostrongylus/; Parasites—Ascariasis http://www.cdc.gov/parasites/ascariasis/; Parasites—Dracunculiasas (also known as guinea worm disease) http://www.cdc.gov/parasites/guineaworm/; Parasites—Hookworm http://www.cdc.gov/parasites/hookworm/; Parasites: Neglected parasitic infections https://www.cdc.gov/parasites/npi/; Parasites— Schistosomiasis http://www.cdc.gov/parasites/schistosomiasis/epi. html; Taeniasis http://www.cdc.gov/parasites/taeniasis/biology. html; Parasites—Toxocariasis (also known as roundworm infection) (http://www.cdc.gov/parasites/toxocariasis/gen_info/faqs.html); Parasites—Trichuriasis (also known as whipworm infection) http:// www.cdc.gov/parasites/whipworm/biology.html.*

species with *Sa. scabiei* var *hominis* infesting people. There are an estimated 300 million cases of scabies per year. In addition, animal scabies mites can be transferred to people with multiple bites but they do not complete a life cycle and die off.

16.6.6.2.2 *DEMODEX* MITES

Two species of mites infest human skin: (1) *Demodex folliculorum* and (2) *Demodex brevis*. These colonize the hair follicles of the eyelash, forehead, nose, cheeks, outer ears, chest, buttocks, and pubic areas (reviewed by Wesolowska et al., 2014). They are small (Table 16.7), wormlike, and live on skin cells and sebum. The mites move over the skin at night for mating. Females

lay their eggs in the sebaceous glands. Transmission is by direct contact with larvae or adult on the skin or with dust containing eggs (reviewed by Wesolowska et al., 2014).

Estimates of the prevalence of *Demodex* mite infestation range from 11% in health professionals aged between 20 and 29 years old (Wesolowska et al., 2014) to 100% in the elderly (Elston, 2010). Mite infestations can be determined by conventional light microscopy or confocal laser scanning microscopy with markedly lower numbers of *Demodex* detected by light microscopy (Jalbert and Rejab, 2015). It is likely that many mite infestations are missed by light microscopy. There is evidence of increased prevalence of *De. brevis* infestations in the eyelid follicles in people with chalazia (blocked sebaceous glands in the eyelid) (Liang et al., 2014). Mites have been detected in almost all contact lens wearers (Jalbert and Rejab, 2015).

Problems, such as dermatitis due to *Demodex* mites, are rare. *Demodex* mites may have a role in rosacea (Jarmuda et al., 2012). Mites are the vector for transmission of pathogens in *Orientia tsutsugamushi* (causing scrub typhus) and *Rickettsia akari* (causing rickettsialpox) (reviewed by Diaz, 2010) (discussed in more detail later in this chapter).

16.6.6.3 Ticks

Ticks might superficially seem to be irrelevant to human well-being. However, ticks acting as disease vectors adversely affect people and livestock. Lyme disease is the most common vector-borne inflammatory disease in the United States with 240,000 to 440,000 new cases per year and costing the U.S. health care system between $712 million and $1.3 billion a year (Johns Hopkins School of Public Health, 2015; Johnson et al., 2011; Zhang et al., 2006).

Bites from some species of ticks can cause the development of a red meat allergy, this ranging in severity from abdominal pain to anaphylaxis. There is a specific immunoglobulin E to an oligosaccharide epitope, galactose-alpha-1,3-galactose.

This is present on glycolipids and glycoproteins including blood groups in nonprimate mammals. Allergies and delayed anaphylactic reactions can be against food containing the epitope including beef, pork, lamb, venison, and milk (reviewed by ACAAI, in press; Commins and Platts-Mills, 2013; Tripathi et al., 2014). Allergies to red meat are associated with bites from lone star ticks (*Amblyomma americanum*) in the southwest of the USA, the castor bean tick (*Ixodes ricinus*) in Europe, and the Australian paralysis tick (*Ixodes holocyclus*) in Australia (reviewed by Commins and Platts-Mills, 2013).

16.7 PARASITES IN POULTRY AND LIVESTOCK

16.7.1 Overview

Parasites have a large economic impact on livestock and poultry production globally. For instance, economic losses to cattle caused by parasites in Brazil were estimated as follows (Grisi et al., 2014):

- gastrointestinal nematodes, US$7.1 billion
- ectoparasites, such as ticks and flies, US$6.9 billion

16.7.2 Single-Celled Parasites and Poultry

16.7.2.1 Coccidiosis

Coccidiosis is a major poultry disease. It is caused by *Eimeria tenella* and other members of the same genus (McDougald, 1998). The methods for preventing coccidiosis are anticoccidial drugs and immunization. The economic impacts of coccidiosis are the following (The Poultry Site, 2013):

- Prevention: globally, over US$3 billion per year ($90 million in the USA)
- Decreased performance, morbidity, and mortality: globally US$300 million

---

Kingdom: Chromista
Superphylum: Alveolata
Phylum: Miozoa
Infraphylum: Apicomplexa
Class: Conoidasida
Order: Eucoccidiorida
Family: Eimeridae
Species: *Eimeria tenella*
(based on Ruggiero et al., 2015).

16.7.2.2 Other Single-Celled Parasites of Poultry

Histomonas meleagridis is the other single-celled parasite of poultry. It is transmitted by the eggs of cecal worm *Heterakis gallinarum* (reviewed by McDougald, 1998). Infection results in morbidity in chickens with mortalities/culling of about 10%. The symptoms are more serious in turkeys with up to 100% mortality (reviewed by McDougald, 1998). Another single-celled parasite of poultry is *Cryptosporidium* with infection resulting in respiratory cryptosporidiosis in turkey poults (reviewed by McDougald, 1998).

Kingdom: Protozoa
Phylum: Metamonada
Class: Parabasalia
Order: Trichomonadida
Family: Monocercomonadidae
Species: *Histomonas meleagridis*

16.7.3 Helminth Parasites in Poultry

Helminths depress poultry production with a disproportionate negative impact in developing countries. It has been known since at least the 1920s that nematode infection reduces growth in chickens (Ackert and Herrick, 1928). The impact of helminth parasites on chickens in a traditional village-type system in Northwest India has been examined by comparing birds receiving or not receiving a broad spectrum anthelmintic (Katoch et al., 2012). Growth rate was markedly lower (by 24%) in chicken not receiving the anthelmintic while mortality was greater (78%) (Table 16.9) (Katoch et al., 2012). As might be expected the load of nematodes was either eliminated or greatly reduced (Table 16.9) (Katoch et al., 2012).

Examples of the prevalence of helminth parasites in chickens in tradition village or backyard systems are as follows:

- Northwest India was 72.0% with gastrointestinal helminths (Katoch et al., 2012).
- Shimoga, a region of Karnataka, India, 73.2% infected with gastrointestinal helminths (Javaregowda et al., 2016).
- Ethiopia, 89.5% (83.0% with cestodes and 58.0% of with nematodes) (Hussen et al., 2012).

- Tanzania, 100% with helminths in both the gastrointestinal tract and trachea (Permin et al., 1997).
- Jordan, 73.1% (three nematode and eight cestode species) (Abdelqader et al., 2008).

16.7.4 Helminth Parasites and Livestock

Charlier et al. (2014) concluded that "helminth infections are a major constraint on efficient livestock production." There is limited information on the economic impact of helminth infections particularly in developing countries but infections with either nematode or trematodes are associated with "tremendous losses" (Fabiyi, 1987). For instance, the impact of gastrointestinal nematodes on cattle production in Brazil is estimated at US$7.11 billion per year (Grisi et al., 2014). The global market for veterinary pharmaceuticals including veterinary anthelmintics (parasiticides) was estimated for 2016 at US$13.3 billion (Marketwired.com, in press).

TABLE 16.9 Effect of a Broad Spectrum Anthelmintic, Fenbendazole, on Growth Rate, Mortality and Helminth Prevalence and Load in Free-Range Chickens in Northwestern India

	Untreated	Fenbendazole
Growth rate over 90 days (g day^{-1})	13.7	18.0
Mortality (%)	32	18
Nematodes in intestines		
Ascaridia galli		
Prevalence	35.3	9.7
Number or load	19.3	1.7
Heterakis gallinarum		
Prevalence	29.4	7.3
Number or load	13.0	1.2
Cestodes in intestines		
Raillietina cesticillus		
Prevalence	17.6	2.4
Number or load	6.9	0.9

Data from Katoch, R., Yadav, A., Godara, R., Khajuria, J.K., Borkataki, S., Sodhi, S.S., 2012. Prevalence and impact of gastrointestinal helminths on body weight gain in backyard chickens in subtropical and humid zone of Jammu, India. J. Parasit. Dis. 36, 49–52.

Helminths influence livestock production by increasing mortality, reducing growth rate, inducing weight loss, depressing reproductive efficiency, and causing economic losses due to liver condemnations because of liver flukes or destruction of the entire carcass (Over et al., 1992). Infection of sheep with the nematode *Haemonchus contortus* causes anemia and then death (Besier, 2006). An indicator of the impact of helminth infection on productivity is seen in a study of the effects of anthelmintic treatment on growth in goats as follows (Githigia et al., 2001):

- control, 1.1 kg gain
- albendazole treatment, 3.1 kg gain

The prevalence of helminths can be high particularly in developing countries with traditional livestock systems. For instance, the prevalence of gastrointestinal helminths was determined in Punjab, Pakistan to be the following (Khan et al., 2010):

- Sheep, 44.2%
- Goats, 40.2%
- Water buffalo, 39.8%
- Cattle, 33.7%

Reports from West Africa indicate high prevalence with 75.5% of goats infected with gastrointestinal helminths in south Nigeria (Owhoeli et al., 2014), while 43.1% of sheep and 55.8% of goats were found to be infected in northeastern Nigeria (Nwosu et al., 2007). Anthelmintic treatment has been highly effective resulting in close to complete elimination of helminth parasites. However, there is increasingly a problem with the parasites gaining resistance to the anthelmintics (reviewed by Kornele et al., 2014).

With the increasing use of outdoor production systems including pastures for pigs in Western Europe and North America, there is an increased possibility of transmission of *Ascaris suum* and *Trichuris suis* (Roepstorff and Murrell, 1997). Moreover, infection with these is likely to impact production and immune responses to vaccines (Roepstorff et al., 2011).

16.8 ECTOPARASITES IN LIVESTOCK

Ectoparasites cause economic losses to livestock and poultry production. For instance, the losses due to ectoparasites in cattle in Brazil were estimated as follows (Grisi et al., 2014):

- cattle tick (*Rhipicephalus microplus*), US$3.2 billion
- horn fly (*Haematobia irritans*), US$2.6 billion
- cattle grub (*Dermatobia hominis*), US$0.4 billion
- New World screwworm fly (*Cochliomyia hominivorax*), $0.3 billion
- stable fly (*Stomoxys calcitrans*), US$0.3 billion

Moreover, losses to the US cattle industry due to the stable fly are estimated at US$2.2 billion (Taylor et al., 2012).

Globally, livestock production is negatively affected by tick-borne diseases particularly in the Indian subcontinent and sub-Saharan Africa (Minjauw and McLeod, 2003). For instance, ticks and tick-borne diseases have been estimated to make up about 80% of all disease costs for cattle production in Uganda (East Africa) (Ocaido et al., 2009). There are high losses to small-holder dairy systems on the Indian subcontinent (Minjauw and McLeod, 2003). The four major tick-borne diseases important to the livestock industry are anaplasmosis, babesiosis, cowdriosis, and theileriosis. Costs include those for control together with losses due to reduced production. In Western countries, there are impacts of ticks with cattle ticks in Queensland (Australia) costing A$2.0 million per year for control and production losses (Jonsson et al., 2001).

16.9 CONCLUSIONS

16.9.1 Introduction

There is some resurgence of Cestode infections in Western countries along with the increased consumption of raw or undercooked fish (Craig, 2012). This is due to the increased popularity of sushi and sashimi (Japanese), ceviche (coming from Peru), together with other dishes with raw fish, such as raw salted or marinated fillets (Baltic and Scandinavian), carpaccio (sliced raw fish in Italian cuisine), *tartare maison* (ground raw salmon), and *poisson du lac façon nordique* (French cuisine) (Craig, 2012). One organism responsible for the increase in parasite infections is the fish tapeworm, *Di. dendriticum* (Kuchta et al., 2013).

16.9.2 Impact of Parasites on Adults in Developing Countries

Associated with parasitic diseases is a poverty trap where sickness prevents work, which leads to lack of money, food, and consequent susceptibility to other pathogens (Kuris, 2012). Moreover, parasitic diseases impede economic development and place a burden on the health systems of developing countries (Bonds et al., 2012; Evans and Jamison, 1994; Sachs and Malaney, 2002).

16.9.3 Impact of Parasites on Children

Parasites have a marked negative effect on children with depressed growth in a metaanalysis (Hall et al., 2008). In individual studies, growth to 3 years old is depressed in children in Kenya infected with hookworm, *Ascaris*, *E. histolytica*, malaria, or *Schistosoma* (LaBeaud et al., 2015). In young children in the Amazonian lowlands, 37% were infected with *Ascaris*, *Trichuris*, and hookworm by the age of 14 months (Gyorkos et al., 2011). Infestation was associated with stunting (Gyorkos et al., 2011). Moreover, after treatment with the parasiticide albendazole, there were increases in appetite and growth in Kenyan school children infected with hookworm or *Trichu. trichiura* or *A. lumbricoides* (Adams et al., 1994). Anemia was present in three-quarters of preschool children associated with parasitic infection and with a heavy hookworm load (Brooker et al., 1999).

16.9.4 Goals

There has been tremendous progress with the treatment of parasitic infections and some are close to eradication (Loker, 2013). An example of a success includes guinea worms (see Section 16.3.4) and this is thanks to the efforts of the WHO and the Carter Center through community education and behavior programs. Globally, life expectancy is increasing (Murray and Lopez, 1997). Programs to eradicate malaria globally between the 1950s and 1980s were largely unsuccessful (Stratton et al., 2008). However, careful analysis suggests that it is feasible to eliminate malaria where "intensity of transmission is low to moderate, and where health systems are strong" (Mendis et al., 2009). An additional issue is the increased resistance of helminths in livestock. New anthelmintics are required along with alternative strategies.

References

ACAAI (American College of Allergy, Asthma and Immunology), in press. Types of meat allergy, 2017. Available from: http://acaai.org/allergies/types/food-allergies/types-food-allergy/meat-allergy.

Abdelqader, A., Gauly, M., Wollny, C.B., Abo-Shehada, M.N., 2008. Prevalence and burden of gastrointestinal helminthes among local chickens, in northern Jordan. Prev. Vet. Med. 85, 17–22.

Ackert, J.E., Herrick, C.A., 1928. Effects of the nematode *Ascaridia lineata* (Schneider) on growing chickens. J. Parasitol. 15, 1–13.

Adams, E.J., Stephenson, L.S., Latham, M.C., Kinoti, S.N., 1994. Physical activity and growth of Kenyan school children with hookworm, *Trichuris trichiura* and *Ascaris lumbricoides* infections are improved after treatment with albendazole. J. Nutr. 124, 1199–1206.

Akinbo, F.O., Omoregie, R., Eromwon, R., Igbenimah, I.O., Airueghiomon, U.-E., 2011. Prevalence of intestinal parasites among patients of a tertiary hospital in Benin City, Nigeria. North Am. J. Med. Sci. 3, 462–464.

Ambrozio, C.L., Nagelm, A.S., Jeske, S., Bragança, G.C., Borsuk, S., Villela, M.M., 2016. Trichomonas vaginalis: prevalence and risk factors for women in Southern Brazil. Rev. Inst. Med. Trop. São Paulo 58, 61.

Bandi, K.M., Saikumar, C., 2013. Sarcoptic mange: a zoonotic ectoparasitic skin disease. J. Clin. Diagn. Res. 7, 156–157.

Berry, A.E., Gardner, M.J., Caspers, G.J., Roos, D.S., Berriman, M., 2004. Curation of the *Plasmodium falciparum* genome. Trends Parasitol. 20, 548–552.

Besier, B., 2006. New anthelmintics for livestock: the time is right. Trends Parasitol. 23, 21–24.

Bitam, I., Dittmar, K., Parola, P., Whiting, M.F., Raoult, D., 2010. Fleas and flea-borne diseases. Int. J. Infect. Dis. 14, e667–e676.

Bloomberg, 2007. The costs of bed bugs. Hotels and rental housing are hit by a resurgence in bedbug infestation and lawsuits are proving it's not a problem that can be swept under the rug. Available from: http://www.bloomberg.com/news/articles/2007-11-08/the-cost-of-bedbugs-businessweek-business-news-stock-market-and-financial-advice.

Bonds, M.H., Dobson, A.P., Keenan, D.C., 2012. Disease ecology, biodiversity, and the latitudinal gradient in income. PLoS Biol. 10, e1001456.

Brooker, S., Peshu, N., Warn, P.A., Mosobo, M., Guyatt, H., Marsh, K., Snow, R.W., 1999. The epidemiology of hookworm infection and its contribution to anaemia among pre-school children on the Kenyan Coast. Trans. R. Soc. Trop. Med. Hyg. 93, 240–246.

Cavalier-Smith, T., 2013. Early evolution of eukaryote feeding modes, cell structural diversity, and classification of the protozoan phyla Loukozoa, Sulcozoa, and Choanozoa. Eur. J. Protistol. 49, 115–178.

CDC, 2017. Parasites. Available from: https://www.cdc.gov/parasites/ including Flukes http://www.cdc.gov/parasites/fasciola/; Parasites—Angiostrongyliasis (also known as angiostrongyliasis infection) http://www.cdc.gov/parasites/angiostrongylus/; Parasites—Ascariasis http://www.cdc.gov/parasites/ascariasis/; Parasites—Dracunculiasas (also known as guinea worm disease) http://www.cdc.gov/parasites/guineaworm/; Parasites—Hookworm http://www.cdc.gov/parasites/hookworm/; Parasites: Neglected parasitic infections https://www.cdc.gov/parasites/npi/; Parasites—Schistosomiasis http://www.cdc.gov/parasites/schistosomiasis/epi.html; Taeniasis http://www.cdc.gov/parasites/taeniasis/biology.html; Parasites—Toxocariasis (also known as roundworm infection) (http://www.cdc.gov/parasites/toxocariasis/gen_info/faqs.html); Parasites—Trichuriasis (also known as whipworm infection) http://www.cdc.gov/parasites/whipworm/biology.html.

Charlier, J., van der Voort, M., Kenyon, F., Skuce, P., Vercruysse, J., 2014. Chasing helminths and their economic impact on farmed ruminants. Trends Parasitol. 30, 361–367.

Chen, S., Ai, L., Zhang, Y., Chen, J., Zhang, W., Li, Y., Muto, M., Morishima, Y., Sugiyama, H., Xu, X., Zhou, X., Yamasaki, H., 2014. Molecular detection of *Diphyllobothrium nihonkaiense* in humans, China. Emerg. Infect. Dis. 20, 315–318.

Cook, G.C., 1994. *Enterobius vermicularis* infection. Gut 35, 1159–1162.

Craig, N., 2012. Fish tapeworm and sushi. Can. Fam. Physician 58, 654–658.

Commins, S.P., Platts-Mills, T.A., 2013. Tick bites and red meat allergy. Curr. Opin. Allergy Clin. Immunol. 13, 354–359.

Del Brutto, O.H., Santibanez, R., Noboa, C.A., Aguirre, R., Diaz, E., Alarcon, T.A., 1992. Epilepsy due to neurocysticercosis: analysis of 203 patients. Neurology 42, 389–392.

Dholakia, S., Buckler, J., Jeans, J.P., Pillai, A., Eagles, N., Dholakia, S., 2014. Pubic lice: an endangered species? Sex. Transm. Dis. 41, 388–391.

Diaz, J.H., 2010. Mite-transmitted dermatoses and infectious diseases in returning travelers. J. Travel Med. 17, 21–31.

Dittmar, K., Mamat, U., Whiting, M., Goldmann, T., Reinhard, K., Guillen, S., 2003. Techniques of DNA-studies on Prehispanic ectoparasites (*Pulex* sp., Pulicidae, Siphonaptera) from animal mummies of the Chiribaya culture, Southern Peru. Mem. Inst. Oswaldo Cruz 98 (Suppl. I), 53–58.

Dupouy-Camet, J., 2000. Trichinellosis: a worldwide zoonosis. Vet. Parasitol. 93, 191–200.

El-Sayed, N.M., Myler, P.J., Blandin, G., Berriman, M., Crabtree, J., Aggarwal, G., Caler, E., Renauld, H., Worthey, E.A., Hertz-Fowler, C., Ghedin, E., Peacock, C., Bartholomeu, D.C., Haas, B.J., Tran, A.N., Wortman, J.R., Alsmark, U.C., Angiuoli, S., Anupama, A., Badger, J., Bringaud, F., Cadag, E., Carlton, J.M., Cerqueira, G.C., Creasy, T., Delcher, A.L., Djikeng, A., Embley, T.M., Hauser, C., Ivens, A.C., Kummerfeld, S.K., Pereira-Leal, J.B., Nilsson, D., Peterson, J., Salzberg, S.L., Shallom, J., Silva, J.C., Sundaram, J., Westenberger, S., White, O., Melville, S.E., Donelson, J.E., Andersson, B., Stuart, K.D., Hall, N., 2005. Comparative genomics of trypanosomatid parasitic protozoa. Science 309, 404–409.

Elston, D.M., 2010. *Demodex* mites: facts and controversies. Clin. Dermatol. 28, 502–504.

EPA, 2007. Joint statement on bed bug control in the United States from the U.S. Centers for Disease Control and Prevention (CDC) and the U.S. Environmental Protection Agency (EPA). Available from: https://www.epa.gov/bedbugs/bed-bugs-public-health-issue.

Evans, D.B., Jamison, D.T., 1994. Economics and the argument for parasitic disease control. Science 264, 1866–1867.

Fabiyi, J.P., 1987. Production losses and control of helminths in ruminants of tropical regions. Int. J. Parasitol. 17, 435–442.

Falagas, M.E., Matthaiou, D.K., Rafailidis, P.I., Panos, G., Pappas, G., 2008. Worldwide prevalence of head lice. Emerg. Infect. Dis. 14, 1493–1494.

Field, N., Clifton, S., Alexander, S., Ison, C.A., Khanom, R., Saunders, P., Hughes, G., Heath, L., Beddows, S., Mercer, C.H., Tanton, C., 2016. *Trichomonas vaginalis* infection is uncommon in the British general population: implications for clinical testing and public health screening. Sex Transm. Infect., (pp.sextrans-2016).

Fournier, P.E., Ndihokubwayo, J.B., Guidran, J., Kelly, P.J., Raoult, D., 2002. Human pathogens in body and head lice. Emerg. Infect. Dis. 8, 1515–1518.

Furtado, J.M., Smith, J.R., Belfort, R., Gattey, D., Kevin, L., Winthrop, K.L., 2011. Toxoplasmosis: a global threat. J. Global Infect. Dis. 3, 281–284.

Gelaw, A., Anagaw, B., Nigussie, B., Silesh, B., Yirga, A., Alem, M., Endris, M., Gelaw, B., 2013. Prevalence of intestinal parasitic infections and risk factors among schoolchildren at the University of Gondar Community School, Northwest Ethiopia: a cross-sectional study. BMC Public Health 13, 304.

G/hiwot, Y., Degarege, A., Erko, B., 2014. Prevalence of intestinal parasitic infections among children under five years of age with emphasis on *Schistosoma mansoni* in Wonji Shoa Sugar Estate, Ethiopia. PLoS One 9, e109793.

Githigia, S.M., Thamsborg, S.M., Munyua, W.K., Maingi, N., 2001. Impact of gastrointestinal helminths on production in goats in Kenya. Small Ruminant Res. 42, 21–29.

Grisi, L., Leite, R.C., de Souza Martins, J.R., Medeiros de Barros, A.T., Andreotti, R., Duarte Canada, P.H., Pérez de León, A.A.P., Pereira, J.B., Silva Villela, H., 2014. Reassessment of the potential economic impact of cattle parasites in Brazil. Braz. J. Vet. Parasitol. 23, 150–156.

Gyorkos, T.W., Maheu-Giroux, M., Casapia, M., Joseph, S.A., Creed-Kanashiro, H., 2011. Stunting and helminth infection in early preschool-age children in a resource-poor community in the Amazon lowlands of Peru. Trans. R. Soc. Trop. Med. Hyg. 105, 204–208.

Hall, A., Hewitt, G., Tuffrey, V., de Silva, N., 2008. A review and meta-analysis of the impact of intestinal worms on child growth and nutrition. Matern. Child Nutr. 1 (4 Suppl.), 118–236.

Harhay, M.O., Horton, J., Olliaro, P.L., 2010. Epidemiology and control of human gastrointestinal parasites in children. Expert Rev. Anti Infect. Ther. 8, 219–234.

Harp, D.F., Chowdhury, I., 2011. Trichomoniasis: evaluation to execution. Eur. J. Obstet. Gynecol. Reprod. Biol. 157, 3–9.

Hopla, C.E., Durden, L.A., Keirans, J.E., 1994. Ectoparasites and classification. Rev. Sci. Tech. 13, 985–1017.

Horton, J., 2003. Global anthelmintic chemotherapy programs: learning from history. Trends Parasitol. 19, 405–409.

Hussen, H., Chaka, H., Deneke, Y., Bitew, M., 2012. Gastrointestinal helminths are highly prevalent in scavenging chickens of selected districts of Eastern Shewa zone, Ethiopia. Pak. J. Biol. Sci. 15, 284–289.

Ibrahim, M.M., 2007. Prevalence and intensity of *Angiostrongylus cantonensis* in freshwater snails in relationship to some ecological and biological factors. Parasite 14, 61–70.

Jalbert, I., Rejab, S., 2015. Increased numbers of *Demodex* in contact lens wearers. Optom. Vis. Sci. 92, 671–678.

Jarmuda, S., O'Reilly, N., Zaba, R., Jakubowicz, O., Szkarad-kiewicz, A., Kavanagh, K., 2012. Potential role of *Demodex* mites and bacteria in the induction of rosacea. J. Med. Microbiol. 61, 1504–1510.

Javaregowda, A.K., Kavitha, Rani, B., Revanna, S.P., Udupa, G., 2016. Prevalence of gastro-intestinal parasites of backyard chickens (*Gallus domesticus*) in and around Shimoga. J. Parasit. Dis. 40, 986–990.

Johns Hopkins School of Public Health, 2015. Lyme disease costs up to $1.3 billion per year to treat, study finds. Available from: http://www.jhsph.edu/news/news-releases/2015/lyme-disease-costs-more-than-one-billion-dollars-per-year-to-treat-study-finds.html.

Johnson, L., Aylward, A., Stricker, R.B., 2011. Healthcare access and burden of care for patients with Lyme disease: a large United States survey. Health Policy 102, 64–71.

Jonsson, N.N., Davis, R., De Witt, M., 2001. An estimate of the economic effects of cattle tick (*Boophilus microplus*) infestation on Queensland dairy farms. Aust. Vet. J. 79, 826–831.

Katoch, R., Yadav, A., Godara, R., Khajuria, J.K., Borkataki, S., Sodhi, S.S., 2012. Prevalence and impact of gastrointestinal helminths on body weight gain in backyard chickens in subtropical and humid zone of Jammu, India. J. Parasit. Dis. 36, 49–52.

Keiser, J., Utzinger, J., 2009. Food-borne trematodiases. Clin. Microbiol. Rev. 22, 466–483.

Khan, M.N., Sajid, M.S., Khan, M.K., Iqbal, Z., Hussain, A., 2010. Gastrointestinal helminthiasis: prevalence and associated determinants in domestic ruminants of district Toba Tek Singh, Punjab, Pakistan. Parasitol. Res. 107, 787–794.

King, C.H., 2011. Schistosomiasis: challenges and opportunities. In: The Causes and Impacts of Neglected Tropical and Zoonotic Diseases: Opportunities for Integrated Intervention Strategies. A12. National Academies Press, Washington, DC.

Korea Centers for Disease Control and Prevention, National Institute of Health, 2013. The 8th National surveys on the prevalence of intestinal parasitic infections. Seoul, Korea, pp. 35–68.

Kornele, M.L., McLean, M.J., O'Brien, A.E., Philippi-Taylor, A.M., 2014. Antiparasitic resistance and grazing livestock in the United States. J. Am. Vet. Med. Assoc. 244, 1020–1022.

Kozek, W.J., D'Alessandro, A., Silva, J., Navarette, S.N., 1982. Filariasis in Colombia: prevalence of mansonellosis in the teenage and adult population of the Colombian bank of the Amazon, Comisaria del Amazonas. Am. J. Trop. Med. Hyg. 31, 1131–1136.

Kuchta, R., Brabec, J., Kubáčková, P., Scholz, T., 2013. Tapeworm *Diphyllobothrium dendriticum* (Cestoda)—neglected or emerging human parasite? PLoS Negl. Trop. Dis. 7, e2535.

Kuchta, R., Serrano-Martínez, M.E., Scholz, T., 2015. Pacific broad tapeworm *Adenocephalus thirdspacificus* as a causative agent of globally reemerging Diphyllobothriosis. Emerg. Infect. Dis. 21, 1697–1703.

Kucik, C.J., Martin, G.L., Sortor, B.V., 2004. Common intestinal parasites. Am. Fam. Physician 69, 1161–1168.

Kuris, A.M., 2012. The global burden of human parasites: who and where are they? How are they transmitted? J. Parasitol. 98, 1056–1064.

LaBeaud, A.D., Singer, M.N., McKibben, M., Mungai, P., Muchiri, E.M., McKibben, E., Gildengorin, G., Sutherland, L.J., King, C.H., King, C.L., Malhotra, I., 2015. Parasitism in children aged three years and under: relationship between infection and growth in rural coastal Kenya. PLoS Negl. Trop. Dis. 9, e0003721.

Langergraber, K.E., Prüfer, K., Rowney, C., Boesch, C., Crockford, C., Fawcett, K., Inoue, E., Inoue-Muruyama, M., Mitani, J.C., Muller, M.N., Robbins, M.M., Schubert, G., Stoinski, T.S., Viola, B., Watts, D., Wittig, R.M., Wrangham, R.W., Zuberbühler, K., Pääbo, S., Vigilant, L., 2012. Generation times in wild chimpanzees and gorillas suggest earlier divergence times in great ape and human evolution. Proc. Natl. Acad. Sci. USA 109, 15716–15721.

Liang, L., Ding, X., Tseng, S.C., 2014. High prevalence of *Demodex brevis* infestation in chalazia. Am. J. Ophthalmol. 157, 342–348.

Lohiya, G.-S., Figueroa, L.-T., Crinella, F.M., Lohiya, S., 2000. Epidemiology and control of enterobiasis in a developmental center. West. J. Med. 172, 305–308.

Loker, E.S., 2013. This de-wormed world? J. Parasitol. 99, 933–942.

Marketwired.com, 2017. Available from: http://www.marketwired.com/press-release/global-veterinary-health-products-market-to-spike-to-over-28-billion-by-2017-1602876.htm.

McDougald, L.R., 1998. Intestinal protozoa important to poultry. Poult. Sci. 77, 1156–1158.

Mendis, K., Rietveld, A., Warsame, M., Bosman, A., Greenwood, B., Wernsdorfer, W.H., 2009. From malaria control to eradication: the WHO perspective. Trop. Med. Int. Health 14, 802–809.

Michael, E., Bundy, D.A.P., Grenfell, B.T., 1996. Re-assessing the global prevalence and distribution of lymphatic filariasis. Parasitology 112, 409–428.

Minjauw, B., McLeod, A., 2003. Tick-borne diseases and poverty. The impact of ticks and tick-borne diseases on the livelihood of small-scale and marginal livestock owners in India and eastern and southern Africa. Research report, DFID Animal Health Programme, Centre for Tropical Veterinary Medicine, University of Edinburgh, UK.

Mourembou, G., Fenollar, F., Lekana-Douki, J.B., Ndjoyi Mbiguino, A., Maghendji Nzondo, S., Matsiegui, P.B.,

Manego, R.Z., Ehounoud, C.H.B., Bittar, F., Raoult, D., Mediannikov, O., 2015. *Mansonella*, including a potential new species, as common parasites in children in Gabon. PLoS Negl. Trop. Dis. 9, e0004155.

Murray, C.J.L., Lopez, A.D., 1997. Alternative projections of mortality and disability by cause 1990–2020: global burden of disease study. Lancet 349, 1498–1504.

Ndimubanzi, P.C., Carabin, H., Budke, C.M., Nguyen, H., Qian, Y.J., Rainwater, E., Dickey, M., Reynolds, S., Stoner, J.A., 2010. A systematic review of the frequency of neurocyticercosis with a focus on people with epilepsy. PLoS Negl. Trop. Dis. 4, e870.

Niyizurugero, E., Ndayanze, J.B., Bernard, K., 2013. Prevalence of intestinal parasitic infections and associated risk factors among Kigali Institute of Education students in Kigali. Rwanda Trop. Biomed. 30, 718–726.

Nwosu, C.O., Madu, P.P., Richards, W.S., 2007. Prevalence and seasonal changes in the population of gastrointestinal nematodes of small ruminants in the semi-arid zone of north-eastern Nigeria. Vet. Parasitol. 144, 118–124.

Ocaido, M., Muwazi, R.T., Opuda, J.A., 2009. Economic impact of ticks and tick-borne diseases on cattle production systems around Lake Mburo National Park in South Western Uganda. Trop. Anim. Health Prod. 41, 731–739.

Otsuji, Y., 2011. History, epidemiology and control of filariasis. Trop. Med. Health 39 (1 Suppl. 2), 3–13.

Over, H.J., Jansen, J., Olm, P.W., 1992. Distribution and impact of helminth diseases of livestock in developing countries. FAO Animal Production and Health Paper 96. Available from: http://www.fao.org/docrep/004/t0584e/T0584E00.htm.

Owhoeli, O., Elele, K., Gboeloh, L.B., 2014. Prevalence of gastrointestinal helminths in exotic and indigenous goats slaughtered in selected abattoirs in Port Harcourt, South-South. Nigeria. Chin. J. Biol. 2014, 435913.

Pani, M.S.P., Kumaraswami, V., Das, L.K., 2005. Epidemiology of lymphatic filariasis with special reference to urogenital-manifestations. Indian J. Urol. 21, 44–49.

Payne, R.J., Turner, L., Morgan, E.R., 2009. Inappropriate measures of population health for parasitic disease? Trends Parasitol. 25, 393–395.

Peacock, C.S., Seeger, K., Harris, D., Murphy, L., Ruiz, J.C., Quail, M.A., Peters, N., Adlem, E., Tivey, A., Aslett, M., Kerhornou, A., Ivens, A., Fraser, A., Rajandream, M.A., Carver, T., Norbertczak, H., Chillingworth, T., Hance, Z., Jagels, K., Moule, S., Ormond, D., Rutter, S., Squares, R., Whitehead, S., Rabbinowitsch, E., Arrowsmith, C., White, B., Thurston, S., Bringaud, F., Baldauf, S.L., Faulconbridge, A., Jeffares, D., Depledge, D.P., Oyola, S.O., Hilley, J.D., Brito, L.O., Tosi, L.R., Barrell, B., Cruz, A.K., Mottram, J.C., Smith, D.F., Berriman, M., 2007. Comparative genomic analysis of three *Leishmania* species that cause diverse human disease. Nat. Genet. 39, 839–847.

Pence, D.B., Kamler, J.F., Ballard, W.B., 2004. Ectoparasites of the Swift Fox in Northwestern Texas. J. Wildlife Dis. 40, 543–547.

Permin, A., Magwisha, H., Kassuku, A.A., Nansen, P., Bisgaard, M., Frandsen, F., Gibbons, L., 1997. A cross-sectional study of helminths in rural scavenging poultry in Tanzania in relation to season and climate. J. Helminthol. 71, 233–240.

Rayan, P., Verghese, S., McDonnell, P.A., 2010. Geographical location and age affects the incidence of parasitic infestations in school children. Indian J. Pathol. Microbiol. 53, 498–502.

Reed, D.L., Light, J.E., Allen, J.M., Churchmen, J.J., 2007. Pair of lice lost or parasites regained: the evolutionary history of anthropoid primate lice. BMC Biol. 5, 7.

Reed, D.L., Smith, V.S., Hammond, S.L., Rogers, A.R., Clayton, D.H., 2004. Genetic analysis of lice supports direct contact between modern and archaic humans. PLoS Biol. 2, e340.

Reid, A.J., Blake, D.P., Ansari, H.R., Billington, K., Browne, H.P., Bryant, J., Dunn, M., Hung, S.S., Kawahara, F., Miranda-Saavedra, D., Malas, T.B., Mourier, T., Naghra, H., Nair, M., Otto, T.D., Rawlings, N.D., Rivailler, P., Sanchez-Flores, A., Sanders, M., Subramaniam, C., Tay, Y.L., Woo, Y., Wu, X., Barrell, B., Dear, P.H., Doerig, C., Gruber, A., Ivens, A.C., Parkinson, J., Rajandream, M.A., Shirley, M.W., Wan, K.L., Berriman, M., Tomley, F.M., Pain, A., 2014. Genomic analysis of the causative agents of coccidiosis in domestic chickens. Genome Res. 24, 1676–1685.

Reid, A.J., Vermont, S.J., Cotton, J.A., Harris, D., Hill-Cawthorne, G.A., Könen-Waisman, S., Latham, S.M., Mourier, T., Norton, R., Quail, M.A., Sanders, M., Shanmugam, D., Sohal, A., Wasmuth, J.D., Brunk, B., Grigg, M.E., Howard, J.C., Parkinson, J., Roos, D.S., Trees, A.J., Berriman, M., Pain, A., Wastling, J.M., 2012. Comparative genomics of the apicomplexan parasites *Toxoplasma gondii* and *Neospora caninum*: Coccidia differing in host range and transmission strategy. PLoS Pathol. 8, e1002567.

Roepstorff, A., Mejer, H., Nejsum, P., Thamsborg, S.M., 2011. Helminth parasites in pigs: new challenges in pig production and current research highlights. Vet. Parasitol. 180, 72–81.

Roepstorff, A., Murrell, K.D., 1997. Transmission dynamics of helminth parasites of pigs on continuous pasture: *Ascaris suum* and *Trichuris suis*. Int. J. Parasitol. 27, 563–572.

Ruggiero, M.A., Gordon, D.P., Orrell, T.M., Bailly, N., Bourgoin, T., Brusca, R.C., Cavalier-Smith, T., Guiry, M.D., Kirk, P.M., 2015. A higher level classification of all living organisms. PLoS One 10, e0119248.

Sachs, J., Malaney, P., 2002. The economic and social burden of malaria. Nature 415, 680–685.

Schantz, P.M., 1983. Trichinellosis in the United States—1947–1981. Food Technol. 37, 83–86.

Sinniah, B., Hassan, A.K.R., Sabaridah, I., Soe, M.M., Ibrahim, Z., Ali, O., 2014. Prevalence of intestinal parasitic infections among communities living in different habitats and its comparison with one hundred and one studies conducted over the past 42 years (1970 to 2013) in Malaysia. Trop. Biomed. 31, 190–206.

Stratton, L., O'Neill, M.S., Kruk, M.E., Bell, M.L., 2008. The persistent problem of malaria: addressing the fundamental causes of a global killer. Soc. Sci. Med. 67, 854–862.

Stoll, N.R., 1947. This wormy world. J. Parasitol. 33, 1–18.

Taylor, D.B., Moon, R.D., Mark, D.R., 2012. Economic impact of stable flies (Diptera: Muscidae) on dairy and beef cattle production. J. Med. Entomol. 49, 198–209.

The Poultry Site, 2013. High cost of Coccidiosis in broilers. Available from: http://www.thepoultrysite.com/poultrynews/28036/high-cost-of-coccidiosis-in-broilers/.

Tree of Life, 2017. Available from: http://www.tolweb.org/Eukaryotes/3.

Tripathi, A., Commins, S.P., Heymann, P.W., Platts-Mills, T.A., 2014. Delayed anaphylaxis to red meat masquerading as idiopathic anaphylaxis. J. Allergy Clin. Immunol Pract. 2, 259–265.

Vaidyanathan, R., Feldlaufer, M.F., 2013. Bed bug detection: current technologies and future directions. Am. J. Trop. Med. Hyg. 88, 619–625.

Varki, A., Altheide, T.K., 2005. Comparing the human and chimpanzee genomes: searching for needles in a haystack. Genome Res. 15, 1746–1758.

Wesolowska, M., Knysz, B., Reich, A., Blazejewska, D., Czarnecki, M., Gladysz, A., Pozowski, A., Misiuk-Hojlo, M., 2014. Prevalence of *Demodex* spp. in eyelash follicles in different populations. Arch. Med. Sci. 10, 319–324.

Whitinga, M.F., Whitinga, A.S., Hastriterb, M.W., Dittmar, K., 2008. A molecular phylogeny of fleas (Insecta: Siphonaptera): origins and host associations. Cladistics 24, 1–31.

WHO, 2015. WHO estimates of the global burden of foodborne diseases. Available from: http://apps.who.int/iris/bitstream/10665/199350/1/9789241565165_eng.pdf?ua=1.

WHO, 2016. Taeniasis/cysticercosis. Available from: http://www.who.int/mediacentre/factsheets/fs376/en/.

WHO, 2017a. Cockroaches. Available from: http://www.who.int/water_sanitation_health/resources/vector-288to301.pdf.

WHO, 2017b. Dracunculiasis (guinea-worm disease). Available from: http://www.who.int/mediacentre/factsheets/fs359/en/.

WHO, 2017c. Available from: Lymphatic filariasis. http://www.who.int/lymphatic_filariasis/epidemiology/en/.

WHO, 2017d. Schistosomiasis. Available from: http://www.who.int/mediacentre/factsheets/fs115/en/.

Wilson, N.O., Hall, R.L., Montgomery, S.P. and Jones, J.L., 2015. Trichinellosis Surveillance—United States, 2008–2012.

CDC Morbidity and Mortality Report. Available from: http://www.cdc.gov/parasites/trichinellosis/biology.html.

World Bank, 2016. Available from: http://www.worldbank.org.

Yaeger, R.G., 1996. Protozoa: structure, classification, growth, and development. In: Medical Microbiology, fourth edition (Chapter 77). Available from: https://www.ncbi.nlm.nih.gov/books/NBK8325/.

Further Reading

Choi, S.-C., Lee, S.-Y., Song, H.-O., Ryu, J.-S., Ahn, M.H., 2014. Parasitic infections based on 320 clinical samples submitted to Hanyang University, Korea (2004–2011). Korean J. Parasitol. 52, 215–220.

Invasive Species

Colin G. Scanes

University of Wisconsin–Milwaukee, Milwaukee, WI, United States

17.1 INTRODUCTION

Invasive species are animals or plants that are nonnative to the region and were introduced by human intervention (Keller et al., 2011). The "100 World's Worst" invaders have been listed by the Invasive Species Specialist Group. Introductions that have resulted in increasing populations and expansion to new geographical areas are viewed by some as "successful" (Wodzicki and Wright, 1980) or as establishing invasive species. If an exotic species survives and is reproducing in the new locale, it is viewed as established. If it spreads widely it is viewed as invasive if it is exerting a "measurable" effect on ecosystems or human activity (Keller et al., 2011). These species can detrimentally influence indigenous biodiversity (both plant and animal) and potentially cause extinctions. The study of the effects of invasive species is difficult as there is unlikely to be data on the region prior to and after the establishment of the invasive species. Effects may be confounded by other changes in the environment (often caused by humans) and it is difficult to definitively quantitate competition for food or nesting sites and/or predation (Grarock et al., 2012).

Examples of invasive species are summarized in Table 17.1. Invasive species can be categorized as one of the following (based on Wodzicki and Wright, 1980):

- Accidentally introduced
  - "Stowaways" on ships [rats (Norway and black) and mice (house mouse) to multiple locales] or planes (brown tree snakes transported to Guam).
  - Escaped from captivity (e.g., gray squirrels in the United Kingdom).
  - Thoughtlessly released (e.g., pet Burmese pythons released into the Everglades, Florida).
- Deliberately introduced
  - On esthetics or other supposed quality of life issues, such as the species being named in Shakespeare plays (e.g., starlings) or to help settlers/immigrants from Europe become comfortable by having familiar animals around them (e.g., North America, South Africa, and Australia).
  - As game species for sport/hunting including wild boar (introduced to the USA), wallabies (New Zealand), deer

TABLE 17.1 Examples of Invasive Species

Species	Category	Location introduced
Mammals		
Camel	Feral	Australia
Cat	Feral	Australia
Deer	Deliberately introduced	Australia, Islands, New Zealand, United Kingdom
Dingos	Accidentally introduced	Australia
European rabbit	Feral	Australia
Gray squirrel	Accidentally introduced	United Kingdom
Horses and donkeys	Feral	North and South America, Australia
Norway or brown rat	Accidentally introduced	Islands
Pacific rat	Deliberately introduced	Polynesian islands
Ship or black rat	Accidentally introduced	Islands
Wild pig	Deliberately introduced + feral	USA
Birds		
Starling	Deliberately introduced	USA
Feral pigeon	Feral	North America, Australia
House sparrow	Deliberately introduced	USA
Common myna	Deliberately introduced	Australia
Reptiles		
Brown tree snake	Accidentally introduced	Guam (Mariana Islands in Pacific Ocean)
Burmese python	Thoughtlessly introduced	USA (Everglades in Florida)
Amphibian		
Cane toad	Deliberately introduced	Australia
Aquatic species		
Alewife	Accidentally introduced	Great Lakes, USA and Canada
Sea lamprey	Accidentally introduced	Great Lakes, USA and Canada
Quagga mussels	Accidentally introduced	Great Lakes, USA and Canada
Zebra mussels	Accidentally introduced	Great Lakes, USA and Canada
Other species (Arthropods)		
Africanized bees	Deliberately introduced	South America

(British Isles, New Zealand, and South America), and game birds, such as pheasants (New Zealand).

- For biological control, such as cane toad and common mynas (introduced to Australia for insect control), hedgehogs (insect control), stoats, ferrets, and weasels (introduced to New Zealand for the control of rabbits), and small Indian mongoose (introduced to over 40 islands for the control of rodents and other pests).

- As utility species for fur or meat, such as possum and rabbits (introduced to New Zealand).
- Feral animals: agricultural or companion animals that have escaped (e.g., camels, cats, dogs/dingos, donkeys, horses, and rabbits).

Given the serious effects that some invasive species cause, the Ecological Society of America made the following recommendations to the US Government about invasive species (Lodge et al., 2006):

1. Use new information and practices to better manage commercial and other pathways to reduce the transport and release of potentially harmful species.
2. Adopt more quantitative procedures for risk analysis and apply them to every species proposed for importation into the country.
3. Use new cost-effective diagnostic technologies to increase active surveillance and sharing of information about invasive species so that responses to new invasions can be more rapid and effective.
4. Create new legal authority and provide emergency funding to support rapid responses to emerging invasions.
5. Provide funding and incentives for cost-effective programs to slow the spread of existing invasive species to protect still uninvaded ecosystems, social and industrial infrastructure, and human welfare.
6. Establish a National Center for Invasive Species Management.

17.2 MAMMALIAN INVASIVE SPECIES

17.2.1 Cats (Free Roaming, Unowned, and Feral)

Free-roaming domestic cats are responsible for killing large numbers of wildlife. According to researchers at the Smithsonian Conservation Biology Institute and the US Fish and Wildlife Service, they may represent "the single greatest source of anthropogenic mortality for US birds and mammals." Cats prefer to hunt small mammals. It is estimated that cats kill the following annually in the contiguous 48 States of the USA (Loss et al., 2013):

- mammals, 12.3 billion
- birds, 2.4 billion
- reptiles, 478 million
- amphibians, 173 million

Free-roaming cats include the following:

- Owned cats allowed to roam.
- Unowned cats including feral cats, semiferal cats/strays (may be fed but not enter homes), and barn cats.

The majority of the wildlife killed are due to so called "unowned cats." Even so, about 30% of the wild birds killed are by owned cats that are allowed to roam. Globally, feral cats are among the "100 of the World's Worst Invasive Alien Species." They are thought to be responsible for 33 modern extinctions of mammals, birds, and reptiles and are threatening many more particularly on islands (Lowe et al., 2000). It is estimated that feral cats on islands caused as much as 14% of all global extinctions of mammals, birds, and reptiles (238 species) (Medina et al., 2011).

Feral cats were successfully eradicated in 1991 from Marion Island; a sub-Antarctic island south of South Africa in the Indian Ocean. Following the eradication, there were rapidly marked improvements of the breeding success of the great-winged petrel (*Pterodroma macroptera*) and the blue petrel (*Halobaena caerulea*) (Cooper et al., 1995).

17.2.2 Dingos and Other Feral Dogs

Based on genomic and archaeological studies, it has been concluded that the wild dog of Australia, the dingo, originated from escaped domestic dogs transported on boats from

East Asia about 5000 years ago (Savolainen et al., 2004). Dingos are found throughout Australia. Damages from dingos and other wild dogs in Australia are reported to be A\$48.5 million (Gong et al., 2009).

17.2.3 Rabbits (*Oryctolagus cuniculus*)

One of the best examples of a damaging invasive alien species is the European wild rabbit, which was introduced into Australia with the first feral population reported in Australia in 1827 (Australian Department of Environment). Damage and losses from rabbits in Australia are estimated at A\$206.0 million (Gong et al., 2009) with damage to grazing for sheep estimated to be between A\$7.1 and A\$38.7 million (Vere et al., 2004).

The myxomatosis virus was introduced to control wild rabbit numbers in Australia in 1950 and also in Europe including the United Kingdom. Although the myxomatosis virus exhibited high lethality killing 99% of rabbits, numbers rebounded as the virus became attenuated and the surviving rabbits showed resistance (Kerr and Best, 1998).

17.2.4 Rodents as Invasive Species

Rats and mice are among the most destructive species to the environment following their introduction by humans, albeit accidentally. Island habitats are especially vulnerable to rats. For example, islands are important breeding sites for seabirds. Rats and mice are threatening both populations and entire species of seabirds. Mitigation programs have been implemented in over 400 islands. Rodent eradication has been successfully carried out on many islands amounting to over 20,000 ha in 2005 (Martins et al., 2006) and extrapolating to 30,000 ha by 2015. Costs increase geometrically with the size of islands (Martins et al., 2006).

Rodents have major economic and social impacts (see Chapter 16). For instance in Asia, rats

have been estimated to consume the equivalent food for 200 million people for an entire year (Singleton, 2003). The economic impact of rats in the USA was estimated in 2000 at \$19 billion year^{-1} many times greater than any other invasive animal species (Pimentel et al. 2000).

17.2.4.1 Black Rat or Ship or Roof Rat (Rattus rattus)

The black rat (Fig. 17.1) originated in the Indian subcontinent and South East Asia (Aplin et al., 2011). Human activity has led to their spreading. There were black rats in the Middle East about 15,000 BP. They reached the areas around the Mediterranean Sea about 6000 BP and the British Isles by CE/AD 400 (Aplin et al., 2011). Accidental introduction from ships has led to its global distribution with the species introduced to North America in the 16th century. The black rat is particularly successful in Tropical areas but tends to be out-competed by the Norway rat (*Rattus norvegicus*) in temperate regions. In the USA, the black rat is found in the Southern States and along the Pacific coast.

According to the Invasive Species Specialist Group (ISSG, 2015), the ship or black rat (*R. rattus*) is the species of rat responsible for many of the island extinctions of birds. It, but not the Norway rat, is included among the 100 of the World's Worse Invasive Species. Recently, it has been reported that logging in Borneo is

FIGURE 17.1   **Black rat (*R. rattus*).** *Source: Courtesy CDC.*

accompanied by large increases in black rats in disturbed areas of a tropic rain forest (Loveridge et al., 2016).

17.2.4.2 Brown or Norway Rat (Rattus norvegicus)

Contrary to its name, the Norway or brown rat (*R. norvegicus*) (Fig. 17.2) does not come from Norway. They originally come from Northern China. They were brought to Europe later than the arrival of the ship or black rat (*R. rattus*) sometime between the AD 800 and AD 1500. Accidental introduction from ships has led to the global distribution of the Norway or brown rat. The species was introduced to North America in about 1750 and spread as humans migrated across the continent. The Norway rat out-competes the ship rat in temperate regions. In the USA, the Norway rat is found throughout the country. Norway rats are found in most of the provinces of Canada. The exception to this is Alberta where there has been a successful series of programs to prevent the western expansion and to eradicate rats.

In around 1800, whaling boats inadvertently brought Norway rats together with house mice to South Georgia, a mountainous island in the South Atlantic Ocean close to the Antarctic with a subarctic climate. Moreover, reindeer

FIGURE 17.2 **Brown rat.** *Source: Courtesy EPA.*

were introduced (Black et al., 2012). At present, there are over 100,000 seabirds that breed on the island (Clarke et al., 2012). South Georgia is the breeding site for 29 seabird species and the most important breeding site for 6 species [Macaroni Penguin (*Eudyptes chrysolophus*), gray-headed albatross (*Thalassarche chrysostoma*), northern giant petrel (*Macronectes halli*), Antarctic prion (*Pachyptila desolata*), white-chinned petrel (*Procellaria aequinoctialis*), and common diving petrel (*Pelecanoides urinatrix*)] and one of the top three sites for 7 other species (Clarke et al., 2012). The rats negatively impact marine bird populations particularly those that nest on the ground on these islands by depredating eggs and chicks (Pye and Bonner, 1980; reviewed by Black et al., 2012). It is estimated that the introduction of the rat led to an over 90% decrease in the number of seabirds nesting on the island (Clarke et al., 2012). A successful rodent eradication program was implemented (Black et al., 2012).

17.2.4.3 Pacific Rat (Rattus exulans)

The Polynesians voyaging across the Pacific and colonizing islands brought with them Pacific rats as a source of food (Matisoo-Smith and Robins, 2004). These escaped leading to the extinction of over 2000 species of birds and are now an important invasive species (Duncan et al., 2013; Steadman, 1995). In part, this was due to the clearing of forests and hence habitat changes, but also due to damage from their accompanying animals, Pacific rats and pigs. The former have had major detrimental effects on seabirds. There is direct observational evidence of predation by Pacific rats of seabirds including adult and chick Laysan albatrosses and reductions in nests (Kepler, 1967).

17.2.4.4 House Mice (Mus musculus)

Until it was studied, it was assumed that the introduction of house mice had little significant effect on ecosystems and particularly birds breeding. However, the effects on birds are supported based on observations on Gough Island.

This is an uninhabited island in the South Atlantic Ocean that it is very useful to study the effects of mice as they are the sole introduced species; there being no rats or feral animals (Wanless et al., 2007). Not only were mice observed by video as killing and eating chicks but they were also demonstrated to be at least one major cause for the poor breeding success in Tristan albatrosses (*Diomedea dabbenena*) and Atlantic petrels (*Pterodroma incerta*) (Wanless et al., 2007).

17.2.4.5 Gray Squirrel (Sciurus carolinensis)

According to the Introduced Species Summary Project (2015) of Columbia University, the gray squirrel (*S. carolinensis*) is the second most damaging introduced species in the British Isles (after the Norway rat). Gray squirrels have also out-competed the native red squirrel (*Sciurus vulgaris*) in much of the United Kingdom leading to the decline and replacement of the latter (Okubo et al., 1989). It is thought that the large population originated from escapees from the London zoo between 1876 and 1929 (Introduced Species Summary Project, 2015) or multiple sites including Woburn Abbey in Bedfordshire (Okubo et al., 1989).

17.2.5 Wild Pigs

Wild pigs (*Sus scrofa*) are also known as wild boar, wild hog, feral pig, feral hog, Old World swine, razorback, Eurasian wild boar, and Russian wild boar. They are derived from escaped domestic pigs or wild boar that were introduced for hunting.

The range of the wild pigs in the USA has been expanding considerably. In 1982, wild pigs were largely restricted to some areas of Texas, Louisiana, and Florida with isolated populations in California, Alabama, Georgia, Louisiana, Tennessee, and Arkansas. By 2015, wild pigs were found throughout Texas, California, Oklahoma, Mississippi, Alabama, Florida, Georgia, South Carolina together with large parts of Oregon, Arizona,

New Mexico, Tennessee, and North Carolina, and extending into Kansas, Missouri, Kentucky, and West Virginia with wild pigs also reported in the Midwest (e.g., Ohio and Michigan), North East (Pennsylvania, Vermont, and New Hampshire), and West (e.g., Washington and Utah) (National Feral Swine Mapping System, 2015).

Losses and damages from wild pigs have been estimated as the following:

- Australia, A\$9.2 million (Gong et al., 2009)
- USA, \$0.8 billion dollars per year (Pimentel et al., 2000, 2005)
- Hawaii, \$4 million including marked soil erosion (Pimentel et al., 2000, 2005)

In addition to damage, wild pigs are a reservoir for pathogens for domestic pigs and for zoonotic diseases for humans.

17.2.6 Mongoose

The small Indian mongoose (*Herpestes javanicus*) was introduced to more than 40 islands including Jamaica (1872), Hawaii (1883), Fiji (1883), Cuba (1886), Okinawa (1910), and the Croatian islands in the Adriatic Sea (1910) with the goal of controlling rat populations (Carnivora: Herpestidae) (Hays and Conant, 2007). It was at best somewhat effective in controlling rats in sugar cane fields but had a negative effect on wildlife particularly birds and reptiles (Hays and Conant, 2007). The mongoose is responsible for \$50 million in losses and damages in the USA, specifically in Hawaii and Puerto Rico (Pimentel et al., 2005).

17.2.7 Horses and Donkeys

The populations of feral donkeys and horses in Australia and the USA are as follows:

- Australia (Australian Department of Environment, 2015)
  - feral donkey 5 million
  - feral horse 400,000

- USA (US Bureau of Land Management, 2016)
  - feral donkey (burros) 10,800
  - feral horse 47,000

The number of feral horses and burros showed a one year 18% increase and now exceeds the appropriate management level by 31,435 (US Bureau of Land Management, 2016). Losses and damages from wild horses and burros in the USA are estimated at $5 million dollars per year (Pimentel et al., 2000, 2005). Losses in Australia are likely much greater.

17.2.8 Other Invasive Mammals in Australia

Other invasive mammals in Australia are feral camels (1 million and damages/losses A$8.9 million), feral goat (2.6 million), feral water buffalo (150,000), and introduced red fox (damage and losses A$21.2 million). Introduced deer (fallow deer, red deer, chital deer, hog deer, rusa deer, and sambar deer) are an emerging problem (Australian Department of Environment, 2015). These invasive species destroy native plants, compete with domestic animals and can cause soil erosion.

17.3 AVIAN INVASIVE SPECIES

In the 19th century, the American Acclimatization Society imported and introduced birds named in Shakespeare's plays to the USA. The group introduced about a hundred European starlings into New York's Central Park in 1890–91 (Mirsky, 2008). They prospered at the expense of native species. Today there are about 200 million starlings across the USA. Another introduced species is the house sparrow. There were multiple introductions of sparrows in the 19th century as part of an effort to ensure that European immigrants had familiar birds around.

17.3.1 Common or European Starling (*Sturnus vulgaris*)

The American Acclimatization Society introduced about a hundred European starlings into New York's Central Park in 1890–91 (Mirsky, 2008). Today there are about 200 million starlings across the USA with populations also in South Africa, New Zealand, and Australia (reviewed by Lintz et al., 2007) (see Chapter 16). Losses and damages from starlings in the USA are estimated at $0.8 billion dollars per year (Pimentel et al., 2000, 2005).

17.3.2 Feral Pigeon (*Columba livia*)

Feral pigeons are derived from domesticated rock pigeons. Pigeons are found in agricultural and urban areas around the world. Urban pigeon populations are growing rapidly with increased urbanization. The highest numbers are found in those urban areas surrounded by agriculture in Poland (Hetmanski et al., 2010). Estimates for some urban pigeon populations include the following:

- Venice and Pisa, 1000–2000 pigeons per km^2 (Giunchi et al., 2012).
- Sheffield, United Kingdom, 15,000 pigeons (Fuller et al., 2008) compared to a total of 550,000 for the entire United Kingdom (Baker et al., 2006).

There are problems with feral pigeons particularly in urban areas including the following (Giunchi et al., 2012):

- Damage to buildings particularly medieval buildings accelerating deterioration (cleaning costs in Europe are estimated at 7–9 euros per pigeon per year) and increasing maintenance costs.
- Public health concerns with pathogens in the excreta and dust from feathers.
- Bird strikes and aircraft.
- Fouling grain in elevators.

- Agricultural damage with losses in Italy estimated between 20 and 43 million euros per year.

Losses and damages from feral pigeons in the USA are estimated at $1.1 billion dollars per year (Pimentel et al., 2000, 2005) (see Chapter 16).

17.3.3 Common Myna (*Acridotheres tristis*)

The common myna was introduced to Australia from India to control insect pests in agriculture (Hone, 1978). The species spread from its initial locus of Melbourne across much of eastern Australia (Hone, 1978). The common myna is one of the world's top 100 worst invaders. Recently, the impact of the common myna on native birds, particularly competition for nesting sites, was reported based on empirical studies prior to and after their arrival in Canberra (Grarock et al., 2012).

17.3.4 Mute Swans (*Cygnus olor*)

Mute swans were introduced in the USA and Canada between 1850 and 1900. Their population around the Great Lakes is increasing at over 10% per year. They are displacing ducks, geese, and loons from wetlands and destroying aquatic vegetation with each swan uprooting about 10 kg of submersed aquatic vegetation daily (Minnesota Department of Natural Resources, in press; Patuxent Wildlife Research Center, in press).

17.3.5 Other Invasive Birds in the USA

Other alien avian species in the USA include the following:

- Eurasian collared-dove (*Streptopelia decaocto*): introduced in Florida, where they have been increasing exponentially since the early 1970s (Patuxent Wildlife Research Center, 2017)

- House sparrow (*Passer domesticus*): the US population of house sparrows (70 million) originated from 16 released house sparrows brought in by European immigrants (Cornell Laboratory of Ornithology, 2017; Moulton et al., 2010).
- Ring-necked pheasant: introduced in the USA for hunting and is now widespread.

17.4 REPTILIAN INVASIVE SPECIES

17.4.1 Burmese Pythons (*Python molurus bivittatus*)

There is growing concern on the detrimental effects of Burmese pythons (Fig. 17.3) on the integrity of the ecosystem of the Everglades (Fig. 17.4) and the Florida Keys. It is thought that a small number of pet Burmese pythons were released into the Everglades in about 1985 (Willson et al., 2011). The numbers have increased based on the number of pythons removed. Also, their distribution is increasing (Fig. 17.5)

The Everglades is biologically vulnerable with endangered species. There has been a decline in the population of mammals based on decreases in the number observed during nocturnal

FIGURE 17.3 **Burmese python (*P. molurus bivittatus*).**
Source: Courtesy National Parks Service.

FIGURE 17.4 **Everglades.** *Source: Courtesy US National Parks Service.*

observation on roads between 2003 and 2013 (Dorcas et al., 2012):

- bobcats decreased 87.5%
- opossums decreased 98.7%
- raccoons decreased 99.3%
- rabbits decreased 100%
- white-tailed deer 94.1%

Moreover, it has been concluded that Burmese pythons are responsible for 77% of mortalities of marsh rabbits (*Sylvilagus palustris*) (McCleery et al., 2015).

17.4.2 Brown Tree Snake (*Boiga irregularis*)

The brown tree snake (Fig. 17.6) is thought to have arrived in Guam in 1949 as a female stowaway on a plane from New Guinea/Solomon Islands (reviewed by Rodda and Savidge, 2007; Wildlife Service, 2011). From a small population, even a single pregnant female, the population grew to over 2 million in 1980 with population densities as high as 100 ha^{-1} (reviewed by Rodda and Savidge, 2007).

Brown tree snakes have the following impacts (reviewed by Rodda and Savidge, 2007):

- To people
  - Injury or death to particularly babies or young children due to snake bites.
  - Injury or death and even consumption of their pets, such as dogs due to snake bites.

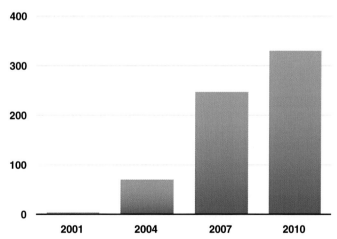

FIGURE 17.5 **Number of Burmese python removed per year from the Everglades.** *Source: Based on data by Dorcas, M.E., Willson, J.D., Reed, R.N., Snow, R.W., Rochford, M.R., Miller, M.A., Meshaka, W.E. Jr., Andreadis, P.T., Mazzotti, F.J., Romagosa, C.M., Hart, K.M. 2012. Severe mammal declines coincide with proliferation of invasive Burmese pythons in Everglades National Park. Proc. Natl. Acad. Sci. USA 109, 2418–2422.*

FIGURE 17.6   **Brown tree snake (*B. irregularis*).** *Source: Courtesy USDA Animal and Plant Health Inspection Service; Photo by R. Anson Eaglin.*

- To human society
  - Electrical system damage estimated at $1.7 billion per year.
  - Losses of poultry.
  - Inspections of exported goods doubling costs.

In contrast in another analysis, the estimated total losses and damages were reported as $11 million per year (Pimentel et al., 2005) (see Chapter 16).

- To the environment
  - Major detrimental effects to the native mammals and birds; decimating

vertebrate fauna (Reed, 2005) resulting in the extinction of the following:
  - 10 of 13 species of endemic forest birds
  - 2 of 3 native mammals
  - 6 endemic lizards
- Decreased transmission of plant seeds
- Reduced negative impact of introduced rats

17.5 AMPHIBIAN INVASIVE SPECIES

17.5.1 Cane Toad (*Rhinella marina*)

According to the Australian Department of Environment (2015), the cane toad is an invasive species with a significant impact. It was initially released in northern Queensland in 1935 as a biological control to eliminate pest beetles.

17.6 AQUATIC INVASIVE SPECIES

Aquatic animals have been introduced accidentally to rivers and lakes. For example, 79 species have been introduced into the Great Lakes (reviewed by Cuhel and Aguilar, 2013). The ecosystems of the Great Lakes of North America (or Laurentian Great Lakes) are new in Geological terms being formed with the retreat/melting of the glacial continental ice sheets from the last Ice Age between 14,000 and 4,000 years ago (reviewed by U.S. Army Corps of Engineers and the Great Lakes Commission, 1999). Of the invasive species in the Great Lakes, four have been demonstrated to be having a major ecological impact. These damaging invasive species are the following: the sea lamprey, the alewife, the zebra mussel, and the quagga mussel (reviewed by Cuhel and Aguilar, 2013).

17.6.1 Alewife (*Alosa pseudoharengus*)

Alewife are also called mulhaden, gray herring, or golden shad. Alewife spread to Lake Erie through the Welland Canal and then to other Great Lakes via the Detroit River and Lake St. Clair (USGS, 2016). The species has spread to reservoirs (e.g., Cherry Merritt Reservoir in Nebraska), lakes (e.g., in Tennessee and New York), and rivers, such as Missouri River and Mississippi River (USGS, 2016). Alefish are problematic as they out-compete with other species for zooplankton and, in die-offs, they foul beaches and negatively impact recreation and local economies (USGS, 2016).

17.6.2 Sea Lamprey (*Petromyzon marinus*)

Sea lampreys are native to the Atlantic Ocean. They spread from Lake Ontario to Lake Erie (1921) again via the Welland Canal then invaded Lake Huron, Lake Michigan, and Lake Superior. Today they are also present in many rivers in Wisconsin, Minnesota, and Michigan (USGS, 2016). Sea lamprey feeds on and/or damage native fish. It is thought that sea lampreys are at least partially responsible for the collapse of commercial fishing on the Great Lakes (USGS, 2016).

17.6.3 Zebra Mussels (*Dreissena bugensis*) and Quagga Mussel (*Dreissena rostriformis bugensis*)

Zebra mussels were first found in the North American Great Lakes in 1986 (reviewed by Mills et al., 1996). It has spread first to all the Great Lakes and then to waterways and lakes throughout the eastern USA and Canada (reviewed by Mills et al., 1996). The quagga mussel was first observed in the Great Lakes in 1989 (reviewed by Mills et al., 1996). The zebra and quagga mussels are native to waterways and lakes around the Black and Caspian seas. It is presumed that these mussels were transported to the Great Lakes in ships from this area.

The invasive mussels, the zebra and quagga mussels, have marked effects on the ecology of the Great Lakes. For instance, their presence has accelerated the decline in populations of native bivalves of the family Unionidae (Burlakova et al., 2014).

17.6.4 Africanized Bees (*Apis mellifera scutellata*) or Killer Bees

African Bees were imported to Brazil in 1956 with the goal of hybridization with the European honeybee to improve the latter's ability to live in tropical environments (reviewed by Kono and Kohn, 2015). Africanized bees escaped an apiary in Rio Claro (São Paulo, Brazil) and spread being first observed in the following places (Ferreira et al., 2012; Lazaneo, 2002; Winston, 1992):

- throughout Brazil by 1979
- reaching Argentina in 1968
- reaching Mexico in 1985
- reaching the USA in 1990; first Texas, then Arizona and in 1994 into California

They have out-competed European honeybees with now over two-thirds of feral bees in San Diego county (California) with Africanized bee mitochondrial DNA (Kono and Kohn, 2015). The spread of Africanized bees has continued with some Africanized bees identified in the San Joaquin or Central Valley, California (Kono and Kohn, 2015).

There are implications for the spread of the Africanized bee. There is improved pollination of both wild and domesticated plants. However, the Africanized bee is much more aggressive to humans stinging people. A single sting may be unpleasant or lead to an allergic reaction (reviewed by Ferreira et al., 2012; Lazaneo, 2002). Multiple stings from an aggressive swarm of Africanized bees can have severe health consequences (reviewed by Ferreira et al., 2012; Lazaneo, 2002).

17.7 CONCLUSIONS

Human introduction of animals or invasive species have disrupted ecosystems globally. In some cases, this is being successfully addressed with eradication programs. However, it is likely that invasive species will continue to be a major problem unless there are concerted efforts by governments and international agencies with science-based policies and careful implementation.

References

Aplin, K.P., Suzuki, H., Chinen, A.A., Chesser, R.T., Ten Have, J., Donnellan, S.C., Austin, J., Frost, A., Gonzalez, J.P., Herbreteau, V., Catzeflis, F., Soubrier, J., Fang, Y.P., Robins, J., Matisoo-Smith, E., Bastos, A.D., Maryanto, I., Sinaga, M.H., Denys, C., Van Den Bussche, R.A., Conroy, C., Rowe, K., Cooper, A., 2011. Multiple geographic origins of commensalism and complex dispersal history of Black Rats. PLoS One 6, e26357.

Australian Department of Environment, 2015. Available from: https://www.environment.gov.au/biodiversity/invasive-species/feral-animals-australia

Baker, H., Stroud, D.A., Aebischer, N.J., Cranswick, P.A., Gregory, R.D., McSorley, C.A., Noble, D.G., Rehfisch, M.M., 2006. Population estimates of birds in Great Britain and the United Kingdom. Br Birds 99, 25–44.

Black, A., Poncet, S., Wolfaardt, A., Peters, D., Hart, T., Wolfaardt, L.-A., Tasker, M., Rexer-Huber, K., 2012. Rodent eradication on South Georgia—preparation and evaluation: a summary report of activities during the 2011/2012 field season. Government of South Georgia and the South Sandwich Islands and UK Overseas Territories Environment Programme. Available from: http://www.gov.gs/docsarchive/Environment/Invasive%20Species/Rodent%20eradication%20preparation%20and%20evaluation%20-OTEP%20report.pdf

Burlakova, L.E., Tulumello, B.L., Karatayev, A.Y., Krebs, R.A., Schloesser, D.W., Paterson, W.L., Griffith, T.A., Scott, M.W., Crail, T., Zanatta, D.T., 2014. Competitive replacement of invasive congeners may relax impact on native species: interactions among zebra, quagga, and native unionid mussels. PLoS One 9, e114926.

Clarke, A., Croxall, J.P., Poncet, S., Martin, A.R., Burton, R., 2012. Important bird areas: South Georgia. British Birds 105, 118–144.

Cooper, J., Marais, A.v.N., Bloomer, J.P., Bester, M.N., 1995. A success story: breeding of burrowing petrels (*Procellariidae*) before and after the eradication of feral cats *Felis catus* at subantarctic Marion Island. Mar. Ornithol. 23, 33–37.

Cornell Laboratory of Ornithology, 2017. All about birds. Available from: https://www.allaboutbirds.org/guide/House_Sparrow/lifehistory

Cuhel, R.L., Aguilar, C., 2013. Ecosystem transformations of the Laurentian Great Lake Michigan by non-indigenous biological invaders. Annu. Rev. Mar. Sci. 5, 289–320.

Dorcas, M.E., Willson, J.D., Reed, R.N., Snow, R.W., Rochford, M.R., Miller, M.A., Meshaka, Jr., W.E., Andreadis, P.T., Mazzotti, F.J., Romagosa, C.M., Hart, K.M., 2012. Severe mammal declines coincide with proliferation of invasive Burmese pythons in Everglades National Park. Proc. Natl. Acad. Sci. USA 109, 2418–2422.

Duncan, R.P., Boyer, A.G., Blackburn, T.M., 2013. Magnitude and variation of prehistoric bird extinctions in the Pacific. Proc. Natl. Acad. Sci. USA 110, 6436–6441.

Ferreira, Jr., R.S., Almeida, R.A., Barraviera, S.R., Barraviera, B., 2012. Historical perspective and human consequences of Africanized bee stings in the Americas. J. Toxicol. Environ. Health B Crit. Rev. 15, 97–108.

Fuller, R.A., Tratalos, J., Gaston, K.J., 2008. How many birds are there in a city of half a million people? Divers. Distrib. 15, 328–337.

Giunchi, D., Albores-Barajas, Y.V., Baldaccini, N.E., Vanni, L., Soldatini, C., 2012. Feral pigeons: problems, dynamics and control methods. Available from: http://cdn.intech-web.org/pdfs/29607.pdf

Gong, W., Sinden, J., Braysher, M., Jones, R., 2009. The economic impacts of vertebrate pests in Australia Invasive Animals Cooperative Research Centre. University of Canberra, Australia. Available from: http://www.pestsmart.org.au/wp-content/uploads/2010/03/IACRC_EconomicImpactsReport.pdf

Grarock, K., Tidemann, C.R., Wood, J., Lindenmayer, D.B., 2012. Is it benign or is it a Pariah? Empirical evidence for the impact of the common myna (*Acridotheres tristis*) on Australian birds. PLoS One 7, e40622.

Hays, W.S.T., Conant, S., 2007. Biology and impacts of Pacific island invasive species. 1. A worldwide review of effects of the small Indian mongoose, *Herpestes javanicus* (Carnivora: *Herpestidae*). Pac. Sci. 61, 3–16.

Hetmanski, T., Bochenski, M., Tryjanowski, P., Skórka, P., 2010. The effect of habitat and number of inhabitants on the population sizes of feral pigeons around towns in northern Poland. Eur. J. Wildlife Res. 57, 421–428.

Hone, J., 1978. Introduction and spread of the common myna in New South Wales. Emu 78, 227–230.

ISSG (Invasive Species Specialist Group), 2015. Available from: http://www.issg.org/database/species/ecology.asp?si=19&fr=1&sts=&lang=EN

Kepler, C.B., 1967. Polynesian rat predation on nesting Laysan albatrosses and other Pacific seabirds. Auk 84, 426–430.

Keller, R.P., Geist, J., Justice, J.M., Kühn, I., 2011. Invasive species in Europe: ecology, status, and policy. Environ. Sci. Eur. 23, 23.

Kerr, P.J., Best, S.M., 1998. Myxoma virus in rabbits. Rev. Sci. Tech. 17, 256–268.

Kono, Y., Kohn, J.R., 2015. Range and frequency of Africanized honey bees in California (USA). PLoS One 10, e0137407.

Lazaneo, V., 2002. Bee alert: Africanized honey bee facts. University of California-Agriculture and Natural Resources; Publication 8068.

Lintz, G.M., Homan, H.J., Gaukler, S.M., Penry, L.B., Bleier, W.J., 2007. European starlings: a review of an invasive species with far reaching impacts. In: Witmer, G.W., Pitt, W.C., Fagerstone, K.A. (Eds.), Managing Vertebrate Invasive Species: Proceedings of an International Symposium. USDA/APHIS/WS, National Wildlife Research Center, Fort Collins, CO, pp. 378–386, Available from: http://www.issg.org/database/species/ecology.asp?si=74&fr=1&sts=&lang=EN.

Lodge, D.M., Williams, S., MacIsaac, H.J., Hayes, K.R., Leung, B., Reichard, S., Mack, R.N., Moyle, P.B., Smith, M., Andow, D.A., Carlton, J.T., McMichael, A., 2006. Biological invasions: recommendations for U.S. policy and management. Ecol. Appl. 16, 2035–2054.

Loss, S.R., Will, T., Marra, P.P., 2013. The impact of free-ranging domestic cats on wildlife of the United States. Nat. Commun. 4, 1396.

Loveridge, R., Wearn, O.R., Vieira, M., Bernard, H., Ewers, R.M., 2016. Movement behavior of native and invasive small mammals shows logging may facilitate invasion in a tropical rain forest. Biotropica 48 (3), 373–380.

Lowe, S., Browne, M., Boudjelas, S., 2000. 100 of the World's Worst Invasive Alien Species: A Selection from the Global Invasive Species Database. Invasive Species Specialist Group, International Union for Conservation of Nature, Auckland, NZ.

Martins, T.L.F., Brooke, M.D.L., Hilton, G.M., Farnsworth, S., Gould, J., Pain, D.J., 2006. Costing eradications of alien mammals from islands. Anim. Conserv. 9, 439–444.

Matisoo-Smith, E., Robins, J.H., 2004. Origins and dispersals of Pacific peoples: evidence from mtDNA phylogenies of the Pacific rat. Proc. Natl. Acad. Sci. USA 101, 9167–9172.

McCleery, R.A., Sovie, A., Reed, R.N., Cunningham, M.W., Hunter, M.E., Hart, K.M., 2015. Marsh rabbit mortalities tie pythons to the precipitous decline of mammals in the Everglades. Proc. Biol. Sci. 282, pii: 20150120.

Medina, F.M., Bonnaud, E., Vidal, E., Tersey, B.R., Zavaleta, E.S., Donlaqn, C.J., Keitt, B.S., Corre, M.l., Horwath, S.V., Nogale, M.S., 2011. A global review of the impacts of invasive cats on island endangered vertebrates. Global Change Biol. 17, 3503–3510.

Mills, E.L., Rosenberg, G., Spidle, A.P., Ludyanskiy, M., Pligin, Y., May, B., 1996. A review of the biology and ecology of the quagga mussel (*Dreissena bugensis*), a second species of freshwater Dreissenid introduced to North America. Am. Zool. 36, 271–286.

Minnesota Department of Natural Resources, 2017. Available from: http://www.dnr.state.mn.us/invasives/terrestrialanimals/muteswan/index.html

Mirsky, S., 2008. Shakespeare to blame for introduction of European starlings to U.S. Scientific American. Available from: http://www.scientificamerican.com/article/call-of-the-reviled/

Moulton, M.P., Cropper, W.P., Avery, M.L., Moulton, L.E., 2010. The earliest house sparrow introductions to North America. Biol. Invasions 12, 2955–2958.

National Feral Swine Mapping System, 2015. Available from: http://swine.vet.uga.edu/nfsms/information/maps/swineDistribution1982.jpg

Okubo, A., Maini, P.K., Williamson, M.H., Murray, J.D., 1989. On the spatial spread of the grey squirrel in Britain. Proc. R. Soc. Lond. B Biol. Sci. 238, 113–125.

Patuxent Wildlife Research Center, 2017. Available from: http://www.pwrc.usgs.gov/bbs/state_of_the_birds_2009.pdf

Pimentel, D., Lach, L., Zuniga, R., Morrison, D., 2000. Environmental and economic costs of non-indigenous species in the United States. BioScience 50, 53–65.

Pimentel, D., Zuniga, R., Morrison, D., 2005. Update on the environmental and economic costs associated with alien-invasive species in the United States. Ecol. Econ. 52, 273–288.

Pye, T., Bonner, W.N., 1980. Feral brown rats, *Rattus norvegicus* in South Georgia (South Atlantic Ocean). J. Zool. 192, 237–255.

Reed, R.N., 2005. An ecological risk assessment of nonnative boas and pythons as potentially invasive species in the United States. Risk Anal. 25, 753–766.

Rodda, G.H., Savidge, J.A., 2007. Biology and impacts of pacific island invasive species. 2. *Boiga irregularis*, the brown tree snake (Reptilia: Colubridae). Pac. Sci. 61, 307–324.

Savolainen, P., Leitner, T., Wilton, A.N., Matisoo-Smith, E., Lundeberg, J., 2004. A detailed picture of the origin of the Australian dingo, obtained from the study of mitochondrial DNA. Proc. Natl. Acad. Sci. USA 101, 12387–12390.

Singleton, G.R., 2003. Impacts of rodents on rice production in Asia. No. 45, IRRI Discussion Paper Series, International Rice Research Institute (IRRI), Los Baños, 30 pp.

Steadman, D.W., 1995. Prehistoric extinctions of Pacific island birds: biodiversity meets zooarchaeology. Science 267, 1123–1131.

U.S. Army Corps of Engineers and the Great Lakes Commission, 1999. About our Great Lakes. Available from: http://www.glerl.noaa.gov/pr/ourlakes/background.html

US Bureau of Land Management, 2016. Wild horse and burro quick facts. Available from: http://www.blm.gov/wo/st/en/prog/whbprogram/history_and_facts/quick_facts.html

USGS (Geological Survey), 2016. Nonindigenous Aquatic Species Database. USGS, Gainesville, FL.

Vere, D.T., Jones, R.E., Saunders, G.R., 2004. The economic benefits of controlling rabbits in Australian temperate pasture systems from the introduction of the rabbit calicivirus disease. Agric. Econ. 30, 143–155.

Wanless, R.M., Angel, A., Cuthbert, R.J., Hilton, G.M., Ryan, P.G., 2007. Can predation by invasive mice drive seabird extinctions? Biol. Lett. 3, 241–244.

Wildlife Service (of APHIS, USDA), 2011. Brown tree snake—an invasive reptile. Available from: https://www.aphis.usda.gov/publications/wildlife_damage/content/printable_version/fs_brown_tree_snake_2011.pdf

Willson, J.D., Dorcas, M.E., Snow, R.W., 2011. Identifying plausible scenarios for the establishment of invasive Bur-mese pythons (*Python molurus*) in Southern Florida. Biol. Invasions 13, 1493–1504.

Winston, M.L., 1992. The biology and management of Africanized honey bees. Annu. Rev. Entomol. 37, 173–193.

Wodzicki, K., Wright, S., 1980. Introduced birds and mammals in New Zealand and their effect on the environment. Tuatara 27, 78–103.

18

Impact of Agricultural Animals on the Environment

Colin G. Scanes

University of Wisconsin–Milwaukee, Milwaukee, WI, United States

18.1 INTRODUCTION

Livestock and poultry production has valuable contributions to the global economy and human welfare (see Chapters 1 and 2), but negatively impact the environment influencing water, air, and soil. Manure or animal waste is the predominant source of concern. Based on (albeit dated) information for the USA from 1992, livestock and poultry produce 120 million metric tons of manure in dry weight (US EPA, 1998). This is 15-fold greater than total human waste (US EPA, 1998). With increases in total production and the efficiency of production, it is likely that manure production in the USA will be similar to figures from 20 years ago. The contribution of animal waste from the livestock and poultry sectors is the following (US EPA, 1998):

- beef cattle, 54%
- dairy cattle, 20%
- poultry, 11%
- pigs, 7%
- horses, 7%

Additional sources of concern are the release of methane from enteric fermentation in ruminants, gases and particulates from animal facilitates, and disposal of mortalities. Livestock and poultry have been increasingly produced in "industrialized systems" in concentrated animal feeding operations (CAFOs) (see box) with large numbers of animals in common facilities and consequently large quantities of manure. This animal waste can impact the environment, specifically water quality, and public health due to the release of antibiotics and pathogens (Burkholder et al., 2007). Livestock and poultry also impact climate change with emissions of greenhouse gas (GHG) (IPCC, 1996; Jun et al., 1996).

The US Environmental Protection Agency (EPA) defines CAFOs in the USA as follows:

- Large CAFO, animal units greater or equal to 1000

Animals and Human Society
http://dx.doi.org/10.1016/B978-0-12-805247-1.00025-3

- Medium CAFO, animal units 300–999 and meets one of the following: "a manmade ditch or pipe that carries manure or wastewater to surface water; or the animals come into contact with surface water."
- Small CAFO, animal units <300

The term CAFO is also synonymous with large CAFO. Small and medium CAFOs can be designated as CAFOs if they are significant contributors of pollutants.

A large CAFO with >999 animal units is equivalent to the following:

- cattle, >999
- cow/calf pairs, >999
- dairy cows, >699
- pigs (25 kg or 55 lb or higher), >2499
- horses, >499
- broiler (meat type) chickens (nonliquid manure system), >124,999
- laying hens, >81,999

18.2 GLOBAL IMPACT OF ANIMAL AGRICULTURE ON WATER USE

All human activity impacts the quantity and quality of freshwater. This is the case whether we consider water for agriculture or specifically animal agriculture, manufacturing, or household use. The impact can be assessed by the water footprint in an analogous manner to the carbon footprint with water footprint being the amount of water used (evaporated or polluted). The total water footprint can be divided into the following (Hoekstra, 2012):

- blue water footprint (surface freshwater from lakes, rivers, and groundwater)
- green water footprint (rainfall)
- gray water footprint (water that has become polluted due to its prior use)

Animal production impacts water globally. Some researchers put the impact of animal agriculture at 27% of human water usage globally for the production of livestock (including water used to grow the feed grains) (Mekonnen and Hoekstra, 2011). Other researchers have much lower estimates, such as 8% (reviewed by Girard, 2012). While the basis of the difference in the estimates is unclear, it is undoubtedly the case that livestock and poultry production uses considerable amounts of global freshwater supplies. Table 18.1 summarizes the amount of water required to produce food. Companies, such as Nestlé, are using water footprints as part of their sustainability audits (Nestle, 2015; White et al., 2015). The following are three pertinent additional points relative to water usage:

- Usage of green water in crop production reduces the requirement for the precious blue water that can be used for drinking.
- Importing grain or meat is the equivalent of importing freshwater and this is known as virtual water imports.
- Improved feed conversion saves water.

18.3 IMPACT OF AGRICULTURAL ANIMALS ON WATER QUALITY (RUN-OFF)

18.3.1 Phosphate and Nitrogen

Phosphate, nitrogen, and potassium are essential plant macronutrients. The application of animal waste or manure to land either by spreading or the animals being on pasture results in the addition of phosphate, nitrogen, and potassium to the soil and to agricultural runoff into watersheds after rainfall, particularly very heavy rains; the phosphate traveling either in solution or bound to soil or manure particles. In the same way, application of fertilizers to

TABLE 18.1   Global Water Footprint for the Production of Food

| Food | Water footprint in 100 gal. lb^{-1} food (L kg^{-1} of food) | | | |
	Green water footprint	Blue water footprint	Gray water footprint	Total water footprint
Vegetables	0.27 (194)	0.06 (43)	0.12 (85)	0.44 (322)
Cereals	1.69 (1,232)	0.31 (228)	0.25 (184)	2.26 (1,644)
Milk	1.19 (863)	0.12 (86)	0.10 (72)	1.40 (1,020)
Eggs	3.56 (2,592)	0.34 (244)	0.59 (429)	4.49 (3,265)
Chicken (meat)	4.88 (3,545)	0.43 (313)	0.64 (467)	5.95 (4,325)
Pork	6.75 (4,907)	0.63 (459)	0.86 (622)	8.24 (5,988)
Beef	19.83 (14,414)	0.76 (550)	0.62 (451)	21.20 (15,415)

Data calculated from Hoekstra, A.Y., 2012. The hidden water resource use behind meat and dairy. Anim. Front. 2, 3–8.

the land can result in the addition of phosphate and nitrogen to runoff into streams. The nitrate or phosphate can lead to eutrophication of waterways (see box) and to the health of people, livestock, and wild animals drinking the water downstream and to oceans into which the river flows (Fig. 18.1). The contribution of various sectors to nitrate levels in the Mississippi River is shown in Table 18.2.

Eutrophication

When waterways have high levels of plant nutrients (usually phosphate), there is a rapid expansion of algae (algal bloom) or cyanobacteria. This is followed by death of freshwater animals due to lack of oxygen.

FIGURE 18.1   **Runoff from livestock facilities can enter watersheds reducing water quality or be a mechanism for transmission of pathogens.** *Source: Courtesy USDA Natural Resource Conservation Service. Photo by Tim McCabe.*

TABLE 18.2   Contribution of Various Sectors to Nitrate in the Mississippi and Hence to Creating a Dead Zone in the Gulf of Mexico (Whittaker et al., 2015)

Nitrogen source	Input to Gulf of Mexico
Fertilizer-soil	50
Animal manure	15
Legumes	11
Municipal and industrial point sources	<2
Atmospheric deposition	24

18.3.1.1 *Phosphate*

Phosphate from livestock and poultry production is moved via surface water to streams, rivers, and oceans. In studies in the USA, it was shown that when anaerobic lagoon liquid is applied to land to meet crop nitrogen needs, phosphate is being overapplied by 2–3 times in high volumes (0.86×10^6 L ha^{-1}), which will likely be carried by the run off (Nelson and Mikkelsen, 2005; Novak et al., 2004). Land application of animal waste can lead to phosphate leakage into groundwater (van Es et al., 2004) with leakage influenced by tillage (Smith et al., 2015). Much of the soluble phosphate is found in runoff after the first rain event (Moore et al., 2011).

In studies in Asia, it is estimated that phosphate is overapplied with a 24% overload and with livestock contributing about 40% to the phosphate application (Gerber et al., 2005) (Fig. 18.2A–B). Application of animal manure providing phosphate at levels higher than the crops utilize can lead to damaging algal blooms and hence eutrophication of surface water including lakes and rivers (Gerber et al., 2005).

18.3.2 Eutrophication Due to Nitrate and/or Phosphate

Just as nitrogen, potassium, and phosphate are essential nutrients for plants on land, they are also nutrients for algae. The rate of algal growth is dependent on the availability of plant nutrients. In freshwater, phosphate is likely to be the limiting nutrient for algal growth. Conversely with the marine environment, nitrogen is limiting. If phosphate concentrations are increased (e.g., due to agricultural runoff), there can be a huge increase in algae and cyanobacteria in lakes, ponds, reservoirs, streams, and rivers leading to eutrophication.

The consequences of eutrophication include the following:

- death of fish
- problems with filtering the water prior to human use
- dirty-looking water that is undesirable for swimming and other recreational activities
- toxins from cyanobacteria (blue–green algae)

18.3.3 Hypoxic Zones in Coastal Regions of Oceans

Phytoplankton growth depends on the availability of plant nutrients with the concentrations of nitrate, phosphate, and iron being rate limiting. Tropical and subtropical marine waters are frequently depleted of nitrogen due to microbial denitrification. The addition of plant nutrients, particularly nitrate from rivers, can lead rapidly to a bloom of phytoplankton (Beman et al., 2005). Nitrate is the principal plant nutrient in the drainage of the Mississippi River basin that is responsible for hypoxia in the Gulf of Mexico (Burkart and James, 1998). The sources of nitrate are the following in order of importance (Burkart and James, 1998):

1. Fertilizer applied agronomically
2. Nitrogen fixation (e.g., soybeans)
3. Manure applied to crop land

Manure was considered as a relatively minor source in the 1990s (Burkart and James, 1998) (Table 18.2). It is likely that this is still the case.

Among the sources of plant nutrients contributing to eutrophication in the Mediterranean Sea is aquaculture (Karydis and Kitsiou, 2011).

As with the situation with freshwater, the growth of phytoplankton in the world's seas and oceans depends on the availability of plant nutrients (Camargo and Alonso, 2006). There can be major increases in marine phytoplankton particularly around large river estuaries. This leads to a zone of low oxygen concentrations (a hypoxic or dead zone) (see Fig. 18.3 for the Mississippi River and the Gulf of Mexico). Another problem is that some phytoplankton species, dinoflagellates, produce toxic chemicals. A bloom of these can result in red or brown tides making the water unappealing not only with the massive numbers of dinoflagellates but also with the dead fish, seabirds, and aquatic mammals. The toxic compounds kill

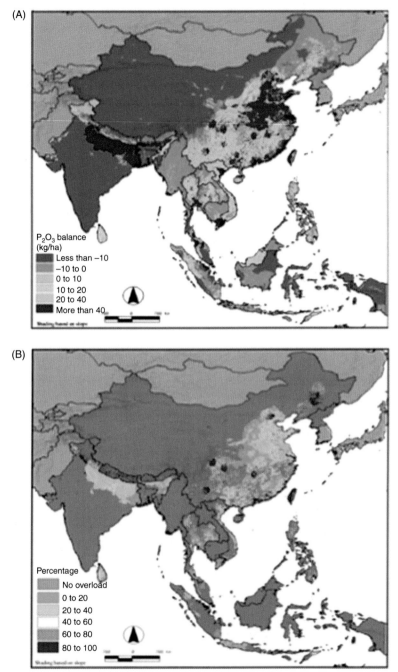

FIGURE 18.2 **Phosphate application to agricultural land in Asia in 1998–2000.** (A) Mass balance of phosphate application. (B) Percentage contribution of livestock to phosphate applied.– *Source: From Gerber, P., Chilonda, P., Franceschini, G., Menzi, H., 2005. Geographical determinants and environmental implications of livestock production intensification in Asia. Bioresour. Technol. 96, 263–276.*

Bottom-water dissolved oxygen—2014

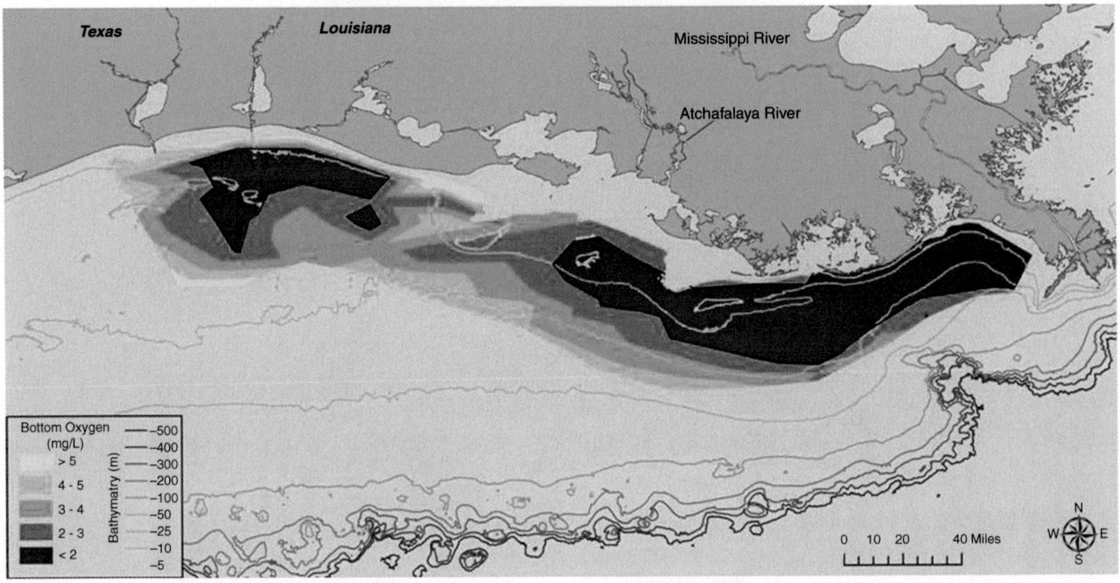

FIGURE 18.3   **Hypoxic region in the Gulf of Mexico 2015.** *Source: From US EPA; http://www.noaanews.noaa.gov/stories2014/ images/map_080414_deadzone.jpg.*

fish, mollusks, mammals, and birds. Examples of algal blooms include the following:

1. The Gulf of California has high phytoplankton production and high loading with nitrogen and phosphate (Kudela et al., 2008).
2. The hypoxic area in the Gulf of Mexico around the inflow of the Mississippi River resulting in a dead zone (Fig. 18.3).
3. The Baltic Sea has potential for eutrophication in its southern part where nitrogen is limiting. The nations surrounding the Baltic Sea have developed the Baltic Sea Action Plan to understand nutrient loading and alleviate the problem (HELCOM, 2009). In a similar manner, the countries around the Mediterranean Sea are working to understand nutrient loading, eutrophication, and provide a sound background for policymakers.
4. There is eutrophication in estuaries around the world, such as the Mae Klong River and Tha Chin River in Thailand (Thaipichitburapa et al., 2010; Thongdonphum et al., 2011).

18.3.4 Reducing Plant Nutrient Loading of Waterways

Approaches to reduce nitrate, for example, in the Mississippi River include the following (Whittaker et al., 2015):

- management of nitrogen fertilizer application
- substituting perennial crops for corn and soybeans
- improving management of animal manure
- creating and restoring wetlands

A good nutrient management plan will reduce the amount of nutrients entering the watershed and conserve them for supporting the growth of crops. Examples of approaches to reduce the impact are the following:

- Soil test to ensure no overloading.
- Inject animal waste into the soil.
- Surface-apply animal waste to fields where runoff is less likely and where nutrients will pass into the soil and be used by growing plants.

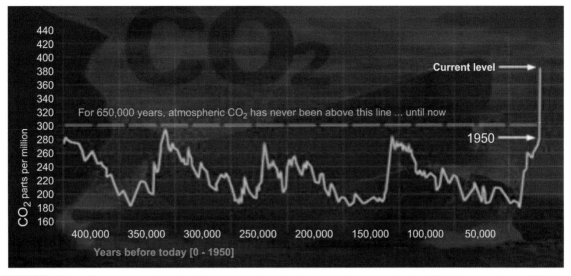

FIGURE 18.4 **Changes in carbon dioxide concentration in the atmosphere over the last 400,000 years.** *Source: NOAA.*

- By planting on contour lines with corn intercropped with a perennial forage crop, it is possible to apply animal waste to the land throughout the spring, summer, and fall (Fig. 18.4).
- Using no-till (no ploughing) or reduced tillage of crop lands reduces soil erosion and phosphate entering waterways bound to soil particles.
- Maintain buffer strips and wetlands around waterways.

18.3.5 Nitrates and Human Health

Nitrate concentrations in water can be a hazard to health with a drinking water standard of >10 mg L^{-1} for NO_3-N (US EPA, 2016). Nitrates can be a risk to the health of babies particularly under 6 months old. If the baby consumes nitrate, such as in baby formula, it is converted to nitrite in the gastrointestinal tract and can bind to hemoglobin impairing oxygen transport. This can be referred to as the "blue baby syndrome" (CDC, 2016a; US EPA, 2016). Nitrate/nitrite can be converted to the chemical messenger, nitric

oxide (NO) (Jones et al., 2015). Although nitrate consumption has been linked to increased miscarriages (Kramer et al., 1996), this is not supported by epidemiological studies (reviewed by Manassaram et al., 2006).

There are examples of nitrate problems in wells and rivers. In the San Joaquin Valley in California (USA), the concentration of nitrate in groundwater, and hence in wells, is related to the proximity to either land application of animal waste from CAFOs (see box for discussion of CAFOs in Section 18.1) or synthetic fertilizer application to fruit and nut crops (Lockhart et al., 2013). Similarly, in the Midwest, USA, nitrate concentration in well water is positively linked to closeness of animal feeding operations (Wheeler et al., 2015). In Iowa (USA), levels of nitrate in the Raccoon River exceed the EPA drinking water standard (US EPA, 2016) and nitrate has to be removed to produce safe drinking water. Concentrations of nitrate in the Raccoon River have been rising over the last 50 years due probably to a shift in crop production with the abandonment of wheat production and its replacement by soybeans (Jayasinghe et al., 2012).

18.3.6 Pathogens in Animal Waste Causing Diseases in People

Animal excreta contain pathogens including the following human pathogenic organisms (Sobsey et al., 2001; Ziemer et al., 2010):

- Viruses: Hepatitis E virus (swine), *Reoviruses*, *Rotaviruses*, *Adenoviruses*, *Caliciviruses*, and influenza viruses (*Orthomyxoviruses*)
- Bacteria: *Campylobacter* spp., *Salmonella* spp. (nontyphoid), *Listeria monocytogenes*, *Escherichia coli* including O157:H7 together with *Aeromonas hydrophila*, *Yersinia enterocolitica*, *Vibrio* spp., and *Leptospira* spp.
- Protozoa: *Cryptosporidium pram*, *Giardia lamblia*, and potentially *Balantidium coli*

A potential source of pathogens could be animal mortalities (dead livestock or poultry or those killed to prevent the disease spreading) from zoonotic diseases. However, these are incinerated or composted or buried to preclude the release of pathogens.

18.3.6.1 Viruses

Manure can be a source of viruses that cause diarrhea in pigs and humans (reviewed by Ziemer et al., 2010). These can be transmitted either by consumption of untreated water or by contamination of foods consumed raw, such as lettuce and salad. Norovirus, sapovirus, and rotavirus A are detected in pig feces (Costantini et al., 2007). Following treatment of manure, norovirus is no longer detected but sapovirus or rotavirus A continue to be found in soil, and in constructed wetlands or following biofiltration (Costantini et al., 2007) potentially leading to the contamination of groundwater.

18.3.6.2 Bacteria

Bacteria pathogenic to humans are found in animal waste and in run-off water. For instance, fecal coliforms and *Campylobacter* spp. have been shown to leach from animal waste into groundwater (Frey et al., 2015; Malik et al., 2004) with run-off into the groundwater at levels of coliformes exceeding US EPA standards (Howell et al., 1994). Runoff from land used for pasture or where manure has been spread can be a nonpoint source of pollution when it exceeds the EPA standards for fecal coliforms (200/100 mL) (US EPA, 1976). *Salmonella* sp. are another potential cause of food-borne diseases in humans. Among the potential sources of *Salmonella* sp. are pig confinement facilities (Davies et al., 1998).

Among the most serious bacteria that can contaminate water are enterohemorrhagic *E. coli* that produce shiga toxin and cause bloody diarrhea. These are called shiga toxin-positive *E. coli* (STEC). There has been a series of disease outbreaks caused by *E. coli* O157:H7, a subset of STEC *E. coli* (Table 18.3). Clearly cattle are responsible for at least a proportion of the O157:H7 outbreaks. Healthy cattle are a reservoir for O157:H7 (reviewed by Lim et al., 2010). The presence of *E. coli* O157:H7 is relatively low but that of other STEC *E. coli* is higher (Table 18.4). Contamination of groundwater with pathogenic bacteria should not be a problem for drinking water if raw water is treated adequately. However, this is not the case globally. When manure

TABLE 18.3   Characteristics of *E. coli* O157:H7 Outbreaks in the USA Between 1982 and 2002

Parameter	No. of outbreaks
Outbreaks	350
Cases	8598
Hospitalizations	1493
Deaths	40
Definitive cause	
Food-borne	52%
Beef	21%
Produce (vegetables)	11%
Waterborne	9%

Based on Rangel, J.M., Sparling, P.H., Crowe, C., Griffin, P.M., Swerdlow, D.L., 2005. Epidemiology of Escherichia coli O157:H7 outbreaks, United States, 1982–2002. Emerg. Infect. Dis. 11, 603–609.

TABLE 18.4 Incidence (%) of O157 and Non-O157 STEC or Enterohemorrhagic *E. coli*

Source	O157 STEC	Non-O157 STEC
Domesticated ruminants	6.6	36.0
Wildlife	1.2	7.4
Soil	0.4	0.4
Produce (leafy greens)	0.00	0.65
Water (ranch[a])	3.0	17.7
Water (farm[b])	0.00	8.2
Water (watershed)	4.0	13.7

STEC, Shiga toxin-positive *Escherichia coli*.
[a] *Ranches with domesticated ruminants.*
[b] *Farms producing leafy greens.*
Based on Cooley, M.B., Jay-Russell, M., Atwill, E.R., Carychao, D., Nguyen, K., Quiñones, B., Patel, R., Walker, S., Swimley, M., Pierre-Jerome, E., Gordus, A.G., Mandrell, R.E., 2013. Development of a robust method for isolation of shiga toxin-positive Escherichia coli (STEC) from fecal, plant, soil and water samples from a leafy greens production region in California. PLoS One 8, e65716.

is used to fertilize, pathogenic bacteria can adhere to vegetables (see Section 18.3.6.2.1). This is a problem when vegetables are consumed raw.

18.3.6.2.1 *E. COLI* IN VEGETABLES

Foods eaten raw can be contaminated with *E. coli* and other bacteria if manure is used as a fertilizer or if there is runoff from fields either where cattle and other livestock are, or where manure has been spread (Table 18.4). For instance, *E. coli* was detected in 19% of raw vegetables in Kurdistan (Saeed et al., 2013). In 2011, there was an outbreak of *E. coli* infections in Europe affecting several thousand people and killing over 20 people. It was suspected that the source of the infection was an organic farm producing beansprouts in Germany (CDC, 2013).

18.3.6.3 *Antibiotic-Resistant Bacteria*

Livestock are a source of antibiotic-resistant bacteria (Economou and Gousia, 2015). Tetracycline antibiotic-resistant *Campylobacter* are found in run-off water after land application of animal waste (Frey et al., 2015). In addition, tylosin-resistant enterococci are found in the manure of pigs receiving the antibiotic tylosin in the diet (Garder et al., 2014).

18.3.6.4 *Protozoa*

18.3.6.4.1 *CRYPTOSPORIDIUM*

Cryptosporidium is responsible for about 20% of childhood diarrhea in developing countries (Putignani and Menichella, 2010) and is a major cause of diarrhea in calves and other livestock. There are multiple species of *Cryptosporidium*:

- *Cryptosporidium parvum* in cattle, humans, and other mammals
- *Cryptosporidium hominis* in humans
- *Cryptosporidium meleagridis* in birds and humans
- *Cryptosporidium bovis* in cattle and sheep
- *Cryptosporidium muris* in rodents
- *Cryptosporidium suis* in pigs

When cryptosporidiosis in humans is caused by *C. parvum*, the infective oocysts come from either contamination of water with cattle or human feces (reviewed by Mosier and Oberst, 2000). One of the worst outbreaks of waterborne disease in the USA was the 1993 outbreak of cryptosporidiosis in Milwaukee, Wisconsin (Kramer et al., 1996). The Milwaukee outbreak encompassed the following (Corso et al., 2003; Kramer et al., 1996):

- 403,000 people becoming ill
- 44,000 people seeking outpatient treatment
- 4,400 hospitalized
- $31.7 million in medical costs
- $64.6 million in lost productivity

The health risks of cryptosporidiosis are increased in the immunocompromised. The case of the outbreak was consumption of the oocytes from the pathogenic protozoa, *Cryptosporidium*, from drinking water that had been inadequately treated (Kramer et al., 1996). Contamination of raw water supplies with *Cryptosporidium* oocysts was reported in 87% of samples in the USA and Canada (LeChevallier et al., 1991) and two-thirds of raw water samples in Ethiopia (Atnafu et al., 2012).

C. parvum oocysts were detected in over two-thirds of surface water samples (Kramer et al., 1996). These must be removed from drinking water by filtration as chemical disinfectants are ineffective (Kramer et al., 1996). Significant outbreaks of waterborne cryptosporidiosis are still occurring in the USA with, for instance, an outbreak sickening 2780 people or 27% of the population in Baker City, Oregon (DeSilva et al., 2016). There was an outbreak of cryptosporidiosis in Lancashire, England, in 2015 (BBC, 2015). This was due to contaminated water with water to 300,000 households suspect (BBC, 2015).

C. parvum is detected in feces from 13% to 78% of preweaning calves (Atwill et al., 1999; Trotz-Williams et al., 2007). There are large numbers of oocysts in feces with calves with diarrhea likely to have >200,000 oocysts per gram of feces (Trotz-Williams et al., 2007). The mean infective dose is estimated to be as low as 132 *C. parvum* oocysts (DuPont et al., 1995). The source of *C. parvum* oocysts in the Oregon outbreak was feces from cattle contaminating the watershed (DeSilva et al., 2016). There are seasonal differences in the percentage of preweaned calves shedding *C. parvum* with 26% in the summer and 11% in the winter (Szonyi et al., 2010). Another source for *C. parvum* is lettuce and other vegetables that are consumed raw and when they are grown where cattle manure is used for fertilization (Ramirez et al., 2004). Another source of *Cryptosporidium* is recreational water for instance when swallowed. Recreational water can be contaminated with *Cryptosporidium* oocysts from sewerage or cattle manure (CDC, 2016b).

18.3.7 Antibiotics in Run-Off From Livestock Operations

Antibiotics are used extensively in pig and poultry production with, for instance, 38.5 million kg used in 2012 in China (Krishnasamy et al., 2015). Antibiotic residues have been detected in ground and surface water close to pig and poultry facilities (Campagnolo et al., 2002). Veterinary antibiotics are similarly detected in farm effluents and rivers in China (Wei et al., 2011). Chlortetracycline and tetracycline are bound to manure solids while tylosin is in the liquid fraction (Frey et al., 2015). Antibiotics are found in run-off water after land application of animal waste (Frey et al., 2015).

18.3.8 Ecotoxicity of Androgenic and Estrogenic Agents

The androgen trenbolone is used to increase growth rate in cattle in the USA. Based on simulation studies rather than direct determinations, trenbolone can be found in soils and stormwater run-off (Webster et al., 2012). When heifers implanted with trenbolone and estradiol were held on pasture at a high density, the levels of the trenbolone metabolite, 17α-trenbolone, in initial run-off exceeded no observed adverse effect levels (Jones et al., 2014a). Leachate from cattle manure contained 17α-trenbolone at concentrations such that dilution of greater than 70-fold is required to achieve no observed adverse effect levels (Jones et al., 2014b). Lagoons receiving waste from cattle implanted with estradiol and trenbolone contained estrone and 17α-trenbolone at levels exceeding the lowest observable effect levels (Khan and Lee, 2012). Thus, as these lagoons are used for irrigation, there is potential for ecotoxicity.

18.4 LIVESTOCK AND POULTRY, GREENHOUSE GASES, AND CLIMATE CHANGE

Climate Change (Global Warming)

There is abundant evidence that human activity in industrialized societies has increased the concentration of GHG, such as carbon dioxide (Fig. 18.5), methane, and nitrous/nitric oxide in the atmosphere. There is also overwhelming evidence of increases in global temperatures (Fig. 18.6). Between 1970 and 2011, atmospheric concentrations of the following GHG increased (Howes et al., 2015):

- carbon dioxide (CO_2) by 40%
- methane (CH_4) by 150%
- nitrous oxide (N_2O) by 20%

It is clear that GHG is the primary cause of increases in temperatures globally. This phenomenon is referred to as anthropogenic (human-caused) climate change or global warming (Hansen et al., 2010; IPCC, 2000; Kerr, 2006a,b; Mann et al., 1999).

Scientific Consensus and Human-Caused (Anthropogenic) Climate Change

A very high percentage (97%) of climate scientists who express an opinion consider that climate change is due to humans (Cook et al., 2013). Similarly, 92% of nonclimate scientists in the USA accept an anthropogenic cause for rising global temperatures (Carlton et al., 2015). Moreover, multiple scientific professional societies, including the American Association for the Advancement of Science, American Medical Association, and the American Geophysical Union endorse the conclusion of anthropogenic climate change (Fig. 18.7) (NASA, 2016a,b).

Public Opinion and Anthropogenic Climate Change

Globally, in 2015, 54% of the public view climate change as a very serious problem and 78% support limiting GHG emissions (Pew Research Center, 2015). There are marked differences by region and country, however, with 74% of people in Latin America but only 18% of Chinese viewing climate change as a very serious problem (Pew Research Center, 2015). In some counties, there are marked differences in views on climate change with political philosophy. For instance in the USA, 20% of Republicans view climate change as a very serious problem while 68% of Democrats do (Pew Research Center, 2015). There was a drop on the public opinion in the USA accepting anthropogenic climate change from 61% in 2007 to 50% in 2010 and then a slight recovery to 57% of people in 2014 believing that "the increases in the earth's temperature is more due to pollution from human activities than to natural changes in the environment not due to human activities" (Gallop, 2014).

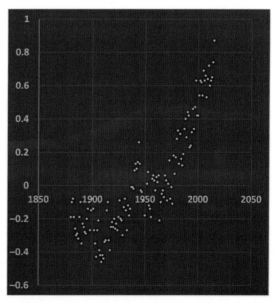

FIGURE 18.5 **Changes in global temperature since 1880.** *Source: Based on Hansen, J., Ruedy, R., Sato, M., Lo, K., 2010. Global surface temperature change. Rev. Geophys. 48, RG4004, updated using data from NASA.*

FIGURE 18.6 **Impact of livestock on rangeland.** (A) Overgrazing (left of fence) compared to well-managed rangeland. (B) Degraded rangeland due to heavy livestock use and wind erosion. *Source: Part A: Courtesy USDA Natural Resource Conservation Service. Photo by Irv Cole. Part B: Courtesy USDA Natural Resource Conservation Service. Photo by Gene Alexander.*

FIGURE 18.7 **Environmental scientists with the USDA Natural Resources Conservation Service have worked with farmers and other agencies to assure the water quality of one of the USA's heritage rivers, the South Fork of the Holston River.** Examples of the approaches employed include terracing, buffers, fencing, and crossings. *Source: Courtesy USDA Natural Resource Conservation Service. Photo by Jeff Vanuga.*

Global Warming Potential is measured in million metric tons of CO_2 equivalents (MMT CO_2 Eq. or teragram CO_2 equivalent or GtCO_2e).

One teragram is equal to 10^{12} g or 1 million metric tons.

18.4.1 Agriculture and Climate Change

Global emissions of GHG were 50.1 million metric tons of CO_2 equivalents with agriculture contributing 11% to global GHG emissions in 2010 (UNEP, 2012). Livestock and poultry make a disproportionately large contribution to global agricultural GHG emissions as is seen from the following (UNEP, 2012):

- animal feed production (crops), 36%
- deforestation and expansion of pasture and feed crops production, 9%

Animal agriculture contribution to GHG emissions (direct contribution):

- enteric, 39%
- manure, 10%

The major sources of GHG emissions directly from animal agriculture are enteric production of methane from ruminants together with GHG emissions from manure. Ruminants (both domesticated and wild) convert about 6% of feed intake to methane. In contrast, pigs convert 0.6% and horses and other equines 2.5% (US EPA, 1997). Global emissions of GHG from enteric fermentation and manure management are summarized in Table 18.5. In addition to enteric and manure GHG, there are GHG emissions

TABLE 18.5 Global Green House Emissions (in Million Metric Tons Carbon Dioxide Equivalents) From Animal Agriculture and Comparison With Meat and Milk Production (in Million Metric Tons) (FAOStats, 2016)

	1990	2005	2012
Total	4.6	4.9	5.4
Enteric	1.9	2.0	2.1
Manure	0.33	0.34	0.36
Meat production	180	256	305
Milk production	544	651	760

from the production of animal feeds (e.g., corn/maize and soybeans) together with processing and transporting of animal products. Somewhat militating against this is the uptake of carbon dioxide by rangelands. There are increases in agricultural GHG emissions globally but disproportionately greater increases in the production of meat and milk (Table 18.5).

Emissions of GHG from agriculture in the USA are summarized in Table 18.6 and represent 9.0% of all US emissions (US EPA, 2015). It might be noted that estimates of the amount

of methane and nitrous oxide produced from pig manure are lower (Park et al., 2006; US EPA, 2003) than those used by the International Panel on Climate Change (IPCC, 2000).

Globally, cattle contribute 61% of the direct GHG emissions from livestock and poultry with the following contributions from the various sectors (UNEP, 2012):

1. beef cattle, 41%
2. dairy cattle, 20%
3. pig, 9%

TABLE 18.6 Contribution of Animal Agriculture to Greenhouse Gas (GHG) Emissions in the USA in 2005 and 2013

Source	GHG emissions (in million metric tons carbon dioxide equivalents)		
	Carbon dioxide	Methane	Nitric/nitrous oxide (NO$_x$)[a]
Total			
2005	6134	708	356
2013	5005 (−18.4%)	636 (−10.2%)	355
Animal agriculture[b]			
2005	<1	225[c]	16.4
2013	<1	226[c]	17.3
Plant agriculture			
2005	<1	9	244
2013	<1	8	264

[a] Predominantly nitrous oxide (N_2O).

[b] Excluding contribution from crop production.

[c] Enteric methane is 164.5 million metric tons carbon dioxide equivalents compared to methane from manure management, which produces 64.6 million metric tons carbon dioxide equivalents.

Data from US EPA, 2015. Inventory of U.S. greenhouse gas emissions and sinks: 1990–2013. Available from: http://www3.epa.gov/climatechange/Downloads/ghgemissions/US-GHG-Inventory-2015-Main-Text.pdf.

4. buffalo, 8%

5. chicken (meat and eggs), 8%

6. small ruminants, 6%

Table 18.7 summarizes the intensity of GHG emission of animal products globally. The intensity of GHG emission is 82% greater for beef from cattle on a grazing system than on a mixed system including grazing and consumption of cereals. Moreover, the intensity of

GHG emission is 11-fold greater for beef than for pork.

Annual percentage increases in GHG emissions by animal agriculture in regions of the world are shown in Table 18.8 with a comparison of changes in the mode of meat and milk production (FAOStats, 2016). The increase in animal products was greatest in Asia with a much smaller increase in GHG emissions. In Oceania, increased production of meat and milk was achieved with a net decrease in GHG emissions. This emphasizes the importance of increasing the efficiency of livestock and poultry production to reduce GHG emissions.

18.4.2 Management Systems and GHG Emissions

Animal waste from operations where large numbers of animals are held together in "industrialized" livestock systems (see box in Section 18.1 for discussion of CAFOs) is generally stored or treated in two ways:

1. Excreta flowing with cleaning water into large anaerobic lagoons (Fig. 18.8D) before land application. The anaerobic bacteria ferment the excreta releasing methane and ammonia.

2. Placing manure in piles or pits (Fig. 18.8C) before land application.

TABLE 18.7   Global Emission Intensity of Animal Products

Product	GHG emissions from animal agriculture (kg CO_2 equivalents per kg product)
Meat	
Beef (cattle meat)	67.6[a]
Pork	6.1
Chicken	5.4
Milk	
Cow	2.6
Buffalo	3.4
Sheep	8.4
Goat	5.2

[a] 102 kg CO_2 equivalents per kg product if grazing and 56 kg CO_2 equivalents per kg product if using a mixed system.

From Gerber et al. (2005).

TABLE 18.8   Annual Rate of Increase (or Decrease) in GHG Emissions From Animal Agriculture in Global Regions and Comparison With Changes in Meat and Milk Production

	Africa	Asia	Americas	Europe	Oceania
Enteric emissions	+2.4	+1.4	+0.8	−2.9	−1.0
Manure	+2.4	+1.5	+0.6	−1.6	+1.3
Animal production					
Meat	+7.8	+12.7	+7.1	−3.1	+2.8
Milk	+9.3	+13.8	+4.5	−2.0	+8.7

Data from FAOStats, 2016. Available from: http://faostat3.fao.org/browse/G1/*/E.

This is irrespective of whether manure is from dairy cattle, beef cattle, or pigs. In contrast, meat chickens are most frequently reared on deep litter (wood shavings). These are then applied to the land.

Globally, enteric fermentation (breakdown of feed by bacteria and protozoa, e.g., in the rumen) produces much more GHG emissions than manure (Table 18.2). However, in a study of dairy cattle in the USA, it was reported that anaerobic lagoons for manure treatment released threefold more methane than enteric-derived methane (Owen and Silver, 2015):

- anaerobic lagoons, 368 kg CH_4 per head per year
- enteric fermentation, 120 kg CH_4 per head per year

Owen and Silver (2015) concluded that "current greenhouse gas emission factors generally underestimate emissions from dairy manure."

18.4.3 Approaches to Reduce GHG Emission

Apart from reducing the consumption of animal products, the most effective approaches to reduce GHG emissions by agriculture are the following:

1. Increasing the efficiency of production. The attractiveness of this approach is exemplified by large increases in meat and milk production coupled with the modest increase in GHG emissions (Tables 18.4 and 18.6). There are trade-offs with this approach. For example, there is improved efficiency with vertical integration but reduced control by farmers (see Chapter 6).
2. GHG emissions being considered per unit meat or milk by policymakers in regulations (Table 18.7).
3. Reducing the use of anaerobic lagoons.

4. Employing management practices that reduce GHG emission. For instance, greater frequency of manure collection decreases methane (CH_4) and nitrous oxide (N_2O) emissions (Lin et al., 2015).
5. More extensive research and outreach programs.

18.5 LIVESTOCK AND POULTRY AND AIR QUALITY

18.5.1 Air Quality in Animal Facilities

Good ventilation is important to ensure good air quality for the animals and people working in the animal facilities. Among the gases that could be problems are the following:

- carbon dioxide (CO_2)
- ammonia (NH_3)
- hydrogen sulfide (H_2S)
- water vapor (H_2O)

Livestock and poultry release carbon dioxide every time they exhale. There is also carbon dioxide coming from the excreta. In animals in confinement facilities, there will be a tendency for carbon dioxide to build up with potentially adverse effects on the animals unless ventilation is sufficient. There is potential build-up of ammonia either from animal waste when livestock are raised over slats or from deep litter in poultry buildings (see Section 18.5.1.1). This potentially adversely affects the animals and people working in the facility. Again ventilation needs to be adequate to remove ammonia.

18.5.1.1 Ammonia

Ammonia is released into the atmosphere from livestock and poultry particularly in concentrated animal feeding operations. Adverse results include the formation of aerosols that cause haze

FIGURE 18.8 **Recommended approaches from the USDA Natural Resources Conservation Service to ameliorate the impacts of agriculture on water quality and decrease soil loss.** (A) Terracing reduces soil losses together with nitrate and phosphate movement into groundwater and then waterways. (B) Run-off of nitrogen, phosphate, and soil into waterways can be reduced by strips of oats and hay interspersed with strips of corn/maize. (C) Animal waste can be stored in a lagoon facilitating loss of nitrogen as ammonia and carbon as methane. The lagoon has a liner to prevent leakage. (D) Manure storage in a concrete pit. (E) Restored wetland at riparian zone with sheep grazing in pasture. (F) A barbed wire fence protects the riparian buffer from cattle in the pasture. (G) Concrete water crossing with associated fencing allowing livestock to cross waterway with minimal erosion or silting. *Source: Part A–B: Courtesy USDA Natural Resources Conservation Service. Photo by Lynn Betts; Part C, F, G: Courtesy USDA Natural Resource Conservation Service. Photo by Jeff Vanuga; Part D: Courtesy USDA Natural Resource Conservation Service. Photo by Tim McCabe; Part E: Courtesy USDA Natural Resources Conservation Service. Photo by Gary Kramer.*

and hence eye irritation, and aerosol particles passing into the lungs and depositing onto the land and groundwater with rainfall (calculated from Arogo et al., 2003). In addition, this ammonia can bind and neutralize sulfur dioxide and nitric oxide released into the atmosphere by coal-fired power plants. This reduces the problem in the air, but with precipitation it becomes acid rain.

Ammonia is released from cattle, pig, and poultry confinement facilities and the associated waste-handling systems. This is due in part to bacterial breakdown of urea or uric acid in the urine of the animals. Dairy cattle waste lagoons are reported to release 2 kg of ammonia per year per animal unit (Grant and Boehm, 2010). The ammonia emission from dairy cattle manure is influenced by levels of dietary protein (Chiavegato et al., 2015). Ammonia is also released from pig facilities and this is influenced by the number/size of animals, the method of manure storage/treatment, and the technique employed for land application (reviewed by Arogo et al., 2003). About two-third of the ammonia emitted from hog operations come from confinement facilities and the rest from the lagoons used for the treatment and storage of manure (Aneja et al., 2000). The range of ammonia emitted from lagoons is about 1.3–100 kg ammonia ha^{-1} day^{-1} (calculated from Arogo et al., 2003). Similarly, a variation of 0–15 kg of NH_3 h^{-1} was reported (Harper et al., 2000). There is a management approach to reduce ammonia release. The amount of urea in the urine increases when the mix of amino-acids in the diet does not match the needs of the animals. Ammonia release can be reduced by matching protein/amino-acids to the needs of the pigs. This improves protein efficiency and reduces nitrogen excretion into the under-floor animal waste moving systems in operations with slatted house ventilation systems (US EPA, 2001).

Ammonia is also a by-product of meat-type chicken production. It has been estimated that 1.7 g of ammonia is released per chicken per day (Siefert et al., 2004). It is estimated that 22% of nitrogen fed to chickens (predominantly as protein) accumulates in the litter (Coufal et al., 2006).

This nitrogen is initially in the form of uric acid but is converted to ammonia by bacteria in the litter. For one major poultry-producing region in the USA, the Delmarva Peninsula, this translates to 25.4×10^3 tons of ammonia per year (Siefert et al., 2004). Moreover, 15% of ammonia is lost to the atmosphere when chicken litter is applied to land by "broadcasting" (mechanically throwing the manure over the cropland) (Moore et al., 2011). This is prevented when the litter is incorporated by "knifing" (placing the manure into narrow hollows created in the cropland) (Moore et al., 2011).

18.5.1.2 *Total Reduced Sulfur Including Hydrogen Sulfide*

Hydrogen sulfide and other reduced sulfur compounds are emitted from CAFOs from the breakdown of excreta, feathers, and mortalities. This can adversely affect animals and humans. For instance, hydrogen sulfide in the atmosphere reduces growth in chickens (Wang et al., 2011). Hydrogen sulfide and other reduced sulfur compounds can smell offensive to neighbors of farms and are potentially hazardous to workers.

Hydrogen Sulfide and Human Health

The Agency for Toxic Substances and Disease Registry is the US federal agency charged with evaluating public health risks from chemicals. This agency has recommended "Minimal Risk Levels for hydrogen sulfide." These were 70 ppb (parts per billion) average exposure for acute exposure and 30 ppb average for long-term exposure (e.g., over a year) (ATSDR, in press). For concentrations of total reduced sulfur including hydrogen sulfide at the center of a feedlot in summer, investigators reported a mean total reduced sulfur of 37 ppb H_2S equivalents but an appreciable number of samples >0.1 ppm concentration ppb H_2S equivalents (Koelsch et al., 2004).

18.5.2 Odor From Animal Facilities

Odors from livestock and poultry can be offensive to neighbors of farms and have the potential to impact human health (Schiffman, 1998; Wing et al., 2014). Offensive odors from swine and other livestock facilities include hydrogen sulfide and volatile organic compounds: the fatty acids (acetic, butanoic, isovaleric, propionic, and valeric), amines (e.g., skatole and trimethyl amine), aromatics (e.g., *p*-cresol, indole, and phenol), and sulfur compounds (e.g., dimethyl disulfide, dimethyl sulfide). Dust particles from feces and feed can absorb odorants at high concentrations. These can be found in the exhaust air and when inhaled are deposited on the olfactory mucosal membranes (Bottcher, 2001). Odorants bind preferentially to small-sized dust particles (Ni et al., 2012); these being spread over larger distances. Odors can be detected by human panels or through various chemical means (Zahn et al., 2001).

18.5.3 Zoonotic Pathogens in Air

Viruses can be transmitted between animal facilities and to humans via aerosol particles (Zhang et al., 2013). Viruses can be detected 2 km downwind of the exhaust from pig facilities (Corzo et al., 2013). The ability of airborne viruses to infect depends on the size of the aerosol particles (Alonso et al., 2015). Manure can also be a source of airborne viruses. There are airborne pathogens aerosolized with center pivot irrigation using diluted wastewater from dairies (Dungan, 2014).

18.6 LIVESTOCK AND POULTRY AND SOIL QUALITY

18.6.1 Soil Degradation and Erosion

Soil degradation includes erosion, salinization (due to freshwater removal), soil loss following erosion after deforestation or overgrazing,

"compaction and crusting (of soils), [which] can be caused by cattle trampling," and waterlogging with impaired water movement (Oldeman et al., 1991). It was estimated in 1991 that there was 1.96 billion ha of land globally with soil degradation caused by human activity. Soil degradation is made up of the following (Oldeman et al., 1991):

- overgrazing of rangeland, 679 million ha globally
- deforestation, 579 million ha globally (also see Chapter 23)
- agricultural mismanagement, 552 million ha globally
- overexploitation, 133 million ha globally

Desertification is defined by the United Nations (1980) as "diminution or destruction of the biological potential of the land which can lead ultimately to desert-like conditions."

While contemporary information is largely lacking, in a series of "classical" studies, the impact of livestock on soil has been demonstrated. It is estimated that the soil of about 75% of rangeland globally was degraded (Dregne and Chou, 1992). Degradation of rangeland can lead to desertification (Dregne and Chou, 1992). Thirty-six years ago, the United Nations (1980) viewed that 35% of the world's land area was at risk of desertification. Poor land management is thought to be responsible for the problem with overgrazing from excessive livestock being one of the causes. The land then becomes more vulnerable to wind and water erosion (reviewed by Nicholson et al., 1998).

18.6.2 Overloading Soil by Application of Manure

The application of animal waste can lead to a build-up of chemicals in the soil. For instance, there are reports of increases in soil phosphate with land application of animal manure. Moreover, where pig waste is treated in a lagoon and the lagoon water applied to the land, there are increases in soil zinc, copper, and phosphate. In a study in North Carolina, application of pig anaerobic lagoon liquid supplied considerable zinc and copper (0.7 kg zinc and 0.6 kg copper ha^{-1} $year^{-1}$). This was followed by large increases >7-fold of zinc and copper concentrations in the soil (Novak et al., 2004). Similarly, copper and zinc levels in the soil were increased by land application of pig manure in China (Xu et al., 2013) and Spain (Berenguer et al., 2008).

18.7 CONCLUSIONS

It is clear that livestock and poultry production have extensive negative impacts on the environment. Caution is expressed about simplistic solutions. According to H.L. Mencken (1880–1956) "there is always a well-known solution to every human problem—neat, plausible, and wrong." What is needed is for policymakers supported by consumers to employ the available evidence and not emotional "knee jerk" responses. Approaches worth considering include the following:

1. Environmental impacts of livestock can be decreased by improved efficiency irrespective of the scale of agriculture if the environmental impacts are carefully considered.
2. Strong research should undergird decisions and application. To quote, "increasing knowledge is the only solution" (Girard, 2012).

References

Alonso, C., Raynor, P.C., Davies, P.R., Torremorell, M., 2015. Concentration, size distribution, and infectivity of airborne particles carrying swine viruses. PLoS One 10, e0135675.

Aneja, V.P., Chauhan, J.P., Walker, J.T., 2000. Characterization of atmospheric ammonia emissions from swine waste storage and treatment lagoons. J. Geophys. Res. 105, 11535–11545.

Arogo, J., Westerman, P.W., Heber, A.J., 2003. A review of ammonia emissions from confined swine feeding operations. Trans. ASAE 46, 805–817.

Atnafu, T., Kassa, H., Keil, C., Frikrie, N., Leta, S., Keil, I., 2012. Presence, viability and determinants *Cryptosporidium* oocysts and *giardia* cysts in the Addis Ababa career supply. Water Qual. Expos. Health 4, 55–65.

ATSDR (Agency for Toxic Substances and Disease Registry), 2017. Available from: http://www.atsdr.cdc.gov/.

Atwill, E.R., Johnson, E., Klingborg, D.J., Veserat, G.M., Marke-gard, G., Jensen, W.A., Pratt, D.W., Delmas, R.E., George, H.A., Forero, L.C., Philips, R.L., Barry, S.J., McDougald, N.K., Gildersleeve, R.R., Frost, W.E., 1999. Age, geographic, and temporal distribution of fecal shedding of *Cryptosporidium parvum* oocyst in cow-calf herds. Am. J. Vet. Res. 60, 420–425.

BBC, 2015. Lancashire water: *Cryptosporidium* "still present" in supply. Available from: http://www.bbc.com/news/uk-england-lancashire-33844758.

Beman, J.M., Arrigo, K.R., Matson, P.A., 2005. Agricultural runoff fuels large phytoplankton blooms in vulnerable areas of the ocean. Nature 434, 211–214.

Berenguer, P., Cela, S., Santiveri, F., Boixadera, J., Lloveras, J., 2008. Copper and zinc soil accumulation and plant concentration in irrigated maize fertilized with liquid swine manure. Agron. J. 100, 1056–1061.

Bottcher, R.W., 2001. An environmental nuisance: odor concentrated and transported by dust. Chem. Senses 26, 327–331.

Burkart, M.R., James, D.E., 1998. Agricultural-nitrogen contributions to hypoxia in the Gulf of Mexico. J. Environ. Qual. 28, 850–859.

Burkholder, J.-A., Libra, B., Weyer, P., Heathcote, S., Kolpin, D., Thorne, P.S., Wichman, M., 2007. Impacts of waste from concentrated animal feeding operations on water quality. Environ. Health Perspect. 115, 308–312.

Camargo, J.A., Alonso, A., 2006. Ecological and toxicological effects of inorganic nitrogen pollution in aquatic ecosystems: a global assessment. Environ. Int. 32, 831–849.

Campagnolo, E.R., Johnson, K.R., Karpati, A., Rubin, C.S., Kolpin, D.W., Meyer, M.T., Esteban, J.E., Currier, R.W., Smith, K., Thu, K.M., McGeehin, M., 2002. Antimicrobial residues in animal waste and water resources proximal to large-scale swine and poultry feeding operations. Sci. Total Environ. 299, 89–95.

Carlton, J.S., Perry-Hill, R., Huber, M., Prokopy, L.S., 2015. The climate change consensus extends beyond climate scientists. Environ. Res. Lett. 10, 094025.

CDC, 2013. Outbreak of *Escherichia coli* O104:H4 infections associated with sprout consumption—Europe and North

America, May–July 2011. MMWR Morb. Mortal. Wkly. Rep. 62, 1029–1031.

CDC (Centers fort Disease Control), 2016a. Available from: http://www.cdc.gov/healthywater/drinking/private/wells/disease/nitrate.html.

CDC (Centers for Disease Control), 2016b. Parasites—*Cryptosporidium* (also known as "Crypto"). Available from: http://www.cdc.gov/parasites/crypto/.

Chiavegato, M.B., Powers, W., Palumbo, N., 2015. Ammonia and greenhouse gas emissions from housed Holstein steers fed different levels of diet crude protein. J. Anim. Sci. 93, 395–404.

Cook, J., Nuccitelli, D., Green, S.A., Richardson, M., Winkler, B., Painting, R., Way, R., Jacobs, P., Skuce, A., 2013. Quantifying the consensus on anthropogenic global warming in the scientific literature. Environ. Res. Lett. 8, 024024.

Corso, P.S., Kramer, M.H., Blair, K.A., Addiss, D.G., Davis, J.P., Haddix, A.C., 2003. Costs of illness in the 1993 waterborne *Cryptosporidium* outbreak, Milwaukee, Wisconsin. Emerg. Infect. Dis. 9, 426–431.

Corzo, C.A., Culhane, M., Dee, S., Morrison, R.B., Torremorell, M., 2013. Airborne detection and quantification of swine influenza a virus in air samples collected inside, outside and downwind from swine barns. PLoS One 8, e71444.

Costantini, V.P., Azevedo, A.C., Li, X., Williams, M.C., Michel, F.C., Saif, L.J., 2007. Effects of different animal waste treatment technologies on detection and viability of porcine enteric viruses. Appl. Environ. Microbiol. 73, 5284–5291.

Coufal, C.D., Chavez, C., Niemeyer, P.R., Carey, J.B., 2006. Nitrogen emissions from broilers measured by mass balance over eighteen consecutive flocks. Poult. Sci. 85, 384–391.

Davies, P.R., Bovee, F.G., Funk, J.A., Morrow, W.E.M., Jones, F.T., Deen, J., 1998. Isolation of Salmonella serotypes from feces of pigs raised in a multiple-site production system. J. Am. Vet. Med. Assoc. 212, 1925–1929.

DeSilva, M.B., Schafer, S., Kendall Scott, M., Robinson, B., Hills, A., Buser, G.L., Salis, K., Gargano, J., Yoder, J., Hill, V., Xiao, L., Roellig, D., Hedberg, K., 2016. Community wide cryptosporidiosis outbreak associated with a surface water-supplied municipal water system—Baker City, Oregon, 2013. Epidemiol. Infect. 144, 274–284.

Dregne, H.E., Chou, N.-T., 1992. Global desertification dimensions and costs. In: Dregne, H.E. (Ed.), Degradation and Restoration of Arid Lands. Texas Technical University, Lubbock, TX.

Dungan, R.S., 2014. Estimation of infectious risks in residential populations exposed to airborne pathogens during center pivot irrigation of dairy wastewaters. Environ. Sci. Technol. 48, 5033–5042.

DuPont, H.L., Chappell, C.L., Sterling, C.R., Okhuysen, P.C., Rose, J.B., Jakubowski, W., 1995. The infectivity of *Cryptosporidium parvum* in healthy volunteers. N. Engl. J. Med. 332, 855–859.

Economou, V., Gousia, P., 2015. Agriculture and food animals as a source of antimicrobial-resistant bacteria. Infect. Drug Resist. 8, 49–61.

FAOStats, 2016. Available from: http://faostat3.fao.org/browse/G1/*/E.

Frey, S.K., Topp, E., Khan, I.U., Ball, B.R., Edwards, M., Gottschall, N., Sunohara, M., Lapen, D.R., 2015. Quantitative *Campylobacter* spp., antibiotic resistance genes, and veterinary antibiotics in surface and ground water following manure application: Influence of tile drainage control. Sci. Total Environ. 532, 138–153.

Gallop, 2014. A steady 57% in U.S. blame humans for global warming. Available from: http://www.gallup.com/poll/167972/steady-blame-humans-global-warming.aspx.

Garder, J.L., Moorman, T.B., Soupir, M.L., 2014. Transport and persistence of tylosin-resistant enterococci, genes, and tylosin in soil and drainage water from fields receiving Swine manure. J. Environ. Qual. 43, 1484–1493.

Gerber, P., Chilonda, P., Franceschini, G., Menzi, H., 2005. Geographical determinants and environmental implications of livestock production intensification in Asia. Bioresour. Technol. 96, 263–276.

Girard, C.L., 2012. Reducing the impact of animal production on the water supply: increasing knowledge is the only solution. Anim. Front. 2, 1–2.

Grant, R.H., Boehm, M.T., 2010. Ammonia emissions from Western livestock waste lagoons. Available from: http://www3.epa.gov/ttn/chief/conference/ei21/session6/rgrant.pdf.

Hansen, J., Ruedy, R., Sato, M., Lo, K., 2010. Global surface temperature change. Rev. Geophys. 48, RG4004.

Harper, L.A., Sharpe, R.R., Parkin, T.B., 2000. Gaseous nitrogen emissions from anaerobic swine lagoons: ammonia, nitrous oxide, and dinitrogen gas. J. Environ. Qual. 29, 1356–1365.

HELCOM, 2009. Eutrophication in the Baltic Sea——an integrated thematic assessment of the effects of nutrient enrichment and eutrophication in the Baltic Sea region: Executive Summary. Baltic Sea Environment Proceedings No. 115A. http://www.helcom.fi/Lists/Publications/BSEP115B.pdf.

Hoekstra, A.Y., 2012. The hidden water resource use behind meat and dairy. Anim. Front. 2, 3–8.

Howell, J.M., Coyne, M.S., Cornelius, P., 1994. Fecal bacteria in agricultural waters of the bluegrass region of Kentucky. J. Environ. Qual. 24, 411–419.

Howes, E.L., Joos, F., Eakin, C.M., Gattuso, J.-P., 2015. An updated synthesis of the observed and projected impacts of climate change on the chemical, physical and biological processes in the oceans. Front. Mar. Sci.

IPCC (International Panel on Climate Change), 1996. Revised 1996 IPCC guidelines for national greenhouse gas inventories, Reference Manual (Revised). v.3.

IPCC, 2000. Good practice guidance and uncertainty management in national greenhouse gas inventories. Available from: http://www.ipcc-nggip.iges.or.jp/public/gp/english/.

Jayasinghe, S., Miller, D., Hatfield, J.L., 2012. Evaluation of variation in nitrate concentration levels in the Raccoon River watershed in Iowa. J. Environ. Qual. 41, 1557–1565.

Jones, G.D., Benchetler, P.V., Tate, K.W., Kolodziej, E.P., 2014a. Trenbolone acetate metabolite transport in rangelands and irrigated pasture: observations and conceptual approaches for agro-ecosystems. Environ. Sci. Technol. 48, 12569–12576.

Jones, G.D., Benchetler, P.V., Tate, K.W., Kolodziej, E.P., 2014b. Surface and subsurface attenuation of trenbolone acetate metabolites and manure-derived constituents in irrigation runoff on agro-ecosystems. Environ. Sci. Process. Impacts 16, 2507–2516.

Jones, J.A., Hopper, A.O., Power, G.G., Blood, A.B., 2015. Dietary intake and bio-activation of nitrite and nitrate in newborn infants. Pediatr. Res. 77, 173–181.

Jun, P., Gibbs, M., Gaffney, K., 1996. CH_4 and N_2O emissions from livestock manure. In: Good Practice Guidance and Uncertainty Management in National Greenhouse Gas Inventories. Available from: http://www.ipcc-nggip.iges.or.jp/public/gp/bgp/4_2_CH4_and_N2O_Livestock_Manure.pdf.

Karydis, M., Kitsiou, D., 2011. Eutrophication and environmental policy in the Mediterranean Sea: a review. Environ. Monit. Assess. 184, 4931–4984.

Kerr, R.A., 2006a. A worrying trend of less ice, higher seas. Science 311, 1698–1701.

Kerr, R., 2006b. Yes, it's been getting warmer in here since the CO_2 began to rise. Science 312, 1854.

Khan, B., Lee, L.S., 2012. Estrogens and synthetic androgens in manure slurry from trenbolone acetate/estradiol implanted cattle and in waste-receiving lagoons used for irrigation. Chemosphere 89, 1443–1449.

Koelsch, R.K., Woodbury, R.L., Stenberg, D.E., Miller, D.N., Schulte, D.D., 2004. Hydrogen sulfide concentrations in vicinity of beef cattle feedlots. Nebraska Beef Cattle Reports. Available from: http://digitalcommons.unl.edu/cgi/viewcontent.cgi?article=1197&context=animalscinbcr.

Kramer, M.H., Herwaldt, B.L., Craun, G.F., Calderon, R.L., Juranek, D.D., 1996. Surveillance for waterborne-disease outbreaks—United States, 1993–1994. MMWR CDC Surveill. Summ. 45, 1–33.

Krishnasamy, V., Otte, J., Silbergeld, E., 2015. Antimicrobial use in Chinese swine and broiler poultry production. Antimicrob. Resist. Infect. Control 4, 17.

Kudela, R.M., Lane, J.Q., Cochlan, W.P., 2008. The potential role of anthropogenically derived nitrogen in the growth of harmful algae in California, USA. Harmful Algae 8, 103–110.

LeChevallier, M.W., Norton, W.D., Lee, R.G., 1991. Occurrence of *Giardia* and *Cryptosporidium* spp. in surface water supplies. Appl. Environ. Microbiol. 57, 2610–2616.

Lim, J.Y., Yoon, J.W., Hovde, C.J., 2010. A brief overview of *Escherichia coli* O157:H7 and its plasmid O157. J. Microbiol. Biotechnol. 20, 5–14.

Lin, Z., Liao, W., Yang, Y., Gao, Z., Ma, W., Wang, D., Cao, Y., Li, J., Cai, Z., 2015. CH_4 and N_2O emissions from China's beef feedlots with ad libitum and restricted feeding in fall and spring seasons. Environ. Res. 138, 391–400.

Lockhart, K.M., King, A.M., Harter, T., 2013. Identifying sources of groundwater nitrate contamination in a large alluvial groundwater basin with highly diversified intensive agricultural production. J. Contam. Hydrol. 151, 140–154.

Malik, Y.S., Randall, G.W., Goyal, S.M., 2004. Fate of Salmonella following application of swine manure to tile-drained clay loam soil. J. Water Health 2, 97–101.

Manassaram, D.M., Backer, L.C., Moll, D.M., 2006. A review of nitrates in drinking water: maternal exposure and adverse reproductive and developmental outcomes. Environ. Health Perspect. 114, 320–327.

Mann, M.E., Bradley, R.S., Hughes, M.K., 1999. Northern hemisphere temperatures during the past millennium: inferences, uncertainties, and limitations. Geophys. Res. Lett. 26, 759–762.

Mekonnen, M.M., Hoekstra, A.Y., 2011. National water footprint accounts: the green, blue and grey water footprint of production and consumption. Value of Water Research Report Series No. 50. UNESCO-IHE, Delft, The Netherlands.

Moore, Jr., P.A., Miles, D., Burns, R., Pote, D., Berg, K., Choi, I.H., 2011. Ammonia emission factors from broiler litter in barns, in storage, and after land application. J. Environ. Qual. 40, 1395–1404.

Mosier, D.A., Oberst, R.D., 2000. Cryptosporidiosis: a global challenge. Ann. N. Y. Acad. Sci. 916, 102–111.

NASA, 2016a. Available from: http://data.giss.nasa.gov/gistemp/graphs_v3/.

NASA, 2016b. Available from: http://climate.nasa.gov/scientific-consensus/.

Nelson, N.O., Mikkelsen, R.L., 2005. Balancing the phosphorus budget of a swine farm: a case study. J. Nat. Res. Life Sci. Educ. 34, 90–95.

Nestle, 2015. The Nestlé sustainability review. Available from: http://www.nestle.com/asset-library/documents/reports/csv%20reports/environmental%20sustainability/sustainability_review_english.pdf.

Ni, J.Q., Robarge, W.P., Xiao, C., Heber, A.J., 2012. Volatile organic compounds at swine facilities: a critical review. Chemosphere 89, 769–788.

Nicholson, S.E., Tucker, C.J., Ba, M.B., 1998. Desertification, drought, and surface vegetation: an example from the West African Sahel. Bull. Am. Meteorol. Soc. 79, 815–829.

Novak, J.M., Watts, D.W., Stone, K.C., 2004. Copper and zinc accumulation, profile distribution, and crop removal in coastal plains soils receiving long-term, intensive application of swine manure. Trans. ASAE 47, 1513–1522.

Oldeman, L.E., Hakkeling, R.T.A., Sombroek, W.G., 1991. Global assessment of soil degradation. International Soil Reference and Information Centre and United Nations Environment Programme. Available from: http://www.isric.org/sites/default/files/ExplanNote_1.pdf.

Owen, J.J., Silver, W.L., 2015. Greenhouse gas emissions from dairy manure management: a review of field-based studies. Global Change Biol. 21, 550–565.

Park, K.-H., Thompson, A.G., Marinier, M., Clark, K., Wagner-Riddle, C., 2006. Greenhouse gas emissions from stored liquid swine manure in a cold climate. Atmos. Environ. 40, 618–627.

Pew Research Center, 2015. Global concern about climate change, broad support for limiting emissions: U.S., China less worried; partisan divides in key countries. Available from: http://www.pewglobal.org/2015/11/05/global-concern-about-climate-change-broad-support-for-limiting-emissions/61%.

Putignani, L., Menichella, D., 2010. Global distribution, public health and clinical Impact of the protozoan pathogen *Cryptosporidium*. Interdiscip. Perspect. Infect. Dis. 2010, 753512.

Ramirez, N.E., Ward, L.A., Sreevatsan, S., 2004. A review of the biology and epidemiology of cryptosporidiosis in humans and animals. Microbes Infect. 6, 773–785.

Saeed, A.Y., Mazin, H., Saadi, A.-A., Hussein, S.O., 2013. Detection of *Escherichia coli* O157 in vegetables. IOSR J. Agric. Vet. Sci. 6, 16–18.

Schiffman, S.S., 1998. Livestock odors: implications for human health and well-being. J. Anim. Sci. 76, 1343–1355.

Siefert, R.L., Scudlark, J.R., Potter, A.G., Simonsen, K.A., Savidge, K.B., 2004. Characterization of atmospheric ammonia emissions from a commercial chicken house on the Delmarva Peninsula. Environ. Sci. Technol. 38, 2769–2778.

Smith, D.R., Francesconi, W., Livingston, S.J., Huang, C.H., 2015. Phosphorus losses from monitored fields with conservation practices in the Lake Erie Basin, USA. Ambio 44 (Suppl. 2), S319–S331.

Sobsey, M.D., Khatib, L.A., Hill, V.R., Alocilja, E., Pillai, S., 2001. Pathogens in animal wastes and impacts of animal waste management practices on their survival, transport and fate. National Center for Manure and Animal Waste Management White Paper. Available from: https://www.cals.ncsu.edu/waste_mgt/natlcenter/whitepapersummaries/pathogens.pdf.

Szonyi, B., Bordonaro, R., Wade, S.E., Mohammed, H.O., 2010. Seasonal variation in the prevalence and molecular epidemiology of *Cryptosporidium* infection in dairy cattle in the New York City Watershed. Parasitol. Res. 107, 317–325.

Thaipichitburapa, P., Meksumpun, C., Meksumpun, S., 2010. Province-based self-remediation efficiency of the Tha Chin river basin, Thailand. Water Sci. Technol. 62, 594–602.

Thongdonphum, B., Meksumpun, S., Meksumpun, C., 2011. Nutrient loads and their impacts on chlorophyll a in the Mae Klong River and estuarine ecosystem: an approach for nutrient criteria development. Water Sci. Technol. 64, 178–188.

Trotz-Williams, L.A., Wayne Martin, S., Leslie, K.E., Duffield, T., Nydam, D.V., Peregrine, A.S., 2007. Calf-level risk factors for neonatal diarrhea and shedding of *Cryptosporidium parvum* in Ontario dairy calves. Prev. Vet. Med. 82, 12–28.

UNEP (United Nations Environment Programme), 2012. The Emissions Gap Report 2012. Available from: http://www.unep.org/pdf/2012gapreport.pdf.

United Nations, 1980. Desertification. In: Biswas, M.K., Biswas, A.K. (Eds.), Associated Case Studies Prepared for the UN Conference on Desertification. Pergamon Press, London.

US EPA, 1976. Fecal coliform limits. Available from: http://www3.epa.gov/npdes/pubs/owm505.pdf.

US EPA, 1997. Enteric fermentation—Greenhouse gases. Available from: http://www3.epa.gov/ttnchie1/ap42/ch14/final/c14s04.pdf.

US EPA, 1998. Environmental impacts of animal feeding operations. U.S. Environmental Protection Agency, Office of Water, Standards and Applied Sciences Division. Available from: http://milk.procon.org/sourcefiles/Impacts_Animal_Feeding_Operations.pdf.

US EPA, 2001. Emissions from animal feeding operations. Draft report. EPA Contract No. 68-D6-0011. U.S. Environmental Protection Agency, Office of Air Quality Planning and Standards, Emissions Standards Division, Research Triangle Park, NC.

US EPA, 2003. Inventory of greenhouse gas emissions and sinks: 1990–2001. Available from: yosemite.epa.gov/OAR/globalwarming.nsf/content/ResourceCenterPublications.

US EPA, 2015. Inventory of U.S. greenhouse gas emissions and sinks: 1990–2013. Available from: http://www3.epa.gov/climatechange/Downloads/ghgemissions/US-GHG-Inventory-2015-Main-Text.pdf.

US EPA, 2016. Available from: http://www.epa.gov/your-drinking-water/table-regulated-drinking-water-contaminants.

van Es, H.M., Schindelbeck, R.R., Jokela, W.E., 2004. Effect of manure application timing, crop, and soil type on phosphorus leaching. J. Environ. Qual. 33, 1070–1080.

Wang, Y., Huang, M., Meng, Q., Wang, Y., 2011. Effects of atmospheric hydrogen sulfide concentration on growth and meat quality in broiler chickens. Poult. Sci. 90, 2409–2414.

Webster, J.P., Kover, S.C., Bryson, R.J., Harter, T., Mansell, D.S., Sedlak, D.L., Kolodziej, E.P., 2012. Occurrence of trenbolone acetate metabolites in simulated confined animal feeding operation (CAFO) runoff. Environ. Sci. Technol. 46, 3803–3810.

Wei, R., Ge, F., Huang, S., Chen, M., Wang, R., 2011. Occurrence of veterinary antibiotics in animal wastewater and surface water around farms in Jiangsu Province, China. Chemosphere 82, 1408–1414.

Wheeler, D.C., Nolan, B.T., Flory, A.R., DellaValle, C.T., Ward, M.H., 2015. Modeling groundwater nitrate concentrations in private wells in Iowa. Sci. Total Environ. 536, 481–488.

White, C., McNeillis, P., Mathews, R. Chapagain, A., 2015. Energising the drops: towards a holistic approach to carbon & water footprint assessment. Available from: http://waterfootprint.org/media/downloads/holistic_approach_carbon__water-1.pdf.

Whittaker, G., Barnhart, B.L., Srinivasan, R., Arnold, J.G., 2015. Cost of areal reduction of gulf hypoxia through agricultural practice. Sci. Total Environ. 505, 149–153.

Wing, S., Lowman, A., Keil, A., Marshall, S.W., 2014. Odors from sewage sludge and livestock: associations with self-reported health. Public Health Rep. 129, 505–515.

Xu, Y., Yu, W., Ma, Q., Zhou, H., 2013. Accumulation of copper and zinc in soil and plant within ten-year application of different pig manure rates. Plant Soil Environ. 59, 492–499.

Zahn, J.A., DiSpirito, A.A., Do, Y.S., Brooks, B.E., Cooper, E.E., Hatfield, J.L., 2001. Correlation of human olfactory responses to airborne concentrations of malodorous volatile organic compounds emitted from swine effluent. J. Environ. Qual. 30, 624–634.

Zhang, H., Li, X., Ma, R., Li, X., Zhou, Y., Dong, H., Li, X., Li, Q., Zhang, M., Liu, Z., Wei, B., Cui, M., Wang, H., Gao, J., Yang, H., Hou, P., Miao, Z., Chai, T., 2013. Airborne spread and infection of a novel swine-origin influenza A (H1N1) virus. Virol. J. 10, 204.

Ziemer, C.J., Bonner, J.M., Cole, D., Vinjé, J., Constantini, V., Goyal, S., Gramer, M., Mackie, R., Meng, X.J., Myers, G., Saif, L.J., 2010. Fate and transport of zoonotic, bacterial, viral, and parasitic pathogens during swine manure treatment, storage, and land application. J. Anim. Sci. 88, E84–E94.

Further Reading

Harper, L.A., Flesch, T.K., Wilson, J.D., 2010. Ammonia emissions from broiler production in the San Joaquin Valley. Poult. Sci. 89, 1802–1814.

US EPA, 2004. National emission inventory—ammonia emissions from animal husbandry operations (Draft Report, January 30, 2004). Available from: http://www3.epa.gov/ttnchie1/ap42/ch09/related/nh3inventorydraft_jan2004.pdf.

19

Human Activity and Habitat Loss: Destruction, Fragmentation, and Degradation

Colin G. Scanes

University of Wisconsin–Milwaukee, Milwaukee, WI, United States

19.1 IMPACT OF AGRICULTURE

19.1.1 Overview

The major form of habitat destruction is deforestation either to develop land for agriculture (70%) or to harvest lumber intensively (World Wildlife Fund, 2015). Two examples of habitat destruction via deforestation will be considered here: (1) deforestation in the Amazon Basin and (2) deforestation in Malaysia, Indonesia, and sub-Saharan Africa from replacement of the natural habitat with oil palm (*Elaeis guineensis*) plantations.

19.1.2 Habitat Destruction and Deforestation in the Amazon Basin

There has been economic pressure to convert the forests of the Amazon to pasture and arable land (Nepstad et al., 2008). The pressures that increase land values include the following (Nepstad et al., 2008):

- Growth of sugarcane for ethanol production as an alternative energy source.
- Growth of palm oil for biodiesel—another biorenewable energy source.
- Corn and soybean production; there being a ready market for corn and soybeans particularly for the growing pig and poultry sector in Brazil but also for exports.
- Pasture for beef cattle.

However, between 2004 and 2010, the rate of deforestation declined by 70% due to the implementation of government policies (Macedo et al., 2012).

Deforestation can be related to other human activities including logging and fires (e.g., in slash and burn subsistence farming) together with such indirect effects as drought (a consequence of decreased rainfall following loss of forests) (Nepstad et al., 2008). The World Wildlife Fund (2015) claims that "Extensive cattle ranching is the number one culprit of deforestation in virtually every Amazon country, and it

Copyright © 2018 Elsevier Inc. All rights reserved.

accounts for 80% of current deforestation." In contrast, refereed quantitative research demonstrates that the major increases in soy production in the Amazon between 2001 and 2010 have been predominantly due to soybean farming on previously cleared pasture areas (Macedo et al., 2012).

19.1.3 Loss of Forests to Oil Palm Plantations in Malaysia, Indonesia, and Sub-Saharan Africa

Palm oil makes up 32% of the global supply of fats and oils (cf. 22% soybean oil) (Palm Oil Research, 2017). Palm oil is used in frying foods including in the fast food industry and in food and nonfood manufacturing. The basis for the increase in the production of palm oil is its production efficiency in comparison with other vegetable oil sources, such as rapeseed and soybean with 1 mt of oil being produced from the following land areas (Palm Oil Research, 2017):

- palm oil from 0.26 ha
- rapeseed oil from 1.5 ha
- soybean oil from 2.2 ha

Production of palm oil is growing (Table 19.1). Native forests have been converted to oil palm plantations in Malaysia and Indonesia. The oil palm originated in West Africa (FAO, 2005). According to Palm Oil Research (2017), 14.2 million ha (35.1 million acres) are used globally for the production of palm oil (compared to 258.9 million ha for agriculture). Today, there are many oil palm plantations particularly in Malaysia and Indonesia. A measure of the importance of oil palms in Malaysia is that 71% of agricultural land is used for palm oil production (Palm Oil Research, 2017). Large oil palm farms can be >500 ha.

Given the increase in palm oil production in Indonesia (Table 19.1), it is not surprising that deforestation is occurring at an accelerating rate (Wich et al., 2011). The Union of Concerned Scientists (Union of Concerned Scientists, in press) argue that the deforestation is having adverse effects on endangered species (Fig. 19.1), such as Bornean orangutans, Sumatran orangutans, pygmy elephants, and tigers and contributing to climate change with the release of carbon dioxide. It has been estimated that "49% of the orangutan distribution will be lost if all forests outside of protected areas and logging concessions are lost" (Wich et al., 2012). There is increasing palm oil production in Africa and with it concerns on the impact on great ape populations and other threatened species (Wich et al., 2014). This is particularly important given the substantial overlap (~60%) between the current oil palm concessions and the range of the great apes in Africa (Wich et al., 2014). There are parallel grave concerns from environmental and conservation groups not only about losses of biodiversity including the great apes and the release of greenhouse gases but also for the indigenous peoples displaced

TABLE 19.1   Increase in Palm Kernel Production in Million Metric Tons (FAOStat, 2015) Destroys Habitats for Animals

Palm kennel production (million metric tons)			
Year	World	Malaysia	Indonesia
1993	4.44	2.27	0.82
2003	8.25	3.63	2.48
2013	15.64	4.86	7.53

FIGURE 19.1   **Orangutan.** Habitat of orangutans is being destroyed with the land going to palm oil plantations. *Source: Image from Shutterstock.*

with the development of oil palm plantations. Such groups include Friends of the Earth (2005), Greenpeace (2014), and zoos. The latter is exemplified by the zoo-based educational campaigns successfully developed at the Melbourne Zoo, Australia with 160,000 people signing an associated petition (Pearson et al., 2014).

19.2 IMPACT OF URBANIZATION

The number and proportion of people living in urban areas are increasing rapidly (Table 19.2). Between 1970 and 2000, there was an increase in urban land area of 5.8 million ha globally (Seto et al., 2011). Between 2000 and 2010, 2.8 million ha was urbanized in China alone (World Bank Group, 2015). The terms urbanization and urban areas are not necessarily consistently employed with an urban area frequently based on administrative boundaries (Liu et al., 2014). Urban areas include parks, yards/gardens, medians on highways, and golf-courses. Another method of quantitating urbanization includes "built-up areas" or "place dominated by the built environment" or "impervious surfaces" (Liu et al., 2014) (Fig. 19.2). Tables 19.3 and 19.4 summarize the present global situation for urban areas, built-up areas, and impervious surfaces globally and in major regions/continents. Akin to urbanization is the increase in human populations close to or in protected areas. Table 19.5 summarizes

FIGURE 19.2 **Impervious surfaces in a city destroying habitat and/or degrading habitat and/or fragmenting habitats.** *Source: Image from Shutterstock.*

TABLE 19.2 The Global Urban Population is Dramatically Increasing

	Urban population (billion)	Percentage of world's population
1950	0.7	29.0
1975	1.5	37.2
2000	2.8	46.7
2030 (projected)	4.9	59.9

Data from UN Department of Economic and Social Affairs Population Division), 2005. World urban population. Available from: http://www.un.org/esa/population/publications/WUP2005/2005WUP_FS1.pdf

TABLE 19.3 Contribution of Urban, Built-Up, and Impervious Areas to Global Land Use

Land use	Percent of global land area[a]	Land area, ha (km²)
Urban areas	3	0.3 billion (3 million)
Built-up areas	0.65	75 million (0.7 million)
Impervious surface (roads, buildings, parking lots, etc.)	0.45	60 million (0.6 million)

a Excluding Greenland and Antarctica.
Data from Liu, Z., He, C., Zhou, Y., Wu, J., 2014. How much of the world's land has been urbanized, really? A hierarchical framework for avoiding confusion. Landscape Ecol. 29, 763–771.

TABLE 19.4   Differences in Urban, Built-Up, and Impervious Areas to Land Use in Different Regions of the World

Region	Area (km²)		
	Urban areas	Built-up areas	Impervious surfaces
Africa	0.8	0.2	0.1
Asia	3.9	0.9	0.6
Europe	3.8	0.9	0.7
Latin America + Caribbean	2.3	0.6	0.3
North America	4.7	0.9	0.8
Oceania	0.6	0.2	0.05

Data from Liu, Z., He, C., Zhou, Y., Wu, J., 2014. How much of the world's land has been urbanized, really? A hierarchical framework for avoiding confusion. Landscape Ecol. 29, 763–771.

TABLE 19.5   Increases in the Number of Housing Units Close to Areas in the USA With High or Relatively High Governmental Protection of Ecosystems

	Area (km²)	Housing units within 50 km in 2000 (compared to 1940)	Housing units within 1 km in 2000 (compared to 1940)
Wilderness (highest level of protection)	191,000	20.5 million (4.4 million)	54,000 (9,400)
National parks (managed for ecosystem protection + recreation)	103,000	6.6 million (1.5 million)	85,000
National forests (managed for sustainable use)	869,000	34.8 million (9.0 million)	1.8 million (0.48 million)

Data from Radeloff, V.C., Stewart, S.I., Hawbaker, T.J., Gimmi, U., Pidgeon, A.M., Flather, C.H., Hammer, R.B., Helmers, D.P., 2010. Housing growth in and near United States protected areas limits their conservation value. Proc. Natl. Acad. Sci. USA 107, 940–945.

the increases in housing units and hence, people close to protected areas in the USA.

Urbanization is one of the primary causes of habitat loss and degradation globally (Fig. 19.3). It has severe impacts on the environment with the following (Grimm et al., 2008; Radeloff et al., 2010; Seto et al., 2011):

- Shifts in land use and cover and consequent habitat loss and fragmentation and hence effects on biodiversity.
- Impacts on hydrological systems including destroyed or disrupted wetlands.
- Polluted air and waterways.
- Effects on local and regional climate.
- Increased predation by mesopredators, pets, and pest rodents.

FIGURE 19.3   **Land being cleared for housing development destroying or degrading habitats.** *Source: Image from Shutterstock.*

19.3 IMPACT OF ROADS ON HABITAT LOSS

19.3.1 Overview

In 2010, there were 25.9 million miles (41.7 million km) of roads globally (Laurance et al., 2014). It is estimated that by 2050, there will be an additional 15.5 million miles (25 million km) of roads with over 90% being in developing countries (Laurance et al., 2014) (Fig. 19.3). New roads are associated with the following negative environmental impacts (Laurance et al., 2014):

- habitat loss (e.g., destruction of wetlands)
- habitat fragmentation
- habitat degradation (e.g., polluted land, water, or air and increased access by human hunters and invasive species)

The effects of roads were categorized by Spellerberg (1998) as follows:

1. Impact of the construction process.
2. Short- and long-term effects (mortalities; increases in carrion-consuming animals; disruption through noise; pollution in run-off, such as lead, cadmium, nickel, and zinc; and emissions including carbon monoxide and NO_x) and increased access of invasive species and humans together with the adverse effects on aquatic environments—adjacent and downstream.
3. Effects of structures associated with roads (clearing rights-of-way), bridges, tunnels, and so on.

Road construction can lead to loss of biodiversity locally or regionally with local extinctions of susceptible species particularly in wetlands with (or without) recolonization (Findlay and Bourdages, 2000).

Roads influence bird populations with the greatest effects being vehicle-caused road mortality and indirect effects of traffic noise (Kociolek et al., 2011). In addition, there are direct effects, such as habitat fragmentation, pollution, and poisoning together with indirect effects,

such as roadway lighting, barriers to movement, and edges (Kociolek et al., 2011). The number and regular breeding of grassland birds is decreased close to roads with the effect greater near roads with heavy traffic (Forman et al., 2002). Similarly, there are lower population densities for stone curlews (*Burhinus oedicnemus*) either relatively close to major roads (1000 m) or in settlements (1500 m) (Clarke et al., 2013).

Road substrate and/or traffic also influences the movement of small mammals and reptiles with many species avoiding paved rural two-lane roads but not dirt and secondary paved roads; small mammals [San Diego pocket mouse (*Chaetodipus fallax*), cactus mouse (*Peromyscus eremicus*), Dulzura kangaroo rat (*Dipodomys simulans*), and deer mouse (*Peromyscus maniculatus*)] and reptiles [western fence lizard (*Sceloporus occidentalis*) and orange-throated whiptail (*Aspidoscelis hyperythra*)] all avoiding paved rural two-lane highways (Brehme et al., 2013).

It is widely held that the mortality of turtles on roads is one of the major causes of decline in turtle populations. This is likely to be a fallacy. There was no difference detected in the population density of painted turtles (*Chrysemys picta*) either very close to or some distance from high-traffic roads (Dorland et al., 2014).

19.3.2 Vehicle Wildlife Collisions

Vehicle wildlife collisions have high impacts with the following reported for deer in the USA in 2014 (including elk and moose) according to State Farm (2015):

- 1.2 million collisions (with likely a similar number of deer fatalities)
- $4 billion in vehicle damage.

The most likely months for these collisions are the mating and hunting seasons (October, November, and December) (State Farm, 2015) together with spring fawning (May and June) when yearlings move to new ranges (WI DNR, 2015).

19.4 IMPACT OF THE ENERGY SECTOR

19.4.1 Impact of Oil, Bitumen, and Gas Extraction

19.4.1.1 Mortalities and Toxicity of Petroleum Oil

Marine birds are exposed to oil following spillages. This is seen in studies using mallard ducks as models for water birds. Petroleum oil decreases egg-laying in ducks and also reduces the fertility of the eggs in mallard ducks (Holmes et al., 1978). Unexpectedly, ducks consuming petroleum oil have reduced circulating concentrations of the stress hormone, corticosterone; this being due to increased catabolism (Gorsline and Holmes, 1981). Thus, not only do oil spills kill birds but also are likely to suppress bird populations due to decreased fertility.

19.4.1.2 Impacts of Extraction and Transportation

The development of extraction and transportation for oil, bitumen, and natural gas requires facilities and wells for extraction, pipelines, and railroads for the transportation, and roads to service the system (reviewed by Muhly et al., 2015). The impacts of oil, gas, and bitumen extraction include the following:

- catastrophic failure, such as explosions
- chronic environmental contamination
- habitat loss and fragmentation

19.4.1.3 Catastrophic Failure and Oil, Bitumen, and Natural Gas Extraction

Catastrophic failure of fossil energy extraction and processing systems can lead to deaths of people and wildlife. For example, the explosion of a train carrying oil in the Lac-Mégantic rail disaster in Quebec resulted in 47 people killed and a 1-km blast radius. The fire in the Magurcherra gas field in Bangladesh (1997) lasted for 3 months destroying 60 ha of natural forest with mortalities including deer, monkeys, and birds (Alam et al., 2010). Moreover, explosions or fires can be followed by the release of oil or bitumen into the environment as in the case of the British Petroleum (BP)-leased Deepwater Horizon offshore oil rig (considered later in more detail). Another disaster as the rupture of the oil tanker *Exxon Valdez*. According to the National Oceanic and Atmospheric Administration (NOAA, 1992), the grounded oil tanker *Exxon Valdez* ruptured releasing about 44 million liters of crude oil into Prince William Sound (Alaska) in 1989 with consequent effects on sea otters, birds, and other marine animal life (Harwell et al., 2012). Other major oil spill disasters occurred with the tankers Torrey Canyon and the Amoco Cadiz.

In the spring of 2010, there was a series of catastrophic incidents including an explosion on the BP-leased Deepwater Horizon offshore oil rig in the Gulf of Mexico (Fig. 19.4) (Levy and Gopalakrishnan, 2010). This resulted in the following:

- The release of 0.76 billion liters of crude oil (Goodbody-Gringley et al., 2013).
- The use of 7 million liters of the chemical dispersant Corexit 9500 (Almeda et al., 2014).

This has major effects on deep and coastal marine ecosystems in the Gulf of Mexico together with the 1600 km (1000 miles) of coastline and threats to wetlands (Fig. 19.4). Components of oil spilt in oceans can undergo evaporation, dissolution, dispersion, emulsification, photooxidation, biodegradation, and sedimentation (Rogowska and Namieśnik, 2010).

There were direct and immediate effects with the deaths of dolphins, turtles, and birds and threats to endangered species including the manatee and brown pelicans (Levy and Gopalakrishnan, 2010). In laboratory studies simulating the contamination, weathered oil greatly increased the mortality of coral larvae (*Porites astreoides* and *Montastraea faveolata*) (Goodbody-Gringley et al., 2013). Bioaccumulation of petroleum

FIGURE 19.4 **An explosion on the BP-leased deepwater horizon offshore oil rig in the Gulf of Mexico resulted in the release of 0.76 billion liters of crude oil and consequent degradation of habitats.** (A) The Deepwater Horizon site. (B) Oil-polluted marine animals including turtles normally associate with sargassum. This sea plant moves with currents along with oil slicks. (C) Massive oil spill resulting in oil emulsion on vegetation in Louisiana, USA. (D) A pod of dolphins following NOAA ship *Pisces* in the Gulf of Mexico after the oil spill. *Source: Part A–D: Courtesy NOAA; part B: photo Carolyn Cole/LA Times.*

hydrocarbons has been demonstrated in fiddler crabs (*Uca minax*) (Chase et al., 2013). In view of the potential for oil to influence nesting birds, it is interesting to note that weathered Gulf oil is toxic to mallard duck embryos but less so than nonweathered oil (Finch et al., 2011). Presumably ducks are more likely to come into contact with weathered oil.

The chemical dispersant Corexit 9500A is highly toxic greatly increasing the mortality of microzooplankton (Almeda et al., 2014) and of coral larvae (Goodbody-Gringley et al., 2013). Moreover, Corexit 9500A increases the toxicity of oil on mallard duck embryos (Finch et al., 2012).

19.4.1.4 Chronic Environmental Contamination

Environmental contamination associated with extractive industries poses risks to wildlife and is viewed as potential habitat degradation. Wildlife is particularly adversely affected when fluids are released, such as the following:

- Oil or bitumen directly released into environment.
- Brine associated with oil and natural gas extraction released into groundwater and hence waterways. These brines have elevated concentrations of chloride with

levels above those suitable for human consumption (Hudak and Wachal, 2001).

- Fracking fluids related to hydraulic fracturing (Vengosha et al., 2013).
- Drilling mud.
- Volatile organic compounds.

The magnitude of the issue is exemplified by the following report about one US state. There were almost 13,000 spills or releases of fluids from oil and gas sites in Oklahoma over 10 years with a median 10 barrels of oil released in each spill or release (Fisher and Sublette, 2005). Leakage can come from active or inactive wells; the latter respective of whether unplugged or inadequately plugged wells or monitored or abandoned. Potential risks from the rupture of oil pipelines also include contamination of wetlands and watersheds (Service et al., 2012).

19.4.1.5 Habitat Loss and Fragmentation From the Development of Oil, Gas, and Bitumen

The ecotype boreal caribou (*Rangifer tarandus caribou*) is threatened and is rapidly declining in Alberta. This has been linked to the development of the Fayetteville shale field for natural gas and the associated increases in roads and edge habitats (Moran et al., 2015). Parenthetically, studies show the caribou crossing under pipelines (Muhly et al., 2015).

19.4.2 Coal Mining

There is a lack of literature quantifying the mortalities or displacement of wildlife by coal mining in the USA (Buehler and Percy, 2002). That is not to say there is no impact. It is estimated that "active, reclaimed, or abandoned mountaintop mines cover ~7% of Central Appalachia" with 6.4 km^2 rock displaced in an 11,500 km^2 region (Ross et al., 2016).

There is regulation of coal mining in the USA by the US Department of the Interior's Office of Surface Mining Reclamation and Enforcement under the Surface Mining Control and Reclamation Act 1977. The industry is required to have reclamation programs for mine sites bringing them to the approximate original contours with revegetation. However, this is not occurring with the land 40% flatter with consequent effects on waterways (Ross et al., 2016). Moreover, contamination, for instance with selenium, has been documented (Ross et al., 2016). Whether the timber has been harvested prior to surface mining, reclamation vegetation will initially be grassland such that the forest habitat is lost and grassland habitat is gained. This will affect forest and grassland species. Reclamation of forests obviously takes much more time. There is a real question as to whether the regulations are working as coal companies can avoid reclamation through bankruptcies (New York Times Editorial Board, 2016).

John Grisham's "Gray Mountain" (2014)

The impact of coal mining in Appalachia is the back-story in "Gray Mountain" (2014) by John Grisham.

19.4.3 Wind Turbines and Habitat Loss/Degradation

The number and size of wind farms has been growing with the shift to alternative renewable energy. This laudable goal is not without impacts. There is strong evidence of sizable mortalities of birds. However, the published research reports have large differences in mortalities. One study in 2009 reported 20,000 birds per year killed by wind power and contrasted this with 330,000 killed by nuclear plants and 14 million killed by fossil fuel power plants (Sovacol, 2013). In other studies, estimates of the number of birds killed by wind turbines in the USA in 2012 were an order of magnitude higher ranging from 234,000 to 573,000 birds (Loss et al., 2013;

Smallwood, 2013). These estimates may not be independent or even unbiased. In a detailed comprehensive and independent analysis, it was estimated that bird mortalities were 0.026 per turbine per day and 0.017 per megawatt per day with half the mortalities in nocturnally migrating passerines (Grodsky et al., 2013). Based on there being 49,000 wind turbines in the USA (American Wind Energy Association, 2016), bird fatalities can be extrapolated to a total of 427,000. The impact of turbines will vary with the species of bird. It is thought that raptors (diurnal carnivorous birds) are particularly sensitive to wind turbines (reviewed by de Lucas et al., 2012; Hoover and Morrison, 2005; Madders and Whitfield, 2006). Smallwood (2013) estimated raptor fatalities in the USA at 83,000 in 2012. In Spain, there is high mortality of Griffon vultures in some wind farms (de Lucas et al., 2008) with a mortality of 0.088 per turbine per year (de Lucas et al., 2012). There is also evidence for the reduced reproductive success of white-tailed eagles (*Haliaeetus albicilla*) with territories within 500 m of wind turbines (Dahl et al., 2012).

Estimates of bat fatalities from wind turbines in the USA are 888,000 bats in 2012 (Smallwood, 2013). No differences were observed whether the turbines were lit or unlit at night (Arnett et al., 2008) or with the rotor diameter, but there was an exponential increase with greater turbine tower height (Barclay et al., 2007).

19.4.4 High-Voltage Electrical Transmission Lines and Possible Habitat Degradation

There is little evidence that the presence of high-voltage electrical transmission lines has any effect on the degradation of the environment and may even have positive effects. There were no discernible effects of the electrical and magnetic fields close to these lines on the biology of kestrels (Costantini et al., 2007; Dell'Omo et al., 2009). Parameters measured included growth rate, body condition score, leukocyte counts, and fledging success together with circulating concentrations of melatonin, carotenoids, reactive oxygen metabolites, and antioxidant capacity (Costantini et al., 2007; Dell'Omo et al., 2009). A positive effect of high voltage electrical transmission lines is that Eurasian kestrels (*Falco tinnunculus*) nest in high-voltage transmission tower/pylons.

19.5 IMPACT OF MINING

19.5.1 Overview of Impact of Mining

Mining and the associated processing of ore can lead to habitat loss irrespective of whether it is surface (strip or "mountain top removal") or subsurface (underground and deep underground) mining. Habitat is lost with the construction of access roads and railroads. Moreover, habitat is lost either permanently or temporarily (with reclamation programs) with surface mining (USFWS, 1983) (Section 19.4.2). Associated with mining are the following problems with waterways and lakes:

- shifts in pH (either acidification or alkalinization)
- silting
- increased turbidity reducing photosynthesis
- metals and other toxicants

Mining can degrade or destroy habitats particularly when environmental regulation is lacking or not adequately enforced (Fig. 19.5). Examples of the effects of mining are given next.

19.5.2 Gold Mining

Gold mining can cause considerable damage to the environment. For example, based on satellite data, gold mining was destroying tropical forests in Peru at the rate of 2166 ha year^{-1} before 2008 but increased to 6145 ha year^{-1} between 2008 and 2012 with over a total of 50,000 ha destroyed (Asner et al., 2013). Gold is extracted from high-grade ores using cyanide.

FIGURE 19.5   **Environmental impact of copper mining destroying habitats.** *Source: Image from Shutterstock.*

Processing of low-grade ore can employ alkaline water containing sodium cyanide. After the extraction of the valuable metal, the tailings or leach residues remain together with liquids used in the extraction. Both of these processes result in retention ponds containing tailings, up to 150 ha, with high concentrations of cyanide. These ponds per se can be hazardous to wildlife especially migratory birds (Donato et al., 2007). Moreover, there is the potential of leakage into groundwater or surface waters or massive loss during flooding affecting the entire watershed (reviewed by Eisler and Wiemeyer, 2004). This is problematic for freshwater animal life. Freshwater fish are sensitive to hydrogen cyanide with the median lethal concentration ranging from 57 to 191 ppb but with juvenile fish more sensitive particularly at low temperatures (Smith et al., 1978). There are multiple other potential toxicants from gold mining. These include arsenic, boron, copper, fluoride, mercury, and zinc; these being released during extraction and processing (reviewed by Eisler and Wiemeyer, 2004). There are elevated arsenic concentrations in water, grasses, invertebrates, and so on, near gold mining operations (reviewed by Eisler, 2004a,b). Gold is also recovered using mercury in gold amalgamation (the formation of an alloy of gold and mercury). This can lead to environmental contamination with elevated

concentrations of mercury in sediments, phytoplankton, zooplankton, fish, and human samples from areas close to gold mines in developing countries, for instance, in Latin America (Marrugo-Negrete et al., 2008) and historic gold mines, for example, in the USA (reviewed by Eisler, 2004b). The environmental impact of the pollutant mercury is magnified as it bioaccumulates along food chains and hence is a particular problem for marine mammals and birds feeding in estuaries (Costa et al., 2009).

19.6 INFLUENCES OF CLIMATE CHANGE ON HABITAT LOSS

19.6.1 Overview

Climate change is considered in Section 18.4 focusing on the causes of climate change and the role of livestock and poultry as sources of greenhouse gases. This section will discuss examples of the habitat-degrading effects of climate change. The examples will be on the impacts on polar regions and oceans.

19.6.2 Arctic Ice Cover Reductions

The extent and thickness of polar sea ice cover is declining (Comiso et al., 2016; IPCC, 2016) particularly the lowest extent of the cover in late summer (Table 19.6).

TABLE 19.6   Nadir of Sea Ice Extent (Late Summer)

Year	Nadir of sea ice extent in million km²
1979–88	7.5
1989–98	7.1
1999–2008	6.0
2007	4.3
2012	3.1

Data calculated from Comiso, J.C., Parkinson, C.L., Markus, T. Cavalieri, D.J., Gersten, R., 2016. Current state of sea ice. cover. Cryosphere Science Research Portal. NASA. Available from: http://neptune.gsfc.nasa.gov/csb/index.php?section=234

19.6.3 Impact of Changes in Polar Ice on Polar Bears

Polar bear (*Ursus maritimus*) populations are decreasing along with increasing ice-free days (Hunter et al., 2010).

There has been a decline in sightings of polar bears (Fig. 19.6) in the southern Beaufort Sea region (north of continental Canada) (Bromaghin et al., 2015). Changes in polar bear numbers in the southern Beaufort Sea region are summarized in Fig. 19.7 with the number declining ($P = 0.01$; adjusted $R^2 = 0.560$; slope = -88.4) (calculated from data by Bromaghin et al., 2015).

Polar bears exhibit changes in behavior and foods consumed to combat the lack of sea ice cover and access to the usual prey species, the primary food of polar bears being ringed (*Pusa hispida*) and bearded (*Erignathus barbatus*) seals. In a study of the Hudson Strait–Northern Hudson Bay Narrows, a low-latitude region of the Canadian Arctic, it was found that polar bear incursions to East Bay Island increased sevenfold from 1997–2001 to 2008–2012 (Iverson et al., 2014). During this time, sea ice coverage was declining (Iverson et al., 2014). Polar bears consumed eggs and destroyed nests of northern common eiders (*Somateria mollissima borealis*) and thick-billed murres (*Uria lomvia*) (Iverson et al., 2014). Parenthetically, polar bears eat garbage when it is available (Lunn and Stirling, 1985). Therefore, human incursions into polar bear-inhabited regions have

FIGURE 19.6 **Polar bears in Beaufort Sea, Alaska with the habitat degrading due to climate age.** *Source: Courtesy NOAA; Photo by Kelley Elliott.*

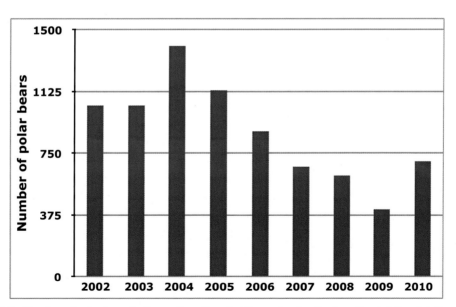

FIGURE 19.7 **Number of polar bears in the southern Beaufort Sea region (north of continental Canada).** *Source: Data from Bromaghin, J.F., McDonald, T.L., Stirling, I., Derocher, A.E., Richardson, E.S., Regehr, E.V., Douglas, D.C., Durner, G.M., Atwood, T., Amstrup, S.C., 2015. Polar bear population dynamics in the southern Beaufort Sea during a period of sea ice decline. Ecol. Appl. 25, 634–651.*

multiple effects: eliminating a pristine environment while augmenting the food availability to polar bears.

19.6.4 Impact of Changes in Polar Ice on Birds

The Arctic Circle lies at 66.3 81 degree N. Since 2005, the Franz Josef Land archipelago (81 degree N and part of the Russian Arctic National Park) has been virtually free of sea ice in the summer.

Little auks (*Alle alle*) lost access to sea ice–associated prey. However, they exhibited plasticity of behavior switching from foraging at distant ice edges to much closer glacier meltwater fronts (Grémillet et al., 2015). There were also no differences in growth rate (Grémillet et al., 2015). In contrast, the survival of adult little auk decreased with the North Atlantic Oscillation index and local summer sea surface temperature in a study of little auk breeding sites close to the West Spitsbergen Current (78 degree N) (Hovinen et al., 2014). This was probably mediated by "change in food quality and/or availability" (Hovinen et al., 2014).

19.6.5 Climate Change and Oceans

19.6.5.1 Ocean Temperatures

Rising ocean temperatures, due to global warming/climate change, per se are likely to have negative and positive effects. For instance, increased ocean temperatures have been reported to increase the growth rate of salmon smolts. However, salmon eggs are susceptible to environmental challenges with a low tolerance to differing temperatures (Elliott and Elliott, 2010). There is decreased oxygen concentration in subsurface oceans due to increased temperatures (reviewed by Doney et al., 2007). Moreover, with increasing ocean temperatures, there are increases in severe weather, for example, hurricanes resulting in mortality of wildlife (Church et al., 2013).

19.6.5.2 Sea Level

Sea level increased due to thermal expansion together with meltwater from polar ice caps (Antarctic and Greenland) and glaciers at a rate of 1.7 mm yr^{-1} between 1900 and 2009 (Church and White, 2011) with a projected further increase of 0.26–0.53 m by 2100 (Church et al., 2013). The rate of sea level increase appears to be accelerating (Haigh et al., 2014). Sea level increases threaten some habitats including the Indo-Pacific mangrove forests (Lovelock et al., 2015) and atolls (Storlazzi et al., 2015).

19.6.5.3 Ocean Carbon Dioxide

A major sink for carbon dioxide from the combustion of fossil fuels is the world's oceans taking up 40% of the anthropogenic-derived carbon dioxide since the beginning of the industrial revolution (Reid et al., 2009). With increasing carbon dioxide dissolved in the oceans (Chapter 18), there are also changes in the pH of these waters. The ocean pH has decreased by 0.1 since prior to the industrial revolution (Royal Society, 2005). The ocean pH is projected to continue to decline by a further 0.3–0.4 by 2100 due to anthropomorphic CO_2 emissions (Royal Society, 2005). Warming of arctic waters may potentially release the powerful greenhouse gas, methane, from hydrates on the ocean floor.

The driving forces for ocean acidification include the following (Doney et al., 2007):

- There is increased concentration of dissolved carbon dioxide (total CO_2 from fossil fuels in the oceans globally = 530 billion tons or gigatons with 1 million tons being added per hour).
- Fossil fuel combustion is associated with sulfur dioxide (0.8 Tmole reactive sulfur per year) and NO_x (2.7 Tmole nitrogen per year) emissions and hence sulfite/sulfate and nitrite/nitrate deposition in oceans particularly in coastal regions.

Ocean pH and Calcium

Changes in the pH in the oceans influence the saturation state for calcium carbonate (Feely et al., 2004). Increased carbon dioxide results in decreased ocean carbonate following this equation:

$$CO_2 + CO_3^= + H_2O \rightarrow 2HCO_3^-$$

FIGURE 19.8 **Coral reef.** *Source: Courtesy of NOAA; Photo by Dr. James P. McVey.*

Increased carbon dioxide partial pressure in oceans is an example of habitat degradation. This has adverse effects on marine animals. Pteropods are particularly vulnerable to ocean pH as their shells are fragile and composed of aragonite, a relatively soluble form of calcium carbonate (Seibel et al., 2012). The decreased pH of the oceans is projected to impact the stability of the external calcium carbonate skeletons in such marine organisms as corals and some plankton (Orr et al., 2005). It has been demonstrated experimentally that increasing the concentrations of carbon dioxide reduces the thickness of aragonite (calcium carbonate, $CaCO_3$) in the shells of mussels (Fitzer et al., 2015). Elevated carbon dioxide partial pressure depresses oxygen consumption in the pteropod *Limacina helicina* (Seibel et al., 2012).

There is great concern about the impact of ocean acidification on coral reefs (Hoegh-Guldberg et al., 2007) (Figs. 19.8 and 19.9). Coral cover decreased by >85% between 1996 and 2002 in four marine reserves in the Tamane Puli Conservation Area, Papua New Guinea (Jones et al., 2004). Calcification rates of *Porites* corals decreased 14.2% in the Great Barrier Reef between 1990 and 2008 (De'ath et al., 2009). This change was unprecedented in the last 400 years (De'ath et al., 2009). The rates of coral growth are declining and with a constant rate of erosion, corals may be close to a "tipping point" (Hoegh-Guldberg et al., 2007). Coral bleaching is increasingly seen as a symptom of adverse

FIGURE 19.9 **Algae growing on destroyed coral in the western Pacific Ocean as the result of climate change.** *Source: Image from Shutterstock.*

health and again it is thought to be anthropogenic (Glynn, 1996).

There are other effects of changes in ocean pH. The ability of clown fish larvae to discriminate by olfaction and avoid water from containing cues from their parents is lost in acidified waters (Munday et al., 2009). Ocean acidification to the level projected for 2100 was simulated by the addition of carbon dioxide (Munday et al., 2009). Acidification impairs settlement of larval fish (Rossi et al., 2015). In studies where carbon dioxide partial pressures are elevated due to natural volcanic vents there is reduced orientation ability of fish larvae due to altered hearing (Rossi et al., 2016). Perhaps surprisingly, lower pH in oceans influences the transmission of sound waves and thereby increases the noise in the ocean (Hester et al., 2008).

19.6.5.4 Adapting to the Effects of Climate Change and Oceans

There is evidence that animals can be resilient or adapt to degraded habitation. For instance, Antarctic sea urchin shells with high-magnesium calcite skeleton appear to be resilient to decreased oceanic pH down decreased carbonate ions (Collard et al., 2015). Climate change-induced changes may also have positive effects with, for instance, evidence that it is responsible for the increase in the population of Chinstrap penguins (Sun and Xie, 2001). Moreover, sea urchins not only survive but also thrive when carbon dioxide partial pressures are elevated close to volcanic vents (Uthicke et al., 2016). Also, there can be shifts in food supply.

19.7 OTHER HUMAN IMPACTS OF HABITATS

There are other effects of human activity from habitat loss, degradation, or fragmentation. Examples of these include the following:

- Dams for hydroelectric power and flood control negatively influence breeding grounds for fish.
- Pollutants degrading habitats. For instance, polychlorinated biphenyls bioaccumulate in marine mammals, such as harbor seals (*Phoca vitulina*), which are at the top of the food chain (Park et al., 2009).
- The ecologically fragile Florida Everglades (Fig. 19.10) is vulnerable to the effects of climate change (Catano et al., 2015) and the invasive species, the Burmese python (discussed in Section 17.4.1).
- The development of a man-made lake and opening of a habitat for human recreation was accompanied by declines in populations of North American wood turtles (*Clemmys insculpta*) (Garber and Burger, 1995).
- Human activity can influence the number of animals by causing disturbance. For instance, the number of sharks was reported to be lower in areas of the Great Barrier Reef that were fished compared with "no entry" or "no take" areas (McCook et al., 2010). It is assumed that this reflects shifts in the number of fish eaten by the sharks (McCook et al., 2010).

19.8 CONCLUSIONS

Clearly, human activity is leading to habitat loss, habitat degradation, and habitat fragmentation. It is questioned whether a tipping point has already been reached or whether we are close to an anthropogenic major extinction event. It is also recognized that even when policymakers base decisions on "good science," concerns about economic development and poverty alleviation may outweigh the importance of healthy ecosystems. The effects of damage to habitats need to be part of the equation. The issue is not about no regulation or no growth but smart regulation with sustainability or smart growth addressing poverty, low quality of life, poor nutrition, and diseases.

FIGURE 19.10 **The Everglades in Florida (USA) are ecologically fragile with habitats is danger of destruction.** (A) Wild turkeys in the Everglades National Park. (B) Alligator in the Everglades National Park. (C) Egret in the Everglades National Park. (D) Seagrass meadows on the western edge of the Everglades providing nutrition to many animal species. *Source: Part A and C: Courtesy US National Parks Service; part B: Courtesy US National Parks Service; photo by Karen Battle-Sanborn; part D: Courtesy US National Parks Service; photo by Justin Campbell.*

References

Alam, J.B., Ahmed, A.A.M., Munna, G.M., Ahmed, A.A.M., 2010. Environmental impact assessment of oil and gas sector: a case study of Magurchara gas field. J. Soil Sci. Environ. Manage. 1, 86–91.

Almeda, R., Hyatt, C., Buskey, E.J., 2014. Toxicity of dispersant Corexit 9500A and crude oil to marine microzooplankton. Ecotoxicol. Environ. Saf. 106, 76–85.

American Wind Energy Association, 2016. U.S. wind energy State facts. Available from: http://www.awea.org/resources/statefactsheets.aspx?itemnumber=890.

Arnett, E.B., Brown, W.K., Erickson, W.P., Fielder, J.K., Hamilton, B.L., Henry, T.H., Jain, A., Johnson, G.D., Kerns, J., Koford, R.R., Nicholson, C.P., O'Connell, T.J., Piorkowski, M.D., Tankersley, Jr., R.D., 2008. Patterns of bat fatalities at wind energy facilities in North America. J. Wildlife Manage. 72, 61–78.

Asner, G.P., Llactayo, W., Tupayachi, R., Ráez Luna, E., 2013. Elevated rates of gold mining in the Amazon revealed through high-resolution monitoring. Proc. Natl. Acad. Sci. USA 110, 18454–18459.

Barclay, R.M.R., Baerwald, E.F., Gruver, J.C., 2007. Variation in bat and bird fatalities at wind energy facilities: assessing the effects of rotor size and tower height. J. Zool. 85, 381–387.

Brehme, C.S., Tracey, J.A., McClenaghan, L.R., Fisher, R.N., 2013. Permeability of roads to movement of scrubland lizards and small mammals. Conserv. Biol. 27, 710–720.

Bromaghin, J.F., McDonald, T.L., Stirling, I., Derocher, A.E., Richardson, E.S., Regehr, E.V., Douglas, D.C., Durner, G.M., Atwood, T., Amstrup, S.C., 2015. Polar bear population dynamics in the southern Beaufort Sea during a period of sea ice decline. Ecol. Appl. 25, 634–651.

Buehler, D.A., Percy, K., 2002. Coal mining and wildlife in the eastern United States: a literature review. University of Tennessee and Appalachian Wildlife. Available from: http://www.appalachianwildlife.com/Coal%20Mining%20and%20Wildlife%20in%20the%20Eastern%20United%20States-final%20draft.pdf.

Catano, C.P., Romañach, S.S., Beerens, J.M., Pearlstine, L.G., Brandt, L.A., Hart, K.M., Mazzotti, F.J., Trexler, J.C., 2015. Using scenario planning to evaluate the impacts of climate change on wildlife populations and communities in the Florida Everglades. Environ. Manage. 55, 807–723.

Chase, D.A., Edwards, D.S., Qin, G., Wages, M.R., Willming, M.M., Anderson, T.A., Maul, J.D., 2013. Bioaccumulation of petroleum hydrocarbons in fiddler crabs (*Uca minax*) exposed to weathered MC-252 crude oil alone and in mixture with an oil dispersant. Sci. Total Environ. 444, 121–127.

Church, J.A., Clark, P.U., Cazenave, A., Gregory, J., Jevrejeva, S., Levermann, A., Merrifield, M.A., Milne, G.A., Nerem, R.S., Nunn, P.D., Payne, A.J., Pfeffer, W.T., Stammer, D., Unnikrishnan, A.S., 2013. Sea level change. In: Climate Change 2013: The Physical Science Basis. Contribution of Working Group I to the Fifth Assessment Report of the Intergovernmental Panel on Climate Change Chapter 13. Cambridge University Press. pp. 1137–1216. Available from: http://www.ipcc.ch/report/ar5/wg1/.

Church, J.A., White, N.J., 2011. Sea-level rise from the late 19th to the early 21st century. Surv. Geophys. 32, 585–602.

Clarke, R.T., Liley, D., Sharp, J.M., Green, R.E., 2013. Building development and roads: implications for the distribution of stone curlews across the Brecks. PLoS One 8, e72984.

Comiso, J.C., Parkinson, C.L., Markus, T. Cavalieri, D.J., Gersten, R., 2016. Current state of sea ice. cover. Cryosphere Science Research Portal. NASA. Available from: http://neptune.gsfc.nasa.gov/csb/index.php?section=234.

Collard, M., De Ridder, C., David, B., Dehairs, F., Dubois, P., 2015. Could the acid-base status of Antarctic sea urchins indicate a better-than-expected resilience to near-future ocean acidification? Glob. Chang Biol. 21, 605–617.

Costa, M.F., Barbosa, S.C., Barletta, M., Dantas, D.V., Kehrig, H.A., Seixas, T.G., Malm, O., 2009. Seasonal differences in mercury accumulation in *Trichiurus lepturus* (Cutlassfish) in relation to length and weight in a Northeast Brazilian estuary. Environ. Sci. Pollut. Res. 16, 423–430.

Costantini, D., Casagrande, S., Dell'Omo, G., 2007. MF magnitude does not affect body condition, pro-oxidants and anti-oxidants in Eurasian kestrel (*Falco tinnunculus*) nestlings. Environ. Res. 104, 361–366.

Dahl, E.L., Bevanger, K., Nygård, T., Røskafta, E., Stokke, B.G., 2012. Reduced breeding success in White-tailed Eagles at Smøla windfarm, western Norway, is caused by mortality and displacement. Biol. Conserv. 145, 79–85.

De'ath, G., Lough, J.M., Fabricius, K.E., 2009. Declining coral calcification on the Great Barrier Reef. Science 323, 116–119.

Dell'Omo, G., Costantini, D., Lucini, V., Antonucci, G., Nonno, R., Polichetti, A., 2009. Magnetic fields produced by power lines do not affect growth, serum melatonin, leukocytes and fledging success in wild kestrels. Comp. Biochem. Physiol. 150, 372–376.

de Lucas, M., Ferrer, M., Janss, G.F.E., 2012. Using wind tunnels to predict bird mortality in wind farms: the case of Griffon Vultures. PLoS One 7, e48092.

de Lucas, M., Janss, G.F.E., Whitfield, P., Ferrer, M., 2008. Collision fatality of raptors in wind farms does not depend on raptor abundance. J. Appl. Ecol. 45, 1695–1703.

Donato, D.B., Nichols, O., Possingham, H., Moore, M., Ricci, P.F., Noller, B.N., 2007. A critical review of the effects of gold cyanide-bearing tailings solutions on wildlife. Environ. Int. 33, 974–984.

Doney, S.C., Mahowald, N., Lima, I., Mackenzie, F.T., Lamarque, J.-F., Rausch, P.J., 2007. Impact of anthropogenic atmospheric nitrogen and sulfur deposition on ocean acidification and the inorganic carbon system. Proc. Natl. Acad. Sci. USA 104, 14,580–114,585.

Dorland, A., Rytwinski, T., Fahrig, L., 2014. Do roads reduce painted turtle (*Chrysemys picta*) populations? PLoS One 9, e98414.

Eisler, R., 2004a. Arsenic hazards to humans, plants, and animals from gold mining. Rev. Environ. Contam. Toxicol. 180, 133–165.

Eisler, R., 2004b. Mercury hazards from gold mining to humans, plants, and animals. Rev. Environ. Contam. Toxicol. 181, 139–198.

Eisler, R., Wiemeyer, S.N., 2004. Cyanide hazards to plants and animals from gold mining and related water issues. Rev. Environ. Contam. Toxicol. 183, 21–54.

Elliott, J.M., Elliott, J.A., 2010. Temperature requirements of Atlantic salmon *Salmo salar*, brown trout *Salmo trutta* and Arctic charr *Salvelinus alpinus*: predicting the effects of climate change. J. Fish. Biol. 77, 1793–1817.

FAO, 2005. Oil palm. Available from: http://www.fao.org/docrep/005/y4355e/y4355e03.htm.

FAOStat, 2015. Production quantities of Palm kernels by country. Available from http://faostat3.fao.org/browse/Q/QC/E.

Feely, R.A., Sabine, C.L., Lee, K., Berelson, W., Kleypas, J., Fabry, V.J., Millero, F.J., 2004. Impact of anthropogenic

CO_2 on the $CaCO_3$ system in the oceans. Science 305, 362–366.

Finch, B.E., Wooten, K.J., Faust, D.R., Smith, P.N., 2012. Embryotoxicity of mixtures of weathered crude oil collected from the Gulf of Mexico and Corexit 9500 in mallard ducks (*Anas platyrhynchos*). Sci. Total Environ. 426, 155–159.

Finch, B.E., Wooten, K.J., Smith, P.N., 2011. Embryotoxicity of weathered crude oil from the Gulf of Mexico in mallard ducks (*Anas platyrhynchos*). Environ. Toxicol. Chem. 30, 1885–1891.

Findlay, C.S., Bourdages, J., 2000. Response time of wetland biodiversity to road construction on adjacent lands. Conserv. Biol. 14, 86–94.

Fisher, J.B., Sublette, K.L., 2005. Environmental releases from exploration and production operations in Oklahoma: type, volume, causes, and prevention. Environ. Geosci. 12, 89–99.

Fitzer, S.C., Vittert, L., Bowman, A., Kamenos, N.A., Phoenix, V.R., Cusack, M., 2015. Ocean acidification and temperature increase impact mussel shell shape and thickness: problematic for protection? Ecol. Evol. 5, 4875–4884.

Forman, R.T., Reineking, B., Hersperger, A.M., 2002. Road traffic and nearby grassland bird patterns in a suburbanizing landscape. Environ. Manage. 29, 782–800.

Friends of the Earth, 2005. The oil for ape scandal: how palm oil is threatening orangutan survival. Available from: https://www.foe.co.uk/sites/default/files/downloads/oil_for_ape_summary.pdf.

Garber, S.D., Burger, J., 1995. A 20-yr study documenting the relationship between turtle decline and human recreation. Ecol. Appl. 5, 1151–1162.

Glynn, P.W., 1996. Coral reef bleaching: facts, hypotheses and implications. Glob. Change Biol. 2, 495–509.

Goodbody-Gringley, G., Wetzel, D.L., Gillon, D., Pulster, E., Miller, A., Ritchie, K.B., 2013. Toxicity of Deepwater Horizon source oil and the chemical dispersant, Corexit 9500, to coral larvae. PLoS One 8, e45574.

Gorsline, J., Holmes, W.N., 1981. Effects of petroleum on adrenocortical activity and on hepatic naphthalene-metabolizing activity in mallard ducks. Arch. Environ. Contam. Toxicol. 10, 765–777.

Greenpeace, 2014. Let's stop the haze. Let's work together to restore and protect Indonesia's forests and peatlands. Available from: http://www.greenpeace.org/international/en/campaigns/forests/asia-pacific/.

Grémillet, D., Fort, J., Amélineau, F., Zakharova, E., Le Bot, T., Sala, E., Gavrilo, M., 2015. Arctic warming: nonlinear impacts of sea-ice and glacier melt on seabird foraging. Glob. Change Biol. 21, 1116–1123.

Grimm, N.B., Faeth, S.H., Golubiewski, N.E., Redman, C.L., Wu, J.G., Bai, X., Briggs, J.M., 2008. Global change and the ecology of cities. Science 319, 756–760.

Grodsky, S.M., Jennelle, C.S., Drake, D., 2013. Bird mortality at a wind-energy facility near a wetland of international importance. Condor 115, 700–711.

Haigh, I.D., Wahl, T., Rohling, E.J., Price, R.M., Pattiaratchi, C.B., Calafat, F.M., Dangendorf, S., 2014. Timescales for detecting a significant acceleration in sea level rise. Nat. Commun. 5, 3635.

Harwell, M.A., Gentile, J.H., Parker, K.R., 2012. Quantifying population-level risks using an individual-based model: sea otters, Harlequin Ducks, and the *Exxon Valdez* oil spill. Integr. Environ. Assess. Manage. 8, 503–522.

Hester, K.C., Peltzer, E.T., Kirkwood, W.J., Brewer, P.G., 2008. Unanticipated consequences of ocean acidification: a noisier ocean at lower pH. Geophys. Res. Lett. 35, L19601.

Hoegh-Guldberg, O., Mumby, P.J., Hooten, A.J., Steneck, R.S., Greenfield, P., Gomez, E., Harvell, C.D., Sale, P.F., Edwards, A.J., Caldeira, K., Knowlton, N., Eakin, C.M., Iglesias-Prieto, R., Muthiga, N., Bradbury, R.H., Dubi, A., Hatziolos, M.E., 2007. Coral reefs under rapid climate change and ocean acidification. Science 318, 1737–1742.

Holmes, W.N., Cavanaugh, K.P., Cronshaw, J., 1978. The effects of ingested petroleum on oviposition and some aspects of reproduction in experimental colonies of mallard ducks (*Anas platyrhynchos*). J. Reprod. Fertil. 54, 335–347.

Hoover, S.L., Morrison, M.L., 2005. Behaviour of red-tailed hawks in a wind turbine development. J. Wildlife Manage. 69, 150–159.

Hovinen, J.E., Welcker, J., Descamps, S., Strøm, H., Jerstad, K., Berge, J., Steen, H., 2014. Climate warming decreases the survival of the little auk (*Alle alle*), a high Arctic avian predator. Ecol. Evol. 4, 3127–3138.

Hudak, P.F., Wachal, D.J., 2001. Effects of brine injection wells, dry holes, and plugged oil/gas wells on chloride, bromide, and barium concentrations in the Gulf Coast Aquifer, southeast Texas, USA. Environ. Int. 26, 497–503.

Hunter, C.M., Caswell, H., Runge, M.C., Regehr, E.V., Amstrup, S.C., Stirling, I., 2010. Climate change threatens polar bear populations: a stochastic demographic analysis. Ecology 91, 2883–2897.

Intergovernmental Panel on Climate Change, 2016. Available from: http://www.ipcc.ch.

Iverson, S.A., Gilchrist, H.G., Smith, P.A., Gaston, A.J., Forbes, M.R., 2014. Longer ice-free seasons increase the risk of nest depredation by polar bears for colonial breeding birds in the Canadian Arctic. Proc. R. Soc. 281, 20133128.

Jones, G.P., McCormick, M.I., Srinivasan, M., Eagle, J.V., 2004. Coral CO_2 decline threatens fish biodiversity in marine reserves. Proc. Natl. Acad. Sci. USA 101, 8251–8253.

Kociolek, A.V., Clevenger, A.P., St Clair, C.C., Proppe, D.S., 2011. Effects of road networks on bird populations. Conserv. Biol. 25, 241–249.

Laurance, W.F., Clements, G.R., Sloan, S., O'Connell, C.S., Mueller, N.D., Goosem, M., Venter, O., Edwards, D.P., Phalan, B., Balmford, A., Van Der Ree, R., Arrea, I.B., 2014. A global strategy for road building. Nature 513, 229–232.

Levy, J.K., Gopalakrishnan, C., 2010. Promoting ecological sustainability and community resilience in the US Gulf Coast after the 2010 deepwater horizon oil spill. J. Nat. Resour. Policy Res. 2, 297–315.

Liu, Z., He, C., Zhou, Y., Wu, J., 2014. How much of the world's land has been urbanized, really? A hierarchical framework for avoiding confusion. Landscape Ecol. 29, 763–771.

Loss, S.R., Will, T., Mara, P.P., 2013. Estimates of bird collisions at wind facilities in the contiguous United States. Biol. Conserv. 168, 201–209.

Lovelock, C.E., Cahoon, D.R., Friess, D.A., Guntenspergen, G.R., Krauss, K.W., Reef, R., Rogers, K., Saunders, M.L., Sidik, F., Swales, A., Saintilan, N., Thuyen le, X., Triet, T., 2015. The vulnerability of Indo-Pacific mangrove forests to sea-level rise. Nature 526, 559–563.

Lunn, N.J., Stirling, I., 1985. The significance of supplemental food to polar bears during the ice-free period of Hudson Bay. Can. J. Zool. 83, 2291–2297.

Macedo, M.N., DeFries, R.S., Morton, D.C., Stickler, C.M., Galford, G.L., Yosio, E., Shimabukuro, Y.E., 2012. Decoupling of deforestation and soy production in the southern Amazon during the late 2000s. Proc. Natl. Acad. Sci. USA 109, 1341–1346.

Madders, M., Whitfield, D.P., 2006. Upland raptors and the assessment of wind farm impacts. IBIS 148, 43–56.

Marrugo-Negrete, J., Benitez, L.N., Olivero-Verbel, J., 2008. Distribution of mercury in several environmental compartments in an aquatic ecosystem impacted by gold mining in northern Colombia. Arch. Environ. Contam. Toxicol. 55, 305–316.

McCook, L.J., Ayling, T., Cappo, M., Choat, J.H., Evans, R.D., De Freitas, D.M., Heupel, M., Hughes, T.P., Jones, G.P., Mapstone, B., Marsh, H., Mills, M., Molloy, F.J., Pitcher, C.R., Pressey, R.L., Russ, G.R., Sutton, S., Sweatman, H., Tobin, R., Wachenfeld, D.R., Williamson, D.H., 2010. Adaptive management of the Great Barrier Reef: a globally significant demonstration of the benefits of networks of marine reserves. Proc. Natl. Acad. Sci. USA 107, 18278–18285.

Moran, M.D., Cox, A.B., Wells, R.L., Benichou, C.C., McClung, M.R., 2015. Habitat loss and modification due to gas development in the Fayetteville shale. Environ. Manage. 55, 1276–1284.

Muhly, T., Serrouya, R., Neilson, E., Li, H., Boutin, S., 2015. Influence of in-situ oil sands development on caribou (*Rangifer tarandus*) movement. PLoS One 10, e0136933.

Munday, P.L., Dixson, D.L., Donelson, J.M., Jones, G.P., Pratchett, M.S., Devitsina, G.V., Døving, K.B., 2009. Ocean acidification impairs olfactory discrimination and homing ability of a marine fish. Proc. Natl. Acad. Sci. USA 106, 1848–1852.

Nepstad, D.C., Stickler, C.M., Soares-Filho, B., Frank Merry, F., 2008. Interactions among Amazon land use, forests and climate: prospects for a near-term forest tipping point. Philos. Trans. R. Soc. 363, 1737–1746.

New York Times Editorial Board, 2016. Will Big Coal pay to clean up its messes? Available from: http://www.nytimes.com/2016/06/10/opinion/will-big-coal-pay-to-clean-up-its-messes.html?_r=0.

NOAA (National Oceanic and Atmospheric Administration) Office of Response and Restoration, 1992. *Exxon Valdez* oil spill. http://response.restoration.noaa.gov/oil-and-chemical-spills/significant-incidents/exxon-valdez-oil-spill.

Orr, J.C., Fabry, V.J., Aumont, O., Bopp, L., Doney, S.C., Feely, R.A., Gnanadesikan, A., Gruber, N., Ishida, A., Joos, F., Key, R.M., Lindsay, K., Maier-Reimer, E., Matear, R., Monfray, P., Mouchet, A., Najjar, R.G., Plattner, G.K., Rodgers, K.B., Sabine, C.L., Sarmiento, J.L., Schlitzer, R., Slater, R.D., Totterdell, I.J., Weirig, M.F., Yamanaka, Y., Yool, A., 2005. Anthropogenic ocean acidification over the twenty-first century and its impact on calcifying organisms. Nature 437, 681–686.

Palm Oil Research, 2017. Untangling the great palm oil debate. Available from: http://www.palmoilresearch.org/statistics.html.

Park, J.S., Kalantzi, O.I., Kopec, D., Petreas, M., 2009. Polychlorinated biphenyls (PCBs) and their hydroxylated metabolites (OH-PCBs) in livers of harbor seals (*Phoca vitulina*) from San Francisco Bay, California and Gulf of Maine. Marine Environ. Res. 67, 129–135.

Pearson, E.L., Lowry, R., Dorrian, J., Litchfield, C.A., 2014. Evaluating the conservation impact of an innovative zoo-based educational campaign: 'Don't Palm Us Off' for orangutan conservation. Zoo Biol. 33, 184–196.

Radeloff, V.C., Stewart, S.I., Hawbaker, T.J., Gimmi, U., Pidgeon, A.M., Flather, C.H., Hammer, R.B., Helmers, D.P., 2010. Housing growth in and near United States protected areas limits their conservation value. Proc. Natl. Acad. Sci. USA 107, 940–945.

Reid, P.C., Fischer, A.C., Lewis-Brown, E., Meredith, M.P., Sparrow, M., Andersson, A.J., Antia, A., Bates, N.R., Bathmann, U., Beaugrand, G., Brix, H., Dye, S., Edwards, M., Furevik, T., Gangstø, R., Hátún, H., Hopcroft, R.R., Kendall, M., Kasten, S., Keeling, R., Le Quéré, C., Mackenzie, F.T., Malin, G., Mauritzen, C., Olafsson, J., Paull, C., Rignot, E., Shimada, K., Vogt, M., Wallace, C., Wang, Z., Washington, R., 2009. Chapter 1. Impacts of the oceans on climate change. Adv. Marine Biol. 56, 1–150.

Rogowska, J., Namieśnik, J., 2010. Environmental implications of oil spills from shipping accidents. Rev. Environ. Contam. Toxicol. 206, 95–114.

Ross, M.R.V., McGlynn, B.L., Bernhardt, E.S., 2016. Deep impact: effects of mountaintop mining on surface topography, bedrock structure, and downstream waters. Environ. Sci. Technol. 50, 2064–2074.

Rossi, T., Nagelkerken, I., Pistevos, J.C., Connell, S.D., 2016. Lost at sea: ocean acidification undermines larval fish orientation via altered hearing and marine soundscape modification. Biol. Lett. 12, pii: 20150937.

Rossi, T., Nagelkerken, I., Simpson, S.D., Pistevos, J.C., Watson, S.A., Merillet, L., Fraser, P., Munday, P.L., Connell, S.D., 2015. Ocean acidification boosts larval fish development but reduces the window of opportunity for successful settlement. Proc. R. Soc. 282, 1954.

Royal Society, 2005. Ocean Acidification Due to Increasing Atmospheric Carbon Dioxide. The Royal Society, London.

Seibel, B.A., Maas, A.E., Dierssen, H.M., 2012. Energetic plasticity underlies a variable response to ocean acidification in the pteropod, *Limacina helicina antarctica*. PLoS One 7, e30464.

Service, C.N., Nelson, T.A., Paquet, P.C., McInnes, W.S.S., Darimont, C.T., 2012. Pipelines and parks: evaluating external risks to protected areas from the proposed Northern Gateway Oil Transport Project. Nat. Areas J. 32, 367–376.

Seto, K.C., Fragkias, M., Güneralp, B., Reilly, M.K., 2011. A meta-analysis of global urban land expansion. PLoS One 6, e23777.

Smallwood, K.S., 2013. Comparing bird and bat fatality-rate estimates among North American wind-energy projects. Wind Energy Wildlife Conserv. 37, 19–33.

Smith, Jr., L.L., Broderius, S.J., Oseid, D.M., Kimball, G.L., Koenst, W.M., 1978. Acute toxicity of hydrogen cyanide to freshwater fishes. Arch. Environ. Contam. Toxicol. 7, 325–337.

Sovacol, B.K., 2013. The avian benefits of wind energy: a 2009 update. Renewable Energy 49, 19–24.

Spellerberg, I.F., 1998. Ecological effects of roads and traffic: a literature review. Global Ecol. Biogeog. Lett. 7, 317–333.

State Farm, 2015. Available from: https://www.statefarm.com/about-us/newsroom/2015/09/14/deer-collision-datafigures for 2014.

Storlazzi, C.D., Elias, E.P., Berkowitz, P., 2015. Many atolls may be uninhabitable within decades due to climate change. Sci. Rep. 5, 14546.

Sun, L., Xie, Z., 2001. Relics: penguin population programs. Sci. Prog. 84, 31–44.

Union of Concerned Scientists, in press. Palm oil. Available from: http://www.ucsusa.org/global_warming/solutions/stop-deforestation/palm-oil-and-forests.html#.Vp6ptporLGg.

U.S. Fish and Wildlife Service, 1983. Practices for protecting and enhancing fish and wildlife on coal surface-mined land in central and southern Appalachia. FWS/OBS-83/08. 208 pp.

Uthicke, S., Ebert, T., Liddy, M., Johansson, C., Fabricius, K.E., Lamare, M., 2016. Echinometra sea urchins acclimatised to elevated pCO_2 at volcanic vents outperform those under present-day pCO_2 conditions. Glob. Change Biol. 22 (7), 2451–2461.

Vengosha, A., Warnera, N., Jacksona, R., Darraha, T., 2013. The effects of shale gas exploration and hydraulic fracturing on the quality of water resources in the United States. Procedia Earth Planet. Sci. 7, 863–866.

Wich, S.A., Garcia-Ulloa, J., Kühl, H.S., Humle, T., Lee, J.S., Koh, L.P., 2014. Will oil palm's homecoming spell doom for Africa's great apes? Curr. Biol. 24, 1659–1663.

Wich, S.A., Gaveau, D., Abram, N., Ancrenaz, M., Baccini, A., Brend, S., Curran, L., Delgado, R.A., Erman, A., Fredriksson, G.M., Goossens, B., Husson, S.J., Lackman, I., Marshall, A.J., Naomi, A., Molidena, E., Nardiyono, Nurcahyo, A., Odom, K., Panda, A., Purnomo, Rafiastanto, A., Ratnasari, D., Santana, A.H., Sapari, I., van Schaik, C.P., Sihite, J., Spehar, S., Santoso, E., Suyoko, A., Tiju, A., Usher, G., Atmoko, S.S., Willems, E.P., Meijaard, E., 2012. Understanding the impacts of land-use policies on a threatened species: is there a future for the Bornean orangutan? PLoS One 7, e49142.

Wich, S., Riswan, Jenson, J., Refisch, J., Nellemann, C., 2011. Orangutans and the economics of sustainable forest management in Sumatra (UNEP/GRASP/PanEco/YEL/ICRAF/GRID-Arendal). Available from: http://www.grida.no/publications/organgutans-sumatra/.

State of Wisconsin Department of Natural Resources, 2015. Vehicle collisions with deer and other wildlife. Available from: http://dnr.wi.gov/topic/wildlifehabitat/cardeer.html.

World Bank Group, 2015. East Asia's changing urban landscape. Measuring a decade of spacial growth. Available from: http://www.worldbank.org/content/dam/Worldbank/Publications/Urban%20Development/EAP_Urban_Expansion_full_report_web.pdf.

World Wildlife Fund (WWF), 2015. Unsustainable cattle ranching. Available from: http://wwf.panda.org/what_we_do/where_we_work/amazon/amazon_threats/unsustainable_cattle_ranching/index.cfm.

Further Reading

Greenpeace, 2016. Coal mining destroying wildlife habitat. Available from: http://www.downtoearth.org.in/news/coal-mining-destroying-wildlife-habitat-greenpeace-38843.

National Wildlife Federation, in press. Habitat loss. Available from: https://www.nwf.org/Wildlife/Threats-to-Wildlife/Habitat-Loss.aspx.

Stein, B.A., Adams, J.S., Kutner, L.S., 2000. Precious Heritage: The Status of Biodiversity in the United States. Oxford University Press, New York.

World Wildlife Federation, in press. Impact of habitat loss on species. Available from: http://wwf.panda.org/about_our_earth/species/problems/habitat_loss_degradation/.

ANIMAL WASTE

C. Mike Williams
NC State University, Raleigh, NC, United States

Humans have consumed animal-sourced food products since the beginning of mankind. These products provide high nutrient density and nutritional quality food options for consumers and currently represent approximately 1/6 and 1/3 of the food energy and food protein, respectively, for humans on a global basis (Bradford, 1999; CAST, 2006). The estimated current annual global production of beef, pork, and poultry meat is approximately 250 metric tons (mt) (USDA, 2015). The societal impacts of producing this amount of food animal products is profound considering that global meat consumption is projected to grow by an annual average of 1.4%, which will result in an additional annual global consumption of approximately 50 million mt by 2024 (OECD/FAO, 2015). These statistics do not include the global egg industry, which represents approximately 5 billion egg-laying hens (International Egg Commission, 2016). In the United States alone approximately 9 billion broiler chickens, 33 million cattle, 240 million turkeys, 2 million sheep and lambs, and 110 million hogs are processed annually (North American Meat Institute, 2016); the United States dairy cattle inventory represents an additional estimated 9 million heads of animals (USDA-ERS, 2016). Using poultry and pork as examples, the United States percent of global production is only approximately 26 and 10%, respectively. The overarching topic of food animals and human society is critically important from environmental, human health, and food security perspectives—it is essential to consider the issues, challenges, and opportunities associated with this topic from a global, not regional, perspective.

The beneficial nutritional contributions of food animals are well recognized and documented in the scientific literature; however coproducts (traditionally referred to as animal waste) associated with the production of these animals can have beneficial, as well as detrimental impacts on society. For a global perspective consider that animal waste manure dry matter (mass after water content is removed) from food animals housed only in the United States is estimated to annually exceed 335 mt (USDA-ARS, 2005). In the United States and elsewhere the management of animal waste in the 21st century continues to predominately follow a model of land application of these coproducts proximate to where the animals are produced—mostly for fertilizer value for targeted crop(s) production. Approved, permitted, and recommended best management practices for land application of animal waste has historically been beneficial and, arguably, is not currently presenting significant environmental or human health threats in many areas utilizing these practices. However, these traditional practices for managing animal waste are unlikely to remain sustainable for many reasons. With the projected increase in global population of humans over the next few decades there will be less farmland available. In addition the future competition between crop and animal agriculture, as well as nonagricultural entities for water resources will be intense (many animal waste management practices currently utilize large volumes of water). Other factors that will impact the future of animal waste management practices include consumer attitudes regarding environmental issues associated with food animal production. These issues, real or perceived, are currently impacting global livestock and poultry production and will only intensify in the future.

An objective assessment of available evidence shows that issues associated with traditional animal waste management practices are geographically specific and include potential discharge of nutrients and pathogenic bacteria to surface and groundwater, and aerial emissions of volatile organic compounds, ammonia, hydrogen sulfide, dust, pathogens, and odor. The consequences of improperly managed animal waste include environmental and human health impacts from elevated concentrations of nitrates, organic matter, sediments, pathogens, heavy metals, hormones, and antibiotics (EPA, 2016). Of particular concern is the emission of gaseous ammonia from animal production housing facilities and associated land application animal waste management practices (EPA, 2004). Atmospheric ammonia (and ammonium) can redeposit via wet and dry deposition processes into aquatic ecosystems that may be nitrogen sensitive resulting in eutrophication (Jones et al., 2013). Ammonia that returns from the atmosphere into certain terrestrial environments may be metabolized by soil bacteria resulting in the release of nitrous oxide, which can potentially impact climate change. Perhaps more concerning is that under certain conditions in the atmosphere ammonia can be converted to ammonium compounds, which comprise a fraction of fine particulate matter (Kwok et al., 2013). Elevated levels of certain fine particulates may have adverse human health impacts (Pope et al., 2009). As with many nonagricultural industrial processes that have benefited society over the past century, animal production agriculture results in externalities due to the resulting environmental emissions to soil, water, and air media. The future of global food security will be dependent upon fair and just solutions regarding the fate and management of animal waste. This will require implementation of proven innovative technology, as well as the development of new food animal agricultural production and management practices.

In summary, it is recognized that nutrients in animal manures can be a valuable source of organic plant nutrients which, if properly managed, reduces the need for chemical fertilizer—a practice that historically and currently is resulting in improved soil conditions and crop production in many parts of the world. It is also recognized that carbon compounds in animal manures can be a source of energy—many areas of the world are benefiting from the utilization of this renewable energy resource. Conceptually, animal waste resources offer holistic animal production opportunities with appropriate technology applications (Shih, 2015). However, economic constraints limit the widespread application of proven technologies in many areas. Policy change, as well as technology development will be required for widespread implementation of new animal agriculture waste management practices (Williams, 2009). A collective global governance and policy initiative providing reasonable incentives for implementing appropriate technologies by providing technical and financial resources can significantly impact environmental and societal issues associated with food animal production agriculture, as well as maintaining the long-term sustainability of this economically important agricultural sector.

References

Bradford, G.E., 1999. Contributions of animal agriculture to meeting global human food demand. Livest. Prod. Sci. 59, 95–112.

Council for Agricultural Science and Technology, 2006. Safety of meat, milk, and eggs from animals fed crops derived from modern biotechnology: animal agriculture's future through biotechnology, Part 5. CAST Issue Paper Number 34, July 2006.

US Environmental Protection Agency, 2004, National emission inventory—ammonia emissions from animal husbandry, January 30, 2004.

US Environmental Protection Agency, 2016. Nutrient pollution—the problem.

International Egg Commission, 2016. The World Egg Industry—a few facts and figures. Available from: https://www.internationalegg.com/corporate/eggindustry/details.asp?id=18.

Jones, L., Nizam, M., Reynolds, B., Bareham, S., Oxley, E., 2013. Upwind impacts of ammonia from an intensive poultry unit. Environ. Pollut. 180, 221–228.

Kwok, R.H.F., Napelenok, S., Baker, K., 2013. Implementation and evaluation of PM2.5 source contribution analysis in a photochemical model. Atmos. Environ. 80, 398–407.

North American Meat Institute, 2016. The United States meat industry at a glance. Available from: https://www.meatinstitute.org/index.php?ht=d/sp/i/47465/pid/47465.

Organization for Economic Co-operation and Development (OECD) and the Food and Agriculture Organization (FAO) of the United Nations, 2015. OECD-FAO Agricultural Outlook 2015.

Pope, III, C., Ezzati, M., Dockery, D., 2009. Fine-particulate air pollution and life expectancy in the United States. N. Engl. J. Med. 360, 376–386.

Shih, J.C.H., 2015. Development of anaerobic digestion of animal waste: from laboratory, research and commercial farms to a value-added new product. In: Fang, H., Zhang, T. (Eds.), Anaerobic Biotechnology: Environmental Protection and Resource Recovery. Imperial College Press, London, UK.

USDA-ERS USDA Economic Research Service, 2016. Dairy. Available from: http://www.ers.usda.gov/topics/animal-products/dairy/background.aspx.

USDA (Foreign Agricultural Service/USDA Office of Global Analysis), 2015. Livestock and Poultry: World Markets and Trade. October 2015.

USDA-ARS (USDA Agricultural Research Service), 2005. FY-2005 Annual report: manure and byproduct utilization—national program 206.

Williams, C.M., 2009. Development of environmentally superior technologies in the US and policy. Bioresour. Technol. 100, 5512–5518.

WILDLIFE AND ENVIRONMENTAL POLLUTION

Barnett A. Rattner

US Geological Survey, Patuxent Wildlife Research Center, Beltsville, MD, United States

The intentional use of poisons can be traced to ancient man (venoms and plant extracts used in hunting and warfare) and the cradle of civilization (Ebers Papyrus, an Egyptian medical document, describes hemlock, aconite, and lead), with the first hint of potential pollution effects in animals associated with the ancient Greek maxim "a bad crow lays a bad egg" (Rattner, 2009). While the term pollution has a broad meaning, encompassing release of naturally occurring or anthropogenic agents (e.g., chemical, pathogens, light, noise, and temperature) that potentially cause harm to the environment, in the present context its use will be limited to chemical stressors. In the 13th century, complaints about smoke from burning of coal in London resulted in a royal proclamation by Edward I prohibiting the use of "sea-coal" due to its annoyance and the concern over "injury of their body health" to man (Mosley, 2014). However, awareness of pollution and wildlife (viz. amphibians, reptiles, birds, and mammals) did not emerge until the late 19th century when popular press reports of pheasant (*Phasianus colchicus*) and waterfowl mortality were related to ingestion of spent lead shot. Once recognized, lead poisoning was considered a common occurrence in waterfowl and other species of birds through the 20th century to the present time. Despite extensive research and regulatory and conservation efforts, intoxication and poisoning associated with accidental ingestion of spent lead ammunition persists today in charismatic species, such as the bald eagle (*Haliaeetus leucocephalus*) and California condor (*Gymnogyps californianus*). Wildlife toxicology is the study of exposure and potential effects of chemical agents (e.g., agrichemicals, industrial compounds, pharmaceuticals, and personal care products) and anthropogenic processes (e.g., irrigation, mining, capture and transport of fossil fuel, and energy generation) on the individual, population, community, and ecosystem. The hazard of spent lead ammunition typifies the discipline. As a component of ecotoxicology, wildlife toxicology has been shaped by chemical use and misuse, ecological mishaps and other catastrophic events, and occasionally by deliberate hypothesis testing, discovery, and accretion of basic knowledge common to the advancement of most other fields of science (Rattner, 2009).

By the 1930s, about 30 pesticides (e.g., pyrethrum, nicotine, arsenicals, mercurial fungicides, and dinitro-*ortho*-cresol herbicide) were in use in the United States, United Kingdom, and elsewhere (Sheail, 1985). Aerial application was commonplace, and potential adverse effects on nontarget wildlife were acknowledged at the Third North American Wildlife Conference (Strong, 1938). The birth of the synthetic pesticide era, including the discovery of the insecticidal properties of DDT (1939), held great promise for agriculture and human health, although premonitions of damage to wildlife appeared in the widely read Atlantic Monthly in 1945 (Wigglesworth, 1945). By the close of the decade, DDT field studies documented harm to passerines (George and Stickel, 1949; Hotchkiss and Pough, 1946; Robbins and Stewart, 1949), and by the mid-1950s organochlorine and organophosphorus insecticides (aldrin, dieldrin, schradan, and parathion) used as sprays and seed-dressings were linked to mortality events in game birds and mammals (Sheail, 1985). Rachel Carson's *Silent Spring* (Carson, 1962) described in lay terms a multitude of pesticide effects (e.g., ecological imbalances, persistence, resistance, and loss of wildlife), which triggered discussion and debate

at all levels of society, and kick-started the environmental movement, new government agencies, research and monitoring programs, and regulatory legislation. Population declines of some fish-eating and raptorial birds, and later other species (e.g., bats), were of great concern. Following the discovery of many damaged peregrine falcon (*Falco peregrinus*) eggs, Moore and Ratcliffe (1962) and others detected organochlorine pesticide residues in eggs, and linked the appearance of eggshell thinning with DDT use (Hickey and Anderson, 1968; Ratcliffe, 1967). Organophosphorus and carbamate insecticides gradually replaced organochlorine pesticides, and while less persistent and seemingly safer, ingestion of pesticide granules and exposed prey caused bird mortality (Grue et al., 1983; Mineau, 1991). Some of the "die-off" events were large (20,000 Swainson's hawks, *Buteo swainsoni*, in Argentina; Hooper et al., 2003), yet subtle effects on food webs may have indirectly had more profound effects on wildlife. While some of these acetylcholinesterase-inhibiting insecticides are still in use, other classes of pesticides (e.g., pyrethroids and neonicotinoids) began to dominate the marketplace. The neonicotinoids, hypothesized by some to cause bee colony collapse (as yet to be rigorously proven, see Fairbrother et al., 2014) may affect granivorous birds consuming pesticide-coated seeds, and likely indirectly affect other wildlife species by reducing invertebrate prey abundance (Gibbons et al., 2015). Despite extensive requirements for their use and registration, it remains difficult to completely characterize exposure, metabolism, adverse effects, and toxicity thresholds of pesticides to provide the means to protect the diversity of wildlife.

Production and large-scale use of the polychlorinated biphenyls (PCBs; 209 congeners) as insulating chemicals, lubricants, and flame retardants (1929–1977 in the US), and their transfer through water, air, and the food chain has resulted in global exposure and often adverse effects in wildlife.

Reproductive problems in waterbirds related to some of the coplanar PCB congeners and other dioxin-like compounds led to the description of the Great Lakes Embryo Mortality, Edema and Deformity Syndrome that met various epidemiological criteria (Gilbertson et al., 1991). Similarly, wasting, edema, immunotoxicity, and reproductive effects were described in mink and marine mammals. While the manufacture, processing, and distribution of PCBs ended in 1979, and remediation efforts have been underway for decades, PCB congeners are routinely detected, and depending on exposure dose, are accompanied by toxic effects in wildlife. Replacement flame retardants, including polybrominated diphenyl ethers, are less toxic but still bioaccumulate (Chen and Hale, 2010), and their use has been phased out in favor of short-lived alternative flame retardants.

Scientific workshops on chemically induced alterations of endocrine and reproductive function, in concert with the publication of *Our Stolen Future* (Colborn et al., 1996), resulted in extensive efforts to identify endocrine disruptive compounds (EDCs) through traditional toxicity tests, high-throughput screening assays, and field studies. While the adage *the dose makes the poison* has been known for centuries (credited to Paracelsus, the founder of toxicology), *the timing* (sic. of exposure) *makes the poison* has been the mantra of EDCs (Gore et al., 2014). Although much has been learned from controlled exposure studies, EDC effects in free-ranging wildlife have been difficult to document. Notable examples include atrazine-induced hermaphroditism in America leopard frogs (*Rana pipiens*) (Hayes et al., 2003) and feminization of alligators (*Alligator mississippiensis*) potentially caused by organochlorine pesticides (Woodward et al., 2011). The complexity of endocrine-mediated effects and interactions with other physiological systems and behavior is challenging to study in free-ranging biota, and thus evidence for causal linkages between EDCs

and effects in wildlife remain incomplete and controversial (Hotchkiss et al., 2008; Murphy et al., 2006).

While some pharmaceuticals and personal care products may be classified as EDCs, drug effects on wildlife can also be of a more direct toxic nature. One such example is the nonsteroidal antiinflammatory drug (NSAID) diclofenac used for the treatment of inflammation, fever, and pain in domestic livestock. Through a remarkable forensic investigation, it was determined that diclofenac was unintentionally ingested by Old World vultures (*Gyps* species) scavenging on carcasses of livestock that had received diclofenac therapeutically shortly before death (Oaks and Watson, 2011; Oaks et al., 2004). Diclofenac is a potent inhibitor of cyclooxygenase-2 and prostaglandin synthetase, and differential sensitivity among avian species is a hallmark of NSAIDs. Mechanistically, it is hypothesized to impair smooth muscle control of the renal portal valve, shunting blood from the renal cortex, causing ischemia, necrosis, visceral gout, and death within a matter of days. Diclofenac appears to have been the principal cause of the unprecedented population crash of *Gyps* vultures in India, Pakistan, and Nepal, and is the only well-documented instance of a veterinary drug resulting in population-level effects and the endangerment of a wildlife species. The decline had implications for humans in the area, including the cost of livestock carcass disposal, cultural and religious impacts, and health implications including the increase in dog populations, which are the principal source of rabies in India (Markandya et al., 2008). While diclofenac use in livestock has been banned in India, and a safe replacement NSAID (meloxicam) is available, its use continues in much of the range of *Gyps* vultures.

Anthropogenic processes, such as irrigation and mining can have profound effects on wildlife. In the 1970s, the potential use of agricultural drainage water for wetland management was being considered in parts of California (Ohlendorf, 2011). In the San Joaquin Valley, natural drainage was inadequate for crop production because saline water accumulated in the plant root zone, but could be ameliorated through subsurface drain tiles to collect and convey saline groundwater from fields. The US Bureau of Reclamation constructed a canal system to dispose of this drainwater; however, the project was not completed as planned and the canal terminus ended at the Kesterson National Wildlife Refuge rather than San Francisco Bay. At this Refuge, irrigation drainwater was impounded in shallow ponds as a beneficial use for wildlife. Monitoring studies revealed that the concentrations of most metals, metalloids, and organochlorine pesticides in mosquito fish (*Gambusia affinis*) from sites receiving drainwater and an untreated reference area were similar; however, levels of selenium in mosquito fish from the drainwater sites were nearly an order of magnitude greater. Field and laboratory studies quickly demonstrated that selenium concentrations in the wildlife food chain at the Refuge were causing epidemic-like problems for waterbirds (up to 40% of nests contained deformed or dead embryos or chicks), and similar selenium-related toxicity in wildlife was found to be widespread in many parts of the western United States. Remediation and management of this situation remains difficult, involving a combination of actions (e.g., source control, drainage reuse, evaporation systems, land retirement, and controlled discharge) (Ohlendorf, 2011).

Mining and smelting of lead, silver, and gold in northern Idaho began in the 1880s and resulted in the deposition of approximately 75 million tons of metal-enriched sediment in Lake Coeur d'Alene (Henny, 2003). Waterfowl collected from this area in 1955 contained high levels of lead, zinc, and copper in their tissues, which was independent of ingestion of spent lead shot. In the decades that followed, annual die-offs of tundra swans (*Cygnus columbianus*) and waterfowl continued.

Through a series of field and laboratory studies examining lead, other metals, and biomarker responses (e.g., inhibition of δ-aminolevulinic acid dehydratase activity, blood protoporphyrin concentration, and presence of renal inclusion bodies), the cause of mortality was linked to the ingestion of "mining-related lead" in sediment and biota rather than the ingestion of spent lead shot from hunting (Beyer et al., 2000; Henny, 2003). These findings resulted in one of the largest natural resource damage assessments of the US Department of the Interior, and the collection of nearly $80 million from mining polluters that was ultimately used for the recovery of wildlife and habitat (US Fish and Wildlife Service, 2009).

Collection and transport of fossil fuels, and the generation of energy, are an integral part of modern civilization, but at times have had devastating effects on wildlife and their supporting habitat. Often thought of as a contemporary issue, petroleum spills accompanied by the oiling and death of waterbirds and other biota have been documented for nearly a century, dating back to maritime events of the 1920s (Phillips and Lincoln, 1930). Highly publicized events (e.g., *Torrey Canyon*, 1967; *Exxon Valdez*, 1989; Gulf War oil spills, 1991; and Deepwater Horizon, 2010), including striking images of petroleum sheen on water, tar-covered habitat, and oil-drenched birds, have captivated public attention and emotion. Interpretation of observations and recovery following petroleum spill events are often controversial, pitting the presumably responsible party against trustees of the natural resource (Wiens, 1996). Extensive rehabilitation efforts of oiled biota, while noble, have yielded mixed results on postrelease survival (Russell et al., 2003).

The meltdown of the Chernobyl nuclear reactor in 1986 released radiocesium and other radionuclides (~3,000,000 trillion Becquerels) resulting in acute and latent effects in humans, plants, and animals at, and a great distance from, the accident site (Eisler, 2003). A nearby pine forest within the "exclusion zone" received a dose of >80 Gy (Gy = 100 rad), died, turned ginger-brown in color, and was subsequently dubbed the Red Forest. There was significant mortality of small mammals in this area; however, the following year rodent population seemed normal, possibly due to immigration from less affected areas. Radioactivity in tissues of game birds near Chernobyl and in large mammals (reindeer, *Rangifer tarandus*; moose *Alces alces*) over 1000 km from Chernobyl, exceeded levels safe for human consumption for years. Several teams of investigators examined genetic effects in small mammals from Sweden and Italy, and at Chernobyl, with findings being inconsistent (Deryabina et al., 2015). A detailed study of the voles (*Microtus* sp.) in the exclusion zone revealed considerable genetic variation, which subsequently was demonstrated to be natural variation among four species of voles (Chesser and Baker, 2006). Although Chernobyl was considered a large-scale disaster, ironically, the subsequent reduced human presence has resulted in abundant populations of wildlife in the exclusion zone (Deryabina et al., 2015).

Man's dependence on and interaction with wildlife can be traced to the beginnings of recorded history. In 1990 just before the 20th anniversary of *Earth Day*, the National Wildlife Federation published a special issue of their magazine entitled *What on Earth Are We Doing?* (National Wildlife Federation, 1990). Dramatic photographs and articles entitled *Innocent World of a Toxic Victim*, *Pollution Knows No Boundaries*, and *Your Contribution of Global Warming* highlighted pollution issues affecting the biosphere, some of our accomplishments, and the many challenges that man and wildlife face. In my lifetime I have tracked societal and regulatory activities that have protected wildlife and habitat. Many of us have witnessed "nature's great experiment" with the recovery of the bald eagle, brown pelican (*Pelecanus occidentalis*), osprey (*Pandion haliaetus*), and peregrine falcon (*Falco peregrinus*) populations that

were devastated by many chlorinated pesticides. There have been advances in green chemistry to replace chemicals that may be persistent, bioaccumulative, or toxic with compounds that are benign by design. Nonetheless, there remains an ongoing conflict between wildlife and man over the habitat we share, with a significant component related to chemical use, agriculture, demand for natural resources, and the generation of energy. Our approach to these conflicts has become holistic, multidisciplinary, and risk-based. The contemporary One Health Initiative that seeks to attain optimal health for people, animals, and the environment, while new, is not really a new concept. Wildlife has time and again served as man's canary in the coal mine, and we need to protect the canary! Clearly, some pollution threats to wildlife have been resolved, "wicked problems" linger, with the unknown looming on the horizon.

Any use of trade, firm, or product names is for descriptive purposes only and does not imply endorsement by the US Government.

References

Beyer, W.N., Audet, D.J., Heinz, G.H., Hoffman, D.J., Day, D., 2000. Relation of waterfowl poisoning to sediment lead concentrations in the Coeur d'Alene River Basin. Ecotoxicology 9, 207–218.

Carson, R.L., 1962. Silent Spring. Houghton Mifflin, Boston, MA, USA.

Chen, D., Hale, R.C., 2010. A global review of polybrominated diphenyl ether flame retardant contamination in birds. Environ. Int. 36, 800–811.

Chesser, R.K., Baker, R.J., 2006. Growing up with Chernobyl. Am. Sci. 94, 542–549.

Colborn, T., Dumandski, D., Myers, J.P., 1996. Our Stolen Future: Are We Threatening Our Fertility, Intelligence and Survival? A Scientific Detective Story. Dutton Publishing, New York, NY, USA.

Deryabina, T.G., Kuchmel, S.V., Nagorskaya, L.L., Hinton, T.G., Beasley, J.C., Lerebours, A., Smith, J.T., 2015. Long-term census data reveal abundant wildlife populations at Chernobyl. Curr. Biol. 25, R811–R826.

Eisler, R., 2003. The Chernobyl nuclear power plant reactor accident: ecological update. In: Hoffman, D.J., Rattner, B.A., Burton, Jr., G.A., Cairns, Jr., J. (Eds.), Handbook of Ecotoxicology. second ed. Lewis Publishing Inc., Boca Raton, FL, USA, pp. 703–736.

Fairbrother, A., Purdy, J., Anderson, T., Fell, R., 2014. Risks of neonicotinoid insecticides to honeybees. Environ. Toxicol. Chem. 33, 719–731.

George, J.L., Stickel, W.H., 1949. Wildlife effects of DDT dust used for tick control on a Texas prairie. Am. Midl. Nat. 42, 228–237.

Gibbons, D., Morrissey, C., Mineau, P., 2015. A review of the direct and indirect effects of neonicotinoids and fipronil on vertebrate wildlife. Environ. Sci. Pollut. Res. 22, 103–118.

Gilbertson, M., Kubiak, T., Ludwig, J., Fox, G., 1991. Great Lakes embryo mortality, edema, and deformities syndrome (GLEMEDS) in colonial fish-eating birds: similarity to chick edema disease. J. Toxicol. Environ. Health 33, 455–520.

Gore, A.G., Crews, D., Doan, L.L., LaMerrill, M., Patisaul, H., Zota, A., 2014. Introduction to endocrine disrupting chemicals (EDCs). Endocrine Society. 76 p. Available from: https://www.motherjones.com/files/introduction_to_endocrine_disrupting_chemicals.pdf.

Grue, C.E., Fleming, W.J., Busby, D.G., Hill, E.F., 1983. Assessing hazards of organophosphate pesticides to wildlife. Trans. North Am. Wildl. Nat. Resour. Conf. 48, 200–220.

Hayes, T., Haston, K., Tsui, M., Hoang, A., Haeffele, C., Vonk, A., 2003. Atrazine-inducted hermaphroditism at 0.1 ppb in American leopard frogs (*Rana pipiens*): laboratory and field evidence. Environ. Health Perspect. 111, 568–575.

Henny, C.J., 2003. Effects of mining lead on birds: a case history of the Coeur d'Alene Basin, Idaho. In: Hoffman, D.J., Rattner, B.A., Burton, Jr., G.A., Cairns, Jr., J. (Eds.), Handbook of Ecotoxicology. second ed. Lewis Publishing Inc., Boca Raton, FL, USA, pp. 755–766.

Hickey, J.J., Anderson, D.W., 1968. Chlorinated hydrocarbons and eggshell changes in raptorial and fish-eating birds. Science 162, 271–273.

Hooper, M.J., Mineau, P., Zaccagnini, M.E., Woodbridge, B., 2003. Pesticides and international migratory bird conservation. In: Hoffman, D.J., Rattner, B.A., Burton, Jr., G.A., Cairns, Jr., J. (Eds.), Handbook of Ecotoxicology. second ed. Lewis Publishing Inc., Boca Raton, FL, USA, pp. 737–754.

Hotchkiss, N., Pough, R.H., 1946. Effect on forest birds of DDT used for gypsy moth control in Pennsylvania. J. Wildl. Manage. 10, 202–207.

Hotchkiss, A.K., Rider, C.V., Blystone, C.R., Wilson, V.S., Hartig, P.C., Ankley, G.T., Foster, P.M., Gray, C.L., Gray, L.E., 2008. Fifteen years after "Wingspread"—environmental endocrine disrupters and human and wildlife health: where we are today and where we need to go. Toxicol. Sci. 105, 235–259.

Markandya, A., Taylor, T., Longo, A., Murty, M.N., Murty, S., Dhavala, K., 2008. Counting the cost of vulture decline—an appraisal of the human health and other benefits of vulture in India. Ecol. Econ. 67, 194–204.

Mineau, P., 1991. Cholinesterase-inhibiting Insecticides: Their Impact on Wildlife and the Environment. Chemicals in Agriculture, Volume 2. Elsevier, New York, NY, USA, 348 pp.

Moore, N.W., Ratcliffe, D.A., 1962. Chlorinated hydrocarbon residues in the egg of a peregrine falcon (*Falco peregrinus*) from Perthshire. Bird Study 9, 242–244.

Mosley, S., 2014. Environmental history of air pollution and protection. In: Agnoletti, M., Serneri, S.N. (Eds.), The Basic Environmental History. Springer International Publishing, New York, NY, USA, pp. 143–169.

Murphy, M.B., Hecker, M., Coady, K.K., Tompsett, A.R., Jones, P.D., DuPreez, L.H., Everson, G.J., Solomon, K.R., Carr, J.A., Smith, E.E., Kendall, R.J., Van Der Kraak, G., Giesy, J.P., 2006. Atrazine concentrations, gonadal gross morphology and histology in ranid frogs collected in Michigan agricultural areas. Aquatic Toxicol. 26, 230–245.

National Wildlife Federation, 1990. What on earth are we doing? National Wildlife February–March Issue. 60 pp.

Oaks, J.L., Gilbert, M., Virani, M.Z., Watson, R.T., Meteyer, C.U., Rideout, B.A., Shivaprasad, H.L., Ahmed, S., Chaudhry, M.J.I., Arshad, M., Mahmood, S., Ali, A., Khan, A.A., 2004. Diclofenac residues as the cause of vulture population decline in Pakistan. Nature 427, 630–633.

Oaks, J.L., Watson, R.T., 2011. South Asian vultures in crisis: environmental contamination with a pharmaceutical. In: Elliott, J.E., Morrissey, C.A., Bishop, C.A. (Eds.), Wildlife Ecotoxicology: Forensic Approaches. Springer Science + Business Media, New York, NY, pp. 412–441.

Ohlendorf, H.M., 2011. Selenium, salty water, and deformed birds. In: Elliott, J.E., Morrissey, C.A., Bishop, C.A. (Eds.), Wildlife Ecotoxicology: Forensic Approaches. Springer Science + Business Media, New York, NY, 325–357.

Phillips, J.C., Lincoln, F.C., 1930. American Waterfowl, Their Present Situation and the Outlook for Their Future. Houghton Mifflin Co., New York, NY, USA.

Ratcliffe, D.A., 1967. Decrease in eggshell weight in certain birds of prey. Nature 215, 208–210.

Rattner, B.A., 2009. History of wildlife toxicology. Ecotoxicology 18, 773–783.

Robbins, C.S., Stewart, R.E., 1949. Effects of DDT on bird population of scrub forest. J. Wildl. Manage. 13, 11–16.

Russell, M., Holcomb, J., Berkner, A., 2003. 30-Years of oiled wildlife response statistics. Proceedings of the 7th International Effects of Oil on Wildlife Conference. Hamburg, Germany. Oct 14–16, 2003.

Sheail, J., 1985. Pesticides and Nation Conservation; the British Experience, 1950–1975. Clarendon Press, Oxford, UK.

Strong, L., 1938. Insect and pest control in relation to wildlife. Transactions of the 3rd North American Wildlife Conference. American Wildlife Institute, Washington, DC, USA, pp. 543–547.

US Fish and Wildlife Service, 2009. Settlement brings millions of dollars for Coeur d'Alene clean-up and restoration. Available from: http://www.fws.gov/news/ShowNews.cfm?ref=settlement-brings-millions-of-dollars-for-coeur-dalene-basin-clean-up-and-&_ID=49.

Wiens, J.A., 1996. Oil, seabirds, and science. BioScience 46, 587–597.

Wigglesworth, V.B., 1945. DDT and the balance of nature. Atlantic Mon. 176, 107–113.

Woodward, A.R., Percival, H.F., Rauschenberger, R.H., Gross, T.S., Rice, K.G., Conrow, R., 2011. Abnormal alligators and organochlorine pesticides in Lake Apopka, Florida. In: Elliott, J.E., Morrissey, C.A., Bishop, C.A. (Eds.), Wildlife Ecotoxicology: Forensic Approaches. Springer Science + Business Media, New York, NY, pp.153–187.

HUMAN–WILDLIFE CONFLICTS

Larry Clark

National Wildlife Research Center, Wildlife Services, Animal and Plant Health Inspection Service, United States Department of Agriculture, Fort Collins, CO, United States

The degree and scope of human–wildlife conflicts is a matter of perspective. Some see the issue through the lens of how human activities and interests have negatively impacted wildlife populations. Others see the issue through the lens of how wildlife negatively impact humans' economic or health interests. In either case, the passions can run deep. The scenarios for solutions have played out differently across the globe (Distefano, 2005; Madden and McQuinn, 2014). Regardless, the issue transcends economic sectors and development status of any given country and is deeply rooted in the cultural practices of agricultural and natural resource sustainability.

In the United States, the issues surrounding human–wildlife conflicts reached a critical juncture in the late 19th and early 20th centuries. Commercial exploitation of wildlife by market hunters was devastating North American wildlife populations (Organ et al., 2012). Alarmed, conservationists and sportsmen began to formalize wildlife management policies into what is now known as the North American wildlife management model, establishing seven principles for the sustainable harvest of wildlife species for the benefit of human consumption, conservation, and recreation (Tober, 1981). At the same time agricultural interests were concerned about the negative impact wildlife had on crop and livestock production. As a result of these concerns the Division of Economic Ornithology and Mammalogy was established under the Department of Agriculture. This division's early purpose was to study food habits of wildlife and educate the agricultural community as to which wildlife species were beneficial or harmful to their interests, and eventually develop programs for wildlife control to prevent damage to crops and livestock. Later, the division took on the task for the prevention of transmission of wildlife disease to human health and economic interests (e.g., rabies) (Hawthorne, 2004).

Both efforts have evolved into well-established and regulated systems of administration and management of wildlife in the United States, yet the inherent conflict of what the social carrying capacity of wildlife should be remains in place (Decker and Purdy, 1988). Fundamentally, wildlife valuation is related to direct experience. A homeowner, farmer, rancher, and so on, generally are less tolerant of wildlife and wants more direct intervention if they have experienced damage to their property or commodity than those who have no such negative experience (Young et al., 2015). Similarly, those who value wildlife for recreation or heuristic reasons are more apt to oppose interventionist management strategies (Reiter et al., 1999). These perspectives have led to some interesting policy debates surrounding methods to resolve conflicts, the economics and effectiveness of the methods, and who should bear the costs of damage should it occur given that, at least in the United States, wildlife are held in public trust (Bulte and Rondeau, 2005; Nyhus et al., 2005). As controversial as some government-sponsored programs might be in terms of compensation, mitigation, or control involvement, when individuals experience conflicts with wildlife, and in the absence of sanctioned and monitored regulated assistance or relief, the consequences of individually or privately based remedies usually are not a favorable outcome for wildlife (Mateo-Tomás et al., 2012). That is to say, individual economic interests will prevail.

By 2050 the world will have 10 billion people, resulting in a rapidly declining per capita availability of arable land. One consequence of this fact is that there will be enormous pressure to extract as much from the land to produce food for an increasingly hungry world (Green et al., 2005).

The growth of the world's population (Phalan et al., 2011), pressures of land and water management (Tilman et al., 2011), food production (Henle et al., 2008), industrialization (Gates et al., 2014), and increased globalization of trade and transport (Hulme, 2009) have all resulted in newer and more intense manifestations of human–wildlife conflicts requiring urgent attention.

Other high-profile examples serve to illustrate the diversity and complexity of human–wildlife conflicts. Each example put forth is a mix of biological processes and human dimensions issues, which include economics, emotions, and value systems. The approaches to solutions range from technical, regulatory, and broader policy. The best cases of addressing and solving any particular human–wildlife conflict has been the use of data and analyses with integration across science and social disciplines, and it demonstrates how successful a willingness of the partners to engage in dialog can be.

Because carnivores eat livestock, and sometimes harm people, human–carnivore conflicts are emblematic of the longstanding economic and emotional issues encountered in the arena of human–wildlife conflicts. Pastoralists and ranchers are closely tied to their animals through culture and economy, thus the loss through predation is monetary and personal (Dickman, 2010). Historically, extirpation of carnivores was a short- and long-term solution. But changes in landscape and ecosystem recovery efforts and targeted species recovery have renewed traditional conflicts. In Europe, in the absence of predation certain management practices, such as livestock protection using guard dogs was lost; where predation remained, those traditions persisted (Kaczensky, 1999). As society values change, the producer must make the economic and societal management decisions on the cost-effectiveness of different practices, that is, lethal damage management control and/or nonlethal methods (which are usually spatially limited and more costly).

Another high-profile area of research and management is aviation safety (DeVault et al., 2011). Keeping birds out of the flight path of aircraft is uniformly seen as a positive goal. In general, two approaches are used. Hazing is the practice of chasing birds away by sound, visual, or other harassment systems. The effects are short-lived with animals rapidly habituating to the methods. Longer-term solutions generally focus on land management practices in and around airports (DeVault et al., 2013). In this case, the goal is to make habitats unattractive to species of birds that pose the highest risks to aircraft, such as flocking species and species with high body density.

Wildlife disease increasingly has become an issue of human–wildlife conflict (Clark, 2014; Karesh et al., 2005; Kock, 2005). The recent visibility of highly pathogenic avian influenza, its impacts on the global poultry industry, and its wildlife origins are good examples of the large-scale magnitude that the conflict can reach (Clark and Hall, 2006). Yet devising the appropriate risk assessments and farm-side management practices to enhance biosecurity against exposure to wildlife is proving to be a challenge that pits agricultural production against wildlife conservation (Graham et al., 2008).

Invasive species may cause conflict because humans value biodiverse communities and invasive species can negatively impact native ecosystems (Olson, 2006). For example, brown tree snakes have been implicated in the extinction of several species of native birds on Guam (Rodda and Savidge, 2007). Interdiction and control efforts are in part used for natural resource recovery, but primarily focused on containment to prevent further environmental disasters by inadvertent transportation routes. But as in most cases, the difficult decision of when to manage an invasive species often depends on resources and perceived attitudes of the threat (Sharp et al., 2011). Because of the large economic and other resource invest-

ments needed for such control programs, there is an imperative to evaluate when an exotic species reaches a threshold where it becomes injurious, hence invasive and thus rises to a situation where action is warranted.

Finally, there is a rich tradition of research on new technical methods and tools used to execute management (Conover, 2001). However, many of these require regulatory approval often not appreciated by researchers and field managers alike. And if manufacture is needed, a business analysis for the scalability to commercial levels, a business plan, intellectual property protection, and technology transfer become filters to getting successful tools into the hands of managers. Few wildlife managers or conservationists are equipped to participate in these arenas. The need for fully integrated partnerships on the research and development and regulatory side are needed. Of course the marketing and success of getting users to accept methodologies is needed as well. The good news is that approaches to methods development have made great strides and there is better integration of a complex network of perspectives, needs, capabilities, and policy than there was a few years ago.

References

Bulte, E.H., Rondeau, D., 2005. Research and management viewpoint: why compensating wildlife damages may be bad for conservation. J. Wildl. Manage. 69, 14–19.

Clark, L., 2014. Disease risks posed by wild birds associated with agricultural landscapes. In: Matthews, K.R., Sapers, G.M., Gerba, C.P. (Eds.), The Produce Contamination Problem. second ed. Academic Press, Elsevier Inc, Boston, MA, pp. 139–165.

Clark, L., Hall, J., 2006. Avian influenza in wild birds: status as reservoirs, and risks to humans and agriculture. Ornithol. Monogr. 11, 3–29.

Conover, M.R., 2001. Resolving human–wildlife conflicts: the science of wildlife damage management. CRC Press, Boca Raton, FL.

Decker, D.J., Purdy, K.G., 1988. Toward a concept of wildlife acceptance capacity in wildlife management. Wildl. Soc. Bull. 16, 53–57.

DeVault, T.L., Belant, J.L., Blackwell, B.F., Seamans, T.W., 2011. Interspecific variation in wildlife hazards to aircraft: implications for airport wildlife management. Wildl. Soc. Bull. 35, 394–402.

DeVault, T.L., Blackwell, B.F., Belant, J.L., 2013. Wildlife in airport environments: preventing animal–aircraft collisions through science-based management. JHU Press.

Dickman, A.J., 2010. Complexities of conflict: the importance of considering social factors for effectively resolving human–wildlife conflict. Anim. Conserv. 13, 458–466.

Distefano, E., 2005. Human–wildlife conflict worldwide: collection of case studies, analysis of management strategies and good practices. Food and Agricultural Organization of the United Nations (FAO), Sustainable Agriculture and Rural Development Initiative (SARDI), Rome, Italy. Available from: FAO Corporate Document repository, http://www.fao.org/documents.

Gates, J.E., Trauger, D.L., Czech, B., 2014. Peak Oil, Economic Growth, and Wildlife Conservation. Springer, New York, pp. 317–339.

Graham, J.P., Leibler, J.H., Price, L.B., Otte, J.M., Pfeiffer, D.U., Tiensin, T., Silbergeld, E.K., 2008. The animal-human interface and infectious disease in industrial food animal production: rethinking biosecurity and biocontainment. Public Health Rep. 123, 282–299.

Green, R.E., Cornell, S.J., Scharlemann, J.P., Balmford, A., 2005. Farming and the fate of wild nature. Science 307, 550–555.

Hawthorne, D., 2004. The history of Federal and cooperative animal damage control. Sheep Goat Res. J. 19, 13–15.

Henle, K., Alard, D., Clitherow, J., Cobb, P., Firbank, L., Kull, T., McCracken, D., Moritz, R.F., Niemelä, J., Rebane, M., Wascher, D., 2008. Identifying and managing the conflicts between agriculture and biodiversity conservation in Europe–A review. Agric. Ecosyst. Environ. 124, 60–71.

Hulme, P.E., 2009. Trade, transport and trouble: managing invasive species pathways in an era of globalization. J. Appl. Ecol. 46, 10–18.

Kaczensky, P., 1999. Large carnivore depredation on livestock in Europe. Ursus 11, 59–71.

Karesh, W.B., Cook, R.A., Bennett, E.L., Newcomb, J., 2005. Wildlife trade and global disease emergence. Emerging Infect. Dis. 11, 1000–1002.

Kock, R.A., 2005. What is this infamous "wildlife/livestock disease interface?" A review of current knowledge for the African continent. Conservation and development interventions at the wildlife/livestock interface: implications for wildlife, livestock and human health,. Osofsky, S.A. (Ed.), Conservation and Development Interventions at the Wildlife/Livestock Interface, 30, IUCN Species Survival Commission, pp. 1–13.

Madden, F., McQuinn, B., 2014. Conservation's blind spot: the case for conflict transformation in wildlife conservation. Biol. Conserv. 178, 97–106.

Mateo-Tomás, P., Olea, P.P., Sánchez-Barbudo, I.S., Mateo, R., 2012. Alleviating human–wildlife conflicts: identifying the causes and mapping the risk of illegal poisoning of wild fauna. J. Appl. Ecol. 49, 376–385.

Nyhus, P.J., Osofsky, S.A., Ferraro, P., Madden, F., Fischer, H., 2005. Bearing the costs of human-wildlife conflict: the challenges of compensation schemes. In: Woodroffe, R., Thirgood, S., Rabinowitz, A. (Eds.), People and Wildlife. Conflict or Coexistence?. Cambridge University Press, London, p. 107.

Olson, L.J., 2006. The economics of terrestrial invasive species: a review of the literature. Agric. Resour. Econ. Rev. 35, 178–194.

Organ, J.F., Geist, V., Mahoney, S.F., Williams, S., Krausman, P.R., Batcheller, G.R., Decker, T.A., Carmichael, R., Nanjappa, P., Regan, R., Medellin, R.A., Cantu, R., McCabe, R.E., Craven, S., Vecellio, G.M., Decker, D.J., 2012. The North American Model of Wildlife Conservation. The Wildlife Society Technical Review 12-04. The Wildlife Society, Bethesda, Maryland, USA, p. 45.

Phalan, B., Onial, M., Balmford, A., Green, R.E., 2011. Reconciling food production and biodiversity conservation: land sharing and land sparing compared. Science 333, 1289–1291.

Reiter, D.K., Brunson, M.W., Schmidt, R.H., 1999. Public attitudes toward wildlife damage management and policy. Wildl. Soc. Bull. 27, 746–758.

Rodda, G.H., Savidge, J.A., 2007. Biology and impacts of Pacific island invasive species. 2. *Boiga irregularis*, the brown tree snake (Reptilia: Colubridae). Pacific Sci. 61, 307–324.

Sharp, R.L., Larson, L.R., Green, G.T., 2011. Factors influencing public preferences for invasive alien species management. Biol. Conserv. 144, 2097–2104.

Tilman, D., Balzer, C., Hill, J., Befort, B.L., 2011. Global food demand and the sustainable intensification of agriculture. Proc. Natl. Acad. Sci. USA 108, 20260–20264.

Tober, J.A., 1981. Who owns the wildlife?: The political economy of conservation in nineteenth-century America (No. 37). ABC-CLIO.

Young, J.K., Ma, Z., Laudati, A., Berger, J., 2015. Human–Carnivore interactions: lessons learned from communities in the American West. Hum. Dimens. Wildl. 20, 349–366.

Animal Welfare and Animal Rights

Clive J.C. Phillips*, Katrina Kluss**

*Centre for Animal Welfare and Ethics, School of Veterinary Science,
University of Queensland, Gatton, QLD, Australia
**TC Beirne School of Law, University of Queensland, Brisbane, QLD, Australia

20.1 INTRODUCTION

As humans have evolved, so too has our relationship with animals. Early humans simply coexisted with animals, using them for food and their bones and skin for clothing, tools, and shelter. Prehistoric art, such as cave paintings, shows that early man was not only reliant on animals, but had respect for the creatures with which they shared the land and other resources (Phillips, 2009). However, the relationship between humans and animals changed with the domestication of wild animals. As humans began to develop closer associations with animals and mastered plant cultivation it became possible to exploit this relationship. Bullocks, horses, and donkeys were trained to work for humans, pulling their goods, cultivating fields, and hauling timber, while the farming of animals such as chickens, sheep, goats, pigs, and cows was intensified to provide a steady, controlled supply of meat, milk, and cheese. Dogs were trained to hunt, protect and provide companionship.

The human–animal relationship further evolved from human dependency on animals to maximizing the benefit in the form of animal agriculture. Initially, local farming provided for villages, but as cities and transport developed, farmers became responsible for feeding larger populations and expanded their enterprises to meet the increased demand.

By the early 19th century, with the use of horse- and bullock-drawn machinery, farmers were able to manage larger farms to supply the growing townships and cities (Reid, 2011). Then, after two world wars that had posed many risks to food security, industrial-scale intensive farming emerged to guarantee supply and reduce cost to consumers. Intensification was facilitated by fertilizers, high-quality feeds, and genetically modified livestock that could produce food at an accelerated rate. Medical advances simultaneously reduced the prevalence of diseases associated with overcrowding, which allowed more intensive housing and reduced the costs of production. In addition, from 1700 to 1980, cultivated land increased by

Animals and Human Society
http://dx.doi.org/10.1016/B978-0-12-805247-1.00030-7

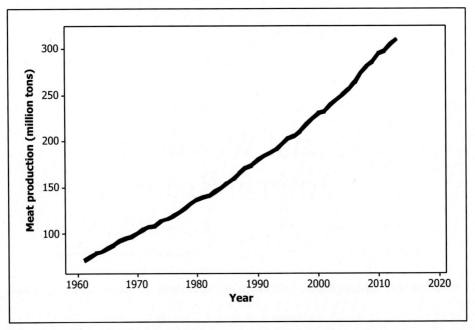

FIGURE 20.1   **Growth in world meat production between 1961 and 2013 (Phillips, 2015, p. 53).**

466% worldwide (Matson et al., 1997). By 1996, while the rate of agricultural land expansion had slowed, the food yielded per area of land had significantly increased and had exceeded global human population growth, from 1 billion in 1800 to 7.4 billion in 2015 (Naylor, 1996).

With increasing prosperity, demand for animal products has increased over time (FAO, 2009). By 2050, global meat and milk production are expected to be 465 and 1043 million tons respectively, from 229 and 58 million tons in 1999/2001. Growth in meat production from 1961–2015 has been increasing at a progressively faster rate (Fig. 20.1). As a result, farmers have been under increasing pressure to produce large amounts of animal products for a reduced market price acceptable to the masses. Although livestock contributes valuable nutrients for crops and provides critical protein and nutrients to human diets, it is also responsible for using one-third of the global freshwater resources, one-third of global cropland for feed, and contributes to nutrient pollution and land

degradation (Herrero et al., 2009). To avoid further environmental problems, the environmental impact per unit of livestock production must be halved (FAO, 2009).

In addition to the development of intensive animal agriculture, society's use of animals also experienced a marked increase over time in the areas of scientific experimentation, entertainment, sport, and breeding. Objectives for the breeding of animals have changed to produce animals we consider aesthetically pleasing or unusual, or to manufacture breeds that produce leaner meats or quicker growth to increase production rates; these may have poor welfare outcomes for the animals involved.

Nevertheless, while society has increasingly exchanged the welfare of some animal species to maximize benefits to humans, other animals have been the subject of increased well-being and legal and societal protection. For example, most countries in the developed world now have legislation in which minimum standards of care are stipulated for companion animals, such as cats and dogs, and breaches of such anticruelty laws result in penalties, sometimes including custodial sentences. This promotes respect for, and protection of, the welfare of certain domestic animals by society. However, welfare standards of domestic animals, as developed by human society, are sometimes placed secondary to economic considerations and market demand (McInerney, 2004).

While animal production has been intensifying, so too has society's concern. This can be seen as an extension of our widening circle of compassion, which started in the Western world with the abolition of slavery at the end of the 18th century, through to the recognition of children's right not to be overworked in the late 19th century, then championing of women's rights in the early 20th century, ethnic minorities rights in the middle of that century, and finally disabled and gay people's rights later in that century (Phillips, 2009). This has led to the emergence of various animal advocate groups, the visions of which are largely founded on jurisprudence and scientific advancements. These groups are frequently divided into two schools of thought: animal welfare and animal rights.

While at first glance, animal welfare and animal rights may appear to be different names for largely the same cause, there is a fundamental difference between the two. On the one hand, animal welfare is based on empirical science related to an animal's quality of life and quantity of experiences, which is operationalized by biological measures, preference testing, and other indices (Phillips, 2009). Therefore, in theory any two people should agree on an evaluation of the welfare of an animal, even though in practice the validity of different welfare assessment methods may mean that there is some disagreement. By contrast, animal rights implies an ethical position, which recognizes that nonhuman animals have rights that extend beyond basic animal welfare considerations and also include some of the rights afforded to humans (Regan, 2004);

whether or not animals without intentionality, such as worms and insects, only have rights as a species, not as individuals per se is a topic of ongoing debate (Scruton, 1998). Species rights could include a right to a biodiverse genotype, sufficient for perpetuation of the species and a right to avoid genetic modification that harms the ability of the species to survive or that impinges on other rights, such as to a good standard of welfare. Others refute this, arguing that an individual animal's rights take precedence over species rights (Regan, 2004). Darwin (1859) recognized that "Man selects only for his own good, Nature only for that of the being which she tends.... Man keeps the natives of many climates in the same country; he seldom exercises each selected character in some peculiar and fitting manner."

The public debate between animal welfare and animal rights has become increasingly vociferous over the past few decades, particularly in the Western world, with many people becoming more interested in the well-being of animals, and our impact on them and the broader environment. Increasing societal concern about the rights attributed to animals reflects both moral developments as a civilization and the recognition that the differences between humans and animals are that of degree, rather than of kind (Francione, 2008).

This recognition has largely been facilitated by scientific advancements that have enabled studies to be conducted, which have revealed higher levels of intelligence, problem-solving skills, and understanding in a variety of animals (e.g., dogs, Colvin and Marino, 2015; calves, Daros et al., 2014).

The human–animal relationship is evolving from one of pure utility to one that more readily encompasses animal welfare considerations and, in more recent times, one that is increasingly concerned with potential rights that ought to be afforded to animals. The animal welfare versus animal rights debate is now, more than ever, relevant to a discussion of the relationship between human society and animals, and is particularly important when looking to the future.

This chapter will explore the differences between animal welfare and animal rights by analyzing both sides of the debate, and how it differs depending upon demographics, such as culture and religion. While often referred to as two opposing fields of thoughts, it is suggested that, although there are certainly principles of difference, animal welfare and animal rights are complementary movements that both drive improvements in the human–animal interaction.

As our understanding of animal capabilities and noncapabilities continues to develop, what ought to guide our attitudes, laws, and policies toward animals? While science provides an undeniably empirical basis on which to guide human attitudes, human moral reasoning has an important role to play. Therefore, a combination of morality and science is guiding human society toward a better relationship with the animals.

Finally, this chapter explores how the relationship between humans and animals may further evolve in the future, which, it is concluded, will largely be determined by the choices made by society today.

20.2 DIFFERENCES BETWEEN ANIMAL WELFARE AND RIGHTS

20.2.1 Animal Welfare Definition and Assessment

The term "welfare" derives from an Old Norse word, velferth, meaning good travel. However, although the etymological derivation of welfare is clear, the concept as understood by the majority of the public remains largely unclear. It is evident that the public hold a naturalness concept of welfare in high regard; that is, what is natural is perceived as also being good. Being close to, and respectful of, nature may have benefited people in the past, as they could find food and locate a safe refuge easily.

Scientists have debated whether animal welfare refers to an animal's ability to cope with its environment (Broom, 1986) or its feelings (Duncan and Fraser, 1997; Fraser et al., 1997; Mason and Veasey, 2010; Spruijt et al., 2001). This divides people into those with a hedonistic focus, based on animals' feelings, and those that recognize a broader and more pragmatic concept of welfare, including an animal's physiology, behavior, production (if kept for this purpose by humans), and even reproduction. The latter approach takes into account the long-term impact of the environment on welfare and acknowledges that short-term hedonistic responses may not be in an animal's long-term interests. However, coping suggests that animals must only be protected from negative experiences, whereas a recent trend has been to include positive experiences in our aims for animals. The position adopted has major implications for the measurement of welfare; the latter, "feelings" approach, suffers from difficulty in assessment and understanding from the limited but growing number of tools at our disposal. An environmental-based assessment, conversely, potentially relates little to an animal's welfare, as perceived by the animal. The dichotomy is linked to an ethical debate, the extent to which humans should determine animals' environment and how they are used. Animal rights protagonists argue for greater agency for animals, including equivalent outcomes when they are used for different purposes. We may consider the "rabbit conundrum" in relation to this, that is, do rabbits have equivalent welfare when kept as pets, used for meat, pelts, research, or existing in the wild, and if not, do we have the right to predetermine their welfare by using them for different purposes?

Welfare considerations are usually restricted to sentient animals, that is, those with conscious awareness. However, sentience is not an all-or-nothing phenomenon; rather, it is best conceptualized as being on a continuum with substantial variation between and within species in terms of

---

BOX 20.1 PROPOSED LEVELS OF AWARENESS, AS PROPOSED BY YOUNG (1994)

1. Phenomenal awareness: the experience of seeing, hearing, touching, and so on.
2. Access awareness: where stored information is brought to mind.
3. Monitoring: including awareness of our own actions and their effects, and monitoring perceptual information for discrepancies with current plans and hypotheses.
4. Executive awareness: awareness of our goals and intentions.

---

their sensory, perceptive, and cognitive awareness (Piggins and Phillips, 1998; Box 20.1). Furthermore, feelings derived from conscious awareness are transient and merely open a window to an animal's longer-term experiences. The relationship between short-term feelings and an animal's long-term welfare is often difficult to deduce, but clearly not always positive. As examples, animals may have positive feelings toward a handler that provides sweet feed that tastes good, but is inappropriate for the their long-term nutrition, or animals may experience short-term pain, in the case of a prophylactic injection for example, that delivers long-term benefit to their welfare.

Since the choices that animals, including humans, make are not always consistent with good welfare states (e.g., pleasurable experiences can be harmful over the long term), it can be useful to create an index of welfare on two dimensions: one that record animals' experiences and another that applies expert human opinion to judge the impact on animal welfare (Gurusamy et al., 2013). This has the disadvantage that "experts" may be individually or collectively inaccurate and/or biased in their assessment of welfare impact. Furthermore, there have been numerous attempts to identify animals' quality of life (Yeates and Main, 2009). This is a concept borrowed from humans nearing the end of their life for whom it is important to determine, just as it is for companion animals, the extent to which care is justified.

An absolute measurement of animal welfare is not agreed by all scientists, but there has been not only extensive research recently on the key components just outlined for many of our managed species, there have also been tools developed that can provide information on an animals' feelings or emotions (Paul et al., 2005). Animal welfare assessment ideally focuses on direct measurements made on the animals themselves, their stress levels, for example, as indicated by heart rate or behavior. However, although this is beneficial for any scientific assessment of welfare, for practical purposes it is sometimes too difficult, time-consuming, or dangerous for effective assessment in animal establishments. On these occasions, it is possible to derive some information about animal welfare from a knowledge of their environment. Measurements of, for example, ambient temperature will provide some information on whether animals are likely to feel too hot or cold.

A key concept when evaluating animal welfare that was initially developed in 1965 was that of the "Five Freedoms" that animals should possess (Brambell Report, 1965). These were later modified (Webster, 2005; Box 20.2), gaining widespread acceptance in regulatory, industry, and advocacy bodies, including incorporation into many codes of practice. Similar concepts have been developed and amplified into a five-domains model (Box 20.3), which focuses on nutrition, environment, health, and behavior, as well as an overarching mental component (Mellor and Beausoleil, 2015).

BOX 20.2 THE FIVE FREEDOMS, AS DEFINED BY WEBSTER (2005)

1. Freedom from hunger or thirst by ready access to fresh water and a diet to maintain full health and vigor.
2. Freedom from discomfort by providing an appropriate environment including shelter and a comfortable resting area.
3. Freedom from pain, injury, or disease by prevention or rapid diagnosis and treatment.
4. Freedom to express (most) normal behavior by providing sufficient space, proper facilities, and company of the animal's own kind.
5. Freedom from fear and distress by ensuring conditions and treatment that avoid mental suffering.

BOX 20.3 THE FIVE DOMAINS MODEL OF MELLOR AND BEAUSOLEIL (2015)

Physical/functional domains							
Survival-related factors						Situation-related factors	
1: Nutrition		2: Environment		3: Health		4: Behavior	
Negative	Positive	Negative	Positive	Negative	Positive	Negative	Positive
Restricted water and food; poor food quality	Enough water and food; balanced and varied diet	Uncomfortable or unpleasant physical features of environment	Physical environment comfortable or pleasant	Disease, injury and/or functional impairment	Healthy, fit, and/or uninjured	Behavioral expression restricted	Able to express rewarding behavior

Affective experience domains						
5: Mental state						
Negative experiences			Positive experiences			
Thirst	Breathlessness	Anger, frustration	Drinking pleasures	Vigor of good health and fitness	Calmness, in control
Hunger	Pain	Boredom, helplessness	Taste pleasures	Reward	Affectionate sociability
Malnutrition malaise	Debility, weakness	Loneliness, depression	Chewing pleasures	Goal-directed engagement	Maternally rewarded
Chilling/overheating	Nausea, sickness	Anxiety, fearfulness	Satiety		Excited playfulness
Hearing discomfort	Dizziness	Panic, exhaustion	Physical comforts		Sexually gratified
Welfare status						

20.2.2 Anthropocentrism Versus Evolutionary Obligation: Is the Consideration of Animal Welfare and Rights a Developed World Privilege? Should All Cultures/Countries Be Held to the Same Obligations?

Traditionally, many Western ethical perspectives are anthropocentric (i.e., human-centered). Anthropocentric theory is often considered to fall into two categories: the "strong interpretation" and the "weak interpretation" (Tanner, 2009, p. 53). The strong interpretation of anthropocentrism holds that humans are the only beings that are owed intrinsic value, while the weak interpretation considers that the significantly greater degree of intrinsic value being afforded to humans, at the expense of nonhumans, is nearly always justified (Tanner, 2009, p. 54). The anthropocentric position largely captures the practical reality of humans' traditional opinion of animals and how the animals fit into a hierarchy with humans at the apex.

Arguably, the first prominent proponent of this view was Aristotle, who considered that "nature has made all things specifically for the sake of man" and that any value attributed to nonhuman things is purely instrumental (Brennan and Lo, 2011). Generally, advocates for the anthropocentric perspective find it difficult to comprehend why it may be wrong to inflict cruelty upon nonhuman things, such as animals, except where it may lead to negative consequences for humans. For example, Immanuel Kant proposed that allowing a person to inflict cruelty on an animal may encourage that person to have a desensitized approach to cruelty toward other humans (Brennan and Lo, 2011). Therefore, animal cruelty is wrong in an instrumental, rather than intrinsic, sense. Similarly, anthropocentrism also acknowledges the instrumentally negative impacts of anthropogenic environmental devastation, in so far as the destruction of the environment may adversely affect the future of humans (Brennan and Lo, 2011). Consequently, anthropocentrism allows for environmental preservation and animal welfare to the extent that they benefit humans.

In recent times, pro-animal arguments have threatened the anthropocentric perspective by challenging the assumed moral superiority of humans and exploring the possibility of rational arguments in favor of affording intrinsic value to the natural environment and its nonhuman inhabitants (Aaltola, 2010, p. 27). Theorists such as Singer and Regan have driven pro-animal arguments to the forefront of philosophical discussion and present consistent, persuasive, and reasoned arguments in support of assigning rights and morality to animals. Advocates of the anthropocentric position have criticized these emerging arguments, suggesting that they fail to recognize the relevance and importance of long-established, common paradigmatic ways of valuing animals (Aaltola, 2010). The moral status of animals is a product of language and "culturally constructed meanings" and therefore cannot be altered solely by moral analyses (Aaltola, 2010). As such, there is no need to provide further justification for anthropocentrism, as the standard animal ethics arguments are defunct

and moral theory is, at the very least, secondary to conventional human paradigms, which exist. It is supported by the fact that in one study, one-third of university students across several disciplines reasoned according to norms (Verrinder et al., 2016). Hence, this is one of the primary ways in which human society continues to justify its attitudes toward different categories of animals.

20.2.3 Animals' Intrinsic or Extrinsic Value

The acceptability of the ways in which we manage animals cannot be evaluated solely from an assessment of their welfare. It must include an assessment of whether our treatment of the animals is morally justifiable. A fundamental issue is whether animals have a moral value just for the benefit that they bring to other species, in particular humans, or whether they have moral value in their own right. There is no doubt that animals bring us enormous benefits; the question is whether this justifies us using them without consideration of their intrinsic value. Their ability to suffer is important here, which relates to negative emotions. Determining this presupposes that we can classify animals' experiences, and hence emotions, as positive or negative. Emotions may be primary (e.g., anger, disgust, fear, and sadness), in which case it is assumed that they evolved to be hard-wired, or secondary, which are more complex emotions that may be constructs of a variety of primary emotions (e.g., shame, guilt, and grief). While it is likely that most primary emotions are felt by most cognitively advanced animals, the extent to which they experience secondary, more complex, emotions is uncertain.

One prominent theory on how to decide whether actions are morally justifiable or not suggests that we should maximize the overall utility of our actions, which in this case means the welfare of all sentient beings. Under this concept of "utilitarianism," good outcomes should be increased and bad outcomes decreased. Thus, it is possible to conceive experiences as positive or negative in their impact on the animal and then add them in the following way (Phillips, 2009):

$$W = g_1G_1 + g_2G_2 \ldots g_nG_n - b_1B_1 - b_2B_2 \ldots b_nB_n$$

where W, welfare of an animal; G_n, extent of good experience; g_n, the number of G_n good experiences; B_n, extent of bad experience; b_n, the number of B_n bad experiences; and n, number 1 to ∞.

Under utilitarian theory, the action taken should be that which maximizes W for all stakeholders. Critics have made objections to this method of evaluating animal welfare; it is not clear how the extent of experiences can be quantified, although some construct of behavioral and physiological indicators may be considered. Second, utilitarianism assumes equal consideration of interests between stakeholders, whereas in reality we consider stakeholders that are genetically closer to ourselves to be more worthy of good welfare. Third, other things than happiness may be valued, such as having a life or having knowledge, which may assist in perpetuating life. Fourth, utilitarianism may inflict all the penalties on some animals, contradicting any idea of egalitarianism. Peter Singer (1995) famously included nonhuman animal species' preferences in a utilitarian equation, which, if they are given equal consideration to humans, has dire consequences for using animals for consumption and research since the number of human beneficiaries is much fewer than the nonhuman animals supporting them.

It is doubtful whether any of us evaluate moral actions in this way in practice, but it is more probable that we learn to make instinctive decisions based on a framework of this nature. Actions taken stem from our moral sensitivity, motivation, judgment, and character (Rest, 1994). People fall broadly into three categories when deciding whether to take moral action: those acting from a perspective of personal gain and interest, those acting according to the law and social norms, and those acting on the basis of universal ethical principles (Rest et al., 1999; Verrinder and Phillips, 2014).

20.2.4 Natural Versus Legal Rights: Should Applied Ethics Result in Legislative Recognition in Some Cases?

Traditionally, rights discussions have been bifurcated into the categories of natural rights and legal rights. Natural rights are those that innately exist simply as a result of a being's natural characteristics (Crowe and Weston-Scheuber, 2013), even though they are still human constructs and as such subject to cultural and other influences. Legal rights are those that exist under, or by virtue of, a legal system (Wenar, 2005). Therefore, natural rights are considered more universal and egalitarian, while the applicability and strength of legal rights vary greatly depending on the jurisdiction in which they apply. Over the last few decades, growing consideration has been given to whether animals, like humans, also possess natural rights by virtue of their nature as sentient beings (Wenar, 2005). Tom Regan purports, in his deontological perspective, that at least some animals possess moral rights that are objective, in the sense that they do not depend on legal recognition to exist (Regan, 2004). In other words, at least some, if not all, categories of animals possess natural rights, simply as a result of their existence as sentient beings. Animal rights may be pursued rapidly, by those advocating abolition of harmful animal use by humans (abolitionists), or gradually, by those who seek incremental change in achieving the same (protectionists).

The majority of theoretical debate, which discusses what rights may be attributed to what beings, largely focuses upon the moral status of the respective beings and the grounds of, or justifications for, that moral status. A being has moral status if its interests matter in a moral sense "for that being's own sake, such that it can be wronged" (DeGrazia, 1996). For example, it may be argued that an animal possesses moral status if its suffering may be perceived as "somewhat morally bad, on account of this animal itself and regardless of the consequences for other beings" (DeGrazia, 1996). Furthermore, "acting unjustifiably against [an animal's] interests is not only wrong, but wrongs the animal" (DeGrazia, 1996). Accordingly, other beings, both human and animal, owe it to this animal to refrain from acting in such a way that may infringe its interests.

Theories of moral status are often argued to exist in degrees, ranging from no-moral status to full moral status (DeGrazia, 1996). No-moral status refers to objects or beings that have no moral interests that matter to the individual, that is, it cannot be wronged, while full moral status is reserved for those deserving the highest degree of morality, on the grounds of their interests (DeGrazia, 1996; Jaworska and Tannenbaum, 2013). As such, it is widely accepted on the grounds of their respective interests that a cognitively unimpaired adult human possesses full moral status while an object such as a rock may be considered to possess no moral status. Animals too must be viewed as the experiencing subjects of a life, with inherent value of their own (Regan, 2004), since the ability to experience similar needs and desires should be valued (Gruen, 2014).

The idea of measuring morality on a continuum is similar to the concept of an awareness spectrum (Box 20.1), which includes the various levels of sentience animals are presumed to possess. While most would expect animals with high levels of sentience would also possess high moral status, by virtue of their similarity to humans, this relationship is not always logical. The public recognize that sentience declines in approximately the following order: human infant > chimpanzee > dolphin > cat > horse > cattle > pig > rat > chicken > octopus > fish (Phillips et al., 2012). Yet there is no anatomical or physiological reason to suspect that a rat is any less sentient than a cat, or that a fish is less sentient than

a horse. However, people's opinion is almost certainly influenced by the fact that we derive companionship from cats and horses, whereas rats and fish are considered less sentient because they are either pests or animals that are objectified because we want to eat them.

When the science of, for example, farm animals' sentience is measured against that of household pets, it is clear that the former possess similar levels of intelligence and pain perception when compared to the latter. However, when one considers the legal protections in place and the common societal opinions held about the two categories of animals, it is evident that companions are protected to a much greater degree than farm animals. As such, it appears that more than sentience and theoretical morality guides the relationship between humans and animals. One of the primary considerations in affording animal welfare, and perhaps animal rights, is that of utility to human society. Where a greater benefit is bestowed on the human race by consuming an animal, it is easier to provide that animal with a level of welfare only so far as it produces a premium food product, and exclude it from any discussion of animal rights.

This notion is equally applicable to some wild animals and endangered species, whom humans feel an obligation to protect, and sometimes an innate need to preserve elements of the natural world as we deem fit. In particular, animals that we perceive as having high levels of human-like intelligence, such as apes and dolphins, will more likely be recipients of basic animal rights in the future. Animals have the level of intelligence that is needed for their survival in their ecosystem; hence, this speciesist approach is to be discouraged.

20.2.5 Virtue Ethics

We suggested earlier that people solve ethical dilemmas in their own mind by one of three main schemas: (1) reasoning from a perspective of personal gain, (2) according to the law, or

(3) using universal principles. Universal principles suggest a high level of virtue, as most religious and other moral guidance organizations advocate similar virtues, such as doing no harm to living beings, do as you would be done by, and belief in the importance of community. Unlike utilitarian or deontological frameworks, deciding how to treat animals on the basis of the type of person that you would like to be has few of the problems of how to evaluate harm and benefit or who or what deserves moral consideration. For example, aspiring to be a good person by treating animals well is something advocated in all the major religions; in a world with declining religious influence (in much of the Western world at least), pursuit of virtue ethics is an important replacement for conventional religious instruction. If people can be persuaded to develop character traits over a long period of time that are relatively stable and benign to other living beings, this approach suggests that moral education can play a major role in improving animal welfare. If the reward is intrinsic benefit to humans that act virtuously to animals, that is, look after animals well because it is good for you, this would still be an acceptable way to advocate better welfare for animals.

20.2.6 A Life Worth Living or a Good Life?

One endpoint that humans should aspire to if they want to utilize animals for their benefit is to make a decision on the quality of animals' lives, both because it is good for them and because animals have interests that people should respect. According to recent debate, this may include whether animals' lives are worth living (Mellor, 2016). However, given that most animals are genetically programmed through a long process of evolution to have reproduction not happiness, interpreted hedonistically, as their major goal in living, it is possible that many animals could have a life that was worth living (in that they fulfilled this objective), but not a good (or presumably happy) life. Bekoff makes this distinction and advocates a compassionate conservation approach (Bekoff, 2016). Mellor postulates a graded scale of quality of life that includes lives worth living and recognizes that lives worth living are not necessarily good (Table 20.1). There is a danger that we accept that animals' lives are worth living, without considering whether it is morally acceptable to keep animals. This concept suffers from some of the same deficiencies as utilitarianism, for example, who decides, what

TABLE 20.1 A Quality of Life Scale Where the Different Categories Are Defined in Terms of the Relative Balance of Positive and Negative Experiences Animals May Have

Category	Description
A good life	The balance of salient positive and negative experiences is strongly positive. Achieved by full compliance with best practice advice well above the minimum requirements of codes of practice or welfare
A life worth living	The balance of salient positive and negative experiences is favorable, but less so. Achieved by full compliance with the minimum requirements of code of practice or welfare that include elements that promote some positive experiences.
Point of balance	The neutral point where salient positive and negative experiences are equally balanced.
A life worth avoiding	The balance of salient positive and negative experiences is unfavorable, but can be remedied rapidly by veterinary treatment or a change in husbandry practices.
A life not worth living	The balance of salient positive and negative experiences is strongly negative and cannot be remedied rapidly so that euthanasia is the only humane alternative.

Adapted from Mellor, D.J., 2016. Updating animal welfare thinking: moving beyond the "Five Freedoms" towards "A Life Worth Living." Animals 6, 21.

is the currency, and deontological theory, for example, how do we compare the moral value of diverse animal species? However, by analogy the existence of pain in animals was the subject of much cynicism just 20 years ago, and it may not be long before we one day have better methods of evaluating animals' quality of life.

It is clear from the discussion that there are many ways of addressing moral issues. If all produced equally valid outcomes, we might adopt a pluralist or moral relativist approach of accepting that there are many theories that differ from our own personal moral norms and these may be equally valid in their context. However, clearly all do not produce equivalent outcomes for animals, even when we relate these to the animals' interests, rather than their actual circumstances. For example, it is illogical to provide the same feed to farm animals as humans, when they have different digestive systems. Cattle and sheep that digest feed in a rumen need more fibrous feed than humans, who digest it initially in their stomach. Thus, using our own personal norm is not an appropriate method to determine how to feed livestock. However, we can aim to satisfy their interests to a common standard by providing adequate nutrients to ensure normal growth patterns. It is therefore acceptable to advocate that some methods of addressing moral issues are better than others. It is hard to discover how people reason morally, but tests have recently

been developed to objectively determine moral sensitivity and reasoning method in relation to animal ethics issues (Verrinder et al., 2016 and Verrinder and Phillips, 2014, respectively). In the test, people read about an ethical dilemma surrounding animals, such as whether to euthanize unwanted pets, and then a number of carefully constructed questions are posed, which demonstrate reasoning based on personal interest, maintaining norms or universal principles. The respondents are asked which questions are most important to them in solving the dilemma, from which it can be deduced exactly how they are reasoning morally.

Students can be educated to adopt reasoning for ethical dilemmas based on universal principles, and this is especially important for those involved in working with animals, such as veterinary students (Verrinder and Phillips, 2014, 2015). Veterinarians have multiple responsibilities, including to their family, their profession, society, the environment, and not least, the animals within their care. These overlap and cannot be considered in isolation (Preston, 2001). Desirable outcomes for animals and others with interests are generally acknowledged to include seeking a good outcome (beneficence), causing no harm (nonmaleficence), respect for autonomy, and justice (using principles of fairness) (Table 20.2; Mepham, 2008, p. 47). A common framework for the evaluation of outcomes of

TABLE 20.2   An Ethical Matrix to Describe the Outcomes for the Principal Interested Parties (Animals, Consumers, and Producers), From the Perspective of Utilitarianism, Autonomy, and Justice as Fairness, When Considering Whether It Is Ethical to Inject Dairy Cows Daily With a Synthetic Hormone That Stimulates Them to Produce More Milk (Mepham, 2008)

	Parties to problem		
Principle	Animal	Consumer	Producer
Utilitarianism	Welfare of cow	Availability of safe food	Producer income and working conditions
Autonomy	Behavioral freedom	Freedom of choice (food labeling?)	Freedom of choice to adopt or not adopt certain practices
Fairness or justice	Equal consideration of interests?	Universally affordable food?	Fair trade?

action on animal ethics issues is an ethical matrix (Mepham, 2008, p. 54), in which the key principles of utilitarianism, autonomy, and justice as fairness are considered for the major stakeholders in a decision-making process. The ethical matrix does not arrive at a conclusion, but it does assist the process by being able to visualize the impact on all parties. Not all parties have equal interests and not all principles have equal meaning to the interested parties. To obtain a decision, a different model is necessary, such as the ethical decision-making model developed by Preston (2001, p. 78). Focusing on respect for life, justice as fairness, and integrity principles, Preston advocates initially analyzing these three basic concepts from the perspectives of the stakeholders. Then he recommends considering possible decisions from a number of different perspectives: what if the decision was universalized, how is the social good served, how does the decision relate to the type of person I ought to be, do any factors warrant greater priority, and finally, is the decision feasible.

20.3 DIFFERENCES IN PEOPLE'S MORAL BEHAVIOR TOWARD ANIMALS

Despite the existing commonality or universalism in people's attitudes to animals, many factors are recognized to have an influence on these attitudes, including culture, religion, and gender. A benign attitude to animals is also linked to attitudes toward other important world issues, such as women's rights and racial equality (Sinclair and Phillips, 2017).

20.3.1 Cultural and Religious Differences

Acknowledging and understanding cultural and religious differences in attitudes to animals is assuming a new importance in an increasingly globalized world, with growing trade links and demands for animal products that cannot be met by local production. A better understanding of these influences will enable education to be targeted where needed, trade in animals between diverse cultures to be facilitated, and multicultural societies to exist in greater harmony.

A Eurasian survey of university students found that culture, as determined by the students' nationality, had a major influence (Phillips et al., 2012). Students in southern European countries had the most benign attitudes to the welfare and rights of animals, and those in a limited number of Asian countries the least. The latter appeared at least partly explicable by the low income of students in Asia, which focuses attention to the welfare of humans, rather than animals. The key religious teachings of the major faiths indicate that there are also significant differences between Christians/Muslims/Jews, who believe that God put animals on the earth for the benefit of humans, who therefore have dominion over them, and Hindus/Buddhists, who believe that deeds in the current life will be reflected in their rebirth, which may be as an animal. The latter belief engenders a great respect for animals, which in the case of Hindus is reflected in their reverence for animals, particularly the cow. Because cow slaughter is illegitimate in Hindu society, surplus animals are offered shelter in gaushallas, which are supported by philanthropy and the Indian government. Such a model for animal caring contrasts markedly with those operating in Western society, in which domesticated animals are destroyed when they no longer provide benefit to humans. Similarly, in Southeast Asia, Hindus and Buddhists are less accepting of killing injured animals or healthy pets than Muslims (Ling et al., 2016).

20.3.2 Gender Differences

Women usually self-report more empathetic attitudes and behavior toward animals than do men (Herzog, 2007). However, this is only the case if women are sufficiently empowered, either within a relationship (Baxter and Kane, 1995) or because they live in a country that is supportive of women's rights (Phillips et al., 2010). Women and men show similar attachment to pets, but women are much more likely to be involved in animal protection and less likely to be involved in perpetrating animal cruelty, including bestiality (Herzog, 2007). In terms of action to improve the welfare of animals, women are more likely to be members of an animal advocacy organization (Phillips and Izmirli, 2012). However, although women are more likely than men to avoid eating meat, this is more likely to be because of perceived health benefits, rather than for animal ethics reasons (Phillips et al., 2010).

20.3.3 Experience With Animals

Adults that have kept pets in childhood have more positive attitudes toward pet animals later (Paul and Serpell, 1993), suggesting that such people have learnt how to respect and look after animals in childhood. This positive attitude generalizes to nonpet animals and even humans. Furthermore, a love of animals and extensive experience with them, particularly pets, are the main factors influencing students to want to become veterinarians (Izmirli et al., 2014; Serpell, 2005). Taken more generally, experience with animals can be taken to include education in animal management. One of the disadvantages of considering virtue ethics to provide the best moral imperative regarding how we should approach animal ethics is that some people have the advantage of having received a virtuous education in how to manage animals, whereas others have not. Acknowledging the merit of pursuing virtue ethics must include recognition of a person's lack of direction, or opportunity, to pursue these ideals in the past.

20.4 THE FUTURE OF ANIMAL WELFARE AND RIGHTS

The growth in affluence and an associated demand for animal products has been particularly rapid in Asia over the last 30 years. This has led to many developing countries there adopting intensive poultry and pig production systems in an attempt to meet the growing demand. In China and major exporting countries, such as Brazil, growth in chicken meat exports has been exponential since the late 1990s (Phillips, 2015, p. xvii). There are now approximately 20 billion chickens produced globally to meet the needs of humans for meat and eggs (FAOSTAT, 2016), almost three for every person on the planet, and most are kept in intensive production systems. This trend is repeated, although not quite so dramatically, for the other major meat- and milk-producing animals: cattle and pigs. Furthermore, as there is growing reliance on chickens for meat production, at the expense of the larger animals, cattle and pigs, this means that more sentient beings are used to produce a given quantity of meat. Taken together, the increase in the number of food-producing animals and the intensity of their management, the focus on small animals, and the transition of production from developed to developing countries, mean that there is a rapidly increasing number of food-producing animals experiencing poor welfare globally. As food-producing animals make up by far the majority of animals that we manage globally, approximately 90% (Phillips, 2009, p. 152), it is evident that our negative impact on animals' welfare globally is increasing. This, together with the inefficiency of resource use, is likely to cause humans to increasingly question the use of scarce feed, water, and land resources for this purpose, particularly in the light of continued human population increases and food shortages (FAO et al., 2014).

By reflecting on past human societies and animals, it is evident that our relationship with animals has evolved from one of necessity, to one that comprises a lesser degree of dependency, but greater intensity of production. While the future of society is unpredictable, one certainty is that, should this intensive production continue to grow at the rapid rate at which it has developed over the last century, the sustainability of the human–animal relationship and the broader environment is likely to be at risk. It may well be that, in the absence of intervening action, humans will steadily lose the choice involved in the present relationship with animals.

Greater human diversity in their use of animals for food is anticipated if prices increase as a result of increasingly scarce feed, water, and land resources on which agricultural producers rely. Meat consumption may no longer be a dietary staple, but confined to the richer echelons of society. Diversity in income is increasing worldwide, leading to a concern that the richer elements of society will utilize a much greater proportion of resources for food production, including for meat, than the poorer elements, for whom staple foods may be increasingly scarce.

It is essential to understand the impact of our actions toward animals today, and the effect such actions may have on the future relationships between animals and human societies, as well as broader environmental sustainability. While it is important for all members of society to consider their position carefully when interacting with animals, whether it be for food, entertainment, or companionship, it is of particular significance for those in the fields of animal science, management, and care to lead by example. Such people are likely to have the most control on the direction in which this relationship takes in the future, not only as consumers but also as agents of science, law, and ethics. After all, it is the actions of today that pave the path for tomorrow.

References

Aaltola, E., 2010. The anthropocentric paradigm and the possibility of animal ethics. Ethics Environ. 15, 27–50.

Baxter, J., Kane, E.W., 1995. Dependence and independence: a cross-national analysis of gender inequality and gender attitudes. Gender Soc. 9, 193–215.

Bekoff, M., 2016. Is "A Life Worth Living" a "Good Life" for Other Animals? Psychology Today, March 26. Available from: https://www.psychologytoday.com/blog/animal-emotions/201603/is-life-worth-living-good-life-other-animals.

Brambell Report, 1965. Report of the Technical Committee to Enquire Into the Welfare of Animals Kept Under Intensive Livestock Husbandry Systems. Her Majesty's Stationery Office, London, UK.

Brennan, A., Lo, Y.S., 2011. Environmental ethics. In: Zalta, E.N. (Ed.), The Stanford Encyclopedia of Philosophy. Available from: http://plato.stanford.edu/archives/fall2011/entries/ethics-environmental/.

Broom, D.M., 1986. Indicators of poor welfare. Br. Vet. J. 142, 524–526.

Colvin, C.M., Marino, L., 2015. Signs of Intelligent Life. Natural History Magazine, Inc, New York, NY.

Crowe, J., Weston-Scheuber, K., 2013. Principles of International Humanitarian Law. Edward Elgar Publishing, Cheltenham, p. 116.

Daros, R.R., Costa, J.H.C., von Keyserlingk, M.A.G., Hoetzel, M.J., Weary, D.M., 2014. Separation from the dam causes negative judgement bias in dairy calves. PLoS One 9 (5), e98429.

Darwin, C., 1859. On the Origin of Species by Means of Natural Selection, first ed. John Murray, London, p. 83.

DeGrazia, D., 1996. Taking Animals Seriously: Mental Life and Moral Status. Cambridge University Press, Cambridge, p. 262.

Duncan, I.J.H., Fraser, D., 1997. Understanding animal welfare. In: Appleby, M.C., Hughes, B.O. (Eds.), Animal Welfare. CABI, Oxford, UK, pp. 19–31.

FAO, 2009. Change in the livestock sector. In: The State of Food and Agriculture: Livestock in the Balance. Available from: http://www.fao.org/docrep/012/i0680e/i0680e.pdf.

FAO, IFAD, and WFP, 2014. The State of Food Insecurity in the World 2014. Strengthening the Enabling Environment for Food Security and Nutrition. FAO, Rome.

FAOSTAT, 2016. World chickens, 2013. Available from: http://faostat3.fao.org/browse/Q/QA/E.

Francione, G., 2008. Animals as Persons: Essays on the Abolition of Animal Exploitation. Columbia University Press, New York, NY, p. 118.

Fraser, D., Weary, D.M., Pajor, E.A., Milligan, B.N., 1997. A scientific conception of animal welfare that reflects ethical concerns. Anim. Welfare 6, 187–205.

Gruen, L., 2014. The moral status of animals. In: Zalta, E.N. (Ed.), The Stanford Encyclopedia of Philosophy (Fall 2014 Edition). Available from: http://plato.stanford.edu/archives/fall2014/entries/moral-animal/.

Gurusamy, V., Tribe, A., Phillips, C.J.C., 2013. Identification by stakeholders of major welfare issues for captive elephant husbandry. Anim. Welfare 23, 11–24.

Herrero, M., Haylik, P., Valin, H., Notenbaert, A., Rufino, M.C., Thornton, P.K., Blummel, M., Weiss, F., Grace, D., Obersteiner, M., 2009. Biomass use, production, feed efficiencies, and greenhouse gas emissions from global livestock systems. Proc. Natl. Acad. Sci. USA 110 (52), 20888–20893.

Herzog, H.A., 2007. Gender differences in human-animal interactions: a review. Anthrozoos 20, 7–21.

Izmirli, S., Yigit, A., Phillips, C.J.C., 2014. The attitudes of Australian and Turkish students of veterinary medicine towards animals and their careers. Soc. Anim. 22, 580–601.

Jaworska, A., Tannenbaum, J., 2013. The grounds of moral status. In: Zalta, E.N. (Ed.), The Stanford Encyclopedia of Philosophy. Available from: http://plato.stanford.edu/archives/sum2013/entries/grounds-moral-status/.

Ling, R.Z., Zulkifli, I., Lampang, P.N., Nhiem, D.V., Wang, Y., Phillips, C.J.C., 2016. Attitudes of students from Southeast and East Asian countries to slaughter and transport of livestock. Anim. Welfare 25, 377–387.

Mason, G.J., Veasey, J.S., 2010. How should the psychological well-being of zoo elephants be objectively investigated? Zoo Biol. 29, 237–255.

Matson, P.A., Parton, W.J., Power, A.G., Swift, M.J., 1997. Agricultural intensification and ecosystem properties. Science 227, 504–509.

McInerney, J., 2004. Animal welfare, economics and policy: report on a study undertaken for the Farm and Animal Health Economics Division of DEFRA. Department of the Environment, Food and Rural Affairs (UK). p. 1.

Mellor, D.J., 2016. Updating animal welfare thinking: moving beyond the "Five Freedoms" towards "A Life Worth Living". Animals 6, 21.

Mellor, D.J., Beausoleil, N.J., 2015. Extending the 'Five Domains' model for animal welfare assessment to incorporate positive welfare states. Anim. Welfare 24, 241–253.

Mepham, B., 2008. Bioethics: An Introduction for the Biosciences. Oxford University Press, Oxford.

Paul, E.S., Harding, E.J., Mendl, M., 2005. Measuring emotional processes in animals: the utility of a cognitive approach. Neurosci. Biobehav. Rev. 29, 469–491.

Naylor, R.L., 1996. Energy and resource constraints on intensive agricultural production. Annu. Rev. Energ. Environ. 21, 99–123.

Paul, E.S., Serpell, J.A., 1993. Childhood pet keeping and humane attitudes in young adulthood. Anim. Welfare 2, 321–337.

Phillips, C.J.C., 2009. The Welfare of Animals: The Silent Majority. Springer, Dordrecht, 9–10.

Phillips, C.J.C., 2015. The Animal Trade. CAB International, Wallingford, Oxford, UK, 208 pp.

Phillips, C.J.C., Izmirli, S., 2012. The effects of levels of support for animal protection organisations on attitudes to the use of animals and other social issues. Anim. Welfare 21, 583–592.

Phillips, C.J.C., Izmirli, S., Aldavood, S.J., Alonso, M., Choe, B.I., Hanlon, A., Handziska, A., Illmann, G., Keeling, L., Kennedy, M., Lee, G.H., Lund, V., Mejdell, C., Pelagic, V.R., Rehn, T., 2010. An international comparison of female and male students' attitudes to the use of animals. Animals 1, 7–26.

Phillips, C.J.C., Izmirli, S., Aldavood, S.J., Alonso, M., Choe, B.I., Hanlon, A., Handziska, A., Illmann, G., Keeling, L., Kennedy, M., Lee, G.H., Lund, V., Mejdell, C., Pelagic, V.R., Rehn, T., 2012. Students' attitudes to animal welfare and rights in Europe and Asia. Anim. Welfare 21, 87–100.

Piggins, D., Phillips, C.J.C., 1998. Awareness in domesticated animals—concepts and definitions. Appl. Anim. Behav. Sci. 57, 181–200, In special issue on Animal Awareness, Concepts and Implications for Domesticated Animals, Piggins D., Phillips, C.J.C. (Eds.).

Preston, N., 2001. Understanding Ethics, second ed. The Federation Press, Sydney.

Regan, T., 2004. The Case for Animal Rights, Updated With a New Preface. University of California Press, Berkeley/Los Angeles, CA, p. 360.

Reid, J.F., 2011. The impact of mechanization on agriculture. Bridge Agric. Inf. Technol. 41, 22–29.

Rest, J., 1994. Background: theory and research. In: Rest, J., Narvaez, D. (Eds.), Moral Development in the Professions: Psychology and Applied Ethics. Lawrence Erlbaum Associates, Hove, UK, pp. 1–26.

Rest, J., Narvaez, D., Bebeau, M., Thoma, S., 1999. A Neo-Kohlbergian approach: the DIT and schema theory. Educ. Psychol. Rev. 11 (4), 291–324.

Scruton, Roger, 1998. Animal Rights and Wrongs. Claridge Press, London.

Serpell, J.A., 2005. Factors influencing veterinary students' career choices and attitudes to animals. J. Vet. Med. Educ. 32, 491–496.

Sinclair, M., Phillips, C.J.C., 2017. The cross-cultural importance of world social issues.

Singer, P., 1995. Animal Liberation. Pimlico, London.

Spruijt, B.M., van den Bos, R., Pijlman, F.T., 2001. A concept of welfare based on reward evaluating mechanisms in the brain: anticipatory behaviour as an indicator for the state of reward systems. Appl. Anim. Behav. Sci. 72, 145–171.

Tanner, J., 2009. The argument from marginal cases and the slippery slope objection. Environ. Values 18, 53.

Verrinder, J.M., Ostini, R., Phillips, C.J.C., 2016. Differences in moral judgment on animal and human ethics issues between university students in animal-related, human medical and ethics programs. PLoS One 11, e0149308.

Verrinder, J.M., Phillips, C.J.C., 2014. Development of a moral judgement measure for veterinary education. J. Vet. Med. Educ. 41, 258–264.

Verrinder, J.M., Phillips, C.J.C., 2015. Assessing veterinary and animal science students' moral judgment development on animal ethics issues. J. Vet. Med. Educ. 42, 206–216.

Webster, J., 2005. Animal Welfare: Limping Towards Eden. Blackwell, Oxford, 12–18.

Wenar, L., 2005. The nature of rights. Philos. Public Aff. 33, 237–239.

Yeates, J., Main, D., 2009. Assessment of companion animal quality of life in veterinary practice and research. J. Small Anim. Pract. 50, 274–281.

Young, A.W., 1994. Neuropschology of awareness. In: Revonsuo, A., Kamppinnen, M. (Eds.), Consciousness in Philosophy and Cognitive Neuroscience. Lawrence Erlbaum, Hillsdale, NJ, (Chapter 8).

21

Animal Extinctions

Samia R. Toukhsati

Honorary Fellow, The University of Melbourne, Parkville, VIC, Australia

The last word in ignorance is the person who says of an animal or plant: what good is it? —*Aldo Leopold 1949*

Extinctions refer to the death of a single or multiple species (or taxon) and are common in the history of life on this planet. Using the fossil record, it has been estimated that 99.9% of the species that existed on earth is now extinct. Extinctions occur when a species fails to meet or adapt to changing environmental forces (such as global warming or cooling, habitat loss, destruction, or fragmentation) or when species origination is low, creating ecological niches for new, better adapted, species. This process of "background" extinction and new species evolution is natural, occurs continuously, and describes the way life diversified and radiated on this planet. However, when extinctions involve vast numbers of species and appear to occur around the same time in many different regions, as may be the case in modern times, they are termed "mass extinctions"; these are much less common, but greatly reduce species diversity. There is much debate and little

consensus as to the cause and timescale of mass extinctions, generally referred to as the "Big Five" extinction events, which mark the point of transition to new geological epochs (Braje and Erlandson, 2013). According to Sodhi et al. (2012), the "Big Five" mass extinction events share the following common features: (1) "a catastrophic loss of global biodiversity," (2) "unfolding rapidly," (3) "taxonomically, their impact was not random," and (4) survivors —"not previously dominant evolutionary groups". The "Big Five" may have been caused by sudden environmental changes due to seismic activity and/or catastrophic events (such as an asteroid impact). Many scientists agree that we are in the middle of a sixth mass extinction event right now; Sodhi et al. (2012), along with many others, consider that humans are playing a direct or indirect role in the "100- to 10,000-fold" increase in the rate of extinctions over background. This chapter will focus on the modern day Holocene–Anthropocene extinction which attribute the possible loss of up to 58,000 species per year to human activities (Dirzo et al., 2014).

Animals and Human Society
http://dx.doi.org/10.1016/B978-0-12-805247-1.00031-9

499

21.1 THE "BIG FIVE" MASS EXTINCTIONS

We know from the fossil record that extinctions are not unusual events in the history of life on this planet. In fact, more that 99% of the species that once lived on this planet is now extinct. Extinctions are an important driver of species evolution; as one species disappears, another species better suited to the environment fills its ecological niche. The timescale for species evolution is vast, occurring over many millions of years, and is inextricably linked to environmental conditions. These so called "natural" or "background" extinctions occur very slowly (such as in response to changing climate conditions associated with earth's changing land mass; Fig. 21.1), but may occur rapidly as a consequence of catastrophic events in earth's history (such as a meteor or asteroid impact).

Five major taxonomic extinctions, known as the "Big Five Mass Extinctions," have traditionally been used to describe the history of life on earth, which occurred at the end of the Ordovician, Devonian, Permian, Triassic, and Cretaceous periods, respectively. An additional period of extinction is likely during the so called "snow-ball" period. There is increasing acceptance that global glaciation extending to the tropic regions occurred in earth's history with two such occurrences happening between 710 and 625 million years ago (Corsetti, 2015). To qualify as one of the so called "Big Five" extinctions requires that the extinction interval is unusual (i.e., higher than usual background extinction rates), sudden, and involves catastrophic loss of life, rather than the slow demise of a given species, as seen in natural extinctions (Kerr, 2001). The "Big Five" were thought to have been caused by sudden environmental changes (such as a relatively sudden drop or increase in temperature or atmospheric changes due to seismic activity) and/or catastrophic events (such as an asteroid impact).

The following are the causes of extinction events (Sodhi et al., 2012):

1. Climatic environmental changes including glaciation, desertification, and sea level changes.
2. Geological events such as earthquakes, volcanic eruptions, and continental drift. An example of such an event with importance was the establishment of a land bridge between South and North America allowing the movement of animals.
3. Anthropogenic causes:
   a. Land use changes (e.g., to agriculture or palm oil production or urbanization or mining) leading to loss of habitat loss and habitat fragmentation.
   b. Invasive species (introduced either purposively or accidentally by humans). Examples include the following:
      - The accidental introduction of the mosquito, *Culex quinquefasciatus*, to Hawaii in 1826 together with the single-celled parasite (*Plasmodium relictum*) led to the spread of malaria to wild birds (reviewed by Sodhi et al., 2012).
      - The purposeful introduction of rats (*Rattus exulans*) to Polynesian islands
      - The accidental introduction of rats (*Rattus rattus* and *Rattus norvegicus*) to islands in the Indian Ocean and around Antarctica.
   c. Overexploitation by hunting and/or consumption of eggs.
4. Population size declining below minimum viable say following decline due to other factors.

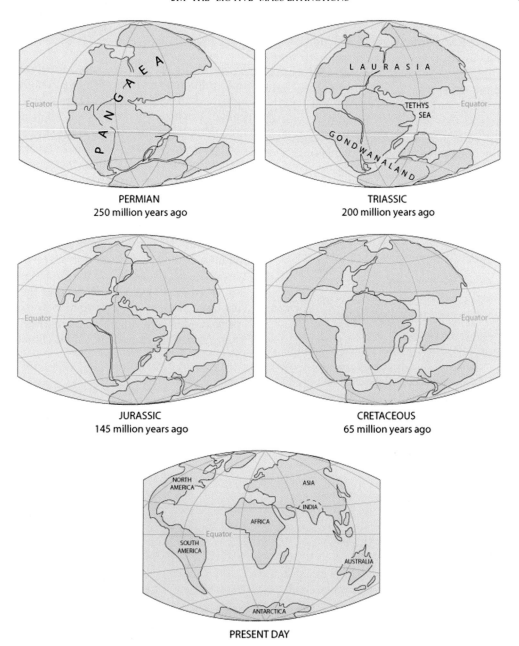

FIGURE 21.1 **Pangaea: http://pubs.usgs.gov/gip/dynamic/historical.html.** *Source: Reprinted with permission from USGS Science Publishing Network.*

Recent evidence appears to challenge the legitimacy of the "Big Five" extinctions, suggesting a slower rate of extinctions, with intermittent phases of increased losses, due to low origination rates (i.e., failure to produce new species) (Braje and Erlandson, 2013; Kerr, 2001). Nonetheless, the "Big Five" provides a useful reference point for exploring species evolution and extinctions throughout earth's history.

Timeline of "Big Five" mass extinction events (Webb, 2013):

1. *444 million years ago: First Mass Extinction—Late Ordovician*
   a. Caused by change in climate conditions.
   b. Loss of up to 25% of marine families and 85% of species of marine life.
2. *407–360 million years ago: Second Mass Extinction—Late Devonian*
   a. Cause uncertain, but probably include change in climate conditions.
   b. Loss of up to 20% of marine families and 80% of all animal species, with marine life impacted most significantly.
3. *266–252 million years ago: Third Mass Extinction—Late Permian (The Great Dying)*
   a. Caused by asteroid/comet impact or volcanic eruptions.
   b. Loss of up to 50% of all families, up to 96% of marine species, and 70% of terrestrial species.
4. *200 million years ago: Fourth Mass Extinction—Late Triassic*
   a. Cause uncertain, but may have included climate change, volcanic eruptions, and lava floods, or asteroid impact.
   b. Loss of 20% of families of marine life and loss of up to 76% of present-day species.
5. *65 million years ago: Fifth Mass Extinction—Late Cretaceous*
   a. Caused by asteroid/comet impact or climate change due to volcanic activity.
   b. Loss of up to 80% of all animal species, including dinosaurs.

21.1.1 First Mass Extinction: Late Ordovician

The Ordovician period lasted for approximately 45 million years (from 488 to 443 million years ago) and describes a period in earth's history known for its thriving diversification of marine invertebrates (such as trilobites and molluscs) and early vertebrates (such as conodonts) and the beginning of the colonization of the supercontinent Gondwana by early terrestrial arthropods and flora (Fig. 21.2). The Ordovician was characterized by two extraordinary events that significantly impacted species biodiversity on this planet. The Great Ordovician Biodiversification Event saw a massive escalation in marine biodiversity, which was suddenly disrupted by the First Mass Extinction (Harper et al., 2014). The end of the Ordovician Extinction saw the loss of up to 85% of marine species and is believed to have occurred over two phases, possibly caused by glacial climate conditions (Harper et al., 2014).

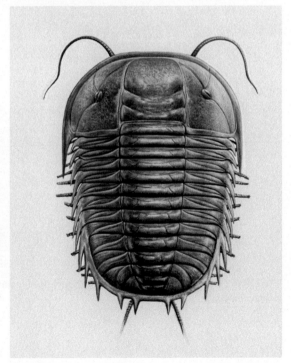

FIGURE 21.2 **Trilobite.** *Source: Courtesy Wikimedia Commons.*

21.1.2 Second Mass Extinction: Late Devonian

The Devonian period, also known as the Age of Fishes, lasted from approximately 420 to 358 million years ago and was characterized by the continued evolution of diverse marine species and the first significant diversification of terrestrial flora and fauna, with the land colonized by plants and insects. Two major clades of vertebrate fish equipped with gills, bladders, and teeth dominated (i.e., the Lobe-finned and Ray-finned fish clades; Fig. 21.3). Marine species flourished, as recorded in the Gogo Formation Lägerstatte, which preserved the three-dimensional fossilized record of reef life in Western Australia, such as the lobe-finned Gogonasus. Lobe-finned fish went on to become the ancestors of aquatic and terrestrial tetrapods (i.e., the first four-limbed vertebrates and their descendants), whereas most modern fish are descended from Ray-finned fish. Large marine predators, such as placoderms and cartilaginous species, such as sharks and rays, flourished in the Devonian period.

The Late Devonian period was also characterized by a series of small extinction events over a period of approximately 20 million years that decimated up to 75% of invertebrate species, particularly marine and freshwater species, such as species of brachiopods and trilobites. This long period of extinctions is thought to have been marked by several major extinction events; the Lower Zilchov, Taghanic, Kellwasser, and Hangenberg events. The cause of the extinctions remains uncertain and controversial, with some prominence of the theory that it was caused by glaciation arising from changes in the availability of atmospheric carbon dioxide attributable to earth's first forests.

21.1.3 Third Mass Extinction: Late Permian

The Permian Mass Extinction, otherwise known as the Great Dying, was "the number one crisis in the history of life" on this planet (Kerr, 2001). Between 299 and 248 million years ago, there was a series of extinction phases that saw the extinction of up to 95% of shallow-water marine species and 70% of terrestrial species, including many families of insects. While the timespan (thought to be somewhere between 60,000 and 15 million years) and cause of the Permian extinction (e.g., rising methane levels

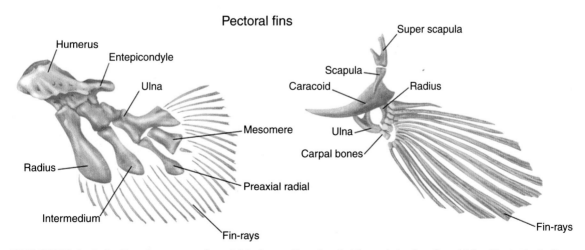

FIGURE 21.3 **Lobe-finned versus ray-finned fish.** *Source: Reproduced with permission from the publisher. Illustration by Karen Carr, www.karencarr.com.*

attributable to methane-producing archaea, resulting in global warming, ocean acidification, and changes in the carbon cycle as one example), remain contentious, with the loss of over half of all taxonomic families, there is agreement that this was the most severe of the "Big Five" Mass Extinctions (Burgess et al., 2014).

21.1.4 Fourth Mass Extinction: Late Triassic

The Triassic mass extinction, occurring over two or three phases between 252 and 201 million years ago, resulted in the loss of up to 76% of all marine and terrestrial species alive at the time, with significant losses to marine species such as radiolarians (protozoa that produce mineral skeletons), ammonoids (marine molluscs), conodonts (primitive marine vertebrates), marine bivalves (such as scallops) and reef-building organisms (such as corals), and terrestrial tetrapods (such as amphibians and reptiles) (Lucas and Tanner, 2008). Recent hypotheses challenge the theory that the Triassic mass extinction occurred rapidly over a period of 10,000 years and suggest a more prolonged period of "elevated extinction rates and low origination rates" extending for 20 to 30 million years (Lucas and Tanner, 2015). Accordingly, there is little agreement as to the cause of the Triassic mass extinction, with recent data suggesting that it may have been caused by atmospheric pollution (such as increased release of CO_2 and/or methane gases) due to intense seismic activity and consequent changes in temperature and ocean acidification, rather than originating from a single event, such as a meteor impact (Lindström et al., 2015; Schootbrugge and Wignall, 2016). For instance, significant fluctuations in environmental temperature, both heating and cooling, may have caused losses in food availability (such as a decrease in surface-dwelling phytoplankton) and consequently impacted marine bioproductivity origination (Lucas and Tanner, 2008). Whatever the cause, the decimation of many species of reptiles of the time, such as phytosaurs (large semiaquatic carnivorous reptiles that

morphologically resembled crocodiles), heralded the beginning of a new era that enabled dinosaurs to dominate the earth.

21.1.5 Fifth Mass Extinction: Late Cretaceous

The Cretaceous period began around 145 million years ago and ended approximately 65 million years ago with a mass extinction event that eliminated up to 80% of all species alive at the time. In the Early Cretaceous, land was predominantly assembled into two continents (i.e., Gondwana and Laurasia); by the end of this period, extending over approximately 79 million years, most present-day continents were separated by large expanses of water, with the exception of Australia, which was still connected to Antarctica. The climate is believed to have been warmer during this period, with the polar regions were dominated by forests rather than ice sheets and glaciers; oceans were correspondingly at their highest known level in earth's history (between 330 and 820 feet higher than present-day sea levels).

The Cretaceous era marks a period of species modernization giving rise to early relatives of present-day species, such as flowering plants (Angiosperms), a substantial increase in the diversification of birds and insects, the evolution and radiation of early species of mammals (including placental and marsupial mammals), abundant marine life productivity, and the last reign of the land-based dinosaurs, such as the apex predator, Tyrannosaurus Rex, and the largest terrestrial animal ever known to have lived, Dreadnoughtus (Fig. 21.4).

The Cretaceous mass extinction, which decimated most large vertebrate species (including the dinosaurs, marine, and flying reptiles) and many marine invertebrates, is thought to have been caused by the catastrophic impact of a massive meteor or asteroid in the Yucatan Peninsula, Mexico (forming the Chicxulub crater), as evidenced by a layer of sediment all over the world believed to comprise high concentrations

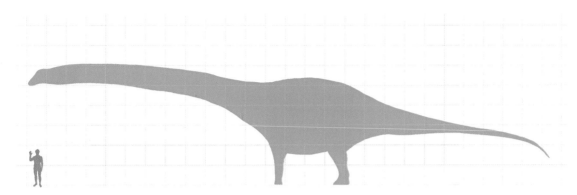

FIGURE 21.4 **Dreadnoughtus.** *Source: Courtesy Wikipedia Commons.*

of elements common in meteorites, such as iridium. In their seminal paper, two American scientists, Luis and Walter Alvarez, proposed that powdered rock debris was ejected into earth's stratosphere following a massive asteroid impact, which cloaked the earth in darkness for several years, disrupting the process of photosynthesis and destroying plant life (including microscopic marine algae), and, ultimately, led to a collapse in the base of marine and terrestrial food chains (Alvarez et al., 1980). This theory, like all other Mass Extinction theories, is subject to ongoing revision with some recent hypotheses proposing that the Cretaceous extinction may have been caused by environmental factors, such as global warming due to increased atmospheric gases caused by volcanic activity, and/or habitat loss due to tectonic plate rearrangement.

21.2 SIXTH MASS EXTINCTION: THE HOLOCENE–ANTHROPOCENE EXTINCTION

Whereas the cause/s for species extinctions throughout earth's history up to recent times are largely thought to be attributable to "gradual or rapid changes in oceanographic, atmospheric, or climate conditions due to a random or cyclical coincidence of causative factors" (Alvarez et al., 1980) that kill off or reduce new species

origination, modern-day extinctions are frequently thought to be attributable to human activities. Presently, the rate of species extinctions is estimated to be approximately 100–1000 times faster than past natural rates of background extinction (Braje and Erlandson, 2013), with the possible loss of up to 58,000 species per year (Dirzo et al., 2014), suggesting that we are experiencing a "current-day biodiversity crisis" (Turvey and Fritz, 2011)—the Sixth Mass Extinction. While there is little agreement as to when this extinction began, there is broad consensus that correlates species extinctions with human dispersal across the globe and consequent human activities, particularly since the industrialized era such as "habitat modification, fragmentation and destruction; overexploitation of species; the spread of invasive species and genes; pollution; and climate change" (Wagler, 2011). For example, whereas primitive humans lived among diverse and abundant large megafauna (e.g., megaherbivores ≥1000 kg and megacarnivores ≥100 kg), losses in these species coincide with the radiation and expansion of human populations across the globe (Malhia et al., 2016). For example, in North America, megafauna such as saber-toothed cats (*Smilodon* spp.), mammoths (*Mammuthus* spp.), and giant ground sloths (*Megalonyx jeffersonii*) became extinct within 2,000 years of the arrival of the first Americans/Native Americans 13,000 years ago (Sodhi et al., 2012) (see also Section 21.4.2).

Humans and the Demise of Megafauna in Australia

In the Late Pleistocene, Australia, Tasmania, and New Guinea were linked due to lower sea level occurring in glaciation periods and, collectively, were known as the continent of Sahul; during this time, 54 species of megafauna (species weighing >40 kg) became extinct (Wroe et al., 2013). The majority of the extinctions occurred prior to the arrival of the indigenous people over 45,000 years ago with extinction of, for instance, the 2.8-ton marsupial herbivore (*Diprotodon optatum*) and the 100- to 130-kg marsupial lion (*Thylacoleo carnifex*) (Wroe et al., 2013). With the arrival of the indigenous peoples, human involvement in extinctions was likely (Roberts and Brook, 2010; Roberts et al., 2001; Turney et al., 2008). For example, there is strong evidence that humans directly contributed to the extinction of the large flightless bird, *Genyornis newtoni*; eggshells from the bird have burn pattern consistent with cooking in fires and then casual disposal (Miller et al., 2016). These have been dated to prior to 45,000 BP but not before human arrival (Miller et al., 2016). It is suggested that humans caused the extinction of this species by reducing its reproductive success (Miller et al., 2016) (Fig. 21.5).

Humans and the Demise of Megafauna in New Zealand

It is estimated that New Zealand lost 36 avian species between the initial settlement by Polynesians and the arrival of European settlers (Holdaway, 1989). Dogs, pigs, and Pacific or Polynesian rats (kiore or *R. exulans*) were brought to New Zealand by the Polynesian settlers (McClone, 1989). The extinction of species of small birds such as some species of kiwis was due to predation and consumption of eggs by kiore (McClone, 1989). Dogs played a role in the extinction of large ground-living birds (McClone, 1989).

FIGURE 21.5 *Genyornis newtoni*. *Source: From Nobu Tamura, 2008, Wikicommons.*

Perhaps beginning 50,000 years ago, it has been estimated that the current-day biotic crisis may eradicate up to 50% of all plant and animal species presently alive on earth, potentially triggering the collapse of ecosystems and food economies (Braje and Erlandson, 2013). It is estimated that up to 33% of all species of vertebrates are presently threatened or already endangered, with similar, albeit less certain, estimates for invertebrates and insects (Dirzo et al., 2014). There is growing consensus among the scientific community that human activities are central to the accelerating rates of extinction of nonhuman species, prompting some scientists to define the current era as "The Anthropocene: The Age of Man" (Nielsen and Nurzynski, 2013; Wagler, 2011). The following sections examine some of the causes and consequences of the present-day extinction era on biodiversity (Tables 21.1 and 21.2).

TABLE 21.1   Examples of Extinctions of Flightless Birds

Region	Species	Extinction date	Cause(s)	References
Australia	Large flightless bird (*G. newtoni*)	>30,000 years ago	Human direct; presumed to be consumption of eggs together with overhunting	Field and Boles (1998)
New Zealand	Moas (Ratite); giant moa (*Dinornis novaezealandiae*; *Dinornis robustus*); stout-legged moa (*Euryapteryx gravis*); heavy-footed moa (*Pachyornis elephantopus*)	Concurrent with Polynesian settlement CE (AD) 1250–1300	Human direct; presumed overhunting	Baker et al. (2005)
	Five species of kiwi (Ratite)	Concurrent with Polynesian settlement CE (AD) 1250–1300	Human indirect; predation and consumption of eggs by rats [kiore (*R. exulans*)] introduced by Polynesians/Maoris	Holdaway (1989); McClone (1989); Weir et al. (2016)
Indian Ocean				
Mauritius (volcanic island)	Dodo (*Rap. cucullatus*) (Columbiformes or related to pigeons and doves)	By CE (AD) 1700	Human (European) indirect and direct effects; destructions of eggs by invasive species such as rats (*R. rattus* and/or *R. norvegicus*) and pigs + overkilling by sailors (<50 people)	Shapiro et al. (2002); Bergman (2005); Hume (2012)
Réunion (volcanic island)	Reunion Solitaire (*Pezophaps solitaria*) (Columbiformes)	By CE (AD) 1700	Human (European); indirect and direct effects as dodo	Shapiro et al. (2002); Hume (2012)
Madagascar	Elephant birds (*Aepyornis maximus*) (Ratite)	By CE (AD) 1700	Human indirect (European) and direct effects probably as dodo	Hume (2012)
Pacific Ocean				
Hawaii (volcanic islands)	Flightless goose (*Thambelochen chauliodo*) (Anseriformes or related to ducks and geese)	Concurrent with Polynesian settlement CE (AD) ~500	Human indirect; introduction of Polynesian rat, or Pacific rat that predated on birds and their eggs	Olson and James (1982a,b)
	Moa-nalos [four species of large flightless birds: *Chelychelynechen quassus* (Kauai), *Thambetochen xanion* (Oahu), *Thambetochen chauliodous* (Molokai, Maui and Lanai) and *Ptaiochen pau* (Maui)] (Anseriformes)	Concurrent with Polynesian settlement CE (AD) ~500	Human indirect; introduction of Polynesian rat, or Pacific rat that predated on birds and their eggs	Sorenson et al. (1999)

(*Continued*)

TABLE 21.1   Examples of Extinctions of Flightless Birds (*cont.*)

Region	Species	Extinction date	Cause(s)	References
New Caledonia	Giant extinct flightless bird (previously referred to as a megapode) (*Sylviornis neocaledoniae*)	Concurrent with Polynesian settlement CE (AD) 200–400	Human; presumed as with New Zealand and Hawaii	Anderson et al. (2010); Worthy et al. (2016)
North America				
California coast and island	California's flightless sea duck (*Chendytes lawi*) (Anseriformes)	~2,400 years ago	Human direct; overhunting by Native American population?	Jones et al. (2008)
Sahul (Australia, Tasmania, and New Guinea conjoined when sea level was much lower due to glaciations	Flightless bird, *G. newtoni*	~46,000 years ago	Human direct; human consumption of eggs (?) by first human settlers—the indigenous (or Aboriginal) Australians	Miller et al. (2016)

TABLE 21.2   Examples of Myths and Extinctions

Nos.	Myth	Reality
1	All Pleistocene extinctions are due to humans (directly or indirectly).	Pleistocene extinctions are attributable to human actions or climatic changes[a] or other shifts in the environment or catastrophic events such as volcanic eruptions.
2	All human-caused extinctions are due to European peoples.	There are examples of extinctions caused (directly and indirectly) by Polynesians, Native Americans, Indigenous Australians, and Europeans.
3	All human-caused extinctions are recent due to industrialized nations.	There are examples of extinctions caused (directly and indirectly) by hunter–gatherers, subsistence farmers, and peoples in modern societies.
4	Different species of animals are equally liable to extinction.	Some species of animals are fragile (particularly vulnerable) to extinction such as flightless birds on islands with no predators.
5	Climate change cannot cause extinctions as animals can move to advantageous environments.	There are numerous examples of climatic changes (e.g., extensive glaciations) being followed by extinctions.
6	Dodos were driven to extinction by overhunting by sailors.	The extinction of dodos was due to the introduction of predator species such as pigs and rats, respectively, purposely and accidentally from ships.
7	Mammoths were driven to extinct by overhunting	The more likely scenario is climatic changes at the end of the last major glaciation. This was probably accompanied by impacts of hunting.[b]

[a] *Wroe et al. (2013).*
[b] *Sodhi et al. (2012).*

21.3 ECOSYSTEM BIODIVERSITY AND ANIMAL EXTINCTIONS

Biodiversity refers to the variety of plant and animal life on earth and offers an essential safeguard for adaption to unpredictable events and uncertainty; the greater the diversity, the more likely that ecosystems will be able to adapt to challenges (Forest et al., 2015). The loss of animal species, without replacement, represents a loss in biodiversity that can have major implications for the functioning of ecosystems, such as the potential impact on global food economies should the decline of insect pollinators continue (Dirzo et al., 2014).

Tropical forested areas are disproportionately impacted by the current biotic crisis (Beaudrot et al., 2016); these biodiversity "hotspots" are areas that feature "exceptional concentrations of endemic species and experience[ing] exceptional loss of habitat" (Myers et al., 2000).

Up to 44% of global plants species and 35% of vertebrate species are proposed to be contained within these hotspots, which now comprise only 1.4% of global land surface; it is argued that conservation investment in these areas yields the most value toward stemming "the mass extinction of species that is now underway" (Myers et al., 2000) (Fig. 21.6).

21.4 THREATS TO BIODIVERSITY

21.4.1 Human Population Growth

While some recent data challenge the role of humans as central to the extinction of terrestrial megafauna (Nielsen and Nurzynski, 2013), there is broad agreement that human activities, particularly postindustrialization, driven by recent and exponential human population growth are the cause of the current-day biotic crisis.

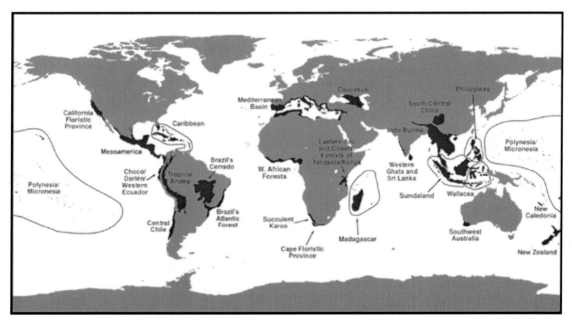

FIGURE 21.6 **The 25 biodiversity hotspots.** The hotspot expanses comprise 30 ± 3% of the red areas. *Source: Myers, N., Mittermeier, R.A., Mittermeier, C.G., da Fonseca, G.A., Kent, J., 2000. Biodiversity hotspots for conservation priorities. Nature 403 (6772), 853–858: Figure 1, reproduced with permission from MacMillian Publishers.*

Modern humans (*Homo sapiens*), thought to have evolved from a common ancestor (*Homo erectus*) approximately 200,000 years ago, are relatively new additions to the history of life on this planet. To use the 24-h clock analogy of the history of the earth, modern humans arrived just a few seconds before midnight. For most of our brief history, human population growth has been extremely slow; short life expectancy (between 30 and 40 years) and high infant mortality kept pre-industrialized global population levels below 1 billion. The advent of manufacturing and coal/steam power during the Industrial Revolution in the 1700s enabled humans to significantly improve their standard of living and to begin a phase of unprecedented population growth. In the last 200 years, human population growth has increased exponentially, reaching 7.3 billion in 2015 and estimated to reach 9.725 billion by 2050, with most growth occurring in developing nations (United Nations, 2015). To meet projected consumption demands, which are largely centralized in developing nations, it has been estimated that the production of an additional billion tons of cereals and 210 million tons of meat will be required annually by 2050, compared to production in 2005/2007 (Bruinsma, 2009). Sustainable realization of these resource needs will require systemic agricultural changes to stem the current biotic crisis.

21.4.2 Animal Overexploitation

Animal overexploitation refers to the harvesting of species at a rate that exceeds their capacity to replenish sufficient numbers to maintain their communities. Although our role in the extinction of megafauna continues to be debated, there are many recent examples of animal overexploitation by humans, such as overfishing and overhunting, which have directly or indirectly led to the extinction of many species of animals, particularly large vertebrates, and threaten ecosystem biodiversity. For example, since modern methods of commercial, mechanized fishing

were introduced to open ocean in the 1950s, many species of marine predators (such as the Atlantic bluefin tuna, groupers, and sea turtles) are under significant threat of extinction (Baum et al., 2003), with up to 80% of fisheries believed to be fully exploited and up to 25% of marine species threatened with extinction (FAO, 2012). Deepwater species (such as Atlantic cod and haddock) were already "fully exploited, overfished or depleted" 20 years ago (FAO, 1993), rendering these species "commercially extinct." Many surface marine species (such as sardines, anchovies, and herring) require careful management if they are to avoid depletion (FAO, 1993). Greater restriction on commercial fisheries, particularly in developed nations, may help to prevent future overfishing and restore depleted species (Sissenwine et al., 2014); efforts to implement sustainable practices also include consumer education (Fig. 21.7).

Wild terrestrial animals are also threatened with extinction as a result of human overhunting. The flightless Dodo bird (*Raphus cucullatus*) from the islands of Mauritius is perhaps one of the most well-known examples of overhunting by humans who played a significant role in taking this species to extinction (Hume, 2012); but there are many others in a shameful history of

FIGURE 21.7 **Overfishing.** About 400 tons of jack mackerel (*Trachurus murphyi*) are caught by a Chilean purse seiner. *Source: C. Ortiz Rojas, http://www.photolib.noaa.gov/htmls/fish2172.htm. Public domain, via Wikimedia Commons.*

unfettered animal overexploitation. For example, unregulated and widespread hunting of the vast population of wild passenger pigeons (*Ectopistes migratorius*) endemic to North America between 1850 and 1900 took this species that once numbered 3–5 billion to extinction in 50 years (Schulz et al., 2014). It is hypothesized that this recent extinction was due to natural cyclical changes in population size coupled with human hunting (Hung et al., 2014).

Humans have long hunted many species of animals for their perceived threat to livestock (such as the Tasmanian tiger, *Thylacinus cynocephalus*, and the Falkland Island wolf, *Dusicyon australis*) (Menzies et al., 2012), for the value of their pelt (such as the flightless great auk, *Pinguinis impennis*, the sea mink, *Neovison macrodon*, and the quagga, *Equus quagga quagga*) (Bourne, 1993; Harley et al., 2009; Sealfon, 2007), for superstitious reasons (such as the Zanzibar leopard, *Panthera pardus adersi*) (Walsh and Goldman, 2012), or for sport (such as the North African Atlas bear, *Ursus arctos crowtheri*). These are but a small number of species that were hunted to extinction by humans; this is an ongoing problem in the contemporary era (such as the African bush meat crisis; Bennett et al., 2007) and one that requires significant investment in local, as well as global, wildlife management strategies (see Section 21.5) (Fig. 21.8).

21.4.3 Habitat Destruction

Habitat loss resulting from exploitation and destruction by humans is one of the principal drivers of global plant and animal extinctions or "defaunation" (Dirzo et al., 2014; Pimm and Raven, 2000). The timeframe for extinctions due to habitat fragmentation and loss may be instantaneous or delayed and depends on the extent to which a given species is also found in unaffected areas (Pimm and Raven, 2000). The "extinction debt" refers to species that are under significant threat of extinction, or will be in the future, as a result of past and present habitat loss (Olivier

FIGURE 21.8 **Passenger pigeon.** *Source: Smith Bennett 1875. Passenger pigeon flock being hunted in Louisiana, https://commons. wikimedia.org/w/index.php?curid=45560761. Public domain.*

et al., 2013). As habitat loss increases, so too does the rate of extinction; this nonlinear, power law is known as the "species–area relationship," which describes the exponentially greater pressure on survival in small areas <10 ha (Tanentzap et al., 2012).

Habitat destruction, such as deforestation, is driven by humans largely for the purposes of clearing land for agriculture and timber harvesting. That is, sustaining the present-day global population requires vast tracts of land to be used in food and other resource production. Beginning in the postindustrialized period, significant advances in agricultural machinery increased farming efficiency and productivity. For instance, industrialized agriculture has seen the departure of subsistence farming to the growth of "agribusiness," which applies modern science and technological innovations at a commercial scale to achieve greater yield and profitability in food production activities. While the intensification of animal husbandry of livestock (such as chickens, beef cattle, pigs, and lamb) has actually seen a reduction in land usage and labor in many developed nations, industrial agriculture presents one of the most significant threats to habitat destruction on this planet due to the use and intensification of land, such as through the growth of food crop monocultures to feed livestock (i.e., up to 91% of the world's farmland is used exclusively for cereal monocultures such as soybeans and corn; Altieri, 2009), and the use of agricultural chemicals (e.g., fertilizers, herbicides, and insecticides) (Wang et al., 2015). At the present time, it is estimated that approximately 40% of land is used for agricultural purposes; tropical regions, such as Brazil, are seeing the faster growth of land clearing and conversion for agricultural cropland and grazing (Galford et al., 2010). The Food and Agricultural Organization of the United Nations (FAO) projects a substantial increase in demand for animal-based foods from the year 2000 to 2030, particularly in developing nations (such as an increase in demand for poultry meat by 850% in India and 725% in South Asia), largely driven by increased per capita demand for the consumption of meat products and population growth (FAO, 2011).

21.4.4 Alien Species

The deliberate, or otherwise, introduction of invasive alien species beyond their native range to naive ecosystems, particularly insular, island ecosystems, by humans is another of the central threats to global biodiversity. In particular, predatory mammalian species [such as domestic cats (*Felis catus*), red foxes (*Vulpes vulpes*), and rats (*R. rattus*)] are perhaps the most damaging of all invasive species, having contributed toward the extinction of many vertebrate species, such as birds, mammals, and reptiles (Doherty et al., 2015); with over 10% of 273 terrestrial mammal species endemic to Australia known to have been lost to predatory invasive species (Woinarski et al., 2015). Others, such as the omnivorous wild boar (*Sus scrofa*), present a significant threat to native species largely through disruption and destruction of the environment and local ecosystem communities (McClure et al., 2015). The wild boar was introduced to North America in 1539 and is one of the most successful invasive vertebrate species worldwide. With a population of approximately 5 million established in 39 states in America, it has been estimated that the economic cost of agricultural losses and population management exceeds $1.5 billion USD annually (USDA, 2013). Wild boar threaten native ecosystem diversity through their opportunistic predation of amphibians, reptiles, rodents, and other small mammals, and disruption of nests/hatchlings of ground-nesting birds, alligators, and turtles (Centner and Shuman, 2015; Krull and Choquenot, 2013).

The spread of plant and animal nonnatives is homogenizing global biota (Vander Zander, 2005) and costs up to $137 billion USD annually in environmental damage and losses (Pimentel et al., 2000). Not all introduced species become invasive or even establish in the wild (Jeschke and Strayer, 2008). Recent research challenges the long-held notion that the success of invasive species is primarily due to their exploitation of opposing characteristics in native species, such as fast versus slow life histories, generalists versus specialists (Jeschke and Strayer, 2008). For example, slow life histories and sexual dimorphism are risk factors for extinction, but the opposite characteristics (a fast life history and sexual monomorphism) do not necessarily predict invasive success; rather, the association of a nonnative species with humans is among the principal predictors of invader success (Jeschke and Strayer, 2008). Advancing the understanding of the determinants of success of nonindigenous invasive species "is among the most interesting and urgent questions in ecology" (Van Kleunen et al., 2010); present-day knowledge does not yet enable effective or efficient management (Chown et al., 2015) (Fig. 21.9).

21.4.5 Climate Change

Climate change has caused mass extinctions in the past (Blois et al., 2013) and there is no doubt that earth's climate is warming; there is consensus among the vast majority of scientists that this is caused by human activities. According to the Intergovernmental Panel on Climate Change, continuing at the current CO_2 emission rates will

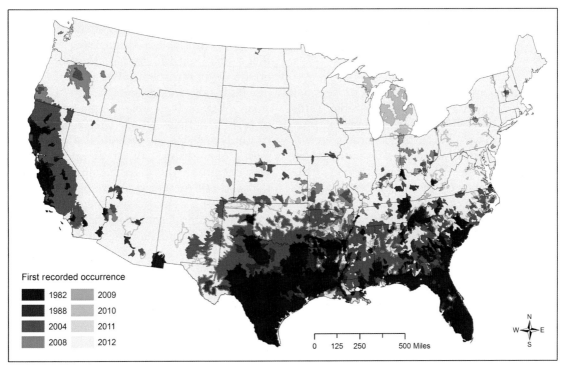

FIGURE 21.9 **Spread of wild pigs in the contiguous United States.** *Source: McClure, M.L., Burdett, C.L., Farnsworth, M.L., Lutman, M.W., Theobald, D.M., Riggs, P.D., et al., 2015. Modeling and mapping the probability of occurrence of invasive wild pigs across the contiguous United States. PLoS One 10 (8), e0133771.*

raise earth's temperature by up to +4.8°C over the next century (IPCC, 2013). A changing climate presents significant challenges to species survival and is one of the leading threats to biodiversity, with ectothermic species (i.e., those dependent on environmental temperatures to regulate their internal temperature) at particular risk of climate change (Bestion et al., 2015). For example, in a novel experiment that modeled future rising temperatures on populations of the common lizard, *Zootoca vivipara*, found that lizard communities exposed to warmer temperatures had a significantly accelerated life cycle and decreased adult survival, with projected extinction in 20 years (Bestion et al., 2015). Yvon-Durocher et al. (2015) undertook a similar experiment to examine the effects of warming on a planktonic marine ecosystem; the authors found that a rise of 4°C produced a significant increase in phytoplankton biomass and biodiversity. Conversely, Seifert et al. (2015) found that high aquatic temperatures promoted extinction rates in temperate marine ectotherms due to significant disruption of food webs such as increased "carnivorous overexploitation" of herbivores and reduced opportunity to replenish populations at high temperatures. These experiments and their divergent outcomes on species extinctions and biodiversity provide "better forecasts of extinction risk and range dynamics" (Fordham, 2015). However, as suggested by Blois et al. (2013, p. 499), "robust predictions about the future require multispecies models that combine long-term insights from the past with specific and shorter-term insights from modern systems" and acknowledge that this presents "a herculean challenge."

21.5 ANIMAL CONSERVATION

Noah got it all wrong. One or two or even a dozen
of each species is not enough *Kaufman, 2012*

Animal extinctions are a natural process that
shape the ongoing evolution of species better
suited to cope with the environmental pressures. However, when extinction rates are unusually high, such as at the present time, the
biodiversity of the planet can be threatened; this
snowball effect can place ever increasing numbers of species at risk. Identifying species at risk
of extinction is a highly complex process fraught
with uncertainty (Forest et al., 2015). The International Union for Conservation of Nature
(IUCN) Red List of Threatened Species provides
the most authoritative list of the world's conservation status of species (IUCN, 2015); it too has
its detractors who suggest that the Red List comprises some of the "more colorful and charismatic" endangered species (Bridgewater, 2016).
These criticisms serve to illustrate the challenge
of selecting species for conservation, since such
decision-making is driven not only by the risk
of extinction, but many other biological (e.g.,
evolutionary distinctiveness), economic, and
human-centered (such as the appeal and popularity of the species to humans) factors. Despite
"uncertainty" in "the information required for
the decision-making in conservation planning,"
current biodiversity losses require urgent intervention (Forest et al., 2015). Based on the assumption that "not all species are equal" (Forest
et al., 2015), one possible way of prioritizing species for conservation is to consider their role in
a given ecosystem, such as "keystone species."

Keystone species uniquely shape the structure
and function of entire ecosystems and their communities, exert an impact that is disproportionate
to their biotic abundance, such as by maintaining
the food web linkages of their community or by
maintaining biodiversity within an ecosystem
(Perry, 2010; Valls et al., 2015). In his seminal
paper, Robert Paine was the first to describe the
keystone species concept, having identified the
pivotal role of predatory starfish "to prevent the
monopolization of the major environmental req-
uisites by one species" (Paine, 1966). Paine found
that a small starfish, *Pisaster ochraceus*, predated
on low-lying mussels, *Mytilus californianus*, during high tide and, despite having a relatively
small biomass, were able to maintain the lower
edge of the mussel bed. Under experimental
conditions, Paine showed that after a 10-year
period of starfish exclusion, the rocky shoreline
was dominated by a monoculture of mussels
that had out-competed other species for space;
starfish were identified as a keystone predator
that significantly enhanced the biodiversity of
the ecosystem and its community by preventing "resource monopolies" by a single species
(Paine, 1966). Since then many other keystone
predators have been identified as having a pivotal impact on ecosystem biodiversity. For example, the predation of sea urchins (*Strongylocentro-
tus polyacanthus*) by sea otters (*Enhydra lutris*) has
been shown to limit their impact on kelp forests,
which could otherwise disappear along with
their associated communities (Estes et al., 1978).

Keystone species need not be predators; there
are many examples of keystone herbivores, pollinators, and habitat modifiers to name a few
(Menge et al., 2013). For example, recent research
undertaken to report on populations of amphibians in the United States identified greater abundance in wetland areas altered by beavers, suggesting that management of this keystone species
could facilitate the recovery of amphibian species in decline (Hossack et al., 2015). The loss of
keystone species, such as through targeted species exploitation by humans, alters the balance of
species interactions within that ecosystem (Eddy
et al., 2014; Valls et al., 2015) and can "trigger a
cascade of secondary extinctions," threatening
the structure and functioning of the ecosystem
(Jonsson et al., 2015).

Despite a "paradigm shift" in the mission of
contemporary zoos from traditional ex situ conservation toward strategies that integrate research,
education, awareness, and advocacy with captive
breeding programs, these attempts are not of a

scale required to protect more than a small fraction of endangered species (Keulartz, 2015). Pressures that have led to species extinction and endangerment, such as habitat destruction, probably mean that "the goal of restoring ecosystems to a desired historical state of "naturalness" is [often] virtually impossible" (Searcy et al., 2016) even if such pressures are managed and minimized. This may undermine the long-term success of in situ conservation strategies, such as taxon substitution, which seeks to replace extinct species through the introduction of nonnative species that serve similar ecological functions (Searcy et al., 2016). The US Endangered Species Act (ESA) of 1973 has been instrumental in protecting species from extinction, with <1% of species identified under the ESA having since gone extinct, such as the successful restoration of the Yellowstone grizzly bear from approximately 136 bears in 1975 to ~700 at the present time (State News Service, 2016). Nonetheless, the ESA has been the subject of criticism for its lack of clarity in criteria for species selection for endangerment, its focus on species protection rather than on ecosystems, and failure to consider public sentiment for investment toward species protection (Greenwald et al., 2013). According to this view, "threatened species are much better seen as symptoms of ecosystem dysfunction, and the imperative for conservation practitioners should be to understand the cause of those symptoms" (Bridgewater, 2016).

Conservation policies, such as the ESA, best serve to limit further harm, but do little to "incentivise stewardship investments" to "align[ing] economic and conservation objectives rather than impose[ing] regulations and restrictions," which is likely to be a particularly relevant consideration for private land-owners, upon which many endangered species and their habitats are found (Male and Donlan, 2015). Contemporary biodiversity conservation strategies will need to do more in this space while also adapting to "new" challenges such as climate change (Bernazzani et al., 2012; Oliver et al., 2016). Since conservation budgets are limited, decision-making about conservation strategies for a given species or ecosystem should

entail a "cost-effectiveness analysis," which identifies the strategy or combination of strategies that provide the greatest return on investment; that is, "the biological impact (benefit) divided by the action's economic cost" (Gjertsen et al., 2014). For example, Wilson et al. (2014) found that the reintroduction of endangered wild orangutans was not a long-term cost-effective strategy compared to the cost of habitat protection. Other conservation strategies, such as the removal of predators of endangered species, appear to compliment the benefits of habitat protection (Melstrom and Horan, 2014). Neoconservationists propose "economic development, poverty alleviation, and corporate partnerships as surrogates or substitutes for endangered species listing, protected areas, and other mainstream conservation tools," with the goal of "enhance[ing] those natural systems that benefit the widest number of people, especially the poor" (Soule, 2013); detractors of this humanitarian approach to conservation warn that it "would hasten ecological collapse globally, eradicating thousands of kinds of plants and animals and causing inestimable harm to humankind in the long run" (Soule, 2013).

References

Altieri, M.A., 2009. Green deserts: monocultures and their impacts on biodiversity. In: Red Sugar, Green Deserts. FIAN International, FIAN Sweden, HIC-AL, and SAL.

Alvarez, L.W., Alvarez, W., Asaro, F., Michel, H.V., 1980. Extraterrestrial cause for the cretaceous-tertiary extinction. Science 208 (4448), 1095–1108.

Anderson, A.S.C., Petchey, F., Worthy, T.H., 2010. Faunal extinction and human habitation in New Caledonia: initial results and implications of new research at the Pindai Caves. J. Pac. Archaeol. 1, 89–109.

Baker, A.J., Huynen, L.J., Haddrath, O., Millar, C.D., Lambert, D.M., 2005. Reconstructing the tempo and mode of evolution in an extinct clade of birds with ancient DNA: the giant moas of New Zealand. Proc. Natl. Acad. Sci. USA 102, 8257–8262.

Baum, J.K., Myers, R.A., Kehler, D.G., Worm, B., Harley, S.J., Doherty, P.A., 2003. Collapse and conservation of shark populations in the Northwest Atlantic. Science 299, 389–392.

Beaudrot, L., Ahumada, J.A., O'Brien, T., Alvarez-Loayza, P., Boekee, K., Campos-Arceiz, A., et al., 2016. Standardized assessment of biodiversity trends in tropical forest protected areas: the end is not in sight. PLoS Biol. 14 (1), e1002357.

Bennett, E.L., Blencowe, E., Brandon, K., Brown, D., Burn, R.W., Cowlishaw, G., Davies, G., Dublin, H., Fa, J.E., Milner-Gulland, E.J., Robinson, J.G., Rowcliffe, J.M., Underwood, F.M., Wilkie, D.S., 2007. Hunting for consensus: reconciling bushmeat harvest, conservation, and development policy in West and Central Africa. Conserv. Biol. 21 (3), 884–887.

Bergman, J., 2005. The history of the dodo bird and the cause of its extinction. Perspect. Sci. Christ. Faith 57, 221–229.

Bernazzani, P., Bradley, B.A., Opperman, J.J., 2012. Integrating climate change into habitat conservation plans under the U. S. Endangered Species Act. Environ. Manage. 49, 1103–1114.

Bestion, E., Teyssier, A., Richard, M., Clobert, J., Cote, J., 2015. Live fast, die young: experimental evidence of population extinction risk due to climate change. PLoS Biol. 13 (10), e1002281.

Blois, J.L., Zarnetske, P.L., Fitzpatrick, M.C., Finnegan, S., 2013. Climate change and the past, present, and future of biotic interactions. Science 341 (6145), 499–504.

Bourne, W.R.P., 1993. The story of the Great Auk *Pinguinis impennis*. Arch. Nat. History 20 (2), 257–278.

Braje, T.J., Erlandson, J.M., 2013. Human acceleration of animal and plant extinctions: a Late Pleistocene, Holocene, and Anthropocene continuum. Anthropocene 4, 14–23.

Bridgewater, P., 2016. The Anthropocene biosphere: do threatened species, Red Lists, and protected areas have a future role in nature conservation? Biodivers. Conserv. 25, 603–607.

Bruinsma, J., 2009. The resource outlook to 2050: by how much do land, water and crop yields need to increase by 2050? In: Paper presented at the FAO Expert Meeting on How to Feed the World in 2050, June 24–26, Rome.

Burgess, S.D., Bowring, S., Shen, S., 2014. High-precision timeline for Earth's most severe extinction. Proc. Natl. Acad. Sci. USA 111 (9), 3316–3321.

Centner, T.J., Shuman, R.M., 2015. Governmental provisions to manage and eradicate feral swine in areas of the United States. Ambio 44, 121–130.

Chown, S.L., Hodgins, K.A., Griffin, P.C., Oakeshott, J.G., Byrne, M., Hoffman, A.A., 2015. Biological invasions, climate change and genomics. Evol. Appl. 8 (1), 23–46.

Corsetti, F.A., 2015. Life during Neoproterozoic Snowball Earth geographically isolated lineages. Geology 43, 559–560.

Dirzo, R., Young, H.S., Galetti, M., Ceballos, G., Isaac, N.J., Collen, B., 2014. Defaunation in the Anthropocene. Science 345 (6195), 401–406.

Doherty, T.S., Dickman, C.R., Nimmo, D.G., Ritchie, E.G., 2015. Multiple threats, or multiplying the threats? Interactions between invasive predators and other ecological disturbances. Biol. Conserv. 190, 60–68.

FAO, 2012. FAO Statistical Yearbook. World Food and Agriculture, Part 3. Feeding the World. FAO, Rome.

Eddy, T.D., Pitcher, T.J., MacDiarmid, A.B., Byfield, T.T., Tam, J.C., Jones, T.T., Bell, J.J., Gardner, J.P.A., 2014. Lobsters as keystone: only in unfished ecosystems? Ecol. Modell. 275, 48–72.

Estes, J.A., Smith, N.S., Palmisano, J.F., 1978. Sea otter predation and community organization in western Aleutian Islands, Alaska. Ecology 59, 822–833.

FAO, 1993. Harvesting Nature's Diversity. FAO, Rome.

FAO, 2011. Mapping supply and demand for animal-source foods to 2030, by T.P. Robinson & F. Pozzi. Animal Production and Health Working Paper No. 2, Rome.

Field, J.H., Boles, W.E., 1998. *Genyornis newtoni* and *Dromaius novaehollandiae* at 30,000 b.p. in central northern New South Wales. Alcheringa 22, 177–188.

Fordham, D.A., 2015. Mesocosms reveal ecological surprises from climate change. PLoS Biol. 13 (12), e1002323.

Forest, F., Crandall, K.A., Chase, M.W., Faith, D.P., 2015. Phylogeny, extinction and conservation: embracing uncertainties in a time of urgency. Philos. Trans. R. Soc. B 370, 20140002.

Galford, G.L., Melillo, J., Mustard, J.F., Cerri, C.E.P., Cerri, C.C., 2010. The Amazon frontier of land-use change: Croplands and consequences for greenhouse gas emissions. Earth Interact. 14, 15.

Gjertsen, H., Squires, D., Dutton, P.H., Eguchi, T., 2014. Cost-effectiveness of alternative conservation strategies with application to the Pacific leatherback turtle. Conserv. Biol. 28 (1), 140–149.

Greenwald, N., Ando, A.W., Butchart, S.H., Tschirhart, J., 2013. Conservation: the Endangered Species Act at 40. Nature 504 (7480), 369–370.

Harley, E.H., Knight, M.H., Lardner, C., Wooding, B., Gregor, M., 2009. The Quagga project: progress over 20 years of selective breeding. J. Wildlife Res. 39 (2), 155–163.

Harper, D.A.T., Hammarlund, E.U., Rasmussen, C.M.O., 2014. End Ordovician extinctions: a coincidence of causes. Gondwana Res. 25, 1294–1307.

Hossack, B.R., Gould, W.R., Patla, D.A., Muths, E., Daley, R., Legg, K., Corn, P.S., 2015. Trends in Rocky Mountain amphibians and the role of beaver as a keystone species. Biol. Conserv. 187, 260–269.

Holdaway, R.N., 1989. New Zealand's pre-human avifauna and its vulnerability. N. Z. J. Ecol. 12 (Suppl.), 11–25.

Hume, J.P., 2012. The Dodo: from extinction to the fossil record. Geol. Today 28 (4), 147–151.

Hung, C.M., Shaner, P.J., Zink, R.M., Liu, W.C., Chu, T.C., Huang, W.S., Li, S.H., 2014. Drastic population fluctuations explain the rapid extinction of the passenger pigeon. Proc. Natl. Acad. Sci. USA 111, 10636–10641.

IPCC, 2013. Climate change 2013: the physical science basis: Working Group I contribution to the fifth assessment report of the Intergovernmental Panel on Climate Change. In: Stocker, T.F., Qin, D., Plattner, G.K., Tignor, M.M.B., Allen, S.K., Boschung, J., et al. (Eds.), Cambridge University Press, Cambridge, United Kingdom/New York, NY.

IUCN, 2015. The IUCN Red List of Threatened Species. Version 2015-4. Available from: http://www.iucnredlist.org.

Jeschke, J.M., Strayer, D.L., 2008. Are threat status and invasion success two sides of the same coin? Ecography 31, 124–130.

Jones, T.L., Porcasi, J.F., Erlandson, J.M., Dallas, Jr., H., Wake, T.A., Schwaderer, R., 2008. The protracted Holocene extinction of California's flightless sea duck (*Chendytes lawi*) and its implications for the Pleistocene overkill hypothesis. Proc. Natl. Acad. Sci. USA 105, 4105–4108.

Jonsson, T., Berg, S., Emmerson, M., Pimenov, A., 2015. The context dependency of species keystone status during food web disassembly. Food Webs 5, 1–10.

Kaufman, L., 2012. Zoos struggle to breed endangered animals. New York Times, July 5.

Kerr, R.A., 2001. Paring down the Big Five Mass Extinctions. Science 284, 2072–2073.

Keulartz, J., 2015. Captivity for conservation? Zoos at a crossroads. J. Agric. Environ. Ethics 28, 335–351.

Krull, C.R., Choquenot, D., 2013. Feral pigs in a temperate rainforest ecosystem: disturbance and ecological impacts. Biol. Invasions 15, 2193–2204.

Lindström, S., Pedersen, G.K., van de Schootbrugge, B., Hansen, K.H., Kuhlmann, N., Thein, J., Johansson, L., Petersen, H.I., Alwmark, C., Dybkjær, K., Weibel, R., Erlström, M., Nielsen, L.H., Oschmann, W., Tegner, C., 2015. Intense and widespread seismicity during the end-Triassic mass extinction due to emplacement of a large igneous province. Geology 43 (5), 387–390.

Lucas, S.G., Tanner, L.H., 2008. Reexamination of the end-Triassic mass extinction. In: Elewa, A.M.T. (Ed.), Mass Extinction. Springer Verlag, New York, NY, pp. 66–103.

Lucas, S.G., Tanner, L.H., 2015. End-Triassic nonmarine biotic events. J. Palaeogeogr. 4 (4), 331–348.

Male, T., Donlan, C.J., 2015. The future of pre-listing conservation programs for wildlife conservation. In: Donlan, C.J. (Ed.), Proactive Strategies for Protecting Species [electronic resource]: Pre-Listing Conservation and the Endangered Species Act. University of California Press, Berkeley, CA.

Malhia, Y., Doughty, C.E., Galettib, M., Smith, F.A., Svenningd, J.-C., Terborgh, J.W., 2016. Megafauna and ecosystem function from the Pleistocene to the Anthropocene. PNAS 113 (4), 838–846.

McClure, M.L., Burdett, C.L., Farnsworth, M.L., Lutman, M.W., Theobald, D.M., Riggs, P.D., et al., 2015. Modeling and mapping the probability of occurrence of invasive wild pigs across the contiguous United States. PLoS One 10 (8), e0133771.

McClone, M.S., 1989. The Polynesian settlement of New Zealand in relationship to environmental and biotic changes. N. Z. J. Ecol. 12 (Supplement), 115–129.

Melstrom, R.T., Horan, R.D., 2014. Interspecies management and land use strategies to protect endangered species. Environ. Resour. Econ. 58, 199–218.

Menge, B.A., Iles, A.C., Freidenburg, T.L., 2013. Keystone species. Levin, S.A. (Ed.), Encyclopedia of Biodiversity, vol. 4, second ed. Academic Press, Waltham, MA, pp. 442–457.

Menzies, B.R., Renfree, M.B., Heider, T., Mayer, F., Hildebrandt, T.B., et al., 2012. Limited genetic diversity preceded extinction of the Tasmanian tiger. PLoS One 7 (4), e35433.

Miller, G., Magee, J., Smith, M., Spooner, N., Baynes, A., Lehman, S., Fogel, M., Johnston, H., Williams, D., Clark, P., Florian, C., Holst, R., DeVogel, S., 2016. Human predation contributed to the extinction of the Australian megafaunal bird *Genyornis newtoni* ~47 ka. Nat. Commun. 7, 10496.

Myers, N., Mittermeier, R.A., Mittermeier, C.G., da Fonseca, G.A., Kent, J., 2000. Biodiversity hotspots for conservation priorities. Nature 403 (6772), 853–858.

Nielsen, R.W., Nurzynski, J., 2013. The Late-Pleistocene extinction of megafauna compared with the growth of human population. Available from: http://arxiv.org/ftp/arxiv/papers/1309/1309.3002.pdf.

Oliver, T.H., Smithers, R.J., Beale, C.M., Watts, K., 2016. Are existing biodiversity conservation strategies appropriate in a changing climate? Biol. Conserv. 193, 17–26.

Olivier, P.I., van Aarde, R.J., Lombard, A.T., 2013. The use of habitat suitability models and species–area relationships to predict extinction debts in coastal forests, South Africa. Diversity Distrib. 19, 1353–1365.

Olson, S.L., James, H.F., 1982a. Fossil birds from the Hawaiian Islands: evidence for wholesale extinction by Man before Western contact. Science 217, 633–635.

Olson, S.L., James, H.F., 1982b. Prodromus of the fossil avifauna of the Hawaiian Islands. Smithsonian Contributions to Zoology No. 365. DOI: https://doi.org/10.5479/si.00810282.365.

Paine, R.T., 1966. Food web complexity and species diversity. Am. Nat. 100 (910), 65–75.

Perry, N., 2010. The ecological importance of species and the Noah's Ark problem. Ecol. Econ. 69, 478–485.

Pimentel, D., Lach, L., Zuniga, R., Morrison, D., 2000. Environmental and economic costs of nonindigenous species in the United States. BioScience 50 (1), 53–65.

Pimm, S.L., Raven, P., 2000. Extinction by numbers. Nature 403, 843–845.

Roberts, R.G., Brook, B.W., 2010. Turning back the clock on the extinction of megafauna in Australia. Quat. Sci. Rev. 29, 593–595.

Roberts, R.G., Flannery, T.F., Ayliffe, L.K., Yoshida, H., Olley, J.M., Prideaux, G.J., Laslett, G.M., Baynes, A., Smith, M.A., Jones, R., Smith, B.L., 2001. New ages for the last Australian megafauna: continent-wide extinction about 46,000 years ago. Science 292, 1888–1892.

Schootbrugge, B.V.D., Wignall, B.P., 2016. A tale of two extinctions: converging end-Permian and end-Triassic scenarios. Geol. Mag. 153 (2), 332–354.

Schulz, J.H., Otis, D.L., Temple, S.A., 2014. 100th Anniversary of the passenger pigeon extinction: lessons for a complex and uncertain future. Wildlife Soc. Bull. 38 (3), 445–450.

Sealfon, R.A., 2007. Dental divergence supports species status of the extinct sea mink (Carnivora: Mustelidae: *Neovison macrodon*). J. Mammal. 88 (2), 371–383.

Searcy, C.A., Rollins, H.B., Shaffer, H.B., 2016. Ecological equivalency as a tool for endangered species management. Ecol. Appl. 26 (1), 94–103.

Seifert, L.I., Weithoff, G., Gaedke, U., Vos, M., 2015. Warming-induced changes in predation, extinction and invasion in an ectotherm food web. Oecologia 178, 485–496.

Shapiro, B., Sibthorpe, D., Rambaut, A., Austin, J., Wragg, G.M., Bininda-Emonds, O.R.P., Lee, P.L.M., Cooper, A., 2002. Flight of the dodo. Science 295, 1683.

Sissenwine, M.M., Mace, P.M., Lassen, H.J., 2014. Preventing over fishing: evolving approaches and emerging challenges. ICES J. Mar. Sci. 71 (2), 153–156.

Sodhi, N.S., Brook, B.W., Bradshaw, C.J.A., 2012. Causes and consequences of species extinctions. In: Levin, S.A., Carpenter, S.R. (Eds.), The Princeton Guide to Ecology. Princeton University Press, Princeton, NJ, pp. 514–520.

Sorenson, M.D., Cooper, A., Paxinos, E.E., Quinn, T.W., James, H.F., Olson, S.L., Fleischer, R.C., 1999. Relationships of the extinct moa-nalos, flightless Hawaiian waterfowl, based on ancient DNA. Proc. R. Soc. Lond. B 266, 2187–2193.

Soule, M., 2013. The "New Conservation". Conserv. Biol. 27 (5), 895–897.

State News Service, 2016. U.S. Fish and Wildlife Service proposes delisting Yellowstone grizzly bear due to recovery decades of conservation lead to Endangered Species Act Success. Available from: https://www.fws.gov/news/.

Tanentzap, A.J., Walker, S., Stephens, R.T.T., Lee, W.G., 2012. A framework for predicting species extinction by linking population dynamics with habitat loss. Conserv. Lett. 5, 149–156.

Turney, C.S., Flannery, T.F., Roberts, R.G., Reid, C., Fifield, L.K., Higham, T.F., Jacobs, Z., Kemp, N., Colhoun, E.A., Kalin, R.M., Ogle, N., 2008. Late-surviving megafauna in Tasmania, Australia, implicate human involvement in their extinction. Proc. Natl. Acad. Sci. USA 105, 12150–12153.

Turvey, S.T., Fritz, S.A., 2011. The ghosts of mammals past: biological and geographical patterns of global mammalian extinction across the Holocene. Phil. Trans R. Soc. B. 366, 2564–2576.

United Nations, Department of Economic and Social Affairs, Population Division, 2015. World Population Prospects: The 2015 Revision. United Nations, New York, NY.

USDA, 2013. Feral Swine: Damage and Disease Threats. Animal and Plant Health Inspection Service Program Aid No. 2086. USDA, Washington, DC.

Valls, A., Coll, M., Christensen, V., 2015. Keystone species: toward an operational concept for marine biodiversity conservation. Ecol. Monogr. 85 (1), 29–47.

Vander Zander, M.J., 2005. The success of animal invaders. Proc. Natl. Acad. Sci. USA 102 (20), 7055–7056.

Van Kleunen, M., Dawson, W., Schlaepfer, D., Jeschke, J.M., Fischer, M., 2010. Are invaders different? A conceptual framework of comparative approaches for assessing determinants of invasiveness. Ecol. Lett. 13 (8), 947–958.

Wagler, R., 2011. The Anthropocene mass extinction: an emerging curriculum theme for science educators. Am. Biol. Teacher 73 (2), 78–83.

Walsh, M., Goldman, H., 2012. Chasing imaginary leopards: science, witchcraft and the politics of conservation in Zanzibar. J. East. Afr. Stud. 6 (4), 727–746.

Wang, S.L., Heisey, P., Schimmelpfennig, D., Ball, E., 2015. Agricultural Productivity Growth in the United States: Measurement, Trends, and DriversERR-189. U.S. Department of Agriculture, Economic Research Service.

Webb, S., 2013. Corridors to extinction and the Australian megafauna. In: Webb, S. (Ed.), Elsevier Science. ProQuest Ebook Central. Available from: http://ebookcentral.proquest.com.ezp.lib.unimelb.edu.au/lib/unimelb/detail.action?docID=1144262.

Weir, J.T., Haddrath, O., Robertson, H.A., Colbourne, R.M., Baker, A.J., 2016. Explosive ice age diversification of kiwi. Proc. Natl. Acad. Sci. USA 113, E5580–E5587.

Wilson, H.B., Meijaard, E., Venter, O., Ancrenaz, M., Possingham, H.P., 2014. Conservation strategies for orangutans: reintroduction versus habitat preservation and the benefits of sustainably logged forest. PLoS One 9 (7), e102174.

Woinarski, J.C.Z., Burbidge, A.A., Harrison, P.L., 2015. Ongoing unraveling of a continental fauna: decline and extinction of Australian mammals since European settlement. Proc. Natl. Acad. Sci. USA 112 (15), 4531–4540.

Worthy, T., Mitri, M., Handley, W., Lee, M., Anderson, A., Sand, C., 2016. Osteology supports a stem-galliform affinity for the giant extinct flightless birds *Sylviornis neocaledoniae* (Sylviornithidae, Galloanseres). PLoS One 11 (3), e0150871.

Wroe, S., Field, J.H., Archer, M., Grayson, D.K., Price, G.J., Louys, J., Faith, J.T., Webb, G.E., Davidson, I., Mooney, S.D., 2013. Climate change frames debate over the extinction of megafauna in Sahul (Pleistocene Australia-New Guinea). Proc. Natl. Acad. Sci. USA 110, 8777–8781.

Yvon-Durocher, G., Allen, A.P., Cellamare, M., Dossena, M., Gaston, K.J., Leitao, M., Montoya, J.M., Reuman, D.C., Woodward, G., Trimmer, M., 2015. Five years of experimental warming increases the biodiversity and productivity of phytoplankton. PLoS Biol. 13 (12), e1002324.

Further Reading

Cardeccia, A., Marchini, A., Occhipinti-Ambrogi, A., Galil, B., Gollasch, S., Minchin, D., Narscius, A., Olenin, S., Ojaveer, H., 2016. Assessing biological invasions in European seas: biological traits of the most widespread non-indigenous species. Estuar. Coast. Shelf Sci., 1–12, doi: 10.1016/j.ecss.2016.02.014.

Mouquet, N., Gravel, D., Massol, F., Calcagno, V., 2012. Extending the concept of keystone species to communities and ecosystems. Ecol. Lett. 16, 1–8.

Index

Printed in the United States
By Bookmasters